D1266958

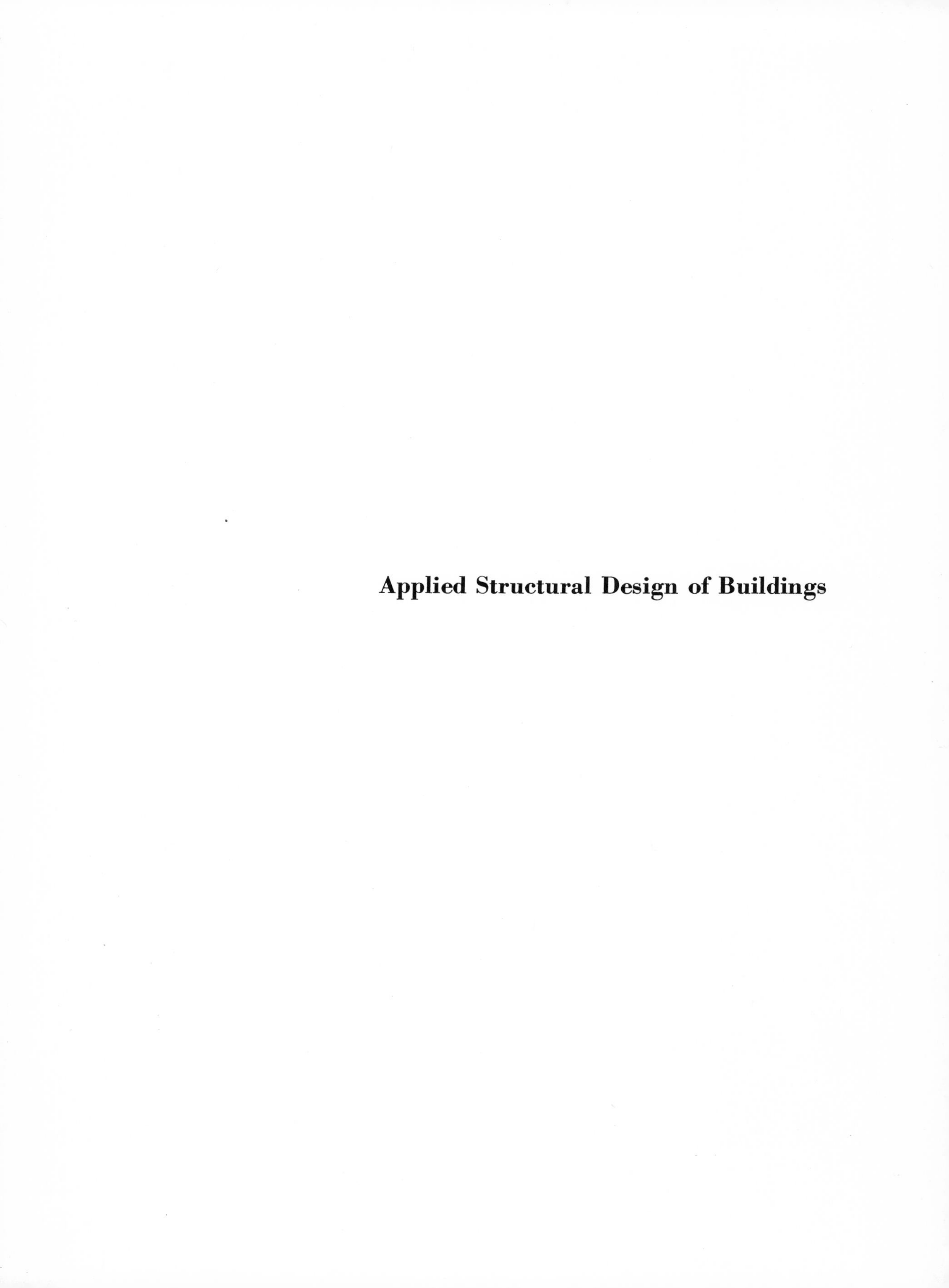

Applied Structural Design of Buildings

Applied Structural Design of Buildings

THOMAS H. McKAIG

B. Arch., C. E.
Consulting Structural Engineer
Buffalo, N.Y.

Third Edition

McGRAW-HILL BOOK COMPANY
New York Toronto London Sydney

APPLIED STRUCTURAL DESIGN OF BUILDINGS

Library of Congress Catalog Card Number 64-21074

45109 ISBN 07-045109-5

3456789101 HDBP 7810798765

Preface

This book had its inception as a series of notes used in the instruction of a group of engineers and architects preparing for state license examinations. The notes were then expanded and published by the author in book form in 1949. In this third edition, the book has been revised to reflect changes in technology, such as the 1963 revision of the ACI Code for reinforced concrete, and also the suggestions for change made by users of the earlier editions.

In its present form the book is intended primarily for the use of architects, structural engineers, plant engineers, license applicants, and draftsmen. It is not strictly a textbook, but rather a handbook, in which the structural engineer will find in simple, practical form the working tools he needs. The presentation is based on the assumption that the person using it has a knowledge of the elements of physics, mechanics, materials of construction, and structural theory. No attempt is therefore made to develop the theory of these subjects, which is obtainable in any textbook. However, the book has been designed to be helpful even if the user's theoretical background was obtained some time ago, and has not been reviewed or brought up to date since.

Certain generally accepted code regulations of various authorities have been used. For structural steel, the AISC Code as adopted in 1963 has been most generally accepted, although some cities still use the AISC 1949 Code (20,000 psi basic stress). For reinforced concrete, the ACI 1963 Code has been used, although again many local codes do not recognize it. Regardless of the suggestions contained in this text as to codes, allowable stresses, and required live loads, the Building Code for the community where the work will be done must be followed. It may be that the recent revisions to the various codes have not yet been adopted by many communities and until they are adopted, the code stresses set up in the local building code will control rather than the recently adopted standard codes.

Although plastic design of steel structures is an accepted part of the 1963 AISC Code, and ultimate strength design of reinforced concrete is similarly a portion of the ACI Code, their use is not described in this volume, nor is the design of prestressed concrete. It is the opinion of the author that these methods as used by a competent and experienced engineer familiar with the limitations and economics of these systems of design will result in some savings, but these methods should only be used by specialists for whom this book is of little value.

Similarly, thin-shell construction and folded plates are not given space in this volume, because although their use is frequently referred to in the architectural and engineering magazines, the actual percentage of use is rather small and the space required for proper solution would be disproportionate to the amount of use.

For most architects, plant engineers and smaller engineering offices, the simple methods of design described herein are sufficient for practically all problems which will be encountered, and the inclusion of the systems mentioned above would probably confuse rather than simplify the application of the structural design theory which formed part of the reader's college course.

Included in this book are tables which the author has most frequently used in his office practice. If the requirements of a specialized problem are beyond the limits of these tables, more complete ones may be found in the various handbooks.

The methods have been taken from good office practice and have proved to be efficient and at the same time acceptable to various building departments and other authorities. Short-cut methods have generally been derived from pure theory. They are recommended, not only to save time, but also to eliminate many sources of error.

Lest it be feared that short cuts sacrifice accuracy, it should be observed that factors of safety both in recommended loads and suggested stresses are more than adequate protection against any lack of refinement in the methods. In addition, the strength of the materials with which the designer is working may vary through a range far greater than the possible variation or error resulting from the short-cut methods.

The author wishes to acknowledge his indebtedness to many sources for his material, including those tables, formulae, and drawings that are reproduced here because they have been especially useful in office practice. Attempt has been made to acknowledge the source of such material wherever it appears, but the following additional acknowledgments, general and specific, are also due:

American Institute of Steel Construction, for much material drawn from the 1963 edition of the Manual of Steel Construction, and other publications.

American Concrete Institute, much of Chapter 5, Art. 7.20, Art. 7.21, Art. 7.40, and Art. 8.11.

Bethlehem Steel Company, particularly for Tables 4.13 and 10.1, and Figs. 9.9 through 9.11.

Engineering News-Record, Table 2.7, Fig. 8.12.

Factory Mutual Fire Insurance Companies, Fig. 10.28.

National Lumber Manufacturers Association, Art. 7.30, Art. 9.30, Art. 1.23, and Art. 6.10.

Portland Cement Association, Figs. 3.24 through 3.27, Fig. 5.1, Fig. 5.7, Art. 7.20, and Tables 8.7 through 8.10, in addition to much material specifically credited.

Practical Builder, Fig. 11.3.

Timber Engineering Company, material credited in Art. 10.12.

Thomas H. McKaig

Contents

6. Timber and Other Materials in Bending

7. Columns

8. Foundations and Walls

9. Connections

10. Complex Structures

11. Office Practice

1

General Principles

1.10 NOMENCLATURE

In earlier editions of this book, an attempt was made to use the same symbol for similar functions of various materials, frequently using subscripts to identify the materials. Many of these symbols had been agreed upon by a committee of the American Society of Civil Engineers and the American Concrete Institute. The recently adopted codes of the American Concrete Institute and the American Institute of Steel Construction vary widely in their use of the various letters. The introduction of a multiplicity of new subscripts makes it advisable to include here only such nomenclature as is used throughout the book, with a similar list in each article where special nomenclature is used.

A	area.
ACI	American Concrete Institute.
AISC	American Institute of Steel Construction.
ASTM	American Society for Testing and Materials.
DS	direct stress.
E	modulus of elasticity.
fem	fixed end moment.
H	horizontal thrust, altitude.
I	American Standard beam.
I	moment of inertia.
k	kips.
ksi	kips per square inch.
ksf	kips per square foot.
L	length in feet; designation for structural angle; design live load.
l	length in inches.
LLO	longer leg outstanding (steel angles).
LLV	longer leg vertical (steel angles).
M	bending moment.
M_s	static moment of section.
o.c.	on center.
P	concentrated load.
psf	pounds per square foot.
psi	pounds per square inch.
r	radius of gyration.
S	section modulus.
SLO	shorter leg outstanding (steel angles).
SLV	shorter leg vertical (steel angles).
W	total uniform load on beam.
w	load per unit length or area.
WF	wide-flange steel section.
XH	extra heavy (pipe).
XXH	double extra heavy (pipe).
Δ	total deformation; deflection.
δ	unit deformation.
ϕ	round bar (reinforcement); angle of repose.
Σ	sum of.

1.11 Definitions

It may be advisable to state a few definitions, not always within the scope of elementary textbooks, that are the result of common usage rather than technical dictation. Some of these may seem elementary, but they have been included either because they have not always been elementary, or because they mean a certain specific thing in this text.

Beam, continuous A beam which is continuous and capable of carrying moment over an interior support.

Beam, restrained A beam of which one or both ends are fixed or continuous over a support.

Beam, simple A beam freely supported at both ends, theoretically with no restraint. The restraint of a wall bearing is considered to be insufficient to change the beam from "simple" to "restrained."

Beam, spandrel A beam from column to column, carrying an exterior wall in a skeleton building.

Girder As distinguished from "beam," the heavier supporting member which carries the load of the beams into the column, as in "beam and girder construction."

Girt A secondary horizontal member in a side wall, designed to resist wind pressure.

Kip A word used to denote a load of 1,000 lb, derived from "kilo-pound."

Lintel A beam especially provided over an opening for a door or window to carry the wall over the opening.

Load, dead The weight of walls, partitions, framing, floors, roofs, and all other permanent stationary construction entering into a building.

Load, live All loads other than dead loads and wind pressure.

Modulus of rupture The unit fiber stress calculated from the beam formula $f = M/S$. In this text it is used as an extreme fiber stress in bending for a material such as plain concrete, which has a different strength in tension and in compression, and for which, because of certain phenomena in change of location of neutral axis, the fiber stress in neither tension nor compression can be used in figuring bending.

Mullion A vertical member between two sections of window sash, usually designed merely to carry wind load and not vertical load (not to be confused with *muntin*, the small member which separates lights of glass within the sash).

Redundant member A member in any frame or truss which may be omitted in the structure without affecting the ability to analyze the frame or truss by ordinary static methods (such as a counterdiagonal in a truss).

Rigid frame Any structure in a plane, made up of beams and columns, so constructed that the joints are rigidly fixed to transmit moment, and thus to reduce moment in other parts of the frame.

Skeleton construction Construction in which the load from all parts of the structure is carried to the foundation by means of beams and columns and not by walls.

Statically determinate structure A structure in which all elements may be computed by ordinary static methods of computation.

Statically indeterminate structure A structure in which all elements cannot be computed by ordinary static methods of computation.

Wall, apron That part of a skeleton building below a window sill and supported on a spandrel beam. (Used principally in reinforced concrete buildings where the entire panel from column to column is filled with window sash.)

Wall, bearing Any wall which supports any vertical load in addition to its own weight.

Wall, curtain As used in this book, a precast or prefabricated wall attached to the building structure.

Wall, nonbearing A wall which carries no load other than its own weight.

Wall, panel A nonbearing wall in skeleton construction, built between columns or piers and wholly supported at each story.

Wall, party A wall used or adapted for joint service between two buildings.

Wall, veneered A wall having a masonry facing which is not attached and bonded to the backing, and so does not form an integral part of the wall for purposes of load bearing and stability.

1.20 PROPERTIES OF MATERIALS

The design of any structural member or detail involves the consideration of two components—one a function of the material and the other a function of the shape and size of the member. Each is independent of the other. A tabulation of the properties of the various materials which go into structures is, therefore, a necessary part of any structural handbook (see Table 1.1).

The properties of materials include weight per cubic foot, allowable working stresses in tension, compression, or bearing, shear and bending, modulus of elasticity, and coefficient of expansion.

1.21 AISC Code

Since structural steel is a basic structural material, it has seemed desirable to include under the general heading of Properties of Materials some discussion of the 1961 AISC "Specifications for Design of Structural Steel for Buildings" or, as it is generally called, the AISC Code. For many years, when an engineer referred to structural steel, it was understood to be what was designated under the ASTM specifications as A7 steel; the allowable minimum yield point was set at 33,000 psi, and maximum basic design stress for beams at 20,000 psi. This basic working stress was established in 1934 and was the generally accepted standard until 1961 when, in keeping with increasing economies afforded by the use of high-strength steel, the old code was rendered obsolete. A new code was adopted which not only drastically revised the stresses which could be used with the various types of steel now produced, but also recognized possible economies inherent in new approaches to construction problems—approaches based on higher strength steel, on the increased use of

Table 1.1 Properties of Materials

Material	Weight, lb, per cu ft	Ultimate stresses except as noted				E (000,000 omitted)	Coef. of expansion per °F
		Tension, f_t	Compression, f_c	Shear, f_v	Bending, f		
Structural steel	490	See Art. 2.13.	See Art. 7.10.	See Art. 4.10.	See Art. 4.10.	29	.0000067
Reinforcing steel	490	See Art. 1.22 for allowable stresses.				29	.0000067
Wrought iron	485	Ultimate 48,000		Ultimate 40,000		29	.0000067
Cast iron	450						.0000059
Aluminum	168–176	Varies with alloy				10.3	.0000128
Brass	534	Varies with alloy				16–17	.0000104–.0000116
Bronze	481–509	Varies with alloy				14–17	.0000102–.0000118
Granite	157–187	600–1,000	7,700–60,000	2,000–4,800	1,430–5,190	5.75–8.2	.0000063–.000009
Marble	165–179	150–2,300	8,000–50,000	1,300–6,500	600–4,900	7.2–14.5	.0000036–.000016
Limestone	117–175	280–890	2,000–28,000	800–4,500	500–2,000	1.5–12.4	.0000042–.000022
Sandstone	168–180	280–500	5,000–20,000	300–3,000	700–2,300	1.9–7.7	.000005–.000012
Concrete	150	See Art. 1.22 for allowable stresses.					.0000055
Lumber (various)	25–59	See Table 1.3 for allowable stresses.					
Brick masonry	120						.0000033

welding, and on research into and acceptance of composite design, of plastic methods of design, of the use of rigid frames, and of many other phases of improved knowledge and design.

The AISC Code recognizes six grades of steel for structural use as follows, using the ASTM designation for each.

A7 —the carbon steel we have used for many years.
A373—a carbon steel developed primarily for welded bridges.
A36 —a carbon steel introduced in 1960 which, because of its greater strength for the same cost, has largely replaced and will increasingly replace A7 steel as our basic structural steel for buildings.
A242—an alloy steel of higher price with high corrosion resistance.
A440—a high-strength alloy steel for bolted or riveted structures.
A441—a high-strength alloy steel for welded structures.

For the purpose of this book from the viewpoint of economics, it is only necessary for us to consider A36 with a yield point of 36,000 psi and A440 or A441 with a basic yield point of 50,000 psi. This latter basic figure may be reduced in heavier sections in recognition of the fact that a higher strength is attained through more passes through the rolls. This factor will be discussed in various design problems in this book.

In the present structural steel market and under the impetus of the new AISC Code, the uses of the various types of steel are changing so rapidly that some of the statements included here may soon be out-of-date. Nevertheless, it seems desirable to set forth the properties and primary uses of these various grades as they now exist, primarily as an indication of the economies which may be effected by a design using perhaps several of these different steels in various parts of the same building. The specific portion of the code which is applicable to a design problem will be cited in connection with the problem solution. In general, the 1961 Code, instead of being set up on a basic working stress of 20,000 psi, is now set up on a basic working stress of a percentage of the yield point as set forth in the various chapters of this book.

The alloy steels, A440 and A441, are available at a premium cost of 17 to 30 percent above carbon steel, but since the allowable stress may be as much as 39 percent higher, it is advisable for the designer to investigate the relative economy of the members under consideration before deciding on the grade of steel to be used. As a simple tool for a preliminary investigation, the AISC has suggested the use of the relationship which they have named the "Material Cost Efficiency," or MCE—the weight ratio of carbon steel to high-strength steel, divided by the cost ratio of high-strength steel to carbon steel, or

$$\text{MCE} = \frac{W_c C_c}{W_h C_h}$$

where the subscript c refers to carbon steel and h to high-strength steel. An MCE of greater than one indicates an economy for high-strength steel; of less than one, for carbon steel.

On the basis of the percentages indicated above, it would appear that the high-strength steel with a strength increase of 40 percent for a cost increase of 20 percent or less must always be the more economical, but this is not necessarily so since other factors may influence the selection of the members. These factors will be referred to in the various articles on beams and columns throughout the book.

Wherever a problem or a table is given without notation as to the grade of steel, it will be assumed that it is based on the use of A36 steel.

1.22 ACI Code

As concrete is a basic structural material, it has seemed desirable to include under the general heading of Properties of Materials those portions of the American Concrete Institute Code which have to do with the design properties of the material. Particular portions of the Code which apply to specific design problems are quoted and referred to in the section to which they apply.

The following material from Chapter 10 of the ACI Code of 1962 may properly be placed in this article.

ALLOWABLE STRESSES—WORKING STRESS DESIGN

1000—*Notation*

f_c = compressive stress in concrete
f_c' = compressive strength of concrete (see Section 301)
n = ratio of modulus of elasticity of steel to that of concrete
v = shear stress
v_c = shear stress carried by the concrete
w = weight of concrete, lb per cu ft

1001—*General*

(*a*) For structures to be designed with reference to allowable stresses, service loads, and the accepted straight-line theory of stress and strain in flexure, the allowable stresses of this chapter shall be used, and designs shall conform to all provisions of this code except Part IV-B.

1002—*Allowable stresses in concrete*

(*a*) The stresses for flexure and bearing on all concrete designed in accordance with Part IV-A shall not exceed the values of Table 1002(a).*

(*b*) The stresses for shear shall not exceed those given in Table 1002(a) except as provided in Chapter 12.

(*c*) The allowable stresses for bond shall not exceed those given in Chapter 13.

* Reproduced as Table 1.2 of this book.

1003—*Allowable stresses in reinforcement*

Unless otherwise provided in this code, steel for concrete reinforcement shall not be stressed in excess of the following limits:

(*a*) In tension

For billet-steel or axle-steel concrete reinforcing bars of structural grade . . . 18,000 psi

For main reinforcement, 3/8 inch or less in diameter, in one-way slabs of not more than 12-ft span, 50 percent of the minimum yield strength specified by the American Society for Testing and Materials for the reinforcement used, but not to exceed . . . . 30,000 psi

For deformed bars with a yield strength of 60,000 psi or more and in sizes #11 and smaller . . . . 24,000 psi

For all other reinforcement . . . . 20,000 psi

(*b*) In compression, vertical column reinforcement spiral columns 40 percent of the minimum yield strength but not to exceed . . . . 30,000 psi

Tied columns, 80 percent of the value for spiral columns but not to exceed . . 24,000 psi

Composite and combination columns:

Structural steel sections

For ASTM A36 Steel . . . . 18,000 psi

For ASTM A7 Steel . . . . 16,000 psi

Steel pipe . . . . see limitations of Section 1406

(*c*) In compression, flexural members. For compression reinforcement in flexural members, see Section 1102.

(*d*) Spirals [yield strength stress for use in Eq. (9-1)]

Hot rolled rods, intermediate grade . . 40,000 psi

Hot rolled rods, hard grade . . . . 50,000 psi

Hot rolled rods, ASTM A432 grade and cold-drawn wire . . . . 60,000 psi

1004—*Allowable stresses—Wind and earthquake forces*

(*a*) Members subject to stresses produced by wind or earthquake forces combined with other loads may be proportioned for stresses 33⅓ percent greater than those specified in Sections 1002 and 1003, provided that the section thus required is not less than that required for the combination of dead and live load.

References made in the above sections of the Code to other sections will be quoted in those particular articles of this book to which they apply.

(Although the ACI Code of 1963 has been adopted by the members of the Institute, it does not necessarily follow that it has become part of all municipal codes, and before proceeding with his design of a reinforced concrete structure, it would be well for the designer to check its acceptability with the building department under which the work is being done.)

1.23 TIMBER

Standard practice for the use of timber has been established by the National Lumber Manufacturers Association and the various inspection bureaus and lumber associations so that stress-grade lumber is now recognized by most authorities in accordance with the NLMA Code. Commercial grades have been established in accordance with the following Code requirements.

102-STRESS-GRADE LUMBER

102-A. *Commercial grades.*

Table 1.2 ACI Table 1002(a)—Allowable Unit Stresses in Concrete

Description		Allowable stresses				
		For any strength of concrete in accordance with Section 502	For strength of concrete shown below			
			$f_c' =$ 2500 psi	$f_c' =$ 3000 psi	$f_c' =$ 4000 psi	$f_c' =$ 5000 psi
Modulus of elasticity ratio: n		$\frac{29,000,000}{w^{1.5} 33\sqrt{f_c'}}$				
For concrete weighing 145 lb per cu ft (see Section 1102)	n		10	9	8	7
Flexure: f_c						
Extreme fiber stress in compression	f_c	$0.45f_c'$	1125	1350	1800	2250
Extreme fiber stress in tension in plain concrete footings and walls	f_c	$1.6\sqrt{f_c'}$	80	88	102	113
Shear: v (as a measure of diagonal tension at a distance d from the face of the support)						
Beams with no web reinforcement*	v_c	$1.1\sqrt{f_c'}$	55*	60*	70*	78*
Joists with no web reinforcement	v_c	$1.2\sqrt{f_c'}$	61	66	77	86
Members with vertical or inclined web reinforcement or properly combined bent bars and vertical stirrups	v	$5\sqrt{f_c'}$	250	274	316	354
Slabs and footings (peripheral shear, Section 1207)*	v_c	$2\sqrt{f_c'}$	100*	110*	126*	141*
Bearing: f_c						
On full area		$0.25f_c'$	625	750	1000	1250
On one-third area or less†		$0.375f_c'$	938	1125	1500	1875

*For shear values for lightweight aggregate concrete see Section 1208.

†This increase shall be permitted only when the least distance between the edges of the loaded and unloaded areas is a minimum of one-fourth of the parallel side dimension of the loaded area. The allowable bearing stress on a reasonably concentric area greater than one-third but less than the full area shall be interpolated between the values given.

102-A-1. Lumber grades shall be specified by commercial grade names.

102-B. *Definitions.*

102-B-1. Stress-grade lumber consists of lumber classifications known as "Beams and Stringers," "Joists and Planks," and "Posts and Timbers," to each grade of which is assigned proper allowable unit stresses.

102-B-2. Beams and Stringers. Lumber of rectangular cross section, 5 or more inches thick and 8 or more inches wide, graded with respect to its strength in bending when loaded on the narrow face.

102-B-3. Joists and Planks. Lumber of rectangular cross section, 2 inches to but not including 5 inches thick and 4 or more inches wide, graded with respect to its strength in bending when loaded either on the narrow face as a joist or on the wide face as a plank.

102-B-4. Posts and Timbers. Lumber of square or approximately square cross section 5 by 5 inches and larger, graded primarily for use as posts or columns carrying longitudinal load but adapted for miscellaneous uses in which strength in bending is not especially important.

The allowable unit stresses for various commercial grades of lumber have been established as given in Table 1.3 of this book. Permissible or required variations of these stresses are established by the following Code sections.

202-A. *Continuously dry conditions.*

202-A-1. The allowable unit stresses in Table 1.3 and adjustments thereof apply to sawn lumber used under conditions continuously dry such as in most covered structures.

202-B. *Wet conditions.*

202-B-1. The allowable unit stresses in Table 1.3, and adjustments thereof, apply to lumber used under conditions where the moisture content of the wood is at or above the fiber saturation point, as when continuously submerged, except that under such conditions of use the allowable unit stresses in compression parallel to grain shall be reduced ten percent (10%), in compression perpendicular to the grain shall be reduced one third (1/3) and the values for modulus of elasticity shall be reduced one-eleventh (1/11).

203-A. *Increase in unit stresses.*

203-A-1. When the duration of the full maximum load does not exceed the period indicated, increase the allowable unit stresses in Tables 1.3, 20 and 21 as follows:

15 percent for 2 months' duration, as for snow.
25 percent for 7 days' duration.
33 1/3 percent for wind or earthquake.
100 percent for impact.

Allowable unit stresses, given in Tables 1.3, 20 and 21 for normal loading conditions, may be used without regard to impact if the stress induced by impact does not exceed the allowable unit stress for normal loading. The above increases are not cumulative. The resulting structural members shall not be smaller than required for a longer duration of loading. The provisions of this section do not apply to modulus of elasticity. These increases apply to mechanical fastenings except as otherwise noted.

204-A. *Decrease in unit stresses.*

204-A-1. Where a member is fully stressed to the maximum allowable stress for many years, either continuously or cumulatively under the condition of maximum design load, use working stresses 90 percent of those in Table 1.3. The provisions of this section do not apply to modulus of elasticity. The provisions of this section apply to mechanical fastenings except as otherwise noted.

202-D. *Fire-retardant treatment.*

202-D-1. The allowable unit stresses in Table 1.3 shall be reduced ten percent for lumber pressure impregnated with fire-retardant chemicals. The resulting stresses are subject to duration of load adjustment as set forth in sections 203 and 204.

209-A. *Surface seasoning.*

209-A-1. The allowable unit stresses for compression perpendicular to the grain in Table 1.3 assume the material will be surface seasoned when installed.

212-A. *Adjustment for seasoning.*

212-A-1. The values for modulus of elasticity in Table 1.3 assume the lumber will be surface seasoned before it is fully loaded to the maximum allowable load. With sawn members thicker than 4 inches, which season slowly, care should be exercised to avoid overload before an appreciable seasoning of the outer fibers has taken place, otherwise the values for modulus of elasticity in Table 1.3 should be reduced one-eleventh (1/11).

1.30 PROPERTIES OF SECTIONS

As has already been stated, the design of any structural member or detail involves the consideration of two components—the shape and size of the member and the material. Some of the functions of shape and size are simple to compute, such as the properties of squares and rectangles, which may be computed from the formulae in Table 1.4. Others, such as the properties of rolled-steel members, are obtainable directly from

Table 1.3 Allowable Unit Stresses for Stress-Grade Lumber

1		2	Allowable unit stresses in pounds per square inch				7
			3	4	5	6	
Species and commercial grade[1]		Rules under which graded	Extreme fiber in bending "f" and tension parallel to grain "t"[3]	Horizontal shear "H"	Compression perpendicular to grain "c ⊥"	Compression parallel to grain "c"	Modulus of elasticity "E"
CEDAR, INCENSE							
Select Dex[10]	Decking	West Coast Lumber Inspection Bureau	1,100	----------	305	--------	990,000
Commercial Dex[10]	Decking		850	----------	305	--------	990,000
CEDAR, INCENSE & WESTERN RED							
Selected Decking[19]	Decking	Western Pine Association	900	----------	240	--------	1,100,000
Commercial Decking[19]	Decking		700	----------	240	--------	1,100,000
CEDAR, WESTERN RED							
Select Dex[10]	Decking	West Coast Lumber Inspection Bureau	900	----------	240	--------	1,100,000
Commercial Dex[10]	Decking		700	----------	240	--------	1,100,000
DOUGLAS FIR, COAST REGION		West Coast Lumber Inspection Bureau					
Dense Select Structural[2]	L.F.[20]		2,050	[5-6-8]120	455	1,500	1,760,000
Select Structural	L.F.[20]		1,900	[5-6-8]120	415	1,400	1,760,000
1750 f Industrial	L.F.[20]		1,750	120	415	1,400	1,760,000
1500 f Industrial	L.F.[20]		1,500	120	390	1,200	1,760,000
1200 f Industrial	L.F.[20]		1,200	95	390	1,000	1,760,000
Dense Select Structural[2]	J.&P.[21]		2,050	[5-6-8]120	455	1,650	1,760,000
Select Structural	J.&P.[21]		1,900	[5-6-8]120	415	1,500	1,760,000
Dense Construction[2]	J.&P.[21]		1,750	[5-7-8]120	455	1,400	1,760,000
Construction	J.&P.[21]		1,500	[5-7-8]120	390	1,200	1,760,000
Standard	J.&P.[21]		1,200	[5-7-8] 95	390	1,000	1,760,000
Dense Select Structural[2]	B.&S.		2,050	[9]120	455	1,500	1,760,000
Select Structural	B.&S.		1,900	[9]120	415	1,400	1,760,000
Dense Construction[2]	B.&S.		1,750	[9]120	455	1,200	1,760,000
Construction	B.&S.		1,500	[9]120	390	1,000	1,760,000
Dense Select Structural[2]	P.&T.		1,900	[9]120	455	1,650	1,760,000
Select Structural	P.&T.		1,750	[9]120	415	1,500	1,760,000
Dense Construction[2]	P.&T.		1,500	[9]120	455	1,400	1,760,000
Construction	P.&T.		1,200	[9]120	390	1,200	1,760,000
Select Dex[10]	Decking		1,500	----------	390	--------	1,760,000
Commercial Dex[10]	Decking		1,200	----------	390	--------	1,760,000
DOUGLAS FIR		Western Pine Association					
Dense Select Structural[2]	L.F.[20]		2,050	[5-6-8]120	455	1,500	1,760,000
Dense Select Structural[2] MC 15	L.F.[20-14]		2,300	[5-6-8]125	455	1,700	1,760,000
Select Structural	L.F.[20]		1,900	[5-6-8]120	415	1,400	1,760,000
Select Structural MC 15	L.F.[20-14]		2,100	[5-6-8]125	415	1,550	1,760,000
1500 f Industrial	L.F.[20]		1,500	120	390	1,200	1,760,000
1500 f Industrial MC 15	L.F.[20-14]		1,750	125	390	1,400	1,760,000
1200 f Industrial	L.F.[20]		1,200	95	390	1,000	1,760,000
1200 f Industrial MC 15	L.F.[20-14]		1,500	110	390	1,200	1,760,000
Dense Select Structural[2]	J.&P.[21]		2,050	[5-6-8]120	455	1,650	1,760,000
Dense Select Structural[2] MC 15	J.&P.[21-14]		2,300	[5-6-8]125	455	1,850	1,760,000
Select Structural	J.&P.[21]		1,900	[5-6-8]120	415	1,500	1,760,000
Select Structural MC 15	J.&P.[21-14]		2,100	[5-6-8]125	415	1,650	1,760,000
Dense Construction[2]	J.&P.[21]		1,750	[5-7-8]120	455	1,400	1,760,000
Dense Construction[2] MC 15	J.&P.[21-14]		2,050	[5-7-8]125	455	1,600	1,760,000
Construction	J.&P.[21]		1,500	[5-7-8]120	390	1,200	1,760,000
Construction MC 15	J.&P.[21-14]		1,750	[5-7-8]125	390	1,400	1,760,000
Standard	J.&P.[21]		1,200	[5-7-8] 95	390	1,000	1,760,000
Standard MC 15	J.&P.[21-14]		1,500	[5-7-8]110	390	1,200	1,760,000
Dense Select Structural[2]	B.&S.		2,050	[9]120	455	1,500	1,760,000
Select Structural	B.&S.		1,900	[9]120	415	1,400	1,760,000
Dense Construction[2]	B.&S.		1,750	[9]120	455	1,200	1,760,000
Construction	B.&S.		1,500	[9]120	390	1,000	1,760,000
Dense Select Structural[2]	P.&T.		1,900	[9]120	455	1,650	1,760,000
Select Structural	P.&T.		1,750	[9]120	415	1,500	1,760,000
Dense Construction[2]	P.&T.		1,500	[9]120	455	1,400	1,760,000
Construction	P.&T.		1,200	[9]120	390	1,200	1,760,000
Selected Decking[19]	Decking		1,500	----------	390	--------	1,760,000
Commercial Decking[19]	Decking		1,200	----------	390	--------	1,760,000

See footnotes at end of table.

Table 1.3 Allowable Unit Stresses for Stress-Grade Lumber (*Continued*)

1		2	Allowable unit stresses in pounds per square inch				7
Species and commercial grade[1]		Rules under which graded	3 Extreme fiber in bending "f" and tension parallel to grain "t"[3]	4 Horizontal shear "H"	5 Compression perpendicular to grain "c ⊥"	6 Compression parallel to grain "c"	Modulus of elasticity "E"
FIR, WHITE							
Select Dex[10]	Decking	West Coast Lumber Inspection Bureau	1,100	—	365	—	1,210,000
Commercial Dex[10]	Decking		850	—	365	—	1,210,000
FIR, WHITE							
Selected Decking[19]	Decking	Western Pine Association	1,100	—	365	—	1,210,000
Commercial Decking[19]	Decking		850	—	365	—	1,210,000
HEMLOCK, EASTERN							
Select Structural	J.&P.[4]–B.&S.[4]	Northern Hardwood & Pine Mfrs. Association	1,300	85	360	850	1,210,000
Prime Structural	J.&P.[14-4]		1,200	60	360	775	1,210,000
Common Structural	J.&P.[14-4]		1,100	60	360	650	1,210,000
Utility Structural	J.&P.[14-4]		950	60	360	600	1,210,000
Select Structural	P.&T.		—	—	360	850	1,210,000
HEMLOCK, WEST COAST							
Select Structural	L.F.[20]	West Coast Lumber Inspection Bureau	1,600	[5-6-12]100	365	1,100	1,540,000
1500 f Industrial	L.F.[20]		1,500	100	365	1,000	1,540,000
1200 f Industrial	L.F.[20]		1,200	80	365	900	1,540,000
Select Structural	J.&P.[21]		1,600	[5-6-12]100	365	1,200	1,540,000
Construction	J.&P.[21]		1,500	[5-11-12]100	365	1,100	1,540,000
Standard	J.&P.[21]		1,200	[5-11-12] 80	365	1,000	1,540,000
Construction	B.&S.		1,500	[13]100	365	1,000	1,540,000
Construction	P.&T.		1,200	[13]100	365	1,100	1,540,000
Select Dex[10]	Decking		1,300	—	365	—	1,540,000
Commercial Dex[10]	Decking		1,000	—	365	—	1,540,000
HEMLOCK, WESTERN							
Select Structural	L.F.[20]	Western Pine Association	1,600	[5-6-12]100	365	1,100	1,540,000
Select Structural MC15	L.F.[20-14]		1,800	[5-6-12]105	365	1,200	1,540,000
1500 f Industrial	L.F.[20]		1,500	100	365	1,000	1,540,000
1500 f Industrial MC 15	L.F.[20-14]		1,650	105	365	1,150	1,540,000
1200 f Industrial	L.F.[20]		1,200	80	365	900	1,540,000
1200 f Industrial MC 15	L.F.[20-14]		1,450	90	365	1,050	1,540,000
Select Structural	J.&P.[21]		1,600	[5-6-12]100	365	1,200	1,540,000
Select Structural MC 15	J.&P.[21-14]		1,800	[5-6-12]105	365	1,300	1,540,000
Construction	J.&P.[21]		1,500	[5-11-12]100	365	1,100	1,540,000
Construction MC 15	J.&P.[21-14]		1,650	[5-11-12]105	365	1,250	1,540,000
Standard	J.&P.[21]		1,200	[5-11-12] 80	365	1,000	1,540,000
Standard MC 15	J.&P.[21-14]		1,450	[5-11-12] 90	365	1,150	1,540,000
Construction	B.&S.		1,500	[13]100	365	1,000	1,540,000
Construction	P.&T.		1,200	[13]100	365	1,100	1,540,000
Selected Decking[19]	Decking		1,300	—	365	—	1,540,000
Commercial Decking[19]	Decking		1,000	—	365	—	1,540,000
LARCH							
Dense Select Structural[2]	L.F.[20]	Western Pine Association	2,050	[5-6-8]120	455	1,500	1,760,000
Dense Select Structural[2] MC 15	L.F.[20-14]		2,300	[5-6-8]125	455	1,700	1,760,000
Select Structural	L.F.[20]		1,900	[5-6-8]120	415	1,400	1,760,000
Select Structural MC 15	L.F.[20-14]		2,100	[5-6-8]125	415	1,550	1,760,000
1500 f Industrial	L.F.[20]		1,500	120	390	1,200	1,760,000
1500 f Industrial MC 15	L.F.[20-14]		1,750	125	390	1,400	1,760,000
1200 f Industrial	L.F.[20]		1,200	95	390	1,000	1,760,000
1200 f Industrial MC 15	L.F.[20-14]		1,500	110	390	1,200	1,760,000
Dense Select Structural[2]	J.&P.[21]		2,050	[5-6-8]120	455	1,650	1,760,000
Dense Select Structural[2] MC 15	J.&P.[21-14]		2,300	[5-6-8]125	455	1,850	1,760,000
Select Structural	J.&P.[21]		1,900	[5-6-8]120	415	1,500	1,760,000
Select Structural MC 15	J.&P.[21-14]		2,100	[5-6-8]125	415	1,650	1,760,000
Dense Construction[2]	J.&P.[21]		1,750	[5-7-8]120	455	1,400	1,760,000
Dense Construction[2] MC 15	J.&P.[21-14]		2,050	[5-7-8]125	455	1,600	1,760,000
Construction	J.&P.[21]		1,500	[5-7-8]120	390	1,200	1,760,000

See footnotes at end of table.

Table 1.3 Allowable Unit Stresses for Stress-Grade Lumber (*Continued*)

1		2	3	4	5	6	7
			Allowable unit stresses in pounds per square inch				
Species and commercial grade[1]		Rules under which graded	Extreme fiber in bending "f" and tension parallel to grain "t"[3]	Horizontal shear "H"	Compression perpendicular to grain "c ⊥"	Compression parallel to grain "c"	Modulus of elasticity "E"
LARCH — Continued							
Construction MC 15	J.&P.[21-14]		1,750	[5-7-8]125	390	1,400	1,760,000
Standard	J.&P.[21]		1,200	[5-7-8] 95	390	1,000	1,760,000
Standard MC 15	J.&P.[21-14]		1,500	[5-7-8]110	390	1,200	1,760,000
Dense Select Structural[2]	B.&S.		2,050	[9]120	455	1,500	1,760,000
Select Structural	B.&S.		1,900	[9]120	415	1,400	1,760,000
Dense Construction[2]	B.&S.		1,750	[9]120	455	1,200	1,760,000
Construction	B.&S.	Western Pine Association	1,500	[9]120	390	1,000	1,760,000
Dense Select Structural[2]	P.&T.		1,900	[9]120	455	1,650	1,760,000
Select Structural	P.&T.		1,750	[9]120	415	1,500	1,760,000
Dense Construction[2]	P.&T.		1,500	[9]120	455	1,400	1,760,000
Construction	P.&T.		1,200	[9]120	390	1,200	1,760,000
Selected Decking[19]	Decking		1,500	----------	390	--------	1,760,000
Commercial Decking[19]	Decking		1,200	----------	390	--------	1,760,000
PINE, SOUTHERN[15]							
Dense Structural 86 KD[2-17-18]	2" thick only		3,000	165	455	2,250	1,760,000
Dense Structural 72 KD[2-17-18]	"		2,500	150	455	1,950	1,760,000
Dense Structural 65 KD[2-17-18]	"		2,250	135	455	1,800	1,760,000
Dense Structural 58 KD[2-17-18]	"		2,050	120	455	1,650	1,760,000
No. 1 Dense KD[2-17-18]	"		2,050	135	455	1,750	1,760,000
No. 1 KD[17]	"		1,750	135	390	1,500	1,760,000
No. 2 Dense KD[2-17-18]	"		1,750	120	455	1,300	1,760,000
No. 2 KD[17]	"		1,500	120	390	1,100	1,760,000
Dense Structural 86[2-18]	2" thick only		2,900	150	455	2,200	1,760,000
Dense Structural 72[2-18]	"		2,350	135	455	1,800	1,760,000
Dense Structural 65[2-18]	"		2,050	120	455	1,600	1,760,000
Dense Structural 58[2-18]	"		1,750	105	455	1,450	1,760,000
No. 1 Dense[2-18]	"		1,750	120	455	1,550	1,760,000
No. 1	"		1,500	120	390	1,350	1,760,000
No. 2 Dense[2-18]	"		1,400	105	455	1,050	1,760,000
No. 2	"		1,200	105	390	900	1,760,000
Dense Structural 86[2-18]	3" & 4" thick	Southern Pine Inspection Bureau	2,900	150	455	2,200	1,760,000
Dense Structural 72[2-18]	"		2,350	135	455	1,800	1,760,000
Dense Structural 65[2-18]	"		2,050	120	455	1,600	1,760,000
Dense Structural 58[2-18]	"		1,750	105	455	1,450	1,760,000
No. 1 Dense SR[2-18]	"		1,750	120	455	1,750	1,760,000
No. 1 SR	"		1,500	120	390	1,500	1,760,000
No. 2 Dense SR[2-18]	"		1,400	105	455	1,050	1,760,000
No. 2 SR	"		1,200	105	390	900	1,760,000
Dense Structural 86[2-18]	5" thick & up		[16]2,400	150	455	1,800	1,760,000
Dense Structural 72[2-18]	"		[16]2,000	135	455	1,550	1,760,000
Dense Structural 65[2-18]	"		[16]1,800	120	455	1,400	1,760,000
Dense Structural 58[2-18]	"		[16]1,600	105	455	1,300	1,760,000
No. 1 Dense SR[2-18]	"		[16]1,600	120	455	1,500	1,760,000
No. 1 SR	"		[16]1,400	120	390	1,300	1,760,000
No. 2 Dense SR[2-18]	"		[16]1,400	105	455	1,050	1,760,000
No. 2 SR	"		[16]1,200	105	390	900	1,760,000
Industrial 86 KD[17]	1", 1¼" and 1½" thick		2,600	165	390	1,950	1,760,000
Industrial 72 KD[17]	"		2,200	150	390	1,650	1,760,000
Industrial 65 KD[17]	"		2,000	135	390	1,550	1,760,000
Industrial 58 KD[17]	"		1,750	120	390	1,400	1,760,000
Industrial 50 KD[17]	"		1,500	120	390	1,100	1,760,000
Industrial 86	"		2,500	150	390	1,900	1,760,000
Industrial 72	"		2,000	135	390	1,550	1,760,000
Industrial 65	"		1,750	120	390	1,350	1,760,000
Industrial 58	"		1,500	105	390	1,250	1,760,000
Industrial 50	"		1,200	105	390	900	1,760,000
Select[10]	Decking		1,750	120	390	1,350	1,760,000
Select No. 1[10]	Decking		1,200	105	390	900	1,760,000
No. 2[10]	Decking		1,200	105	390	900	1,760,000

See footnotes at end of table.

Table 1.3 Allowable Unit Stresses for Stress-Grade Lumber (*Continued*)

1		2	Allowable unit stresses in pounds per square inch				7
			3	4	5	6	
Species and commercial grade[1]		Rules under which graded	Extreme fiber in bending "f" and tension parallel to grain "t"[3]	Horizontal shear "H"	Compression perpendicular to grain "c ⊥"	Compression parallel to grain "c"	Modulus of elasticity "E"
PINE, NORWAY							
Prime Structural	J.&P.[14-4]	Northern Hardwood & Pine Mfrs. Association	1,200	75	360	900	1,320,000
Common Structural	J.&P.[14-4]		1,100	75	360	775	1,320,000
Utility Structural	J.&P.[14-4]		950	75	360	650	1,320,000
PINE (IDAHO WHITE, LODGEPOLE, PONDEROSA and SUGAR)							
Selected Decking[19]	Decking	Western Pine Association	900	----------	305	---------	1,100,000
Commercial Decking[19]	Decking		700	----------	305	---------	1,100,000
REDWOOD							
Dense Structural[2]	J.&P.[4]–B.&S.[4]	Redwood Inspection Service	1,700	110	320	1,450	1,320,000
Heart Structural	J.&P.[4]–B.&S.[4]		1,300	95	320	1,100	1,320,000
Dense Structural[2]	P.&T.		----------	----------	320	1,450	1,320,000
Heart Structural	P.&T.		----------	----------	320	1,100	1,320,000
SPRUCE, EASTERN							
1450 f Structural Grade	J.&P.[4]	Northeastern Lumber Manufacturers Association, Inc.	1,450	110	300	1,050	1,320,000
1300 f Structural Grade	J.&P.[4]		1,300	95	300	975	1,320,000
1200 f Structural Grade	J.&P.[4]		1,200	95	300	900	1,320,000
SPRUCE, ENGELMANN							
Selected Decking[19]	Decking	Western Pine Association	750	----------	215	---------	1,100,000
Commercial Decking[19]	Decking		600	----------	215	---------	1,100,000
SPRUCE, SITKA							
Select Dex[10]	Decking	West Coast Lumber Inspection Bureau	1,100	----------	305	---------	1,320,000
Commercial Dex[10]	Decking		850	----------	305	---------	1,320,000

[1] Abbreviations: (For description of classification of material, see section 102-B) J&P = Joists and Planks; B&S = Beams and Stringers; P&T = Posts and Timbers; LF = Light Framing; KD = See Note 17; SR = Stress Rated.

[2] These grades meet the requirements for density.

[3] In tension members the slope of grain limitations applicable to the middle portion of the length of the Joist and Plank and Beam and Stringer grades used shall apply throughout the length of the piece. This note does not apply to Southern Pine as indicated in Note 15.

[4] The allowable unit stresses for tension parallel to grain "t" and for compression parallel to grain "c" given for these Joist and Plank and Beam and Stringer grades are applicable when the following additional provisions are applied to the grades:

The sum of the sizes of all knots in any 6 inches of the length of the piece shall not exceed twice the maximum permissible size of knot. Two knots of maximum permissible size shall not be within the same 6 inches of length of any face.

[5] Value applies to pieces used as planks.

[6] Value applies to 2″ thick pieces of Select Structural grade used as joists.

[7] For 2″ thick pieces of Construction, Standard, Structural and Standard Structural grades used as joists:
- H = 120 when length of split is approximately equal to ½ the width of piece.
- H = 100 when length of split is approximately equal to the width of piece.
- H = 70 when length of split is approximately equal to 1½ times width of piece.

[8] For 3″ thick pieces of Select Structural, Construction, Standard, and Standard Structural Grades used as joists:
- H = 120 when length of split is approximately 2¼″.
- H = 80 when length of split is approximately 4½″, and for 4″ thick pieces of Select Structural Construction, Standard and Standard Structural grades used as joists:
- H = 120 when length of split is approximately 3″.
- H = 80 when length of split is approximately 6″.

[9] For Beams and Stringers and for Posts and Timbers:
- H = 120 when length of split is equal to ½ the nominal narrow face dimension.
- H = 100 when length of split is equal to the nominal narrow face dimension.
- H = 80 when length of split is equal to 1½ times the nominal narrow face dimension.

NOTE: Values for lengths of split other than those given in Notes 7, 8 and 9 are proportionate.

[10] These grades cover 3-inch and 4-inch nominal thickness, double tongued and grooved material of 6-inch nominal width. Stresses recommended are limited to where material is used as planks.

[11] For 2″ thick pieces of Construction and Standard Grades used as joists:
- H = 100 when length of split is approximately equal to ½ the width of piece.
- H = 80 when length of split is approximately equal to the width of piece.
- H = 60 when length of split is approximately equal to 1½ times width of piece.

[12] For 3″ thick pieces of Select Structural, Construction and Standard grades used as joists:
- H = 100 when length of split is approximately 2¼″.
- H = 70 when length of split is approximately 4½″ and for 4″ thick pieces of Select Structural, Construction and Standard grade used as joists:
- H = 100 when length of split is approximately 3″.
- H = 70 when length of split is approximately 6″.

[13] For Beams and Stringers and for Posts and Timbers:
- H = 100 when length of split is equal to ¾ the nominal narrow face dimension.
- H = 90 when length of split is equal to the nominal narrow face dimension.
- H = 70 when length of split is equal to 1½ times the nominal narrow face dimension.

Table 1.3 Allowable Unit Stresses for Stress-Grade Lumber (*Continued*)

NOTE: Value for lengths of splits other than those given in Notes 11, 12, and 13 are proportionate.

[14] These grades applicable to 2″ thickness only.

[15] All stress-grades under the 1960 Standard Grading Rules are all-purpose grades and apply to all sizes. Pieces so graded may be cut to shorter lengths without impairment of the stress rating of the shorter pieces.

Grade restrictions provided by the 1960 Standard Grading Rules apply to the entire length of the piece, and each piece is suitable for use in continuous spans, over double spans or under concentrated loads without regrading for special shear or other special stress requirements.

The following variations apply to the provisions of paragraph 202-B for lumber in service under wet conditions or where the moisture content is at or above fiber saturation point, as when continuously submerged, (a) the allowable unit stresses in bending, tension parallel to grain and horizontal shear shall be limited in all thicknesses to the stresses indicated for thicknesses of 5″ and up; (b) the allowable unit stresses for compression parallel to grain shall be limited to the stresses indicated for thicknesses of 5″ and up reduced by 10%; (c) the allowable unit stresses for compression perpendicular to grain shall be reduced one-third; and (d) the values for modulus of elasticity shall be reduced one-eleventh.

[16] These stresses apply for loading either on narrow face or on wide face, which is an exception to sections 102-B-1 and 205-B.

[17] KD = Kiln dried in accordance with the provisions of paragraphs 219 and 220 of the 1960 Standard Grading Rules.

[18] Longleaf may be specified by substituting "Longleaf" for "Dense" in the grade name, and when so specified the same allowable stresses shall apply.

[19] The grades apply for tongued and grooved material 2″ to 4″ in nominal thickness and 6″ or more in nominal width. Stresses recommended are limited to where the lumber is used as planks.

[20] The allowable unit stresses listed apply to lumber 2″ to 4″ thick and 3″ to 4″ wide.

[21] The allowable unit stresses listed apply to lumber 2″ to 4″ thick and 6″ wide and wider.

Tables 1.6 to 1.11. Still others must be computed by the more complex methods presented in Art. 1.31.

The elements or properties of sections are certain mathematical functions of the dimensions and shape of a homogeneous section, commonly used in structural calculations. The functions usually listed as elements or properties are (1) area, (2) position of neutral axis or center of gravity, (3) moment of inertia, (4) section modulus, and (5) radius of gyration. The properties of sections commonly used are shown in Table 1.4, taken from the handbook of the American Institute of Steel Construction (1947 edition).

1.31 Properties of compound sections

It is frequently necessary in structural design to compute the functions of compound sections. This consists of the application of several simple formulae to the elements of simple sections and is always best performed by means of a tabulation.

a The computation of position of neutral axis or center of gravity of a composite cross section, such as an angle, may be found from the following formula (Fig. 1.1):

$$\bar{y} = \frac{A_1 y_1 + A_2 y_2 + A_3 y_3 + \cdots A_n y_n}{A_1 + A_2 + A_3 + \cdots A_n} \tag{1.1}$$

where A_1, A_2, A_3, etc., are the component areas involved, and y_1, y_2, y_3, etc., are the distances from the various neutral axes of the component areas to a known reference line (such as the bottom leg of an angle). Then the value of $\bar{y}$ calculated from this formula is the distance from this *same reference line* to the center of gravity of the compound section.

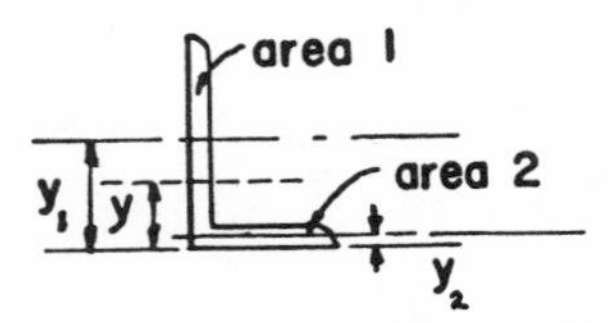

Fig. 1.1 Center of Gravity of Angles

Problem Find the distance of the center of gravity from the back of the short leg of a 6 × 4 × ½-in. *L*, disregarding all fillets.

Area A	y	Ay
6 × ½ = 3.00	3.00	9.00
3½ × ½ = 1.75	0.25	0.44
4.75		9.44

$$\bar{y} = \frac{9.44}{4.75} = 1.99 \text{ in.}$$

This checks exactly the figure y for this angle from Table 1.11.

b Ordinarily the center of gravity required is much more complex than indicated by the above problem, and it may involve the use of negative as well as positive areas. In using the above formula for $\bar{y}$, any hole is treated as a minus area; that is, area that is not present, area subtracted from the whole. A simple application of this principle would be to the problem just given. Instead of two small positive areas as shown, the cross section might be assumed to consist of one large (6 × 4-in.) *positive* rectangle and one smaller (3½ × 5½-in.) *negative* rectangle, yielding the same result. Since the position of the reference line is an arbitrary one, some areas may extend below it, in which case y distances to the neutral axes of these areas are minus. It should be noted that the conventional rules for multiplication of signs are to be followed in computing the products A_1y_1, A_2y_2, etc.; that is, minus A_1 times minus y_1 distance becomes plus A_1y_1, etc.

Problem To illustrate the use of Formula (1.1) in a more complex case, locate the center of gravity of the crane girder section shown in Fig. 1.2. As a reference axis, we shall use a horizontal line through the center of the I beam.

Item	A	y	Ay
1. Beam	+24.71	0	0
2. Channel	+9.90	+12.96	+128.3
3. Bottom plate	+6.00	−13.6	−81.6
4. Top holes	−1.8	+13.24	−23.8
5. Bottom holes	−1.97	−13.29	+26.2
	+36.84		+49.1

Therefore, $\bar{y} = \dfrac{+49.1}{+36.84} = +1.33$ in. (above reference line).

Explanation of terms:

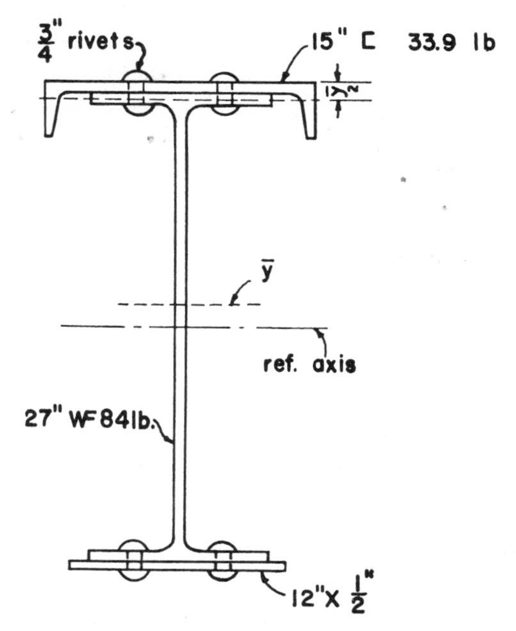

Fig. 1.2 Center of Gravity—Compound Section

If a problem in an examination is so worded that specific mention is made of "net section," the holes should be taken as negative areas, as has been done in the last paragraph. Otherwise gross section may be used.

$$y_2 = \frac{\text{beam depth}}{2} + (\text{web of channel} - x \text{ of channel}) = 13.35 + 0.4 - 0.79$$

$$y_3 = \frac{\text{beam depth}}{2} + \frac{\text{plate thickness}}{2} = 13.35 + 0.25$$

$$y_4 = \frac{\text{beam depth}}{2} + \text{web of channel} - \frac{\text{rivet length}}{2} = 13.75 - 0.51$$

$$y_5 = \frac{\text{beam depth}}{2} + \text{plate thickness} - \frac{\text{rivet length}}{2} = 13.85 - 0.56$$

$A_4 = 2 \times 0.875 \times 1.03 = 1.80$ sq in. (see Art. 2.13*a*)
$A_5 = 2 \times 0.875 \times 1.13 = 1.97$ sq in. (see Art. 2.13*a*)

Therefore, the center of gravity of the compound section is 1.33 in. *above* the center of the 27-in. beam.

c Formerly in computing properties of cover plated beams and plate girders, the AISC Code required that net section be used. Although this Code has now been revised so that properties are based on gross section, this provision has not been generally accepted. This fact should be looked into before computing properties in any problem.

There are many problems, such as those involving combined footings, in which it is necessary to find the center of gravity of several loads or pressures instead of areas. The principle is the same except that weights or pressures are substituted for areas. Problems of this type are given in Art. 8.12.

d The computation of the moment of inertia of a compound section about any reference axis is an additive process, taking into account all component areas. The moment of inertia of any single component area about the reference axis is the sum of two terms: the moment of inertia of the area about its own centroidal axis parallel to the reference line; and a "transfer" term equal to Ad^2, in which A is the area of the component part and d is the distance between the reference axis and the centroidal axis of the area in question. This summation including all components is expressed algebraically:

$$I = (I_1 + A_1d_1^2) + (I_2 + A_2d_2^2) + \cdots (I_n + A_nd_n^2) \quad (1.2)$$

e The moment of inertia may be taken about any line as an axis, but ordinarily the only moment of inertia in which we are interested is about the centroidal or center of gravity axis. The moment of inertia about the gravity axis of any cross section is the least moment of inertia for that section, and may be determined by application of the fundamental principles expressed by Formulae (1.1) and (1.2). The center of gravity is first determined as in the preceding problem, then a horizontal line through this point is arbitrarily taken as the reference axis, and Formula (1.2) is applied as shown in the following section.

f The solution of an unsymmetrical member, a rather common type of problem frequently given in state license examinations, is given below. Because of the complexity of the problem, it should always be tabulated following the procedure outlined above.

Problem The top chord of a bridge truss is made up of two 15-in. channels 40, each with a side plate 12 × 3/8 in., and with one cover plate 22 × 1/2 in. Disregarding rivet holes, compute the moment of inertia through the center of gravity. (From a state license examination. Refer to Art. 1.32 for properties of steel beams and see Fig. 1.3.)

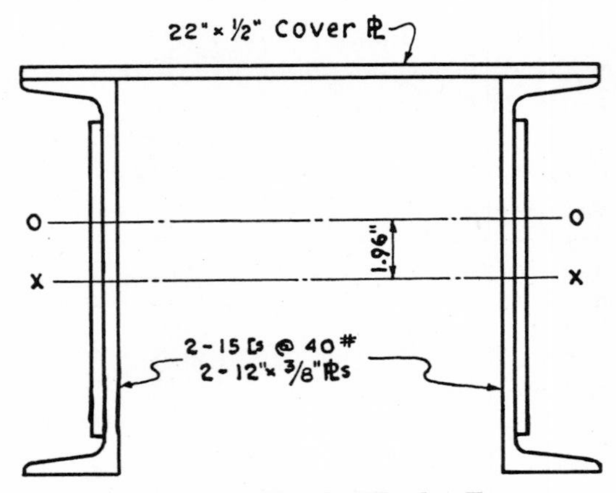

Fig. 1.3 Top Chord of Bridge Truss

The problem resolves itself into two steps, the determination of the center of gravity of the compound section, and the computation of the moment of inertia, first through a given reference axis, and then translated to the computed gravity axis.

The entire problem may be set up in one tabulation. We will use the center line of the 15-in. channels as the given reference axis.

Item	Area A	y	Product Ay	I of member	Ay^2
Two 15-in. channels 40	23.4	0	0	692.6	0
Two plates 12 × 3/8 in.	9.0	0	0	108	0
Cover plate 22 × 1/2 in.	11.0	7.75	85.25	0.23	660.69
	43.4		85.25	800.83	660.69

$$\bar{y} = +\frac{85.25}{43.4} = +1.96$$

$$\begin{aligned} \text{Total } I \text{ about reference axis} &= 660.69 + 800.83 = 1{,}461.52 \\ \text{Transfer term } Ay^2 &= 85.25 \times 1.96 = \phantom{0{,}}167.09 \\ I - A\bar{y}^2 &= 1{,}294.43 \end{aligned}$$

Table 1.4 Properties of Sections

SQUARE

Axis of moments through center

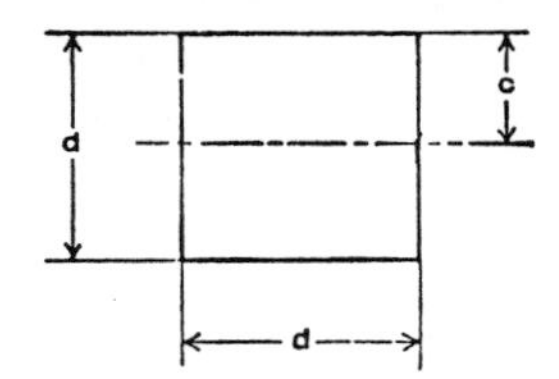

$A = d^2$

$c = \frac{d}{2}$

$I = \frac{d^4}{12}$

$S = \frac{d^3}{6}$

$r = \frac{d}{\sqrt{12}} = .288675\ d$

SQUARE

Axis of moments on base

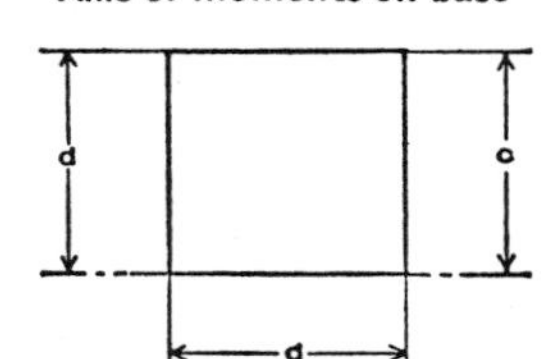

$A = d^2$

$c = d$

$I = \frac{d^4}{3}$

$S = \frac{d^3}{3}$

$r = \frac{d}{\sqrt{3}} = .577350\ d$

SQUARE

Axis of moments on diagonal

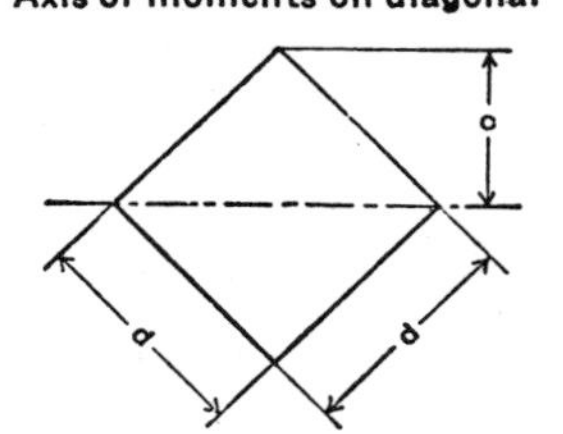

$A = d^2$

$c = \frac{d}{\sqrt{2}} = .707107\ d$

$I = \frac{d^4}{12}$

$S = \frac{d^3}{6\sqrt{2}} = .117851\ d^3$

$r = \frac{d}{\sqrt{12}} = .288675\ d$

RECTANGLE

Axis of moments through center

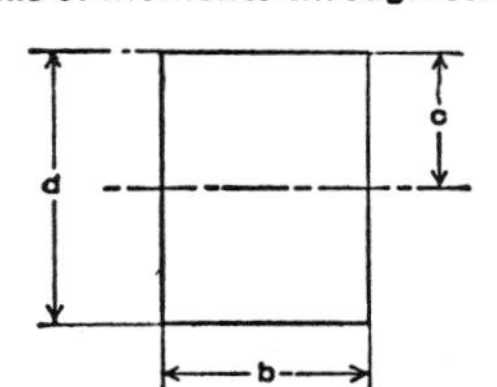

$A = bd$

$c = \frac{d}{2}$

$I = \frac{bd^3}{12}$

$S = \frac{bd^2}{6}$

$r = \frac{d}{\sqrt{12}} = .288675\ d$

Table 1.4 Properties of Sections (*Continued*)

RECTANGLE

Axis of moments on base

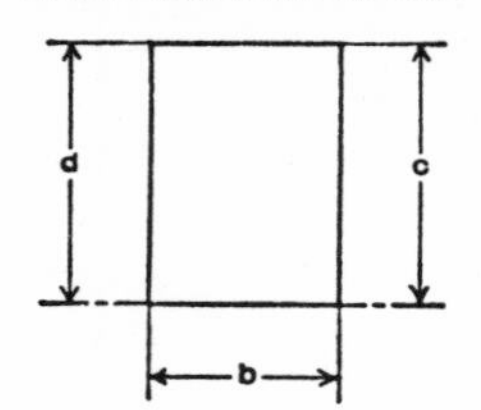

$A = bd$

$c = d$

$I = \frac{bd^3}{3}$

$S = \frac{bd^2}{3}$

$r = \frac{d}{\sqrt{3}} = .577350\ d$

RECTANGLE

Axis of moments on diagonal

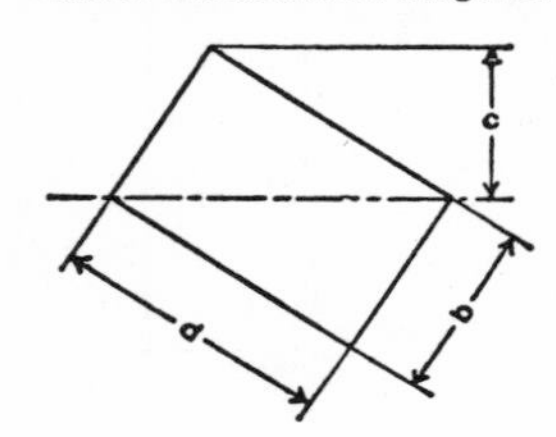

$A = bd$

$c = \frac{bd}{\sqrt{b^2 + d^2}}$

$I = \frac{b^3d^3}{6\,(b^2 + d^2)}$

$S = \frac{b^2d^2}{6\sqrt{b^2 + d^2}}$

$r = \frac{bd}{\sqrt{6\,(b^2 + d^2)}}$

RECTANGLE

Axis of moments any line through center of gravity

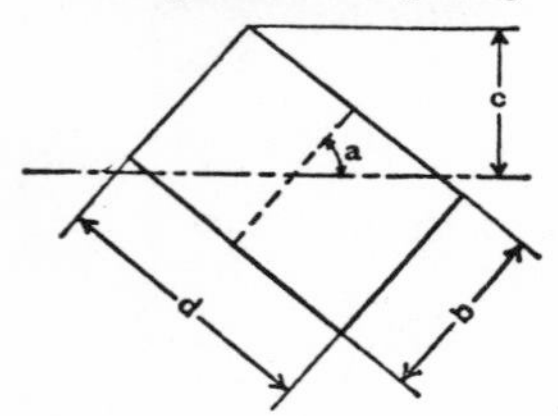

$A = bd$

$c = \frac{b \sin a + d \cos a}{2}$

$I = \frac{bd\,(b^2 \sin^2 a + d^2 \cos^2 a)}{12}$

$S = \frac{bd\,(b^2 \sin^2 a + d^2 \cos^2 a)}{6\,(b \sin a + d \cos a)}$

$r = \sqrt{\frac{b^2 \sin^2 a + d^2 \cos^2 a}{12}}$

HOLLOW RECTANGLE

Axis of moments through center

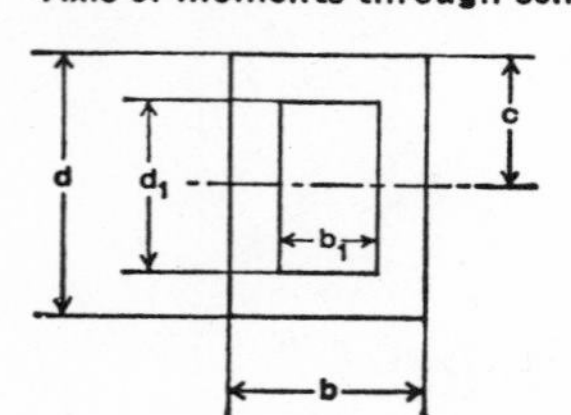

$A = bd - b_1d_1$

$c = \frac{d}{2}$

$I = \frac{bd^3 - b_1d_1^3}{12}$

$S = \frac{bd^3 - b_1d_1^3}{6d}$

$r = \sqrt{\frac{bd^3 - b_1d_1^3}{12\,A}}$

Table 1.4 Properties of Sections (*Continued*)

EQUAL RECTANGLES

Axis of moments through center of gravity

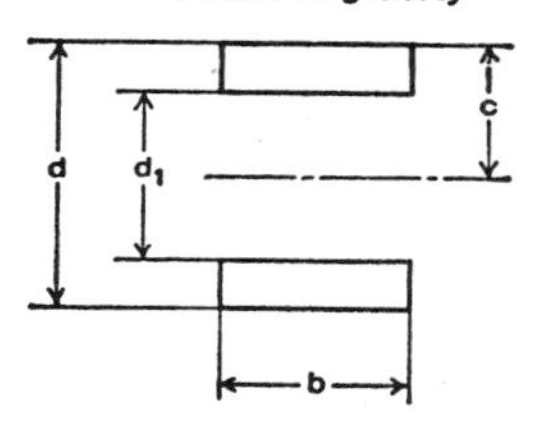

$$A = b(d - d_1)$$

$$c = \frac{d}{2}$$

$$I = \frac{b(d^3 - d_1^3)}{12}$$

$$S = \frac{b(d^3 - d_1^3)}{6d}$$

$$r = \sqrt{\frac{d^3 - d_1^3}{12(d - d_1)}}$$

UNEQUAL RECTANGLES

Axis of moments through center of gravity

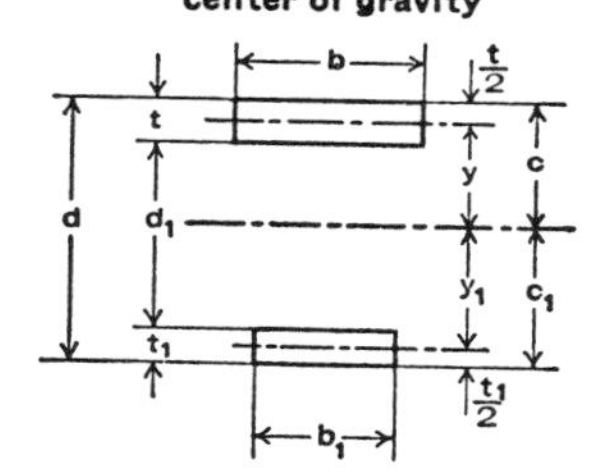

$$A = bt + b_1t_1$$

$$c = \frac{\frac{1}{2}bt^2 + b_1t_1(d - \frac{1}{2}t_1)}{A}$$

$$I = \frac{bt^3}{12} + bty^2 + \frac{b_1t_1^3}{12} + b_1t_1y_1^2$$

$$S = \frac{I}{c} \qquad S_1 = \frac{I}{c_1}$$

$$r = \sqrt{\frac{I}{A}}$$

TRIANGLE

Axis of moments through center of gravity

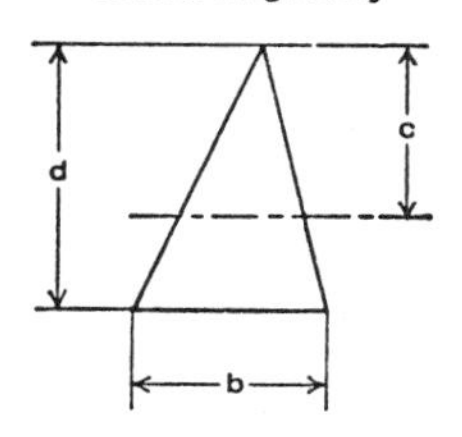

$$A = \frac{bd}{2}$$

$$c = \frac{2d}{3}$$

$$I = \frac{bd^3}{36}$$

$$S = \frac{bd^2}{24}$$

$$r = \frac{d}{\sqrt{18}} = .235702\ d$$

TRIANGLE

Axis of moments on base

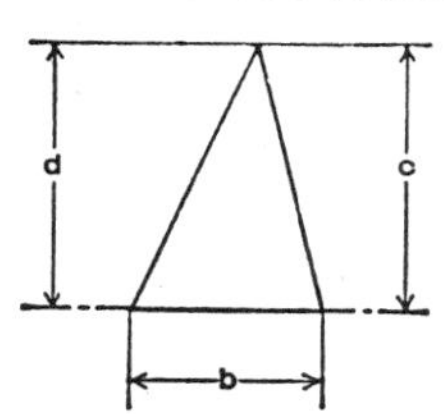

$$A = \frac{bd}{2}$$

$$c = d$$

$$I = \frac{bd^3}{12}$$

$$S = \frac{bd^2}{12}$$

$$r = \frac{d}{\sqrt{6}} = .408248\ d$$

Table 1.4 Properties of Sections (*Continued*)

TRAPEZOID

Axis of moments through center of gravity

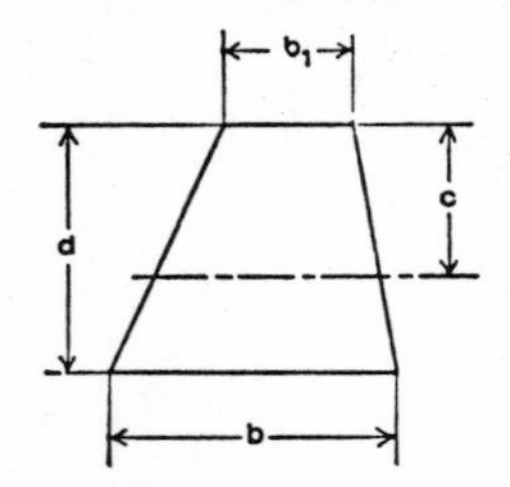

$$A = \frac{d(b + b_1)}{2}$$

$$c = \frac{d(2b + b_1)}{3(b + b_1)}$$

$$I = \frac{d^3 (b^2 + 4bb_1 + b_1^2)}{36 (b + b_1)}$$

$$S = \frac{d^2 (b^2 + 4bb_1 + b_1^2)}{12 (2b + b_1)}$$

$$r = \frac{d}{6(b + b_1)} \sqrt{2 (b^2 + 4bb_1 + b_1^2)}$$

CIRCLE

Axis of moments through center

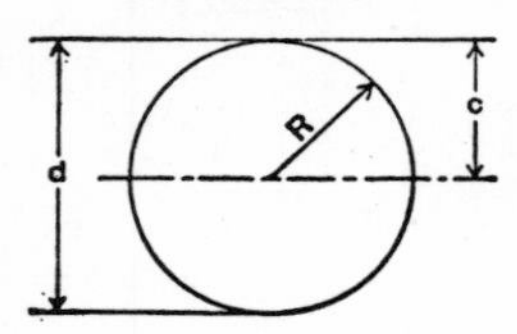

$$A = \frac{\pi d^2}{4} = \pi R^2 = .785398\ d^2 = 3.141593\ R^2$$

$$c = \frac{d}{2} = R$$

$$I = \frac{\pi d^4}{64} = \frac{\pi R^4}{4} = .049087\ d^4 = .785398\ R^4$$

$$S = \frac{\pi d^3}{32} = \frac{\pi R^3}{4} = .098175\ d^3 = .785398\ R^3$$

$$r = \frac{d}{4} = \frac{R}{2}$$

HOLLOW CIRCLE

Axis of moments through center

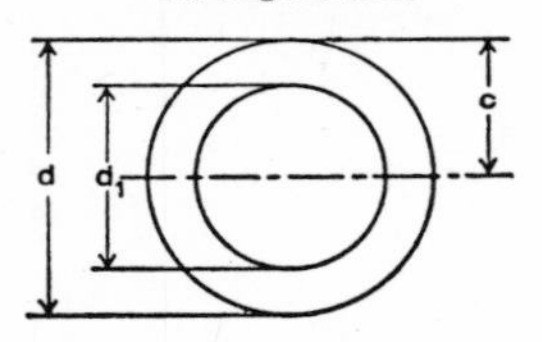

$$A = \frac{\pi(d^2 - d_1^2)}{4} = .785398\ (d^2 - d_1^2)$$

$$c = \frac{d}{2}$$

$$I = \frac{\pi(d^4 - d_1^4)}{64} = .049087\ (d^4 - d_1^4)$$

$$S = \frac{\pi(d^4 - d_1^4)}{32d} = .098175\ \frac{d^4 - d_1^4}{d}$$

$$r = \frac{\sqrt{d^2 + d_1^2}}{4}$$

HALF CIRCLE

Axis of moments through center of gravity

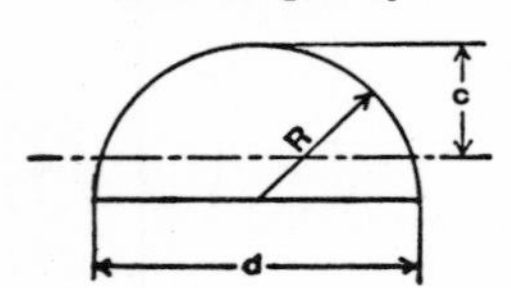

$$A = \frac{\pi R^2}{2} = 1.570796\ R^2$$

$$c = R\left(1 - \frac{4}{3\pi}\right) = .575587\ R$$

$$I = R^4\left(\frac{\pi}{8} - \frac{8}{9\pi}\right) = .109757\ R^4$$

$$S = \frac{R^3}{24}\ \frac{(9\pi^2 - 64)}{(3\pi - 4)} = .190687\ R^3$$

$$r = R\,\frac{\sqrt{9\pi^2 - 64}}{6\pi} = .264336\ R$$

Table 1.4 Properties of Sections (*Continued*)

PARABOLA

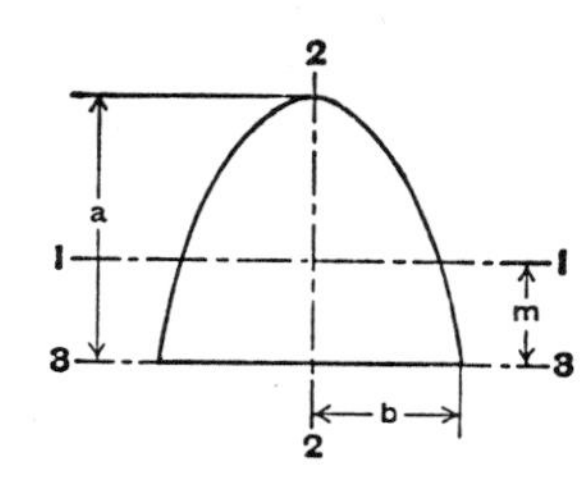

$A = \frac{4}{3}ab$

$m = \frac{2}{5}a$

$I_1 = \frac{16}{175}a^3b$

$I_2 = \frac{4}{15}ab^3$

$I_3 = \frac{32}{105}a^3b$

HALF PARABOLA

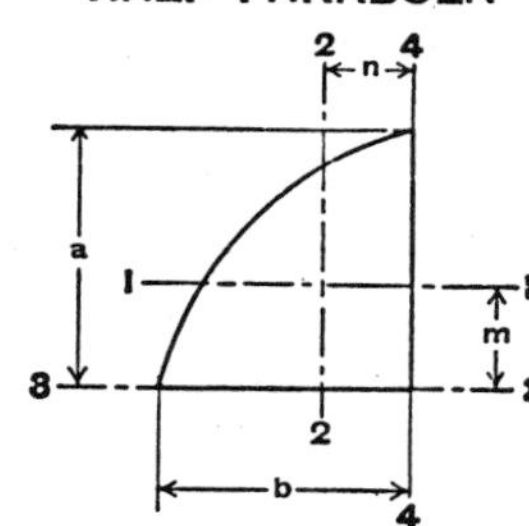

$A = \frac{2}{3}ab$

$m = \frac{2}{5}a$

$n = \frac{3}{8}b$

$I_1 = \frac{8}{175}a^3b$

$I_2 = \frac{19}{480}ab^3$

$I_3 = \frac{16}{105}a^3b$

$I_4 = \frac{2}{15}ab^3$

COMPLEMENT OF HALF PARABOLA

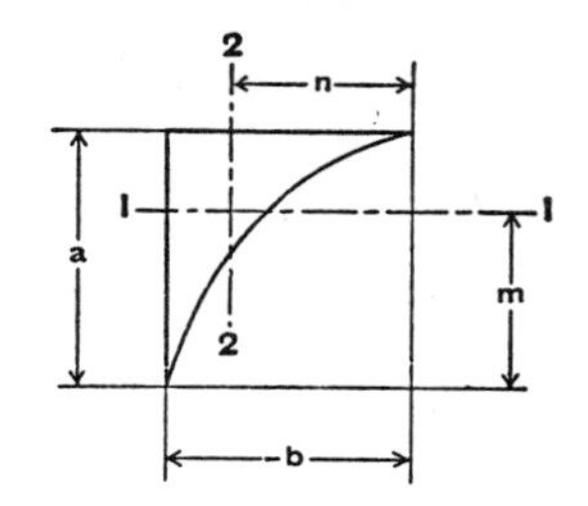

$A = \frac{1}{3}ab$

$m = \frac{7}{10}a$

$n = \frac{3}{4}b$

$I_1 = \frac{37}{2100}a^3b$

$I_2 = \frac{1}{80}ab^3$

PARABOLIC FILLET IN RIGHT ANGLE

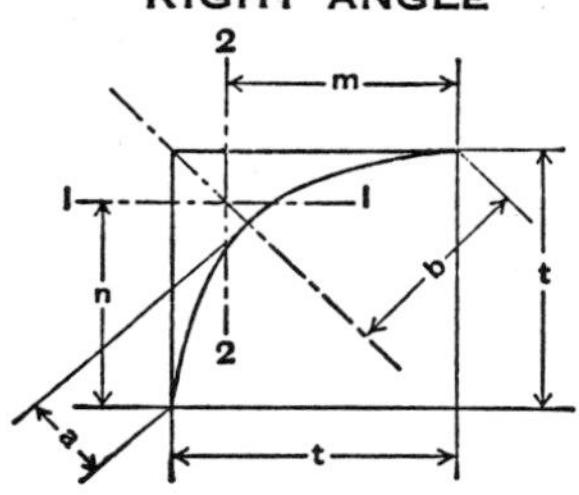

$a = \frac{t}{2\sqrt{2}}$

$b = \frac{t}{\sqrt{2}}$

$A = \frac{1}{6}t^2$

$m = n = \frac{4}{5}t$

$I_1 = I_2 = \frac{11}{2100}t^4$

Table 1.4 Properties of Sections (*Continued*)

REGULAR POLYGON

Axis of moments through center

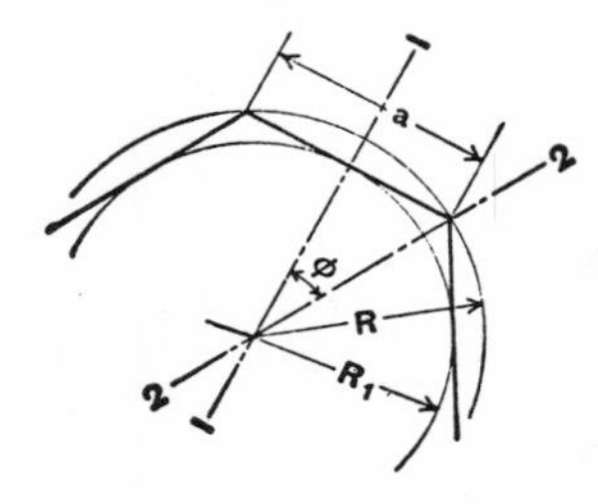

n = Number of sides

$\phi = \frac{180°}{n}$

$a = 2\sqrt{R^2 - R_1^2}$

$R = \frac{a}{2 \sin \phi}$

$R_1 = \frac{a}{2 \tan \phi}$

$A = \frac{1}{4} na^2 \cot \phi = \frac{1}{2} nR^2 \sin 2\phi = nR_1^2 \tan \phi$

$I_1 = I_2 = \frac{A(6R^2 - a^2)}{24} = \frac{A(12R_1^2 + a^2)}{48}$

$r_1 = r_2 = \sqrt{\frac{6R^2 - a^2}{24}} = \sqrt{\frac{12R_1^2 + a^2}{48}}$

ANGLE

Axis of moments through center of gravity

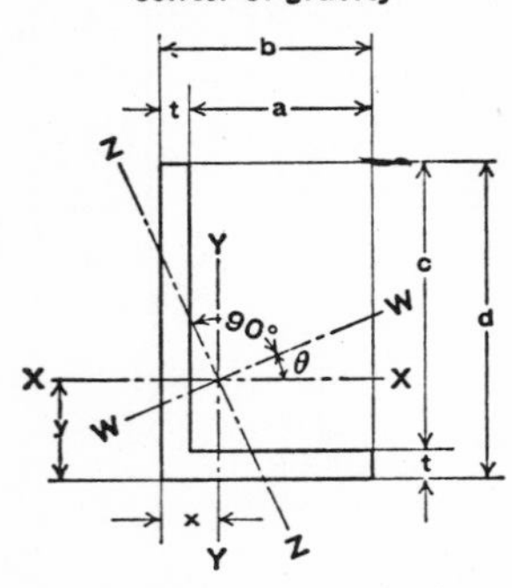

Z-Z is axis of minimum I

$\tan 2\theta = \frac{2K}{I_Y - I_X}$

$A = t(b + c) \quad x = \frac{b^2 + ct}{2(b + c)} \quad y = \frac{d^2 + at}{2(b + c)}$

K = Product of Inertia about X-X & Y-Y

$= \mp \frac{abcdt}{4(b + c)}$

$I_X = \frac{1}{3}\left(t(d - y)^3 + by^3 - a(y - t)^3 \right)$

$I_Y = \frac{1}{3}\left(t(b - x)^3 + dx^3 - c(x - t)^3 \right)$

$I_Z = I_X \sin^2\theta + I_Y \cos^2\theta + K \sin 2\theta$

$I_W = I_X \cos^2\theta + I_Y \sin^2\theta - K \sin 2\theta$

K is negative when heel of angle, with respect to c. g., is in 1st or 3rd quadrant, positive when in 2nd or 4th quadrant.

BEAMS AND CHANNELS

Transverse force oblique through center of gravity

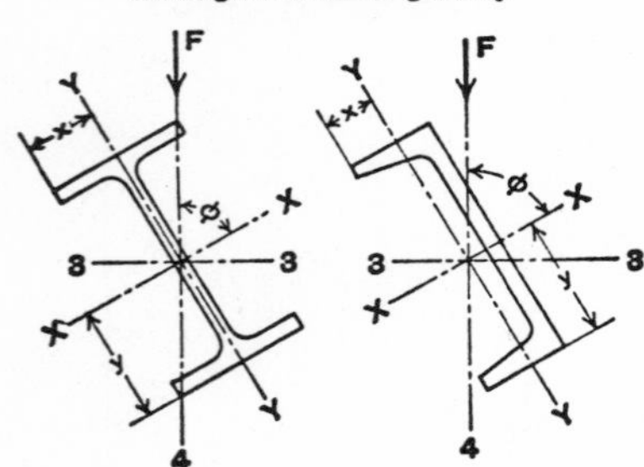

$I_3 = I_X \sin^2\phi + I_Y \cos^2\phi$

$I_4 = I_X \cos^2\phi + I_Y \sin^2\phi$

$f = M \left(\frac{y}{I_X} \sin\phi + \frac{x}{I_Y} \cos\phi \right)$

where M is bending moment due to force F.

g The section modulus is a property of the section used in the computation of a beam. Its definition is best stated by formula

$$S = \frac{I}{c}$$

in which c is the distance from the neutral axis to the extreme outermost fiber, the fiber farthest from the neutral axis. In a completely symmetrical beam this distance c is half the depth.

For the preceding problem, c is either

$$7.5 + 1.96 = 9.46$$

or

$$15.5 - 9.46 = 6.04$$

Table 1.5 Properties of Pipe and Tubing

(*a*) Pipe sections for use as columns

Designation	OD, in.	Wall thickness, in.	Weight per ft, lb	A	I	S	r	Designation	OD, in.	Wall thickness, in.	Weight per ft, lb	A	I	S	r
2½″ Std	2.875	.203	5.79	1.70	1.53	1.06	.95	4″ XXH	4.5	.674	27.54	8.10	15.28	6.82	1.37
2½″ XH	2.875	.276	7.66	2.25	1.92	1.34	.92	5″ Std	5.563	.258	14.62	4.30	15.16	6.46	1.88
2½″ XXH	2.875	.552	13.70	4.03	2.87	2.00	.84	5″ XH	5.563	.375	20.78	6.11	20.67	7.44	1.84
3″ Std	3.5	.216	7.58	2.23	3.02	1.72	1.16	5″ XXH	5.563	.750	38.55	11.34	33.64	8.50	1.72
3″ XH	3.5	.300	10.25	3.02	3.89	2.22	1.14	6″ Std	6.625	.280	18.97	5.58	28.14	8.51	2.25
3″ XXH	3.5	.600	18.58	5.47	5.99	3.38	1.05	6″ XH	6.625	.432	28.57	8.41	40.49	12.30	2.20
3½″ Std	4	.226	9.11	2.68	4.79	2.39	1.34	6″ XXH	6.625	.864	53.16	15.64	66.33	20.10	2.06
3½″ XH	4	.318	12.51	3.68	6.28	3.14	1.31	8″ Std	8.625	.277	24.70	7.27	63.35	15.45	2.95
3½″ XXH	4	.636	22.85	6.72	9.85	4.92	1.21	8″ Std	8.625	.322	28.55	8.40	72.49	16.80	2.94
4″ Std	4.5	.237	10.79	3.17	7.24	3.22	1.51	8″ XH	8.625	.500	43.39	12.76	105.70	24.50	2.88
4″ XH	4.5	.337	14.98	4.41	9.61	4.26	1.48	8″ XXH	8.625	.875	72.42	21.30	162.00	37.70	2.76

(*b*) Properties of square tubing

Size, in.	Wall, in.	Weight per ft, lb	A	I	S	r	Size, in.	Wall, in.	Weight per ft, lb	A	I	S	r
1 × 1	.095	1.09	.321	.042	.084	.362	5 × 5	.188	11.86	3.49	13.21	5.28	1.95
	.133	1.41	.415	.048	.097	.342		.250	15.42	4.54	16.60	6.64	1.91
2 × 2	.110	2.69	.791	.457	.457	.760		.375	21.94	6.45	21.95	8.78	1.84
	.125	3.04	.893	.508	.508	.754		.500	27.68	8.14	25.52	10.21	1.77
	.154	3.65	1.08	.591	.591	.742	6 × 6	.188	14.41	4.24	23.50	7.83	2.35
	.188	4.31	1.27	.667	.667	.725		.250	18.82	5.54	29.85	9.95	2.32
2½ × 2½	.141	4.32	1.27	1.15	.920	.951		.375	27.04	7.95	40.44	13.48	2.25
	.188	5.59	1.64	1.42	1.14	.930		.500	34.48	10.14	48.38	16.13	2.18
	.250	7.10	2.09	1.68	1.35	.898	7 × 7	.188	16.85	4.96	37.70	10.77	2.76
3 × 3	.155	5.78	1.70	2.25	1.50	1.15		.250	22.04	6.48	48.05	13.73	2.72
	.188	6.86	2.02	2.60	1.73	1.13		.375	31.73	9.33	65.54	18.73	2.65
	.250	8.80	2.59	3.15	2.10	1.10		.500	40.55	11.93	78.91	22.55	2.57
3½ × 3½	.156	6.88	2.02	3.71	2.12	1.35	8 × 8	.250	25.44	7.48	73.38	18.35	3.13
	.188	8.14	2.39	4.29	2.45	1.34		.375	36.83	10.83	101.46	25.37	3.06
	.250	10.50	3.09	5.28	3.02	1.31		.500	47.35	13.93	124.08	31.02	2.98
	.313	12.69	3.73	6.08	3.48	1.28	10 × 10	.250	32.23	9.48	147.89	29.58	3.95
4 × 4	.188	9.31	2.74	6.47	3.23	1.54		.375	47.03	13.83	208.31	41.64	3.88
	.250	12.02	3.54	7.99	3.99	1.50		.500	60.95	17.93	259.81	51.96	3.81
	.313	14.52	4.27	9.20	4.60	1.47							
	.375	16.84	4.95	10.15	5.08	1.43							

Therefore, $c = 9.46$, the greater of the two, and the section modulus

$$S = \frac{1{,}294.43}{9.46} = 136.83$$

h The radius of gyration is a property of the section used in computing its stiffness as a column and may be defined by the formula

$$r = \sqrt{\frac{I}{A}}$$

For the section about the x axis in the preceding problem

$$r = \sqrt{\frac{1{,}294.43}{43.4}} = \sqrt{29.82} = 5.46$$

Table 1.5 Properties of Pipe and Tubing (*Continued*)

(*c*) Properties of rectangular steel tubing

Size, in. Axis y-y	Size, in. Axis x-x	Wall, in.	Weight per ft, lb	A	I_y	S_y	r_y	I_x	S_x	r_x
3	2	.141	4.32	1.27	1.50	1.00	1.08	.80	.80	.79
		.1875	5.59	1.64	1.86	1.24	1.06	.98	.98	.77
		.250	7.10	2.09	2.20	1.47	1.03	1.15	1.15	.74
4	2	.155	5.78	1.70	3.35	1.67	1.40	1.12	1.12	.81
		.1875	6.86	2.02	3.87	1.93	1.38	1.28	1.28	.80
		.250	8.80	2.59	4.69	2.34	1.35	1.53	1.53	.77
4	3	.156	6.88	2.02	4.52	2.26	1.49	2.89	1.93	1.20
		.1875	8.14	2.39	5.23	2.61	1.48	3.34	2.23	1.18
		.250	10.50	3.09	6.45	3.22	1.45	4.10	2.73	1.15
		.3125	12.69	3.73	7.43	3.72	1.41	4.70	3.13	1.12
5	3	.1875	9.31	2.74	8.86	3.55	1.80	4.01	2.67	1.21
		.250	12.02	3.54	10.95	4.38	1.76	4.92	3.28	1.18
		.375	16.84	4.95	13.91	5.56	1.68	6.16	4.10	1.11
6	4	.1875	11.86	3.49	17.16	5.72	2.22	9.20	4.60	1.62
		.250	15.42	4.54	21.57	7.19	2.18	11.51	5.75	1.59
		.375	21.94	6.45	28.55	9.52	2.10	15.10	7.55	1.53
7	5	.1875	14.41	4.24	29.38	8.39	2.63	17.55	7.02	2.04
		.250	18.82	5.54	37.34	10.67	2.60	22.24	8.90	2.00
		.375	27.04	7.95	50.65	14.47	2.52	29.99	11.99	1.94
		.500	34.48	10.14	60.64	17.33	2.45	35.69	14.28	1.88
8	4	.1875	14.41	4.24	34.83	8.71	2.87	11.92	5.96	1.68
		.250	18.82	5.54	44.23	11.06	2.83	15.03	7.51	1.65
		.375	27.04	7.95	59.86	14.97	2.74	20.04	10.02	1.59
		.500	34.48	10.14	71.48	17.87	2.65	23.57	11.78	1.52
8	6	.1875	16.85	4.96	45.77	11.44	3 04	29.55	9.85	2.44
		.250	22.04	6.48	58.36	14.59	3.00	37.61	12.54	2.41
		.375	31.73	9.33	79.64	19.91	2.92	51.14	17.05	2.34
		.500	40.55	11.93	95.92	23.98	2.84	61.37	20.46	2.27
10	6	.250	25.44	7.48	100.35	20.07	3.66	45.88	15.29	2.48
		.375	36.83	10.83	138.69	27.74	3.58	63.03	21.01	2.41
		.500	47.35	13.93	169.48	33.90	3.49	76.54	25.51	2.34

Table 1.6

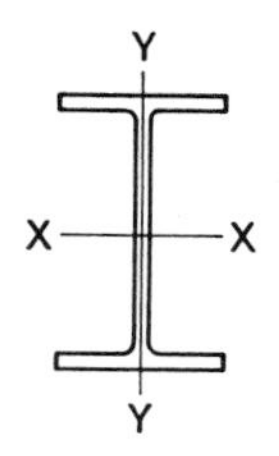

WF SHAPES
Properties for designing

Nominal Size	Weight per Foot	Area	Depth	Flange Width	Flange Thickness	Web Thickness	$\frac{d}{A_f}$	AXIS X-X I	AXIS X-X S	AXIS X-X r	AXIS Y-Y I	AXIS Y-Y S	AXIS Y-Y r
In.	Lb.	In.^{2}	In.	In.	In.	In.		In.^{4}	In.^{3}	In.	In.^{4}	In.^{3}	In.
36 × 16½	300	88.17	36.72	16.655	1.680	.945	1.31	20290.2	1105.1	15.17	1225.2	147.1	3.73
	280	82.32	36.50	16.595	1.570	.885	1.40	18819.3	1031.2	15.12	1127.5	135.9	3.70
	260	76.56	36.24	16.555	1.440	.845	1.52	17233.8	951.1	15.00	1020.6	123.3	3.65
	245	72.03	36.06	16.512	1.350	.802	1.62	16092.2	892.5	14.95	944.7	114.4	3.62
	230	67.73	35.88	16.475	1.260	.765	1.73	14988.4	835.5	14.88	870.9	105.7	3.59
36 × 12	194	57.11	36.48	12.117	1.260	.770	2.39	12103.4	663.6	14.56	355.4	58.7	2.49
	182	53.54	36.32	12.072	1.180	.725	2.55	11281.5	621.2	14.52	327.7	54.3	2.47
	170	49.98	36.16	12.027	1.100	.680	2.73	10470.0	579.1	14.47	300.6	50.0	2.45
	160	47.09	36.00	12.000	1.020	.653	2.94	9738.8	541.0	14.38	275.4	45.9	2.42
	150	44.16	35.84	11.972	.940	.625	3.19	9012.1	502.9	14.29	250.4	41.8	2.38
	‡135	39.70	35.55	11.945	.794	.598	3.75	7796.1	438.6	14.01	207.1	34.7	2.28
33 × 15¾	240	70.52	33.50	15.865	1.400	.830	1.51	13585.1	811.1	13.88	874.3	110.2	3.52
	220	64.73	33.25	15.810	1.275	.775	1.65	12312.1	740.6	13.79	782.4	99.0	3.48
	200	58.79	33.00	15.750	1.150	.715	1.82	11048.2	669.6	13.71	691.7	87.8	3.43
33 × 11½	152	44.71	33.50	11.565	1.055	.635	2.74	8147.6	486.4	13.50	256.1	44.3	2.39
	141	41.51	33.31	11.535	.960	.605	3.01	7442.2	446.8	13.39	229.7	39.8	2.35
	130	38.26	33.10	11.510	.855	.580	3.36	6699.0	404.8	13.23	201.4	35.0	2.29
	‡118	34.71	32.86	11.484	.738	.554	3.88	5886.9	358.3	13.02	170.3	29.7	2.22
30 × 15	210	61.78	30.38	15.105	1.315	.775	1.53	9872.4	649.9	12.64	707.9	93.7	3.38
	190	55.90	30.12	15.040	1.185	.710	1.69	8825.9	586.1	12.57	624.6	83.1	3.34
	172	50.65	29.88	14.985	1.065	.655	1.87	7891.5	528.2	12.48	550.1	73.4	3.30
30 × 10½	132	38.83	30.30	10.551	1.000	.615	2.87	5753.1	379.7	12.17	185.0	35.1	2.18
	124	36.45	30.16	10.521	.930	.585	3.08	5347.1	354.6	12.11	169.7	32.3	2.16
	116	34.13	30.00	10.500	.850	.564	3.36	4919.1	327.9	12.00	153.2	29.2	2.12
	108	31.77	29.82	10.484	.760	.548	3.74	4461.0	299.2	11.85	135.1	25.8	2.06
	‡ 99	29.11	29.64	10.458	.670	.522	4.23	3988.6	269.1	11.70	116.9	22.4	2.00

‡ Non-compact shape in A242, A440, and A441.

Table 1.6 (*Continued*)

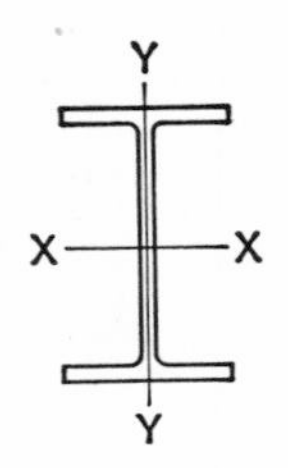

WF SHAPES
Properties for designing

Nominal Size	Weight per Foot	Area	Depth	Flange Width	Flange Thickness	Web Thickness	$\frac{d}{A_f}$	AXIS X-X I	AXIS X-X S	AXIS X-X r	AXIS Y-Y I	AXIS Y-Y S	AXIS Y-Y r
In.	Lb.	In.²	In.	In.	In.	In.		In.⁴	In.³	In.	In.⁴	In.³	In.
27 × 14	177	52.10	27.31	14.090	1.190	.725	1.63	6728.6	492.8	11.36	518.9	73.7	3.16
	160	47.04	27.08	14.023	1.075	.658	1.80	6018.6	444.5	11.31	458.0	65.3	3.12
	145	42.68	26.88	13.965	.975	.600	1.97	5414.3	402.9	11.26	406.9	58.3	3.09
27 × 10	114	33.53	27.28	10.070	.932	.570	2.91	4080.5	299.2	11.03	149.6	29.7	2.11
	102	30.01	27.07	10.018	.827	.518	3.27	3604.1	266.3	10.96	129.5	25.9	2.08
	94	27.65	26.91	9.990	.747	.490	3.61	3266.7	242.8	10.87	115.1	23.0	2.04
	‡ 84	24.71	26.69	9.963	.636	.463	4.21	2824.8	211.7	10.69	95.7	19.2	1.97
24 × 14	160	47.04	24.72	14.091	1.135	.656	1.55	5110.3	413.5	10.42	492.6	69.9	3.23
	145	42.62	24.49	14.043	1.020	.608	1.71	4561.0	372.5	10.34	434.3	61.8	3.19
	‡130	38.21	24.25	14.000	.900	.565	1.93	4009.5	330.7	10.24	375.2	53.6	3.13
24 × 12	120	35.29	24.31	12.088	.930	.556	2.16	3635.3	299.1	10.15	254.0	42.0	2.68
	110	32.36	24.16	12.042	.855	.510	2.34	3315.0	274.4	10.12	229.1	38.0	2.66
	‡100	29.43	24.00	12.000	.775	.468	2.58	2987.3	248.9	10.08	203.5	33.9	2.63
24 × 9	94	27.63	24.29	9.061	.872	.516	3.07	2683.0	220.9	9.85	102.2	22.6	1.92
	84	24.71	24.09	9.015	.772	.470	3.47	2364.3	196.3	9.78	88.3	19.6	1.89
	76	22.37	23.91	8.985	.682	.440	3.90	2096.4	175.4	9.68	76.5	17.0	1.85
	‡ 68	20.00	23.71	8.961	.582	.416	4.55	1814.5	153.1	9.53	63.8	14.2	1.79
21 × 13	142	41.76	21.46	13.132	1.095	.659	1.49	3403.1	317.2	9.03	385.9	58.8	3.04
	127	37.34	21.24	13.061	.985	.588	1.65	3017.2	284.1	8.99	338.6	51.8	3.01
	‡112	32.93	21.00	13.000	.865	.527	1.87	2620.6	249.6	8.92	289.7	44.6	2.96
21 × 9	96	28.21	21.14	9.038	.935	.575	2.50	2088.9	197.6	8.60	109.3	24.2	1.97
	82	24.10	20.86	8.962	.795	.499	2.93	1752.4	168.0	8.53	89.6	20.0	1.93
21 × 8¼	73	21.46	21.24	8.295	.740	.455	3.46	1600.3	150.7	8.64	66.2	16.0	1.76
	68	20.02	21.13	8.270	.685	.430	3.73	1478.3	139.9	8.59	60.4	14.6	1.74
	62	18.23	20.99	8.240	.615	.400	4.15	1326.8	126.4	8.53	53.1	12.9	1.71
	‡ 55	16.18	20.80	8.215	.522	.375	4.85	1140.7	109.7	8.40	44.0	10.7	1.65

‡ Non-compact shape in A242, A440 and A441.

Table 1.6 (*Continued*)

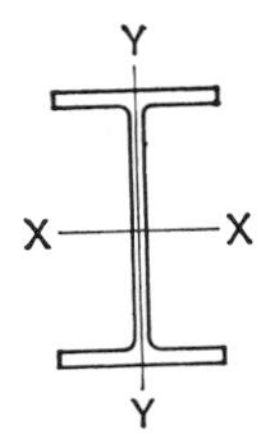

WF SHAPES
Properties for designing

Nominal Size	Weight per Foot	Area	Depth	Flange		Web Thickness	$\frac{d}{A_f}$	AXIS X-X			AXIS Y-Y		
				Width	Thickness			I	S	r	I	S	r
In.	Lb.	In.^{2}	In.	In.	In.	In.		In.^{4}	In.^{3}	In.	In.^{4}	In.^{3}	In.
18 × 11¾	114	33.51	18.48	11.833	.991	.595	1.58	2033.8	220.1	7.79	255.6	43.2	2.76
	105	30.86	18.32	11.792	.911	.554	1.71	1852.5	202.2	7.75	231.0	39.2	2.73
	96	28.22	18.16	11.750	.831	.512	1.86	1674.7	184.4	7.70	206.8	35.2	2.71
18 × 8¾	85	24.97	18.32	8.838	.911	.526	2.28	1429.9	156.1	7.57	99.4	22.5	2.00
	77	22.63	18.16	8.787	.831	.475	2.49	1286.8	141.7	7.54	88.6	20.2	1.98
	70	20.56	18.00	8.750	.751	.438	2.74	1153.9	128.2	7.49	78.5	17.9	1.95
	64	18.80	17.87	8.715	.686	.403	2.99	1045.8	117.0	7.46	70.3	16.1	1.93
18 × 7½	60	17.64	18.25	7.558	.695	.416	3.48	984.0	107.8	7.47	47.1	12.5	1.63
	55	16.19	18.12	7.532	.630	.390	3.82	889.9	98.2	7.41	42.0	11.1	1.61
	50	14.71	18.00	7.500	.570	.358	4.22	800.6	89.0	7.38	37.2	9.9	1.59
	‡45	13.24	17.86	7.477	.499	.335	4.79	704.5	78.9	7.30	31.9	8.5	1.55
16 × 11½	96	28.22	16.32	11.533	.875	.535	1.62	1355.1	166.1	6.93	207.2	35.9	2.71
	88	25.87	16.16	11.502	.795	.504	1.77	1222.6	151.3	6.87	185.2	32.2	2.67
16 × 8½	78	22.92	16.32	8.586	.875	.529	2.17	1042.6	127.8	6.74	87.5	20.4	1.95
	71	20.86	16.16	8.543	.795	.486	2.38	936.9	115.9	6.70	77.9	18.2	1.93
	64	18.80	16.00	8.500	.715	.443	2.63	833.8	104.2	6.66	68.4	16.1	1.91
	58	17.04	15.86	8.464	.645	.407	2.91	746.4	94.1	6.62	60.5	14.3	1.88
16 × 7	50	14.70	16.25	7.073	.628	.380	3.66	655.4	80.7	6.68	34.8	9.8	1.54
	45	13.24	16.12	7.039	.563	.346	4.07	583.3	72.4	6.64	30.5	8.7	1.52
	40	11.77	16.00	7.000	.503	.307	4.54	515.5	64.4	6.62	26.5	7.6	1.50
	‡36	10.59	15.85	6.992	.428	.299	5.30	446.3	56.3	6.49	22.1	6.3	1.45

‡ Non-compact shape in A242, A440 and A441.

Table 1.6 (*Continued*)

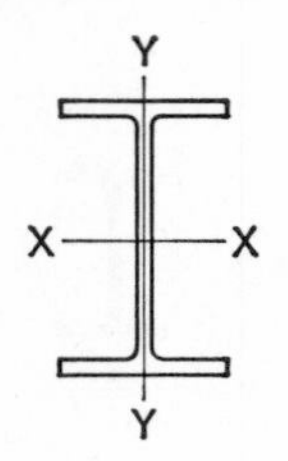

WF SHAPES
Properties for designing

Nominal Size	Weight per Foot	Area	Depth	Flange		Web Thickness	$\frac{d}{A_f}$	AXIS X-X			AXIS Y-Y		
				Width	Thickness			I	S	r	I	S	r
In.	Lb.	In.²	In.	In.	In.	In.		In.⁴	In.³	In.	In.⁴	In.³	In.
14 × 16	426	125.25	18.69	16.695	3.033	1.875	.369	6610.3	707.4	7.26	2359.5	282.7	4.34
	398	116.98	18.31	16.590	2.843	1.770	.388	6013.7	656.9	7.17	2169.7	261.6	4.31
	370	108.78	17.94	16.475	2.658	1.655	.410	5454.2	608.1	7.08	1986.0	241.1	4.27
	342	100.59	17.56	16.365	2.468	1.545	.435	4911.5	559.4	6.99	1806.9	220.8	4.24
	314	92.30	17.19	16.235	2.283	1.415	.464	4399.4	511.9	6.90	1631.4	201.0	4.20
	287	84.37	16.81	16.130	2.093	1.310	.498	3912.1	465.5	6.81	1466.5	181.8	4.17
	264	77.63	16.50	16.025	1.938	1.205	.531	3526.0	427.4	6.74	1331.2	166.1	4.14
	246	72.33	16.25	15.945	1.813	1.125	.562	3228.9	397.4	6.68	1226.6	153.9	4.12
	237	69.69	16.12	15.910	1.748	1.090	.580	3080.9	382.2	6.65	1174.8	147.7	4.11
	228	67.06	16.00	15.865	1.688	1.045	.597	2942.4	367.8	6.62	1124.8	141.8	4.10
	219	64.36	15.87	15.825	1.623	1.005	.618	2798.2	352.6	6.59	1073.2	135.6	4.08
	211	62.07	15.75	15.800	1.563	.980	.638	2671.4	339.2	6.56	1028.6	130.2	4.07
	202	59.39	15.63	15.750	1.503	.930	.660	2538.8	324.9	6.54	979.7	124.4	4.06
	193	56.73	15.50	15.710	1.438	.890	.686	2402.4	310.0	6.51	930.1	118.4	4.05
	184	54.07	15.38	15.660	1.378	.840	.713	2274.8	295.8	6.49	882.7	112.7	4.04
	176	51.73	15.25	15.640	1.313	.820	.743	2149.6	281.9	6.45	837.9	107.1	4.02
	167	49.09	15.12	15.600	1.248	.780	.777	2020.8	267.3	6.42	790.2	101.3	4.01
	158	46.47	15.00	15.550	1.188	.730	.812	1900.6	253.4	6.40	745.0	95.8	4.00
	150	44.08	14.88	15.515	1.128	.695	.850	1786.9	240.2	6.37	702.5	90.6	3.99
	142	41.85	14.75	15.500	1.063	.680	.895	1672.2	226.7	6.32	660.1	85.2	3.97
	320	94.12	16.81	16.710	2.093	1.890	.481	4141.7	492.8	6.63	1635.1	195.7	4.17
14 × 14½	136	39.98	14.75	14.740	1.063	.660	.941	1593.0	216.0	6.31	567.7	77.0	3.77
	127	37.33	14.62	14.690	.998	.610	.997	1476.7	202.0	6.29	527.6	71.8	3.76
	‡119	34.99	14.50	14.650	.938	.570	1.06	1373.1	189.4	6.26	491.8	67.1	3.75
	‡111	32.65	14.37	14.620	.873	.540	1.13	1266.5	176.3	6.23	454.9	62.2	3.73
	†103	30.26	14.25	14.575	.813	.495	1.20	1165.8	163.6	6.21	419.7	57.6	3.72
	† 95	27.94	14.12	14.545	.748	.465	1.30	1063.5	150.6	6.17	383.7	52.8	3.71
	† 87	25.56	14.00	14.500	.688	.420	1.40	966.9	138.1	6.15	349.7	48.2	3.70
14 × 12	‡ 84	24.71	14.18	12.023	.778	.451	1.52	928.4	130.9	6.13	225.5	37.5	3.02
	‡ 78	22.94	14.06	12.000	.718	.428	1.63	851.2	121.1	6.09	206.9	34.5	3.00
14 × 10	74	21.76	14.19	10.072	.783	.450	1.80	796.8	112.3	6.05	133.5	26.5	2.48
	68	20.00	14.06	10.040	.718	.418	1.95	724.1	103.0	6.02	121.2	24.1	2.46
	‡ 61	17.94	13.91	10.000	.643	.378	2.16	641.5	92.2	5.98	107.3	21.5	2.45

† Non-compact shape in A36, A242, A440 and A441.
‡ Non-compact shape in A242, A440 and A441.

Table 1.6 (*Continued*)

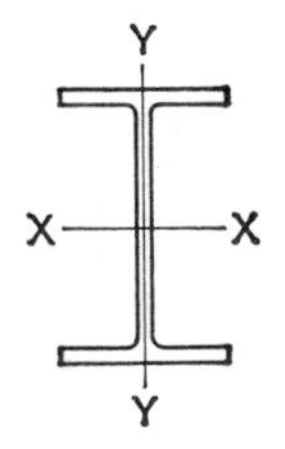

WF SHAPES
Properties for designing

Nominal Size	Weight per Foot	Area	Depth	Flange Width	Flange Thickness	Web Thickness	$\frac{d}{A_f}$	AXIS X-X I	AXIS X-X S	AXIS X-X r	AXIS Y-Y I	AXIS Y-Y S	AXIS Y-Y r
In.	Lb.	In.²	In.	In.	In.	In.		In.⁴	In.³	In.	In.⁴	In.³	In.
14 × 8	53	15.59	13.94	8.062	.658	.370	2.63	542.1	77.8	5.90	57.5	14.3	1.92
	48	14.11	13.81	8.031	.593	.339	2.90	484.9	70.2	5.86	51.3	12.8	1.91
	‡ 43	12.65	13.68	8.000	.528	.308	3.24	429.0	62.7	5.82	45.1	11.3	1.89
14 × 6¾	38	11.17	14.12	6.776	.513	.313	4.06	385.3	54.6	5.87	24.6	7.3	1.49
	‡ 34	10.00	14.00	6.750	.453	.287	4.58	339.2	48.5	5.83	21.3	6.3	1.46
	‡ 30	8.81	13.86	6.733	.383	.270	5.37	289.6	41.8	5.73	17.5	5.2	1.41
12 × 12	190	55.86	14.38	12.670	1.736	1.060	.654	1892.5	263.2	5.82	589.7	93.1	3.25
	161	47.38	13.88	12.515	1.486	.905	.746	1541.8	222.2	5.70	486.2	77.7	3.20
	133	39.11	13.38	12.365	1.236	.755	.875	1221.2	182.5	5.59	389.9	63.1	3.16
	120	35.31	13.12	12.320	1.106	.710	.963	1071.7	163.4	5.51	345.1	56.0	3.13
	106	31.19	12.88	12.230	.986	.620	1.07	930.7	144.5	5.46	300.9	49.2	3.11
	99	29.09	12.75	12.190	.921	.580	1.14	858.5	134.7	5.43	278.2	45.7	3.09
	92	27.06	12.62	12.155	.856	.545	1.21	788.9	125.0	5.40	256.4	42.2	3.08
	‡ 85	24.98	12.50	12.105	.796	.495	1.29	723.3	115.7	5.38	235.5	38.9	3.07
	‡ 79	23.22	12.38	12.080	.736	.470	1.39	663.0	107.1	5.34	216.4	35.8	3.05
	† 72	21.16	12.25	12.040	.671	.430	1.52	597.4	97.5	5.31	195.3	32.4	3.04
	† 65	19.11	12.12	12.000	.606	.390	1.67	533.4	88.0	5.28	174.6	29.1	3.02
12 × 10	‡ 58	17.06	12.19	10.014	.641	.359	1.90	476.1	78.1	5.28	107.4	21.4	2.51
	‡ 53	15.59	12.06	10.000	.576	.345	2.09	426.2	70.7	5.23	96.1	19.2	2.48
12 × 8	50	14.71	12.19	8.077	.641	.371	2.35	394.5	64.7	5.18	56.4	14.0	1.96
	45	13.24	12.06	8.042	.576	.336	2.60	350.8	58.2	5.15	50.0	12.4	1.94
	‡ 40	11.77	11.94	8.000	.516	.294	2.89	310.1	51.9	5.13	44.1	11.0	1.94
12 × 6½	36	10.59	12.24	6.565	.540	.305	3.45	280.8	45.9	5.15	23.7	7.2	1.50
	31	9.12	12.09	6.525	.465	.265	3.98	238.4	39.4	5.11	19.8	6.1	1.47
	‡ 27	7.97	11.96	6.500	.400	.240	4.60	204.1	34.1	5.06	16.6	5.1	1.44

†Non-compact shape in A36, A242, A440 and A441.
‡Non-compact shape in A242, A440 and A441.

Table 1.6 (*Continued*)

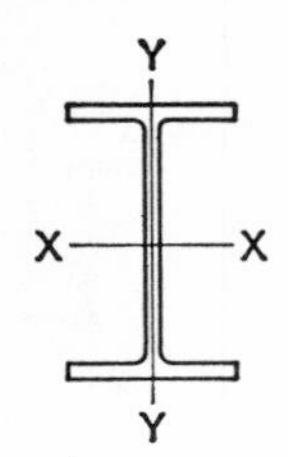

WF SHAPES
Properties for designing

Nominal Size	Weight per Foot	Area	Depth	Flange Width	Flange Thickness	Web Thickness	$\frac{d}{A_f}$	AXIS X-X I	AXIS X-X S	AXIS X-X r	AXIS Y-Y I	AXIS Y-Y S	AXIS Y-Y r
In.	Lb.	In.²	In.	In.	In.	In.		In.⁴	In.³	In.	In.⁴	In.³	In.
10 × 10	112	32.92	11.38	10.415	1.248	.755	.876	718.7	126.3	4.67	235.4	45.2	2.67
	100	29.43	11.12	10.345	1.118	.685	.961	625.0	112.4	4.61	206.6	39.9	2.65
	89	26.19	10.88	10.275	.998	.615	1.06	542.4	99.7	4.55	180.6	35.2	2.63
	77	22.67	10.62	10.195	.868	.535	1.20	457.2	86.1	4.49	153.4	30.1	2.60
	72	21.18	10.50	10.170	.808	.510	1.28	420.7	80.1	4.46	141.8	27.9	2.59
	66	19.41	10.38	10.117	.748	.457	1.37	382.5	73.7	4.44	129.2	25.5	2.58
	‡ 60	17.66	10.25	10.075	.683	.415	1.49	343.7	67.1	4.41	116.5	23.1	2.57
	‡ 54	15.88	10.12	10.028	.618	.368	1.63	305.7	60.4	4.39	103.9	20.7	2.56
	† 49	14.40	10.00	10.000	.558	.340	1.79	272.9	54.6	4.35	93.0	18.6	2.54
10 × 8	45	13.24	10.12	8.022	.618	.350	2.04	248.6	49.1	4.33	53.2	13.3	2.00
	‡ 39	11.48	9.94	7.990	.528	.318	2.36	209.7	42.2	4.27	44.9	11.2	1.98
	† 33	9.71	9.75	7.964	.433	.292	2.83	170.9	35.0	4.20	36.5	9.2	1.94
10 × 5¾	29	8.53	10.22	5.799	.500	.289	3.52	157.3	30.8	4.29	15.2	5.2	1.34
	25	7.35	10.08	5.762	.430	.252	4.08	133.2	26.4	4.26	12.7	4.4	1.31
	‡ 21	6.19	9.90	5.750	.340	.240	5.07	106.3	21.5	4.14	9.70	3.4	1.25
8 × 8	67	19.70	9.00	8.287	.933	.575	1.16	271.8	60.4	3.71	88.6	21.4	2.12
	58	17.06	8.75	8.222	.808	.510	1.32	227.3	52.0	3.65	74.9	18.2	2.10
	48	14.11	8.50	8.117	.683	.405	1.53	183.7	43.2	3.61	60.9	15.0	2.08
	40	11.76	8.25	8.077	.558	.365	1.83	146.3	35.5	3.53	49.0	12.1	2.04
	‡ 35	10.30	8.12	8.027	.493	.315	2.05	126.5	31.1	3.50	42.5	10.6	2.03
	† 31	9.12	8.00	8.000	.433	.288	2.31	109.7	27.4	3.47	37.0	9.2	2.01
8 × 6½	28	8.23	8.06	6.540	.463	.285	2.66	97.8	24.3	3.45	21.6	6.6	1.62
	‡ 24	7.06	7.93	6.500	.398	.245	3.07	82.5	20.8	3.42	18.2	5.6	1.61
8 × 5¼	20	5.88	8.14	5.268	.378	.248	4.08	69.2	17.0	3.43	8.50	3.2	1.20
	‡ 17	5.00	8.00	5.250	.308	.230	4.95	56.4	14.1	3.36	6.72	2.6	1.16

† Non-compact shape in A36, A242, A440 and A441.
‡ Non-compact shape in A242, A440 and A441.

1.32 Steel-beam tables

The steel-beam tables, Tables 1.6 to 1.9, combine in one series the properties of rolled-steel beam, column, and channel sections in which the designer is interested, but they omit some of the dimensions pertaining primarily to steel detailing. There are a number of other useful figures which are constant for each individual beam section and which it has seemed advisable to include in Chapter 4, since they are not actually properties of the steel section but are dependent on these properties.

I and S are used in steel beams and plate-girder design—S usually for simple beam design as described in Art. 4.10, and I for plate-girder design as described in Art. 4.20. The radius of gyration about axis x-x and axis y-y are used when these sections act as struts or columns, as described in Art. 7.10.

1.40 SELECTION OF SYSTEM

a The selection of the various systems of construction to be used is dependent on a number of factors. Frequently the whole matter is dictated to the structural designer, but certain phases of the design may be left for his recommendation, and it is well to consider the various factors which would influence the selection. Among these are:

1. Location
    - Building code limitations
    - General nature of neighborhood
2. Economics
    - First cost of building
    - Relation of building to land value
    - Maintenance cost
    - Insurance cost
3. Use
    - Building code limitations
    - Effect of loads
4. Advertising value of the building
5. General policy of the owner

b Perhaps the best primary classification for types of construction is the generally accepted code classification. The Building Code of the National Board of Fire Underwriters makes a standard classification as follows:

"Fireproof construction," as applied to buildings, means that in which the structural members are of approved noncombustible construction having the necessary strength and stability and having the fire resistance ratings of not less than 4 hours for exterior nonbearing walls and for wall panels, for columns, and for wall-supporting girders, and trusses, and not less than 3 hours for floors, for roofs, and for floor and roof supporting beams, girders and trusses; and in which exterior bearing walls and interior bearing walls, if any, are of approved masonry or of reinforced concrete.

This type of construction corresponds generally with that sometimes called "fully protected" or "fire-resistive" construction. Because of the long, well-established, and almost universal use of the term "fireproof," it is thought best to retain that term for the type of construction here defined.

"Semifireproof construction," as applied to buildings, means that in which the structural members are of approved noncombustible construction having the necessary strength and stability, and having fire resistance ratings of not less than 4 hours for exterior walls and for wall panels; not less than 3 hours for columns, and for wall-supporting girders and trusses; not less than 2 hours for floors, for roofs, and for floor and roof supporting beams, girders and trusses; and in which exterior bearing walls and interior bearing walls, if any, are of approved masonry or reinforced concrete.

This type of construction corresponds generally with that sometimes called "protected" or "fire safe."

"Heavy timber construction," as applied to buildings, means that in which walls are of approved masonry or reinforced concrete; and in which the interior structural elements, including columns, floors and roof construction, consist of heavy timbers with smooth, flat surfaces assembled to avoid thin sections, sharp projections and concealed or inaccessible spaces; and in which all structural members which support masonry walls shall have a fire resistance rating of not less than 3 hours; and other structural members of steel or reinforced concrete, if used in lieu of timber construction, shall have a fire resistance rating of not less than 1 hour.

This type of construction is the same as that called "mill construction" for many years, and sometimes called "slow burning construction."

"Ordinary construction," as applied to build-

Table 1.7 Miscellaneous Sections

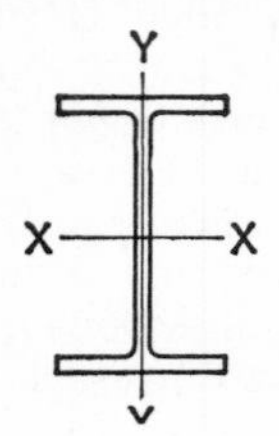

H BEARING PILES
Properties for designing

Section Number and Nominal Size	Weight per Foot	Area	Depth	Flange		Web Thickness	$\frac{d}{A_f}$	AXIS X-X			AXIS Y-Y		
				Width	Thickness			I	S	r	I	S	r
In.	Lb.	In.²	In.	In.	In.	In.		In.⁴	In.³	In.	In.⁴	In.³	In.
BP 14	†117	34.44	14.23	14.885	.805	.805	1.19	1228.5	172.6	5.97	443.1	59.5	3.59
14 × 14½	†102	30.01	14.03	14.784	.704	.704	1.35	1055.1	150.4	5.93	379.6	51.3	3.56
	† 89	26.19	13.86	14.696	.616	.616	1.53	909.1	131.2	5.89	326.2	44.4	3.53
	† 73	21.46	13.64	14.586	.506	.506	1.85	733.1	107.5	5.85	261.9	35.9	3.49
BP 12	† 74	21.76	12.12	12.217	.607	.607	1.63	566.5	93.5	5.10	184.7	30.2	2.91
12 × 12	† 53	15.58	11.78	12.046	.436	.436	2.24	394.8	67.0	5.03	127.3	21.2	2.86
BP 10	† 57	16.76	10.01	10.224	.564	.564	1.74	294.7	58.9	4.19	100.6	19.7	2.45
10 × 10	† 42	12.35	9.72	10.078	.418	.418	2.31	210.8	43.4	4.13	71.4	14.2	2.40
BP 8	† 36	10.60	8.03	8.158	.446	.446	2.21	119.8	29.9	3.36	40.4	9.9	1.95
8 × 8													

† Non-compact shape in A36, A242, A440 and A441.

Table 1.7 Miscellaneous Sections (*Continued*)

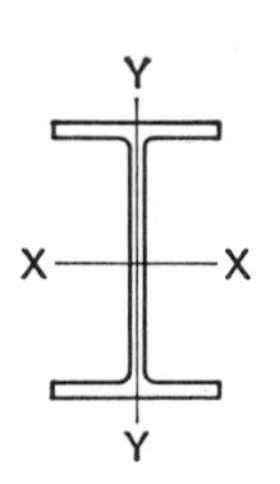

LIGHT WF COLUMNS
MISCELLANEOUS COLUMNS (M)
and
JUNIOR BEAMS AND CHANNELS
Properties for designing

Nominal Size and Designation	Weight per Foot	Area	Depth	Flange Width	Flange Average Thickness	Web Thickness	$\frac{d}{A_f}$	Axis X-X I	Axis X-X S	Axis X-X r	Axis Y-Y I	Axis Y-Y S	Axis Y-Y r
In.	Lb.	In.²	In.	In.	In.	In.		In.⁴	In.³	In.	In.⁴	In.³	In.
LIGHT WF AND MISCELLANEOUS COLUMNS													
8 × 8 M	‡34.3	10.09	8.00	8.000	.459	.375	2.18	115.5	28.9	3.40	35.1	8.8	1.87
	‡32.6	9.59	8.00	7.938	.459	.313	2.20	112.8	28.2	3.45	34.2	8.6	1.90
6 × 6 WF	25.0	7.37	6.37	6.080	.456	.320	2.30	53.5	16.8	2.69	17.1	5.6	1.52
6 × 6 M	25.0	7.35	6.00	5.938	.481	.313	2.10	47.0	15.7	2.53	14.9	5.0	1.43
	‡22.5	6.62	6.00	6.063	.380	.375	2.60	41.0	13.7	2.49	12.2	4.0	1.36
6 × 6 WF	‡20.0	5.90	6.20	6.018	.367	.258	2.81	41.7	13.4	2.66	13.3	4.4	1.50
6 × 6 M	‡20.0	5.88	6.00	5.938	.380	.250	2.66	38.8	12.9	2.57	11.4	3.8	1.39
6 × 6 WF	†15.5	4.62	6.00	6.000	.269	.240	3.72	30.3	10.1	2.56	9.69	3.2	1.45
5 × 5 M	18.9	5.56	5.00	5.000	.417	.313	2.40	23.8	9.5	2.08	7.85	3.1	1.20
5 × 5 WF	18.5	5.45	5.12	5.025	.420	.265	2.43	25.4	9.94	2.16	8.89	3.5	1.28
	16.0	4.70	5.00	5.000	.360	.240	2.78	21.3	8.53	2.13	7.51	3.0	1.26
4 × 4 WF	13.0	3.82	4.16	4.060	.345	.280	2.97	11.3	5.45	1.72	3.76	1.9	.99
4 × 4 M	13.0	3.82	4.00	3.937	.372	.250	2.73	10.4	5.20	1.65	3.39	1.7	.94
JUNIOR BEAMS													
12 × 3 JR	‡11.8	3.45	12.00	3.063	.225	.175	17.4	72.2	12.0	4.57	.98	.64	.53
10 × 2¾ JR	‡ 9.0	2.64	10.00	2.688	.206	.155	18.1	39.0	7.8	3.85	.61	.45	.48
8 × 2¼ JR	6.5	1.92	8.00	2.281	.189	.135	18.6	18.7	4.7	3.12	.34	.30	.42
7 × 2⅛ JR	5.5	1.61	7.00	2.078	.180	.126	18.7	12.1	3.5	2.74	.25	.24	.39
6 × 1⅞ JR	4.4	1.30	6.00	1.844	.171	.114	19.0	7.3	2.4	2.37	.17	.18	.36

JUNIOR CHANNELS

Nominal Size and Designation	Weight per Foot	Area	Depth	Flange Width	Flange Average Thickness	Web Thickness	$\frac{d}{A_f}$	Axis X-X I	Axis X-X S	Axis X-X r	Axis Y-Y I	Axis Y-Y S	Axis Y-Y r	Axis Y-Y x
In.	Lb.	In.²	In.	In.	In.	In.		In.⁴	In.³	In.	In.⁴	In.³	In.	In.
12 × 1½ JR ⊏	10.6	3.12	12.00	1.500	.309	.190	25.9	55.8	9.3	4.23	.39	.32	.35	.27
10 × 1½ JR ⊏	8.4	2.47	10.00	1.500	.280	.170	23.8	32.3	6.5	3.61	.33	.28	.37	.29
10 × 1⅛ JR ⊏	6.5	1.91	10.00	1.125	.201	.150	44.2	22.1	4.4	3.47	.12	.13	.25	.19

† Non-compact shape in A36, A242, A440 and A441.
‡ Non-compact shape in A242, A440 and A441.

Table 1.7 Miscellaneous Sections (*Continued*)

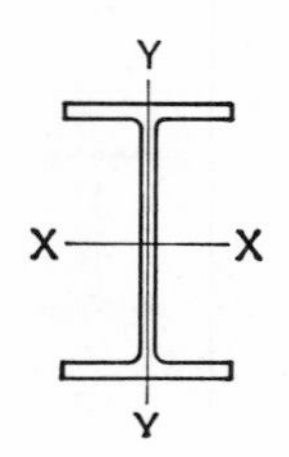

MISCELLANEOUS SHAPES (M) and LIGHT BEAMS (B) Properties for designing

I

Nominal Size and Designation	Weight per Foot	Area	Depth	Flange		Web Thickness	$\frac{d}{A_f}$	AXIS X-X			AXIS Y-Y		
				Width	Average Thickness			I	S	r	I	S	r
In.	Lb.	In.^{2}	In.	In.	In.	In.		In.^{4}	In.^{3}	In.	In.^{4}	In.^{3}	In.
16 × 5½ B	31	9.12	15.84	5.525	.442	.275	6.49	372.5	47.0	6.39	11.57	4.19	1.13
	‡26	7.65	15.65	5.500	.345	.250	8.25	298.1	38.1	6.24	8.71	3.17	1.07
14 × 5 B	26	7.65	13.89	5.025	.418	.255	6.61	242.6	34.9	5.63	8.26	3.29	1.04
	‡22	6.47	13.72	5.000	.335	.230	8.19	197.4	28.8	5.52	6.40	2.56	.99
14 × 4 B	‡17.2	5.05	14.00	4.000	.272	.210	12.9	147.3	21.0	5.40	2.65	1.32	.72
12 × 4 B	22	6.47	12.31	4.030	.424	.260	7.20	155.7	25.3	4.91	4.55	2.26	.84
	19	5.62	12.16	4.010	.349	.240	8.69	130.1	21.4	4.81	3.67	1.83	.81
	‡16.5	4.86	12.00	4.000	.269	.230	11.2	105.3	17.5	4.65	2.79	1.39	.76
	†14	4.14	11.91	3.970	.224	.200	13.4	88.2	14.8	4.61	2.25	1.13	.74
10 × 5¾ M	‡29.1	8.55	9.88	5.935	.389	.425	4.28	131.5	26.6	3.92	11.2	3.7	1.14
	‡22.9	6.73	9.88	5.750	.389	.240	4.42	116.6	23.6	4.16	9.9	3.5	1.22
	‡21	6.10	9.90	5.750	.338	.240	5.09	104.4	21.1	4.14	9.2	3.2	1.22
10 × 4 B	19	5.61	10.25	4.020	.394	.250	6.47	96.2	18.8	4.14	4.19	2.08	.86
	17	4.98	10.12	4.010	.329	.240	7.67	81.8	16.2	4.05	3.45	1.72	.83
	‡15	4.40	10.00	4.000	.269	.230	9.29	68.8	13.8	3.95	2.79	1.39	.80
	†11.5	3.39	9.87	3.950	.204	.180	12.3	51.9	10.5	3.92	2.01	1.02	.77
8 × 6½ M	‡28	8.23	8.00	6.650	.398	.390	3.02	90.1	22.5	3.31	17.73	5.33	1.47
	‡24	7.06	8.00	6.500	.398	.240	3.09	83.8	21.0	3.45	16.52	5.08	1.53
8 × 5¼ M	‡22.5	6.61	8.00	5.395	.352	.375	4.21	68.3	17.1	3.23	7.5	2.8	1.08
	‡20	5.88	8.00	5.360	.305	.350	4.89	60.7	15.2	3.22	6.6	2.46	1.06
	‡18.5	5.44	8.00	5.250	.352	.230	4.33	62.1	15.5	3.38	6.9	2.6	1.13
	‡17	5.00	8.00	5.250	.305	.240	5.00	56.0	14.0	3.35	6.16	2.35	1.11
8 × 4 B	15	4.43	8.12	4.015	.314	.245	6.44	48.0	11.8	3.29	3.30	1.65	.86
	‡13	3.83	8.00	4.000	.254	.230	7.87	39.5	9.88	3.21	2.62	1.31	.83
	†10	2.95	7.90	3.940	.204	.170	9.83	30.8	7.79	3.23	1.99	1.01	.82
6 × 4 B	16	4.72	6.25	4.030	.404	.260	3.84	31.7	10.1	2.59	4.32	2.14	.96
	12	3.53	6.00	4.000	.279	.230	5.38	21.7	7.24	2.48	2.89	1.44	.90
	† 8.5	2.50	5.83	3.940	.194	.170	7.63	14.8	5.07	2.43	1.89	.96	.87

† Non-compact shape in A36, A242, A440 and A441.
‡ Non-compact shape in A242, A440 and A441.

Table 1.8

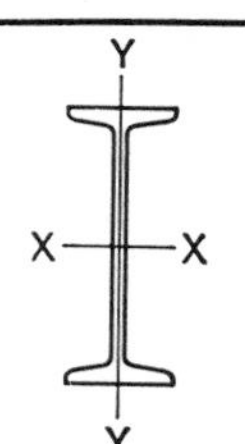

AMERICAN STANDARD BEAMS
Properties for designing

I

Nominal Size	Weight per Foot	Area	Depth	Flange		Web Thickness	$\frac{d}{A_f}$	AXIS X-X			AXIS Y-Y		
				Width	Average Thickness			I	S	r	I	S	r
In.	Lb.	In.^{2}	In.	In.	In.	In.		In.^{4}	In.^{3}	In.	In.^{4}	In.^{3}	In.
24 × 7⅞	120.0	35.13	24.00	8.048	1.102	.798	2.71	3010.8	250.9	9.26	84.9	21.1	1.56
	105.9	30.98	24.00	7.875	1.102	.625	2.76	2811.5	234.3	9.53	78.9	20.0	1.60
24 × 7	100.0	29.25	24.00	7.247	.871	.747	3.81	2371.8	197.6	9.05	48.4	13.4	1.29
	90.0	26.30	24.00	7.124	.871	.624	3.87	2230.1	185.8	9.21	45.5	12.8	1.32
	79.9	23.33	24.00	7.000	.871	.500	3.94	2087.2	173.9	9.46	42.9	12.2	1.36
20 × 7	95.0	27.74	20.00	7.200	.916	.800	3.03	1599.7	160.0	7.59	50.5	14.0	1.35
	85.0	24.80	20.00	7.053	.916	.653	3.09	1501.7	150.2	7.78	47.0	13.3	1.38
20 × 6¼	75.0	21.90	20.00	6.391	.789	.641	3.97	1263.5	126.3	7.60	30.1	9.4	1.17
	65.4	19.08	20.00	6.250	.789	.500	4.05	1169.5	116.9	7.83	27.9	8.9	1.21
18 × 6	70.0	20.46	18.00	6.251	.691	.711	4.17	917.5	101.9	6.70	24.5	7.8	1.09
	54.7	15.94	18.00	6.000	.691	.460	4.35	795.5	88.4	7.07	21.2	7.1	1.15
15 × 5½	50.0	14.59	15.00	5.640	.622	.550	4.28	481.1	64.2	5.74	16.0	5.7	1.05
	42.9	12.49	15.00	5.500	.622	.410	4.39	441.8	58.9	5.95	14.6	5.3	1.08
12 × 5¼	50.0	14.57	12.00	5.477	.659	.687	3.32	301.6	50.3	4.55	16.0	5.8	1.05
	40.8	11.84	12.00	5.250	.659	.460	3.47	268.9	44.8	4.77	13.8	5.3	1.08
12 × 5	35.0	10.20	12.00	5.078	.544	.428	4.34	227.0	37.8	4.72	10.0	3.9	.99
	31.8	9.26	12.00	5.000	.544	.350	4.41	215.8	36.0	4.83	9.5	3.8	1.01
10 × 4⅝	35.0	10.22	10.00	4.944	.491	.594	4.12	145.8	29.2	3.78	8.5	3.4	.91
	25.4	7.38	10.00	4.660	.491	.310	4.37	122.1	24.4	4.07	6.9	3.0	.97
8 × 4	23.0	6.71	8.00	4.171	.425	.441	4.51	64.2	16.0	3.09	4.4	2.1	.81
	18.4	5.34	8.00	4.000	.425	.270	4.71	56.9	14.2	3.26	3.8	1.9	.84
7 × 3⅝	20.0	5.83	7.00	3.860	.392	.450	4.62	41.9	12.0	2.68	3.1	1.6	.74
	15.3	4.43	7.00	3.660	.392	.250	4.88	36.2	10.4	2.86	2.7	1.5	.78
6 × 3⅜	17.25	5.02	6.00	3.565	.359	.465	4.69	26.0	8.7	2.28	2.3	1.3	.68
	12.5	3.61	6.00	3.330	.359	.230	5.02	21.8	7.3	2.46	1.8	1.1	.72
5 × 3	14.75	4.29	5.00	3.284	.326	.494	4.67	15.0	6.0	1.87	1.7	1.0	.63
	10.0	2.87	5.00	3.000	.326	.210	5.11	12.1	4.8	2.05	1.2	.82	.65
4 × 2⅝	9.5	2.76	4.00	2.796	.293	.326	4.89	6.7	3.3	1.56	.91	.65	.58
	7.7	2.21	4.00	2.660	.293	.190	5.13	6.0	3.0	1.64	.77	.58	.59
3 × 2⅜	7.5	2.17	3.00	2.509	.260	.349	4.60	2.9	1.9	1.15	.59	.47	.52
	5.7	1.64	3.00	2.330	.260	.170	4.95	2.5	1.7	1.23	.46	.40	.53

Table 1.9

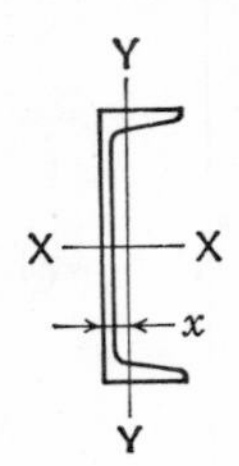

AMERICAN STANDARD CHANNELS
Properties for designing

Nominal Size	Weight per Foot	Area	Depth	Flange		Web Thick-ness	$\frac{d}{A_f}$	AXIS X-X			AXIS Y-Y			
				Width	Avg. Thick-ness			I	S	r	I	S	r	x
In.	Lb.	In.^{2}	In.	In.	In.	In.		In.^{4}	In.^{3}	In.	In.^{4}	In.^{3}	In.	In.
*18 × 4	58.0	16.98	18.00	4.200	.625	.700	6.86	670.7	74.5	6.29	18.5	5.6	1.04	.88
	51.9	15.18	18.00	4.100	.625	.600	7.02	622.1	69.1	6.40	17.1	5.3	1.06	.87
	45.8	13.38	18.00	4.000	.625	.500	7.20	573.5	63.7	6.55	15.8	5.1	1.09	.89
	42.7	12.48	18.00	3.950	.625	.450	7.29	549.2	61.0	6.64	15.0	4.9	1.10	.90
15 × 3⅜	50.0	14.64	15.00	3.716	.650	.716	6.21	401.4	53.6	5.24	11.2	3.8	.87	.80
	40.0	11.70	15.00	3.520	.650	.520	6.56	346.3	46.2	5.44	9.3	3.4	.89	.78
	33.9	9.90	15.00	3.400	.650	.400	6.79	312.6	41.7	5.62	8.2	3.2	.91	.79
12 × 3	30.0	8.79	12.00	3.170	.501	.510	7.56	161.2	26.9	4.28	5.2	2.1	.77	.68
	25.0	7.32	12.00	3.047	.501	.387	7.68	143.5	23.9	4.43	4.5	1.9	.79	.68
	20.7	6.03	12.00	2.940	.501	.280	8.15	128.1	21.4	4.61	3.9	1.7	.81	.70
10 × 2⅝	30.0	8.80	10.00	3.033	.436	.673	7.56	103.0	20.6	3.42	4.0	1.7	.67	.65
	25.0	7.33	10.00	2.886	.436	.526	7.95	90.7	18.1	3.52	3.4	1.5	.68	.62
	20.0	5.86	10.00	2.739	.436	.379	8.37	78.5	15.7	3.66	2.8	1.3	.70	.61
	15.3	4.47	10.00	2.600	.436	.240	8.82	66.9	13.4	3.87	2.3	1.2	.72	.64
9 × 2½	20.0	5.86	9.00	2.648	.413	.448	8.23	60.6	13.5	3.22	2.4	1.2	.65	.59
	15.0	4.39	9.00	2.485	.413	.285	8.77	50.7	11.3	3.40	1.9	1.0	.67	.59
	13.4	3.89	9.00	2.430	.413	.230	8.97	47.3	10.5	3.49	1.8	.97	.67	.61
8 × 2¼	18.75	5.49	8.00	2.527	.390	.487	8.12	43.7	10.9	2.82	2.0	1.0	.60	.57
	13.75	4.02	8.00	2.343	.390	.303	8.75	35.8	9.0	2.99	1.5	.86	.62	.56
	11.5	3.36	8.00	2.260	.390	.220	9.08	32.3	8.1	3.10	1.3	.79	.63	.58
8 × 1⅞	8.5	2.49	8.00	1.875	.321	.180	13.29	23.7	5.9	3.08	.65	.46	.51	.45
7 × 2⅛	14.75	4.32	7.00	2.299	.366	.419	8.32	27.1	7.7	2.51	1.4	.79	.57	.53
	12.25	3.58	7.00	2.194	.366	.314	8.72	24.1	6.9	2.59	1.2	.71	.58	.53
	9.8	2.85	7.00	2.090	.366	.210	9.15	21.1	6.0	2.72	.98	.63	.59	.55
6 × 2	13.0	3.81	6.00	2.157	.343	.437	8.11	17.3	5.8	2.13	1.1	.65	.53	.52
	10.5	3.07	6.00	2.034	.343	.314	8.60	15.1	5.0	2.22	.87	.57	.53	.50
	8.2	2.39	6.00	1.920	.343	.200	9.11	13.0	4.3	2.34	.70	.50	.54	.52
5 × 1¾	9.0	2.63	5.00	1.885	.320	.325	8.29	8.8	3.5	1.83	.64	.45	.49	.48
	6.7	1.95	5.00	1.750	.320	.190	8.93	7.4	3.0	1.95	.48	.38	.50	.49
4 × 1⅝	7.25	2.12	4.00	1.720	.296	.320	7.86	4.5	2.3	1.47	.44	.35	.46	.46
	5.4	1.56	4.00	1.580	.296	.180	8.55	3.8	1.9	1.56	.32	.29	.45	.46
3 × 1½	6.0	1.75	3.00	1.596	.273	.356	6.89	2.1	1.4	1.08	.31	.27	.42	.46
	5.0	1.46	3.00	1.498	.273	.258	7.34	1.8	1.2	1.12	.25	.24	.41	.44
	4.1	1.19	3.00	1.410	.273	.170	7.79	1.6	1.1	1.17	.20	.21	.41	.44

* Car and shipbuilding channel: not an American Standard.

AMERICAN INSTITUTE OF STEEL CONSTRUCTION

ings, means that in which exterior walls and bearing walls are of approved masonry or reinforced concrete, and in which the structural elements are wholly or partly of wood of smaller dimensions than required for heavy timber construction, or of steel or iron not protected as required for fireproof construction or semi-fireproof construction.

The term "ordinary construction" corresponds generally with that variously called "nonfireproof," "masonry walls and wooden joists," or "ordinary masonry" construction.

"Light noncombustible construction," as applied to buildings, means that in which all structural members including walls, floors, roofs and their supports are of steel, iron, concrete, or of other noncombustible materials, and in which the exterior enclosure walls are of masonry or concrete, or are of other fire resistive materials or assemblies of materials which have not less than 2-hour fire resistance ratings.

Structures similar to all-metal gasoline service stations would be included under this classification.

"Frame construction," as applied to buildings, means that in which walls and interior construction are wholly or partly of wood.

Buildings of exterior masonry veneer, metal, or stucco on wooden frame, constituting wholly or in part the structural supports of the building or its loads, are frame buildings within the meaning of this definition.

"Unprotected metal construction," as applied to buildings, means that in which the structural supports are unprotected metal in which the roofing and walls or other enclosures are of sheet metal, or of other noncombustible materials, or of masonry deficient in thickness or otherwise not conforming to approved masonry.

c The second classification of construction is between skeleton and wall-bearing. Wall-bearing buildings are limited in height because of the thickness of walls required in buildings of over three stories. Even in buildings of less than three stories it may be advisable to use skeleton construction on some occasions. Skeleton construction permits greater overall speed of construction, because the framework can be completely erected for walls and interior construction at the same time instead of slowing down steel erection to keep pace with the masonry walls. In factory buildings it is often necessary to use skeleton construction even for a one-story building, sometimes to provide for crane runways and sometimes to allow sufficient window area.

The chief advantages of wall-bearing construction, except for buildings using wood columns and beams, are slightly lower cost and freedom from breaks in the inside wall line, which are needed to cover up wall columns.

In any area subject to earthquakes, it is extremely advisable to use skeleton construction.

d The third classification, and the one subject to most argument, distinguishes between structural-steel frame and reinforced concrete. Without going into too much detail, these general facts may be given pro and con:

1. For high buildings, structural steel has the advantage because of the lesser column areas and the greater simplicity of wind bracing.

2. For buildings carrying heavy loads, the cost advantage is decidedly with reinforced concrete. No other form of construction for heavy loads can approach flat-slab reinforced concrete in cost when the spans are right.

3. For any type of factory where it will be necessary to change operations from time to time or where structural changes will be required, steel is better because it is easier to connect to.

4. Steel is usually better for long-span construction, although since the advent of rigid-frame construction the reinforced concrete advocates can dispute this point.

5. Structural steel is more foolproof because the main supporting parts are shop-fabricated, and it is not so subject to weather changes, human failings, etc., as reinforced concrete.

The preference for one material over the other in many types of structures where the requirement is not self-evident is largely a question of minor factors like the current local market, the contractors available, and the attitude of the owner.

e It is permissible to change from one

type of construction to another in various parts of the same building; for instance, in a two-story school where the structure in general is wall-bearing reinforced concrete, skeleton structural steel may be used over the auditorium because of the long span and light walls. However, it is inadvisable to use several systems of construction in the same job. It tends to increase the cost and slow down the work.

1.41 Floor systems

The selection of a floor system is largely independent of the various general systems of construction discussed in Art. 1.40. For instance, short-span concrete floors may be used whether a building is wall-bearing or skeleton, fireproof or semifireproof, with a structural-steel or reinforced concrete frame.

The number of floor systems using patented or specially manufactured materials is far too great to cover in any but a superficial manner. The general types of systems, however, may be grouped briefly as to spans, materials, and theory of design:

1. Short-span systems, suitable for spans from 3 to 10 ft
   - *a.* Concrete slabs
   - *b.* Gypsum systems
   - *c.* Cellular steel systems
2. Long-span one-way systems, suitable from 9 to about 30 ft
   - *a.* Concrete joists cast in place
     - Tile fillers
     - Metal fillers
   - *b.* Precast concrete plank or double-T slabs
   - *c.* Steel joists or junior beams
   - *d.* Cellular steel systems
3. Two-way systems
   - *a.* Two-way slabs on steel or concrete beams
     - Solid concrete
     - Concrete with tile fillers (Shuster system)
     - Concrete with concrete-block fillers (Republic system)
     - Concrete with metal-dome fillers (grid system)
   - *b.* Flat slab

Reference was made in the Preface to several types of concrete roof construction which are not mentioned above. Several of the above-mentioned systems of concrete construction lend themselves economically to a lift-slab system of construction, designed and constructed by competent engineers and contractors.

The various factors which might influence the selection of the floor system to be used are as follows:

1. Use of floor
   - *a.* Loading
     - Light or heavy
     - Uniform or concentrated
   - *b.* Span
     - Long or short
     - Regularity of support arrangement
   - *c.* Freedom from obstruction
     - Columns
     - Beams or girders
     - Incorporation of pipes, conduits, or vent shafts
   - *d.* Rigidity and resistance to:
     - Machinery shocks
     - Outside vibrations
     - Wind stresses or earthquakes
   - *e.* Adaptability to alteration
     - Main members
     - Secondary members
   - *f.* Conformity to code
     - City or other local codes
     - Fire or other insurance codes
   - *g.* Insulation
     - Sound
     - Heat or cold
   - *h.* Acoustical properties
2. Cost of floor
   - *a.* First cost
   - *b.* Maintenance cost
   - *c.* Indirect cost (affecting remainder of building)
     - Ability to "tie in" to structure
     - Difference in weight and story height as affecting size and cost of supporting columns and foundations
   - *d.* Royalty costs of patented types
   - *e.* Salvage of form work

3. Appearance of floor
   a. Floor finish
      Integral concrete
      Any finish applied later
   b. Ceiling finish
      Exposed and untreated
      Plaster
      Paint
4. Local building conditions
   a. Material
      Supply, including freight rate
      Handling on the job
      Storage space
   b. Labor
      Type available
      Possibility of importation
      Labor restrictions
   c. Special conditions of bidding by contractors
   d. Climatic conditions
5. Time requirements
   a. Speed of construction as affecting revenue of the building
   b. Delays caused by purchase of special materials not available locally
6. Supervision
   a. Normal
   b. Extra rigid, due to unusual or new system involved
   c. Ease of repairing the result of slight errors or omissions, without reducing factors of safety

It is advisable to avoid the use of any system of construction which requires that the rest of the building be built to fit the floor system. What advantage is there in saving 3¢ per sq ft in the floor construction and spending 3½¢ extra in electric and heating contracts?

Further discussion which may influence the selection of a floor system will be found in the sections dealing with the design of floors, as follows:

1. Short-span systems
   Concrete slabs—Art. 5.40
   Gypsum systems—Art. 5.60
2. Long-span one-way systems
   Concrete joists—Art. 5.41
   Steel joists—Art. 4.30
3. Two-way systems
   Two-way slabs—Art. 5.50
   Flat slabs—Art. 5.51

1.42 Fire-resistance ratings

In Art. 1.40 (*a*), under the heading of "2. Economics," reference is made to "Insurance cost." Insurance rates vary for different localities and it is impossible to discuss them intelligently in such a book as this. Any local insurance broker can give you figures which will enable you to weigh the economics of this item; but assuming a typical case with an 80 percent co-insurance clause, the figures would approximate this ratio:

For a frame building—$1.10/$100
For a brick building—$.55/$100
For an incombustible building—$.24/$100
For a fire-resistive building—$.11/$100

One of the chief criteria for the use of any construction system or detail of construction is based on the fire-resistance rating, or as it is frequently called, the fire rating. This rating is being adopted by building codes throughout the country as the standard of measurement of construction required. Ratings are based on standard fire tests as set up by the American Standards Association. The generally recognized authority for fire ratings is the "Technical Report on Building Materials 44" (commonly called "TRBM-44") and pamphlet BMS-92 of the National Bureau of Standards. The generally accepted standards are listed in the pamphlet "Fire Resistance Ratings" published by the National Board of Fire Underwriters.

1.50 DESIGN LOADS

a Every part of a building or structure must be designed to support safely a prescribed live load plus the full dead load.

The dead load is the entire load of construction, including all carrying members, slabs, fill and finish, ceilings, and partitions. Live loads include all superimposed loads, machinery, furniture, moving loads, snow loads on roofs, occupants, and stored material. Unless it is known that the use of the

Table 1.10

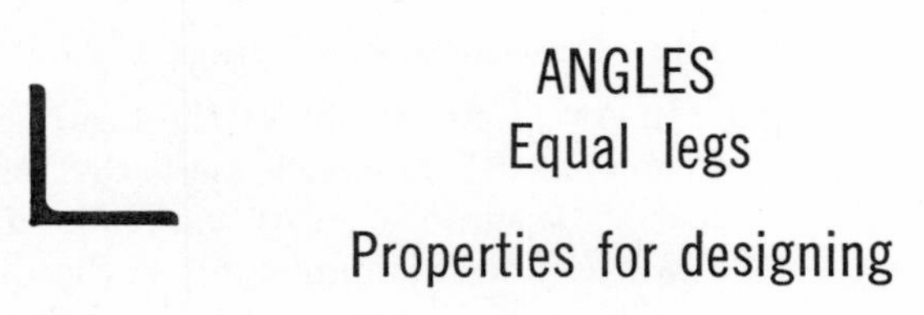

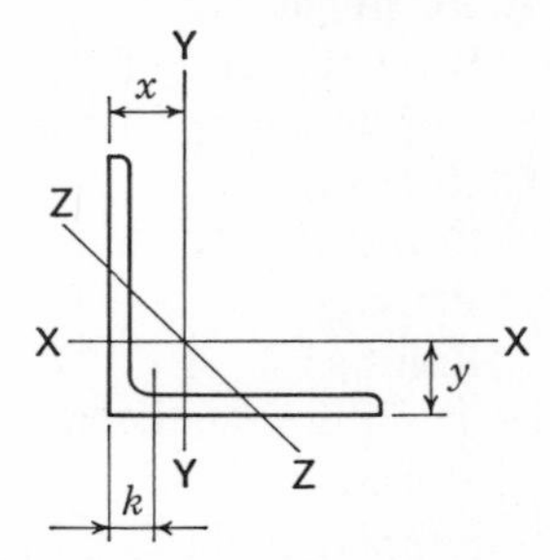

Size	Thickness	k	Weight per Foot	Area	AXIS X-X AND AXIS Y-Y				AXIS Z-Z
					I	S	r	x or y	r
In.	In.	In.	Lb.	In.²	In.⁴	In.³	In.	In.	In.
8 × 8	1⅛	1¾	56.9	16.73	98.0	17.5	2.42	2.41	1.55
	1	1⅝	51.0	15.00	89.0	15.8	2.44	2.37	1.56
	⅞	1½	45.0	13.23	79.6	14.0	2.45	2.32	1.57
	¾	1⅜	38.9	11.44	69.7	12.2	2.47	2.28	1.57
	⅝	1¼	32.7	9.61	59.4	10.3	2.49	2.23	1.58
	9/16	1 3/16	29.6	8.68	54.1	9.3	2.50	2.21	1.58
	½	1⅛	26.4	7.75	48.6	8.4	2.50	2.19	1.58
6 × 6	1	1½	37.4	11.00	35.5	8.6	1.80	1.86	1.16
	⅞	1⅜	33.1	9.73	31.9	7.5	1.81	1.82	1.17
	¾	1¼	28.7	8.44	28.2	6.7	1.83	1.78	1.17
	⅝	1⅛	24.2	7.11	24.2	5.7	1.84	1.73	1.18
	9/16	1 1/16	21.9	6.43	22.1	5.1	1.85	1.71	1.18
	½	1	19.6	5.75	19.9	4.6	1.86	1.68	1.18
	7/16	15/16	17.2	5.06	17.7	4.1	1.87	1.66	1.19
	⅜	⅞	14.9	4.36	15.4	3.5	1.88	1.64	1.19
	5/16	13/16	12.4	3.65	13.0	3.0	1.89	1.62	1.20
5 × 5	⅞	1⅜	27.2	7.98	17.8	5.2	1.49	1.57	.96
	¾	1¼	23.6	6.94	15.7	4.5	1.51	1.52	.97
	⅝	1⅛	20.0	5.86	13.6	3.9	1.52	1.48	.97
	½	1	16.2	4.75	11.3	3.2	1.54	1.43	.98
	7/16	15/16	14.3	4.18	10.0	2.8	1.55	1.41	.98
	⅜	⅞	12.3	3.61	8.7	2.4	1.56	1.39	.99
	5/16	13/16	10.3	3.03	7.4	2.0	1.56	1.37	.99
4 × 4	¾	1⅛	18.5	5.44	7.7	2.8	1.19	1.27	.77
	⅝	1	15.7	4.61	6.7	2.4	1.20	1.23	.77
	½	⅞	12.8	3.75	5.6	2.0	1.22	1.18	.78
	7/16	13/16	11.3	3.31	5.0	1.8	1.23	1.16	.78
	⅜	¾	9.8	2.86	4.4	1.5	1.23	1.14	.79
	5/16	11/16	8.2	2.40	3.7	1.3	1.24	1.12	.79
	¼	⅝	6.6	1.94	3.0	1.0	1.25	1.09	.79

AMERICAN INSTITUTE OF STEEL CONSTRUCTION

Table 1.10 (*Continued*)

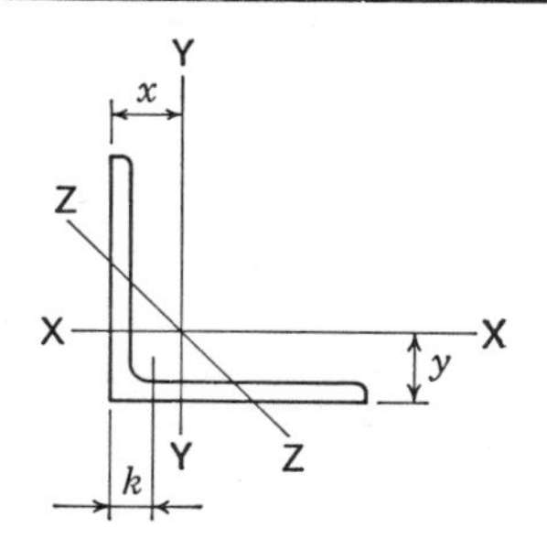

ANGLES
Equal legs
Properties for designing

Size	Thick-ness	k	Weight per Foot	Area	AXIS X-X AND AXIS Y-Y				AXIS Z-Z
					I	S	r	x or y	r
In.	In.	In.	Lb.	In.²	In.⁴	In.³	In.	In.	In.
3½ × 3½	½	⅞	11.1	3.25	3.6	1.5	1.06	1.06	.68
	7/16	13/16	9.8	2.87	3.3	1.3	1.07	1.04	.68
	⅜	¾	8.5	2.48	2.9	1.2	1.07	1.01	.69
	5/16	11/16	7.2	2.09	2.5	.98	1.08	.99	.69
	¼	⅝	5.8	1.69	2.0	.79	1.09	.97	.69
3 × 3	½	13/16	9.4	2.75	2.2	1.1	.90	.93	.58
	7/16	¾	8.3	2.43	2.0	.95	.91	.91	.58
	⅜	11/16	7.2	2.11	1.8	.83	.91	.89	.58
	5/16	⅝	6.1	1.78	1.5	.71	.92	.87	.59
	¼	9/16	4.9	1.44	1.2	.58	.93	.84	.59
	3/16	½	3.71	1.09	.96	.44	.94	.82	.59
2½ × 2½	½	¾	7.7	2.25	1.2	.73	.74	.81	.47
	⅜	⅝	5.9	1.73	.98	.57	.75	.76	.48
	5/16	9/16	5.0	1.47	.85	.48	.76	.74	.49
	¼	½	4.1	1.19	.70	.39	.77	.72	.49
	3/16	7/16	3.07	.90	.55	.30	.78	.69	.49
2 × 2	⅜	9/16	4.7	1.36	.48	.35	.59	.64	.39
	5/16	½	3.92	1.15	.42	.30	.60	.61	.39
	¼	7/16	3.19	.94	.35	.25	.61	.59	.39
	3/16	⅜	2.44	.71	.28	.19	.62	.57	.40
	⅛	5/16	1.65	.48	.19	.13	.63	.55	.40
1¾ × 1¾	¼	7/16	2.77	.81	.23	.19	.53	.53	.34
	3/16	⅜	2.12	.62	.18	.14	.54	.51	.35
	⅛	5/16	1.44	.42	.13	.10	.55	.48	.35
1½ × 1½	¼	⅜	2.34	.69	.14	.13	.45	.47	.29
	3/16	5/16	1.80	.53	.11	.10	.46	.44	.29
	⅛	¼	1.23	.36	.078	.072	.46	.42	.30
1¼ × 1¼	¼	⅜	1.92	.56	.077	.091	.37	.40	.24
	3/16	5/16	1.48	.43	.061	.071	.38	.38	.24
	⅛	¼	1.01	.30	.044	.049	.38	.35	.25
1 × 1	¼	⅜	1.49	.44	.037	.056	.29	.34	.19
	3/16	5/16	1.16	.34	.030	.044	.30	.32	.19
	⅛	¼	.80	.23	.022	.031	.31	.30	.20

AMERICAN INSTITUTE OF STEEL CONSTRUCTION

Table 1.11

ANGLES
Unequal legs
Properties for designing

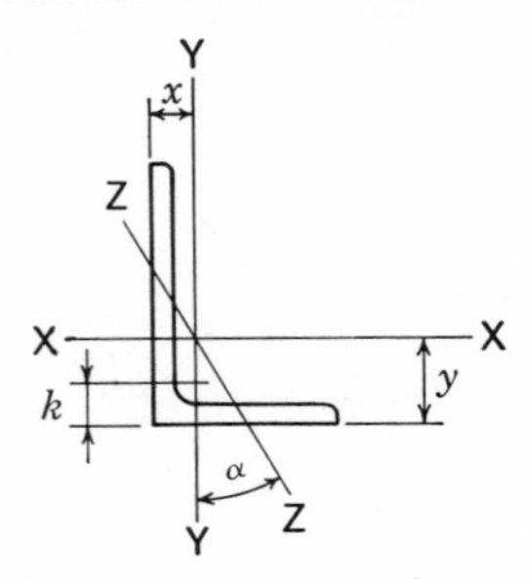

Size	Thick-ness	k	Weight per Foot	Area	AXIS X-X				AXIS Y-Y				AXIS Z-Z	
					I	S	r	y	I	S	r		r	Tan α
In.	In.	In.	Lb.	In.^{2}	In.^{4}	In.^{3}	In.	In.	In.^{4}	In.^{3}	In.	In.	In.	
*9 × 4	1	1½	40.8	12.00	97.0	17.6	2.84	3.50	12.0	4.0	1.00	1.00	.83	.203
	⅞	1⅜	36.1	10.61	86.8	15.7	2.86	3.45	10.8	3.6	1.01	.95	.84	.208
	¾	1¼	31.3	9.19	76.1	13.6	2.88	3.41	9.6	3.1	1.02	.91	.84	.212
	⅝	1⅛	26.3	7.73	64.9	11.5	2.90	3.36	8.3	2.6	1.04	.86	.85	.216
	9/16	1 1/16	23.8	7.00	59.1	10.4	2.91	3.33	7.6	2.4	1.04	.83	.85	.218
	½	1	21.3	6.25	53.2	9.3	2.92	3.31	6.9	2.2	1.05	.81	.85	.220
8 × 6	1	1½	44.2	13.00	80.8	15.1	2.49	2.65	38.8	8.9	1.73	1.65	1.28	.543
	⅞	1⅜	39.1	11.48	72.3	13.4	2.51	2.61	34.9	7.9	1.74	1.61	1.28	.547
	¾	1¼	33.8	9.94	63.4	11.7	2.53	2.56	30.7	6.9	1.76	1.56	1.29	.551
	⅝	1⅛	28.5	8.36	54.1	9.9	2.54	2.52	26.3	5.9	1.77	1.52	1.29	.554
	9/16	1 1/16	25.7	7.56	49.3	9.0	2.55	2.50	24.0	5.3	1.78	1.50	1.30	.556
	½	1	23.0	6.75	44.3	8.0	2.56	2.47	21.7	4.8	1.79	1.47	1.30	.558
	7/16	15/16	20.2	5.93	39.2	7.1	2.57	2.45	19.3	4.2	1.80	1.45	1.31	.560
8 × 4	1	1½	37.4	11.00	69.6	14.1	2.52	3.05	11.6	3.9	1.03	1.05	.85	.247
	⅞	1⅜	33.1	9.73	62.5	12.5	2.53	3.00	10.5	3.5	1.04	1.00	.85	.253
	¾	1¼	28.7	8.44	54.9	10.9	2.55	2.95	9.4	3.1	1.05	.95	.85	.258
	⅝	1⅛	24.2	7.11	46.9	9.2	2.57	2.91	8.1	2.6	1.07	.91	.86	.262
	9/16	1 1/16	21.9	6.43	42.8	8.4	2.58	2.88	7.4	2.4	1.07	.88	.86	.265
	½	1	19.6	5.75	38.5	7.5	2.59	2.86	6.7	2.2	1.08	.86	.86	.267
	7/16	15/16	17.2	5.06	34.1	6.6	2.60	2.83	6.0	1.9	1.09	.83	.87	.269
7 × 4	⅞	1⅜	30.2	8.86	42.9	9.7	2.20	2.55	10.2	3.5	1.07	1.05	.86	.318
	¾	1¼	26.2	7.69	37.8	8.4	2.22	2.51	9.1	3.0	1.09	1.01	.86	.324
	⅝	1⅛	22.1	6.48	32.4	7.1	2.24	2.46	7.8	2.6	1.10	.96	.86	.329
	9/16	1 1/16	20.0	5.87	29.6	6.5	2.24	2.44	7.2	2.4	1.11	.94	.87	.332
	½	1	17.9	5.25	26.7	5.8	2.25	2.42	6.5	2.1	1.11	.92	.87	.335
	7/16	15/16	15.8	4.62	23.7	5.1	2.26	2.39	5.8	1.9	1.12	.89	.88	.337
	⅜	⅞	13.6	3.98	20.6	4.4	2.27	2.37	5.1	1.6	1.13	.87	.88	.339

* Rolled by Bethlehem Steel Company and U. S. Steel Corp.

Table 1.11 (***Continued***)

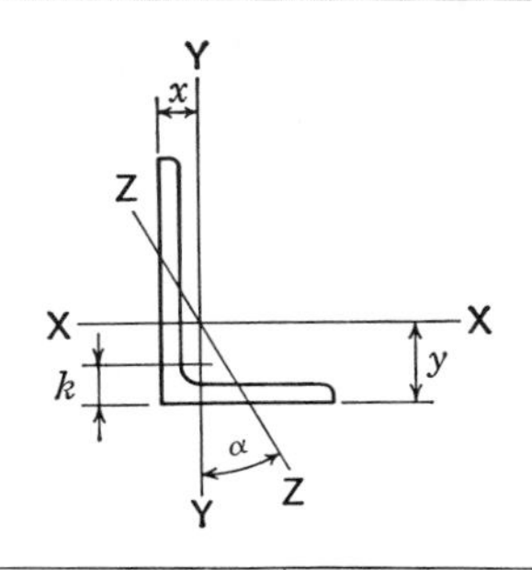

ANGLES
Unequal legs
Properties for designing

L

Size	Thick-ness	k	Weight per Foot	Area	AXIS X-X				AXIS Y-Y				AXIS Z-Z	
					I	S	r	y	I	S	r	x	r	Tan α
In.	In.	In.	Lb.	In.^{2}	In.^{4}	In.^{3}	In.	In.	In.^{4}	In.^{3}	In.	In.	In.	
6 × 4	7/8	1 3/8	27.2	7.98	27.7	7.2	1.86	2.12	9.8	3.4	1.11	1.12	.86	.421
	3/4	1 1/4	23.6	6.94	24.5	6.3	1.88	2.08	8.7	3.0	1.12	1.08	.86	.428
	5/8	1 1/8	20.0	5.86	21.1	5.3	1.90	2.03	7.5	2.5	1.13	1.03	.86	.435
	9/16	1 1/16	18.1	5.31	19.3	4.8	1.90	2.01	6.9	2.3	1.14	1.01	.87	.438
	1/2	1	16.2	4.75	17.4	4.3	1.91	1.99	6.3	2.1	1.15	.99	.87	.440
	7/16	15/16	14.3	4.18	15.5	3.8	1.92	1.96	5.6	1.9	1.16	.96	.87	.443
	3/8	7/8	12.3	3.61	13.5	3.3	1.93	1.94	4.9	1.6	1.17	.94	.88	.446
	5/16	13/16	10.3	3.03	11.4	2.8	1.94	1.92	4.2	1.4	1.17	.92	.88	.449
	1/4	3/4	8.3	2.44	9.3	2.3	1.95	1.89	3.4	1.1	1.18	.89	.89	.449
6 × 3 1/2	1/2	1	15.3	4.50	16.6	4.2	1.92	2.08	4.3	1.6	.97	.83	.76	.344
	3/8	7/8	11.7	3.42	12.9	3.2	1.94	2.04	3.3	1.2	.99	.79	.77	.350
	5/16	13/16	9.8	2.87	10.9	2.7	1.95	2.01	2.9	1.0	1.00	.76	.77	.352
	1/4	3/4	7.9	2.31	8.9	2.2	1.96	1.99	2.3	0.85	1.01	.74	.78	.355
5 × 3 1/2	3/4	1 3/16	19.8	5.81	13.9	4.3	1.55	1.75	5.6	2.2	.98	1.00	.75	.464
	5/8	1 1/16	16.8	4.92	12.0	3.7	1.56	1.70	4.8	1.9	.99	.95	.75	.472
	1/2	15/16	13.6	4.00	10.0	3.0	1.58	1.66	4.1	1.6	1.01	.91	.75	.479
	7/16	7/8	12.0	3.53	8.9	2.6	1.59	1.63	3.6	1.4	1.01	.88	.76	.482
	3/8	13/16	10.4	3.05	7.8	2.3	1.60	1.61	3.2	1.2	1.02	.86	.76	.486
	5/16	3/4	8.7	2.56	6.6	1.9	1.61	1.59	2.7	1.0	1.03	.84	.76	.489
	1/4	11/16	7.0	2.06	5.4	1.6	1.61	1.56	2.2	.83	1.04	.81	.76	.492
5 × 3	1/2	7/8	12.8	3.75	9.5	2.9	1.59	1.75	2.6	1.1	.83	.75	.65	.357
	7/16	13/16	11.3	3.31	8.4	2.6	1.60	1.73	2.3	1.0	.84	.73	.65	.361
	3/8	3/4	9.8	2.86	7.4	2.2	1.61	1.70	2.0	.89	.84	.70	.65	.364
	5/16	11/16	8.2	2.40	6.3	1.9	1.61	1.68	1.8	.75	.85	.68	.66	.368
	1/4	5/8	6.6	1.94	5.1	1.5	1.62	1.66	1.4	.61	.86	.66	.66	.371

Table 1.11 (*Continued*)

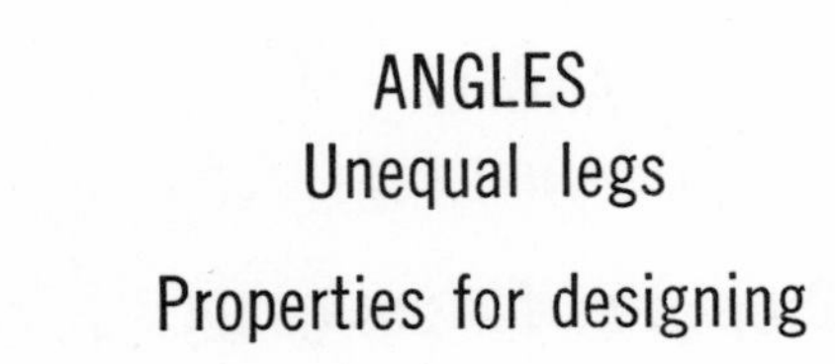

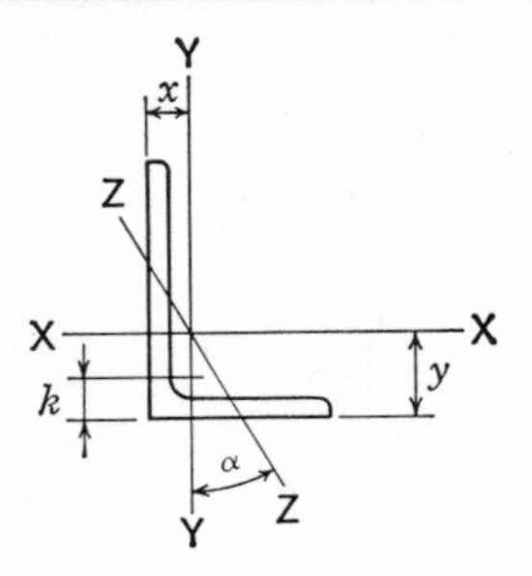

Size	Thickness	k	Weight per Foot	Area	AXIS X-X				AXIS Y-Y				AXIS Z-Z	
					I	S	r	y	I	S	r	x	r	Tan α
In.	In.	In.	Lb.	In.²	In.⁴	In.³	In.	In.	In.⁴	In.³	In.	In.	In.	
4 × 3½	⅝	1	14.7	4.30	6.4	2.4	1.22	1.29	4.5	1.8	1.03	1.04	.72	.745
	½	⅞	11.9	3.50	5.3	1.9	1.23	1.25	3.8	1.5	1.04	1.00	.72	.750
	7/16	13/16	10.6	3.09	4.8	1.7	1.24	1.23	3.4	1.4	1.05	.98	.72	.753
	⅜	¾	9.1	2.67	4.2	1.5	1.25	1.21	3.0	1.2	1.06	.96	.73	.755
	5/16	11/16	7.7	2.25	3.6	1.3	1.26	1.18	2.6	1.0	1.07	.93	.73	.757
	¼	⅝	6.2	1.81	2.9	1.0	1.27	1.16	2.1	.81	1.07	.91	.73	.759
4 × 3	⅝	1	13.6	3.98	6.0	2.3	1.23	1.37	2.9	1.4	.85	.87	.64	.534
	½	⅞	11.1	3.25	5.1	1.9	1.25	1.33	2.4	1.1	.86	.83	.64	.543
	7/16	13/16	9.8	2.87	4.5	1.7	1.25	1.30	2.2	1.0	.87	.80	.64	.547
	⅜	¾	8.5	2.48	4.0	1.5	1.26	1.28	1.9	.87	.88	.78	.64	.551
	5/16	11/16	7.2	2.09	3.4	1.2	1.27	1.26	1.7	.73	.89	.76	.65	.554
	¼	⅝	5.8	1.69	2.8	1.0	1.28	1.24	1.4	.60	.90	.74	.65	.558
3½ × 3	½	⅞	10.2	3.00	3.5	1.5	1.07	1.13	2.3	1.1	.88	.88	.62	.714
	7/16	13/16	9.1	2.65	3.1	1.3	1.08	1.10	2.1	.98	.89	.85	.62	.718
	⅜	¾	7.9	2.30	2.7	1.1	1.09	1.08	1.9	.85	.90	.83	.62	.721
	5/16	11/16	6.6	1.93	2.3	.95	1.10	1.06	1.6	.72	.90	.81	.63	.724
	¼	⅝	5.4	1.56	1.9	.78	1.11	1.04	1.3	.59	.91	.79	.63	.727
3½ × 2½	½	13/16	9.4	2.75	3.2	1.4	1.09	1.20	1.4	.76	.70	.70	.53	.486
	7/16	¾	8.3	2.43	2.9	1.3	1.09	1.18	1.2	.68	.71	.68	.54	.491
	⅜	11/16	7.2	2.11	2.6	1.1	1.10	1.16	1.1	.59	.72	.66	.54	.496
	5/16	⅝	6.1	1.78	2.2	.93	1.11	1.14	.94	.50	.73	.64	.54	.501
	¼	9/16	4.9	1.44	1.8	.75	1.12	1.11	.78	.41	.74	.61	.54	.506
3 × 2½	½	13/16	8.5	2.50	2.1	1.0	.91	1.00	1.3	.74	.72	.75	.52	.667
	7/16	¾	7.6	2.21	1.9	.93	.92	.98	1.2	.66	.73	.73	.52	.672
	⅜	11/16	6.6	1.92	1.7	.81	.93	.96	1.0	.58	.74	.71	.52	.676
	5/16	⅝	5.6	1.62	1.4	.69	.94	.93	.90	.49	.74	.68	.53	.680
	¼	9/16	4.5	1.31	1.2	.56	.95	.91	.74	.40	.75	.66	.53	.684

Table 1.11 (*Continued*)

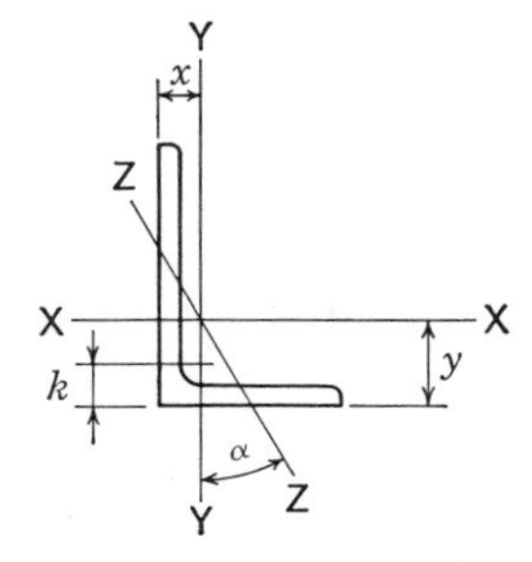

ANGLES
Unequal angles
Properties for designing

L

Size	Thickness	k	Weight per Foot	Area	AXIS X-X				AXIS Y-Y				AXIS Z-Z	
					I	S	r	y	I	S	r	x	r	Tan α
In.	In.	In.	Lb.	In.^{2}	In.^{4}	In.^{3}	In.	In.	In.^{4}	In.^{3}	In.	In.	In.	
3 × 2	1/2	13/16	7.7	2.25	1.9	1.0	.92	1.08	.67	.47	.55	.58	.43	.414
	7/16	3/4	6.8	2.00	1.7	.89	.93	1.06	.61	.42	.55	.56	.43	.421
	3/8	11/16	5.9	1.73	1.5	.78	.94	1.04	.54	.37	.56	.54	.43	.428
	5/16	5/8	5.0	1.47	1.3	.66	.95	1.02	.47	.32	.57	.52	.43	.435
	1/4	9/16	4.1	1.19	1.1	.54	.96	.99	.39	.26	.57	.49	.43	.440
	3/16	1/2	3.07	.90	.84	.41	.97	.97	.31	.20	.58	.47	.44	.446
2½ × 2	3/8	5/8	5.3	1.55	.91	.55	.77	.83	.51	.36	.58	.58	.42	.614
	5/16	9/16	4.5	1.31	.79	.47	.78	.81	.45	.31	.58	.56	.42	.620
	1/4	1/2	3.62	1.06	.65	.38	.78	.79	.37	.25	.59	.54	.42	.626
	3/16	7/16	2.75	.81	.51	.29	.79	.76	.29	.20	.60	.51	.43	.631
2½ × 1½	5/16	1/2	3.92	1.15	.71	.44	.79	.90	.19	.17	.41	.40	.32	.349
	1/4	7/16	3.19	.94	.59	.36	.79	.88	.16	.14	.41	.38	.32	.357
	3/16	3/8	2.44	.72	.46	.28	.80	.85	.13	.11	.42	.35	.33	.364
2 × 1½	1/4	7/16	2.77	.81	.32	.24	.62	.66	.15	.14	.43	.41	.32	.543
	3/16	3/8	2.12	.62	.25	.18	.63	.64	.12	.11	.44	.39	.32	.551
	1/8	5/16	1.44	.42	.17	.13	.64	.62	.085	.075	.45	.37	.33	.558
2 × 1¼	1/4	7/16	2.55	.75	.30	.23	.63	.71	.089	.097	.34	.33	.27	.372
	3/16	3/8	1.96	.57	.23	.18	.64	.69	.071	.075	.35	.31	.27	.378
	1/8	5/16	1.33	.40	.17	.13	.65	.65	.050	.052	.36	.28	.27	.396
1¾ × 1¼	1/4	7/16	2.34	.69	.20	.18	.54	.60	.085	.095	.35	.35	.27	.486
	3/16	3/8	1.80	.53	.16	.14	.55	.58	.069	.075	.36	.33	.27	.496
	1/8	5/16	1.23	.36	.11	.094	.56	.56	.049	.052	.37	.31	.27	.506

building will require the use of heavier loads, the live load shall be in accordance with code requirements. It is well to check all city and state codes covering the requirements, but in the absence of any such code, the requirements of "Minimum Live Loads Allowable for Use in the Design of Buildings" compiled by the U.S. Department of Commerce (Tables 1.12, 1.13, and 1.14) are accepted as good engineering practice.

b The dead load of the floor slab is made up of

1. Finish
2. Fill
3. Slab
4. Ceiling
5. Partition

Some codes require the load of the partition to be spread out uniformly over the entire slab as a means of providing for possible relocation of partitions, as is frequently done in office buildings.

The New York City Code requires for office and public buildings, where partitions are apt to be moved to different locations, that an allowance of 20 psf shall be made as additional dead load on the floors in place of partition load. The U.S. Army requires for hospitals and similar structures, using short-span systems, that a load of 300 lb per lin ft be applied centrally concentrated on any slab, or on any beam or girder.

The weights of materials entering into the dead weight are given in Table 1.15.

The weight of the beam and haunching does not need to be taken into the slab, but to arrive at the design load for the beam it must be added in.

If there is no haunching, the weight of a steel beam does not add much, and the beam can be estimated as 1 in. depth for 1.75 ft span, using the next commercial size above this; for a 17-ft span, 10 WF 21. If the beam is haunched, assume the beam as above, and estimate the weight of beam and haunching per foot to be $d(b + 4) + \frac{2}{3}w'$, where d is the depth of steel beam, b the width of flange of beam, and w' the weight of beam estimated. Thus a 10 WF 21 would be

$$10(5.75 + 4) + (\tfrac{2}{3} \times 21) = 97.5 + 14 = 111.5 \text{ psf}$$

This allows for the weight of stone or gravel concrete.

If the beams are spaced 7 ft o.c. this will add to the uniform floor load 111.5/7 = 16 psf. Ordinarily, however, this weight of beam and haunching is merely added into the load per linear foot of beam for the design.

The load on the roof slab is made up of

1. Roofing
2. Fill (if required)
3. Slab
4. Ceiling

Spandrel beams carry their share of the interior load, and in addition they carry wall load. Solid brick wall may be estimated to weigh 10 psf per inch of nominal thickness—an 8-in. wall, 80 psf. Four-inch brick with 8-in. block backing, which is the common spandrel material for skeleton buildings, weighs 80 psf with plaster. Window openings are deducted, no allowance being made for the window itself. Loads from lintels are considered as spread uniformly over the piers supporting them to the beam below. Lintel beams in high brick walls are figured to carry a triangular load of masonry equal in height to the span of the lintel.

c For buildings over three stories high it is often permissible to reduce column loads as follows:

Roof and top story—full design load

Second story down—full dead load plus 95 per cent of live load

Third story down—full dead load plus 90 per cent of live load

Continue this reduction of 5 percent per story till 50 percent of the live load is reached, and carry this same 50 percent reduction through the balance of the floors to the foundations. The U.S. Department of Commerce Code, which has been used by many cities, uses the full design load for the top story, then reduces by 10 percent jumps to the 50 percent limitation. This load reduction is not usually permitted in warehouses and other buildings where the full load may be on all stories at the same time.

The approximate weight of the column and covering for any given story may be estimated by the equation $w = 60 + 0.4P/1{,}000$, where w = weight per foot of height, and P is the total weight down to the section of column under consideration.

d The American Standard Building Code Requirements for Minimum Design Loads in Buildings and Other Structures, sponsored by the National Bureau of Standards and approved 1955 by the American Standards Association, requires the minimum uniformly distributed live loads listed in Table 1.12. The following are extracts from these code requirements.

2.2 WEIGHT OF FIXED SERVICE EQUIPMENT. In estimating dead loads for purposes of design, the weight of fixed service equipment, such as plumbing stacks and risers, electrical feeders, heating, ventilating and air-conditioning systems, shall be included, whenever it is carried by structural members.

2.3 PROVISION FOR PARTITIONS. In office build-

Table 1.12 Minimum Uniformly Distributed Live Loads

Occupancy or use	Live load lb per sq ft
Apartments (*see* Residential)	
Armories and drill rooms	150
Assembly halls and other places of assembly:	
Fixed seats	60
Movable seats	100
Balcony (exterior)	100
Bowling alleys, poolrooms, and similar recreational areas	75
Corridors:	
First floor	100
Other floors, same as occupancy served except as indicated	
Dance halls	100
Dining rooms and restaurants	100
Dwellings (*see* Residential)	
Garages (passenger cars)	100
Floors shall be designed to carry 150 percent of the maximum wheel load anywhere on the floor.	
Grandstands (*see* Reviewing stands)	
Gymnasiums, main floors and balconies	100
Hospitals:	
Operating rooms	60
Private rooms	40
Wards	40
Hotels (*see* Residential)	
Libraries:	
Reading rooms	60
Stack rooms	150
Manufacturing	125
Marquees	75
Office buildings:	
Offices	80
Lobbies	100
Penal institutions:	
Cell blocks	40
Corridors	100
Residential:	
Multifamily houses:	
Private apartments	40
Public rooms	100
Corridors	60
Dwellings:	
First floor	40
Second floor and habitable attics	30
Uninhabitable attics	20
Hotels:	
Guest rooms	40
Public rooms	100
Corridors serving public rooms	100
Public corridors	60
Private corridors	40
Reviewing stands and bleachers	100
Schools:	
Classrooms	40
Corridors	100
Sidewalks, vehicular driveways, and yards, subject to trucking	250
Skating rinks	100
Stairs, fire escapes, and exitways	100
Storage warehouse, light	125
Storage warehouse, heavy	250
Stores:	
Retail:	
First floor, rooms	100
Upper floors	75
Wholesale	125
Theaters:	
Aisles, corridors, and lobbies	100
Orchestra floors	60
Balconies	60
Stage floors	150
Yards and terraces, pedestrians	100

ings or other buildings where partitions might be subject to erection or rearrangement, provision for partition weight shall be made, whether or not partitions are shown on the plans, unless the specified live load exceeds 80 pounds per square foot.

3.2 CONCENTRATED LOADS. Floors shall be designed to support safely the uniformly distributed live loads prescribed in Table 1.12 or the concentrated load in pounds given in Table 1.13, whichever produces the greater stresses. Unless otherwise specified, the indicated concentration shall be assumed to occupy an area 2½ feet square and shall be so located as to produce the maximum stress conditions to the structural members.

3.2.1 *Roof trusses.* Any panel point of the lower chord of roof trusses or any point of other primary structural members supporting roofs over garage, manufacturing, and storage floors shall be capable of carrying safely a suspended concentrated load of not less than 2000 pounds.

3.4 IMPACT LOADS. The live loads specified shall be assumed to include adequate allowance for ordinary impact conditions. Provision shall be made in the structural design for uses and loads which involve unusual vibration and impact forces.

3.4.1 *Elevators.* All moving elevator loads shall be increased 100 percent for impact, and the structural supports shall be designed within the limits of deflection prescribed by the American Standard Safety Code for Elevators, Dumbwaiters, and Escalators, A17.1-1937, and American Standard Inspection of Elevators (Inspector's Manual), A17.2-1945, or the latest revisions thereof approved by the American Standards Association, Incorporated.

3.4.2 *Heavy machinery.* For the purpose of design, the weight of heavy machinery and moving loads shall be increased not less than 25 percent for impact, unless otherwise specified.

3.4.3 *Craneways.* All craneways shall be designed to resist a horizontal transverse force equal to 25 percent of the crane capacity plus the weight of the trolley applied one-half at the top of each runway rail for impact; and a horizontal longitudinal force equal to 12½ percent of the total of the maximum wheel loads applied at the top of each rail.

3.5 REDUCTION IN LIVE LOADS

3.5.1 *Roof live loads.* No reduction shall be applied to the roof live load.

3.5.2 *Live loads 100 pounds per square foot or less.* For live loads of 100 pounds or less per square foot, the design live load on any member supporting 150 square feet or more may be reduced at the rate of 0.08 percent per square foot of area supported by the member, except that no reduction shall be made for areas to be occupied as places of public assembly. The reduction shall exceed neither R as determined by the following formula, nor 60 percent:

$$R = 100 \times \frac{D + L}{4.33L}$$

in which

R = reduction in percent
D = dead load per square foot of area supported by the member
L = design live load per square foot of area supported by the member

3.5.3 *Live loads exceeding 100 pounds per square foot.* For live loads exceeding 100 pounds per square foot, no reduction shall be made, except that the design live loads on columns may be reduced 20 percent.

Table 1.13 Concentrated Loads

Location	Load, lb
Elevator machine room grating (on area of 4 sq in.)	300
Finish light floor plate construction (on area of 1 sq in.)	200
Office floors	2,000
Scuttles, skylight ribs, and accessible ceilings	200
Sidewalks	8,000
Stair treads (on center of treads)	300

3.8 MINIMUM ROOF LOADS

3.8.1 *Flat, pitched, or curved roofs.* Ordinary roofs, either flat, pitched, or curved, shall be designed for a load of not less than 20 pounds per square foot of horizontal projection in addition to the dead load, and in addition to either the wind or the earthquake load, whichever produces the greater stresses.

NOTE: The unit load recommended in 3.8.1 is a minimum. It is intended to provide for loads incidental to construction and repair, for sleet loads and minor snow loads, and to insure reasonable stiffness to the roof. In preparing a local code, the snow records of the nearest U. S. Weather Bureau station or, if Weather Bureau records are not available, the snow-load map shown in Fig. A1 in the Appendix should be consulted and the indicated unit snow load for the locality, if larger, substituted for the minimum. In such a case, provision should be made in the local code to the effect that any excess over 20 pounds per square foot may be reduced for each degree of pitch over 20 degrees by $S/40 - \frac{1}{2}$, where S is the total snow load in pounds per square foot.

3.8.2 *Special conditions.* (*a*) When the effect

Table 1.14 Supplementary List of Recommended Uniformly Distributed Live Loads for Special Conditions

	lb per sq ft		lb per sq ft
Air-conditioning (machine space)	200*	Hangars	150§
Amphitheater:		Incinerator charging floor	100
Fixed seats	60	Kitchens, other than domestic	150*
Movable seats	100	Laboratories, scientific	100
Amusement park structure	100*	Laundries	150*
Attic:		Libraries, corridors	100†
Nonstorage	25	Manufacturing, ice	300
Storage	80†	Morgue	125
Bakery	150	Office buildings:	
Balcony:		Files (*see* File room)	
Exterior	100	Business machine equipment	100*
Interior (fixed seats)	60	Printing plants:	
Interior (movable seats)	100	Composing rooms	100
Boathouse, floors	100†	Linotype rooms	100
Boiler room, framed	300*	Press rooooms	150*
Broadcasting studio	100	Paper storage	‖
Catwalks	25	Public rooms	100
Ceiling, accessible furred	10	Railroad tracks	#
Cold storage:		Ramps, driveway (*see* Garages)	
No overhead system	250‡	Ramps, pedestrian (*see* Sidewalks and Corridors in Table 1.12)	
Overhead system—		Ramps, seaplane (*see* Hangars)	
Floor	150	Rest rooms	60
Roof	250	Rink, ice skating	250
Dormitories:		Roof (*see* 3.8 following)	
Partitioned	40	Storage, hay or grain	300*
Nonpartitioned	80	Telephone exchange	150*
Driveways (*see* Garages)		Theaters:	
Elevator machine room	150*	Dressing rooms	40
Fan room	150*	Grid-iron floor or fly gallery—	
File room:		Grating	60
Letter	80†	Well beams, 250 lb per lin ft per pair	
Card	125†	Header beams, 1,000 lb per lin ft	
Addressograph	150†	Pin rail, 250 lb per lin ft	
Foundries	600*	Projection room	100
Fuel rooms, framed	400	Toilet rooms	60
Garages:		Transformer rooms	200*
Trucks, with load, 3 to 10 tons	150§	Vaults, in offices	250†
Trucks, with load, above 10 tons	200§		
Greenhouses	150		

* Use weight of actual equipment when greater.
† Increase when occupancy exceeds this amount.
‡ Plus 150 lb for trucks.
§ Also subject to not less than 125 percent maximum axle load.
‖ Paper storage 50 lb per ft of clear story height.
As required by railroad company.

Table 1.15 Weights of Building Materials

Materials	Weight Lb. per Sq. Ft.
CEILINGS	
Gypsum ceiling block, 2″ thick, unplastered	10
Plaster board, unplastered	3
Plaster, ¾″, and wood lath	8
Plaster, ¾″, and metal lath	8
Plaster, on tile or concrete	5
Suspended, metal lath and plaster	10
FLOORS	
Hardwood flooring, ⅞″ thick	4
Sheathing, white, red and Oregon pine, spruce or hemlock, ⅞″ thick	2½
Sheathing, yellow pine, 1″ thick	4
Wood block, creosoted, 3″ thick	15
Cement finish, per inch thick	12
Cinder concrete, per inch thick	9
Cinder concrete fill, per inch thick	5
Terrazzo, Tile, Mastic, Linoleum, per inch thick, including base	12
Gypsum slab, per inch thick	5
ROOFS	
Corrugated metal	Page 143
Roofing felt, 3 ply and gravel	5½
Roofing felt, 5 ply and gravel	6½
Roofing felt, 3 ply and slag	4½
Roofing felt, 5 ply and slag	5½
3-ply ready roofing	1
Shingles, wood	2
Tile or slate	5-20
PARTITIONS	
Channel studs, metal lath, cement plaster, solid, 2″ thiqk	20
Studs, 2″ x 4″, wood or metal lath, ¾″ plaster both sides	18
Studs, 2″ x 4″ plaster board, ½″ plaster both sides	18
Plaster, ½″, on gypsum block or clay tile (one side)	4
Hollow clay tile, 2″	13
Hollow clay tile, 3″	16
Hollow clay tile, 4″	18
Hollow clay tile, 5″	20
Hollow clay tile, 6″	25
Hollow clay tile, 8″	30
Hollow clay tile, 10″	35
Hollow gypsum block, 3″	10
Hollow gypsum block, 4″	13
Hollow gypsum block, 5″	15½
Hollow gypsum block, 6″	16½
Solid gypsum block, 2″	9½
Solid gypsum block, 3″	13
Steel partitions	4
WALLS	
Brick, 9″ thick	84
Brick, 13″ thick	121
Brick, 18″ thick	168
Brick, 22″ thick	205
Brick, 26″ thick	243
Wall tile, 6″ thick	30
Wall tile, 8″ thick	33
Wall tile, 10″ thick	40
Wall tile, 12″ thick	45
Brick 4″, tile backing 4″	60
Brick 4″, tile backing 8″	75
Brick 9″, tile backing 4″	100
Brick 9″, tile backing 8″	115
Limestone 4″, brick 9″	140
Limestone 4″, brick 13″	175
Limestone 4″, tile 8″	90
Limestone 4″, tile 12″	100
Windows, glass, frame and sash	8

of the shape of roof structure as determined by actual test or experience indicates lesser or greater snow-retention value than specified herein, the roof load shall be modified as directed or approved by the building official.

(*b*) When valleys are formed by a multiple series of roofs, special provision shall be made for the increased load at the intersections.

3.8.3 *Special-purpose roofs.* When used for incidental promenade purposes, roofs shall be designed for a minimum live load of 60 pounds per square foot; and 100 pounds per square foot when designed for roof-garden or assembly uses. Roofs to be used for other special purposes shall be designed for appropriate loads as directed or approved by the building official.

The subject of roof loads is discussed further in "Housing Research Paper 19" of the Housing and Home Finance Agency, entitled "Snow Load Studies," in which the following recommendations are made:

Flat Roofs Minimum Uniformly Distributed Design Vertical Live Roof Loads

Flat roofs and roofs having a rise of 3 inches per ft or less	Pounds per sq ft of horizontal projection
Southern States	20
Central States	25
Northern States	30
Great Lakes, New England, and Mountain areas	40

Great Lakes and New England areas include northern portions of Minnesota, Wisconsin, Michigan, New York, and Massachusetts and the states of Vermont, New Hampshire, and Maine. Mountain areas include Appalachians above 2,000 ft elevation, Pacific Coastal ranges above 1,000 ft elevation, and Rocky Mountains above 4,000 ft elevation.

Sloping roofs

Vertical live loads for roofs having a rise of more than 3 inches per ft, and so designed as to have lesser snow retention values, may be progressively reduced from the values given above for flat roofs to a minimum operational or sleet load of 10 pounds per sq ft of horizontal projection of roof area for roofs having a slope of 12 inches or more per ft. Suggested design vertical live loads for sloping roofs are approximately as follows:

Sloping Roofs Minimum Uniformly Distributed Design Vertical Live Roof Loads

Slope	Pounds per sq ft of horizontal projection			
	3 in 12 or less	6 in 12	9 in 12	12 in 12 or more
Southern States	20	15	12	10
Central States	25	20	15	10
Northern States	30	25	17	10
Great Lakes, New England, and Mountain areas	40	30	20	10

A3.1.2 *Stairway and balcony railings.* Stairway and balcony railings, both exterior and interior, should be designed to resist a horizontal thrust of 50 pound per linear ft applied at the top of the railing.

1.51 Heavy concentrated loads

a A frequent problem concerns the distribution of a heavy concentrated load on a floor slab, such as an automobile or a truck, or of a heavy safe. The selection of the proper design load may be left up to the engineer or it may be specified by code. The Syracuse Building Code requires garages with doors not over 8 ft high to be designed to carry a 1,500-lb wheel load (presumably because heavy trucks require over 8 ft clearance); and garages with doors over 8 ft high to carry an 8,000-lb wheel concentration. This 8,000-lb wheel load represents a total truck load of 24,000 lb. This should be considered not only for garage floors, but for loading docks and sidewalk slabs. Moreover, modern codes require office-building floors to be designed to carry a load of 2,000 lb on an area 2.5 ft square of otherwise unloaded floor. This would seem to indicate a floor-load requirement of 320 psf instead of the 50-lb requirement elsewhere specified.

In neither of the above instances, however, is the concentration as serious as it might seem to be. The manner in which this load may be distributed should be investigated to determine whether the concentration will have any effect on the design load of any part of the floor system.

b Probably the most commonly accepted method of distributing any concentrated load is the method used by the highway departments for distributing loads on bridge floor slabs. The load is considered as concentrated along a line at right angles to the span and along a width E. For the common condition, with a single load centered on the span (Fig. 1.4),

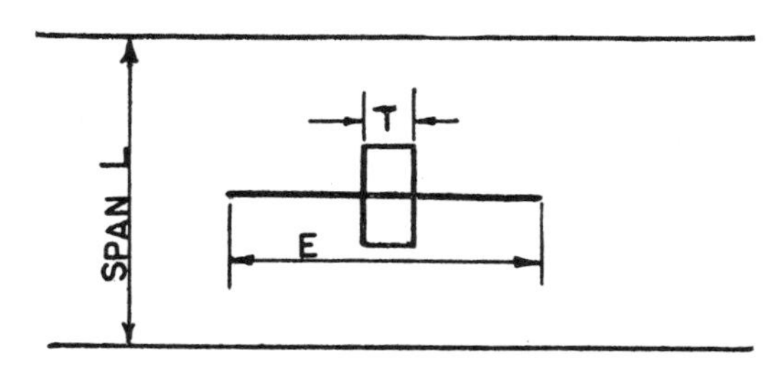

Fig. 1.4 Distribution of Concentrated Load

$$E = 0.7(L + T)$$

where T is the width of tire or lesser dimension of the load. The centrally concentrated load per unit width is then P/E. Where the

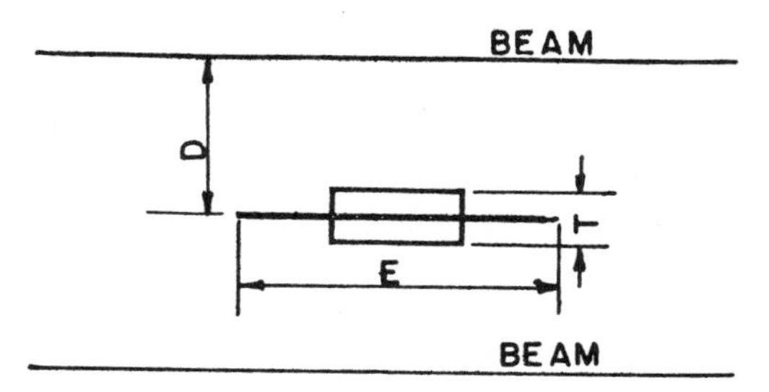

Fig. 1.5 Alternative Distribution of Concentrated Load

load is placed at some given point on the span, as the concentrated load of a machine base, the width is (Fig. 1.5)

$$E = 0.7(2D + T)$$

where D is the distance to the closest supporting beam.

Problem Design and check an office-building floor panel with the following conditions.

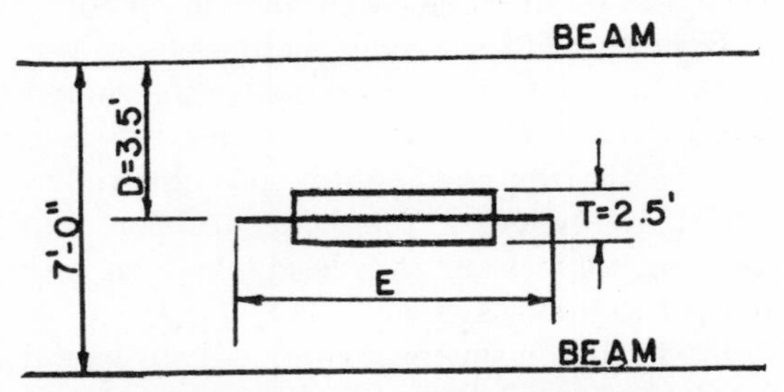

Fig. 1.6 Distribution of Concentrated Load

Assume a beam and slab floor; beams on 7-ft spacing; live load 50 psf, or a 2,000-lb safe on a 2½-ft square area (Fig. 1.6).

$$E = 0.7(2D + T)$$
$$E = 0.7(7 + 2.5)$$
$$E = 0.7 \times 9.5 = 6.65$$

Centrally concentrated load = 2,000/6.65 = 302 lb per linear foot. Therefore the design for live load must be based on 302 lb per linear foot centrally concentrated load, or 350 lb per foot of width of slab uniform load.

From the formulae given in Art. 3.11, for concentrated load,

$$M_{LL} = \frac{302 \times 7 \times 12}{8} = 3{,}180 \text{ in.-lb}$$

and for uniform load,

$$M_{LL} = \frac{350 \times 7 \times 12}{12} = 2{,}450 \text{ in.-lb}$$

Therefore the controlled load of the safe would be the controlling factor. It will be found that for this commonly used code, the design for weight of safe will control the slab design for a span up to 8.1 ft, and the load of 50 psf beyond this slab span.

Similarly for a truck-wheel load of 8,000 lb against a uniform load of 100 psf, the concentration of the truck wheel will control for spans up to 7.6 ft, and the uniform load for greater spans. Where two loads overlap, as in adjoining traffic lanes, the maximum permissible width to design for is the traffic lane width of 9 ft.

c Some years ago M. Hirschthal, Concrete Engineer for the Lackawanna R.R., published in *Engineering News-Record* the method used by his office in the design of overhead bridges, which approximate a two-way slab in their shape. While most of their

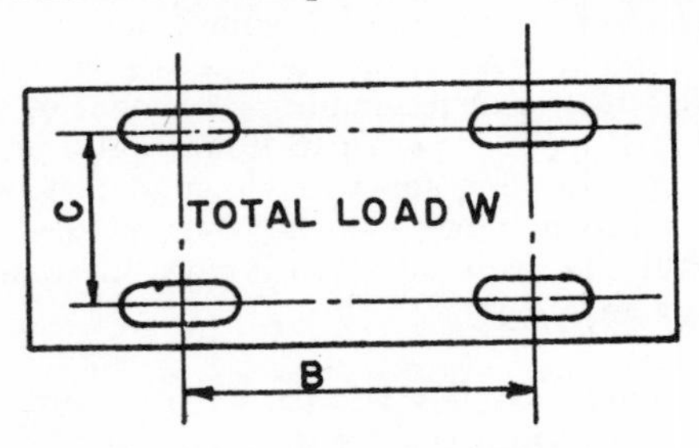

Fig. 1.7 Truck Load Diagram

designs are for a 20-ton truck (H-20 highway loading) with ⅔ of the load on the rear axle, the method given is general.

Using the nomenclature shown in Figs. 1.7, 1.8, and 1.9, for longitudinal distribution the

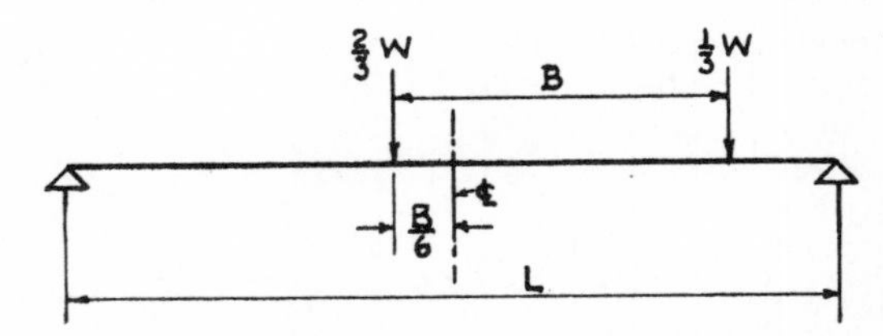

Fig. 1.8 Longitudinal Distribution of Truck Load

maximum moment occurs when the load is placed as shown in Fig. 1.8 and the equivalent uniform load longitudinally is

$$w = \frac{W}{3L^3}(3L^2 - 2bL + \tfrac{1}{3}b^2) \qquad (1.3)$$

With two trucks placed at a distance apart

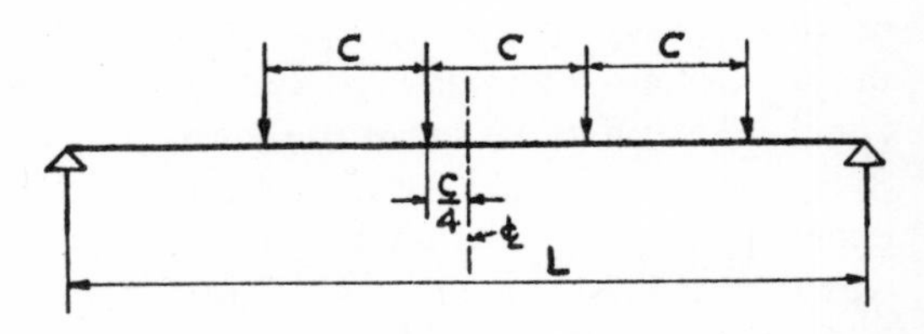

Fig. 1.9 Transverse Distribution of Truck Load

approximately equal to the transverse wheel base, the maximum transverse moment occurs when the load is placed as shown in Fig. 1.9 and the equivalent uniform load, referring to the uniform load w already found for the longitudinal distribution, is

$$w_1 = \frac{8w}{L_1^3}(L_1^2 - 2cL_1 + \tfrac{1}{4}c^2) \qquad (1.4)$$

Thus w_1 is the uniform load to design for, and in the design of a two-way slab should be distributed in accordance with the method given in Art. 5.50. While the method and the formula apply to two-way reinforced slabs, the method may be applied to one-way slabs for any width decided on for distribution.

Problem Assume a 20-ton truck, wheel base $b = 12$ ft, $c = 6$ ft; panel, 30 ft long by 28 ft wide. From Formula (1.3) above,

$$w = \frac{40{,}000}{3 \times 30^3}(3 \times 30^2 - 24 \times 30 + \tfrac{1}{3} \times 144)$$
$$= 0.494 \times 2{,}028 = 1{,}002$$
$$w_1 = \frac{8 \times 1{,}002}{28^3}(28^2 - 2 \times 6 \times 28 + \tfrac{1}{4} \times 6^2)$$
$$= 0.365 \times 457 = 167 \text{ psf equivalent live load}$$

1.60 MATHEMATICAL PROBLEMS

There are several problems in structural design which require the use of geometry, algebra, or other mathematics which the engineer may not have used since he left college, or the use of some special geometric section. Probably the most important of these involves the parabola, and it is discussed here at some length.

a Because the parabola is required for uniform-load–moment diagrams, two simple methods of construction are given here, a purely graphical method and a direct or ordinate method which involves a small amount of computation. In the graphical method, shown in Fig. 1.10, lay out the base line and the center height or maximum moment. Then divide the height into any number of equal parts, marking these points on a side ordinate. Divide the space from the center line of the parabola (the origin) to the side line, into the same number of equal spaces, and draw vertical lines through each of these points to the base, parallel with the side line. Then from the point of origin, draw rays to the points on the side lines. The point where ray 1 intersects vertical line 1 is a point on the parabola, etc., thus giving any required number of points.

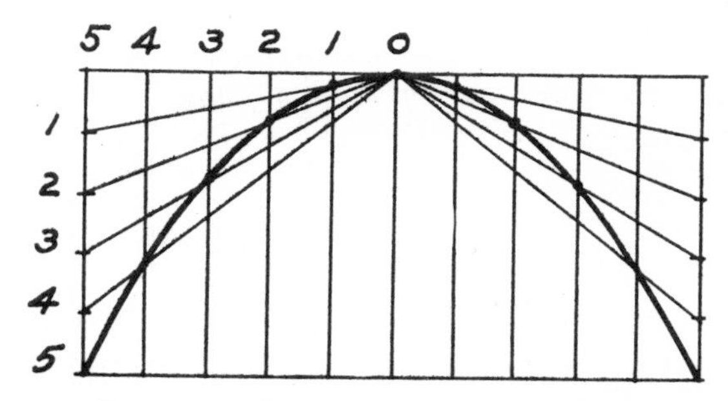

Fig. 1.10 Construction of Parabola—Ray Method

The ordinate method is a correct mathematical method of construction, as shown in Fig. 1.11. Divide the base into an even number of equal parts and number them from left to right and again from right to left as shown. Then the height of the ordinate next to the center is $(4 \times 6)/(5 \times 5)$ of the center ordinate, the second ordinate is $(3 \times 7)/(5 \times 5)$ of the center ordinate, the third ordinate is $(2 \times 8)/(5 \times 5)$ of the center ordinate. The method is identical, regardless of how many sections may be used, the denominator being the square of the center ordinate. Figure 1.12 indicates the method of laying out a parabola on part of the span only, as mentioned in Art. 3.11*c*.

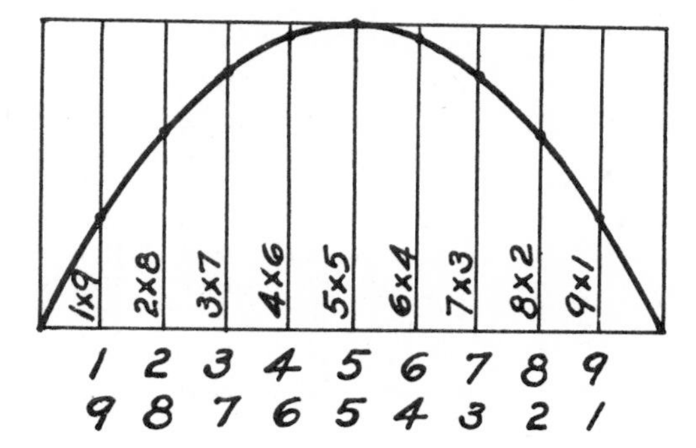

Fig. 1.11 Construction of Parabola—Ordinate Method

For simplicity in determining the area of any fractional part of a parabola, which is required in deflection problems by the moment-area theorem (Art. 2.31), Table 1.16 will be found useful.

Table 1.16 Altitude and Area of Section of Parabola

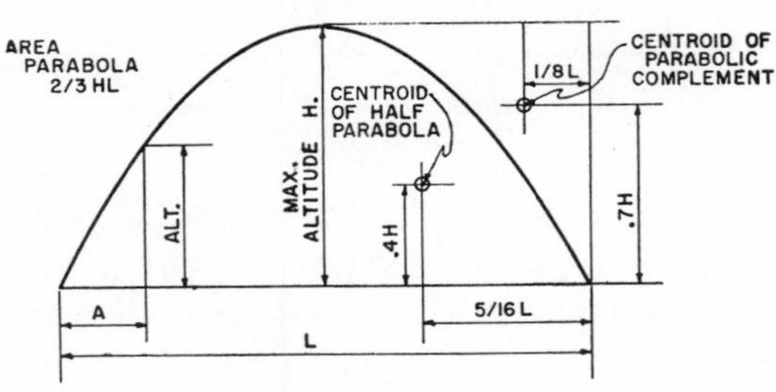

Area of parabola = $\frac{2}{3}HL$

For given altitude (expressed in table as decimal fraction of maximum altitude H), area of section equals decimal fraction given of rectangle HL.

A/L	Altitude (× H)	Area (× HL)	A/L	Altitude (× H)	Area (× HL)
.01	.0396	.000198	.26	.7696	.111754
.02	.0784	.000788	.27	.7884	.119544
.03	.1164	.001762	.28	.8064	.127518
.04	.1536	.003116	.29	.8236	.135668
.05	.19	.004834	.30	.84	.143986
.06	.2256	.006912	.31	.8556	.152464
.07	.2604	.009342	.32	.8704	.161094
.08	.2944	.012116	.33	.8844	.169868
.09	.3276	.015228	.34	.8976	.178778
.10	.36	.018666	.35	.91	.187816
.11	.3916	.022424	.36	.9216	.196974
.12	.4224	.026494	.37	.9324	.206244
.13	.4524	.030868	.38	.9424	.215618
.14	.4816	.035538	.39	.9516	.225088
.15	.51	.040496	.40	.96	.234636
.16	.5376	.045734	.41	.9676	.244274
.17	.5644	.051244	.42	.9744	.253984
.18	.5904	.057018	.43	.9804	.263758
.19	.6156	.063048	.44	.9856	.273588
.20	.64	.069326	.45	.99	.283466
.21	.6636	.075844	.46	.9936	.293384
.22	.6864	.082594	.47	.9964	.303334
.23	.7084	.089568	.48	.9984	.313308
.24	.7296	.096758	.49	.9996	323298
.25	.75	.104156	.50	1.0	.333333

Either Table 1.16 or the method of Fig. 1.11 may be used to determine the moment at any given point on a simple uniformly loaded beam. To determine from Table 1.16 the moment at a point 3.5 ft from the

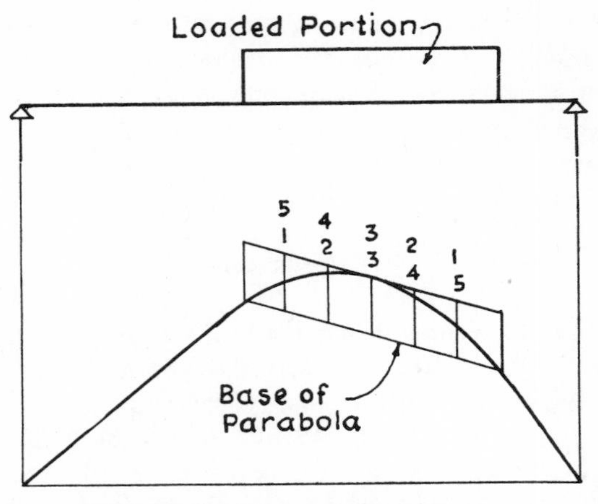

Fig. 1.12 Parabola on Part of Span

end of a beam 17 ft long,

$$\frac{A}{L} = \frac{3.5}{17} = 0.206$$

By interpolation between 0.6636 and 0.64, M = 0.654 of the center moment. By the method of Fig. 1.11,

$$M = \frac{3.5 \times 13.5}{8.5 \times 8.5} = 0.654 \text{ of the center moment}$$

b It is occasionally necessary to find the area of a parabolic segment such as the one shown in Fig. 1.13. Given the span l, the rise r, and the segmental length c,

$$A = \frac{2}{3} \times \frac{r}{l^2} \times c^3$$

c Another common geometrical problem in structural design is the solution of a pressure trapezoid. If the load on any wall, buttress, footing, etc., is applied off center, the pressure is not uniform across the bearing,

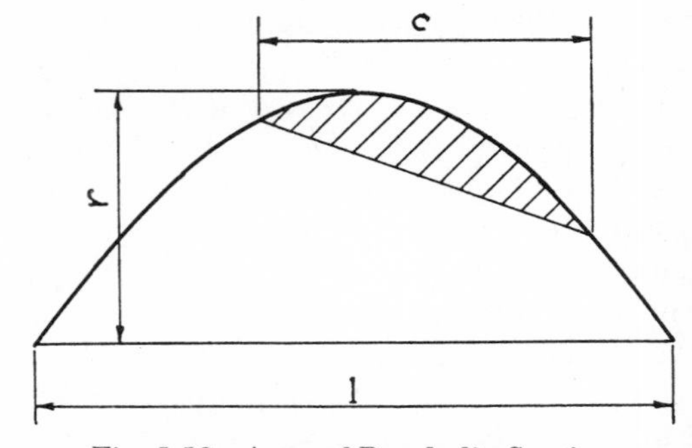

Fig. 1.13 Area of Parabolic Section

but varies uniformly from a maximum at the edge nearer the point of load to a minimum at the opposite edge.

Graphically we may represent the resultant pressures by a trapezoid, as shown in Fig. 1.14. A unit length is represented by

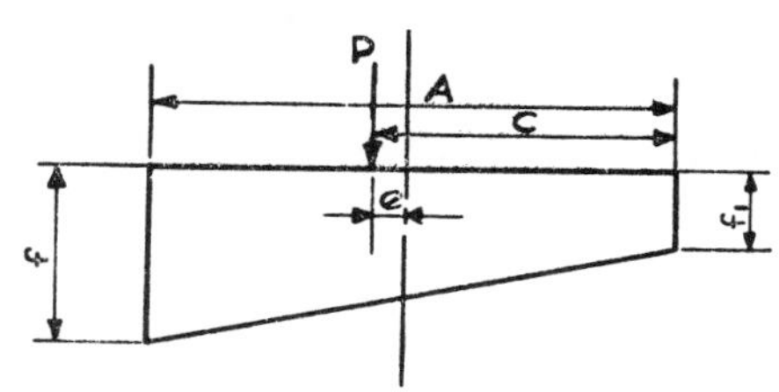

Fig. 1.14 Pressure Trapezoid

area A, for the sake of simplifying the problem. The pressure f increases and f_1 decreases very rapidly as the eccentricity of the load increases, until the eccentricity is $\frac{1}{6}A$, at which point f_1 becomes zero and f is double the average load. The unit pressure at the edge is obtained from the formula,

$$f = \frac{P}{A}\left(1 \pm \frac{6e}{A}\right) \tag{1.5}$$

in which plus is used to compute f and minus to compute f_1.

Another typical trapezoidal problem is the design of a trapezoidal footing, as described in Art. 8.12*e*. In this instance the two ends of the trapezoid are the widths of the footing required. The eccentricity is determined by solving for center of gravity of the loads, as described in Art. 1.31*a*. Then referring to the properties of the trapezoid in Art. 1.30, knowing the required area of footing A, and the available length of footing d, the factor $(b + b_1)$ in the first equation may be determined. Referring to the second equation, c is known, d is known, and the sum of $b + b_1$ is known, from which $2b + b_1$ may be computed. By solving the two simultaneous equations, b and b_1 may be obtained. For example, assume the maximum permissible length of footing to be 20 ft, with columns 18 ft o.c., each 1 ft from the end of the footing. The load at one end is 300 kips and at the other end 250 kips, and the allowable net soil pressure 5 kips per sq ft. The required $A = {}^{550}\!/_{5} = 110$ sq ft. From Table 1.4, the first equation is

$$110 = \frac{20(b + b_1)}{2} = 10(b + b_1)$$
$$b + b_1 = 11$$

The center of gravity of the column loads from the heavy-load column is $250 \times {}^{18}\!/_{550} = 8.18$ ft from the column, or 9.18 ft from the wall, and $c = 9.18$.

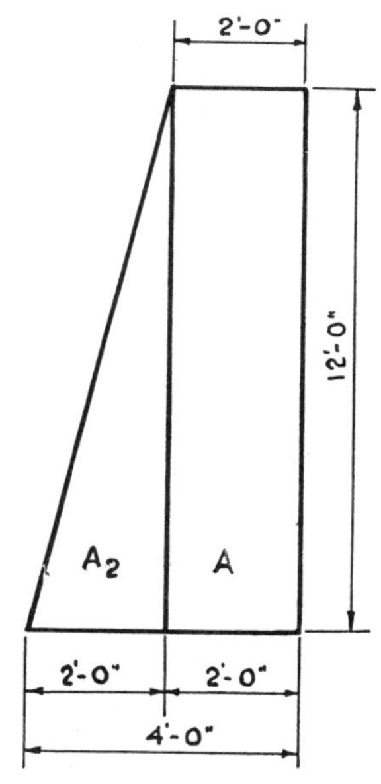

Fig. 1.15 Trapezoidal Retaining Wall

From the second equation,

$$9.18 = \frac{20(2b + b_1)}{3 \times 11}$$

from which

$$\begin{aligned} 2b + b_1 &= 15.1 \\ b + b_1 &= 11 \\ \hline b \quad &= 4.1 \\ b_1 \quad &= 6.9 \end{aligned}$$

Therefore the footing is 20 ft long, 4.1 ft wide at the end nearest the light load, and 6.9 ft at the heavy end.

Perhaps a simple solution would be to solve by application of Formula (1.5). In

Table 1.17 Trigonometric Formulae

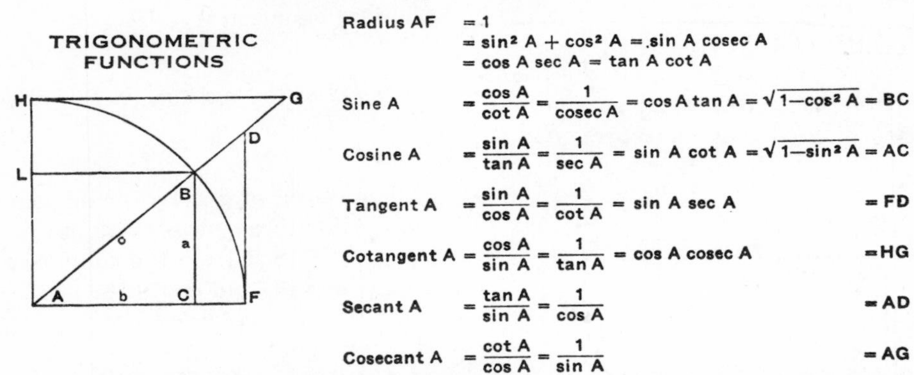

Radius AF $= 1$
$= \sin^2 A + \cos^2 A = \sin A \operatorname{cosec} A$
$= \cos A \sec A = \tan A \cot A$

Sine A $= \dfrac{\cos A}{\cot A} = \dfrac{1}{\operatorname{cosec} A} = \cos A \tan A = \sqrt{1-\cos^2 A} = BC$

Cosine A $= \dfrac{\sin A}{\tan A} = \dfrac{1}{\sec A} = \sin A \cot A = \sqrt{1-\sin^2 A} = AC$

Tangent A $= \dfrac{\sin A}{\cos A} = \dfrac{1}{\cot A} = \sin A \sec A = FD$

Cotangent A $= \dfrac{\cos A}{\sin A} = \dfrac{1}{\tan A} = \cos A \operatorname{cosec} A = HG$

Secant A $= \dfrac{\tan A}{\sin A} = \dfrac{1}{\cos A} = AD$

Cosecant A $= \dfrac{\cot A}{\cos A} = \dfrac{1}{\sin A} = AG$

RIGHT ANGLED TRIANGLES

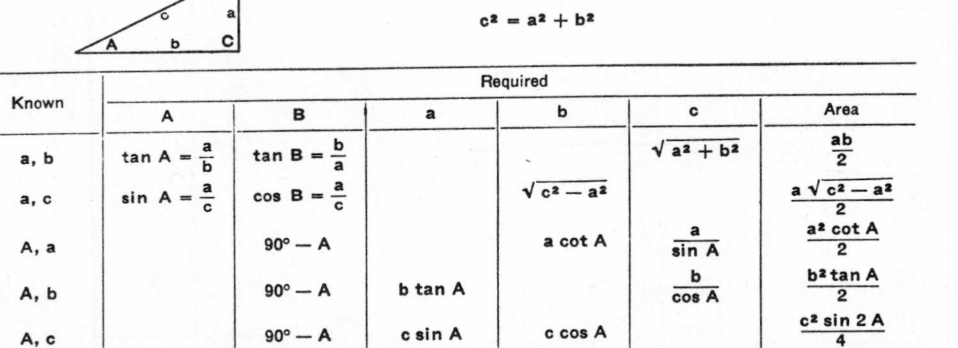

$a^2 = c^2 - b^2$

$b^2 = c^2 - a^2$

$c^2 = a^2 + b^2$

Known	Required: A	B	a	b	c	Area
a, b	$\tan A = \frac{a}{b}$	$\tan B = \frac{b}{a}$			$\sqrt{a^2+b^2}$	$\frac{ab}{2}$
a, c	$\sin A = \frac{a}{c}$	$\cos B = \frac{a}{c}$		$\sqrt{c^2-a^2}$		$\frac{a\sqrt{c^2-a^2}}{2}$
A, a		$90° - A$		$a \cot A$	$\frac{a}{\sin A}$	$\frac{a^2 \cot A}{2}$
A, b		$90° - A$	$b \tan A$		$\frac{b}{\cos A}$	$\frac{b^2 \tan A}{2}$
A, c		$90° - A$	$c \sin A$	$c \cos A$		$\frac{c^2 \sin 2A}{4}$

OBLIQUE ANGLED TRIANGLES

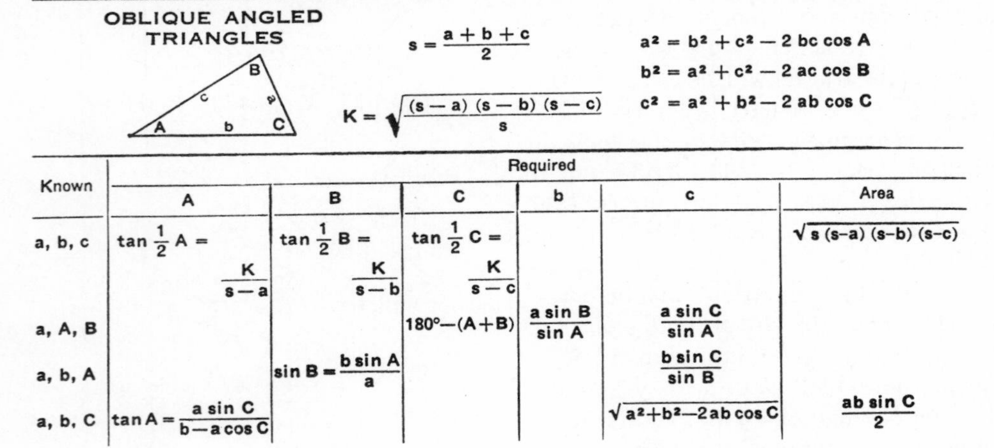

$s = \dfrac{a+b+c}{2}$

$K = \sqrt{\dfrac{(s-a)(s-b)(s-c)}{s}}$

$a^2 = b^2 + c^2 - 2bc \cos A$

$b^2 = a^2 + c^2 - 2ac \cos B$

$c^2 = a^2 + b^2 - 2ab \cos C$

Known	Required: A	B	C	b	c	Area
a, b, c	$\tan \frac{1}{2} A = \frac{K}{s-a}$	$\tan \frac{1}{2} B = \frac{K}{s-b}$	$\tan \frac{1}{2} C = \frac{K}{s-c}$			$\sqrt{s(s-a)(s-b)(s-c)}$
a, A, B			$180° - (A+B)$	$\frac{a \sin B}{\sin A}$	$\frac{a \sin C}{\sin A}$	
a, b, A		$\sin B = \frac{b \sin A}{a}$			$\frac{b \sin C}{\sin B}$	
a, b, C	$\tan A = \frac{a \sin C}{b - a \cos C}$				$\sqrt{a^2+b^2-2ab \cos C}$	$\frac{ab \sin C}{2}$

Table 1.18 Properties of the Circle

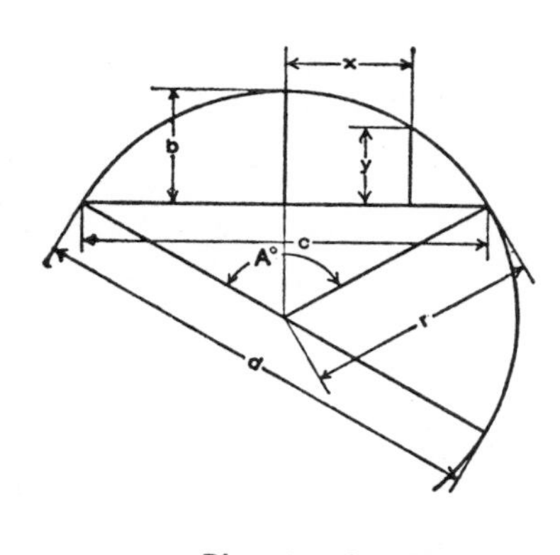

Circumference $= 6.28318\ r = 3.14159\ d$

Diameter $= 0.31831$ circumference

Area $= 3.14159\ r^2$

Arc $a = \dfrac{\pi r A^\circ}{180^\circ} = 0.017453\ r A^\circ$

Angle $A^\circ = \dfrac{180^\circ a}{\pi r} = 57.29578 \dfrac{a}{r}$

Radius $r = \dfrac{4b^2 + c^2}{8b}$

Chord $c = 2\sqrt{2br - b^2} = 2r \sin\dfrac{A}{2}$

Rise $b = r - \tfrac{1}{2}\sqrt{4r^2 - c^2} = \dfrac{c}{2}\tan\dfrac{A}{4}$

$= 2r\sin^2\dfrac{A}{4} = r + y - \sqrt{r^2 - x^2}$

$y = b - r + \sqrt{r^2 - x^2}$

$x = \sqrt{r^2 - (r + y - b)^2}$

Diameter of circle of equal periphery as square = 1.27324 side of square
Side of square of equal periphery as circle = 0.78540 diameter of circle
Diameter of circle circumscribed about square = 1.41421 side of square
Side of square inscribed in circle = 0.70711 diameter of circle

---

CIRCULAR SECTOR

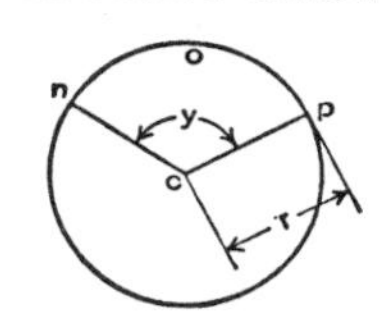

r = radius of circle y = angle ncp in degrees

Area of Sector ncpo = ½ (length of arc nop × r)

$= \text{Area of Circle} \times \dfrac{y}{360}$

$= 0.0087266 \times r^2 \times y$

CIRCULAR SEGMENT

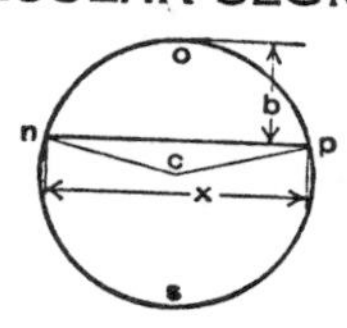

r = radius of circle x = chord b = rise

Area of Segment nop = Area of Sector ncpo − Area of triangle ncp

$= \dfrac{(\text{Length of arc nop} \times r) - x(r - b)}{2}$

Area of Segment nsp = Area of Circle − Area of Segment nop

VALUES FOR FUNCTIONS OF π

$\pi = 3.14159265359$, log = 0.4971499

$\pi^2 = 9.8696044$, log = 0.9942997 $\dfrac{1}{\pi} = 0.3183099$, log = $\bar{1}.5028501$ $\sqrt{\dfrac{1}{\pi}} = 0.5641896$, log = $\bar{1}.7514251$

$\pi^3 = 31.0062767$, log = 1.4914496 $\dfrac{1}{\pi^2} = 0.1013212$, log = $\bar{1}.0057003$ $\dfrac{\pi}{180} = 0.0174533$, log = $\bar{2}.2418774$

$\sqrt{\pi} = 1.7724539$, log = 0.2485749 $\dfrac{1}{\pi^3} = 0.0322515$, log = $\bar{2}.5085500$ $\dfrac{180}{\pi} = 57.2957795$, log = 1.7581226

this instance, the eccentricity is $10 - 9.18 = 0.82$. Then

$$\text{Width} = \frac{110}{20}\left[1 \pm 6 \times \frac{0.82}{20}\right]$$
$$= 5.5(1 \pm 0.246)$$
$$(= 1.246 \quad \text{or} \quad 0.754)$$

Then the width at the light end is $0.754 \times 5.5 = 4.1$, and at the heavy end $1.246 \times 5.5 = 6.9$ ft, which checks with the previous solution.

Problem A concrete wall is 12 ft high, 4 ft thick at the base, and 2 ft thick at the top. One face is vertical. What are the maximum and minimum pressures under the base due to the weight of concrete? (From a state license examination. See Fig. 1.15.)

The center of gravity of the wall, from the back face, is,

$$A_1x_1 = 24 \times 1 \quad = 24$$
$$A_2x_2 = \underline{12} \times 2.67 = \underline{32}$$
$$A = 36 \qquad Ax = 56$$
$$x = \tfrac{56}{36} = 1.555$$
$$e = 2 - 1.555 = 0.445$$

$$P = 36 \times 150 = 5{,}400$$

$$f_1 = \frac{5{,}400}{4}\left(1 - 6 \times \frac{0.445}{4}\right) = \text{450 psf at front of wall}$$

$$f = \frac{5{,}400}{4}\left(1 + 6 \times \frac{0.445}{4}\right) = \text{2,250 psf at back of wall}$$

With the two end pressures known, determining the eccentricity is a problem in determining the center of gravity of a trapezoid. Referring to Fig. 1.14, the formula becomes

$$c = \frac{A\,(2f + f_1)}{3\,(f + f_1)}$$

$$e = c - \frac{A}{2}$$

d The solution of a simple quadratic equation is required in some cases in locating the point of zero shear in a beam with a uniformly increasing load, such as the design of a basement wall against earth pressure (see Art. 3.10*o*). For this problem in structural design, given the equation $x^2 + ax = b$,

$$x = \sqrt{b + \frac{a^2}{2}} - \frac{a}{2}$$

2

Simple Stresses and Elastic Theory

2.10 SIMPLE STRESSES

Throughout this chapter, the following nomenclature is used in addition to that already listed in Art. 1.10.

C	coefficient.
F	unit stress, psi.
F_t	allowable unit tensile stress, psi.
F_v	unit shear stress, psi.
F_{vH}	unit horizontal shear stress, psi.
K	fractional multiplier, slab length.
M_s	statical moment.
N	$\frac{E_1}{E_2}$.
P	total stress.
V	total vertical shear.

By simple stresses we mean tension, compression, or shear in the simple form, that is, without combination with bending, torsion, or any other stresses, and applied to a homogeneous material such as steel or plain concrete. This is the simplest form of construction computation and may be solved by some variation of the formula

$$F = \frac{P}{A} \tag{2.1}$$

2.11 Simple shear

a Vertical shear comes in almost entirely as part of some more complex design problem, such as the design of a concrete beam, a riveted joint, or a plate girder, and will be discussed under these headings. In any instance, the problem is solved by a direct application of Formula (2.1),

$$P = AF_v$$

Problem (from a state license examination) Compute the capacity of a ¾-in. power-driven rivet in single shear if the allowable stress is 15,000 psi.

$$P = AF_v = 0.4418 \times 15{,}000 = 6.627 \text{ lb}$$

b Horizontal shear stress is caused by the tendency of horizontal layers or fibers to slip on each other because of the shortening of the fibers on the compression side of a beam and the lengthening of the fibers on the tension side as the beam undergoes bending. It is therefore a more complex type of stress to determine.

Problems in horizontal shear are found in the design of riveting between flange angles and cover plates of a steel plate girder (see Art. 4.20) and in various wood-beam problems (see Art. 6.10).

The formula for horizontal shear stress at any fiber distance from the neutral axis of any cross section is

$$F_{vH} = \frac{VM_s}{I} \tag{2.2}$$

In this formula, V is the total vertical

shear force at the point along the span under consideration, M_s is the statical moment of the cross-sectional area beyond the fiber in question, taking the neutral axis as an axis of moments. In the denominator, I is the moment of inertia of the entire area about the neutral axis.

Problem Derive an expression for the horizontal shear stress at the neutral axis of a rectangular cross section. From Fig.

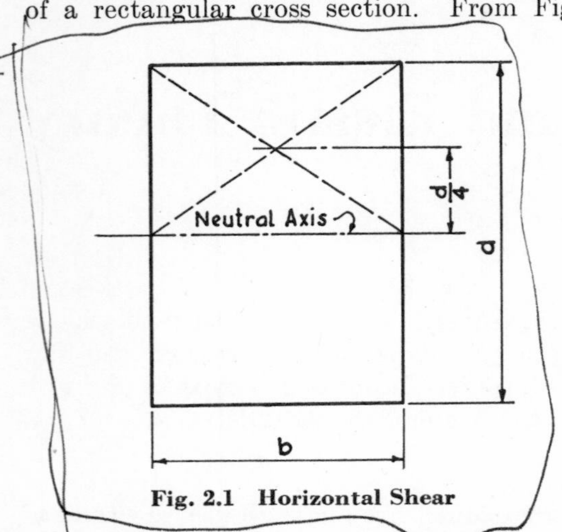

Fig. 2.1 Horizontal Shear

2.1 the statical moment of the area above the neutral axis is

$$\frac{bd}{2} \times \frac{d}{4} = \frac{bd^2}{8}$$

From Table 1.3, the moment of inertia of a rectangle is $\frac{bd^3}{12}$. Substituting in formula (2.2), the shear per linear inch is

$$F_{vH} = \frac{V\frac{bd^2}{8}}{\frac{bd^3}{12}} = \frac{1.5V}{d}$$

from which the shear in psi may be derived by dividing by b, or

$$F_{vH} = \frac{1.5V}{bd} = 1.5\frac{V}{A}$$

It might be noted that the value derived is 1.5 times the vertical shear stress acting across the entire cross section. For fibers other than those at the neutral axis, different values will be found, varying parabolically from a maximum at the neutral axis to zero at the outermost fibers.

Another problem of this type is given in Art. 4.22.

2.12 Direct compression

Compression problems in the use of steel will be treated under the subject of columns in Art. 7.10, since they usually are not simple stress problems but are dependent on length and on secondary bending. Rivet holes are not deducted in the solution of compression problems, since the rivet or bolt is supposed to fill the hole sufficiently.

Except for the design of details which may be classed rather as bearing problems, compression in timber is ordinarily a problem in column design and will be considered in Art. 7.30. In compression problems using timber, remember that timber has two primary compression unit stresses (see Table 1.1)—on the side grain, or perpendicular to the grain, and on the end grain, or parallel with the grain—and any number of stresses on the diagonal grain, which will be treated in Art. 9.30.

Problems in bearing area may be classed as compression problems, as in the following example.

Problem The base plate of a column is 16 in. square. What is the unit bearing stress on the top of a concrete footing under a concentric load of 125,000 lb?

From Formula (2.1),

$$F = \frac{P}{A} = \frac{125{,}000 \text{ lb}}{256 \text{ sq in.}} = 488 \text{ psi}$$

Similar problems involve the determination of footing and foundation area (Arts. 8.10 and 8.12) and masonry piers (Art. 8.20).

2.13 Tension

a Problems in tension, although they may occur in other materials, are usually problems in steel design. In reinforced concrete design, the tension value of the concrete is disregarded. In wood problems, because of the difficulty of obtaining proper connection, tension is seldom a controlling factor. Wood tension is discussed in Arts. 9.30 and 10.12.

Direct tension in building steel, as in hangers, truss members, and steel details is subject to the following limits in the AISC Code.

Section 1.5.1.1, on the net section, except at pin holes, $F_t = 0.60F_y$. (Inasmuch as pin connections are practically never used in building construction, the remainder of this section would only confuse the designer.)

Other stresses in tension are given in Section 1.5.1.4, *Bending*, and will be quoted under the subject matter of Chapter 4.

In Section 1.5.2.1, the allowable tension on rivets and bolts is given as follows in (psi):

A141 hot-driven rivets	20,000
A195 and A406 hot-driven rivets	27,000
A307 bolts and threaded parts of A7 and A373 steel	14,000
Threaded parts of other steel	$0.40F_y$
A325 bolts	40,000
A354 Grade BC bolts	50,000

In Sec. 1.8.4, the slenderness ratio, l/r, of tension members other than rods, preferably should not exceed

For main members,	240
For bracing and other secondary members,	300

Unless otherwise specified, tension members shall be designed on the basis of net section.

Sec. 1.14.3 *Net Section*

In the case of a chain of holes extending across a part of any diagonal or zigzag line, the net width of the part shall be obtained by deducting from the gross width the sum of the diameters of all the holes in the chain, and adding, for each gage space in the chain, the quantity

$$\frac{s^2}{4g}$$

where s = longitudinal spacing (pitch, in inches) of any two consecutive holes
g = transverse spacing (gage, in inches) of the same two holes

The critical net section of the part is obtained from that chain which gives the least net width; however, the net section taken through a hole shall in no case be considered as more than 85 percent of the corresponding gross section (see Fig. 2.4).

In determining the net section across plug or slot welds, the weld metal shall not be considered as adding to the net area.

Sec. 1.14.4 *Angles*

For angles, the gross width shall be the sum of the widths of the legs less the thickness. The gage for holes in opposite legs shall be the sum of the gages from back of angles less the thickness.

Sec. 1.14.5 *Size of Holes*

In computing net area the diameter of a rivet or bolt hole shall be taken as ⅛ inch greater than the nominal diameter of the rivet or bolt.

Sec. 1.15.1 *Minimum Connections*

Connections carrying calculated stresses, except for lacing, sag bars, and girts, shall be designed to support not less than 6,000 pounds.

b If a rod is upset and threaded, the least section is in the body of the rod, since the area at the root of the thread is in excess of the area of the rod. The cost of work involved in upsetting, however, will ordinarily make it more economical to use a plain rod, threaded. With such rods, the area used in tension problems is the area at root of thread (see Table 2.4). It is not good policy to use any threaded rod of less than ⅝-in. diameter, because smaller rods may be easily overstressed or even fractured at the root of the thread when the nut is pulled up. In building work, rods of diameters larger than 1¾ in. are seldom used. A satisfactory upset rod may be made by welding a threaded rod the size of upset to a base rod of the required size. This is only economical if the rod is long enough so that the resultant saving in the base rod is sufficient to offset the cost of the welding. In order to develop the strength of the rod, special design may be necessary in the weld detail. If stresses require more area than this, angles or flats are ordinarily used.

Problem A rod hanger is required to carry a load of 11,000 lb. Using a maximum tensile fiber stress of 22,000 psi, what size standard rod will be required? The rod will not be upset.

From Formula (2.1),

$$A = \frac{11{,}000}{22{,}000} = 0.50 \text{ sq in.}$$

Table 2.4 shows that a 1-in. round rod has an area of 0.55 sq in. at root of thread.

If no allowable stress is mentioned, it is advisable to use the AISC Code, which allows 22,000 psi on the net section of a member of A36 steel.

c If a plate or angle is used, the holes are

to be deducted. For bolts or rivets, the hole is ordinarily punched 1/16 in. larger than the nominal size of the bolt or rivet to allow easy insertion, and an additional allowance of 1/16-in. diameter is deducted to allow for deformation of the plate or angle due to punching. Thus a 3/4-in. rivet requires a 13/16-in. hole and an allowance of 7/8 in. will be deducted from the width in computing the net area of the plate or angle.

Problem Assume a flat bar hanger 4 in. wide with the required number of 3/4-in. rivets in a vertical row to carry a load of 20,000 lb. What thickness will be required, allowing a unit fiber stress of 22,000 psi? (See Fig. 2.2.)

From Formula (2.1), $A = \dfrac{20{,}000}{22{,}000} = 0.9$

Net width of bar $= 4 - 0.875 = 3.125$

Thickness required $= \dfrac{0.9}{3.125} = 0.29$ in.

Use 4 × 5/16-in. bar.

Further problems in tension on net sections of bars will be found in Art. 9.10, Riveting.

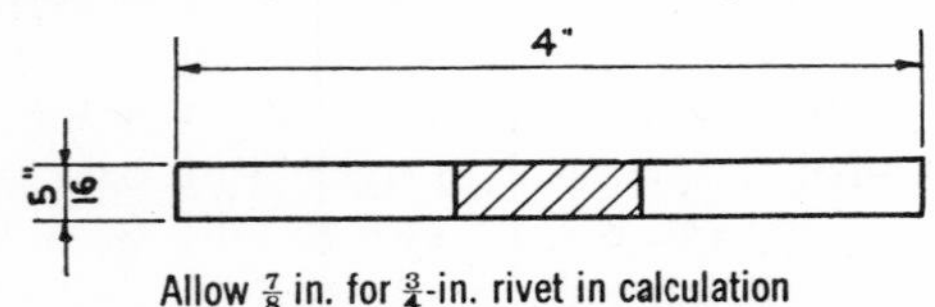

Allow 7/8 in. for 3/4-in. rivet in calculation

Fig. 2.2 Rivet Hole in Flat Bar

d The design of angle hangers is in accordance with the AISC Code, Section 1.5.1.1, quoted in Art. 2.13*a*.

The former Code considered the "effective net area" of an angle to be the net area of the connected leg plus one-half the area of the unconnected leg, and the tables in the former Manual of the AISC were laid out on this basis. The new Code allows the full value of both legs of the angle less only the area of rivet or bolt holes.

Table 2.2 will be found useful for the design of angle hangers or of truss members in that the areas are listed in ascending sequence from the smallest to the largest desirable size. Because of the increasing use of welding, the sequence is based on A_1, the gross area, but to assist the designer in selecting a riveted or bolted angle section, the area of a single hole for 3/4-in. rivet or bolt (a 7/8-in. hole) is deducted in column A_2 of this table. It will be noted that the sequence variation in column A_2 may vary slightly from the sequence listed under A_1. Since A_1 is gross area, it is also the sequence for weight of member.

Variations such as the use of 7/8-in. rivets (1-in. holes) or several rivets in the same cross-sectional area may be applied to this table as may be required.

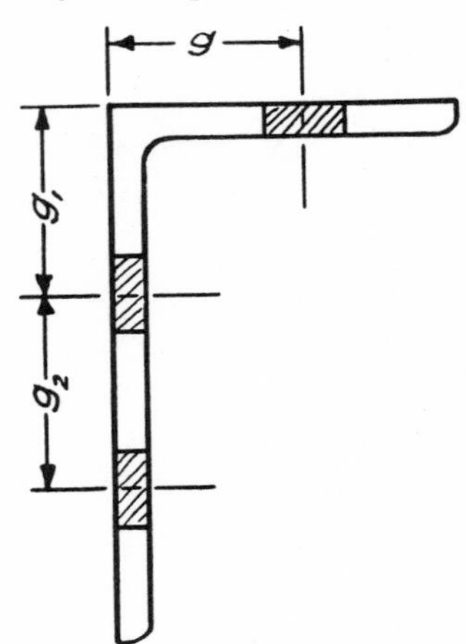

Fig. 2.3 Gage Lines in Angle Legs

Problem Find the lightest angle which may be used to provide a net area of 2.75 sq in. with one hole out for a 3/4-in. bolt for a member with a minimum radius of gyration of 0.90 in. From Table 2.2, a 5 × 5 × 5/16-in. angle has a net area of 2.76 sq in. and a radius of gyration of 0.99 and may therefore be used. A 6 × 4 × 5/16-in. angle will provide the same net area but with a smaller radius of gyration, less than the required.

e The usual gages for angles are given in Table 2.1 (Fig. 2.3). If two or more of the

Table 2.1 Gage for Angle Legs

Leg	8 in.	7 in.	6 in.	5 in.	4 in.	3½ in.	3 in.	2½ in.	2 in.	1¾ in.	1½ in.	1⅜ in.	1¼ in.	1 in.
g	4½	4	3½	3	2½	2	1¾	1⅜	1⅛	1	⅞	⅞	¾	⅝
g_1	3	2½	2¼	2										
g_2	3	3	2½	1¾										

Table 2.2 Area of Steel Angles in Sequence

Size of angle	r	A_1	A_2	Size of angle	r	A_1	A_2
$3 \times 2 \times \frac{1}{4}$	.43	1.19	.97	$5 \times 3\frac{1}{2} \times \frac{1}{2}$	.75	4.00	3.56
$3 \times 2\frac{1}{2} \times \frac{1}{4}$	.53	1.31	1.09	$5 \times 5 \times \frac{7}{16}$	.98	4.18	3.80
$3 \times 3 \times \frac{1}{4}$	.59	1.44	1.22	$6 \times 4 \times \frac{7}{16}$	.87	4.18	3.80
$3\frac{1}{2} \times 2\frac{1}{2} \times \frac{1}{4}$	.54	1.44	1.22	$4 \times 3\frac{1}{2} \times \frac{5}{8}$	.72	4.30	3.75
$3 \times 2 \times \frac{5}{16}$	.43	1.47	1.20	$6 \times 6 \times \frac{3}{8}$	1.19	4.36	4.03
$3\frac{1}{2} \times 3 \times \frac{1}{4}$	.63	1.56	1.34	$6 \times 3\frac{1}{2} \times \frac{1}{2}$	.75	4.50	4.06
$3 \times 2\frac{1}{2} \times \frac{5}{16}$	.53	1.62	1.35	$4 \times 4 \times \frac{5}{8}$	.78	4.61	4.06
$3\frac{1}{2} \times 3\frac{1}{2} \times \frac{1}{4}$	.69	1.69	1.47	$7 \times 4 \times \frac{7}{16}$	.88	4.62	4.24
$4 \times 3 \times \frac{1}{4}$	.65	1.69	1.47	$5 \times 5 \times \frac{1}{2}$	.98	4.75	4.31
$3 \times 2 \times \frac{3}{8}$	.43	1.73	1.40	$6 \times 4 \times \frac{1}{2}$	.87	4.75	4.31
$3 \times 3 \times \frac{5}{16}$	.59	1.78	1.51	$5 \times 3\frac{1}{2} \times \frac{5}{8}$	.75	4.92	4.37
$3\frac{1}{2} \times 2\frac{1}{2} \times \frac{5}{16}$	.54	1.78	1.51	$6 \times 6 \times \frac{7}{16}$	1.19	5.06	4.68
$4 \times 3\frac{1}{2} \times \frac{1}{4}$	.73	1.81	1.59	$8 \times 4 \times \frac{7}{16}$	.87	5.06	4.68
$3 \times 2\frac{1}{2} \times \frac{3}{8}$	.52	1.92	1.59	$7 \times 4 \times \frac{1}{2}$	.87	5.25	4.81
$3\frac{1}{2} \times 3 \times \frac{5}{16}$	.83	1.93	1.66	$6 \times 4 \times \frac{9}{16}$	.87	5.31	4.82
$4 \times 4 \times \frac{1}{4}$	.80	1.94	1.72	$4 \times 4 \times \frac{3}{4}$	.78	5.44	4.78
$3 \times 2 \times \frac{7}{16}$	.43	2.00	1.62	$6 \times 6 \times \frac{1}{2}$	1.18	5.75	5.31
$3\frac{1}{2} \times 3\frac{1}{2} \times \frac{5}{16}$	.69	2.09	1.82	$8 \times 4 \times \frac{1}{2}$	.86	5.75	5.31
$4 \times 3 \times \frac{5}{16}$	.65	2.09	1.82	$5 \times 3\frac{1}{2} \times \frac{3}{4}$	.75	5.81	5.15
$3 \times 3 \times \frac{3}{8}$	.58	2.11	1.78	$5 \times 5 \times \frac{5}{8}$	.98	5.86	5.31
$3\frac{1}{2} \times 2\frac{1}{2} \times \frac{3}{8}$	.54	2.11	1.78	$6 \times 4 \times \frac{5}{8}$	.86	5.86	5.31
$3 \times 2\frac{1}{2} \times \frac{7}{16}$	.52	2.21	1.83	$7 \times 4 \times \frac{9}{16}$	.87	5.87	5.35
$3 \times 2 \times \frac{1}{2}$	.43	2.25	1.81	$8 \times 6 \times \frac{7}{16}$	1.31	5.93	5.55
$4 \times 3\frac{1}{2} \times \frac{5}{16}$	.73	2.25	1.98	$9 \times 4 \times \frac{1}{2}$	.85	6.25	5.81
$3\frac{1}{2} \times 3 \times \frac{3}{8}$	.62	2.30	1.97	$6 \times 6 \times \frac{9}{16}$	1.18	6.43	5.94
$4 \times 4 \times \frac{5}{16}$	.79	2.40	2.13	$8 \times 4 \times \frac{9}{16}$	.86	6.43	5.94
$5 \times 3 \times \frac{5}{16}$	.65	2.40	2.13	$7 \times 4 \times \frac{5}{8}$	.86	6.48	5.93
$3 \times 3 \times \frac{7}{16}$	.58	2.43	2.05	$8 \times 6 \times \frac{1}{2}$	1.30	6.75	6.31
$3\frac{1}{2} \times 2\frac{1}{2} \times \frac{7}{16}$	.54	2.43	2.05	$5 \times 5 \times \frac{3}{4}$	.97	6.94	6.28
$3\frac{1}{2} \times 3\frac{1}{2} \times \frac{3}{8}$	.69	2.48	2.15	$6 \times 4 \times \frac{3}{4}$	.86	6.94	6.28
$4 \times 3 \times \frac{3}{8}$	.64	2.48	2.15	$9 \times 4 \times \frac{9}{16}$	.85	7.00	6.51
$3 \times 2\frac{1}{2} \times \frac{1}{2}$	.52	2.50	2.06	$6 \times 6 \times \frac{5}{8}$	1.18	7.11	6.56
$5 \times 3\frac{1}{2} \times \frac{5}{16}$	.77	2.56	2.29	$8 \times 4 \times \frac{5}{8}$	.86	7.11	6.56
$3\frac{1}{2} \times 3 \times \frac{7}{16}$	.62	2.65	2.27	$8 \times 6 \times \frac{9}{16}$	1.30	7.56	7.07
$4 \times 3\frac{1}{2} \times \frac{3}{8}$	.73	2.67	2.34	$7 \times 4 \times \frac{3}{4}$	.86	7.69	7.03
$3 \times 3 \times \frac{1}{2}$	.58	2.75	2.31	$9 \times 4 \times \frac{5}{8}$	.85	7.75	7.18
$3\frac{1}{2} \times 2\frac{1}{2} \times \frac{1}{2}$	.53	2.75	2.31	$8 \times 8 \times \frac{1}{2}$	1.59	7.75	7.31
$4 \times 4 \times \frac{3}{8}$	.79	2.86	2.53	$5 \times 5 \times \frac{7}{8}$	.97	7.98	7.21
$5 \times 3 \times \frac{3}{8}$	.65	2.86	2.53	$6 \times 4 \times \frac{7}{8}$	.86	7.98	7.21
$3\frac{1}{2} \times 3\frac{1}{2} \times \frac{7}{16}$	.68	2.87	2.49	$8 \times 6 \times \frac{5}{8}$	1.29	8.36	7.81
$4 \times 3 \times \frac{7}{16}$	.64	2.87	2.49	$6 \times 6 \times \frac{3}{4}$	1.17	8.44	7.78
$6 \times 3\frac{1}{2} \times \frac{5}{16}$	.77	2.87	2.60	$8 \times 4 \times \frac{3}{4}$	.85	8.44	7.78
$3\frac{1}{2} \times 3 \times \frac{1}{2}$	.62	3.00	2.56	$8 \times 8 \times \frac{9}{16}$	1.58	8.68	8.19
$5 \times 5 \times \frac{5}{16}$	.99	3.03	2.76	$7 \times 4 \times \frac{7}{8}$	.86	8.86	8.09
$6 \times 4 \times \frac{5}{16}$	.88	3.03	2.76	$9 \times 4 \times \frac{3}{4}$	.84	9.19	8.53
$5 \times 3\frac{1}{2} \times \frac{3}{8}$	.76	3.05	2.72	$8 \times 8 \times \frac{5}{8}$	1.58	9.61	9.06
$4 \times 3\frac{1}{2} \times \frac{7}{16}$	.72	3.09	2.71	$6 \times 6 \times \frac{7}{8}$	1.17	9.73	8.96
$3\frac{1}{2} \times 3\frac{1}{2} \times \frac{1}{2}$	.68	3.25	2.81	$8 \times 4 \times \frac{7}{8}$	.85	9.73	8.96
$4 \times 3 \times \frac{1}{2}$	.64	3.25	2.81	$8 \times 6 \times \frac{3}{4}$	1.29	9.94	9.28
$4 \times 4 \times \frac{7}{16}$	.78	3.31	2.93	$9 \times 4 \times \frac{7}{8}$	.84	10.61	9.84
$6 \times 3\frac{1}{2} \times \frac{3}{8}$	.77	3.42	3.09	$6 \times 6 \times 1$	1.17	11.00	10.12
$4 \times 3\frac{1}{2} \times \frac{1}{2}$	.72	3.50	3.06	$8 \times 4 \times 1$	.85	11.00	10.12
$5 \times 3\frac{1}{2} \times \frac{7}{16}$	.76	3.53	3.15	$8 \times 8 \times \frac{3}{4}$	1.57	11.44	10.78
$5 \times 5 \times \frac{3}{8}$	.99	3.61	3.28	$8 \times 6 \times \frac{7}{8}$	1.28	11.48	10.71
$6 \times 4 \times \frac{3}{8}$	.88	3.61	3.28	$9 \times 4 \times 1$	.83	12.00	11.12
$6 \times 6 \times \frac{5}{16}$	1.20	3.65	3.38	$8 \times 6 \times 1$	1.28	13.00	12.12
$4 \times 4 \times \frac{1}{2}$	.78	3.75	3.31	$8 \times 8 \times \frac{7}{8}$	1.57	13.23	12.46
$5 \times 3 \times \frac{1}{2}$	.65	3.75	3.31	$8 \times 8 \times 1$	1.56	15.00	14.12
$4 \times 3 \times \frac{5}{8}$	.64	3.98	3.43	$8 \times 8 \times 1\frac{1}{8}$	1.56	16.73	15.75
$7 \times 4 \times \frac{3}{8}$	.88	3.98	3.65				

Table 2.3 Reduction of Area for Rivet Holes

Area in square inches = assumed diameter of hole by thickness of metal.
For computation purposes rivet holes shall be taken at the
nominal diameter of the rivet plus ⅛ in.

Thickness of Metal Inches	Diameter of Hole, Inches					
	⅝	¾	⅞	1	1⅛	1¼
3/16	.117	.141	.164	.188	.211	.234
¼	.156	.188	.219	.250	.281	.313
5/16	.195	.234	.273	.313	.352	.391
⅜	.234	.281	.328	.375	.422	.469
7/16	.273	.328	.383	.438	.492	.547
½	.313	.375	.438	.500	.563	.625
9/16	.352	.422	.492	.563	.633	.703
⅝	.391	.469	.547	.625	.703	.781
11/16	.430	.516	.602	.688	.773	.859
¾	.469	.563	.656	.750	.844	.938
13/16	.508	.609	.711	.813	.914	1.016
⅞	.547	.656	.766	.875	.984	1.094
15/16	.586	.703	.820	.938	1.055	1.172
1	.625	.750	.875	1.000	1.125	1.250
1 1/16	.664	.797	.930	1.063	1.195	1.328
1⅛	.703	.844	.984	1.125	1.266	1.406
1 3/16	.742	.891	1.039	1.188	1.336	1.484
1¼	.781	.938	1.094	1.250	1.406	1.563
1 5/16	....	.984	1.148	1.313	1.477	1.641
1⅜	....	1.031	1.203	1.375	1.547	1.719
1 7/16	....	1.078	1.258	1.438	1.617	1.797
1½	....	1.125	1.313	1.500	1.688	1.875
1 9/16	....	1.172	1.367	1.563	1.758	1.953
1⅝	....	1.219	1.422	1.625	1.828	2.031
1 11/16	....	1.266	1.477	1.688	1.898	2.109
1¾	....	1.313	1.531	1.750	1.969	2.188
1 13/16	....	....	1.586	1.813	2.039	2.266
1⅞	....	....	1.641	1.875	2.109	2.344
1 15/16	....	....	1.695	1.938	2.180	2.422
2	....	....	1.750	2.000	2.250	2.500
2 1/16	....	....	1.805	2.063	2.320	2.578
2⅛	....	....	1.859	2.125	2.391	2.656
2 3/16	....	....	1.914	2.188	2.461	2.734
2¼	....	....	1.969	2.250	2.531	2.813
2 5/16	....	....	2.023	2.313	2.602	2.891
2⅜	....	....	2.078	2.375	2.672	2.969
2 7/16	....	....	2.133	2.438	2.742	3.047
2½	....	....	2.188	2.500	2.813	3.125
2⅝	....	....	2.297	2.625	2.953	3.281
2¾	....	....	2.406	2.750	3.094	3.438
2⅞	....	....	2.516	2.875	3.234	3.594
3	....	....	2.625	3.000	3.375	3.750

holes are in a line, their net area may be taken directly from Table 2.2. However, occasionally the holes are staggered, or in a diagonal or zigzag line, and the net width must be determined by means of the chart and problem indicating the use of the formula as given in Fig. 2.4.

2.20 SIMPLE STRESS-STRAIN PROBLEMS

a The fundamental formula for stress-strain problems is derived from the definition of the modulus of elasticity,

$$E = \frac{F}{\delta} \quad \text{or} \quad \frac{\text{Unit stress}}{\text{Unit deformation}} \tag{2.3}$$

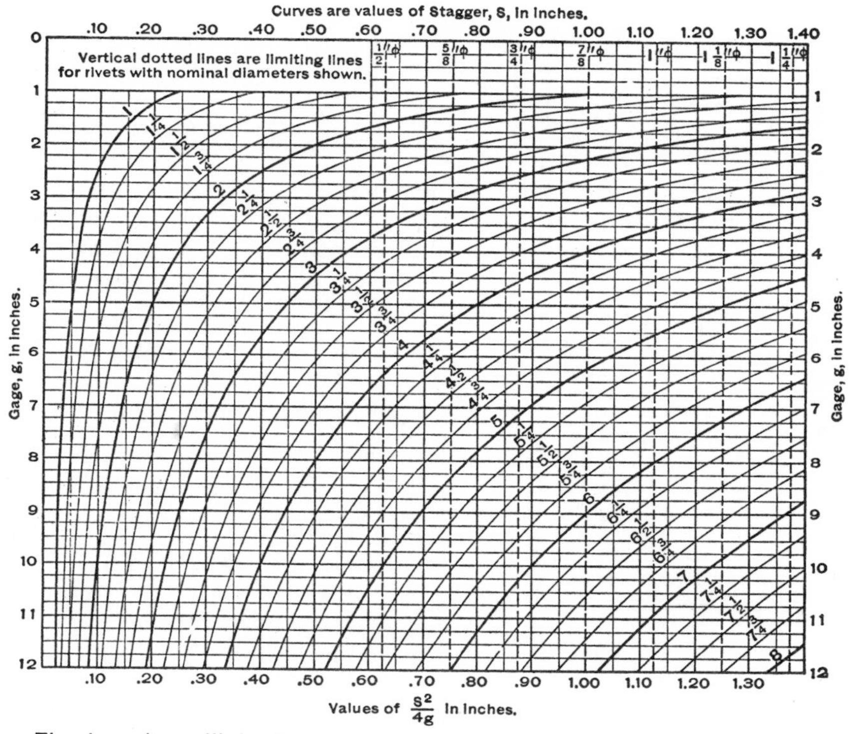

The above chart will simplify the application of the rule for net width, Section 19, Pars. (c) and (d) of the Institute Specifications. Entering the chart at left or right with the gauge "g" and proceeding horizontally to intersection with the curve for the pitch "s", thence vertically to top or bottom, the value of s²/4g may be read directly.

The example below illustrates the application of the rule, and the use of the chart.

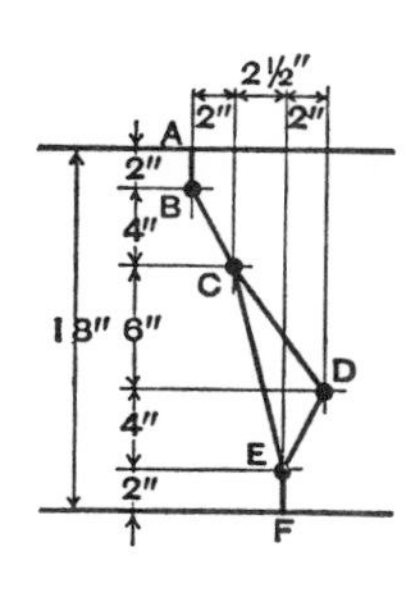

Chain A B C E F		
Deduct for 3 holes @ (¾ + ⅛)	=	—2.625
BC, g = 4, s = 2; add s²/4g	=	+0.25
CE, g = 10, s = 2½; add s²/4g	=	+0.16
Total Deduction	=	—2.215"
Chain A B C D E F		
Deduct for 4 holes @ (¾ + ⅛)	=	—3.50
BC, as above, add	=	+0.25
CD, g = 6, s = 4½; add s²/4g	=	+0.85
DE, g = 4, s = 2; add s²/4g	=	+0.25
Total Deduction	=	—2.15"

Net Width = 18.0 — 2.215 = 15.785".

¾" Rivets

In comparing the path CDE with the path CE, it is seen that if the sum of the two values of s²/4g for CD and DE exceed the single value of s²/4g for CE, by more than the deduction for one hole, then the path CDE is not critical as compared with CE.

Evidently if the value of s²/4g for one leg CD of the path CDE is greater than the deduction for one hole, the path CDE cannot be critical as compared with CE. The vertical dotted lines in the chart serve to indicate, for the respective rivet diameters noted at the top thereof, that any value of s²/4g to the right of such line is derived from a non-critical chain which need not be further considered.

Fig. 2.4 Net Section of Riveted Tension Members

Table 2.4 Screw Threads

American National Form
American Standard, B 1.1—1935.

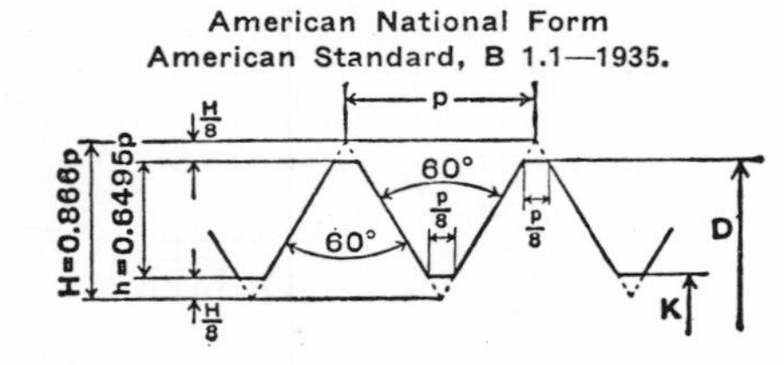

DIAMETER		AREA		Number of Threads per Inch
Total D In.	Net K In.	Total Dia., D Sq. In.	Net Dia., K Sq. In.	
1/4	.185	.049	.027	20
3/8	.294	.110	.068	16
1/2	.400	.196	.126	13
5/8	.507	.307	.202	11
3/4	.620	.442	.302	10
7/8	.731	.601	.419	9
1	.838	.785	.551	8
1 1/8	.939	.994	.693	7
1 1/4	1.064	1.227	.890	7
1 3/8	1.158	1.485	1.054	6
1 1/2	1.283	1.767	1.294	6
1 3/4	1.490	2.405	1.744	5
2	1.711	3.142	2.300	4 1/2
2 1/4	1.961	3.976	3.021	4 1/2
2 1/2	2.175	4.909	3.716	4
2 3/4	2.425	5.940	4.619	4

Since $F = \dfrac{P}{A}$ and $\delta = \dfrac{\Delta}{l}$, the above formula may be written $E = \dfrac{Pl}{\Delta A}$.

Most stress-strain problems require either the deformation due to a given load or stress or the stress from a given deformation. The value of the modulus of elasticity of various materials is given in Table 1.1. If the problem is an examination problem, both the total length and the gage length of the member may be given. The gage length is the length to use here.

b Probably the commonest form of stress-strain problem is that which requires the deformation under a given stress. This is a favorite type of problem in professional engineer's license examinations.

Problem What is the elongation of a steel rod 1/2 in. in diameter, 40 ft long, carrying a total suspended load of 6,000 lb?

For the solution of this problem, Formula (2.3) may be written

$$\Delta = \frac{Pl}{EA} =$$

$$\frac{6{,}000 \text{ lb} \times 40 \text{ ft} \times 12 \text{ in. per ft}}{30{,}000{,}000 \text{ psi} \times 0.1963 \text{ sq in.}} = 0.49 \text{ in.}$$

c The above problem brings up another point—the units in any problem must cancel out properly. Note that in the problem the load is expressed in pounds, the length in feet multiplied by 12 in. per ft. The numerator above is therefore

$$\text{Pounds} \times \text{feet} \times \frac{\text{inches}}{\text{feet}}$$

which cancels out to pounds times inches. In the denominator, E is expressed in pounds per square inch and the area in square inches, so the denominator becomes

$$\frac{\text{Pounds}}{\text{Square inches}} \times \text{square inches} = \text{pounds}$$

The problem is therefore

$$\Delta = \frac{\text{pounds} \times \text{inches}}{\text{pounds}} = \text{inches}$$

In complex problems, it is frequently advisable to set down and cancel out units to be sure of the final unit. Otherwise we are apt to run into the error of forgetting to reduce lengths or areas from feet to inch units, one of the most fruitful sources of error in engineering design.

d A very common stress-strain problem is the computation of temperature stresses. In this type of problem, neither total length nor area is necessary. It is necessary to compute what the change in unit length would be if the member were free to expand or contract—then to translate this directly into a unit stress.

Problem A piece of medium steel 40 ft long is rigidly fixed so that it cannot expand. If temperature increases 30°F, what change in unit stress results? From formula (2.3)

$$F = E\delta = 30{,}000{,}000 \text{ psi} \times 0.0000067 \text{ in. per deg} \times 30° = 6{,}000 \text{ psi}$$

Refer to Table 1.1 for the properties of steel.

e Occasionally a problem may be encountered which combines a stress-strain problem with some other factor such as thermal expansion. The following problem of this type was taken from a professional engineer's license examination.

Problem A heavy steel strut 20 ft long between two walls undergoes a temperature rise of 40°F. If the walls each yield 0.01 in., what will be the unit stress in the strut?

If the member were free to expand, its increase in length would be $\Delta = 40 \times 0.0000067 \times 20 \times 12 = 0.064$ in. The actual expansion measured by the yielding of the walls is 0.02 in. The strain is therefore $0.064 - 0.02 = 0.044$. Therefore

$$F = \frac{E\Delta}{l} = \frac{30{,}000{,}000 \times 0.044}{12 \times 20} = 5{,}500 \text{ psi}$$

2.21 Nonhomogeneous members

a Formula (2.3) may be written in the form

$$\delta = \frac{F}{E}$$

Therefore if two materials work together to resist the same type of stress in unison,

$$\delta = \frac{F_1}{E_1} = \frac{F_2}{E_2} \quad \text{or} \quad F_1 = \frac{E_1}{E_2}F_2$$

If we designate the ratio E_1/E_2 by n, the formula becomes $F_1 = nF_2$. In other words, if two materials having different moduli of elasticity are used together, they do not both work to the limit of their stress, but each carries stress in the proportion given by the above formula, up to the capacity of the weaker material. This principle is used in several common design problems, such as the design of flitch plate (or wood and steel) beams, and composite (or rolled-steel and concrete) beams, and will be referred to elsewhere.

Problem What is the carrying capacity of a short wood post 11½ in. square, with two steel side plates (one each side), each being 11 × ½ in.? Assume 1,000 psi on the wood with $E = 1{,}200{,}000$; 22,000 psi on the steel with $E = 29{,}000{,}000$.

From the above,

$$n = \frac{29{,}000{,}000}{1{,}200{,}000} = 24.2$$

The combination of stresses could be either

$$F_s = nF_w = 24.2 \times 1.000 = 24{,}200 \text{ and } F_w = 1{,}000$$

or

$$F_w = \frac{F_s}{n} = \frac{22{,}000}{24.2} = 910 \quad \text{and } F_s = 22{,}000$$

The controlling stresses therefore stay within the limits of the second combination and are 22,000 psi on steel and 910 psi on wood, and the capacity would be

$$\begin{aligned} 22{,}000 \times 2 \times 11 \times \tfrac{1}{2} &= 242{,}000 \\ 910 \times 11.5 \times 11.5 &= 120{,}350 \\ \text{Total capacity} &= 362{,}350 \end{aligned}$$

2.30 DEFLECTION

a Deflection may or may not be of importance in a structure. If the structure carries a suspended plaster ceiling, that part of the load which is applied after the plaster has set may cause sufficient deflection to crack plaster. Likewise, any beam carrying a brick wall may deflect enough to crack the brickwork. Any floor which carries vibratory machinery, crowds of people, or small groups marching in unison (such as church floors or lodges) should be designed with deflection in mind. Deflection and vibration go together, and even though the deflection is not enough to crack plaster, the vibration may be sufficient to affect the reputation of the building for safety—particularly in a church, a school, or a theater. In a house, an apartment house, or an office building, the effect of vibration is not so important. As a matter of actual safety, however, deflection does not indicate weakness. A short beam may be overloaded to the verge of collapse without an apparent deflection, while, on the other hand, a long-span shallow beam may deflect far beyond the so-called allowable deflection and still be perfectly safe.

If deflection is to be considered, the computed deflection in inches should not exceed 1/360 of the span or span in feet/30. With structural steel, using a fiber stress of 24,000 psi, it is not necessary to investigate for deflection unless the span of the beam in feet exceeds 1.33 times the depth of the beam in inches. In wood, the ratio of span in feet to depth in inches is about 1.25, depending on the allowable fiber stress and the modulus of elasticity of the lumber. With reinforced concrete, there are so many factors which influence the design, such as continuity, tension taken in the concrete, etc., that deflection is more or less uncertain; but reinforced concrete in general is less subject to deflection than steel or timber. It is however subject to plastic flow over a period of time, and in shallow long-span members provision must be made to prevent excessive permanent set in the concrete. This subject is discussed in Chapter 5 in connection with the design of concrete in bending.

Although no allowance is usually made for the fact, we know that a poured-concrete floor slab bonds to a steel beam in such a way as to form to some extent a T beam, whether designed as a composite beam or not. The effect of this T-beam action is to raise the neutral axis and give the effect of a beam having twice the depth of the beam below the neutral axis.

Bar joists in general follow the same rules of deflection as steel beams except that the shallow-end construction (see Fig. 4.33) increases the deflection. Load tests on identical joists with different types of top support—all loaded to the tabular load capacity of the joist as listed in Art. 4.13—gave the following results as compared with the computed deflection of a bare steel beam of uniform equal depth.

A joist system with a 2½-in. top slab deflected 35 percent of the computed deflection.

A joist system with a 1½-in. top slab deflected 47 percent of the computed deflection.

A joist system with metal deck tack welded to the joists deflected 110 percent of the computed deflection.

The results of the above tests indicate the effect of a properly bonded concrete slab in reducing deflection and the increase in deflection as the result of shallow ends.

b The deflection in inches of a uniformly loaded rolled-steel beam stressed to either 24,000 psi or 30,000 psi may be readily obtained by dividing the coefficient given in Table 2.5 by the depth of the beam in inches. For a span in excess of those tabulated, divide the required span by 2 and multiply the coefficient by 4. Thus to determine the deflection of a 10-in. beam stressed to 30,000 psi on a span of 16 ft,

$$\Delta = \frac{\text{coefficient}}{\text{depth}} = \frac{7.945}{10} = 0.7945 \text{ in.}$$

The deflection coefficient for stresses not listed is in direct proportion to the stress so that the deflection may be readily computed.

Table 2.5 Deflection Coefficients

Span, ft	F_s = 24,000	F_s = 30,000	Span, ft	F_s = 24,000	F_s = 30,000
1	0.026	0.035	26	16.784	20.980
2	0.099	0.124	27	18.099	22.624
3	0.224	0.279	28	19.466	24.333
4	0.398	0.496	29	20.880	26.100
5	0.621	0.776	30	22.346	27.932
6	0.894	1.117	31	23.859	29.824
7	1.217	1.521	32	25.424	31.780
8	1.589	1.986	33	27.038	33.797
9	2.012	2.514	34	28.701	35.876
10	2.483	3.104	35	30.414	38.018
11	3.003	3.755	36	32.177	40.221
12	3.575	4.469	37	33.989	42.486
13	4.196	5.245	38	35.852	44.815
14	4.866	6.083	39	37.763	47.204
15	5.586	6.984	40	39.725	49.656
16	6.356	7.945	41	41.735	52.219
17	7.175	8.969	42	43.796	54.745
18	8.045	10.055	43	45.906	57.383
19	8.963	11.204	44	48.066	60.082
20	9.932	12.414	45	50.276	62.845
21	10.949	13.686	46	52.535	65.674
22	12.017	15.021	47	54.843	68.554
23	13.134	16.418	48	57.203	71.504
24	14.301	17.876	49	59.612	74.515
25	15.518	19.396	50	62.069	77.586

c If the load is applied to the beam in any other manner than as a uniform load, a correction factor must be applied to obtain maximum deflection in accordance with Table 2.6.

Thus, to determine the deflection of a 10-in. beam on a 17-ft span with a centrally concentrated load, stressed to 17,500 psi,

Table 2.6 Deflection Factors for Types of Loading

Diagram (Fig. 3.13)	Load conditions	Deflection factor
1	Uniformly distributed	1.0
2	Load increasing uniformly to one end	.97
3	Load increasing uniformly to center	.96
7	Centrally located concentrated load	.8
12	Fixed at one end, freely supported at other, uniformly distributed	.42
13	Fixed at one end, freely supported at other, centrally located concentrated load	.48
15	Uniformly distributed load on fixed end beam	.3
16	Centrally concentrated load on fixed end beam	.4
18	Cantilever—load increasing uniformly to fixed end	1.92
19	Cantilever—uniformly loaded	2.4
22	Cantilever—concentrated load at free end	3.2

$$\Delta = \frac{17.5}{24} \times 0.8 \times \frac{7.175}{10} = 0.419$$

Other examples are given in Art. 4.10 (*e, f,* and *g*). If the load condition does not come under any of the cases for which the diagrams are listed, the method of approach is by means of the moment-area method, Art. 2.31.

Another simple approach to the problem of deflection is given in the "Manual of Steel Construction" of the AISC and is included as Table 2.8 of this book.

d The deflection of a wood beam or joist for a uniform load is

$$\Delta = \frac{15F1^2}{Ec}$$

Because the fiber stress and modulus of elasticity vary for different kinds of lumber, no simple deflection table like Table 2.5 can be set up. The various modifications given in Table 2.6 for different conditions of loading will modify the above formula for timber also.

e The tables of coefficients for beam deflections in Table 2.7 were published in *Engineering News-Record* by Robins Fleming and will prove quite helpful for determining the deflection of any point on a beam under a number of various loading conditions. The coefficient C is the multiplier to be used for PL^3 for concentrated loads and WL^3 for uniform loads in the formulae of Fig. 3.13.

f A footnote to Section 909 of the ACI Code makes the following statement:

Deflections in reinforced concrete structures depend on the elastic and inelastic properties of concrete and steel, as well as on shrinkage and creep which in turn are influenced by temperature and humidity, curing conditions, age of concrete at the time of loading and other factors. All simple methods for computation of deflections are therefore necessarily approximate. The methods specified in (c) and (d) of Sec. 909 of the ACI Code (Art. 5.20 of this book) may be considered satisfactory for guarding against excessive deflections in structures of common types and sizes.

The generally accepted methods of computing deflection in a reinforced concrete beam or slab are based on a number of assumptions of variables, so that at best the result is an approximation probably far on the safe side. A simple, rather direct approach is based on the theory that, for a computed stress in the steel, the deflection is the same as for a steel beam with the same type of load and fixity, and of a depth equal to twice the distance from the neutral axis to the plane of the tensile steel. Since reinforced concrete beams or slabs are usually continuous or semicontinuous, the factor for continuity taken from Table 2.6 must be applied—0.3 for a fully continuous uniformly loaded beam or 0.42 for a semicontinuous end condition. Thus

$$\Delta = \frac{CD}{2d(1-k)} \times \frac{A_s \text{ required}}{A_s \text{ furnished}}$$

where C is the deflection coefficient for the span taken from Table 2.5, D is the deflection factor from Table 2.6, and d, k, and A_s have the meaning usually applied to them in reinforced concrete formulae.

Table 2.7 Coefficients for Beam Deflection

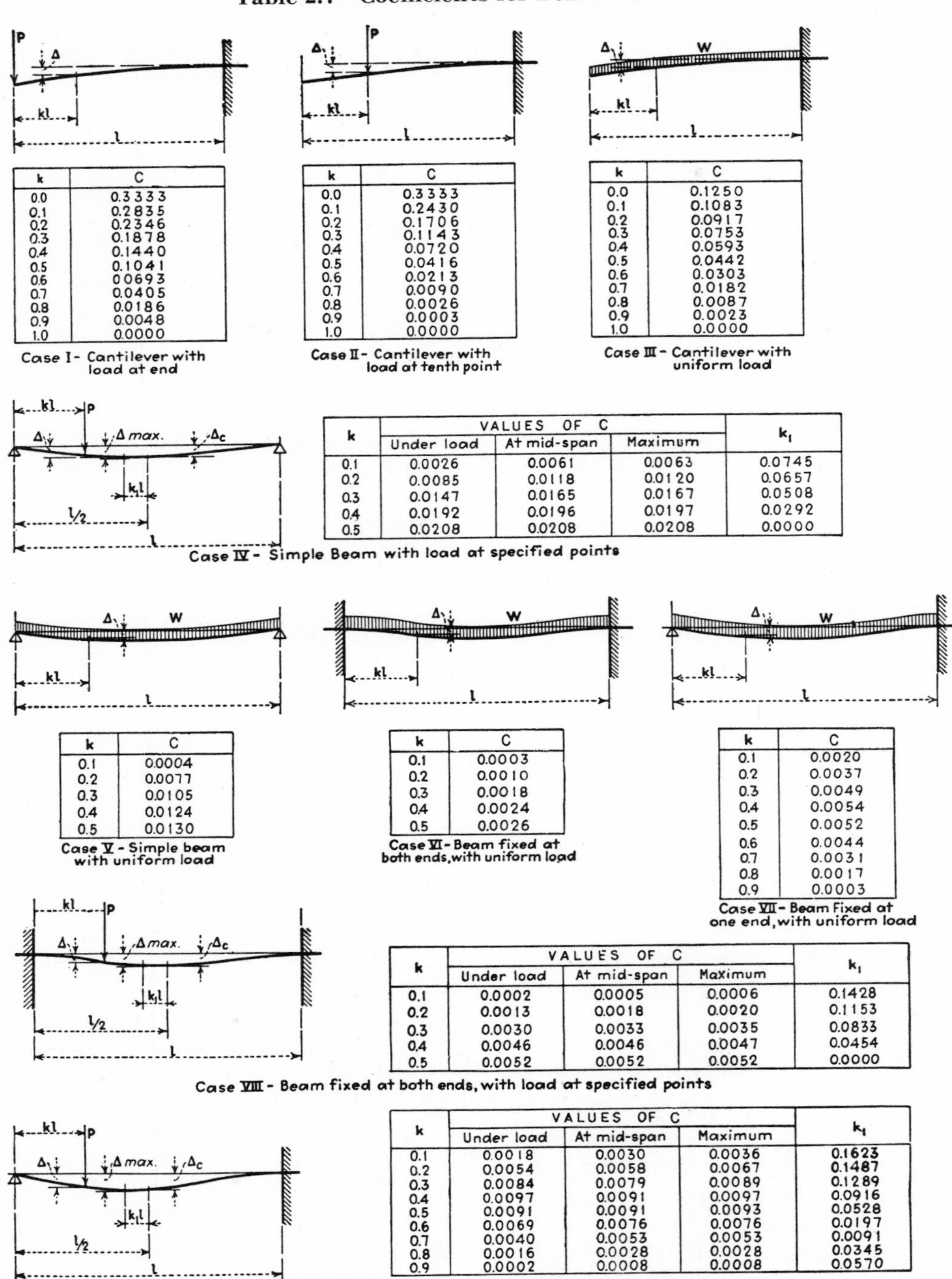

k	C
0.0	0.3333
0.1	0.2835
0.2	0.2346
0.3	0.1878
0.4	0.1440
0.5	0.1041
0.6	00693
0.7	0.0405
0.8	0.0186
0.9	0.0048
1.0	0.0000

Case I - Cantilever with load at end

k	C
0.0	0.3333
0.1	0.2430
0.2	0.1706
0.3	0.1143
0.4	0.0720
0.5	0.0416
0.6	0.0213
0.7	0.0090
0.8	0.0026
0.9	0.0003
1.0	0.0000

Case II - Cantilever with load at tenth point

k	C
0.0	0.1250
0.1	0.1083
0.2	0.0917
0.3	0.0753
0.4	0.0593
0.5	0.0442
0.6	0.0303
0.7	0.0182
0.8	0.0087
0.9	0.0023
1.0	0.0000

Case III - Cantilever with uniform load

k	VALUES OF C			k_1
	Under load	At mid-span	Maximum	
0.1	0.0026	0.0061	0.0063	0.0745
0.2	0.0085	0.0118	0.0120	0.0657
0.3	0.0147	0.0165	0.0167	0.0508
0.4	0.0192	0.0196	0.0197	0.0292
0.5	0.0208	0.0208	0.0208	0.0000

Case IV - Simple Beam with load at specified points

k	C
0.1	0.0004
0.2	0.0077
0.3	0.0105
0.4	0.0124
0.5	0.0130

Case V - Simple beam with uniform load

k	C
0.1	0.0003
0.2	0.0010
0.3	0.0018
0.4	0.0024
0.5	0.0026

Case VI - Beam fixed at both ends, with uniform load

k	C
0.1	0.0020
0.2	0.0037
0.3	0.0049
0.4	0.0054
0.5	0.0052
0.6	0.0044
0.7	0.0031
0.8	0.0017
0.9	0.0003

Case VII - Beam Fixed at one end, with uniform load

k	VALUES OF C			k_1
	Under load	At mid-span	Maximum	
0.1	0.0002	0.0005	0.0006	0.1428
0.2	0.0013	0.0018	0.0020	0.1153
0.3	0.0030	0.0033	0.0035	0.0833
0.4	0.0046	0.0046	0.0047	0.0454
0.5	0.0052	0.0052	0.0052	0.0000

Case VIII - Beam fixed at both ends, with load at specified points

k	VALUES OF C			k_1
	Under load	At mid-span	Maximum	
0.1	0.0018	0.0030	0.0036	0.1623
0.2	0.0054	0.0058	0.0067	0.1487
0.3	0.0084	0.0079	0.0089	0.1289
0.4	0.0097	0.0091	0.0097	0.0916
0.5	0.0091	0.0091	0.0093	0.0528
0.6	0.0069	0.0076	0.0076	0.0197
0.7	0.0040	0.0053	0.0053	0.0091
0.8	0.0016	0.0028	0.0028	0.0345
0.9	0.0002	0.0008	0.0008	0.0570

Case IX - Beam fixed at one end, with load at specified points

Table 2.8 Camber and Deflection Coefficients

Given the simple span length, the depth of a beam, girder or truss, and the design unit bending stress, the center deflection in inches may be found by multiplying the span length in feet by the tabulated coefficients given in the following table.

For the unit stress values not tabulated, multiply the factor given for 10,000 psi by the ratio of the given unit bending stress to 10,000.

The maximum fiber stresses listed in this table correspond to the allowable unit stresses as provided in Section 1.5.1.4 of the AISC Specification for steels having yield points ranging between 33,000 psi and 50,000 psi.

The table values, as given, assume a uniformly distributed load. For a single load at center span multiply these factors by 0.80; for two equal concentrated loads at third points, multiply by 1.02. Likewise, for three equal concentrated loads at quarter points multiply by 0.95.

The tabulated factors are correct for beams of constant cross section; reasonably accurate for cover plated beams and girders; and approximate for trusses.

Ratio of Depth/Span	Maximum Fiber Stress in Lbs. Per Sq. Inch									
	10,000	20,000	22,000	24,000	25,000	27,500	28,000	30,000	30,500	33,000
1/4	.0034	.0069	.0076	.0083	.0086	.0095	.0097	.0103	.0105	.0114
1/5	.0043	.0086	.0095	.0103	.0108	.0119	.0121	.0129	.0131	.0142
1/6	.0052	.0103	.0114	.0124	.0129	.0142	.0145	.0155	.0158	.0171
1/7	.0060	.0121	.0133	.0145	.0151	.0166	.0169	.0181	.0184	.0199
1/8	.0069	.0138	.0152	.0166	.0172	.0190	.0193	.0207	.0210	.0228
1/9	.0078	.0155	.0171	.0186	.0194	.0213	.0217	.0233	.0237	.0256
1/10	.0086	.0172	.0190	.0207	.0216	.0237	.0241	.0259	.0263	.0284
1/11	.0095	.0190	.0209	.0228	.0237	.0261	.0266	.0284	.0289	.0313
1/12	.0103	.0207	.0228	.0248	.0259	.0284	.0290	.0310	.0316	.0341
1/13	.0112	.0224	.0247	.0269	.0280	.0308	.0314	.0336	.0342	.0370
1/14	.0121	.0241	.0266	.0290	.0302	.0332	.0338	.0362	.0368	.0398
1/15	.0129	.0259	.0284	.0310	.0323	.0356	.0362	.0388	.0394	.0427
1/16	.0138	.0276	.0303	.0331	.0345	.0379	.0386	.0414	.0421	.0455
1/17	.0147	.0293	.0322	.0352	.0366	.0403	.0410	.0440	.0447	.0484
1/18	.0155	.0310	.0341	.0372	.0388	.0427	.0434	.0466	.0473	.0512
1/19	.0164	.0328	.0360	.0393	.0409	.0450	.0459	.0491	.0500	.0541
1/20	.0172	.0345	.0379	.0414	.0431	.0474	.0483	.0517	.0526	.0569
1/21	.0181	.0362	.0398	.0434	.0453	.0498	.0507	.0543	.0552	.0597
1/22	.0190	.0379	.0417	.0455	.0474	.0522	.0531	.0569	.0578	.0626
1/23	.0198	.0397	.0436	.0476	.0496	.0545	.0555	.0595	.0605	.0654
1/24	.0207	.0414	.0455	.0497	.0517	.0569	.0579	.0621	.0631	.0683

AMERICAN INSTITUTE OF STEEL CONSTRUCTION

For example, referring to the concrete slab designed in Art. 5.40*c*, compute the deflection of the slab noted. For a 10½-ft slab, $C = 2.20$ (the coefficient from Table 2.5 for 21, divided by 4), $D = 0.3$ (from Table 2.6), A_s required $= 0.458$, A_s furnished $=$ 0.48.

$$\Delta = \frac{2.20 \times 0.3}{2 \times 3 \times 0.625} \times \frac{0.458}{0.48} = 0.168$$

2.31 Moment-area method

a For irregular loading conditions or conditions for which the deflection cannot be readily computed by the methods described in Art. 2.30, there are several methods for determining deflections. One will be described here—the conjugate-beam, or moment-area, method. It takes the name from certain similarities between the method of calculation and the method of calculating a complex beam. Without explaining the mathematical derivation, the steps in the solution are:

1. Set down the beam diagram, compute the reactions and moments, and select the beam as described in Art. 4.10.

2. Lay out a moment diagram of the beam as described in Art. 3.11. This moment diagram becomes the equivalent of the beam-load diagram in an ordinary beam problem, and the areas of the moment diagram are the applied loads.

3. Divide the moment diagram into triangles, and other shapes of which the area and center of gravity may be readily obtained. Compute the area of these various parts of the moment diagrams and compute the reactions, as described in Art. 3.10. These reactions are the product $EI\theta$, which gives the slope deflection θ at the support.

4. Locate the point of zero shear. This is the point of maximum deflection.

5. Compute the moment at the point of zero shear. This moment is the product $EI\Delta$, from which the deflection Δ may be computed.

The above method may be used to determine moment at any intermediate point. For simplicity in determining the area of any fraction of a parabola, refer to Table 1.16.

Problem Compute the deflection of a steel beam under the following load conditions, and locate the point of maximum deflection (see Fig. 2.5). (See Art. 4.10 for method of computation of beam.)

$$M = 8 \times 9.6 = 76.8$$
$$S = 0.5 \times 76.8 = 38.4$$

Use 14 WF 30 ($S = 41.8$).

By the method of Art. 3.11*c*, draw the moment diagram Fig. 2.6. Divide Fig. 2.6 into three parts, two triangles and a parabola. The moment at 6 ft from R_1 is (Fig. 2.6)

$$12.4 \times 6 = 74.4$$
$$-(6 \times 3) = \underline{18.0}$$
$$56.4$$

The altitude of the triangle at this point is $76.8/2 = 38.4$. The altitude of the parabola therefore is $56.4 - 38.4 = 18.0$ (which would be the moment on a uniformly loaded beam carrying 1 kip per ft in a 12-ft span).

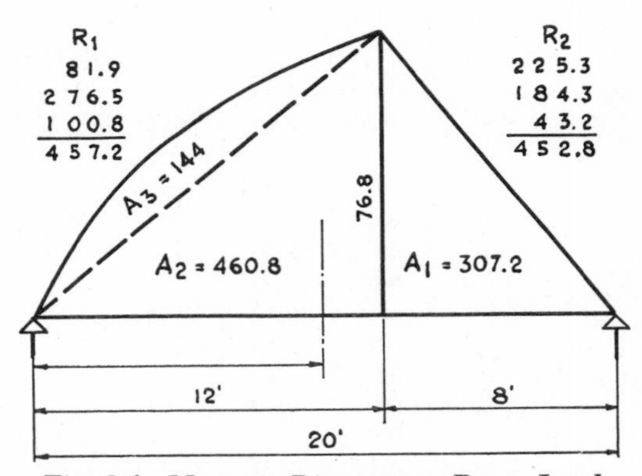

Fig. 2.6 Moment Diagram as Beam Load

We may compute the areas and apply them as loads to the beam, and determine the reactions and point of zero shear. Referring to Table 1.16, let us compute the area at the midpoint of the beam. Starting from R_2,

Triangle A_1 $= 307.2$

$2/12$ of $76.8 = 12.8 =$ small triangle,

$12.8 \times 2/2$ $= 12.8$

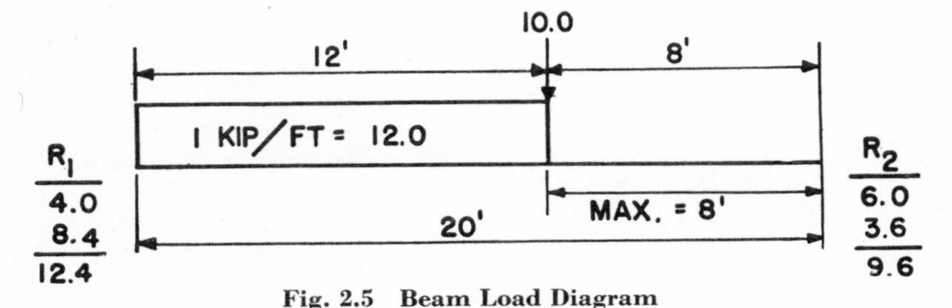

Fig. 2.5 Beam Load Diagram

Small rectangle, 76.8 − 12.8 = 64 × 2 = 128.0
Parabola, $\frac{2}{12}$ = 0.167. From Table 1.16, 0.05 × 18 × 12 = 10.8

458.8

R_2 = 452.8, so the point of zero shear is just less than 10 ft from R_2.

$$\text{Height at 10 ft} = 64 + (0.55 \times 18) = 73.9$$
$$458.8 - 452.8 = 6$$
$$\frac{6}{73.9} = 0.081$$

Therefore the point of zero shear of the conjugate beam, which is point of maximum deflection of the real beam, is approximately 10 − 0.08 = 9.92 ft from R_2.

From this point of zero shear, the maximum moment may be computed:

$$M = 452.8 \times 9.92 = 4{,}491.8$$

$$-307.2 \times 4.59 = 1{,}410.0$$
$$-\frac{1.92}{12} \times 76.8 \times \frac{1.92}{2} \times \frac{1.92 \times 2}{3} = 15.1$$
$$-64.5 \times 1.92 \times 0.96 = 118.9$$
$$-8.64 \times 1.92 \times 0.375 = 6.2$$

−1,550.2

2,941.6

Then $EI\Delta$ = 2,941.6 in ft³-kip units. Therefore

$$\Delta = \frac{2{,}941{,}600 \times 1{,}728}{29{,}000{,}000 \times 289.6} = 0.606 \text{ in.}$$

(It is interesting to note that the computed deflection for a uniformly loaded beam for equal section modulus on this span is 0.651.)

3

Moments

3.10 SIMPLE BEAMS

a Beams may be divided into two primary classes, simple, and restrained or continuous. Because of the method of construction, steel and wood beams are ordinarily computed as simple beams and concrete beams as continuous or restrained. There are several conditions under which this general rule is varied. A cantilever beam is obviously restrained. Sometimes, too, steel beams are rigidly connected for the sake of wind bracing, or for other design reasons. The 1961 AISC Code recognizes semirigid and rigid framing, and establishes code requirements for them.

b If any structure is to be sustained in equilibrium, the sum of the moments about any point on it must be algebraically equal to zero. Therefore, if an imaginary section could be cut at any given point within the beam, the sum of all external moments at either side of the section cut must be balanced by an internal moment or resisting moment within the beam. Although this principle is elementary mechanics, it may be well to recall it as the foundation for the solution of all beam problems. If we could concentrate all internal tension at a point and all internal compression at a point, as is indicated in Fig. 3.1, the stress couple thus obtained would be a measure of the external moment at this point. This stress-couple idea is useful in some aspects of design, and there will be occasion to refer to this formula later on:

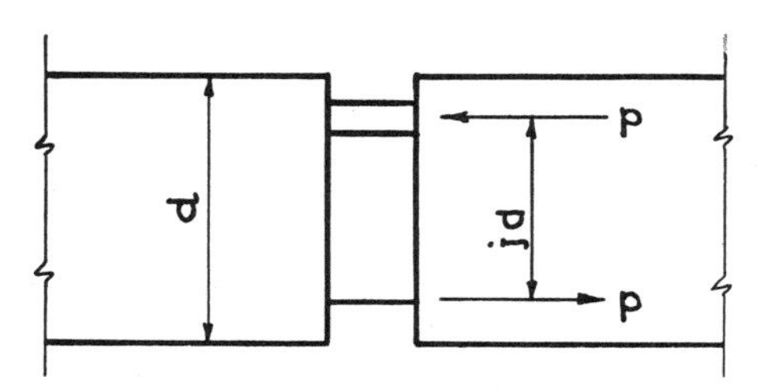

Fig. 3.1 Moment Couple Method

$$M = Pjd \qquad \text{or} \qquad P = \frac{M}{jd} \tag{3.1}$$

c Ordinary beam problems, however, involving steel or wood beams, are most readily solved by the use of the section-modulus formula, the fundamental beam formula,

$$M = F\frac{I}{c} \quad \text{or} \quad M = FS \quad \text{or} \quad S = \frac{M}{F} \tag{3.2}$$

d Practically all engineers have been taught the general method of beam design, as given in most mechanics or materials of construction textbooks. In general, these methods will be given, but they will be by-passed by various short cuts and simplifications.

e After the beam has been described, with a diagram to show load conditions if necessary, the first step is to obtain the reactions. The general method of finding reactions is to take moments about one end. The reaction at the left end for each load is the product of the load multiplied by its distance from the right end and divided by the total length. The total reaction is the sum of all reactions at that end, the balance of the load going to the opposite end. Most textbooks state the formula in this manner:

$$R_1 = \frac{P_1L_1 + P_2L_2 + P_3L_3 + \cdots P_nL_n}{L}$$

distances being taken from the right end. But each one can better be taken separately:

$$R_1 = \frac{P_1L_1}{L} + \frac{P_2L_2}{L} + \frac{P_3L_3}{L} + \cdots \frac{P_nL_n}{L} \qquad (3.3)$$

setting down the reactions from each load separately. The advantages of this method are (1) if any load is changed it is easier to change the reaction accordingly, and (2) frequently the relation of L_n/L will give an even fraction which can be solved by observation without the use of a slide rule. Obviously, for any symmetrically loaded beam the reactions are each half the total load on the beam.

f The general rule for obtaining the bending moment at any section of a simple beam may be stated as follows: Multiply one end reaction by the length from the end of the section. From the moment thus obtained, subtract the sum of the moments of the loads lying between the section cut and the chosen reaction, using as moment arms each length from the center of gravity of the load to the section cut. The general formula then becomes

$$M_a = R_1a - P_1a_1 - P_2a_2 - P_3a_3 - \cdots P_na_n \qquad (3.4)$$

where a is the distance from the reaction to the section cut, and a_1, a_2, etc., the distance from each load to the section cut.

g Ordinarily the only M in which we are interested is the moment at the dangerous or critical section. Without going into the proof of the axiom, it will be stated that the moment is maximum at the point on the beam where the shear is zero. The shear diagrams in Fig. 3.13 will permit solving the shear diagram of any such problem. It is not necessary to set down a shear diagram for each beam problem but it is necessary to have a mental picture of the diagram in order to locate this point of zero shear. The simple method of locating the point of zero shear is to subtract the various loads successively from the reaction until the result becomes zero. A little experience will show which reaction to select (see Art. 3.11).

h For the sake of simplicity in discussion, simple beam problems may be placed in one of five classifications.

Class I—Uniformly distributed load ($M = WL/8$).

Class II—Equal, concentrated loads equally spaced.

Class III—Simple triangular, trapezoidal, or cantilever load condition.

Class IV—Combination of any two or more of the preceding three classes.

Class V—Irregular or complex load conditions.

A glance at the beam will permit placing it in one of the above classifications. Classes I and II do not require any sketch and their calculation is quite simple. Classes III and V will require load diagrams; Class V may become quite complicated. Class IV may be simple or complicated, depending on the combinations used. Only the computation of the moment will be considered here. Beam selection will be left to other articles.

The beam diagrams and formulae given in Fig. 3.13 give shear and moment diagrams for various general conditions of loading. The formula for maximum moment is derived in all instances from the general formula in Art. 3.10*f*. The equivalent tabular load for each of the various load conditions is the ratio of load which, applied to a beam uniformly loaded on this span, would give the same maximum moment. For example, for Case 7 (Fig. 3.13), a centrally concentrated load, the equivalent tabular load is $2P$, or the effect of this type of loading causes the same moment as twice the same load if applied uniformly as in Case 1. Thus a load of 10,000 lb centrally concentrated on a 16-ft span would cause 40 ft-kips moment; but for a uniform load on a 16-ft span it would require 20,000 lb to cause 40 ft-kips moment.

i *Class II, equal concentrated loads, equally spaced,* is common for girders, or main supporting beams. The general formula may be written $M = CWL$, where C is

For one load,	.25
For two or three loads,	.167
For four or five loads,	.15
For six or seven loads,	.143
For eight or nine loads,	.139

j *Class III, simple triangular, trapezoidal, or cantilever load condition,* groups certain beam loadings which are encountered occasionally and which may best be solved by means of the formulae in Fig. 3.13. Cases 2 and 3 are found in hip and valley roof design. A beam with trapezoidal load, as indicated in Fig. 3.2, may be part of a roof construction, but it is more frequently the condition found in basement walls. It may be solved with an

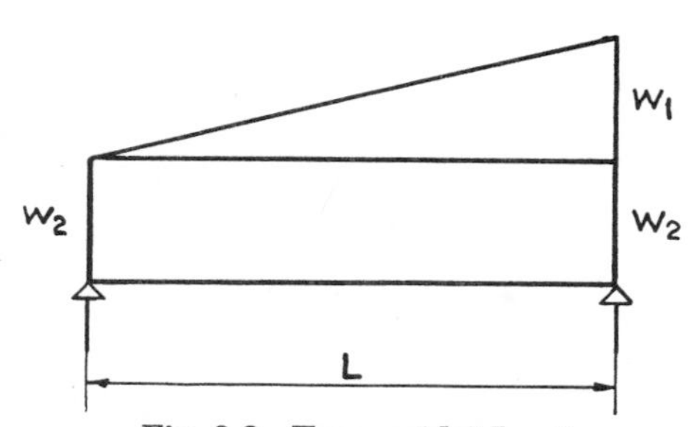

Fig. 3.2 Trapezoidal Load

error of not over 3 percent by dividing it between uniform and triangular load, as shown in the sketch, and adding the results, or it may be solved accurately by the general method similar to the problem in Art. 3.10*o*.

Problem A state license examination has this question: A vertical surface 40 ft high and 40 ft wide is exposed to a wind pressure of 30 psf. What is the total overturning moment? What is the total horizontal shear? This problem comes under Fig. 3.13, Case 19, cantilever beam, uniformly distributed load.

The shear $V = W = wL$.

$$V = 40 \times 40 \times 30 = 48{,}000 \text{ lb maximum}$$

$$M_{\max} = \frac{wL^2}{2} = \frac{48{,}000 \times 40}{2} = 960{,}000 \text{ ft-lb}$$

k *Class IV, combined loadings.* Any two symmetrical systems of loadings may be combined, and this is frequently the simplest method of solution. For instance, a beam carries a uniform load of 1,000 lb per ft on 15 ft and a centrally concentrated load of 6,000 lb. What is the total moment? From Fig. 3.13, Case 1,

$$M = \frac{WL}{8} = \frac{15{,}000 \times 15}{8} = 28{,}125$$

From Fig. 3.13, Case 7,

$$M = \frac{PL}{4} = \frac{6{,}000 \times 15}{4} = 22{,}500$$

Total $M = 50{,}625$ ft-lb

A simpler method would be to double the concentrated load and treat it as uniform (because of the ratio of denominators of the above two formulae). Thus, $15{,}000 + 12{,}000 = 27{,}000$.

$$M = \frac{27{,}000 \times 15}{8} = 50{,}625 \text{ ft-lb}$$

(See Art. 4.10*c*).

Obviously a symmetrical loading cannot be combined with an unsymmetrical loading by adding moments, since the point of maximum bending does not occur at the same point. In this case, it is advisable to treat the problem as indicated in the next paragraph.

l *Class V, irregular or complex load conditions,* is a general method for moments. If the problem does not come within the limits of one of the above classes, it must fall into this class. A sketch is necessary. It is preferable to divide the problem into two distinct steps as follows:

1. Compute the reaction from each load at each end. This may often be done by observation, as in the beam sketched in Fig. 3.3. The load of 10,000 lb is at the ¼ point—therefore ¼ × 10,000 goes to the opposite end—2,500. The balance of this load goes to the left end. The center of gravity of the 6,000-lb load is 3 ft out, or ⅙ span; therefore ⅙ × 6,000 = 1,000 goes to the left end, and 5,000 to the right end.

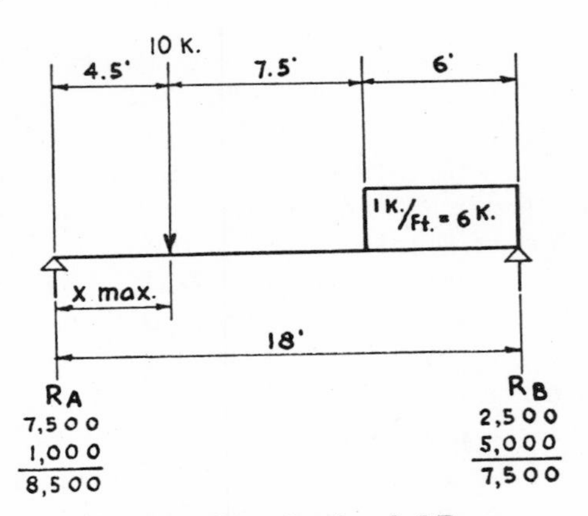

Fig. 3.3 Complex Loaded Beam

2. With the total end reactions known, subtracting from either end, the point where the shear becomes zero is readily found—the critical section. For this purpose it is advisable to have a mental picture of the various shear diagrams, Fig. 3.13.

3. Compute M, the difference between all clockwise and counterclockwise moments to one side or the other of this critical section. The sides will be identical, so select the simpler problem of the two. In the problem above, $M = 8{,}500 \times 4.5 = 38{,}250$ ft-lb.

Problem This question is from a state license examination: The left half of a simple beam of 20 ft span is loaded with 2,000 lb per lin ft. Where is the point of maximum moment? What is its value? (See Fig. 3.4.)

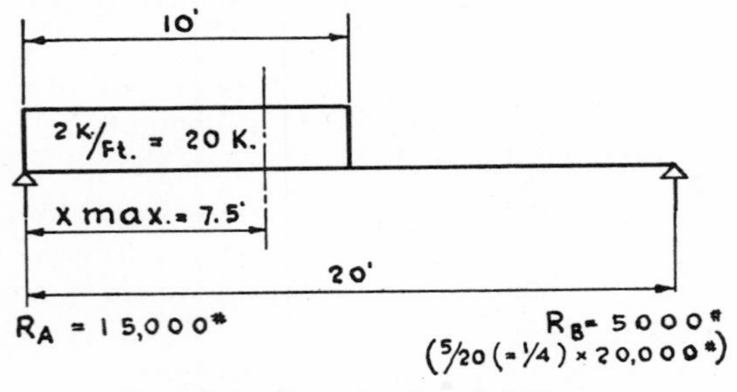

Fig. 3.4 Complex Loaded Beam

The computation of end reactions R_A and R_B and the computation of critical section are indicated. The method of computation of moment then becomes

$$\begin{aligned} M = \quad & 15{,}000 \times 7.5 = 112{,}500 \\ & -15{,}000 \times 3.75 = -56{,}250 \\ & \qquad\qquad 56{,}250 \end{aligned}$$

m This brings out a little short cut in a problem of this kind, where there is no change in uniform load condition between the critical section and the end of the beam. Since the counterclockwise moment is exactly half the clockwise moment, the final computation might have been written

$$M = 15{,}000 \times \frac{7.5}{2} = 56{,}250$$

Note that this applies only when the reaction is entirely used up by uniform load.

n The problem shown in Fig. 3.5 is selected because it may clarify the location of

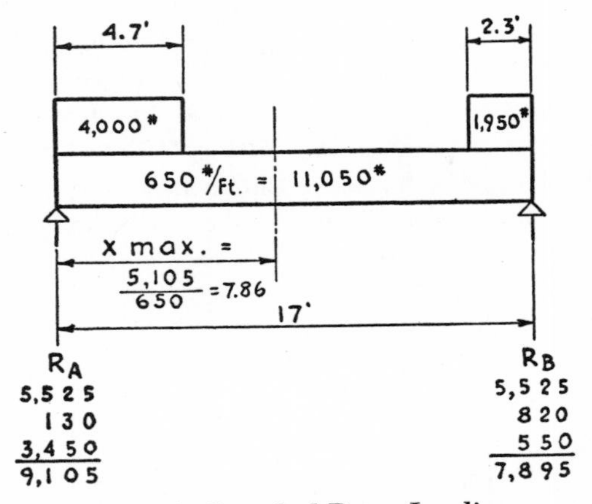

Fig. 3.5 Spandrel Beam Loading

point of zero shear. From R_A deduct the 4,000-lb load on the 4.7-ft length, leaving 5,105. This load must be used up in uniform load at 650 lb per ft, or $x = 5{,}105/650 = 7.86$.

$$\begin{aligned} M &= 9{,}105 \times 7.86 = 71{,}700 \\ -4{,}000 &\times 5.51 = 22{,}040 \\ -5{,}105 &\times 3.93 = 20{,}100 \\ & \qquad -42{,}140 \\ & \qquad 29{,}560 \text{ maximum moment} \end{aligned}$$

o Probably the most complicated of beams coming under this classification is one involving a trapezoidal load over part of the span, in combination with loads of other types. The point of zero shear must be found by means of a quadratic equation. The reactions are computed and set down in the usual manner, as has been done in Fig. 3.6. From inspection it may be seen that the dangerous section lies between the

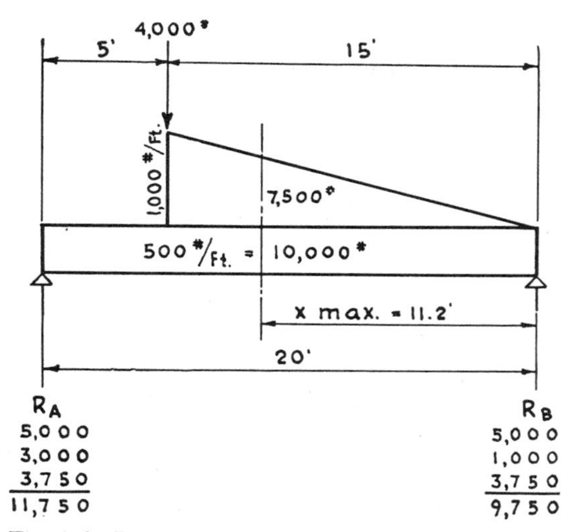

Fig. 3.6 Beam with Uniformly Increasing Load

4,000-lb load and R_B. The triangular load increases from R_B so that at any point x ft from R_B, by similar triangles,

$$\frac{w_x}{1{,}000} = \frac{x}{15} \quad \text{or} \quad w_x = \frac{1{,}000x}{15}$$

The area of the triangle to the right of this would be

$$\frac{w_x x}{2} \quad \text{or from similar triangles} \quad \frac{1{,}000x^2}{30} = 33.3x^2$$

The area of the rectangle to the right would be $500x$. Therefore the shear crosses the zero line where $33.3x^2 + 500x = 9.750$.

$$x^2 + 15x = 293$$
$$x + 7.5 = \sqrt{349.25} = 18.7$$
$$x = 18.7 - 7.5 = 11.2$$

From this point the computation is simple.

$$M = 9{,}750 \times 11.2 = 109{,}200$$
$$-500 \times 11.2 \times 5.6 = 31{,}360$$
$$-\frac{11.2}{15} \times 1{,}000 \times \frac{11.2}{2} \times \frac{11.2}{3} = 15{,}610$$

$$-46{,}950$$
$$62{,}250 \text{ maximum moment}$$

This method may be used for solving accurately any trapezoidal load beam as noted in Art. 3.10*j*.

p Attention should be called to several points with regard to cantilever beams which may be overlooked. In Fig. 3.13 for the beams shown in Cases 18, 19, 21, and 22, the load diagrams and the consequent reactions

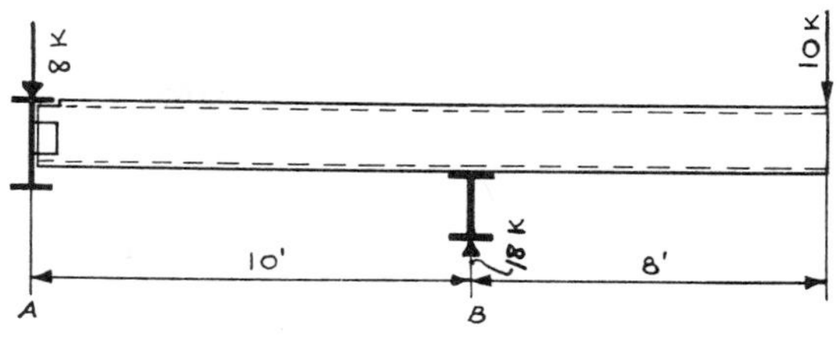

Fig. 3.7 Cantilever Beam

given in the table would indicate that the reactions are equal to P or to W. This is true so far as the load of the cantilever beam itself is concerned. However, cantilever beams are not simply fixed at a point as indicated in Fig. 3.13 but are supported by an uplift load, or a countermoment on a beam or a wall at the end against overturning. The reaction then becomes the sum of the downward load plus the uplift load.

Assume a cantilever beam as shown in Fig. 3.7. The load of 10 kips on a lever arm of 8 ft produces a cantilever moment of 80 ft-kips. The tendency to rotate is resisted by the connection to the beam at A; and to balance the overturning moment of 80 ft-kips would require a load of 8 kips on the lever arm of 10 ft, thus putting a total reaction of 18 kips on beam B instead of merely the cantilever load. Perhaps a more direct approach to the problem of obtaining the reaction would be to take moments about

A. Then

$$10R_B = 18 \times 10$$
$$R_B = 18$$

A problem involving a combined simple and cantilever loading may be rather confusing. Figure 3.8 indicates a problem of this type. It will be noted that there are two points of zero shear, one within the simple span, 8.375 ft from the end, where the moment is 28.0 ft-kips, and the cantilever

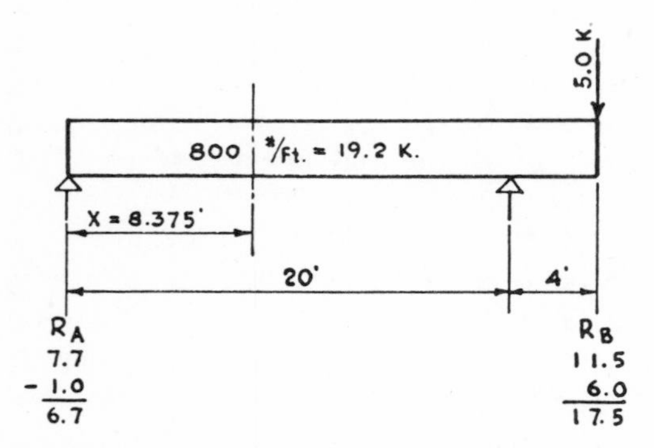

Fig. 3.8 Cantilever Beam—Fully Loaded

moment, 26.4 ft-kips. In order to design the beam it is necessary to compute the moment at both points.

In connection with this beam, it is well to note that the simple beam or uplift end of the cantilever must be checked against the possibility of the simple span having full live and dead load, while the cantilever span supports dead load only. Let us assume in the preceding problem that the total load consists of 50 percent dead load and 50 percent live load. In this case the load condition and reactions are as indicated in Fig. 3.9. Although the

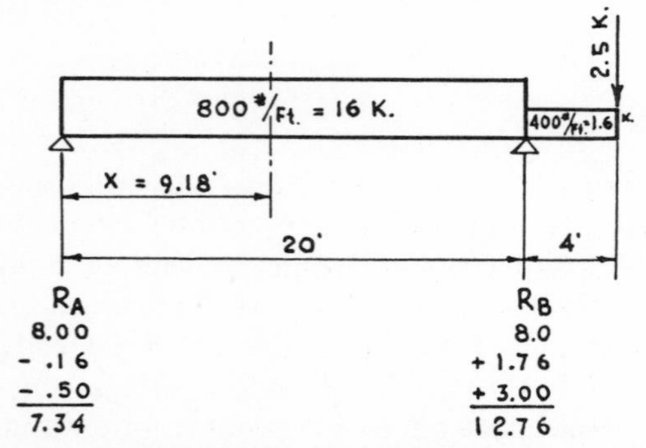

Fig. 3.9 Cantilever Beam—Partially Loaded

cantilever moment is only half the original cantilever moment, R_A is increased so that the point of zero shear moves further out into the span, and the maximum moment becomes $7.34 \times 4.59 = 33.69$ ft-kips.

If the uplift arm is relatively short or if the load available to resist uplift is very low, it becomes necessary also to investigate the uplift with the cantilever fully loaded and the uplift span with the dead load only.

q *Moving loads.* Moving loads on two wheels, such as the load of a traveling crane on a crane girder or a road roller or a truck on a bridge, present a special problem in that the position of the load to produce maximum moment must be determined. The general statement for the position of loads on a simple span is as follows: The maximum bending moment in a simple beam with a moving load occurs when the heavier load is as far from one end of the beam as the center of gravity of all the moving loads is from the other end of the beam. Obviously, with a single moving load, this moment occurs when the load is at the center of the span.

If two wheels carry equal loads, such as on the end truck of a traveling crane, and the wheel base a, the distance between the wheels, is less than $0.586L$ (Fig. 3.10),

$$M_{\max} = \frac{P(L - a/2)^2}{2L} \qquad (3.5)$$

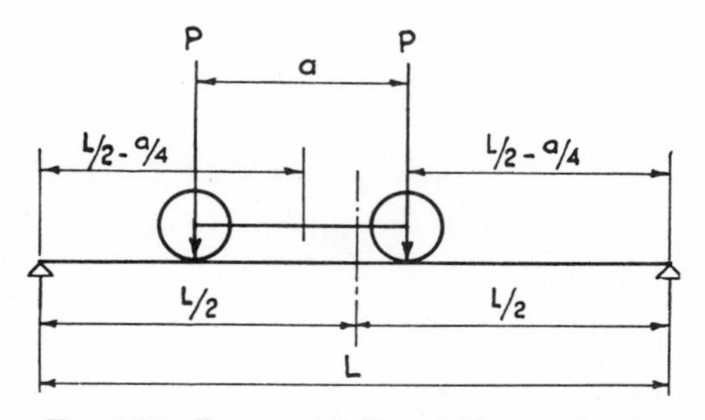

Fig. 3.10 Beam with Equal Moving Loads

When the wheel base exceeds $0.586L$, $M_{\max} = PL/4$. M will occur twice on the span as the load moves along. The maximum shear V is at one end when the first wheel is about to leave the span.

Problem A 10-ton crane on a steel-beam runway having a 20-ft span has a wheel base

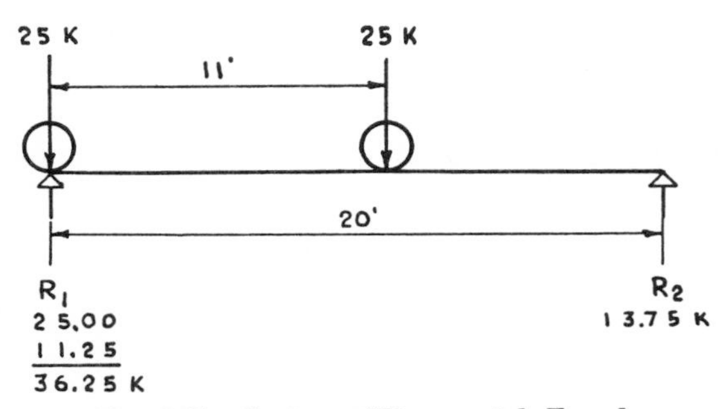

Fig. 3.11 Design of Beam with Equal Moving Loads

of 11 ft and a wheel load of 25,000 lb. Figure maximum moment and shear.

$$M_{max} = \frac{25(20 - 11/2)^2}{2 \times 20} = 131.4 \text{ ft-kip}$$

The maximum shear is the maximum reaction when the load is in the position indicated in Fig. 3.11, 36.25 kips.

If two unequal rolling loads are a fixed distance apart (Fig. 3.12), the maximum moment is under the heavier load in the position computed by the formula

$$y = \frac{P_A a}{2(P_A + P_B)} \tag{3.6}$$

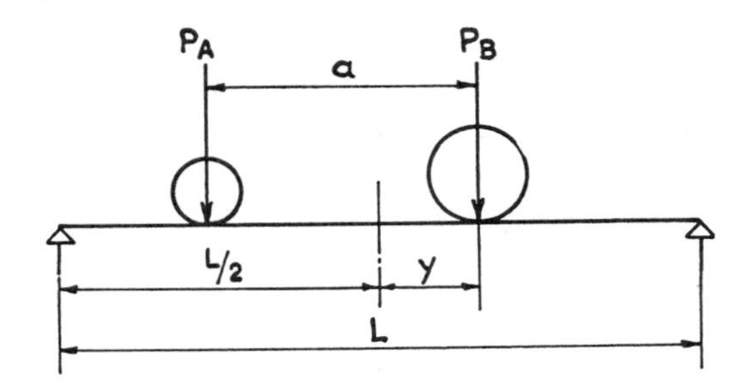

Fig. 3.12 Beam with Unequal Moving Loads

The maximum shear occurs as the heavier wheel is about to leave the span.

Problem How shall a truck be placed on a span of 30 ft to produce the greatest bending moment? The loads are 5 tons on the front axle and 15 tons on the rear axle. The distance between the axles is 14 ft.

$$y = \frac{5 \times 14}{2 \times 20} = 1.75 \text{ ft}$$

The greatest moment occurs when the rear wheel is $15 - 1.75 = 13.25$ ft on the span, and the front is $30 - (13.25 + 14) = 2.75$ ft from the other end.

3.11 Shear and moment diagrams

Although usually shear and moment diagrams are not necessary for the solution of beams, the engineer should know their construction. The shear diagram provides the means of locating the point of maximum moment on a beam (see Art. 3.10), and although it is usually unnecessary to go to the trouble of drawing it, for irregularly loaded beams it is at least necessary to have a mental picture of the diagram. The shear diagram is also used in computation of stirrups in a concrete beam (see Art. 5.30).

The shear diagram represents graphically the actual vertical shear at all points along a beam, and the moment diagram represents the moment at all points along the beam. Figure 3.13 shows the general shape of shear and moment diagrams for the various types of uniformly loaded and concentrated load beams.

a The shear diagram is drawn by laying out to scale—a vertical scale to represent load and a horizontal scale to represent length—a beam diagram showing all upward loads drawn as straight lines up (reactions) and downward loads downward, a straight drop for a concentrated load and a slant line downward for a uniform load. Reference to Fig. 3.13 will indicate the following facts about shear diagrams:

1. Concentrated loads are indicated by a straight drop downward.
2. Uniform loads are indicated by a sloping line downward.
3. Uniform loads that vary on several parts of a beam are indicated by sloping lines which change slope at points where the ratio of uniform load changes.
4. Uniformly increasing or decreasing loads are represented by curved lines.
5. Through sections where no loads either uniform or concentrated are applied, the line of the shear diagram is parallel with the base line.
6. A properly drawn shear diagram is a good check on the reactions of a beam.

1. SIMPLE BEAM—UNIFORMLY DISTRIBUTED LOAD

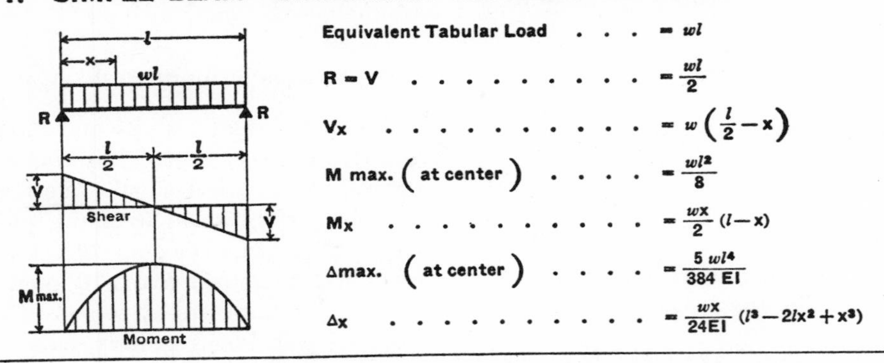

2. SIMPLE BEAM—LOAD INCREASING UNIFORMLY TO ONE END

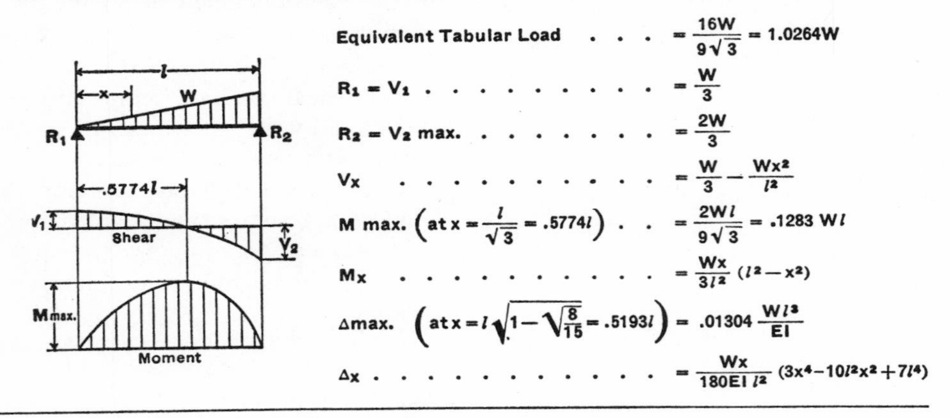

3. SIMPLE BEAM—LOAD INCREASING UNIFORMLY TO CENTER

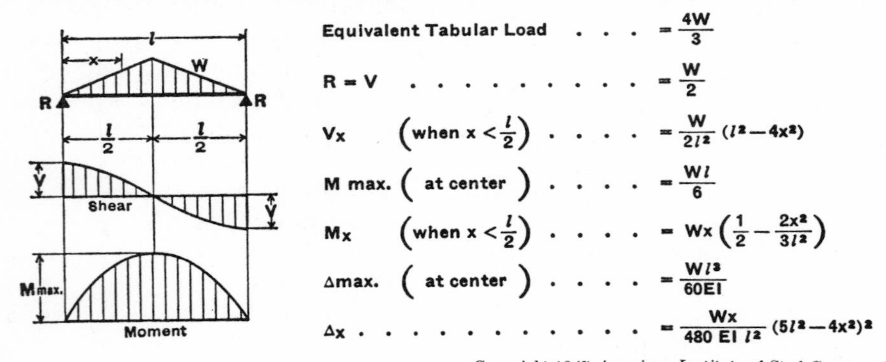

Fig. 3.13 Beam Diagrams and Formulae

b The moment diagram is useful for solving several problems in structural design and is referred to under the following headings:

Art. 2.31—Computation of deflection by the use of moment-area method.

Art. 3.21—Finding contraflexure points, and positive moment in continuous beams.

Art. 4.20—Determination of length of cover plates on a plate girder.

c The moment diagrams for problems involving concentrated loads only are simple to draw. It is only necessary to compute the

4. SIMPLE BEAM—UNIFORM LOAD PARTIALLY DISTRIBUTED

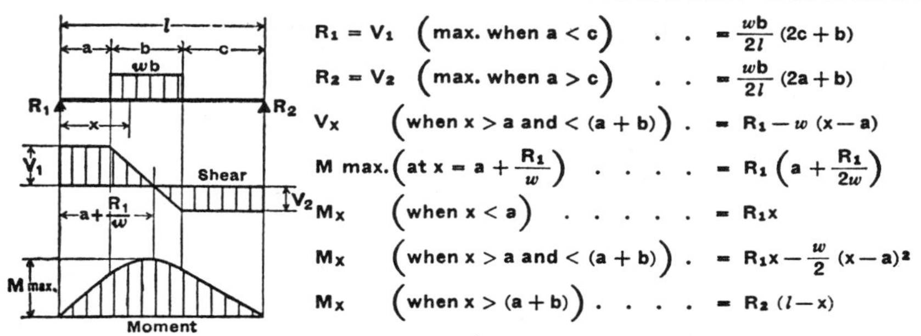

$R_1 = V_1 \left(\text{max. when } a < c\right) \quad . \; . \quad = \frac{wb}{2l}(2c+b)$

$R_2 = V_2 \left(\text{max. when } a > c\right) \quad . \; . \quad = \frac{wb}{2l}(2a+b)$

$V_x \quad \left(\text{when } x > a \text{ and } < (a+b)\right) \; . \quad = R_1 - w(x-a)$

$M\ \text{max.}\left(\text{at } x = a + \frac{R_1}{w}\right) \quad . \; . \; . \; . \quad = R_1\left(a + \frac{R_1}{2w}\right)$

$M_x \quad \left(\text{when } x < a\right) \quad . \; . \; . \; . \; . \quad = R_1 x$

$M_x \quad \left(\text{when } x > a \text{ and } < (a+b)\right) \; . \quad = R_1 x - \frac{w}{2}(x-a)^2$

$M_x \quad \left(\text{when } x > (a+b)\right) \; . \; . \; . \; . \quad = R_2(l-x)$

5. SIMPLE BEAM—UNIFORM LOAD PARTIALLY DISTRIBUTED AT ONE END

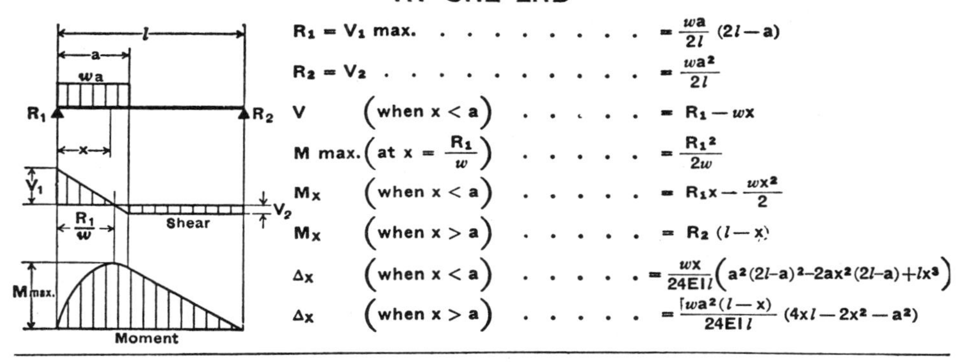

$R_1 = V_1\ \text{max.} \; . \; . \; . \; . \; . \; . \; . \; . \; . \quad = \frac{wa}{2l}(2l-a)$

$R_2 = V_2 \; . \; . \; . \; . \; . \; . \; . \; . \; . \; . \quad = \frac{wa^2}{2l}$

$V \quad \left(\text{when } x < a\right) \; . \; . \; . \; . \; . \quad = R_1 - wx$

$M\ \text{max.}\left(\text{at } x = \frac{R_1}{w}\right) \; . \; . \; . \; . \; . \quad = \frac{R_1^2}{2w}$

$M_x \quad \left(\text{when } x < a\right) \; . \; . \; . \; . \; . \quad = R_1 x - \frac{wx^2}{2}$

$M_x \quad \left(\text{when } x > a\right) \; . \; . \; . \; . \; . \quad = R_2(l-x)$

$\Delta x \quad \left(\text{when } x < a\right) \; . \; . \; . \; . \; . \quad = \frac{wx}{24EIl}\left(a^2(2l-a)^2 - 2ax^2(2l-a) + lx^3\right)$

$\Delta x \quad \left(\text{when } x > a\right) \; . \; . \; . \; . \; . \quad = \frac{wa^2(l-x)}{24EIl}(4xl - 2x^2 - a^2)$

6. SIMPLE BEAM—UNIFORM LOAD PARTIALLY DISTRIBUTED AT EACH END

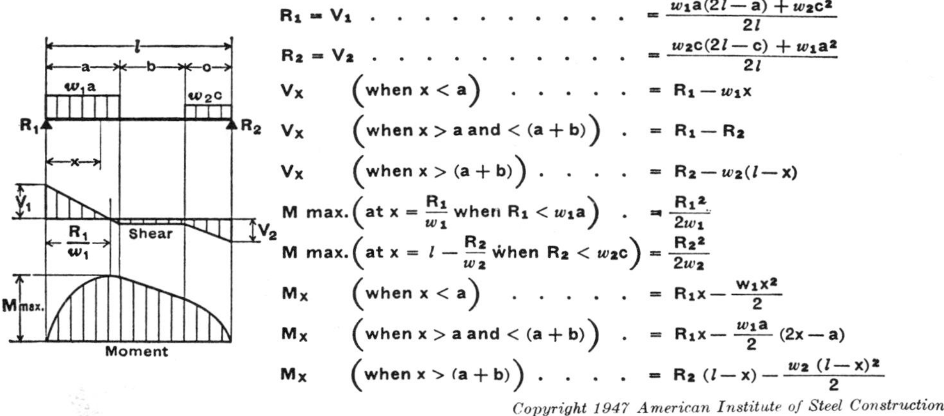

$R_1 = V_1 \; . \; . \; . \; . \; . \; . \; . \; . \; . \; . \quad = \frac{w_1 a(2l-a) + w_2 c^2}{2l}$

$R_2 = V_2 \; . \; . \; . \; . \; . \; . \; . \; . \; . \; . \quad = \frac{w_2 c(2l-c) + w_1 a^2}{2l}$

$V_x \quad \left(\text{when } x < a\right) \; . \; . \; . \; . \; . \quad = R_1 - w_1 x$

$V_x \quad \left(\text{when } x > a \text{ and } < (a+b)\right) \; . \quad = R_1 - R_2$

$V_x \quad \left(\text{when } x > (a+b)\right) \; . \; . \; . \; . \quad = R_2 - w_2(l-x)$

$M\ \text{max.}\left(\text{at } x = \frac{R_1}{w_1} \text{ when } R_1 < w_1 a\right) \; . \quad = \frac{R_1^2}{2w_1}$

$M\ \text{max.}\left(\text{at } x = l - \frac{R_2}{w_2} \text{ when } R_2 < w_2 c\right) = \frac{R_2^2}{2w_2}$

$M_x \quad \left(\text{when } x < a\right) \; . \; . \; . \; . \; . \quad = R_1 x - \frac{w_1 x^2}{2}$

$M_x \quad \left(\text{when } x > a \text{ and } < (a+b)\right) \; . \quad = R_1 x - \frac{w_1 a}{2}(2x-a)$

$M_x \quad \left(\text{when } x > (a+b)\right) \; . \; . \; . \; . \quad = R_2(l-x) - \frac{w_2(l-x)^2}{2}$

Fig. 3.13 Beam Diagrams and Formulae (Continued)

moment at points of load, lay them out to scale above the base line, and connect the points with straight lines.

Moment diagrams involving uniform loads, even through part of the span, require the construction of parabolas (see Table 1.16). For a beam carrying a uniform load throughout its full length plus one or more concentrated loads, the simplest method is to construct the two separate moment diagrams, one above the line, the other beneath the line. If the use of the moment diagram requires a

7. SIMPLE BEAM—CONCENTRATED LOAD AT CENTER

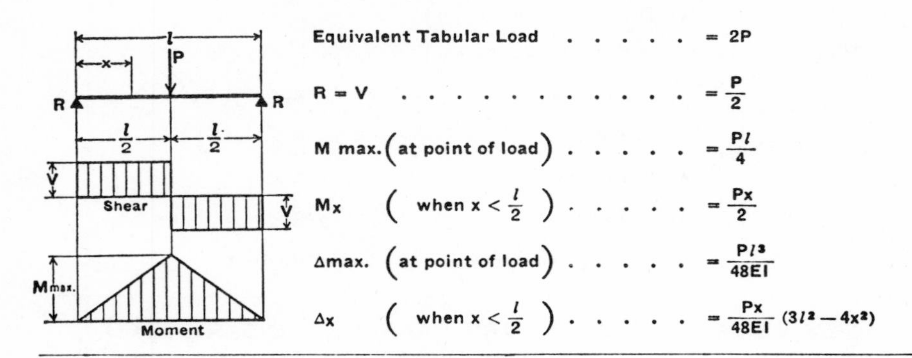

8. SIMPLE BEAM—CONCENTRATED LOAD AT ANY POINT

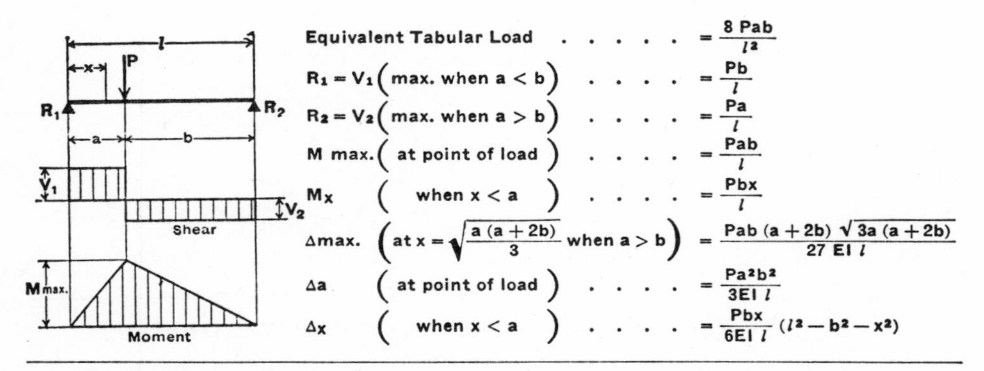

9. SIMPLE BEAM—TWO EQUAL CONCENTRATED LOADS SYMMETRICALLY PLACED

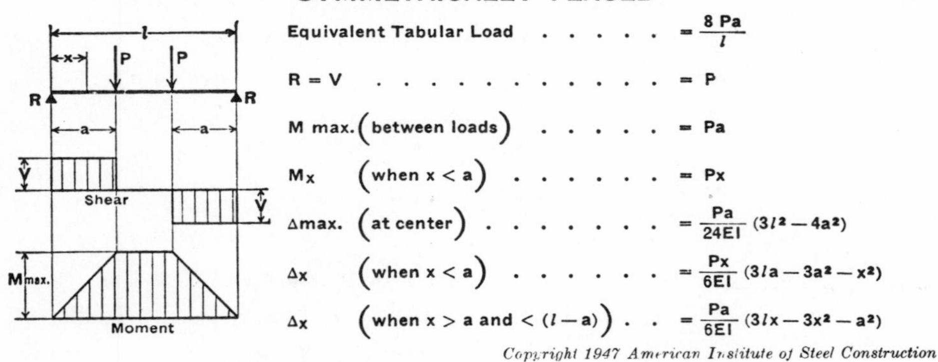

Fig. 3.13 Beam Diagrams and Formulae (Continued)

level base line, reconstruct the combined diagram above the base line by adding the length of ordinate below the base line to each ordinate on top of the parabola. Then construct a smooth curve through the various points. If the beam carries a uniform load through one section only, compute the moment at the beginning and the end of the load, then draw a base line through these two points and construct a parabola on the new base line, the height of the parabola to be determined by the moment at the center of the uniform-load area. Properly drawn, the parabola at its ends should be tangent to the straight lines (see Fig. 3.13, Case 4).

The construction and use of moment dia-

10. SIMPLE BEAM—TWO EQUAL CONCENTRATED LOADS UNSYMMETRICALLY PLACED

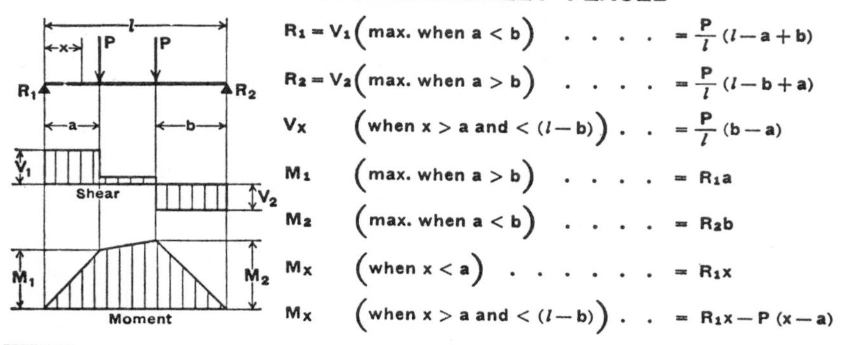

11. SIMPLE BEAM—TWO UNEQUAL CONCENTRATED LOADS UNSYMMETRICALLY PLACED

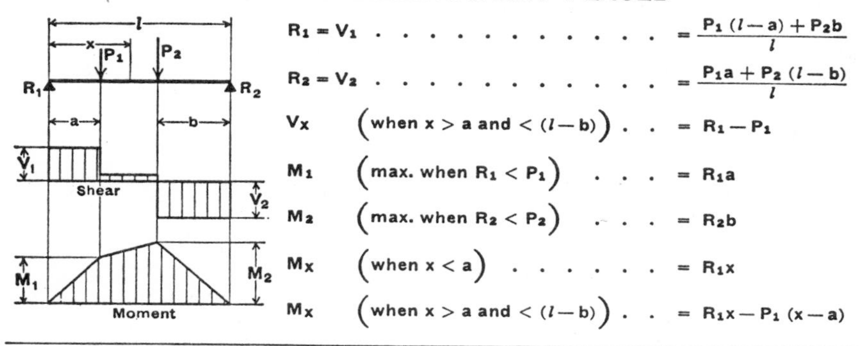

12. BEAM FIXED AT ONE END, SUPPORTED AT OTHER—UNIFORMLY DISTRIBUTED LOAD

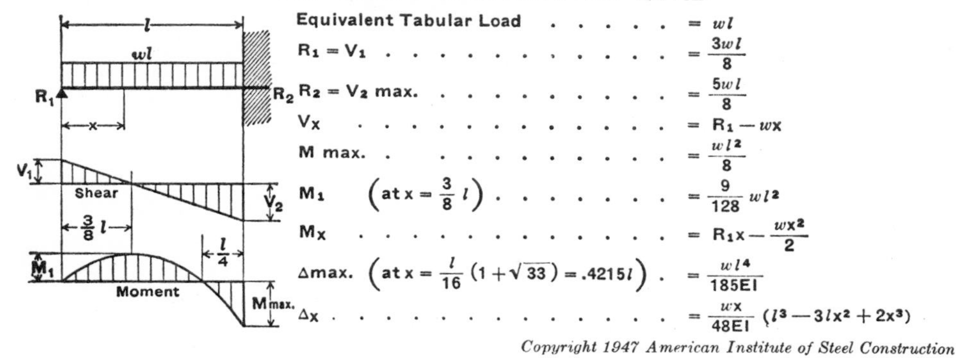

Fig. 3.13 Beam Diagrams and Formulae (Continued)

grams for continuous beams is discussed in Art. 3.21 on the Hardy Cross method of moment distribution.

3.20 MOMENTS—CONTINUOUS BEAMS

a *Continuous beams,* that is, beams over three or more supports, are statically indeterminate structures and are not subject to analysis by simple beam methods. The continuous-beam diagrams in Fig. 3.13 give the moment and the shear resulting from (1) a uniformly distributed load on equal spans, with beams of constant moment of inertia, and (2) concentrated loads under several different conditions on equal spans, with beams

13. BEAM FIXED AT ONE END, SUPPORTED AT OTHER—CONCENTRATED LOAD AT CENTER

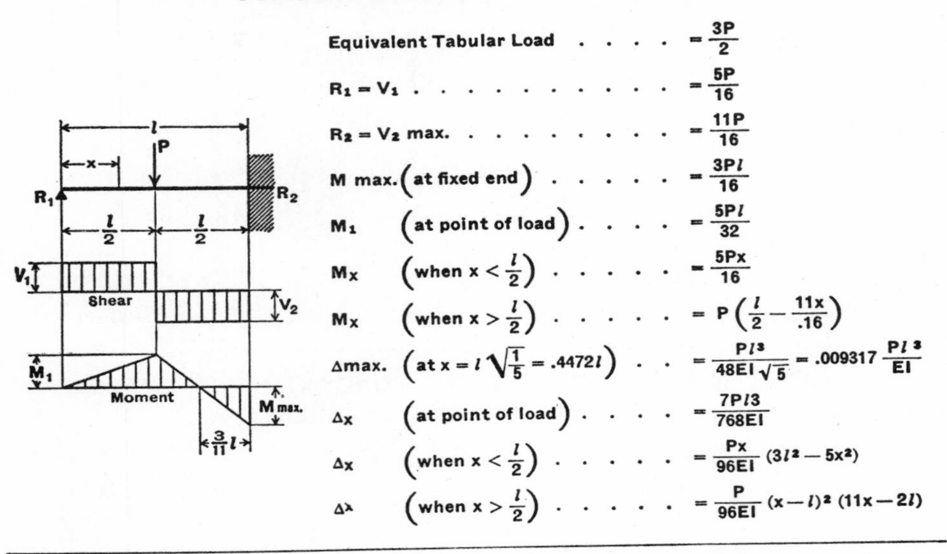

Equivalent Tabular Load . . . . $= \frac{3P}{2}$

$R_1 = V_1$ . . . . $= \frac{5P}{16}$

$R_2 = V_2$ max. . . . . $= \frac{11P}{16}$

M max. (at fixed end) . . . . . . $= \frac{3Pl}{16}$

M_1 (at point of load) . . . . $= \frac{5Pl}{32}$

M_x $\left(\text{when } x < \frac{l}{2}\right)$ . . . . . $= \frac{5Px}{16}$

M_x $\left(\text{when } x > \frac{l}{2}\right)$ . . . . . $= P\left(\frac{l}{2} - \frac{11x}{.16}\right)$

Δmax. $\left(\text{at } x = l\sqrt{\frac{1}{5}} = .4472l\right)$ . . $= \frac{Pl^3}{48EI\sqrt{5}} = .009317\frac{Pl^3}{EI}$

Δx (at point of load) . . . . $= \frac{7Pl^3}{768EI}$

Δx $\left(\text{when } x < \frac{l}{2}\right)$ . . . . . $= \frac{Px}{96EI}(3l^2 - 5x^2)$

Δx $\left(\text{when } x > \frac{l}{2}\right)$ . . . . . $= \frac{P}{96EI}(x - l)^2(11x - 2l)$

14. BEAM FIXED AT ONE END, SUPPORTED AT OTHER—CONCENTRATED LOAD AT ANY POINT

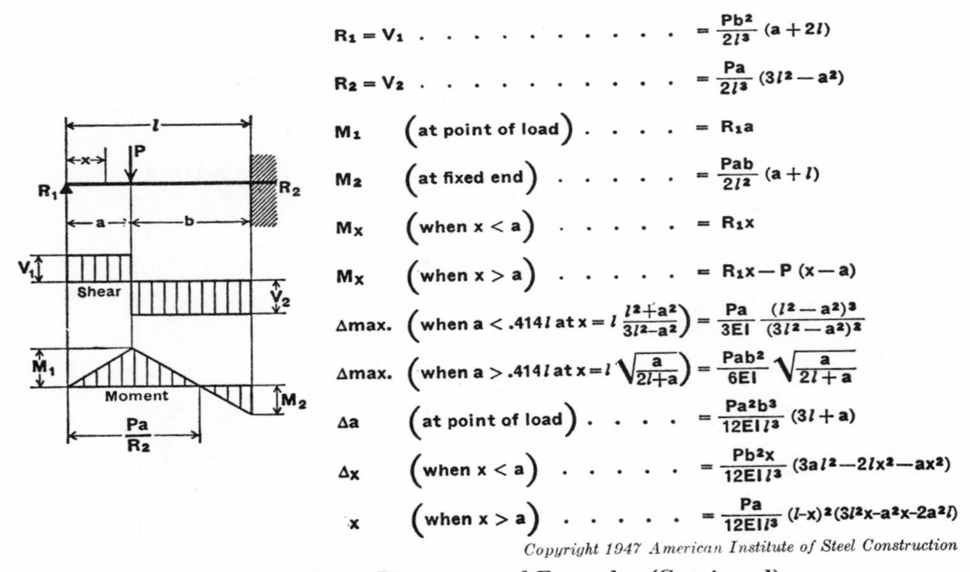

$R_1 = V_1$ . . . . $= \frac{Pb^2}{2l^3}(a + 2l)$

$R_2 = V_2$ . . . . $= \frac{Pa}{2l^3}(3l^2 - a^2)$

M_1 (at point of load) . . . . $= R_1 a$

M_2 (at fixed end) . . . . . $= \frac{Pab}{2l^2}(a + l)$

M_x (when $x < a$) . . . . . $= R_1 x$

M_x (when $x > a$) . . . . . $= R_1 x - P(x - a)$

Δmax. $\left(\text{when } a < .414l \text{ at } x = l\frac{l^2 + a^2}{3l^2 - a^2}\right) = \frac{Pa}{3EI}\frac{(l^2 - a^2)^3}{(3l^2 - a^2)^2}$

Δmax. $\left(\text{when } a > .414l \text{ at } x = l\sqrt{\frac{a}{2l + a}}\right) = \frac{Pab^2}{6EI}\sqrt{\frac{a}{2l + a}}$

Δa (at point of load) . . . . $= \frac{Pa^2b^3}{12EIl^3}(3l + a)$

Δx (when $x < a$) . . . . . $= \frac{Pb^2x}{12EIl^3}(3al^2 - 2lx^2 - ax^2)$

x (when $x > a$) . . . . . . $= \frac{Pa}{12EIl^3}(l - x)^2(3l^2x - a^2x - 2a^2l)$

Fig. 3.13 Beam Diagrams and Formulae (Continued)

of constant moment of inertia. For rolled steel beams which are naturally of a constant moment of inertia, these tables are quite satisfactory. Since welding and the advantages of continuity are becoming more generally accepted, they may save time for the designer. For concrete beams or for built-up steel beams in which the moment of inertia may be varied by adding or omitting flange area, these charts do not give the maximum results to be designed for. The maximum negative moment at any support results when the two adjacent members are fully loaded and all other panels have dead load

15. BEAM FIXED AT BOTH ENDS—UNIFORMLY DISTRIBUTED LOADS

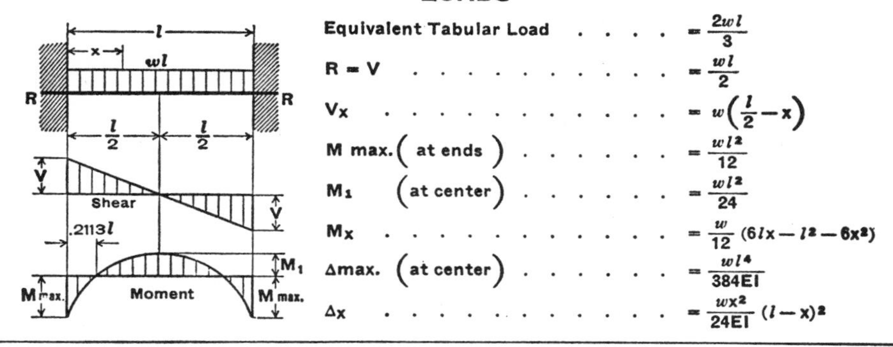

16. BEAM FIXED AT BOTH ENDS—CONCENTRATED LOAD AT CENTER

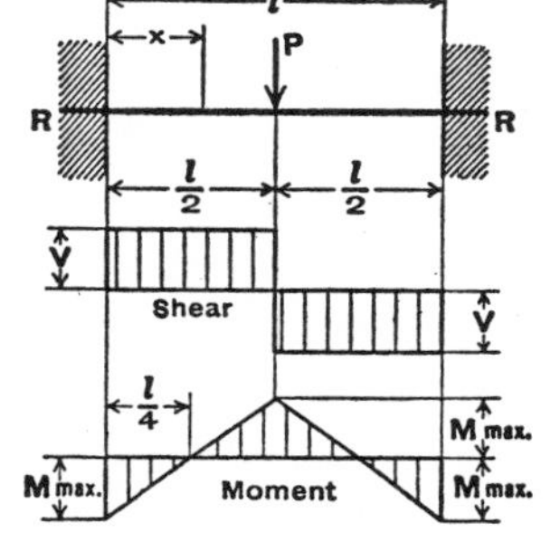

Equivalent Tabular Load . . . . $= P$

$R = V$ . . . . $= \frac{P}{2}$

M max. (at center and ends) . . . $= \frac{Pl}{8}$

M_x $\left(\text{when } x < \frac{l}{2}\right)$ . . . . . $= \frac{P}{8}(4x - l)$

Δmax. (at center) . . . . . . $= \frac{Pl^3}{192EI}$

Δx . . . . $= \frac{Px^2}{48EI}(3l - 4x)$

17. BEAM FIXED AT BOTH ENDS—CONCENTRATED LOAD AT ANY POINT

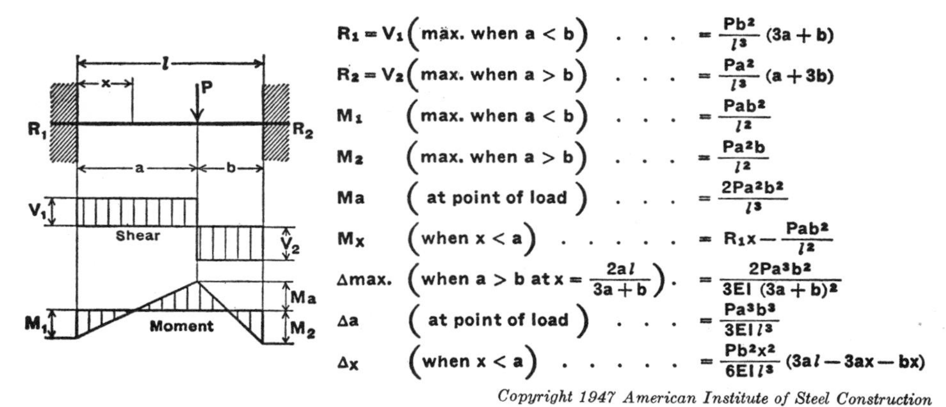

Fig. 3.13 Beam Diagrams and Formulae (Continued)

only. The maximum positive moment results when the beam or slab in question is loaded, the spans on either side of the span in question carry dead load only, and the balance of the spans are alternately loaded and unloaded. Obviously the dead load is always applied to all spans.

The charts (Fig. 3.13) are useful in indicating the general type of moment resulting; or, since under normal conditions the maximum moment is the negative moment, the charts give an approximation of the maximum moment, from which the size of beam may be determined.

18. CANTILEVER BEAM—LOAD INCREASING UNIFORMLY TO FIXED END

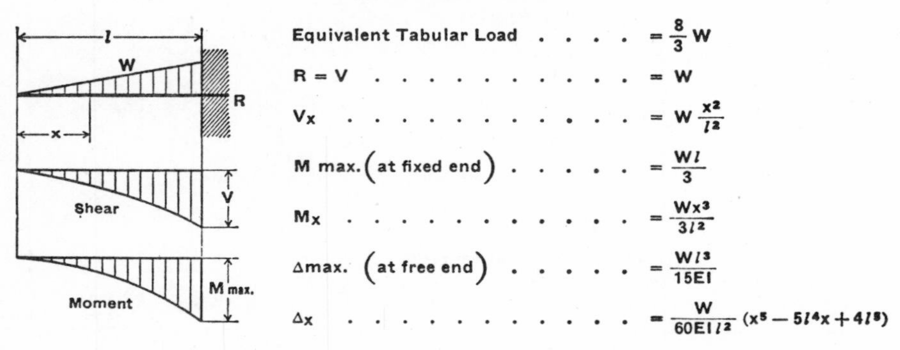

19. CANTILEVER BEAM—UNIFORMLY DISTRIBUTED LOAD

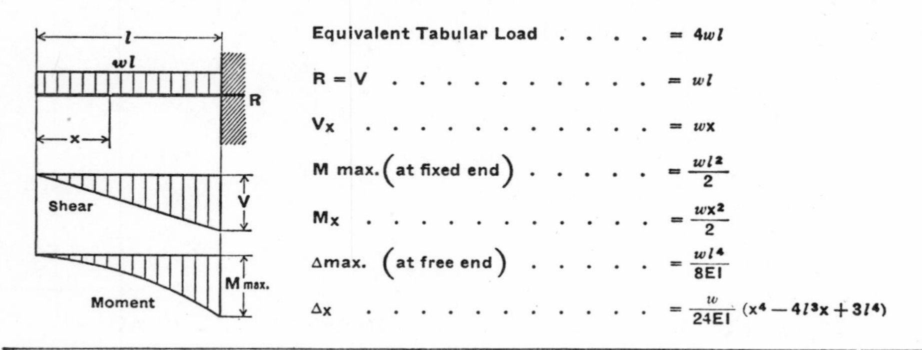

20. BEAM FIXED AT ONE END, FREE BUT GUIDED AT OTHER—UNIFORMLY DISTRIBUTED LOAD

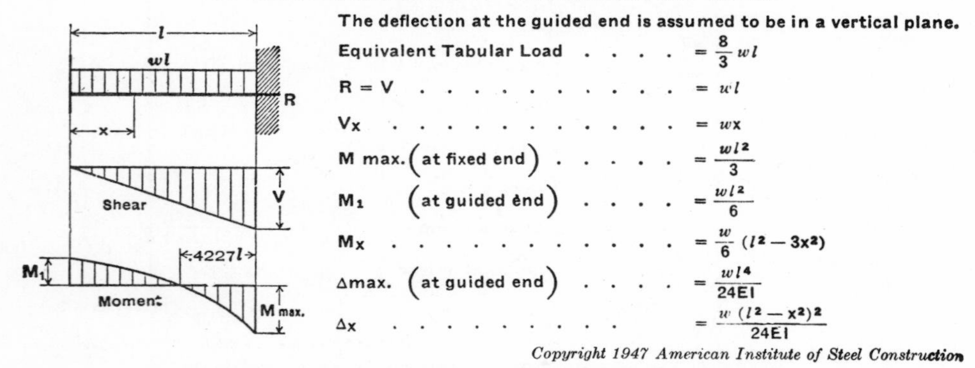

Fig. 3.13 Beam Diagrams and Formulae (Continued)

Occasionally a problem is encountered which may be solved by the use of the coefficients found in Fig. 3.13. Below is a typical problem of this type.

Problem A floor is made up of 2-in. boards (actually 1⅝ in. thick) 20 ft long and is supported on joists 4 ft apart. With a fiber stress of 1,600 psi, what is the maximum total load per square foot which may be put on the floor?

This indicates five equal spans. Reference to the continuous beam diagrams in Fig. 3.13 for five spans indicates that the maximum moment is at the first support from the end, and amounts to $(304/2{,}888)wL^2$, or $0.105WL$. The section modulus per foot

21. CANTILEVER BEAM—CONCENTRATED LOAD AT ANY POINT

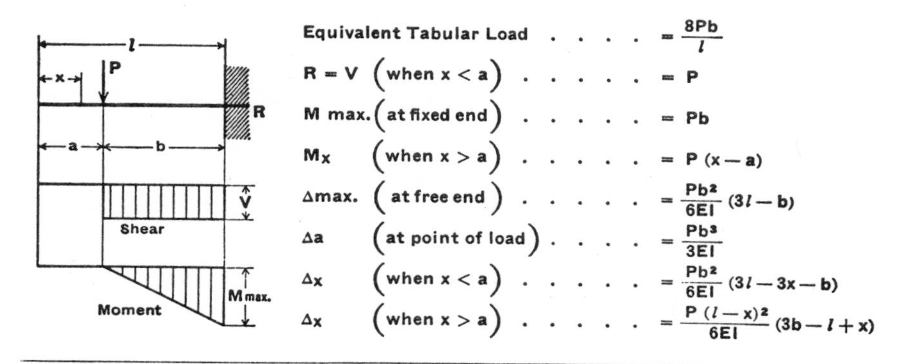

22. CANTILEVER BEAM—CONCENTRATED LOAD AT FREE END

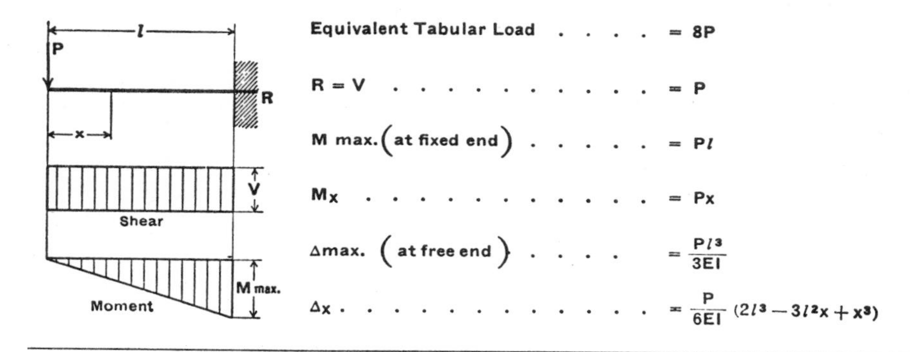

23. BEAM FIXED AT ONE END, FREE BUT GUIDED AT OTHER—CONCENTRATED LOAD AT GUIDED END

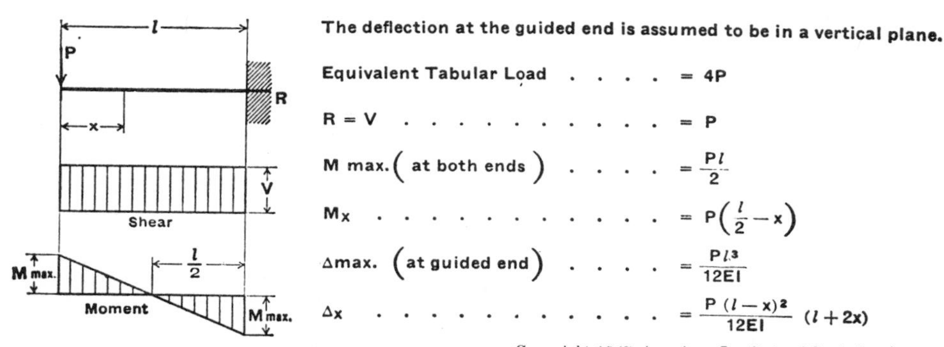

Fig. 3.13 Beam Diagrams and Formulae (Continued)

width is $\frac{1}{6}bd^2$, or

$$S = \frac{12 \times 1.625^2}{6} = 5.28 \text{ cu in.}$$

We will assume a trial load of 100 psf on the floor.

$$W = 100 \times 4 = 400 \text{ lb}$$
$$M = 0.105 \times 400 \times 4 \times 12 = 2016 \text{ in.-lb}$$

$$f = \frac{2,016}{5.28} = 382 \text{ psi}$$

Therefore if 100 psf causes a stress of 382 psi, the allowable load per square foot for 1,600 psi stress will be

$$\frac{1,600}{382} \times 100 = 419 \text{ psf}$$

24. BEAM OVERHANGING ONE SUPPORT—UNIFORMLY DISTRIBUTED LOAD

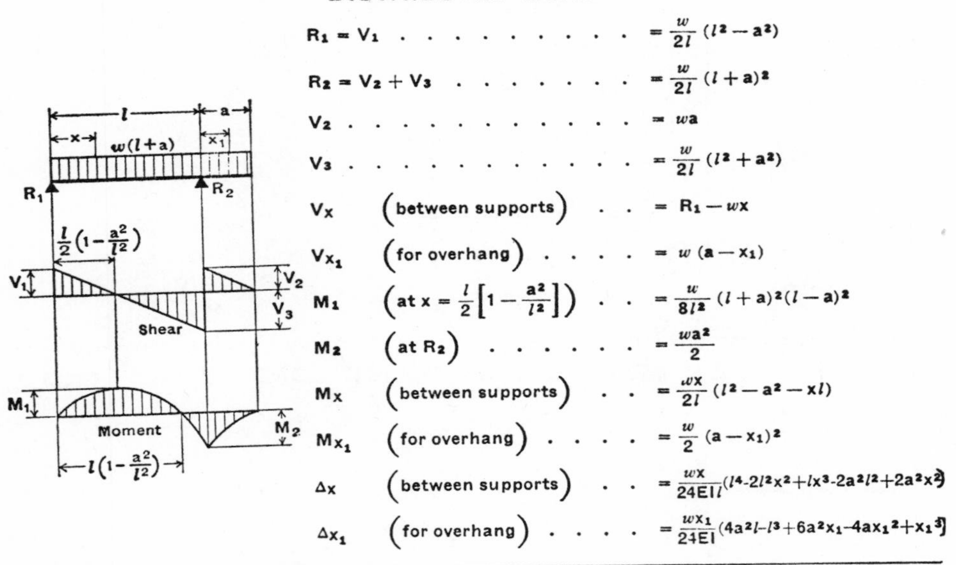

25. BEAM OVERHANGING ONE SUPPORT—UNIFORMLY DISTRIBUTED LOAD ON OVERHANG

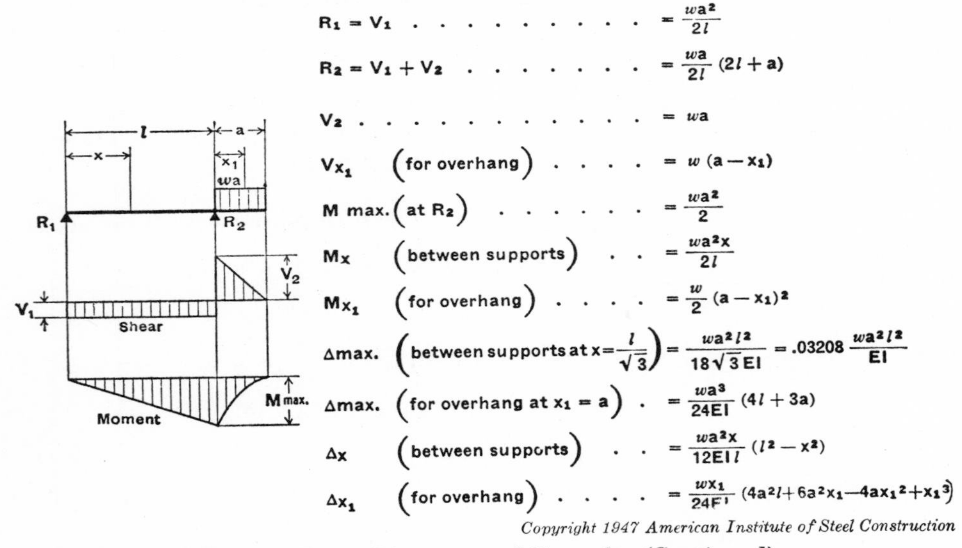

Fig. 3.13 Beam Diagrams and Formulae (Continued)

b Most continuous-beam problems do not give any assurance of constant load, so that instead of using the charts for an exact solution, some other method must be used. For spans which vary by not over 20 percent, with approximately equal uniform loads, it is accurate enough to use the ordinary moment coefficients as follows:

1. Beams and slabs of one span, positive moment near center (simple beam):

$$M = \frac{wL^2}{8}$$

2. Beams and slabs continuous for two spans only, positive moment near center (semicontinuous):

26. BEAM OVERHANGING ONE SUPPORT—CONCENTRATED LOAD AT END OF OVERHANG

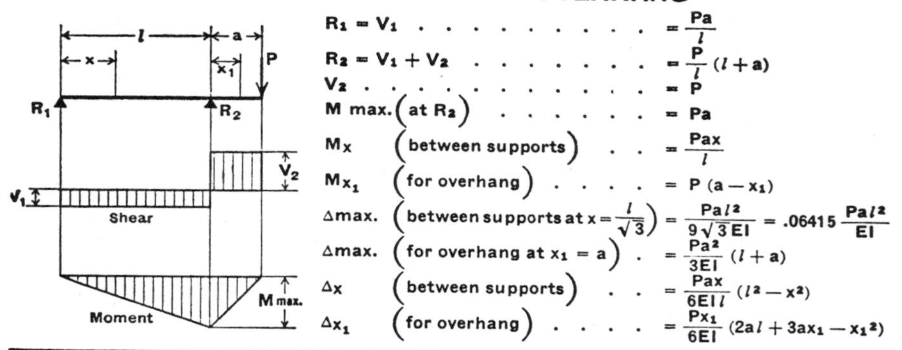

$R_1 = V_1 \ldots = \frac{Pa}{l}$

$R_2 = V_1 + V_2 \ldots = \frac{P}{l}(l + a)$

$V_2 \ldots = P$

M max. (at R_2) $\ldots = Pa$

M_x (between supports) $\ldots = \frac{Pax}{l}$

M_{x_1} (for overhang) $\ldots = P(a - x_1)$

Δmax. $\left(\text{between supports at } x = \frac{l}{\sqrt{3}}\right) = \frac{Pal^2}{9\sqrt{3}EI} = .06415\frac{Pal^2}{EI}$

Δmax. (for overhang at $x_1 = a$) $\ldots = \frac{Pa^2}{3EI}(l + a)$

Δx (between supports) $\ldots = \frac{Pax}{6EIl}(l^2 - x^2)$

Δx_1 (for overhang) $\ldots = \frac{Px_1}{6EI}(2al + 3ax_1 - x_1^2)$

27. BEAM OVERHANGING ONE SUPPORT—UNIFORMLY DISTRIBUTED LOAD BETWEEN SUPPORTS

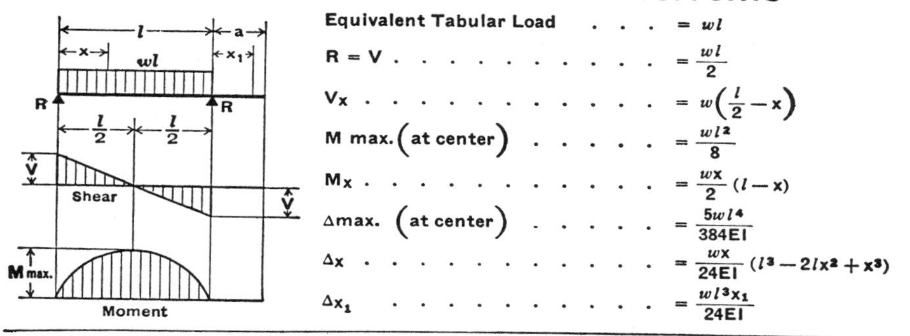

Equivalent Tabular Load $\ldots = wl$

$R = V \ldots = \frac{wl}{2}$

$V_x \ldots = w\left(\frac{l}{2} - x\right)$

M max. (at center) $\ldots = \frac{wl^2}{8}$

$M_x \ldots = \frac{wx}{2}(l - x)$

Δmax. (at center) $\ldots = \frac{5wl^4}{384EI}$

$\Delta x \ldots = \frac{wx}{24EI}(l^3 - 2lx^2 + x^3)$

$\Delta x_1 \ldots = \frac{wl^3x_1}{24EI}$

28. BEAM OVERHANGING ONE SUPPORT—CONCENTRATED LOAD AT ANY POINT BETWEEN SUPPORTS

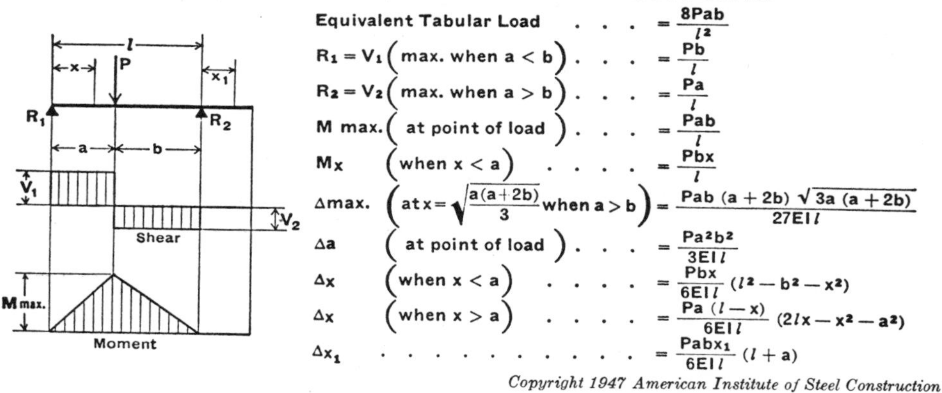

Equivalent Tabular Load $\ldots = \frac{8Pab}{l^2}$

$R_1 = V_1$ (max. when $a < b$) $\ldots = \frac{Pb}{l}$

$R_2 = V_2$ (max. when $a > b$) $\ldots = \frac{Pa}{l}$

M max. (at point of load) $\ldots = \frac{Pab}{l}$

M_x (when $x < a$) $\ldots = \frac{Pbx}{l}$

Δmax. $\left(\text{at } x = \sqrt{\frac{a(a+2b)}{3}} \text{ when } a > b\right) = \frac{Pab(a + 2b)\sqrt{3a(a + 2b)}}{27EIl}$

Δa (at point of load) $\ldots = \frac{Pa^2b^2}{3EIl}$

Δx (when $x < a$) $\ldots = \frac{Pbx}{6EIl}(l^2 - b^2 - x^2)$

Δx (when $x > a$) $\ldots = \frac{Pa(l - x)}{6EIl}(2lx - x^2 - a^2)$

$\Delta x_1 \ldots = \frac{Pabx_1}{6EIl}(l + a)$

Fig. 3.13 Beam Diagrams and Formulae (Continued)

$$M = \frac{wL^2}{14}$$

Negative moment over interior support:

$$M = \frac{wL^2}{9}$$

3. Beams and slabs continuous for more than two spans (fully continuous), positive moment near center of interior spans:

$$M = \frac{wL^2}{16}$$

29 CONTINUOUS BEAM—TWO EQUAL SPANS—UNIFORM LOAD ON ONE SPAN

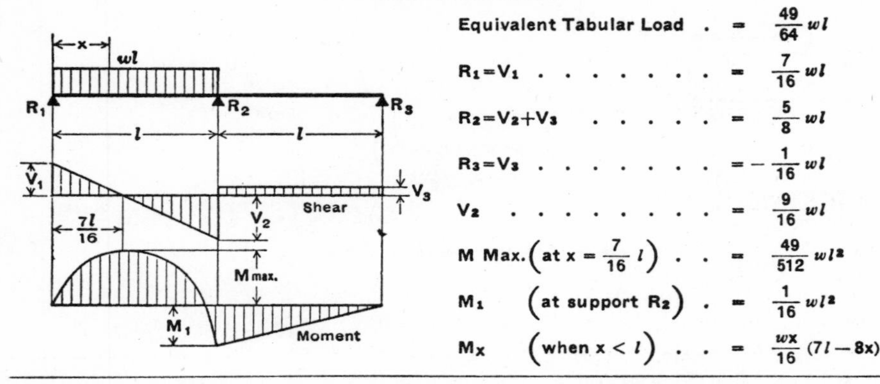

Equivalent Tabular Load . = $\frac{49}{64} wl$

$R_1 = V_1$ . . . . = $\frac{7}{16} wl$

$R_2 = V_2 + V_3$ . . . . . = $\frac{5}{8} wl$

$R_3 = V_3$ . . . . = $-\frac{1}{16} wl$

V_2 . . . . = $\frac{9}{16} wl$

M Max. $\left(\text{at } x = \frac{7}{16} l\right)$ . . = $\frac{49}{512} wl^2$

M_1 (at support R_2) . = $\frac{1}{16} wl^2$

M_x (when $x < l$) . . = $\frac{wx}{16}(7l - 8x)$

30. CONTINUOUS BEAM—TWO EQUAL SPANS—CONCENTRATED LOAD AT CENTER OF ONE SPAN

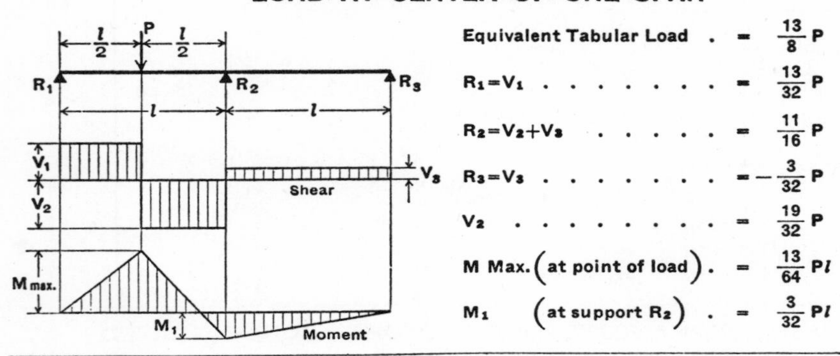

Equivalent Tabular Load . = $\frac{13}{8} P$

$R_1 = V_1$ . . . . = $\frac{13}{32} P$

$R_2 = V_2 + V_3$ . . . . . = $\frac{11}{16} P$

$R_3 = V_3$ . . . . = $-\frac{3}{32} P$

V_2 . . . . = $\frac{19}{32} P$

M Max. (at point of load) . = $\frac{13}{64} Pl$

M_1 (at support R_2) . = $\frac{3}{32} Pl$

31. CONTINUOUS BEAM—TWO EQUAL SPANS—CONCENTRATED LOAD AT ANY POINT

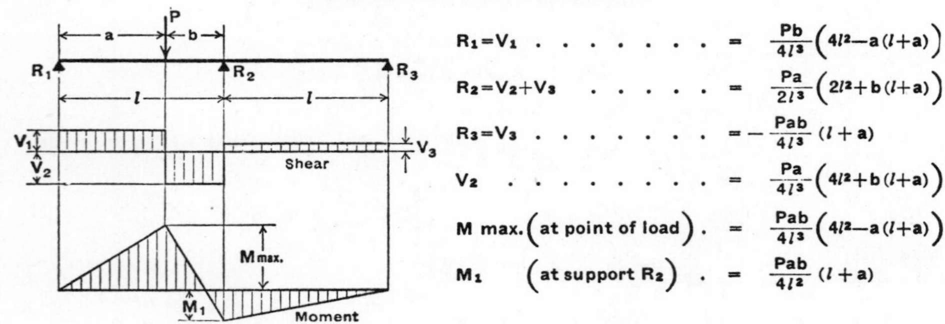

$R_1 = V_1$ . . . . = $\frac{Pb}{4l^3}\left(4l^2 - a(l+a)\right)$

$R_2 = V_2 + V_3$ . . . . . = $\frac{Pa}{2l^3}\left(2l^2 + b(l+a)\right)$

$R_3 = V_3$ . . . . = $-\frac{Pab}{4l^3}(l+a)$

V_2 . . . . = $\frac{Pa}{4l^3}\left(4l^2 + b(l+a)\right)$

M max. (at point of load) . = $\frac{Pab}{4l^3}\left(4l^2 - a(l+a)\right)$

M_1 (at support R_2) . = $\frac{Pab}{4l^2}(l+a)$

Fig. 3.13 Beam Diagrams and Formulae (Continued)

Negative moment at first interior support:

$$M = \frac{wL^2}{10}$$

Negative moment at other interior supports:

$$M = \frac{wL^2}{12}$$

Negative moment at end supports for cases 1, 2, and 3:

$$M = \text{not less than } \frac{wL^2}{24}$$

32. SIMPLE BEAM—ONE CONCENTRATED MOVING LOAD

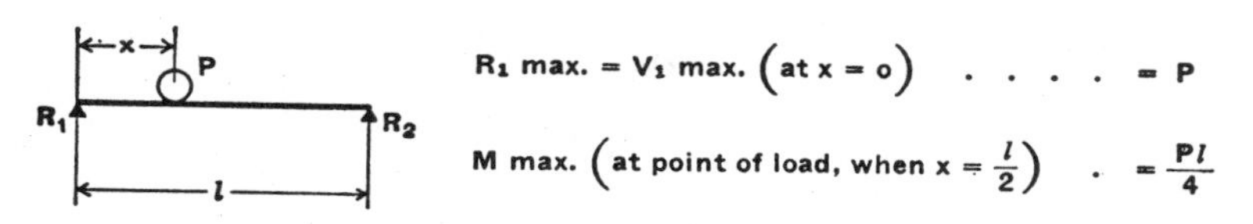

33. SIMPLE BEAM—TWO EQUAL CONCENTRATED MOVING LOADS

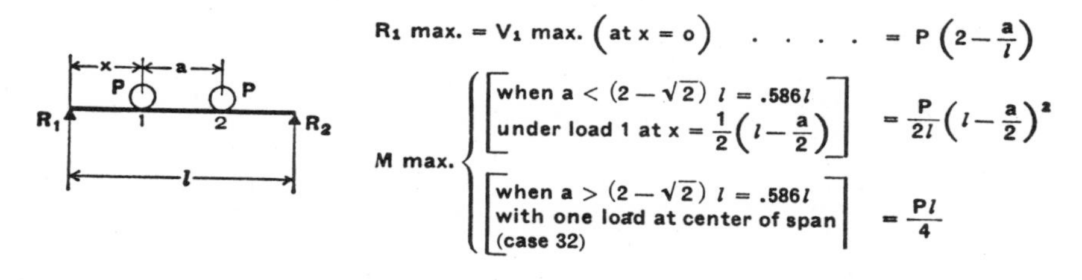

34. SIMPLE BEAM—TWO UNEQUAL CONCENTRATED MOVING LOADS

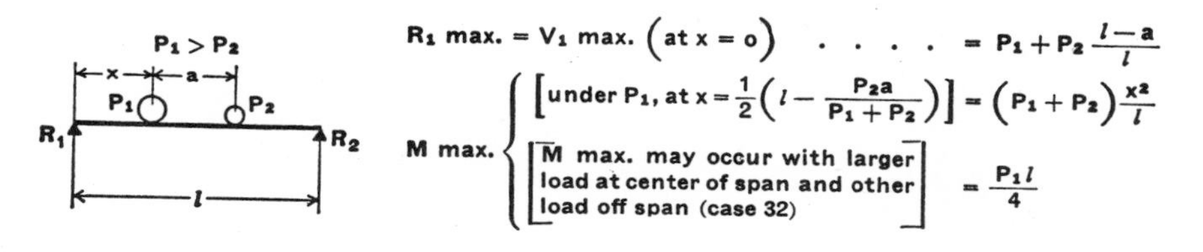

GENERAL RULES FOR SIMPLE BEAMS CARRYING MOVING CONCENTRATED LOADS

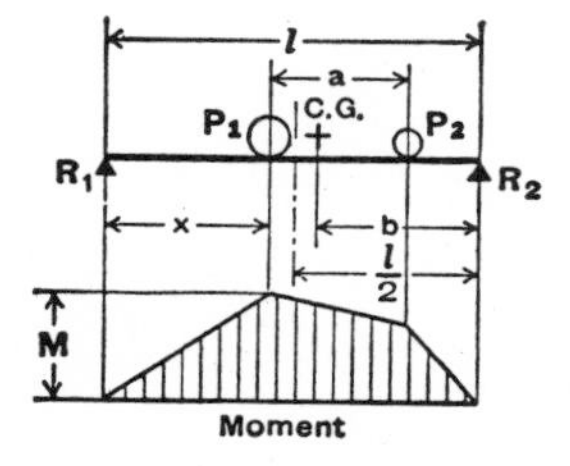

The maximum shear due to moving concentrated loads occurs at one support when one of the loads is at that support. With several moving loads, the location that will produce maximum shear must be determined by trial.

The maximum bending moment produced by moving concentrated loads occurs under one of the loads when that load is as far from one support as the center of gravity of all the moving loads on the beam is from the other support.

In the accompanying diagram, the maximum bending moment occurs under load P_1 when x = b. It should also be noted that this condition occurs when the center line of the span is midway between the center of gravity of loads and the nearest concentrated load.

Fig. 3.13 Beam Diagrams and Formulae (Continued)

c The most modern practice in concrete recommends a complete analysis of the beams, or even an analysis combining beams and columns. The method of approach to such problems is to determine by one of several methods the statically indeterminate moments at the supports or at the joints, and apply these indeterminate moments to the ordinary static-moment diagrams. When the moments have been determined, the reactions, points of contraflexure, and other necessary information may be obtained. In simple beams the necessary information is dependent only on load, span, and conditions

CONTINUOUS BEAM DIAGRAMS

EQUAL SPANS, SIMILARLY LOADED, CONSTANT MOMENT OF INERTIA
SUPPORTS AT SAME LEVEL

UNIFORMLY DISTRIBUTED LOAD

Reaction and shear coefficients of wl Moment coefficients of wl^2

w = Load per unit length

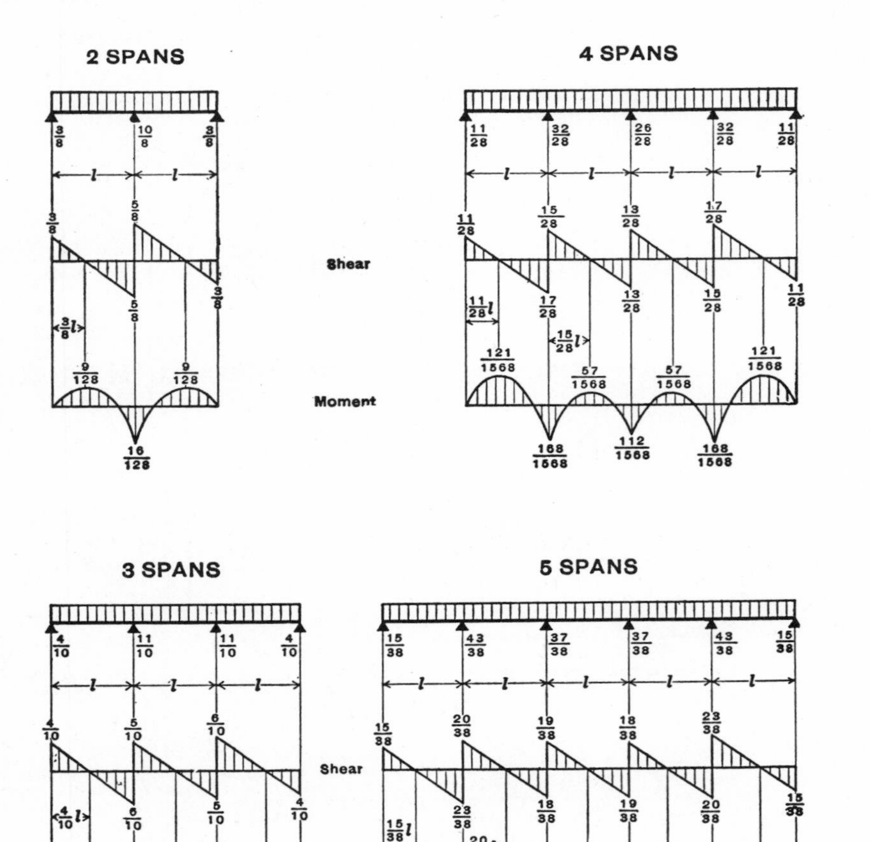

Fig. 3.13 Beam Diagrams and Formulae (Continued)

of loading. From this information a moment is obtainable from which, for any given material and fiber stress, the design may be made. In continuous beams or rigid frames, however, the stiffness factor K, which is the ratio I/L of each member respectively, controls the distribution of the indeterminate moments and enters into each of the various methods of solution. Since I/L is a factor for each individual member, it becomes a ratio between the various members; therefore it makes no difference if L is taken in feet and I in inches, or vice versa. The main idea is to keep the mathematics in the simplest possible terms. In any case, the units must be kept the same for the stiffness factors of all members in a series.

Until a few years ago the common method

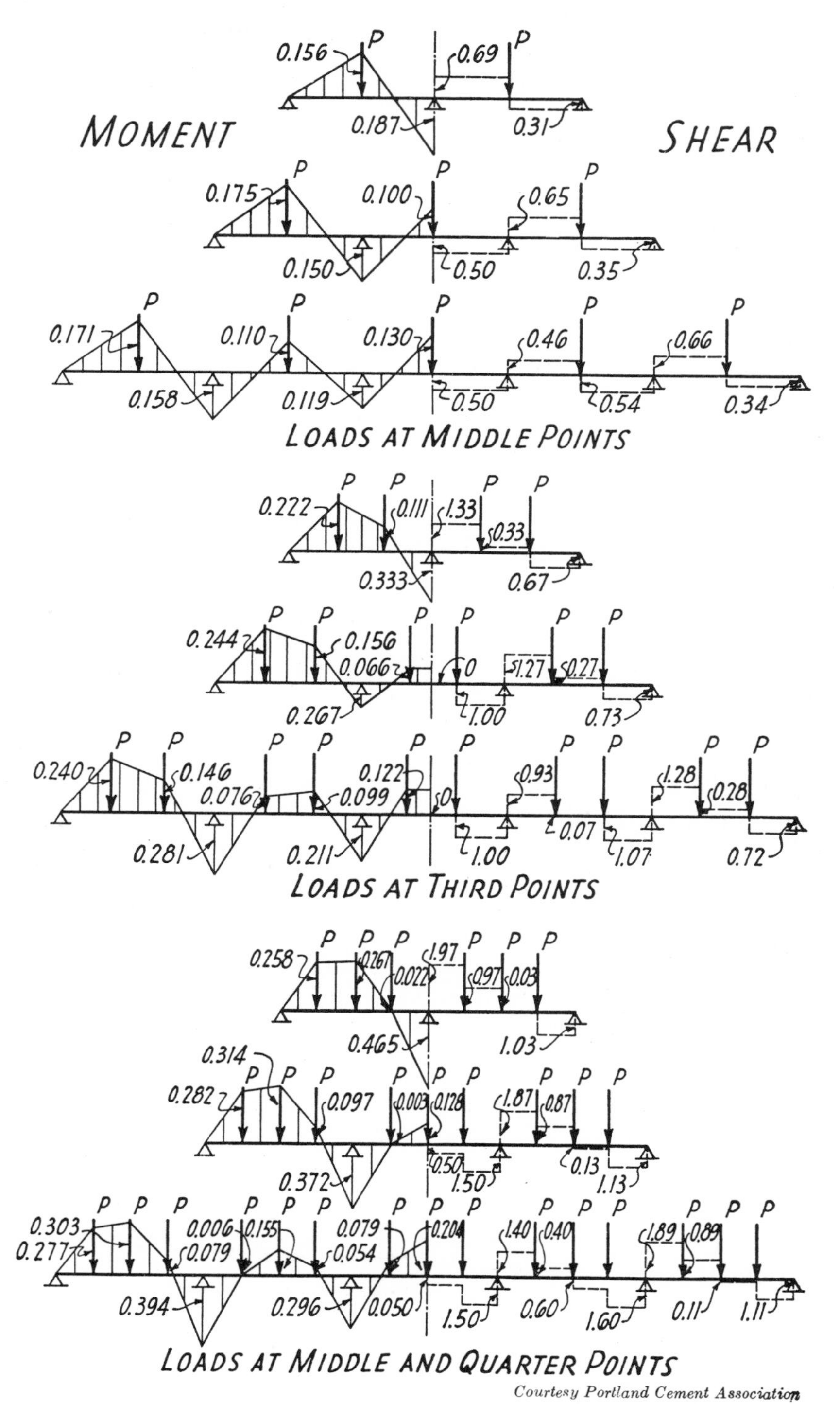

Courtesy Portland Cement Association

Fig. 3.13 Beam Diagrams and Formulae (Continued)

of approach to the problem of continuous-beam moments was the rather cumbersome three-moment theorem. This has been replaced in most offices by the Hardy Cross method of moment distribution—a close approximation devised by Prof. Hardy Cross, given in Art. 3.21. No other method of solution for continuous beams is given because no other method is so usable in actual design.

d The shear at the end of a continuous beam is determined by the sum of the shear in the beam simply supported, with a correction plus or minus due to the difference between the end moments. This correction is found by $\frac{(-M_1) - (-M_2)}{L}$, the correction being added to the end with the greater negative moment and subtracted from the end with the lesser negative moment. Since the negative moments may vary depending on adjacent spans, it is advisable in computing shears to add the full correction but to subtract only ⅔ of the full correction. This correction is not usually large for interior bays but may be an appreciable amount in an end bay (see Fig. 3.13).

3.21 Method of moment distribution

a The Hardy Cross method of moment distribution starts by assuming each span as a fixed beam, similar to Case 15, 16, or 17 of the beam diagrams in Fig. 3.13. However, since the moment must be the same on both sides of a support—that is, since the slope of both beams must be tangent to the same line over a support—the moment on both sides of the support must be the same. The Hardy Cross method is a means of adjusting the assumed moments to an approximately correct result—correct within all necessary limits.

Fixed end moments, frequently designated as "fem," for all normal conditions are given in the series of charts in Fig. 3.24.

For any combination of load conditions, add the fixed end moments. Thus for a trapezoidal load, add Case 5 and Case 12 of Fig. 3.24.

To eliminate some cumbersome calculation, coefficients of influence are given in Table 3.1 for Case 15, and in Tables 3.2 and 3.3 for Cases 14 and 11.

b The various steps in the design are as follows, referring to Fig. 3.14 as typical:

1. Compute for each span a stiffness factor $K = I/L$. Since this is a ratio only, I may be taken as 1 for each if the beam is of the same section over all supports. In Fig. 3.14 all spans are alike and the beam is continuous, so $K = 1$ for all spans. Put this figure in a circle over each span.

2. Compute the fixed end moment in each span from Fig. 3.24. To distinguish between positive and negative moment, ordinarily bending upward is considered negative, downward positive. It will simplify this problem, however, if the convention is forgotten until the corrected moment has been obtained, and bending in a clockwise direction about a joint is considered positive, counterclockwise negative. The computed moments are set down under the ends of the beam as has been done in Fig. 3.14.

3. If the slope on each side of the support is to be tangent to the same line, the algebraic sum of the moments about a joint must equal zero. At point B there is an unbalanced moment of -20. At A there is an unbalanced moment of $+20$. Since end A is freely supported, the moment on both sides of the joint must equal zero.

The moments may be balanced in turn by adding and subtracting in proportion to the stiffness ratio K till each joint comes to an

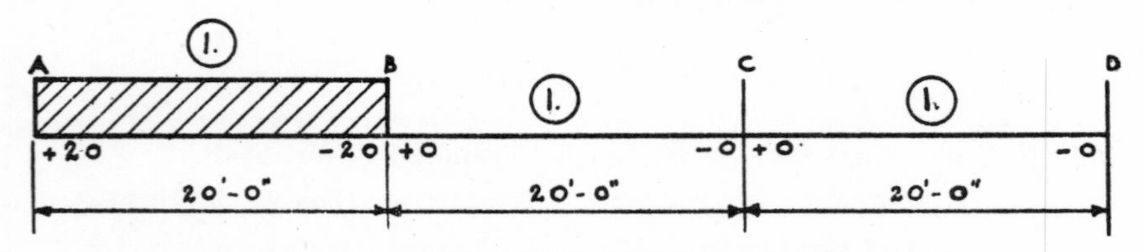

Fig. 3.14 First Step in Moment-Distribution Problem

algebraic total of zero. At the free end A, since the adjacent span has no stiffness, the full moment must be distributed on one side. Set down the balancing moment and draw a line under it as in Fig. 3.15.

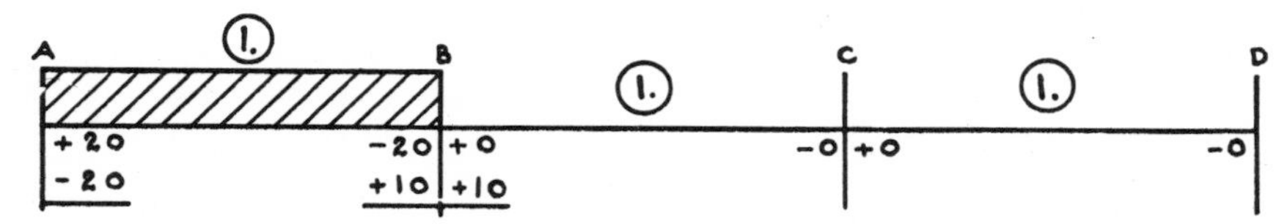

Fig. 3.15 First Distribution in Moment-Distribution Problem

4. This process of balancing, however, sets up a "carry-over" moment in the opposite end of the same beam span, equal to one-half the balancing moment and of the same sign. (This carry-over factor is one-half only when the beam span is of uniform section throughout. For instance, concrete beams bracketed at the columns have a different carry-over factor. See Art. 3.21*h*.) Set down the carry-over moment as in Fig. 3.16.

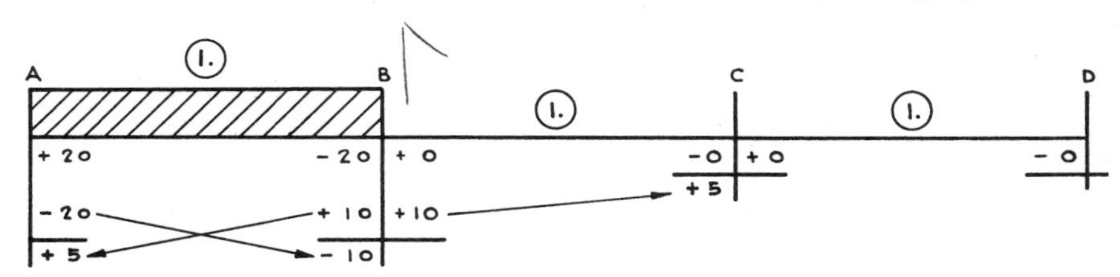

Fig. 3.16 First Carry-over in Moment-Distribution Problem

This carry-over moment again unbalances the joints, and they must again be balanced, carried over, balanced again, carried over again, etc., until sufficient accuracy is reached, as in Fig. 3.17.

5. The final M is obtained by adding all moments algebraically (original, balancing, and carry-over moments), and these will balance both sides of any joint.

6. The procedure for application of positive moments and obtaining shears is shown in Arts. 3.21*e* and 3.21*f*.

c With a free end like A or D in the preceding problem, the computation can be greatly simplified by multiplying K for the end span by ¾, balancing the end moment once, and avoiding the end span in any further calculations. If K is not the same in all spans, the percentage of stiffness is put down each side of the joint as in Fig. 3.18 where this short-cut method has been applied to the problem of the last paragraph. The figure 0.43 at B represents 0.75/1.75, and 0.57 is 1/1.75. This calculation has been carried to a finer conclusion than in the earlier paragraph, and is probably more accurate.

d A typical problem from actual design calculations is shown in Figs. 3.19 and 3.20. In determination of K, I has been assumed at 10 throughout, so that in span BC, $K = 10/15 = 0.667$. In the two end spans, the multiplier 0.75 has been used as described in the preceding paragraph.

e In accordance with the statement made in Art. 3.20*c*, the actual moments throughout, including maximum positive moments, are found by computing the simple moment and plotting it, then applying the indeterminate moments to the static-moment diagrams.

To compute the positive moments of each span in the problem of Figs. 3.19 and 3.20:

$$M_1 = \frac{10{,}000 \times 20}{8} = 25{,}000$$

$$M_2 = \frac{13{,}500 \times 15}{8} = 25{,}300$$

$$M_3 = \frac{10{,}800 \times 18}{8} = 24{,}300$$

Laying the moments out graphically permits obtaining the maximum positive and negative moments.

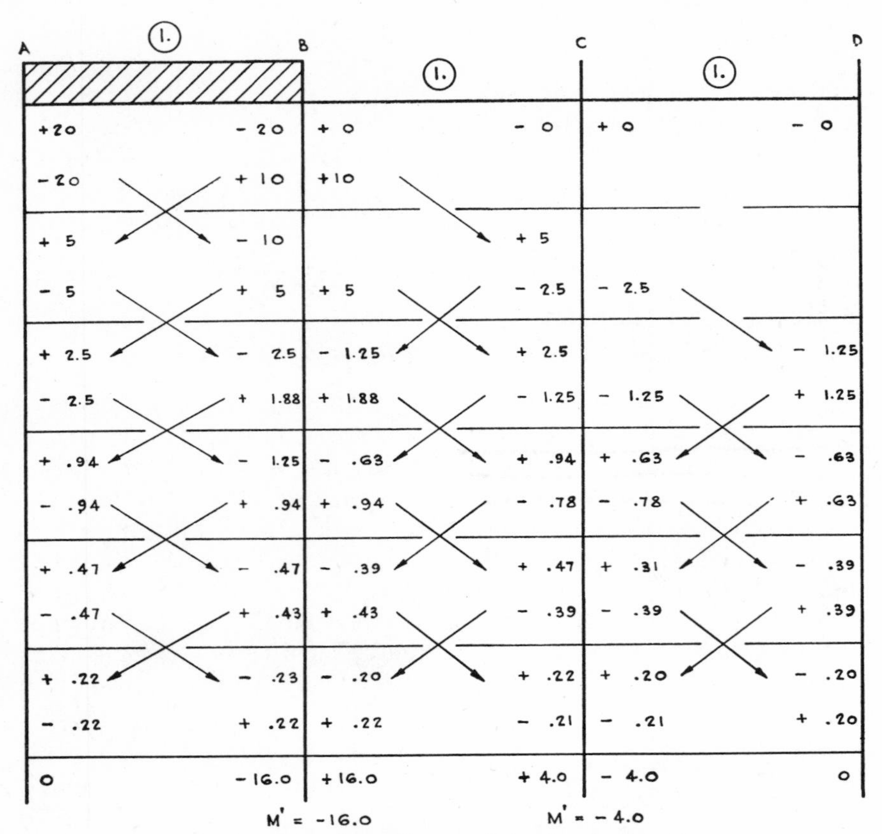

Fig. 3.17 Complete Moment-Distribution Problem

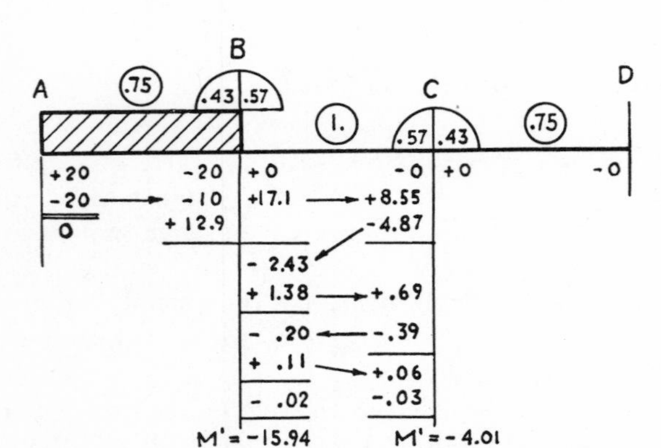

Fig. 3.18 Short Cut in Moment-Distribution Problem

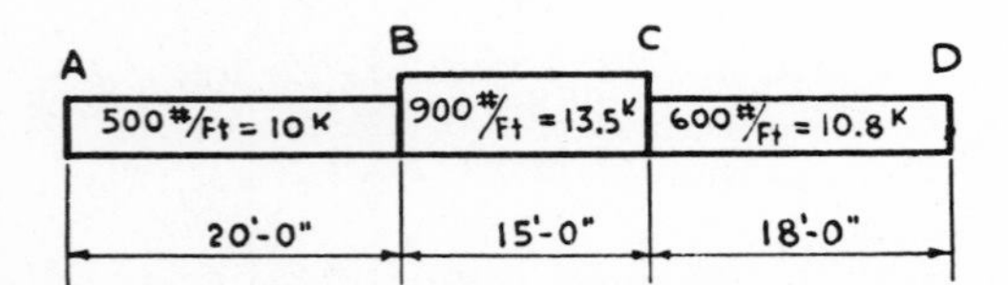

Fig. 3.19 Complex Moment-Distribution Problem

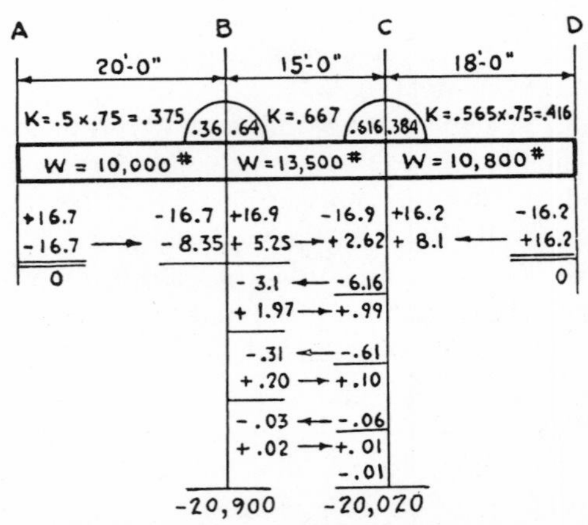

Fig. 3.20 Solution of Moment-Distribution Problem

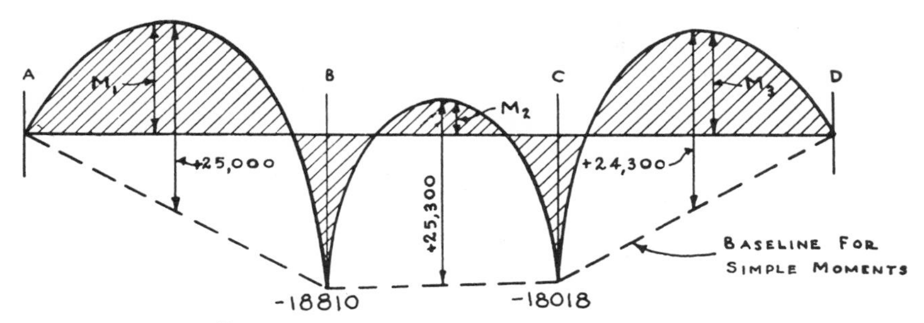

Fig. 3.21 Determination of Final Moments

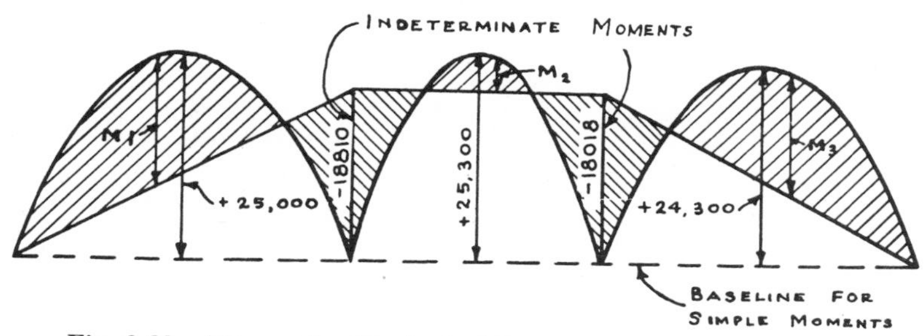

Fig. 3.22 Alternative Method of Determining Final Moments

Instead of constructing the moment diagrams as shown in Fig. 3.21, they may be constructed on a straight base line as shown in Fig. 3.22, and the indeterminate moments laid off above the line to obtain actual moments.

The 1961 AISC Code makes a concession to the theory of limit design in making an arbitrary adjustment in positive and negative moment values. Under most conditions, the maximum moment obtained by computation is the negative moment over the support, but test results show that before failure takes place, the positive moment increases until some figures approaching a balance of positive and negative moments result. The AISC Code therefore allows a reduction of 10 percent in the negative moments, with a resultant increase in the adjacent positive moments. In the above problem, as indicated in Figs. 3.20, 3.21, and 3.22, the indeterminate negative moments may be corrected so that at B, it becomes 20,900 − 2,090 = 18,810, and at C, it becomes 20,020 − 2,002 = 18,018. The adjusted positive moments therefore become

$$M_1 = 25{,}000 - \frac{18{,}810}{2} = +15{,}595$$

$$M_2 = 25{,}300 - \frac{18{,}810 + 18{,}018}{2} = +6{,}886$$

$$M_3 = 24{,}300 - \frac{18{,}018}{2} = +15{,}291$$

f Having obtained the maximum moments, the next step is to compute reactions. The simplest method is to compute the reactions as a simple beam, which for a uniformly loaded beam may be done by observation. Then apply a correction to each reaction, equal to the difference in end moments on a span, divided by the span length. This correction is added to the reaction where the actual negative end moment is the greater, and subtracted from the reaction where the actual negative end moment is the lesser. Applying this method to the problem in Art. 3.21*e*, the reactions for the first span are 5,000 lb. The correction is

$$\frac{18{,}810}{20} = 940 \text{ lb}$$

$$R_a = 5{,}000 - 940 = 4{,}060 \text{ lb}$$

$$R_b = 5{,}000 + 940 = 5{,}940 \text{ lb}$$

For the second span, the normal reactions are 6,750 at each end, and the corrections

$$\frac{792}{15} = 53$$

$$R_b = 6{,}750 + 53 = 6{,}803 \text{ lb}$$
$$R_c = 6{,}750 - 53 = 6{,}697 \text{ lb}$$

For the third span, the normal reactions are 5,400. The correction is

$$\frac{18{,}018}{18} = 1{,}001$$

$$R_c = 5{,}400 + 1{,}001 = 6{,}401 \text{ lb}$$
$$R_d = 5{,}400 - 1{,}001 = 4{,}399 \text{ lb}$$

The above corrections are based on all spans being fully loaded, a condition which may or may not exist. A conservative method of application would be to apply the full additive correction, but at the opposite end to subtract only in proportion of live load to total load but not to exceed half of the computed correction for full live plus dead load.

g *The indirect Hardy Cross method.* There are certain problems which cannot be solved directly, but which require an indirect approach: certain moments are assumed, the structure is balanced using these moments, and correction from the resulting shears is obtained. We approach the method by putting forth two axioms: (1) for equal deflections the moment varies as I/L^2 or as K/L, and (2) the horizontal thrust varies as I/L^3 or as K/L^2.

Problem Assume a crane column subject to a horizontal thrust, of 60 kips, with unequal section above and below load point.

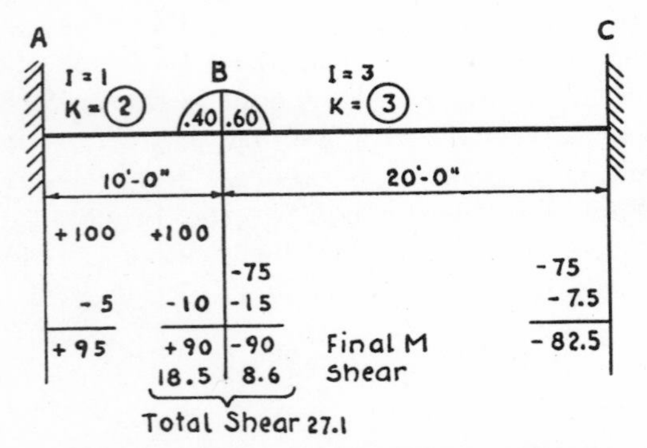

Fig. 3.23 Indirect Hardy-Cross Method

Treating the column as a beam, the method is shown in Fig. 3.23. Assume in span AB a moment of 100, the sign being the same at both ends of the span. In order to satisfy axiom 1, the moment in the span BC would be $\frac{3}{4} \times 100 = 75$; but observation indicates that the bending would be of the opposite sign. If we balance the center joint and carry over, we arrive at the final moments indicated. The shear in each span is obtained by dividing the sum of the two end moments by the span.

$$V_{AB} = \frac{95 + 90}{10} = 18.5$$

$$V_{BC} = \frac{90 + 82.5}{20} = 8.6$$

This gives a total shear of 27.1, corresponding to the assumed moment of 100. But the actual shear is 60, so the moments corresponding to this shear would be 60/27.1 of the computed M in Fig. 3.23; or using the ordinary moment conventions, $M_A = -213$, $M_B = 199$, $M_C = -183$.

h *Continuity in beams and columns with variable moment of inertia.* The preceding discussion of the Hardy Cross method has been based on the assumption that the members have a constant moment of inertia. If members have a variable moment of inertia, as is the case with concrete beams with haunched ends, or with gusset-braced steel beams, the stiffness coefficient K, the carry-over factor C, and the fixed-end moment will vary depending on a number of factors.

For prismatic members $K = 4EI/L$, but since $4E$ is constant in any structure, K varies as I/L, and this is the stiffness factor we have previously used in this section. In Figs. 3.25 and 3.26 the stiffness factor K is selected from the charts and divided by 4. Thus when the relation of minimum d to maximum d equals 1, as in the beams considered previously, $K = 4I/4L$. For any other combination of a with minimum d over maximum d, the value of K may be selected from charts No. 1 of Figs. 3.25 and 3.26 and the K we use is $K/4$. The carry-over factor C may be similarly selected, directly from charts No. 2 of Figs. 3.25 and 3.26.

Charts No. 3 of Figs. 3.25 and 3.26, giving uniform load fem coefficient, may be readily understood when it is noted that for the minimum to maximum depth ratio of 1, the factor is 0.083, or $\frac{1}{12}WL$. In other words,

Table 3.1 Value of Coefficient C_1 for Single Concentrated Load on Beam

Case 15 of Fig. 3.24; fem $= C_1WL$

C_1	+.00	.01	.02	.03	.04	.05	.06	.07	.08	.09
.0		.010	.019	.028	.037	.045	.053	.061	.068	.075
.1	.081	.087	.093	.098	.104	.108	.113	.117	.121	.125
.2	.128	.131	.134	.136	.139	.141	.142	.144	.145	.146
.3	.147	.148	.148	.148	.148	.148	.147	.147	.146	.145
.4	.144	.143	.141	.140	.138	.136	.134	.132	.130	.127
.5	.125	.122	.120	.117	.114	.111	.108	.105	.102	.099
.6	.096	.093	.090	.086	.083	.080	.076	.073	.070	.066
.7	.063	.060	.056	.053	.050	.047	.044	.041	.038	.035
.8	.032	.029	.027	.024	.022	.019	.017	.015	.013	.011
.9	.009	.007	.006	.005	.003	.002	.002	.001	.000	.000

(Application: For load at $.26L$ from near end, $.74L$ from far end, $C_1 = .142$ at near end and .050 at far end.)
For uniformly spaced equal loads the following may be used:

For 1 load: fem $= .125WL$ For 2 loads: fem $= .111WL$
For 3 loads: fem $= .104WL$ For 4 loads: fem $= .10\ WL$

$$\text{For } n \text{ loads: fem} = \frac{n+2}{12\,(n+1)}\,WL$$

Table 3.2 Value of Coefficient C_2 for Partial Uniform Load on Beam for Loaded End

Case 14 of Fig. 3.24; fem $= C_2WL$

C_2	+.00	.01	.02	.03	.04	.05	.06	.07	.08	.09
.0		.005	.010	.014	.019	.023	.028	.032	.036	.040
.1	.044	.047	.051	.054	.058	.061	.064	.067	.070	.073
.2	.075	.078	.080	.083	.085	.087	.089	.091	.093	.095
.3	.097	.098	.100	.101	.103	.104	.105	.106	.107	.108
.4	.109	.110	.111	.112	.112	.113	.113	.114	.114	.114
.5	.115	.115	.115	.115	.115	.115	.115	.115	.115	.114
.6	.114	.114	.113	.113	.112	.112	.111	.111	.110	.110
.7	.109	.108	.108	.107	.106	.105	.105	.104	.103	.102
.8	.101	.100	.100	.099	.098	.097	.096	.095	.094	.093
.9	.092	.091	.090	.089	.089	.088	.087	.086	.085	.084

The coefficient at the loaded end of the beam is given by the formula: $C_2 = \dfrac{a\,(6 - 8a + 3a^2)}{12}$

Table 3.3 Value of Coefficient C_3 for Partial Uniform Load on Beam for Unloaded End

Case 14 of Fig. 3.24; fem $= C_3WL$

C_3	+.00	.01	.02	.03	.04	.05	.06	.07	.08	.09
.0		.000	.000	.000	.001	.001	.001	.002	.002	.003
.1	.003	.004	.004	.005	.006	.007	.008	.008	.009	.010
.2	.011	.011	.013	.015	.016	.017	.018	.019	.021	.022
.3	.023	.025	.026	.027	.029	.030	.032	.033	.034	.036
.4	.037	.039	.040	.042	.043	.045	.046	.048	.049	.051
.5	.052	.054	.055	.056	.058	.059	.061	.062	.063	.065
.6	.066	.067	.069	.070	.071	.072	.073	.074	.076	.077
.7	.078	.079	.079	.080	.081	.082	.083	.084	.084	.085
.8	.085	.086	.086	.087	.087	.087	.088	.088	.088	.088
.9	.088	.088	.087	.087	.087	.086	.086	.085	.085	.084

The coefficient at the unloaded end of the beam is given by the formula: $C_3 = \dfrac{a^2\,(4 - 3a)}{12}$

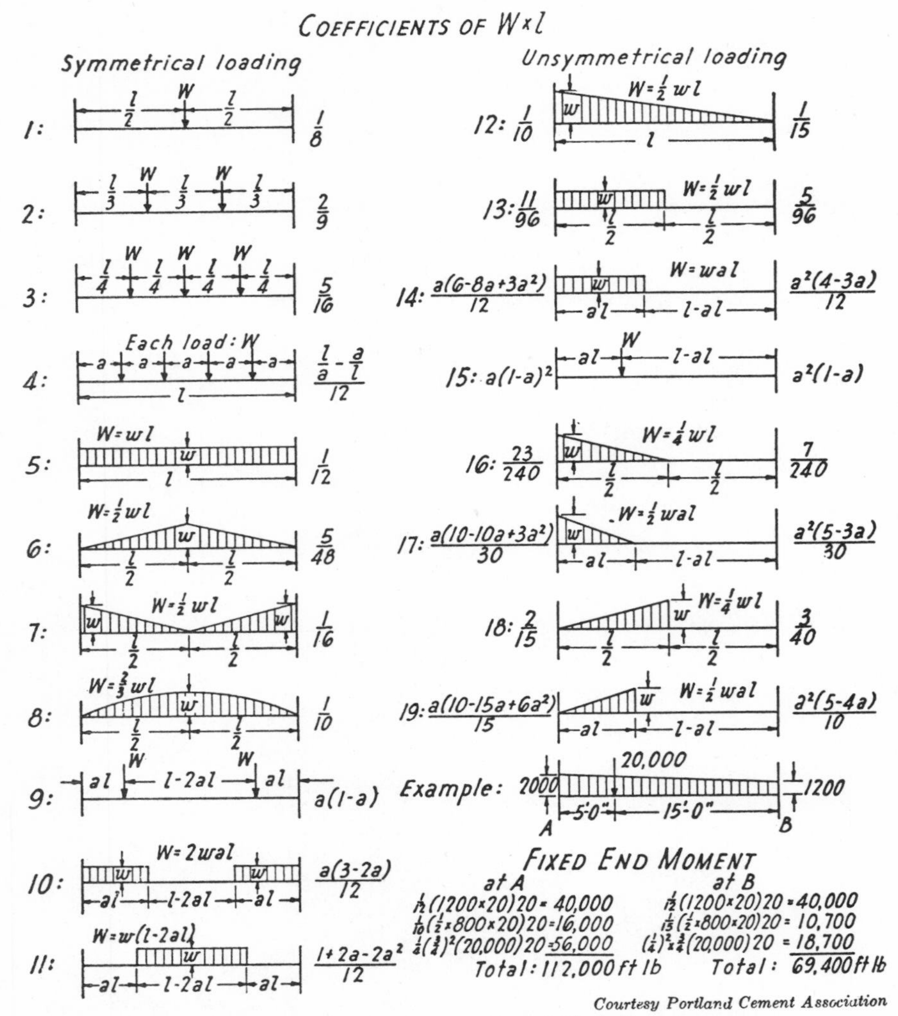

Courtesy Portland Cement Association

Fig. 3.24 Determination of Fixed End Moments

the fem coefficient for uniformly loaded beams of variable moment of inertia is not $\frac{1}{12}WL$, but fWL, the factor f being selected from Table 3 on each chart. For a concentrated load, the coefficient f corresponds to C in charts No. 4 of Figs. 3.25 and 3.26. Note that the end of the beam is designated by the reading at the top and the bottom of the chart.

It is also well to note that these charts are laid out for either a ratio minimum d to maximum d or a factor b, which represents the ratio minimum I to maximum I. The former would probably be used for concrete beams, the latter for steel beams.

i The conditions for which the engineer must design are not normally constant load conditions as described in this section. The maximum positive and negative moments and the maximum reactions result from the following combinations of live load (dead load always being present).

1. When alternate spans are loaded, there is maximum positive moment in the loaded spans and maximum negative moment in the unloaded spans.

2. When two adjacent spans are loaded, with alternate spans beyond them unloaded, there is maximum negative moment, maximum shears, and maximum reactions at the support between the two loaded spans.

Therefore, to design for all maximum con-

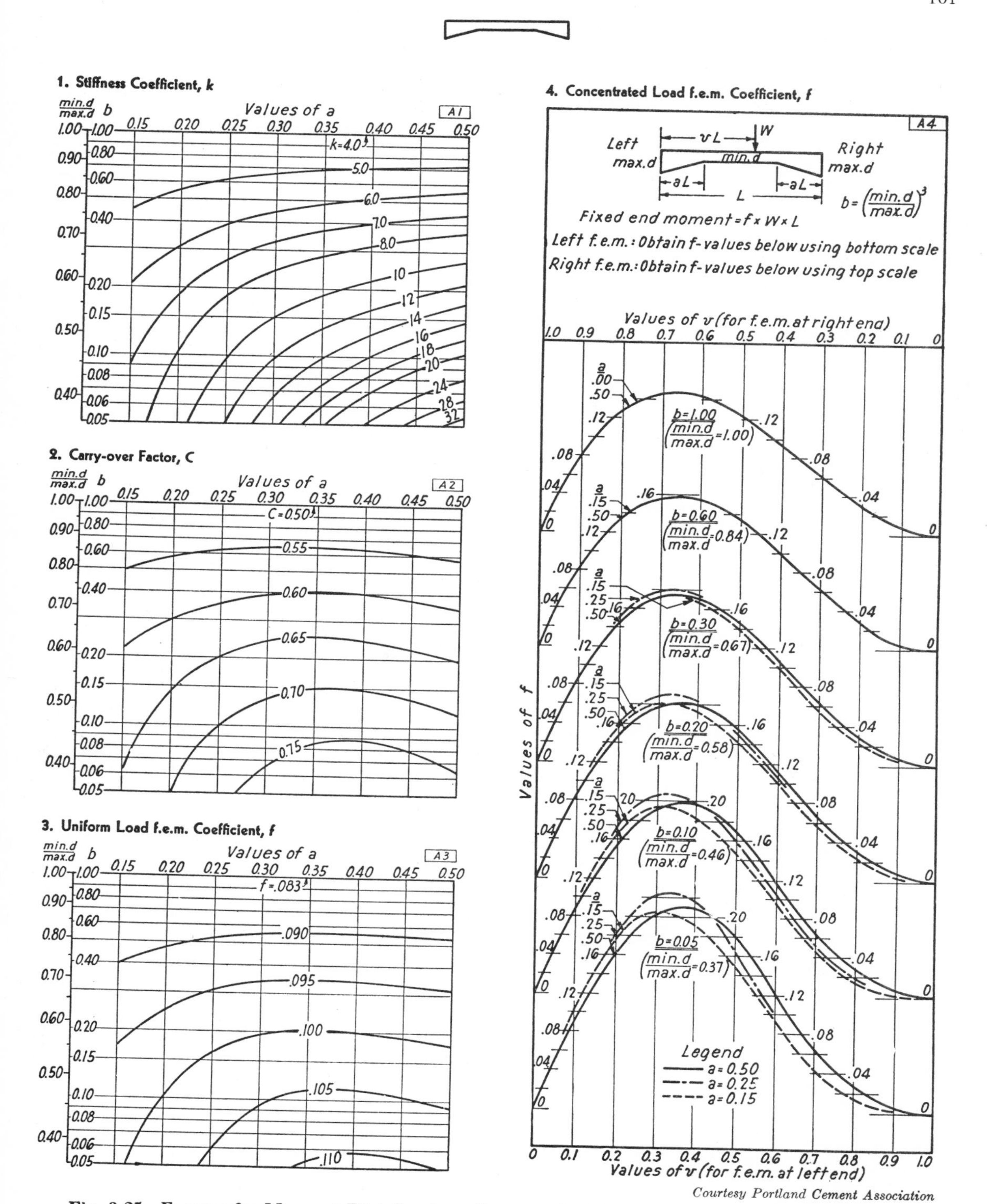

Courtesy Portland Cement Association

Fig. 3.25 Factors for Moment Distribution—Symmetrical Members with Straight Haunches

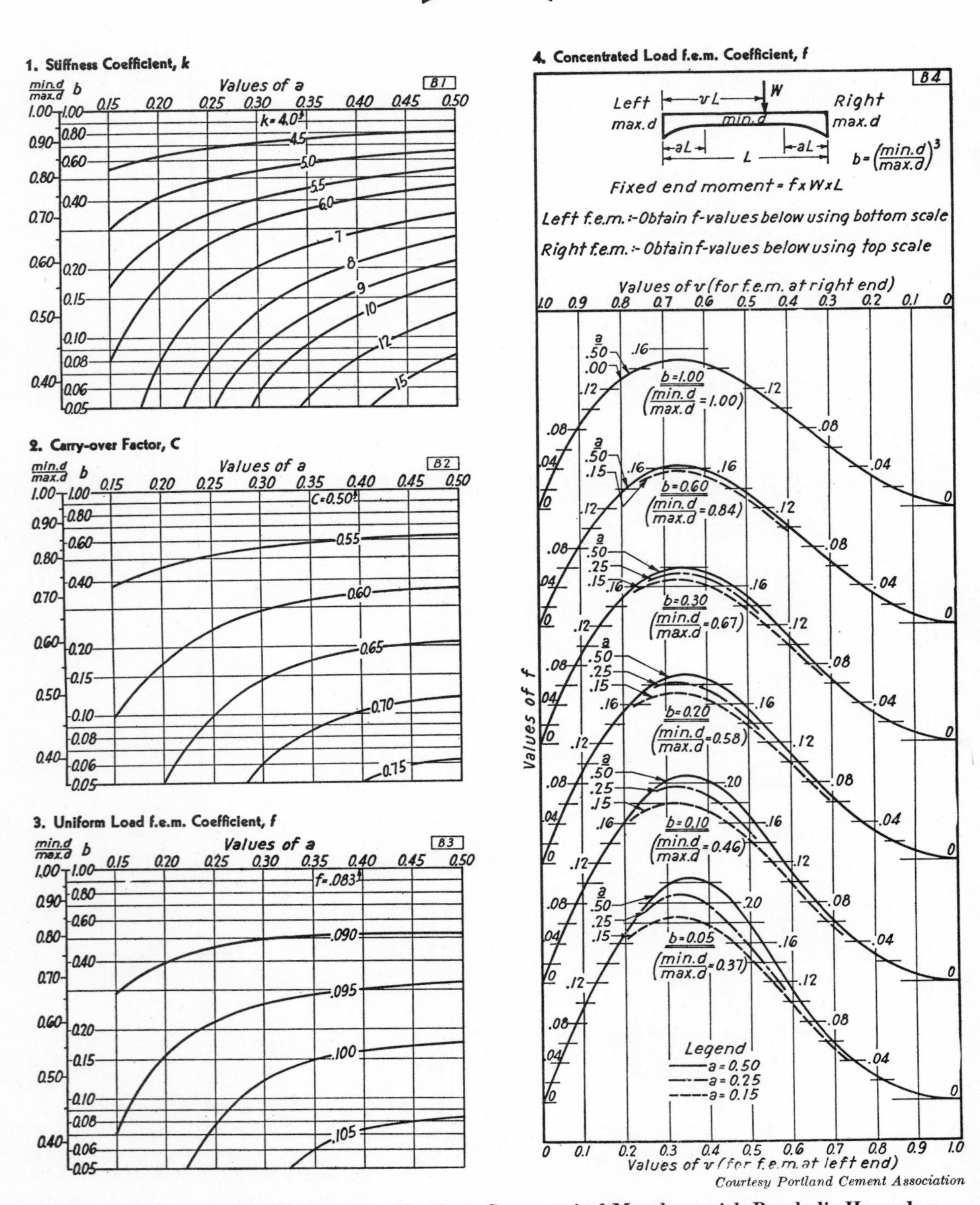

Courtesy Portland Cement Association

Fig. 3.26 Factors for Moment Distribution—Symmetrical Members with Parabolic Haunches

ditions over each interior support, two designs must be made—for load condition 1 and for load condition 2 (see Art. 5.41*d*).

j The American Concrete Institute Code recommends that the moment of inertia to be used for stiffness computation should be based on the gross section of the T beam, allowing for the effect of the flange, but with no allowance for the reinforcement. The curves shown in Fig. 3.27 give a method of computation of the I of the gross section for various conditions. These curves are from the publications of the Portland Cement Association.

ments which lie between the limits of uniform and equivalent central concentrated loads. By equivalent load is meant the resultant load from the same uniform load per running foot of frame. For example, a 40-ft frame with 1 kip per ft uniform load has a total uniform load of 40 kips. The equivalent central concentrated load is 20 kips, the equivalent load applied at the three quarter-points is 10 kips, and so on.

Rigid frames of this type must be designed for the following factors:

1. Size of main members

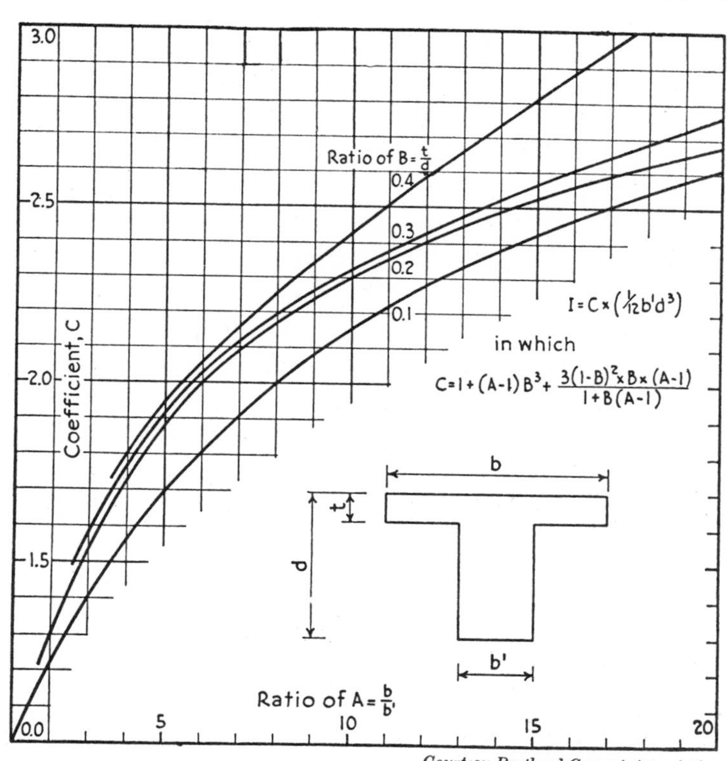

Courtesy Portland Cement Association

Fig. 3.27 Moment of Inertia for T-Beams

3.22 Gable-end rigid frame

a The entire subject of rigid frames is too large and important to cover with any degree of thoroughness in such a work as this. However, the gable-end rigid frame in Fig. 3.28 is by far the most commonly used frame in building construction, and it may be solved using formulae given with the figure.

Other loading conditions, such as three, five, or seven equivalent loads, develop mo-

2. Moment at the eave for design of joint
3. Moment at the peak for design of joint
4. Horizontal thrust at base

There are certain relationships between moments and horizontal thrusts for frames under vertical loads, and an understanding of these relationships will assist greatly in eliminating error in design. These relationships are sketched in Fig. 3.28. A basic axiom for these relationships is that vertical load on a

RIDGE AND RECTANGULAR FRAMES

(RECTANGULAR WHEN Q = 0)

* In formulas for M_x, "x" is always measured from point B.

General Formulas:

$K = \frac{I_2 h}{I_1 m}$ $\qquad Q = \frac{f}{h}$

$N = 4(K + 3 + 3Q + Q^2)$

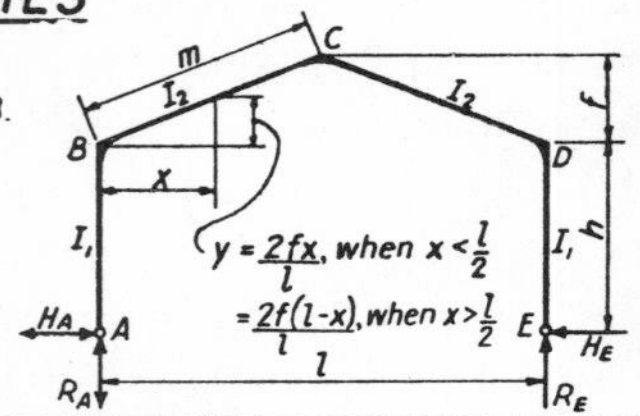

Plus sign (+) denotes moments which cause tension on the inside of the frame when the vertical loads act downward and the horizontal loads, applied to the left side of the frame, act toward the right. The direction of the reactions, and the signs of all terms in the moment formulas, are shown correctly for this condition. When the direction of the loads (but not their position) is reversed the direction of the reactions may, and the signs for all moments will, be reversed.

CASE I - UNIFORMLY DISTRIBUTED VERTICAL LOAD ENTIRE SPAN

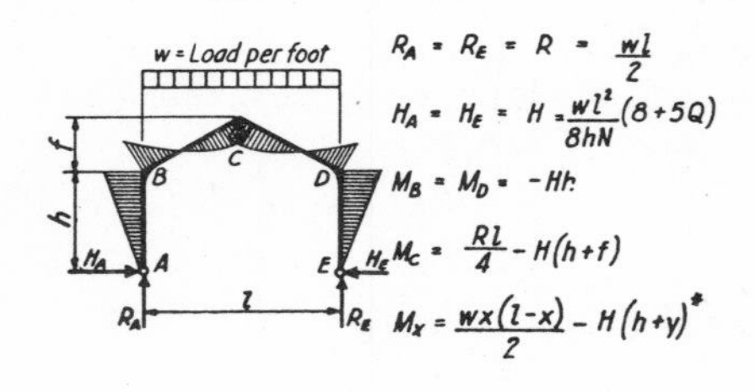

$R_A = R_E = R = \frac{wl}{2}$

$H_A = H_E = H = \frac{wl^2}{8hN}(8 + 5Q)$

$M_B = M_D = -Hh$

$M_C = \frac{Rl}{4} - H(h+f)$

$M_x = \frac{wx(l-x)}{2} - H(h+y)$ *

CASE IA - UNIFORMLY DISTRIBUTED VERTICAL LOAD HALF SPAN (LEFT)

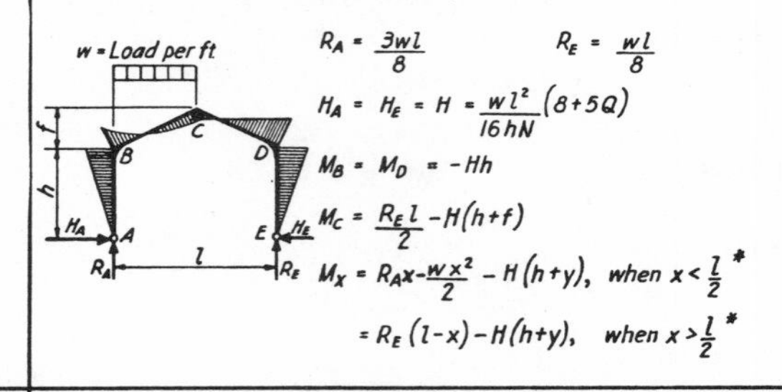

$R_A = \frac{3wl}{8}$ $\qquad R_E = \frac{wl}{8}$

$H_A = H_E = H = \frac{wl^2}{16hN}(8 + 5Q)$

$M_B = M_D = -Hh$

$M_C = \frac{R_E l}{2} - H(h+f)$

$M_x = R_A x - \frac{wx^2}{2} - H(h+y)$, when $x < \frac{l}{2}$ *

$= R_E(l-x) - H(h+y)$, when $x > \frac{l}{2}$ *

CASE II - ONE CONCENTRATED ROOF LOAD AT ANY POSITION

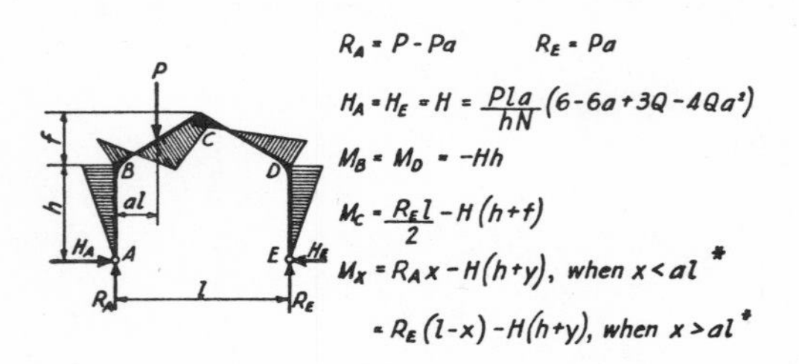

$R_A = P - Pa$ $\qquad R_E = Pa$

$H_A = H_E = H = \frac{Pla}{hN}(6 - 6a + 3Q - 4Qa^2)$

$M_B = M_D = -Hh$

$M_C = \frac{R_E l}{2} - H(h+f)$

$M_x = R_A x - H(h+y)$, when $x < al$ *

$= R_E(l-x) - H(h+y)$, when $x > al$ *

CASE III - BRACKET LOAD ON ONE COLUMN (LEFT COLUMN)

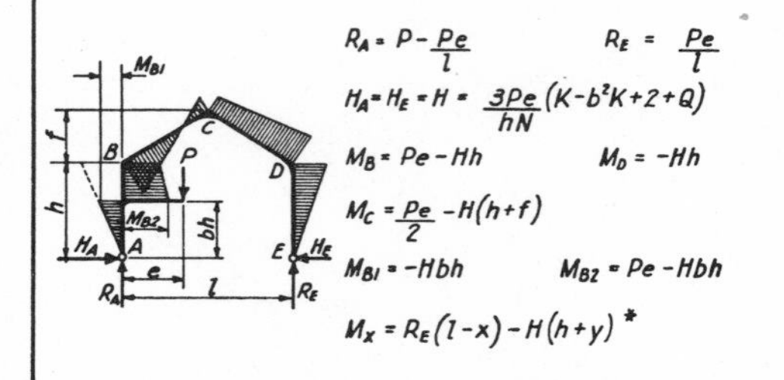

$R_A = P - \frac{Pe}{l}$ $\qquad R_E = \frac{Pe}{l}$

$H_A = H_E = H = \frac{3Pe}{hN}(K - b^2K + 2 + Q)$

$M_B = Pe - Hh$ $\qquad M_D = -Hh$

$M_C = \frac{Pe}{2} - H(h+f)$

$M_{B1} = -Hbh$ $\qquad M_{B2} = Pe - Hbh$

$M_x = R_E(l-x) - H(h+y)$ *

CASE IVA - UNIFORM HORIZONTAL LOAD INCLINED (ROOF) PORTION ONLY

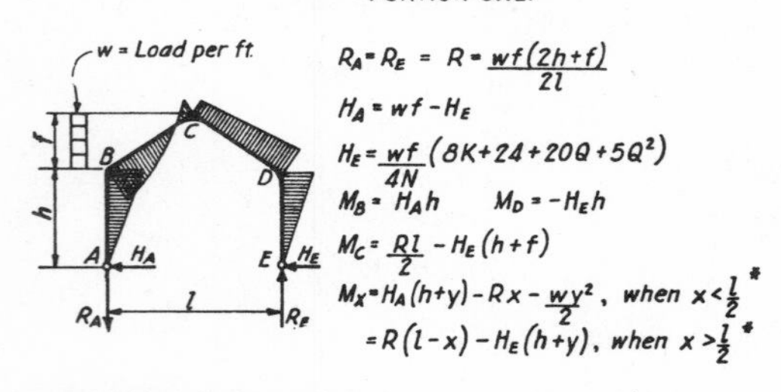

$R_A = R_E = R = \frac{wf(2h+f)}{2l}$

$H_A = wf - H_E$

$H_E = \frac{wf}{4N}(8K + 24 + 20Q + 5Q^2)$

$M_B = H_A h$ $\qquad M_D = -H_E h$

$M_C = \frac{Rl}{2} - H_E(h+f)$

$M_x = H_A(h+y) - Rx - \frac{wy^2}{2}$, when $x < \frac{l}{2}$ *

$= R(l-x) - H_E(h+y)$, when $x > \frac{l}{2}$ *

CASE IVB - UNIFORM HORIZONTAL LOAD VERTICAL (COLUMN) PORTION ONLY

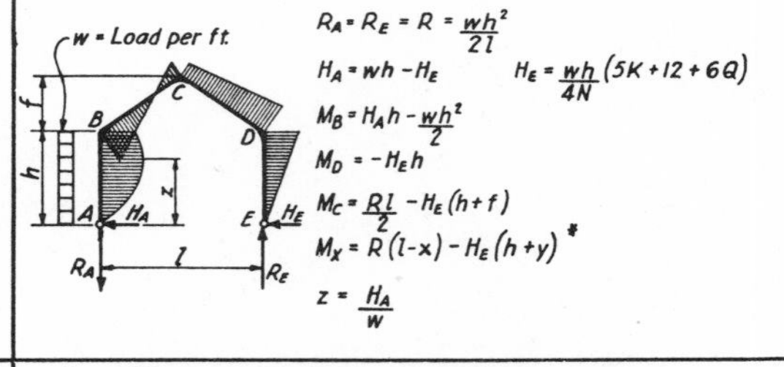

$R_A = R_E = R = \frac{wh^2}{2l}$

$H_A = wh - H_E$ $\qquad H_E = \frac{wh}{4N}(5K + 12 + 6Q)$

$M_B = H_A h - \frac{wh^2}{2}$

$M_D = -H_E h$

$M_C = \frac{Rl}{2} - H_E(h+f)$

$M_x = R(l-x) - H_E(h+y)$ *

$z = \frac{H_A}{w}$

CASE V - UNIFORM HORIZONTAL LOAD ON BOTH INCLINED AND VERTICAL SURFACES

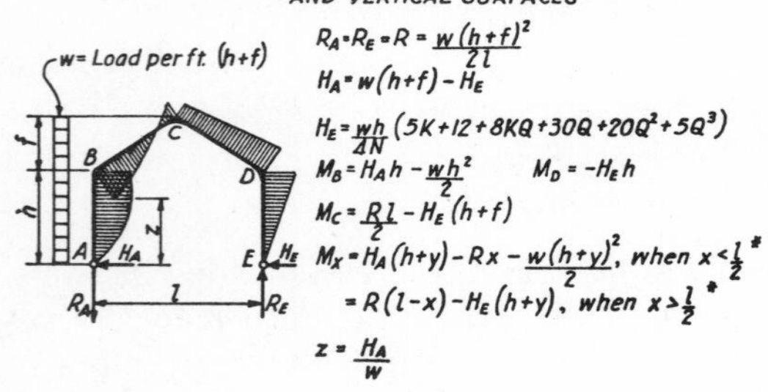

$R_A = R_E = R = \frac{w(h+f)^2}{2l}$

$H_A = w(h+f) - H_E$

$H_E = \frac{wh}{4N}(5K + 12 + 8KQ + 30Q + 20Q^2 + 5Q^3)$

$M_B = H_A h - \frac{wh^2}{2}$ $\qquad M_D = -H_E h$

$M_C = \frac{Rl}{2} - H_E(h+f)$

$M_x = H_A(h+y) - Rx - \frac{w(h+y)^2}{2}$, when $x < \frac{l}{2}$ *

$= R(l-x) - H_E(h+y)$, when $x > \frac{l}{2}$ *

$z = \frac{H_A}{w}$

CASE VI - ONE HORIZONTAL CONCENTRATED LOAD AT ANY POSITION ON COLUMN ($b \leqq 1.0$)

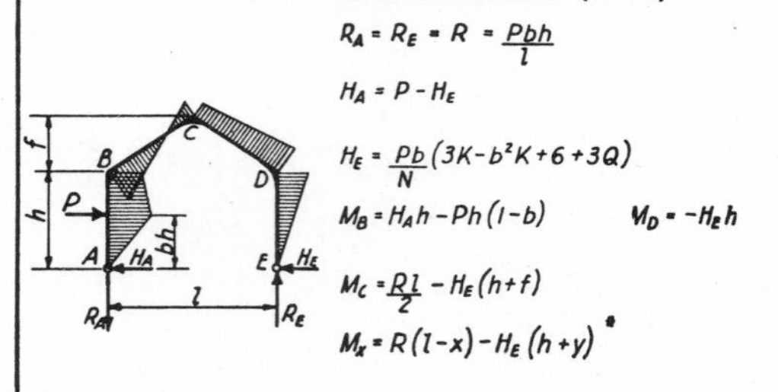

$R_A = R_E = R = \frac{Pbh}{l}$

$H_A = P - H_E$

$H_E = \frac{Pb}{N}(3K - b^2K + 6 + 3Q)$

$M_B = H_A h - Ph(1-b)$ $\qquad M_D = -H_E h$

$M_C = \frac{Rl}{2} - H_E(h+f)$

$M_x = R(l-x) - H_E(h+y)$ *

Fig. 3.28 Gable-End Rigid Frames

frame induces a horizontal thrust at the base of the frame, or a tendency to spread; and the moment at any point in the frame is the ordinary vertical load moment derived by the methods of Arts. 3.10 and 3.11, corrected by the negative moment resulting from the horizontal thrust.

Reference to Fig. 3.28, Case I, will illustrate this axiom. The triangle on the base line AB is the moment diagram for a cantilever moment about B due to a horizontal thrust H applied at the base A. This cantilever moment at B is carried around the corner and applied as a negative moment at the end of a parabola. From B to C to D the moment diagram is a parabola derived as in Fig. 3.13, Case 1, corrected by a negative moment equal to the product of thrust H times the total height to the point in question. The moment at the eave is always negative for this type of load, but it may be either positive or negative at the peak. Stating these moments in formulae,

$$M_B\ (= M_D) = -Hh$$

$$M_C = +M - H(h + f) = +M - M_B\left(1 + \frac{f}{h}\right)$$

where M is the simple beam moment at the peak.

b Although a gable-end rigid frame is seldom uniformly loaded, the formulae for uniformly loaded frames may be used as a basis of calculation, and corrections applied to these results for the load condition applicable. It will be found that the moment at the eave is greater for uniform load than for any other type of equivalent roof load, while the moment at the peak is greater for equivalent central concentrated load.

c The American Institute of Steel Construction has published an excellent pamphlet on "Single Span Rigid Frames in Steel," from which the method of design for roof loads applied as a series of equivalent concentrated loads, with modifications, is as follows. Compute the horizontal thrust H_E from Fig. 3.29 for uniform load, using Chart I. Modify H_E: For three loads at quarter points, each equal to $\frac{1}{4}wL$,

$$H_3 = 0.94H_E$$

For five loads at one-sixth points, each equal to $\frac{1}{6}wL$,

$$H_5 = 0.975H_E$$

Beyond five loads, the result so closely approaches that for a uniform load that H_E may be used without correction. The above figures will vary slightly for different pitch, ratio of height to span, and other factors, but the result is accurate within 2 percent. If the designer prefers greater accuracy, the sum of the horizontal thrusts obtained from Chart II of Fig. 3.29 may be used.

d Because of the allowable increase in the unit stresses for wind loads, it is frequently unnecessary to make any revision in the design for wind. However, it is advisable to check the design on the basis of moments obtained from Fig. 3.30.

e Solution of the various formulae in Figs. 3.29 and 3.30 requires either an assumption of size of column and beam from which I_B/I_C may be solved or an assumption of the relationship. The controlling moment for size of beam and column is the negative moment at the eave. A rough approximation from which this ratio may be determined is an assumption of $M_B = M_D = \frac{1}{16}Wl^2$, or one-half the uniform load moment on a simple supported beam on the total span. With this moment and the direct load on the column, a beam and column may be selected from which the relationships may be assumed, and the final size usually computed after one trial calculation.

The adjustment of maximum negative and positive moments mentioned in Art. 3.21*e* for continuous beams permits a 10 percent reduction in the negative moments and a corresponding increase in the positive moments in rigid frames also.

Problem Assume the load conditions indicated in Fig. 3.31, using wide-flange A36 steel sections and no increase in depth of section at the knee (from AISC pamphlet

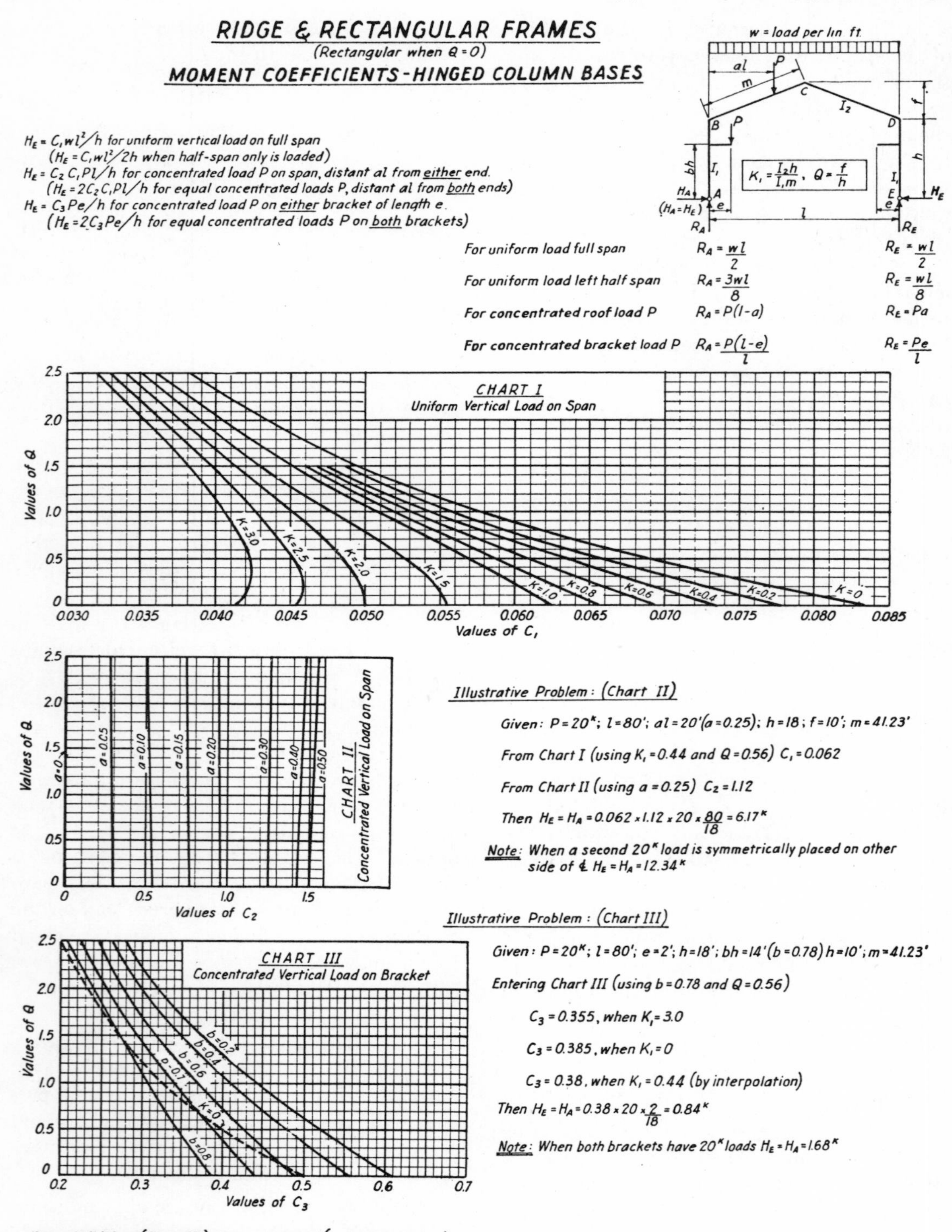

Fig. 3.29 Moment Coefficients—Gable-End Rigid Frames, Vertical Loads

RIDGE & RECTANGULAR FRAMES

(Rectangular when $Q = 0$)

MOMENT COEFFICIENTS - HINGED COLUMN BASES

$H_E = C_4 wh$, for uniform load against roof.
$H_E = C_5 wh$, for uniform load against vertical side.
$H_E = C_6 wh$, for uniform load against total height.
$H_E = Pb(C_7 - b^2 C_8)$, for concentrated load P on one vertical side ($b \leq 1.0$).
($H_E = P$, for concentrated load P on both vertical sides at same elevation.

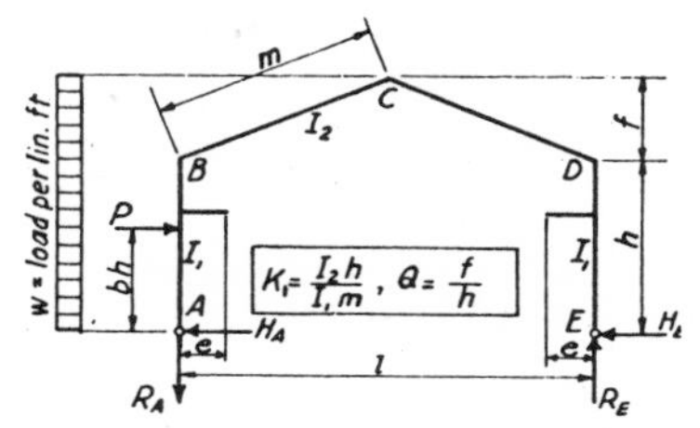

For concentrated horizontal load P	$H_A = P - H_E$	$R_A = R_E = \frac{Pbh}{l}$
For uniform load against roof	$H_A = wf - H_E$	$R_A = R_E = \frac{wf(2h+f)}{2l}$
For uniform load against vertical side	$H_A = wh - H_E$	$R_A = R_E = \frac{wh^2}{2l}$
For uniform load against total height	$H_A = w(h+f) - H_E$	$R_A = R_E = \frac{w(h+f)^2}{2l}$

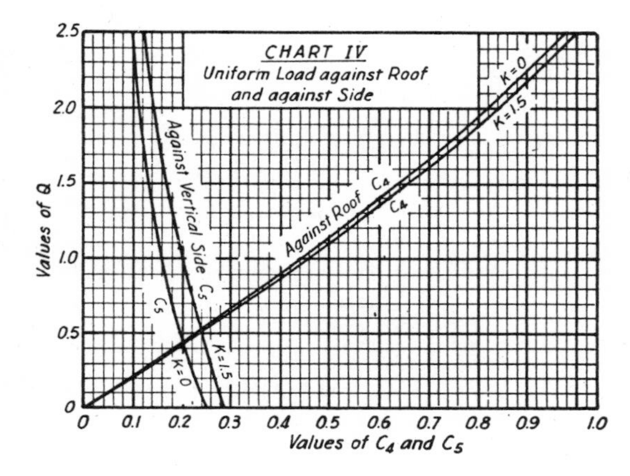

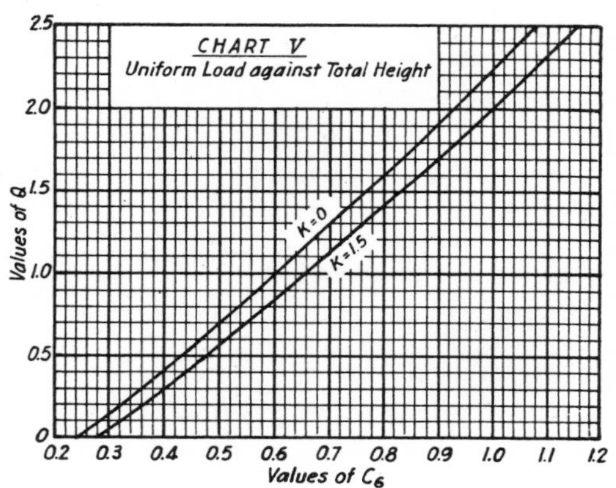

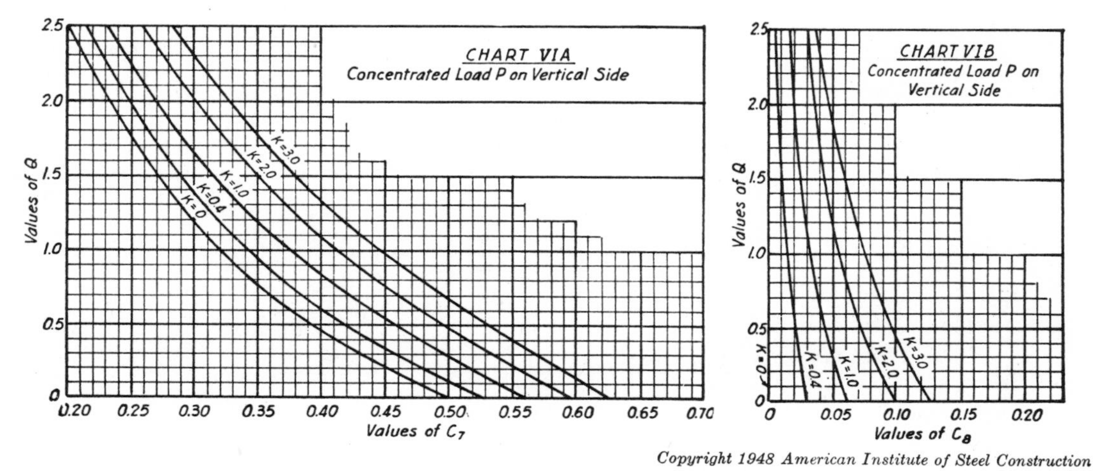

Fig. 3.30 Moment Coefficients—Gable-End Rigid Frames, Horizontal Loads

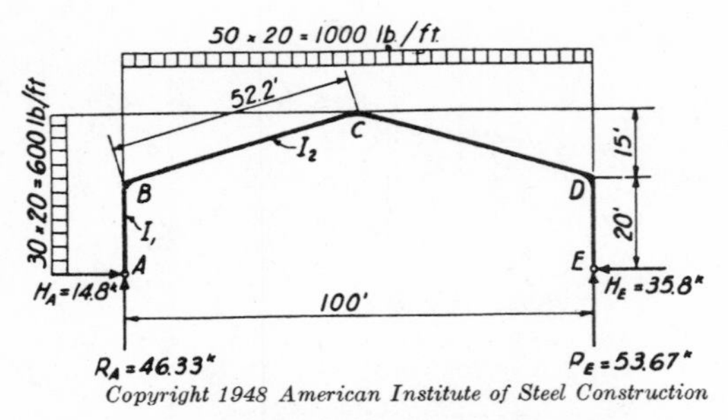

Fig. 3.31 Load Conditions for Rigid-Frame Problem

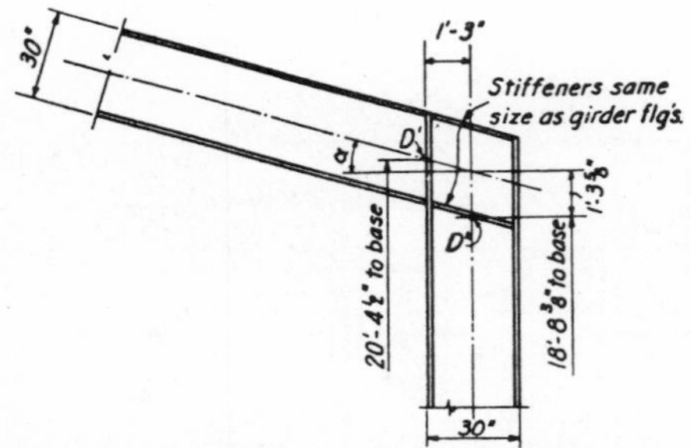

Fig. 3.32 Detail at Knee of Rigid Frame

referred to above). In problems of this type, the load is normally brought into the frame by purlins spaced 6 to 9 ft apart, and the wind load by means of girts. Therefore in accordance with paragraph *c* above, the effect of the load is practically the same as if it were brought into the frame as a uniformly distributed load.

Assume $I_2/I_1 = 1.0$.

$$Q = \frac{15}{20} = 0.75$$

$$K = \frac{1}{1} \times \frac{20}{52.2} = 0.38$$

From Chart I, Fig. 3.29,

$$C_1 = 0.059$$

$$H_E = 0.059 \times 1.0 \times \frac{100^2}{20} = 29.5$$

From Chart V, Fig. 3.30,

$$C_6 = 0.525$$

$$H_E = 0.525 \times 0.6 \times 20 \quad = \underline{6.3}$$

$$\text{Total } H_E = 35.8 \text{ kips}$$

For the girder the moment at D' (Fig. 3.32) with wind is

$$35.8 \times 20.38 = -730$$

$$1.0 \times \frac{1.25^2}{2} = -1$$

$$53.67 \times 1.25 = +\underline{67}$$

$$-664 \text{ ft-kips}$$

and without wind it is

$$29.5 \times 20.38 = -601$$

$$1.0 \times \frac{1.25^2}{2} = -1$$

$$50 \quad \times 1.25 = +\underline{63}$$

$$-539 \text{ ft-kips}$$

The critical condition is without wind, since stress with wind may be increased 33⅓ percent.

The axial force at D' is

$$P = 29.5 \cos \alpha + 50 \sin \alpha = 42.6 \text{ kips}$$

The moment at the peak is

$$\frac{1 \times 100^2}{8} = +1250$$

$$-29.5 \times 35 = \underline{-1032}$$

$$+218 \text{ ft-kips}$$

The moment on the column at D' (Fig. 3.32) is

$$29.5 \times 18.70 = 552 \text{ ft-kips}$$

Art. 1.5.1.4.1 of the AISC Code contains this concession to our knowledge of the behavior of a structure before failure:

> Beams which are rigidly framed to columns may be proportioned for 9/10 of the negative moments which are maximum at the points of support provided that for such members the maximum positive moment shall be increased by one tenth of the maximum negative moments.

The effect of this redistribution of moments is shown in Fig. 3.33, and the adjustment applied to the moments on this frame

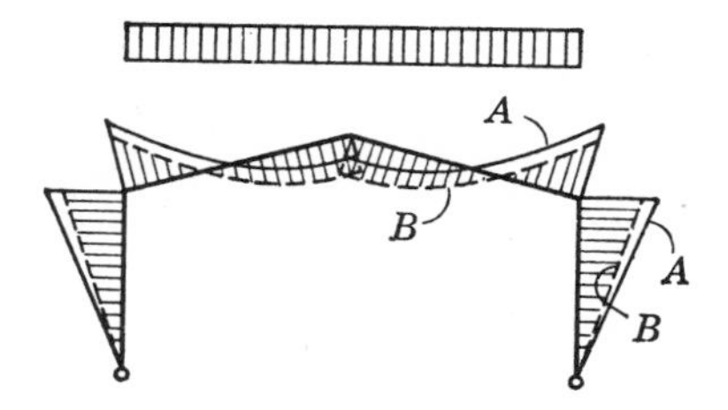

A = **Actual moment diagram**
B = **Modified diagram corresponding to 10 percent moment reduction allowance at interior supports**

Fig. 3.33 Modification of Moment

gives the following adjusted results:

At the peak, the moment becomes

$$+218 + 55 = +273 \text{ ft-kips}$$

At the knee, the moment becomes

$$-552 + 55 = -497 \text{ ft-kips}$$

At the critical point, this moment becomes

$$\frac{20.0 - 1.8}{20} \text{ (see Fig. 3.32)} \times 497 = 452 \text{ ft-kips}$$

Try 30 WF 99 for the beam.

The bracing is assumed to be 7.5 ft (purlin spacing) for the beam, and 5.0 ft (girt spacing) for the column.

$$l/r = \frac{7.5 \times 12}{2.00} = 45$$

From Table 7.1, $F_a = 18.78$

As a compact section, $F_b = 24.00$

$$\%a + \%b = \frac{42.6}{29.11 \times 18.78} + \frac{452 \times 12}{269.1 \times 24} = 0.078 + 0.84 = 0.918$$

This is less than unity and is therefore satisfactory. The next lighter section would be the 27 WF 94 with an $r = 2.04$ and an $S = 242.8$, correcting the above computations for the 27-in. section would give

$$\%a + \%b = \frac{42.6}{27.65 \times 18.85} + \frac{455 \times 12}{242.8 \times 24} = 0.082 + 0.936 = 1.018$$

Although slightly in excess of unity, this is within the ordinarily acceptable tolerance.

For the column, the axial force is 50.0 and assuming the 27 WF 94,

$$l/r = 5 \times 12 = 29.4, \quad F_a = 19.98$$

$$\%a + \%b = \frac{50}{27.65 \times 19.98} + \frac{455 \times 12}{242.8 \times 24} = 0.091 + 0.936 = 1.027$$

Therefore since both beam and column exceed the allowable stress, it is advisable to use the 30 WF 99.

4

Steel in Bending

4.10 STEEL BEAMS

a It is well to repeat here two fundamental statements. First, the strength of a structural member is dependent upon two factors, the material used, and its shape and size. Second, with the moment induced in a beam by any system of loading known, there are two methods of computing the internal resisting moment in the beam, the moment-couple method, indicated in Fig. 3.1, and the section-modulus method. Because steel beams are homogeneous and because the section modulus is easily tabulated for standard beam sections, this method is generally used.

To select a steel beam, it is therefore necessary to compute the moment as described in Art. 3.10. Reduce this moment to inch-pound units, and divide by the allowable fiber stress to obtain the section modulus. Then from Table 4.1 or Tables 1.5 to 1.10 select the beam which will furnish the required section modulus and otherwise satisfy the required conditions.

The various handbooks have tables of maximum bending moment in foot-pounds and allowable uniform load in kips on different spans, but the simplest and best way is to use the section-modulus method for all types of beam loading.

Usually the two governing factors in the selection of a steel beam are section modulus required and economy of weight. Occasionally shallower depth, deflection, lateral deflection, shear, or buckling of the web will control, but these will be discussed in later paragraphs.

If the section modulus required is greater than that obtainable in the wide-flange sections listed in Table 1.5, it is necessary to use either a cover-plated beam as treated in Art. 4.22 or, for still greater strength, a plate girder as treated in Art. 4.20.

b Until the Gray Mill was used, about 1908, the only sections available for I beams were the American Standard beams, listed as such in the handbooks and in Table 1.7. Now the major steel companies are rolling a uniform list of wide-flange sections and light beams. The wide-flange sections require a slight increase over base cost, usually about 5 percent, but the better range of sections available, the economy of weight of about 10 percent for equal section modulus, the more uniform strength due to the method of rolling, and the wider flange for the same strength beam—have caused the wide-flange sections largely to displace the old American Standard beam sections. The best designation for wide-flange beams is WF, instead of B as used by Bethlehem or CB as used by United States Steel. Thus we would note 10 WF 21 (or 10 WF 21) instead of 10 B 21 or

10 CB 21. For light-weight beams the designation is B, as 10 B 15, for either United States Steel or Bethlehem sections.

If American Standard beams are used it is best policy to use the lightest section in each group listed—3 I 5.7, 4 I 7.7, etc., including both 12 I 31.8 and 12 I 40.8, 15 I 42.9, 15 I 60.8, etc. (Note that American Standard beams are always indicated as I.) The reason for selecting the lightest beam only is that if others are called for, a long delay may be encountered while waiting for a rolling.

American Standard channels are frequently used for a special purpose where one flat side is desirable, or occasionally for other purposes, as will be noted later. They are not ordinarily used as floor beams, however, because of their narrow flange, lower efficiency in bending for a given depth, and the difficulty in supporting concrete forms from a channel.

Angles are not economical as beams, although they are used as lintels and will be discussed under that heading. T's are not economical and are only used in making up such things as stair landings, marquees, etc. Z bars are now practically off the market. Do not use car or ship sections; they are hard to get and are not economical.

c In accordance with the 1963 AISC Code, the nomenclature used throughout this chapter will be as follows:

A_f area of compression flange.
A_{st} cross-sectional area of stiffener or pair of stiffeners.
A_w area of girder web.
F_b bending stress permitted in the absence of axial stress.
F'_b allowable bending stress in compression flange of plate girders as reduced because of large web depth-to-thickness ratio.
F_p allowable bearing stress.
F_t allowable tensile stress.
F_v allowable shear stress.
F_y specified minimum yield point of the type of steel being used (pounds per square inch unless otherwise noted).
L span length, in feet.
M moment.
M_1 smaller end moment on unbraced length of beam-column.
M_2 larger end moment on unbraced length of beam-column.
M_D moment produced by dead load.
M_L moment produced by live load.
N length of bearing of applied load.
P applied load.
V statical shear on beam.
Y ratio of yield point of web steel to yield point of stiffener steel.
a clear distance between transverse stiffeners.
a' distance required at ends of welded partial length cover plate to develop stress.
d depth of beam or girder. Also diameter of roller or rocker bearing.
f_b computed bending stress.
f_t computed tensile stress.
f_v computed sheer stress, in pounds per square inch.
f_{vs} shear between girder web and transverse stiffeners, in pounds per linear inch of single stiffener or pair of stiffeners.
h clear distance between flanges of a beam or girder.
l actual unbraced length, in inches.
t girder or beam web thickness.
t_f flange thickness.

d The following sections from the AISC specifications apply to the design of steel beams.

SECTION 1.2 TYPES OF CONSTRUCTION

Three basic types of construction and associated design assumptions are permissible under the respective conditions stated hereinafter, and each will govern in a specific manner the size of members and the types and strength of their connections.

Type 1, commonly designated as "rigid-frame" (continuous frame), assumes that beam-to-column connections have sufficient rigidity to hold virtually unchanged the original angles between intersecting members.

Type 2, commonly designated as "conventional" or "simple" framing (unrestrained, free-ended), assumes that the ends of beams and girders are connected for shear only, and are free to rotate under load.

Type 3, commonly designated as "semi-rigid framing" (partially restrained), assumes that the connections of beams and girders possess a dependable and known moment capacity intermediate in degree between the complete rigidity of Type 1 and the complete flexibility of Type 2.

The design of all connections shall be consistent with the assumptions as to type of construction called for on the design drawings.

Type 1 construction is unconditionally permitted under this Specification. Two different methods of design are recognized. Within the limitations laid down in Sect. 2.1, members of continuous frames, or continuous portions of frames, may be proportioned, on the basis of their maximum predictable strength, to resist the specified design loads multiplied by the prescribed load factors. Otherwise Type 1 construction shall be designed, within the limitations of Sect. 1.5, to resist the stresses produced by the specified design loads, assuming moment distribution in accordance with the elastic theory.

Type 2 construction is permitted under this specification, subject to the stipulations of the following paragraph wherever applicable. Beam-to-column connections with seats for the reactions and with top clip angles for lateral support only are classed under Type 2.

In tier buildings, designed in general as Type 2 construction (that is, with beam-to-column connections other than wind connections flexible), the distribution of the wind moments between the several joints of the frame may be made by a recognized empirical method provided that either:

1. The wind connections, designed to resist the assumed moments, are adequate to resist the moments induced by the gravity loading and the wind loading at the increased unit stresses permitted therefor, or
2. The wind connections, if welded and if designed to resist the assumed wind moments, are so designed that larger moments induced by the gravity loading under the actual condition of restraint will be relieved by deformation of the connection material without over-stress in the welds.

Type 3 (semi-rigid) construction will be permitted only upon evidence that the connections to be used are capable of furnishing, as a minimum, a predictable proportion of full end restraint. The proportioning of main members joined by such connections shall be predicated upon no greater degree of end restraint than this minimum.

Types 2 and 3 construction may necessitate some nonelastic but self-limiting deformation of a structural steel part.

SECTION 1.12 SIMPLE AND CONTINUOUS SPANS

1.12.1 *Simple spans*

Beams, girders and trusses shall ordinarily be designed on the basis of simple spans whose effective length is equal to the distance between centers of gravity of the members to which they deliver their end reactions.

1.12.2 *End restraint*

When designed on the assumption of full or partial end restraint, due to continuous, semi-continuous or cantilever action, the beams, girders and trusses, as well as the sections of the members to which they connect, shall be designed to carry the shears and moments so introduced, as well as all other forces, without exceeding at any point the unit stresses prescribed in Sect. 1.5.1; except that some nonelastic but self-limiting deformation of a part of the connection may be permitted when this is essential to the avoidance of overstressing of fasteners.

1.5.1.4 *Bending*

1.5.1.4.1 Tension and compression on extreme fibers of laterally supported compact rolled shapes and compact built-up members having an axis of symmetry in the plane of loading

$$F_b = 0.66F_y$$

(In order to qualify as a compact section the width-thickness ratio of projecting elements of the compression flange shall not exceed $1600/\sqrt{F_y}$ except that for rolled shapes an upward variation of 3% may be tolerated. The width-thickness ratio of flange plates in box sections and flange cover plates included between longitudinal lines of rivets, high strength bolts or welds shall not exceed $6000/\sqrt{F_y}$. The depth-thickness ratio of the web, d/t, shall not exceed $13{,}300/\sqrt{F_y}$. When subjected to combined axial force and bending moment d/t shall not exceed $13{,}300 \left(1 - 1.43 \frac{f_a}{F_a}\right) \Big/ \sqrt{F_y}$ except that it need not be less than $8000/\sqrt{F_y}$. Flanges of compact built-up sections shall be continuously connected to the web or webs. Such members are deemed to be supported laterally when the distance, in inches, between points of support of the compression flange does not exceed $2400\ b_f/\sqrt{F_y}$ nor $20{,}000{,}000\ A_f/dF_y$.)

Beams and girders which meet the requirements of the preceding paragraph and are continuous over supports or are rigidly framed to columns by means of rivets, high strength bolts or welds, may be proportioned for $\frac{9}{10}$ of the negative moments produced by gravity loading which are maximum at points of support, provided that, for such members, the maximum positive moment shall be increased by $\frac{1}{10}$ of the average negative moments. This reduction shall not apply to moments produced by loading on cantilevers. If the negative moment is resisted by a column rigidly framed to the beam or girder, the $\frac{1}{10}$ reduction may be used in proportioning the column for the combined axial and bending loading, provided that the unit stress, f_a, due to any concurrent axial load on the member, does not exceed $0.15F_a$.

The reference in Sec. 1.5.1.4.1 to compact sections may be simplified by the statement that all wide flange beams may be considered as Compact Sections with the exception of the following "Non-Compact Sections":

Non-Compact Sections

6 B 8.5	8 WF 31	14 WF 87
8 B 10	10 WF 33	14 WF 95
10 B 11.5	10 WF 49	14 WF 103
12 B 14	12 WF 65	8 M 20
6 WF 15.5	12 WF 72	

In addition to the above list, other non-compact sections in certain grades of steel are noted in Table 4.1.

1.5.1.4.2 Tension and compression on extreme fibers of unsymmetrical members, except channels, supported in the region of compression stress as in Sect. 1.5.1.4.1

$$F_b = 0.60F_y$$

1.5.1.4.3 Tension and compression on extreme fibers of box-type members whose proportions do not meet the provisions of a compact section, but do conform to the provisions of Sect. 1.9

$$F_b = 0.60F_y$$

1.5.1.4.4 Tension on extreme fibers of other rolled shapes, built-up members and plate girders

$$F_b = 0.60F_y$$

1.5.1.4.5 Compression on extreme fibers of rolled shapes, plate girders and built-up members having an axis of symmetry in the plane of their web (other than box-type beams and girders), the larger value computed by Formulas (4) and (5), but not more than $0.60F_y$

$$F_b = \left[1.0 - \frac{(l/r)^2}{2C_c^2C_b}\right]0.60F_y{}^* \qquad (4)$$

$$F_b = \frac{12{,}000{,}000}{ld/A_f} \qquad (5)$$

where l is the unbraced length of the compression flange; r is the radius of gyration of a tee section comprising the compression flange plus one-sixth of the web area, about an axis in the plane of the web; A_f is the area of the compression flange; C_c is defined in Sect. 1.5.1.3, and C_b, which can conservatively be taken as unity, is equal to

$$C_b = 1.75 - 1.05\left(\frac{M_1}{M_2}\right) + 0.3\left(\frac{M_1}{M_2}\right)^2$$

but not more than 2.3, where M_1 is the smaller and M_2 the larger bending moment at the ends of the unbraced length, taken about the strong axis of the member, and where M_1/M_2, the ratio of end moments, is positive when M_1 and M_2 have the same sign (single curvature bending) and negative when they are of opposite signs (reverse curvature bending). When the bending moment at any point within an unbraced length is larger than that at both ends of this length, the ratio M_1/M_2 shall be taken as unity. See Sect. 1.10 for further limitation in plate girder flange stress.

* Where l/r is less than 40, stress reduction according to Formula (4) may be neglected.

1.5.1.4.6 Compression in extreme fibers of channels, the value computed by Formula (5), but not more than

$$F_b = 0.60F_y$$

1.5.1.4.7 Tension and compression on extreme fibers of large pins

$$F_b = 0.90F_y$$

1.5.1.4.8 Tension and compression on extreme fibers of rectangular bearing plates

$$F_b = 0.75F_y$$

In accordance with the Code, the AISC has set up the following numerical simplifications for carbon steel A36 and for high-strength steel A440 and A141 (for 50,000-psi yield strength).

e The general formula for moment is $M = FS$, or $S = M/F_b$ [Formula (3.2)]. In steel-beam problems, M is ordinarily figured in foot-pounds or foot-kips and S in inch units. M as computed from Art. 3.10 must, therefore, be reduced to inches before finding the section modulus.

The AISC Code, which is the generally accepted design code for steel, uses a stress in bending for compact sections of 24,000 psi for A36 steel, or 33,000 for high-strength steels A440 and A441. Unless otherwise noted, problems in this chapter are figured

Table 4.1

S_x ELASTIC SECTION MODULUS TABLE
For shapes used as beams

Elastic Modulus	Shape
1105.1	**36 WF 300**
1031.2	**36 WF 280**
951.1	**36 WF 260**
892.5	**36 WF 245**
835.5	**36 WF 230**
811.1	33 WF 240
740.6	**33 WF 220**
669.6	**33 WF 200**
663.6	**36 WF 194**
649.9	30 WF 210
621.2	**36 WF 182**
586.1	30 WF 190
579.1	**36 WF 170**
541.0	**36 WF 160**
528.2	30 WF 172
502.9	**36 WF 150**
492.8	27 WF 177
486.4	33 WF 152
446.8	**33 WF 141**
444.5	27 WF 160
438.6	**‡36 WF 135**
413.5	24 WF 160
404.8	**33 WF 130**
402.9	27 WF 145
379.7	30 WF 132
372.5	24 WF 145
358.3	**‡33 WF 118**
354.6	30 WF 124
330.7	‡24 WF 130
327.9	**30 WF 116**
317.2	21 WF 142
299.2	**30 WF 108**
299.2	27 WF 114
299.1	24 WF 120
284.1	21 WF 127
274.4	24 WF 110
269.1	**‡30 WF 99**
266.3	27 WF 102
263.2	12 WF 190
250.9	24 I 120
249.6	‡21 WF 112
248.9	‡24 WF 100

Elastic Modulus	Shape
242.8	**27 WF 94**
234.3	24 I 105.9
222.2	12 WF 161
220.9	**24 WF 94**
220.1	18 WF 114
216.0	14 WF 136
211.7	**‡27 WF 84**
202.2	18 WF 105
202.0	14 WF 127
197.6	21 WF 96
197.6	24 I 100
196.3	**24 WF 84**
189.4	‡14 WF 119
185.8	24 I 90
184.4	18 WF 96
182.5	12 WF 133
176.3	‡14 WF 111
175.4	**24 WF 76**
173.9	24 I 79.9
168.0	21 WF 82
166.1	16 WF 96
163.6	†14 WF 103
163.4	12 WF 120
160.0	20 I 95
156.1	18 WF 85
153.1	**‡24 WF 68**
151.3	16 WF 88
150.7	21 WF 73
150.6	†14 WF 95
150.2	20 I 85
144.5	12 WF 106
141.7	18 WF 77
139.9	**21 WF 68**
138.1	†14 WF 87
134.7	12 WF 99
130.9	‡14 WF 84
128.2	18 WF 70
127.8	16 WF 78
126.4	**21 WF 62**
126.3	20 I 75
126.3	10 WF 112
125.0	12 WF 92
121.1	‡14 WF 78
117.0	18 WF 64
116.9	20 I 65.4
115.9	16 WF 71
115.7	‡12 WF 85
112.4	10 WF 100
112.3	14 WF 74

Elastic Modulus	Shape
109.7	**‡21 WF 55**
107.8	18 WF 60
107.1	‡12 WF 79
104.2	16 WF 64
103.0	14 WF 68
101.9	18 I 70
99.7	10 WF 89
98.2	**18 WF 55**
97.5	†12 WF 72
94.1	16 WF 58
92.2	‡14 WF 61
89.0	**18 WF 50**
88.4	18 I 54.7
88.0	†12 WF 65
86.1	10 WF 77
80.7	**16 WF 50**
80.1	10 WF 72
78.9	**‡18 WF 45**
78.1	‡12 WF 58
77.8	14 WF 53
74.5	*18 [58
73.7	10 WF 66
72.4	**16 WF 45**
70.7	‡12 WF 53
70.2	14 WF 48
69.1	*18 [51.9
67.1	‡10 WF 60
64.7	12 WF 50
64.4	**16 WF 40**
64.2	15 I 50
63.7	*18 [45.8
62.7	‡14 WF 43
61.0	*18 [42.7
60.4	‡10 WF 54
60.4	8 WF 67
58.9	15 I 42.9
58.2	12 WF 45
56.3	**‡16 WF 36**
54.6	14 WF 38
54.6	†10 WF 49
53.6	*15 [50
52.0	8 WF 58
51.9	‡12 WF 40
50.3	12 I 50
49.1	10 WF 45
48.5	**‡14 WF 34**

Table 4.1 (*Continued*)

ELASTIC SECTION MODULUS TABLE
For shapes used as beams

S_x

Elastic Modulus	Shape
47.0	**16 B 31**
46.2	*15 [40
45.9	12 WF 36
44.8	12 I 40.8
43.2	8 WF 48
42.2	‡10 WF 39
41.8	**‡14 WF 30**
41.7	*15 [33.9
39.4	12 WF 31
38.1	**‡16 B 26**
37.8	12 I 35
36.0	12 I 31.8
35.5	8 WF 40
35.0	†10 WF 33
34.9	**14 B 26**
34.1	‡12 WF 27
31.1	‡ 8 WF 35
30.8	10 WF 29
29.2	10 I 35
28.8	**‡14 B 22**
28.9	‡ 8 M 34.3
28.2	‡ 8 M 32.6
27.4	† 8 WF 31
26.9	*12 [30
26.6	‡10 M 29.1
26.4	**10 WF 25**
25.3	**12 B 22**
24.4	10 I 25.4
24.3	8 WF 28
23.9	*12 [25
23.6	‡10 M 22.9
22.5	‡ 8 M 28
21.5	**‡10 WF 21**
21.4	**12 B 19**
21.4	*12 [20.7
21.1	‡10 M 21
21.0	‡ 8 M 24

Elastic Modulus	Shape
21.0	**‡14 B 17.2**
20.8	‡ 8 WF 24
20.6	*10 [30
18.8	10 B 19
18.1	*10 [25
17.5	**‡12 B 16.5**
17.1	‡ 8 M 22.5
17.0	8 WF 20
16.8	6 WF 25
16.2	10 B 17
16.0	8 I 23
15.7	*10 [20
15.7	6 M 25
15.5	‡ 8 M 18.5
15.2	‡ 8 M 20
14.8	**†12 B 14**
14.2	8 I 18.4
14.1	‡ 8 WF 17
14.0	‡ 8 M 17
13.8	‡10 B 15
13.7	‡ 6 M 22.5
13.5	* 9 [20
13.4	*10 [15.3
13.4	‡ 6 WF 20
12.9	‡ 6 M 20
12.0	**‡12 JR 11.8**
12.0	7 I 20
11.8	8 B 15
11.3	* 9 [15
10.9	* 8 [13.75
10.5	**†10 B 11.5**
10.5	* 9 [13.4
10.4	7 I 15.3
10.1	† 6 WF 15.5
10.1	6 B 16
9.9	‡ 8 B 13
9.9	5 WF 18.5
9.5	5 M 18.9

Elastic Modulus	Shape
9.3	* **12 JR [10.6**
9.0	* 8 [13.75
8.7	6 I 17.25
8.5	5 WF 16
8.1	* 8 [11.5
7.8	**‡10 JR 9**
7.8	† 8 B 10
7.7	* 7 [14.75
7.3	6 I 12.5
7.2	6 B 12
6.9	* 7 [12.25
6.5	* **10 JR [8.4**
6.0	* 7 [9.8
6.0	5 I 14.75
5.8	* 6 [13
5.4	4 WF 13
5.2	4 M 13
5.1	† 6 B 8.5
5.0	* 6 [10.5
4.8	5 I 10
4.7	**8 JR 6.5**
4.4	*10 JR [6.5
4.3	* 6 [8.2
3.5	**7 JR 5.5**
3.5	* 5 [9
3.3	4 I 9.5
3.0	* 5 [6.7
3.0	4 I 7.7
2.4	**6 JR 4.4**
2.3	* 4 [7.25
1.9	* 4 [5.4
1.9	3 I 7.5
1.7	3 I 5.7
1.4	* 3 [6
1.2	* **3 [5**
1.1	* **3 [4.1**

†Identifies non-compact shapes for which bending stress F_b may not exceed 0.60 F_y in A36, A242, A440, and A441 steels (Specification Section 1.5.1.4.1).

‡Identifies non-compact shapes for which bending stress F_b may not exceed 0.60 F_y in A242, A440, and A441 steels (Specification Section 1.5.1.4.1).

*Bending stress F_b may not exceed 0.60 F_y (Specification Section 1.5.1.4.6).

Shapes subjected to combined axial force and bending moment may not be compact under Section 1.5.1.4.1, of the AISC Specification. Check all shapes for compliance with this Section.

Table 4.2

SECTION 1.5 ALLOWABLE UNIT STRESSES A36 STEEL

1.5.1 Structural Steel

1.5.1.1 Tension

Tension on net section, except at pin holes. . . . . F_t = 22,000 psi
Tension on net section at pin holes. . . . . F_t = 16,000 psi

1.5.1.2 Shear

Shear on gross section (see Table 3-36 for reduced values for girder webs). . . . . F_v = 14,500 psi

1.5.1.3 Compression

$$C_c = 126.1$$

For values of F_a given by Formulas (1), (2) and (3) see Table 1-36.

1.5.1.4 Bending

1.5.1.4.1 Tension and compression for compact, adequately braced beams having an axis of symmetry in the plane of loading. . . . . F_b = 24,000 psi
1.5.1.4.2 Tension and compression for unsymmetrical rolled shapes continuously braced in the region under compression stress. . . . . F_b = 22,000 psi
1.5.1.4.3 Tension and compression for box-type members not included in Sect. 1.5.1.4.1. . . . . F_b = 22,000 psi
1.5.1.4.4 Tension for other rolled shapes, built-up members and plate girders. . . . . F_b = 22,000 psi
1.5.1.4.5 Compression, except as provided by Sect. 1.5.1.4.1, 1.5.1.4.2, 1.5.1.4.3, 1.5.1.4.7 and 1.5.1.4.8: the larger value given by Formulas (4) and (5).

$$F_b = 22{,}000 - \frac{0.692}{C_b}\left(\frac{l}{r}\right)^2 \quad (4)$$

$$F_b = \frac{12{,}000{,}000}{ld/A_f} \leq 22{,}000 \text{ psi} \quad (5)$$

on the basis of A36 steel. If we use M in foot-kips, with a 24,000-psi fiber stress, the basic formula for a uniform-load moment of 1 ft-kip becomes

$$S = \frac{12 \times 1{,}000M}{24{,}000}, \text{ or } S = 0.5M$$

Expressing it as a statement rather than a formula, the section modulus is one-half of the moment in foot-kips.

This approach combined with the formula for moments given in Fig. 3.13 will greatly simplify the computation of steel beams. For instance, for a uniformly distributed load on a simple beam (Case 1),

$$M = \frac{wL^2}{8} = \frac{WL}{8} \quad (4.1)$$

$$S = 0.5M = \frac{0.5WL}{8} = \frac{WL}{16} = 0.0625WL \quad (4.2)$$

Thus we may go directly from the combination of total load and length to section modulus.

Similarly, combining the formula for Case 7, centrally concentrated load,

$$S = \frac{PL}{8} = 0.125PL \quad (4.3)$$

For the beams listed in Art. 3.10, Class II,

Table 4.2 (*Continued*)

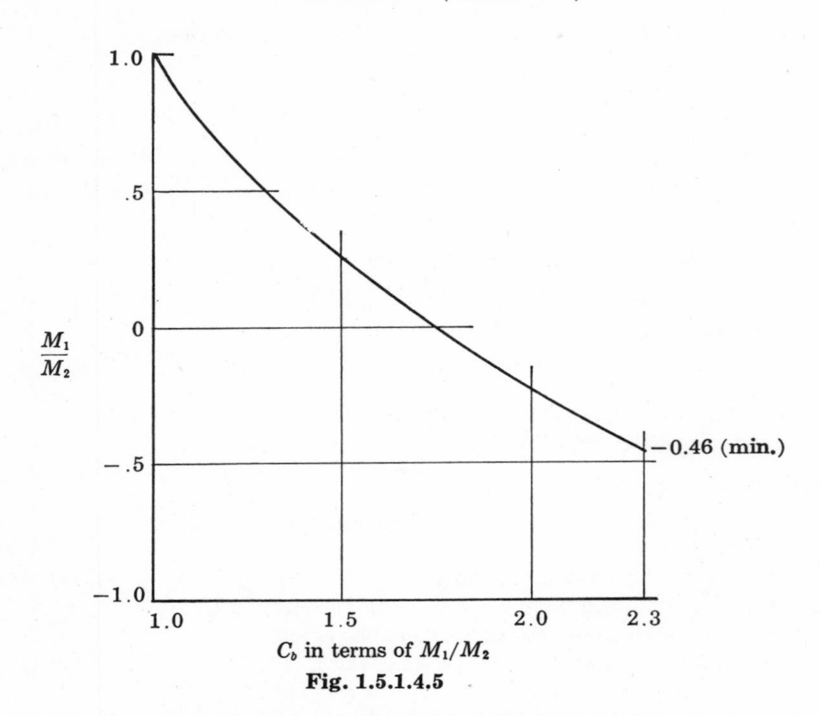

C_b in terms of M_1/M_2

Fig. 1.5.1.4.5

1.5.1.4.6 Compression for channels: Use Formula (5) above.
1.5.1.4.7 Tension and compression for large pins. . . . . . F_b = 33,000 psi
1.5.1.4.8 Tension and compression for rectangular bearing plates. . . . . F_b = 27,000 psi

1.5.1.5 Bearing

1.5.1.5.1 On milled surfaces and pins in reamed, drilled or bored holes. . . . . F_p = 33,000 psi
1.5.1.5.2 On finished stiffeners. . . . . F_p = 30,000 psi
1.5.1.5.3 On expansion rockers and rollers (in pounds per linear inch). . . . . F_p = 760*d*

$S = C(WL/8)$ using C as given in the paragraph noted.

Referring to the problem in Art. 3.10*k*,

$$S = 0.5 \times 50.625 = 25.3$$

From Table 4.1, select 12 WF 27. This could have been arrived at directly from the problem set up in Art. 3.10*k*.

$$15{,}000 + 12{,}000 = 27{,}000$$

$$S = \frac{27.0 \times 15}{16} = 25.3$$

f A fairly common type of problem encountered in practice is the determination of permissible load on a beam—the capacity of an existing structure. This may be approached directly by use of the formula $S = 0.0625WL$, writing in S and L and solving for W. It is preferable to compute the section modulus required to carry a load of 100 psf, and by prorating the actual against required section modulus, determine the load.

Problem Floor beams in a building are spaced 7 ft 6 in. apart on a 19-ft span. The beams are 18 WF 50. What total floor load per square foot will the beams carry?

$$W = 7.5 \times 19 \times 100 = 14{,}250$$

$$S = 0.0625 \times 14.25 \times 19 = 16.9$$

$$\text{Allowable } W = \frac{89.0}{16.9} \times 100 = 473 \text{ psf}$$

Table 4.3

SECTION 1.5 ALLOWABLE UNIT STRESSES $F_y = 50{,}000$ psi

1.5.1 Structural Steel

1.5.1.1 Tension

Tension on net section, except at pin holes. . . . . $F_t = 30{,}000$ psi
Tension on net section at pin holes. . . . . $F_t = 22{,}500$ psi

1.5.1.2 Shear

Shear on gross section (see Table 3-50 for reduced values for girder webs). . . . . $F_v = 20{,}000$ psi

1.5.1.3 Compression

$$C_c = 107.0$$

For values of F_a given by Formulas (1), (2) and (3) see Table 1-50.

1.5.1.4 Bending

1.5.1.4.1 Tension and compression for compact,* adequately** braced beams having an axis of symmetry in the plane of loading. . . . . $F_b = 33{,}000$ psi

1.5.1.4.2 Tension and compression for unsymmetrical rolled shapes except channels continuously braced in the region under compression stress. . . . . $F_b = 30{,}000$ psi

1.5.1.4.3 Tension and compression for box-type members not included in Sect. 1.5.1.4.1 . . . . $F_b = 30{,}000$ psi

1.5.1.4.4 Tension for other rolled shapes, built-up members and plate girders. . . . . $F_b = 30{,}000$ psi

1.5.1.4.5 Compression, except as provided by Sect. 1.5.1.4.1, 1.5.1.4.2, 1.5.1.4.3, 1.5.1.4.7 and 1.5.1.4.8: the larger value given by Formulas (4) and (5).

$$F_b = 30{,}000 - \frac{1.310}{C_b}\left(\frac{l}{r}\right)^2 \tag{4}$$

$$F_b = \frac{12{,}000{,}000}{ld/A_f} \leq 30{,}000 \text{ psi} \tag{5}$$

* $b_f/2t_f \leq 7$; $d/t \leq (60 - 85\, f_a/F_a)$, but not less than 36.
** $l_b \leq 11b_f$ and $400\, A_f/d$.

g Deflection of a steel beam. Occasionally the selection of a beam is modified by deflection. Sometimes the depth available is reduced by architectural limitations, or with light loads on a long span, the economical beam would deflect beyond the allowable limit. The allowable live load deflection is ordinarily taken at $1/360$ of the span, or

$$\Delta = \frac{L}{30} = \text{deflection in inches}$$

Live load deflection beyond this limit may crack plastered ceilings, and beams deflected beyond this point will give appreciable vibration under moving load.

The deflection of a steel beam on a simple span carrying a uniformly distributed load throughout its entire length may be obtained by dividing the coefficient from Table 2.5 for the required span by the depth of the beam in inches, and multiplying by the ratio S required/S furnished.

Table 4.3 (*Continued*)

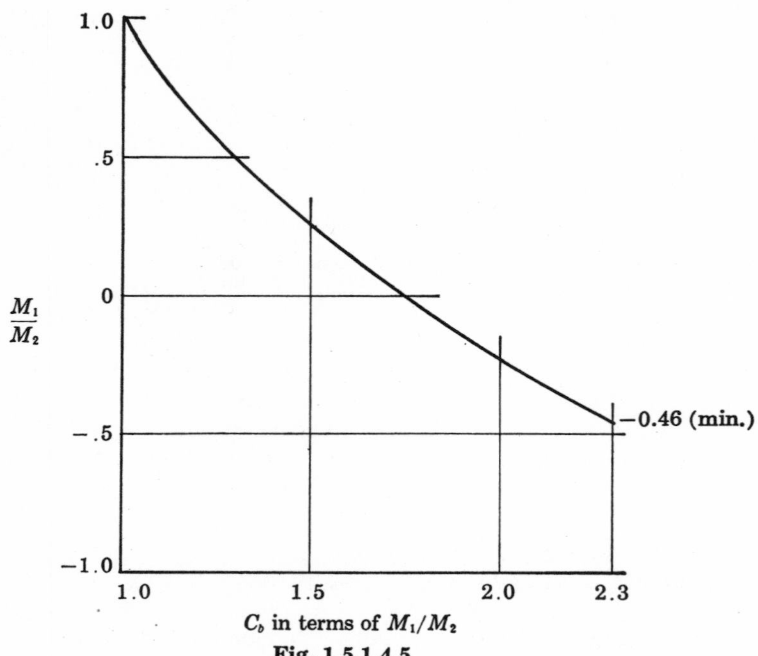

Fig. 1.5.1.4.5

1.5.1.4.6 Compression for channels: Use formula (5) above.
1.5.1.4.7 Tension and compression for large pins . . . . . . F_b = 45,000 psi
1.5.1.4.8 Tension and compression for rectangular bearing plates . . . . F_b = 37,500 psi

1.5.1.5 Bearing

1.5.1.5.1 On milled surfaces and pins in reamed, drilled or bored holes . . . . F_p = 45,000 psi
1.5.1.5.2 On finished stiffeners . . . . F_p = 40,000 psi
1.5.1.5.3 On expansion rockers and rollers (in pounds per linear inch) . . . . F_p = 1,220*d*

Problem Select a beam and compute the deflection in a steel beam carrying 1,500 lb per running foot on a span of 20 ft—no limitations as to depth.

$$W = 1.5 \times 20 = 30 \text{ kips}$$

From Formula (4.2)

$$S = 0.0625 \times 30 \times 20 = 37.5$$

From Table 4.1 select 16 B 26 ($S = 38.1$). From Table 2.5

$$\Delta = \frac{37.5}{38.1} \times \frac{9.60}{15.65} = 0.602$$

In Table 2.6 the "relative deflection" is given for different types of loading. This means the ratio of deflection of a beam loaded in accordance with the diagrams in Fig. 3.13 to the deflection under equal fiber stress of a simple supported uniformly loaded beam. In other words, for a regularly loaded beam in any of these classifications, compute the deflection as described above and multiply by the relative deflection.

Problem Compute the beam required and the deflection in a simple steel beam on a 20-ft span carrying a centrally concentrated load of 25 kips and properly supported laterally.

From Formula (4.3)

$$S = 0.125 \times 25 \times 20 = 62.5$$

From Table 4.1 select 16 WF 40 ($S = 64.4$). From Tables 2.5 and 2.6

$$\Delta = 0.8 \times \frac{62.5}{64.4} \times \frac{9.6}{16} = 0.466$$

In order to bring the beam within the deflection limits, it is not necessary to use a deeper beam. A reduction of the fiber stress by increasing S proportionately will accomplish the same thing. All these deflection tables are based on fiber stress of 24,000 psi, and in accordance with Hooke's law, the deflection is directly proportional to the fiber stress.

Problem Let us assume a uniformly distributed load of 1,500 lb per ft on 20 ft

$$W = 30{,}000$$

At 24,000-psi fiber stress,

$$S = 0.0625 \times 30 \times 20 = 37.5$$

From Table 4.1 we may use 16 B 26, 14 WF 30, 12 WF 31, 10 WF 39, or 8 WF 48. The allowable deflection would be $^{20}/_{30}$, or 0.667 in. The actual deflection for each of the above beams, allowing for actual fiber stress and using the subscript for the beam size, would be

$$\Delta_{16} = \frac{37.5}{38.1} \times \frac{9.6}{15.65} = 0.602 \text{ in.}$$

$$\Delta_{14} = \frac{37.5}{41.8} \times \frac{9.6}{13.86} = 0.625 \text{ in.}$$

$$\Delta_{12} = \frac{37.5}{39.4} \times \frac{9.6}{12} = 0.762 \text{ in.}$$

$$\Delta_{10} = \frac{37.5}{42.2} \times \frac{9.6}{9.94} = 0.842 \text{ in.}$$

$$\Delta_{8} = \frac{37.5}{43.2} \times \frac{9.6}{8.5} = 0.98 \text{ in.}$$

It will be seen that the 8-, 10-, and 12-in. beams are beyond the deflection limit. In order to use beams of these three depths, the section modulus will have to be increased approximately thus,

$$S_{12} = \frac{0.762}{0.667} \times 39.4 = 45.0$$

Use 12 WF 36.

$$S_{10} = \frac{0.842}{0.667} \times 42.2 = 53.3$$

Use 10 WF 49.

$$S_{8} = \frac{0.98}{0.667} \times 43.2 = 63.5$$

The latter value is beyond the S furnished by the heaviest 8 WF rolled, therefore 8 in. cannot be used.

It will be found that for any of the loads ordinarily encountered, uniformly distributed or concentrated on simple beams, stressed to 24,000 psi, the deflection limit will not be exceeded until L in feet is greater than $1.39d$ in inches. Therefore, a glance at the beam length and depth will tell whether it is necessary to check deflection. Thus on a span of 16 ft, using a 12-in. beam, $L = 1.33d$, and it is unnecessary to check for deflection.

h *Lateral deflection of beams.* The top flange of a simple beam is a compression member, and if the full unit fiber stress is to be utilized, the top flange must be braced against lateral deflection.

The AISC Code, Section 1.5.1.4.5 with Formulae (4) and (5) set up a formidable method of application of this lateral deflection, but the use of Tables 4.4 and 4.5 together with the charts in the 1963 AISC *Manual of Steel Construction,* pages 2–48 through 2–53, simplify this section to usable form. If the unsupported length of the top flange does not exceed the figure L_c in the tables referred to above, the full fiber stress of 24,000 psi for A36 steel or 33,000 psi for A440 or A441 steel may be used. This represents the length to which the beam may be used as a compact section. Beyond L_c and up to the value of L_u, the fiber strength must be reduced to $0.6F_y$ or 22,000 psi for A36 and 30,000 psi for high-strength steels. Beyond the listed limits given in the table, the capacity of the beam drops off rapidly and the beam required may be selected from the curved portion of the charts in the AISC Manual.

Table 4.4 Supplementary Table of Properties for Beam Sections—A36 Steel

Section	Weight per ft	S	k	V	R	R_i	N_e	L_c	L_u
36 WF	300	1,105.1	2.81	503	161	25.5	16.9	18.0	34.7
36 × 16½	280	1,031.2	2.69	468	148	23.9	16.9	18.0	32.5
	260	951.1	2.56	444	138	22.8	16.9	17.9	29.9
	245	892.5	2.44	419	129	21.7	16.9	17.9	28.1
	230	835.5	2.38	398	121	20.7	16.9	17.8	26.3
36 × 12	194	663.6	2.13	407	117	20.8	17.5	13.1	19.0
	182	621.2	2.06	382	109	19.6	17.4	13.1	17.8
	170	579.1	1.94	357	100	18.4	17.5	13.0	16.7
	160	541.0	1.88	341	95	17.6	17.5	13.0	15.5
	150	502.9	1.81	325	90	16.9	17.4	13.0	14.2
	135	438.6	1.63	308	83	16.1	17.5	....	12.1
33 WF	240	811.1	2.44	403	133	22.4	15.6	17.2	30.1
33 × 15¾	220	740.6	2.31	374	122	20.9	15.5	17.1	27.5
	200	669.6	2.19	342	110	19.3	15.5	17.1	25.0
33 × 11½	152	486.4	1.88	308	92	17.1	16.1	12.5	16.6
	141	446.8	1.75	292	86	16.3	16.1	12.5	15.1
	130	404.8	1.69	278	81	15.7	16.1	12.5	13.5
	118	358.3	1.56	264	76	15.0	16.1	....	11.7
30 WF	210	649.9	2.31	341	122	20.9	14.0	16.4	29.7
30 × 15	190	586.1	2.19	310	109	19.2	14.0	16.3	26.9
	172	528.2	2.06	284	98	17.7	14.0	16.2	24.3
30 × 10½	132	379.7	1.69	270	86	16.6	14.6	11.4	15.8
	124	354.6	1.63	256	81	15.8	14.6	11.4	14.8
	116	327.9	1.56	245	77	15.2	14.5	11.4	13.5
	108	299.2	1.50	237	74	14.8	14.5	11.4	12.2
	99	269.1	1.38	224	69	14.1	14.5	....	10.7
27 WF	177	492.8	2.13	287	110	19.6	12.5	15.3	27.9
27 × 14	160	444.5	2.06	258	99	17.8	12.5	15.2	25.3
	145	402.9	1.94	234	88	16.2	12.5	15.1	23.1
27 × 10	114	299.2	1.63	225	79	15.4	13.0	10.9	15.6
	102	266.3	1.56	203	71	14.0	13.0	10.9	13.9
	94	242.8	1.44	191	65	13.2	13.0	10.8	12.6
	84	211.7	1.38	179	61	12.5	13.0	10.8	10.8
24 WF	160	413.5	2.00	235	97	17.7	11.3	15.3	29.3
24 × 14	145	372.5	1.88	216	88	16.4	11.3	15.2	26.6
	130	330.7	1.75	199	80	15.3	11.3	15.2	23.6
24 × 12	120	299.1	1.69	196	78	15.0	11.4	13.1	21.0
	110	274.4	1.63	179	71	13.8	11.3	13.0	19.4
	100	248.9	1.56	163	64	12.6	11.3	13.0	17.6
24 × 9	94	220.9	1.44	182	69	13.9	11.6	9.8	14.8
	84	196.3	1.38	164	62	12.7	11.6	9.8	13.1
	76	175.4	1.25	153	56	11.9	11.6	9.7	11.7
	68	153.1	1.19	143	53	11.2	11.5	9.7	10.0

Table 4.4 Supplementary Table of Properties for Beam Sections—A36 Steel (*Continued*)

Section	Weight per ft	S	k	V	R	R_i	N_e	L_c	L_u
21 WF	142	317.2	1.88	205	96	17.8	9.6	14.2	30.5
21 × 13	127	284.1	1.75	181	83	15.9	9.7	14.1	27.5
	112	249.6	1.63	160	73	14.2	9.7	14.1	24.3
21 × 9	96	197.6	1.56	176	79	15.5	9.8	9.8	18.2
	82	168.0	1.44	151	67	13.5	9.8	9.7	15.5
21 × 8¼	73	150.7	1.31	140	59	12.3	10.1	9.0	13.1
	68	139.9	1.25	132	55	11.6	10.1	9.0	12.2
	62	126.4	1.19	122	51	10.8	10.1	8.9	11.0
	55	109.7	1.06	113	46	10.1	10.1	8.9	9.4
18 WF	114	220.1	1.69	159	83	16.1	8.2	12.8	28.8
18 × 11¾	105	202.2	1.63	147	77	15.0	8.2	12.8	26.6
	96	184.4	1.50	135	69	13.8	8.3	12.7	24.4
18 × 8¾	85	156.1	1.50	140	71	14.2	8.3	9.6	19.9
	77	141.7	1.38	125	63	12.8	8.4	9.5	18.3
	70	128.2	1.31	114	57	11.8	8.4	9.5	16.6
	64	117.0	1.25	104	52	10.9	8.3	9.4	15.2
18 × 7½	60	107.8	1.19	110	53	11.2	8.6	8.2	13.1
	55	98.2	1.13	102	49	10.5	8.6	8.2	11.9
	50	89.0	1.06	93	44	9.7	8.6	8.1	10.8
	45	78.9	1.00	87	41	9.0	8.6	8.1	9.5
16 WF	96	166.1	1.63	127	74	14.4	7.1	12.5	28.1
16 × 11½	88	151.3	1.50	118	68	13.6	7.2	12.5	25.7
16 × 8½	78	127.8	1.50	125	71	14.3	7.3	9.3	20.9
	71	115.9	1.38	114	64	13.1	7.3	9.3	19.1
	64	104.2	1.31	103	58	12.0	7.3	9.2	17.3
	58	94.1	1.25	94	52	11.0	7.3	9.2	15.6
16 × 7	50	80.7	1.13	90	47	10.3	7.6	7.7	12.4
	45	72.4	1.06	81	43	9.3	7.6	7.6	11.2
	40	64.4	1.00	71	37	8.3	7.6	7.6	10.0
	36	56.3	0.94	69	36	8.1	7.6	7.6	8.6
14 WF	119	189.4	1.56	120	78	15.4	6.2	15.9	42.9
14 × 14½	111	176.3	1.50	113	73	14.6	6.2	15.8	40.2
	103	163.6	1.44	102	66	13.4	6.2	....	37.9
	95	150.6	1.38	95	61	12.6	6.2	....	35.0
	87	138.1	1.31	85	55	11.3	6.2	....	32.5
14 × 12	84	130.9	1.38	93	59	12.2	6.2	13.0	29.9
	78	121.1	1.31	87	56	11.6	6.2	13.0	27.9
14 × 10	74	112.3	1.38	93	59	12.2	6.2	10.9	25.3
	68	103.0	1.31	85	54	11.3	6.2	10.9	23.3
	61	92.2	1.25	76	48	10.2	6.2	10.8	21.0
14 × 8	53	77.8	1.25	75	47	10.0	6.2	8.7	17.3
	48	70.2	1.19	68	43	9.2	6.2	8.7	15.7
	43	62.7	1.13	61	38	8.3	6.2	8.7	14.0

Table 4.4 Supplementary Table of Properties for Beam Sections— A36 Steel (*Continued*)

Section	Weight per ft	S	k	V	R	R_i	N_e	L_c	L_u
14 × 6¾	38	54.6	1.00	64	38	8.5	6.6	7.3	11.2
	34	48.5	0.94	58	34	7.7	6.6	7.3	9.9
	30	41.8	0.88	54	32	7.3	6.6	7.3	8.5
12 WF 12 × 12	85	115.7	1.38	90	65	13.4	5.3	13.1	35.2
	79	107.1	1.31	84	61	12.7	5.3	13.1	32.7
	72	97.5	1.25	76	55	11.6	5.3	....	29.9
	65	88.0	1.19	69	49	10.5	5.3	....	27.2
12 × 10	58	78.1	1.25	63	46	9.7	5.3	10.8	23.9
	53	70.7	1.19	60	44	9.3	5.3	10.8	21.7
12 × 8	50	64.7	1.25	66	48	10.0	5.3	8.8	19.3
	45	58.2	1.19	59	43	9.1	5.3	8.7	17.5
	40	51.9	1.13	51	37	7.9	5.3	8.7	15.7
12 × 6½	36	45.9	0.94	54	37	8.2	5.6	7.1	13.2
	31	39.4	0.88	46	31	7.2	5.6	7.1	11.4
	27	34.1	0.81	42	28	6.5	5.6	7.0	9.9
10 WF 10 × 10	66	73.7	1.25	69	59	12.3	4.3	11.0	33.2
	60	67.1	1.19	62	53	11.2	4.3	10.9	30.5
	54	60.4	1.13	54	46	9.9	4.3	10.9	27.9
	49	54.6	1.06	49	42	9.2	4.3	....	25.4
10 × 8	45	49.1	1.13	51	44	9.5	4.3	8.7	22.3
	39	42.2	1.06	46	39	8.6	4.3	8.7	19.3
	33	35.0	0.94	41	35	7.9	4.3	....	16.1
10 × 5¾	29	30.8	0.88	43	34	7.8	4.6	6.3	12.9
	25	26.4	0.81	37	29	6.8	4.6	6.2	11.1
	21	21.5	0.69	34	27	6.5	4.6	6.2	9.0
8 WF 8 × 8	35	31.1	0.88	37	37	8.5	3.5	8.7	22.2
	31	27.4	0.81	33	34	7.8	3.5	....	19.7
8 × 6½	28	24.3	0.81	33	33	7.7	3.5	7.1	17.1
	24	20.8	0.81	28	29	6.6	3.4	7.0	14.8
8 × 5½	20	17.0	0.69	29	28	6.7	3.7	5.7	11.1
	17	14.1	0.63	27	26	6.2	3.7	5.7	9.2
B 16 × 5½	31	47.0	0.94	63.2	32.9	7.4	7.6	6.0	7.0
	26	38.1	0.81	56.7	29.1	6.8	7.6	....	5.5
B 14 × 5	26	34.9	0.88	51.4	30.1	6.9	6.6	5.4	6.9
	22	28.8	0.81	45.8	26.8	6.2	6.6	5.4	5.6
B 14 × 4	17.2	21.0	0.56	42.6	23.0	5.7	7.0	....	3.5

Table 4.4 Supplementary Table of Properties for Beam Sections—A36 Steel (*Continued*)

Section	Weight per ft	S	k	V	R	R_i	N_e	L_c	L_u
B									
12 × 4	22	25.3	0.75	46.4	29.8	7.0	5.9	4.4	6.3
	19	21.4	0.69	42.3	27.1	6.5	5.8	4.3	5.2
	16.5	17.5	0.63	40.0	25.6	6.2	5.8	....	4.1
	14	14.8	0.56	34.5	21.9	5.4	5.8	....	3.4
M									
10 × 5¾	29.1	26.6	0.81	60.9	49.5	11.5	4.5	6.4	10.6
	22.9	23.6	0.81	34.4	27.9	6.5	4.5	6.2	10.3
	21	21.7	0.75	34.5	27.5	6.5	4.6	6.2	9.2
B									
10 × 4	19	18.8	0.69	37.2	28.3	6.8	4.8	4.4	7.0
	17	16.2	0.63	35.2	26.7	6.5	4.8	4.3	5.9
	15	13.8	0.56	33.4	25.2	6.2	4.8	4.3	4.9
	11.5	10.5	0.50	25.8	19.4	4.9	4.8	....	3.7
M									
8 × 6½	28	22.5	0.88	45.2	46.1	10.5	3.4	7.2	15.1
	24	21.0	0.88	27.8	28.4	6.5	3.4	7.0	14.7
M									
8 × 5¼	22.5	17.1	0.75	43.5	43.0	10.1	3.5	5.8	10.8
	20	15.2	0.69	40.6	39.6	9.5	3.6	5.8	9.3
	18.5	15.5	0.75	26.7	26.4	6.2	3.5	5.7	10.5
	17	14.0	0.69	27.8	27.1	6.5	3.6	5.7	9.1
B									
8 × 4	15	11.8	0.63	28.8	27.3	6.6	3.7	4.3	7.1
	13	9.88	0.56	26.7	25.2	6.2	3.7	4.3	5.8
	10	7.79	0.50	19.5	18.4	4.6	3.7	....	4.6
B									
6 × 4	16	10.1	0.69	23.6	29.4	7.0	2.7	4.4	11.8
	12	7.24	0.56	20.0	25.2	6.2	2.7	4.3	8.4
	8.5	5.07	0.44	14.4	18.1	4.6	2.7	....	6.0
Jr. Beams									
12 × 3	11.8	12.0	0.50	30.5	18.9	4.7	5.9	....	2.6
10 × 2¾	9.0	7.8	0.44	22.5	16.5	4.2	4.9	....	2.5
8 × 2¼	6.5	4.7	0.38	15.7	14.1	3.6	3.9	....	2.4
7 × 2⅛	5.5	3.5	0.38	12.8	13.2	3.4	3.4	2.3	2.4
6 × 1⅞	4.4	2.4	0.38	9.9	11.9	3.1	2.8	2.0	2.4
American Standard Beams									
24 × 7⅞	120	250.9	1.94	278	117	21.5	11.0	8.7	16.8
	105.9	234.3	1.94	218	92	16.9	11.0	8.5	16.5
24 × 7	100	197.6	1.63	260	103	20.2	11.3	7.9	11.9
	90	185.8	1.63	217	86	16.8	11.3	7.7	11.7
	79.9	173.9	1.63	174	69	13.5	11.3	7.6	11.5

Table 4.4 Supplementary Table of Properties for Beam Sections—A36 Steel (*Continued*)

Section	Weight per ft	S	k	V	R	R_i	N_e	L_c	L_u
20 × 7	95	160.0	1.75	232	113	21.6	9.0	7.8	15.0
	85	150.2	1.75	189	93	17.6	9.0	7.6	14.7
20 × 6¼	75	126.3	1.56	186	88	17.3	9.2	6.9	11.4
	65.4	116.9	1.56	145	68	13.5	9.2	6.8	11.2
18 × 6	70	101.9	1.38	186	94	19.2	8.3	6.8	10.9
	54.7	88.4	1.38	120	61	12.4	8.3	6.5	10.4
15 × 5½	50	64.2	1.25	120	71	14.9	6.8	6.1	10.6
	42.9	58.9	1.25	89	53	11.1	6.8	6.0	10.4
12 × 5¼	50	50.3	1.31	120	89	18.5	5.1	5.9	13.7
	40.8	44.8	1.31	80	60	12.4	5.1	5.7	13.1
12 × 5	35	37.8	1.13	74	53	11.6	5.3	5.5	10.5
	31.8	36.0	1.13	61	44	9.5	5.3	5.4	10.3
10 × 4⅝	35.0	29.2	1.00	86	72	16.0	4.4	5.4	11.0
	25.4	24.4	1.00	45	38	8.4	4.4	5.0	10.4
8 × 4	23.0	16.0	0.88	51	52	11.9	3.4	4.5	10.1
	18.4	14.2	0.88	31	32	7.3	3.4	4.3	9.7
7 × 3⅝	20.0	12.0	0.81	46	52	12.2	2.9	4.2	9.8
	15.3	10.4	0.81	25	29	6.8	2.9	4.0	9.3
6 × 3⅜	17.25	8.7	0.75	40.5	53.4	12.6	2.5	3.9	9.7
	12.5	7.3	0.75	20.0	26.4	6.2	2.5	3.6	9.1
5 × 3	14.75	6.0	0.69	35.8	55.9	13.3	2.0	3.6	9.7
	10.0	4.8	0.69	15.2	23.7	5.7	2.0	3.3	8.9
4 × 2⅝	9.5	3.3	0.63	18.9	36.3	8.8	1.5	3.0	9.3
	7.7	3.0	0.63	11.0	21.2	5.1	1.5	2.9	8.9
3 × 2⅜	7.5	1.9	0.56	15.2	38.3	9.4	1.0	2.7	9.9
	5.7	1.7	0.56	7.4	18.6	4.6	1.0	2.5	9.2
American Standard Channels									
18 × 4	58	74.5	1.31	183	91	18.9	8.4	....	6.6
	51.9	69.1	1.31	157	78	16.2	8.4	....	6.5
	45.8	63.7	1.31	131	65	13.5	8.4	....	6.3
	42.7	61.0	1.31	117	58	12.2	8.4	....	6.2
15 × 3⅜	50	53.6	1.31	156	93	19.3	6.7	....	7.3
	40	46.2	1.31	113	68	14.0	6.7	....	6.9
	33.9	41.7	1.31	87	52	10.8	6.7	....	6.7
12 × 3	30	26.9	1.06	89	63	13.8	5.4	....	6.0
	25	23.9	1.06	67	48	10.4	5.4	....	5.9
	20.7	21.4	1.06	49	34	7.6	5.4	....	5.6

Table 4.4 Supplementary Table of Properties for Beam Sections—A36 Steel (*Continued*)

Section	Weight per ft	S	k	V	R	R_i	N_e	L_c	L_u
10 × 2⅝	30	20.6	0.94	98	81	18.2	4.4	....	6.0
	25	18.1	0.94	76	63	14.2	4.4	....	5.7
	20	15.7	0.94	55	45	10.2	4.4	....	5.4
	15.3	13.4	0.94	35	29	6.5	4.4	....	5.2
9 × 2½	20	13.5	0.88	58	53	12.1	4.0	....	5.5
	15	11.3	0.88	37	34	7.7	4.0	....	5.2
	13.4	10.5	0.88	30	27	6.2	4.0	....	5.1
8 × 2¼	18.75	10.9	0.81	56	57	13.1	3.5	....	5.6
	13.75	9.0	0.81	35	35	8.2	3.5	....	5.2
	11.5	8.1	0.81	26	26	5.9	3.5	....	5.0
7 × 2⅛	14.75	7.7	0.81	43	49	11.3	2.9	....	5.5
	12.25	6.9	0.81	32	37	8.5	2.9	....	5.2
	9.8	6.0	0.81	21	24	5.7	2.9	....	5.0
6 × 2	13.0	5.8	0.75	38.0	50.1	11.8	2.5	....	5.6
	10.5	5.0	0.75	27.3	36.0	8.5	2.5	....	5.3
	8.2	4.3	0.75	17.4	23.0	5.4	2.5	....	5.0
5 × 1¾	9.0	3.5	0.69	23.6	36.7	8.8	2.0	....	5.5
	6.7	3.0	0.69	13.8	21.5	5.1	2.0	....	5.1
4 × 1⅝	7.25	2.3	0.63	18.6	35.6	8.6	1.5	....	5.8
	5.4	1.9	0.63	10.4	20.0	4.9	1.5	....	5.3
3 × 1½	6.0	1.4	0.63	15.5	39.6	9.6	1.0	....	6.6
	5.0	1.2	0.63	11.2	28.7	7.0	1.0	....	6.2
	4.1	1.1	0.63	7.4	18.9	4.6	1.0	....	5.8
Jr. Channels									
12 × 1½	10.6	9.3	0.63	33.1	21.2	5.1	5.8	....	1.8
10 × 1½	8.4	6.5	0.50	24.7	18.4	4.6	4.9	....	1.9
10 × 1⅛	6.5	4.4	0.38	21.8	15.7	4.1	5.0	....	1.1

Table 4.5 Supplementary Table of Properties for Beam Sections—A440 and A441 Steel

Section	Weight per ft	S	k	V	R	R_i	N_e	L_c	L_u
36 WF 36 × 16½	300	1,105.1	2.81	639	204	32.4	21.5	16.0	27.7
	280	1,031.2	2.69	594	188	30.4	21.5	15.9	25.9
	260	951.1	2.56	564	175	29.0	21.5	15.9	23.8
	245	892.5	2.44	532	164	27.6	21.5	15.8	22.4
	230	835.5	2.38	505	154	26.3	21.5	15.8	21.0
36 × 12	194	663.6	2.13	562	161	28.7	24.2	11.1	13.9
	182	621.2	2.06	527	150	27.0	24.1	11.1	13.1
	170	579.1	1.94	493	138	25.4	24.2	11.0	12.2
	160	541.0	1.88	471	131	24.3	24.2	11.0	11.3
	150	502.9	1.81	449	124	23.3	24.1	....	10.4
	135	438.6	1.63	260	104	20.1	24.2	....	8.9
33 WF 33 × 15¾	240	811.1	2.44	512	169	28.4	19.8	15.2	24.0
	220	740.6	2.31	474	155	26.5	19.7	15.2	22.0
	200	669.6	2.19	437	140	24.5	19.7	15.1	19.9
33 × 11½	152	486.4	1.88	425	127	23.6	22.2	10.6	12.2
	141	446.8	1.75	403	119	22.5	22.2	10.6	11.1
	130	404.8	1.69	384	112	21.7	22.2	....	9.9
	118	358.3	1.56	330	95	18.8	20.1	....	8.6
30 WF 30 × 15	210	649.9	2.31	471	168	28.8	19.3	13.9	21.8
	190	586.1	2.19	428	150	26.4	19.3	13.8	19.7
	172	528.2	2.06	392	135	24.4	19.3	13.7	17.8
30 × 10½	132	379.7	1.69	373	119	22.9	20.1	9.7	11.6
	124	354.6	1.63	353	112	21.8	20.1	9.6	10.8
	116	327.9	1.56	338	106	21.0	20.0	9.6	9.9
	108	299.2	1.50	327	102	20.4	20.0	....	8.9
	99	269.1	1.38	281	86	17.6	18 1	....	7.9
27 WF 27 × 14	177	492.8	2.13	396	152	27.0	17.3	12.9	20.4
	160	444.5	2.06	356	137	24.6	17.3	12.9	18.5
	145	402.9	1.94	323	121	22.4	17.3	12.8	16.9
27 × 10	114	299.2	1.63	311	109	21.3	17.9	9.2	11.5
	102	266.3	1.65	280	98	19.3	17.9	9.2	10.2
	94	242.8	1.44	264	90	18.2	17.9	9.2	9.2
	84	211.7	1.38	224	76	15.6	16.3	....	7.9
24 WF 24 × 14	160	413.5	2.00	324	134	24.4	15.6	12.9	21.5
	145	372.5	1.88	298	121	22.6	15.6	12.9	19.5
	130	330.7	1.75	249	100	19.1	14.1	....	17.3
24 × 12	120	299.1	1.69	270	108	20.7	15.7	11.1	15.4
	110	274.4	1.63	247	98	17.4	15.6	11.0	14.2
	100	248.9	1.56	204	80	15.8	14.1	....	12.9
24 × 9	94	220.9	1.44	251	95	19.2	16.0	8.3	10.9
	84	196.3	1.38	226	86	17.5	16.0	8.3	9.6
	76	175.4	1.25	211	77	16.4	16.0	8.2	8.5
	68	153.1	1.19	179	66	13.5	14.4	....	7.3

Table 4.5 Supplementary Table of Properties for Beam Sections—A440 and A441 Steel (*Continued*)

Section	Weight per ft	S	k	V	R	R_i	N_e	L_c	L_u
21 WF 21 × 13	142	317.2	1.88	283	132	24.6	13.2	12.0	22.4
	127	284.1	1.75	250	115	21.9	13.4	12.0	20.2
	112	249.6	1.63	200	91	17.8	12.1	....	17.8
21 × 9	96	197.6	1.56	242	109	21.9	13.5	8.3	13.3
	82	168.0	1.44	208	92	18.6	13.5	8.2	11.4
21 × 8¼	73	150.7	1.31	193	81	17.4	13.9	7.6	9.6
	68	139.9	1.25	182	76	16.0	13.9	7.6	8.9
	62	126.4	1.19	168	70	14.9	13.9	7.6	8.0
	55	109.7	1.06	141	58	12.6	12.6	....	6.9
18 WF 18 × 11¾	114	220.1	1.69	219	115	22.2	11.2	10.8	21.1
	105	202.2	1.63	203	106	20.7	11.2	10.8	19.5
	96	184.4	1.50	186	95	19.0	11.3	10.8	17.9
18 × 8¾	85	156.1	1.50	193	98	19.6	11.3	8.1	14.6
	77	141.7	1.38	173	87	17.7	11.4	8.1	13.4
	70	128.2	1.31	157	79	16.3	11.4	8.0	12.2
	64	117.0	1.25	144	72	15.0	11.3	8.0	11.1
18 × 7½	60	107.8	1.19	152	73	15.5	11.9	6.9	9.6
	55	98.2	1.13	141	68	14.5	11.9	6.9	8.7
	50	89.0	1.06	128	61	13.4	11.9	6.9	7.9
	45	78.9	1.00	109	51	11.3	10.8	....	7.0
16 WF 16 × 11½	96	166.1	1.63	127	102	19.9	9.8	10.6	20.6
	88	151.3	1.50	118	94	18.8	9.7	10.5	18.8
16 × 8½	78	127.8	1.50	173	98	19.7	10.1	7.9	15.4
	71	115.9	1.38	157	88	18.1	10.1	7.8	14.0
	64	104.2	1.31	142	80	16.6	10.1	7.8	12.7
	58	94.1	1.25	130	72	14.2	10.1	7.8	11.5
16 × 7	50	80.7	1.13	124	65	14.2	10.5	6.5	9.1
	45	72.4	1.06	112	59	12.8	10.5	6.5	8.2
	40	64.4	1.00	98	51	11.5	10.5	6.4	7.3
	36	56.3	0.94	86	45	10.1	9.5	....	6.3
14 WF 14 × 14½	119	189.4	1.56	150	98	19.3	7.8	....	31.4
	111	176.3	1.50	144	91	18.3	7.8	....	29.5
	103	163.6	1.44	128	83	16.8	7.8	....	27.8
	95	150.6	1.38	119	76	15.8	7.8	....	25.6
	87	138.1	1.31	106	69	14.1	7.8	....	23.8
14 × 12	84	130.9	1.38	116	74	15.3	7.8	....	21.9
	78	121.1	1.31	109	70	14.4	7.8	....	20.4
14 × 10	74	112.3	1.38	128	81	16.8	8.6	9.2	18.5
	68	103.0	1.31	117	75	15.6	8.6	9.2	17.1
	61	92.2	1.25	95	60	12.8	7.2	....	15.4

Table 4.5 Supplementary Table of Properties for Beam Sections—A440 and A441 Steel (*Continued*)

Section	Weight per ft	S	k	V	R	R_i	N_e	L_c	L_u
14 × 8	53	77.8	1.25	104	65	13.8	8.6	7.4	12.7
	48	70.2	1.19	94	59	12.7	8.6	7.4	11.5
	43	62.7	1.13	76	48	10.4	7.2	....	10.3
14 × 6¾	38	54.6	1.00	88	52	11.7	9.1	6.2	8.2
	34	48.5	0.94	73	43	9.6	8.3	....	7.3
	30	41.8	0.88	68	40	9.1	8.3	....	6.2
12 WF 12 × 12	85	115.7	1.38	113	81	16.8	6.6	....	25.8
	79	107.1	1.31	105	76	15.9	6.6	....	24.0
	72	97.5	1.25	95	69	14.5	6.6	....	21.9
	65	88.0	1.19	86	61	13.1	6.6	....	20.0
12 × 10	58	78.1	1.25	79	58	12.1	6.6	....	17.5
	53	70.7	1.19	75	55	11.6	6.6	....	15.9
12 × 8	50	64.7	1.25	91	66	13.8	7.3	7.4	14.2
	45	58.2	1.19	81	59	12.6	7.3	7.4	12.8
	40	51.9	1.13	64	46	9.9	6.6	....	11.5
12 × 6½	36	45.9	0.94	68	51	11.3	7.7	6.0	9.7
	31	39.4	0.88	63	43	9.9	7.7	6.0	8.4
	27	34.1	0.81	53	35	8.1	7.0	....	7.2

There is no clearly defined rule as to what constitutes sufficient lateral support and the designer must make his own assumption. It is the practice of the author's office to use the following rules:

1. For a concrete slab, haunching or setting of the slab forms so that they come below the top flange as shown in Fig. 6.6 or Fig. 4.1 will obviously furnish ample support. The use of a composite system as described in Art. 6.50 and shown in Fig. 6.7 will also provide support. If the slab rests on the beam but does not grip the top flange, a steel form welded to the flange and left in place is also acceptable. A slab resting on the beam with nothing but frictional resistance or the bond of the concrete to the steel beam is not satisfactory lateral support.

2. In the case of steel joists, tack welds to each joist will brace the beam. Steel clips will not, because they may loosen up.

3. Where gypsum roof deck is used, the welding of the tees will provide proper support.

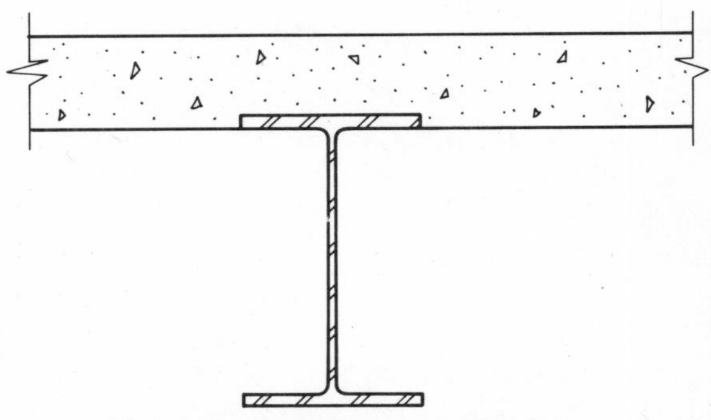

Fig. 4.1 Lateral Support of Beam Normal Slab Construction

4. Where precast concrete is used, in long span planks such as double tee or Flexicore slabs, welding clips may be cast into the slabs for tack welds to the steel beams. Where short-span concrete plank systems are used, properly attached to the concrete and to the beam flanges by means of clips as indicated in Fig. 6.7, they will furnish satisfactory support. Clips attached in one direction only will not.

5. In order to provide proper rigidity with

wood joists, a nailer must be bolted (not clipped) to the top flange of the beam to which the joists may be nailed. Resting the joist on the beam and driving and bending nails around the beam flange will not provide rigidity since the wood will shrink and permit movement of the flange.

i Web shear and buckling. Web shear or buckling of the web will control the selection of the beam infrequently; in general, when:

The beam is short but very heavily loaded.

A heavy load occurs very close to one end of the beam.

A column is carried on a beam or on a grillage.

A heavily loaded beam bears over the top of another beam and is placed at right angles to it.

Under any of these conditions, the beam selected should be investigated for web shear and for buckling before being chosen. Certain applications of this statement will be given later. The AISC Code, Section 1.5.1.2, sets the allowable shear on the gross section of beam webs as

$$F_v = 0.40F_y$$

which for A36 steel is 14,500 psi and for A440 and A441 steel, 20,000 psi. These values for the beams normally used are listed in Tables 4.4 and 4.5 in the column headed V.

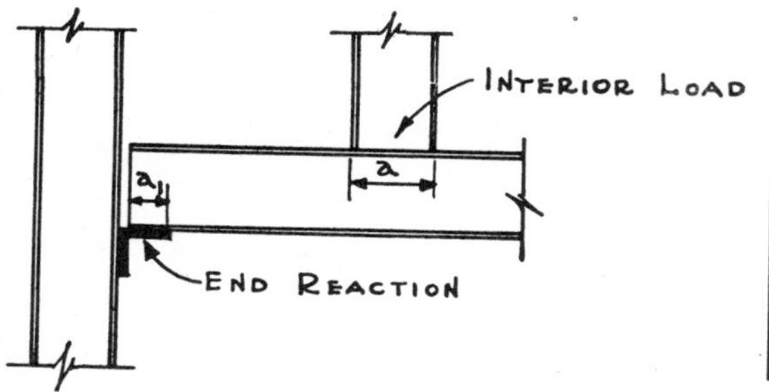

Fig. 4.2 Web Shear and Buckling

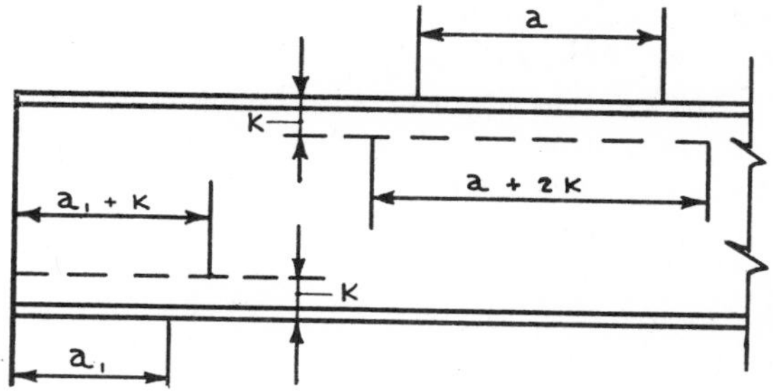

Fig. 4.3 Distribution for Web Shear and Buckling

Web buckling or crippling as indicated in Fig. 4.2 and Fig. 4.3 is limited in the AISC Code as follows:

1.10.10 *Web crippling*

1.10.10.1 Webs of beams and welded plate girders shall be so proportioned that the compressive stress at the web toe of the fillets, resulting from concentrated loads not supported by bearing stiffeners, shall not exceed the value of $0.75F_y$ pounds per square inch allowed in Sect. 1.5.1; otherwise, bearing stiffeners shall be provided. The governing formulas shall be:

For interior loads,

$$\frac{R}{t(N + 2k)} = \text{not over } 0.75F_y \text{ pounds per square inch} \quad (13)$$

For end-reactions,

$$\frac{R}{t(N + k)} = \text{not over } 0.75F_y \text{ pounds per square inch} \quad (14)$$

where R = concentrated load or reaction, in pounds
t = thickness of web, in inches
N = length of bearing in inches (not less than k for end reactions)
k = distance from outer face of flange to web toe of fillet, in inches

This section is not required for the design of the beam itself, but for details only. If the figures for R and R_i in the aforesaid tables are exceeded, the web can be strengthened against crippling by the use of stiffeners as described in Art. 4.20.

4.11 Design of steel details

While it is obvious that the design of steel details is as important as the design of the member itself, these details are ordinarily selected by the steel company detailer from standards set up by the American Institute of Steel Construction. Thus, while it is the duty of the designer to check the shop drawings (see Art. 11.16), it is not necessary to repeat all the detail information given in the

AISC Manual relative to all of the standard types of connections. The AISC Manual, Part 4, covers the entire subject of connections and contains complete tables of framed beam connections, both for bolted or riveted and for welded connections. Where dictated by economy, the fabricator may sometimes choose to weld shop connections and to bolt field connections.

4.12 Special steel details

a *Reduced end depth of beam.* This problem is frequently encountered in detailing beams. It is shown in Fig. 4.4. The solution involves a problem in moment, shear,

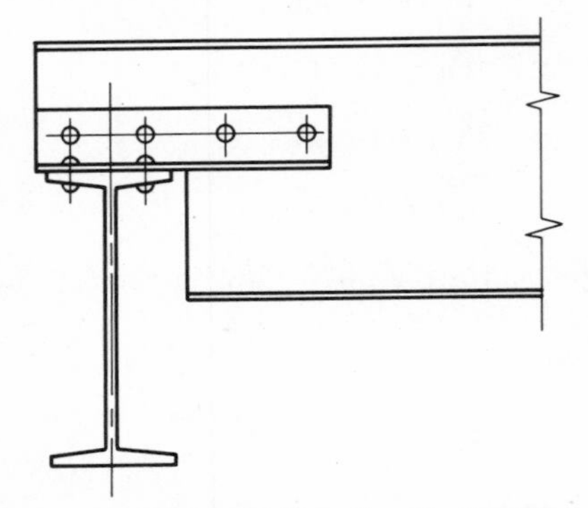

Fig. 4.4 Reduced Depth—End of Beam

properties of complex sections, and riveting. Assume a 16 WF 36 of A36 steel bearing on an 18 WF 55, carrying a reaction of 20,000 lb, and with half the web of the 16-in. beam cut away. From the table for shearing value of beams Table 4.4, the shear on a 16 WF 36 is 69 kips for the full web, or 34.5 kips for the half web. The reaction of 20 kips must be developed in the rivets on either side of the cut. The beam has a web thickness of 0.299 in., and from the table for ¾-in. rivets in double shear on enclosed bearing (Table 9.1), each rivet is good for 10.9 kips. This would require the use of two rivets. These can be located as shown in Fig. 4.4 and would require a 3½-in. upstanding leg to take them.

The moment required at the cut would be 20.0×4 in. = 80.0 in.-kips. This would require a section modulus of $^{80}/_{22} = 3.64$. From observation, the least angle which should be used is $3\frac{1}{2} \times 3\frac{1}{2} \times \frac{5}{16}$ in., which would give a beam of greater section than an 8 WF 17, which has an S of 14.1. Therefore it is not necessary to compute the S of the section, which is made up of two angles $3\frac{1}{2} \times 3\frac{1}{2} \times \frac{5}{16}$ in. and half of a 16 WF 36. However, in some problems it might be necessary to compute the section modulus of a compound section as described in Art. 1.31.

b *Angle legs in bending—fixed ends.* When an angle forms part of a wind-bracing bracket, a hanger, or another similar mem-

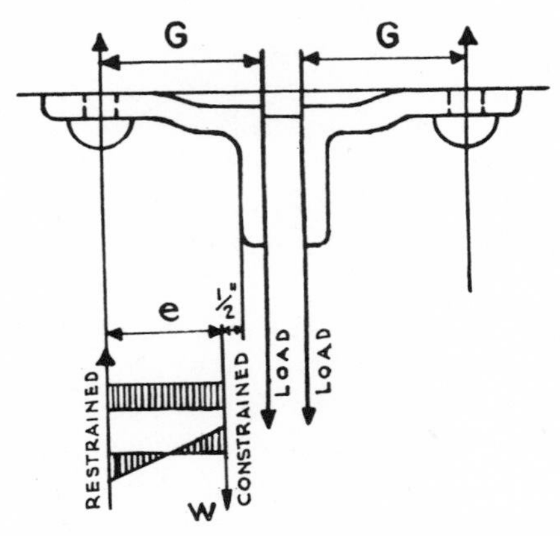

Fig. 4.5 Angle Legs in Bending

ber where the faces are held at right angles but are caused to bend as indicated in Fig. 4.5, the angle leg or T leg is computed as a cantilever with the load applied on half the lever arm e. Figures 4.6 and 4.7 show practical applications of this condition.

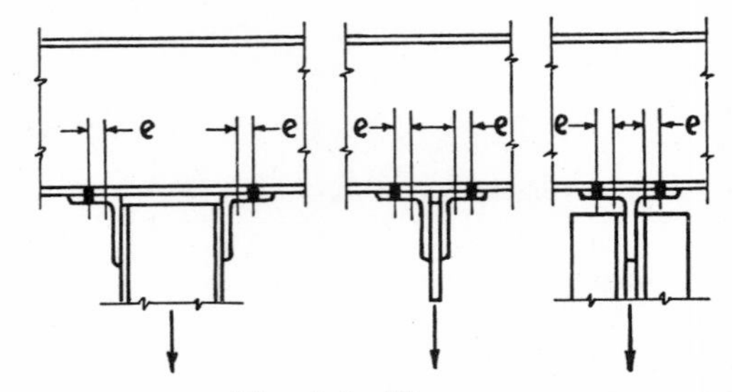

Fig. 4.6 Hangers

For a direct hanger load on two angles or one T, the total load per inch of length of angle or T for a fiber stress of 22,000 psi is $P = 14.67t^2/e$. The allowable load in kips per inch on two angles or a structural T according to this formula and the suggested method of solution is given in Table 4.6, all

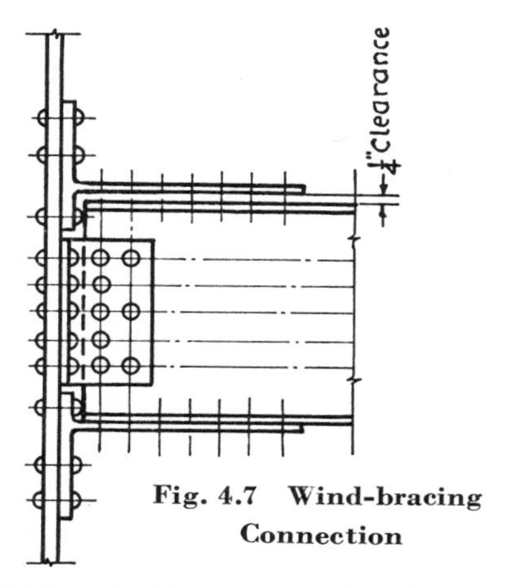

Fig. 4.7 Wind-bracing Connection

of which is reproduced from the AISC Manual.

Where the pull is one-sided as in an angle wind-bracing detail similar to Fig. 4.7, but with cap and seat angles, the allowable load reduces to half as much, since only one side is resisting the tension; hence $P = 7.33t^2/e$.

Problem A problem in the design of a wind bracing connection will indicate the use of Table 4.6.

Assume a wind-bracing connection designed to take wind moments in either direction of 250 ft-kips, allowing normal working

Table 4.6 Hanger and Bracket Connections Structural Tee or Double Angle

Allowable loads in kips per inch

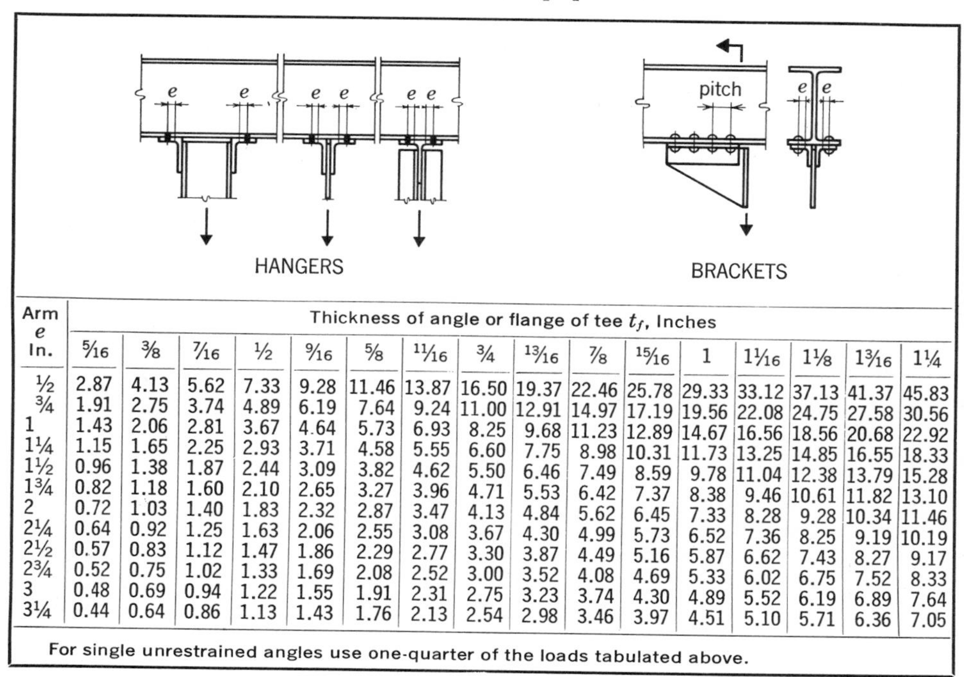

Arm e In.	Thickness of angle or flange of tee t_f, Inches															
	5/16	3/8	7/16	1/2	9/16	5/8	11/16	3/4	13/16	7/8	15/16	1	1 1/16	1 1/8	1 3/16	1 1/4
1/2	2.87	4.13	5.62	7.33	9.28	11.46	13.87	16.50	19.37	22.46	25.78	29.33	33.12	37.13	41.37	45.83
3/4	1.91	2.75	3.74	4.89	6.19	7.64	9.24	11.00	12.91	14.97	17.19	19.56	22.08	24.75	27.58	30.56
1	1.43	2.06	2.81	3.67	4.64	5.73	6.93	8.25	9.68	11.23	12.89	14.67	16.56	18.56	20.68	22.92
1 1/4	1.15	1.65	2.25	2.93	3.71	4.58	5.55	6.60	7.75	8.98	10.31	11.73	13.25	14.85	16.55	18.33
1 1/2	0.96	1.38	1.87	2.44	3.09	3.82	4.62	5.50	6.46	7.49	8.59	9.78	11.04	12.38	13.79	15.28
1 3/4	0.82	1.18	1.60	2.10	2.65	3.27	3.96	4.71	5.53	6.42	7.37	8.38	9.46	10.61	11.82	13.10
2	0.72	1.03	1.40	1.83	2.32	2.87	3.47	4.13	4.84	5.62	6.45	7.33	8.28	9.28	10.34	11.46
2 1/4	0.64	0.92	1.25	1.63	2.06	2.55	3.08	3.67	4.30	4.99	5.73	6.52	7.36	8.25	9.19	10.19
2 1/2	0.57	0.83	1.12	1.47	1.86	2.29	2.77	3.30	3.87	4.49	5.16	5.87	6.62	7.43	8.27	9.17
2 3/4	0.52	0.75	1.02	1.33	1.69	2.08	2.52	3.00	3.52	4.08	4.69	5.33	6.02	6.75	7.52	8.33
3	0.48	0.69	0.94	1.22	1.55	1.91	2.31	2.75	3.23	3.74	4.30	4.89	5.52	6.19	6.89	7.64
3 1/4	0.44	0.64	0.86	1.13	1.43	1.76	2.13	2.54	2.98	3.46	3.97	4.51	5.10	5.71	6.36	7.05

For single unrestrained angles use one-quarter of the loads tabulated above.

The following method of design using a maximum bending stress of 22,000 psi is recommended. Angles and structural tees are considered restrained in cases shown above or in similar cases. Point of critical moment is assumed at tangent of fillet of outstanding leg of angle or tee.

$$M = \frac{P}{2} \times \frac{e}{2} = \frac{22t_f^2}{6} \qquad {}^*P = \frac{88t^2}{6e} = \frac{14.67t_f^2}{e}$$

where P = Allowable load on two angles or structural tee, in kips per linear inch.

e = Distance from tangent of fillet of angle or tee to center of rivet, in inches. ($e/2$ is the lever arm used to determine moment, because angles and tees are considered restrained.)

t_f = Thickness of angle or flange of tee, in inches.

For brackets as shown above, divide the tension in the two top rivets by the rivet pitch to obtain the load per linear inch of two angles.

* P is for fitting only; for prying action on fastener see Section 1.5.2.1 of Commentary on AISC Specification.

stresses to be increased by 33 percent for wind stresses. Assuming that wind rather than live and dead loads governs the beam selection, the allowable working stress on the beam is $24{,}000 \times 1.33 = 32{,}000$ psi. The S required for the beam would be

$$S = \frac{250 \times 12}{32{,}000} = 93.75$$

Use 21 WF 55.

Under all conditions the moment will be transferred into the column by means of a moment couple through the top and bottom rivets—compression on one group, tension on the other. The direct stress resulting from moment is

$$\frac{\text{Moment in foot-kips}}{\text{Depth of beam in feet}} = \frac{250}{1.75} = 142.5 \text{ kips}$$

(See Art. 3.10*b*.) Using ⅞-in. rivets of A141 steel in single shear, would require

$$\frac{142.9}{9.02 \times 1.33} = 11.9 = 12 \text{ rivets}$$

The same number would be required from the T to the column flange.

Because of the way in which the T is applied, it may be assumed that if failure occurred by tearing the T from the face of the column, all 12 rivets would fail simultaneously. Therefore we may consider the load of 8 rivets to be concentrated along the line just above the top flange of the beam or just below the bottom flange. The stress on this line is therefore $\frac{8}{12} \times 142.9 = 95.3$ kips. For any such wind stress it is safe to assume that the column at this point is a 14-in. column, so the stress per inch width is $95.3/14 = 6.8$ kips. The center gage line on the T is 5½ in., and assuming ⅜-in. web and ½-in. fillets, the e distance is $\frac{5\frac{1}{2} - 1\frac{3}{8}}{2} = 2\frac{1}{16}$ in. Since Table 4.6 is based on 22,000-psi fiber stress and we are permitted a 33⅓ percent increase, we may use the table by dividing 6.8 by 1.33, or use 5.1. This, however, would be for two sides, while the load we used is from one side only. We must therefore use $5.1 \times 2 = 10.2$. With $e = 2\frac{1}{16}$, $t = 1\frac{3}{16}$ in.

Only the 27 WF 177, 30 WF 190, 33 WF 220, 36 WF 182, and all heavier sections will furnish this thickness of flange together with enough web length to make the connection. Of these, the 27 WF 177 is the lightest but not the most economical. Minimum rivet spacing on the flange of the beam indicates that a minimum length of 15 in. is required and a 27 WF will provide two tees only 13½ in. long. The economical beam then must be 30 in. or more in depth. Also, in order to get two rows of rivets above the web, the minimum flange must be 13 in., which would eliminate the 36 WF 182. Of the remaining two sections, the 30 WF 190 is the more economical and will be used.

c *Stair strings*, or stringers, constitute a special design problem frequently encountered—not so much in the design of the details as in the fact that a wide variety of standard channels, junior channels, steel plates, and pressed-steel channels are used as stair strings. Table 4.7 is a table of standard stair strings.

d It is frequently necessary with a cantilever beam to use a system of framing which permits the beam to occupy the same space vertically as the member which supports it, so as to avoid the greater depth required by the framing shown in Fig. 3.7. Since this requirement is encountered so frequently in connection with the design of stair-supporting members, it has come to be known as a "stair cantilever" detail. It consists of converting the moment into a moment couple taken through top and bottom plates, at the same time taking the shear through a standard end connection.

Let us assume the condition indicated in Fig. 4.8. From the computations, the simple beam carries a moment of 33.69 ft-kips, and $S = 0.5 \times 33.69 = 16.8$. The cantilever moment is 24.0 ft-kips, and $S = 0.5 \times 24 = 12.0$. For stair cantilevers, although it is not necessary to use the same beam for the simple beam and the cantilever, it is advisable to use a beam of the same depth; so from Table 4.1 we shall select for the simple beam 10 B 19 and for the cantilever 10 B 15. Thus from the moment couple

Table 4.7 Stair Loads, Strings, and Risers

Total allowable uniform loads in kips and deflections in inches for stair strings

	1/8"		1/2"		1/2"		2.6"		2.94"		3/16"		1/4"		5/16"		3/16"		1/4"		1/2"		1/2"	
Depth	10		10		12		10		12		10		10		10		12		12		10		12	
Weight	6.5		8.4		10.6		15.3		20.7		6.53		8.5		10.63		7.84		10.2		8.03		9.33	
Web	.15		.17		.19		.24		.28		3/16		1/4		5/16		3/16		1/4		3/16		3/16	
Area	1.91		2.47		3.12		4.47		6.03		1.88		2.5		3.13		2.25		3.0		2.3		2.68	
I	22.1		32.3		55.8		66.9		128.1		15.6		20.8		26.0		27.0		36.0		25.9		41.9	
S	4.4		6.5		9.3		13.4		21.4		3.12		4.16		5.2		4.5		6.0		5.2		7.0	
Span ft.	Load	Defl.	Load	Defl.	Load	Defl.	Load	Defl.	Load	Defl.	Load	Defl.	Load	Defl.	Load	Defl.	Load	Defl.	Load	Defl.	Load	Defl.	Load	Defl.
3	21.7	.021	31.6	.021	45.4	.018	65.3	.021	98.5	.018	15.3	.021	20.4	.021	25.5	.021	22.0	.018	30.4	.018	25.5	.021	34.2	.018
4	16.2	.036	23.7	.036	34.1	.031	49.1	.036	78.3	.031	11.4	.036	15.3	.036	19.1	.036	16.5	.031	22.0	.031	19.1	.036	25.6	.031
5	13.0	.057	18.9	.057	27.3	.047	39.3	.057	62.6	.047	9.1	.057	12.2	.057	15.3	.057	13.2	.047	17.6	.047	15.3	.057	20.5	.047
6	10.8	.081	15.7	.081	22.8	.068	32.8	.081	52.1	.068	7.6	.081	10.2	.081	12.8	.081	11.0	.068	14.6	.068	12.8	.081	17.1	.068
7	9.2	.111	13.5	.111	19.5	.092	27.9	.111	44.8	.092	6.5	.111	8.7	.111	10.9	.111	9.5	.092	12.7	.092	10.9	.111	14.6	.092
8	8.1	.145	11.8	.145	17.1	.121	24.5	.145	39.2	.121	5.7	.145	7.6	.145	9.6	.145	8.3	.121	11.0	.121	9.6	.145	12.9	.121
9	7.3	.185	10.6	.185	15.2	.154	21.8	.185	34.9	.154	5.1	.185	6.8	.185	8.5	.185	7.4	.154	9.8	.154	8.5	.185	11.4	.154
10	6.5	.228	9.5	.228	13.6	.189	19.7	.228	31.2	.189	4.6	.228	6.2	.228	7.6	.228	6.6	.189	8.8	.189	7.6	.228	10.2	.189
11	5.9	.275	8.6	.275	12.4	.230	17.8	.275	28.5	.230	4.2	.275	5.5	.275	6.9	.275	6.1	.230	8.1	.230	6.9	.275	9.4	.230
12	5.4	.328	7.9	.328	11.3	.273	16.4	.328	26.2	.273	3.9	.328	5.1	.328	6.4	.328	5.5	.273	7.4	.273	6.4	.328	8.6	.273
13	5.0	.385	7.3	.385	10.5	.320	15.1	.385	24.1	.320	3.5	.385	4.7	.385	5.8	.385	5.1	.320	6.8	.320	5.8	.385	7.9	.320
14	4.6	.447	6.7	.447	9.8	.372	14.1	.447	22.3	.372	3.3	.447	4.4	.447	5.5	.447	4.7	.372	6.3	.372	5.5	.447	7.4	.372
15	4.3	.513	6.3	.513	9.1	.427	13.1	.513	20.9	.427	3.1	.513	4.1	.513	5.1	.513	4.4	.427	5.8	.427	5.1	.513	6.8	.427
16	4.1	.583	5.9	.583	8.5	.485	12.2	.583	19.6	.485	2.9	.583	3.9	.583	4.7	.583	4.2	.485	5.6	.485	4.7	.583	6.4	.485
17	3.9	.658	5.6	.658	8.0	.548	11.4	.658	18.5	.548	2.6	.658	3.6	.658	4.5	.658	3.9	.548	5.2	.548	4.5	.658	6.1	.548
18	3.6	.737	5.3	.737	7.6	.615	10.9	.737	17.4	.615	2.5	.737	3.4	.737	4.3	.737	3.6	.615	4.8	.615	4.3	.737	5.7	.615
19	3.4	.822	5.0	.822	7.2	.684	10.2	.822	16.5	.684	2.4	.822	3.2	.822	4.1	.822	3.5	.684	4.7	.684	4.1	.822	5.4	.684
20	3.2	.911	4.7	.911	6.8	.759	9.8	.911	15.6	.759	2.3	.911	3.1	.911	3.9	.911	3.3	.759	4.4	.759	3.9	.911	5.2	.759
Dimensions, in. T	9.228		8.934		10.664		8.125		9.875												9.25		11.25	
M	.261		.38		.418		.633		.723												3/16		3/16	
N	.14		.18		.20		.24		.28												3/8		3/8	
K	.386		.63		.668		.938		1.06												3/16		3/16	
R	.125		.25		.25		.34		.38												.26		.23	
Y	.19		.29		.27		.64		.70															

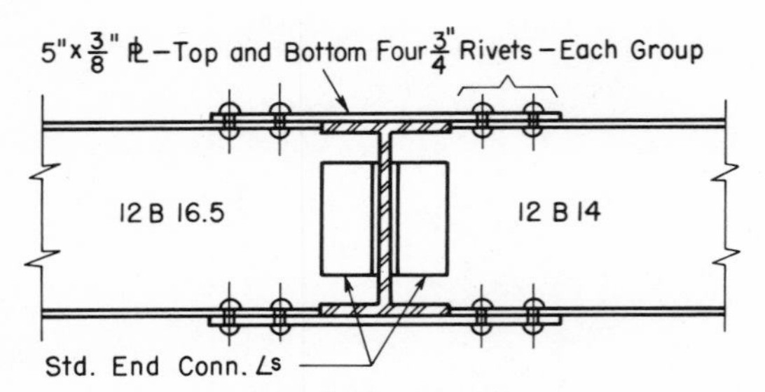

Fig. 4.8 Stair Cantilever—Beams of Equal Depth

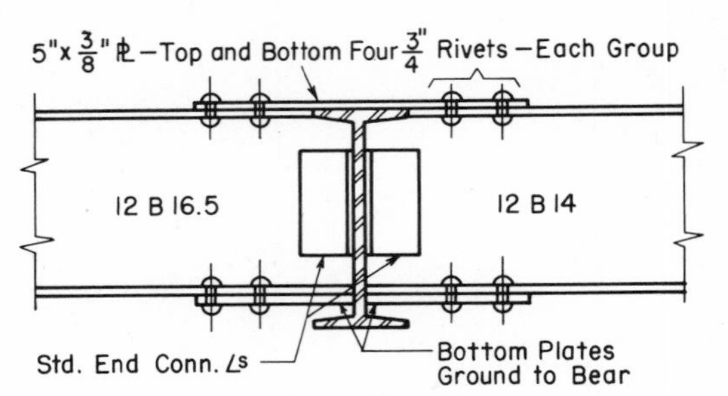

Fig. 4.9 Stair Cantilever—Beams of Unequal Depth

10 B 15. Thus from the moment couple

$$P = \frac{24 \times 12}{10} = 28.8 \text{ kips}$$

This requires a net area of $28.8/22 = 1.3$ sq in. in the plate or, allowing for two ¾-in. rivets in the width of a 5-in. plate, $t = 1.3/(5 - 1.75) = 0.4$. Use $\frac{7}{16} \times 5$-in. plate.

The flange thickness of a 10 B 15 (Table 1.6) is 0.269 and of a 10 B 19, 0.394. From Table 9.1, single shear at 6.63 will control the rivet value, and the number of rivets is $28.8/6.63 = 4.4$. Use six ¾-in. rivets. Figure 4.8 shows a detail using beams of equal depth, and Fig. 4.9, the same detail with a supporting beam of greater depth.

4.13 Beam-bearing plates

When a beam is supported by a masonry wall or pier, it is essential that the beam reaction be distributed over an area sufficient to keep the average pressure on the masonry within the allowable limits. Steel bearing plates are generally used for this pressure distribution. The following method of design taken from the AISC Manual, pages 2-44 and 2-45, is recommended.

Example

An 18 W 50 beam, ASTM A36, for which web thickness = .358 inch, $k = 1\frac{1}{16}$ inches and $F_y = 36.0$ ksi, has a reaction of 49 kips and is to be supported by a masonry wall with an allowable $F_p = 250$ psi. The length of bearing, C, is limited to 10 inches. Using ASTM A36 steel,

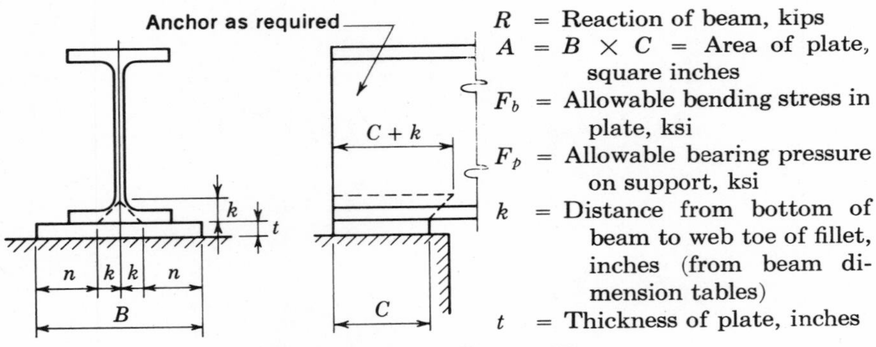

R = Reaction of beam, kips
$A = B \times C$ = Area of plate, square inches
F_b = Allowable bending stress in plate, ksi
F_p = Allowable bearing pressure on support, ksi
k = Distance from bottom of beam to web toe of fillet, inches (from beam dimension tables)
t = Thickness of plate, inches

Fig. 4.10 Beam Bearing Plate

The beam reaction, R, is assumed to be uniformly distributed to the plate over the area $C \times 2k$. The bearing plate is assumed to distribute this load uniformly over the masonry support.

1. Establish F_b, ksi and F_p, ksi.
2. Determine the required area, $A = R/F_p$, square inches.
3. Establish C and solve for $B = A/C$. The length of bearing, C, is usually governed by the available wall thickness or some other structural consideration. B and C should preferably be in full inches, and B rounded off so that $B \times C \geq A$.
4. Determine the actual bearing pressure, $F_p = R/(B \times C)$.
5. Determine $n = (B/2) - k$ and, using the actual F_p, solve for t in the formula:

$$t = \sqrt{\frac{3F_p n^2}{F_b}}$$

6. Check web crippling on length $C + k$.

$$\frac{R}{(C + k) \times \text{web thickness}} \leq 0.75F_y$$

F_b = 27.0 ksi (AISC Specification, Sect. 1.5.1.4.8), design a bearing plate for the beam.

$$A \text{ (req'd.)} = 49/.250 = 196 \text{ sq. in.}$$

$$B = 196/10 = 19.6 \text{ in.; use 20 in.}$$
$$A = 10 \times 20 = 200 \geq 196$$

$$F_p \text{ (actual)} = 49/(10 \times 20) = .245 \text{ ksi}$$

$$n = 20/2 - 1.06 = 8.94 \text{ in.}$$

$$t = \sqrt{\frac{3 \times .245 \times 8.94^2}{27.0}}$$
$$= 1.48 \text{ inches; use } 1\tfrac{1}{2} \text{ in.}$$

Check for web crippling:
$.75\ F_y$ = 27.0 ksi (allowable)

$$\frac{49}{(10 + 1.06) \times .358} = 12.38 \text{ ksi}$$

$$12.38 \leq 27.0 \text{ o.k.}$$

Use: Bearing plate 10 × 1½ × 1′-8

Steel bearing plates are usually shipped separately and grouted in place prior to erection of the beam. Beams are generally not attached to the bearing plates, but should be properly anchored to the wall. Recommended anchorage details are shown on page 4-77.

In the event light loads or high allowable bearing pressures reduce the bearing area required sufficiently to permit support of the beam without bearing plates, it is recommended that the beam flange be investigated for bending by the formula,

$$F_b \text{ (allowable)} = \frac{3F_p n^2}{t^2}$$

in which

t = thickness of flange, inches
n = (flange width/2) − k, inches
$F_b = 0.75F_y$, ksi, allowable bending stress in flange when acting as a bearing plate

Despite the fact that the beam flange may furnish sufficient width to carry the load without the use of a bearing plate, the author has found that it is good construction practice to bed a thin steel plate, usually ¼-in. thick and slightly greater than the beam flange in width, so that the beam may be set accurately without holding up the erection crane and the time of a crew of erectors. The cost of the plate and masons' time is more than offset by the time saved in erection.

4.14 Unsymmetrical bending in purlins

a Any beam the main axis of which is not horizontal—as found most frequently in roof purlins on a sloping roof—must be analyzed for the effect of this unsymmetrical bending. Some textbooks and handbooks discuss this subject at length, working out "S polygons" and otherwise complicating a simple problem. The simplest approach is to make a vector analysis of the applied loads, apply the two components parallel to the two axes of the beam, and compute the unit stresses obtained by dividing the vector moment by the section modulus in each direction.

However, since the bending parallel with the plane of the roof must be resisted by the top flange only, the section modulus to be used must be half the section modulus listed for the weaker way of the beam.

Inasmuch as the component parallel with the slope of the roof is usually greater than can be carried on the full span, it is common practice to divide the span by putting in one or more rows of sag rods in the length of the span. The rods serve as a support the weak way of the beam, carrying the thrust load up to the peak of the roof and balancing it against a corresponding load from the other side of the roof. This is an important reason why the vector analysis of loads is a simpler method of approach to the problem than the S-polygon method.

Problem Select a purlin for a span of 21 ft and a uniform load of 420 lb per running foot, the roof to have ⅕ pitch (see Fig. 4.11).

The pitch of a roof is stated in terms of a ratio of center height to total span, so ⅕ pitch means a slope of 1 on 2½. The total load on one beam is

$$W = 21 \times 420 = 8{,}820$$

From the triangle relationship,

$$1^2 + 2.5^2 = 7.25$$
$$\sqrt{7.25} = 2.7$$

Therefore,

$$W_p = \frac{1}{2.7} \times 8{,}820 = 3{,}267$$

$$W_n = \frac{2.5}{2.7} \times 8{,}820 = 8{,}167$$

Using the normal load only and disregarding deflection,

$$S = \frac{8.167 \times 21}{16} = 10.7$$

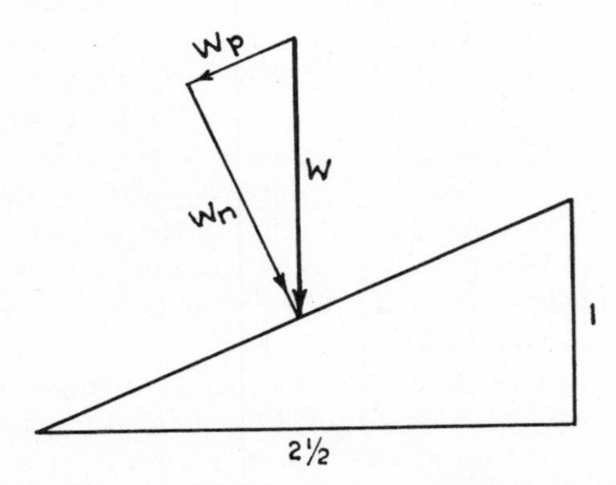

Fig. 4.11 Load Distribution on Sloping Beam

Try 8 WF 17.

$$M_n = \frac{8{,}167 \times 21 \times 12}{8} = 257{,}300 \text{ in.-lb}$$

$$M_p = \frac{3{,}267 \times 21 \times 12}{8} = 102{,}900 \text{ in.-lb}$$

$$f_n = \frac{257{,}300}{14.1} = 18{,}250$$

$$f_p = \frac{102{,}900}{1.3} = \frac{79{,}160}{97{,}411}$$

Although f_p could be reduced greatly by using sag rods, it is obvious that the 8 WF 17 will not do. Try 10 WF 21.

$$f_n = \frac{257{,}300}{21.5} = 11{,}970$$

$$f_p = \frac{102{,}900}{1.7} = \frac{60{,}520}{71{,}590}$$

This is still much too high without sag rods.

The effect of introducing sag rods is to make the beam a continuous beam so far as W_p is concerned, and thus introduce different coefficients (see Art. 3.20*a*) and divide the fiber stress by the square of the number of parts.

The combined effect for introducing sag rods is as follows:

For a single row of sag rods, the divisor is 4.

For two rows of sag rods, the divisor is $1.25 \times 9 = 11.25$.

For three rows of sag rods, the divisor is $1.25 \times 16 = 20$.

If two rows of sag rods are used with a 10 WF 21, the calculation becomes

$$f_n = 11{,}970$$

$$f_p = \frac{60{,}520}{11.25} = \frac{5{,}390}{17{,}360} \text{ psi}$$

Thus, using two rows of sag rods, the 10 WF 21 may be used.

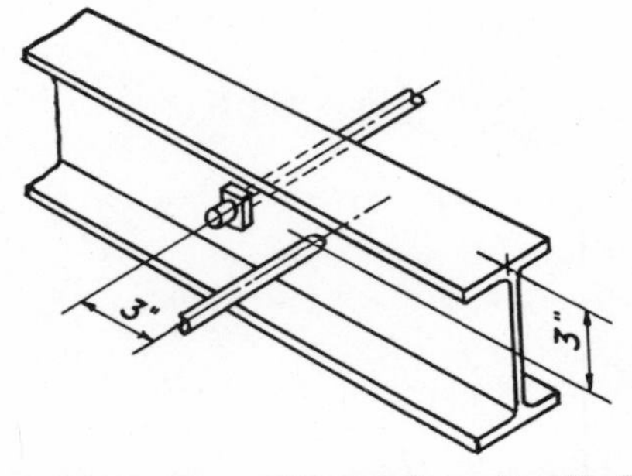

Fig. 4.12 Use of Tie Rods or Sag Rods in Purlin

b The area of sag rods required must be computed also. The rods are ordinarily run from beam to beam, 3 in. below the top of the beam and staggered laterally 3 in. The load for which they are designed is the cumulative W or thrust from the eave to the peak.

Problem Assume that the purlin designed above is used on a roof which is 10 beam spaces wide. Thus there will be 5 beams either side of the peak purlin. The eave purlin will be assumed to carry half the purlin load.

From the preceding problem, $W = 3{,}267$ for full span or 1,090 for one-third span as carried by a row of sag rods. Then

$$P = 4.5 \times 1{,}090 = 4{,}905$$

$$A = \frac{4{,}905}{22{,}000} = 0.223 \text{ sq in. net}$$

From Table 2.4 a ¾-in. round rod would be required just below the peak. Since the rod is run in short lengths from beam to beam, the size may be decreased as the load drops off. A tabulation of the load and size is as follows: ⅝-in. rod is good for $0.202 \times 22{,}000 = 4{,}444$, or $4 \times 1{,}090$. Thus it may be used below the upper bay.

Rods of less than $\frac{5}{8}$-in. diameter should not be used. Therefore, it will be necessary to use $\frac{3}{4}$-in. sag rods for one beam space each side of the peak, then $\frac{5}{8}$-in. from this beam down to the eaves.

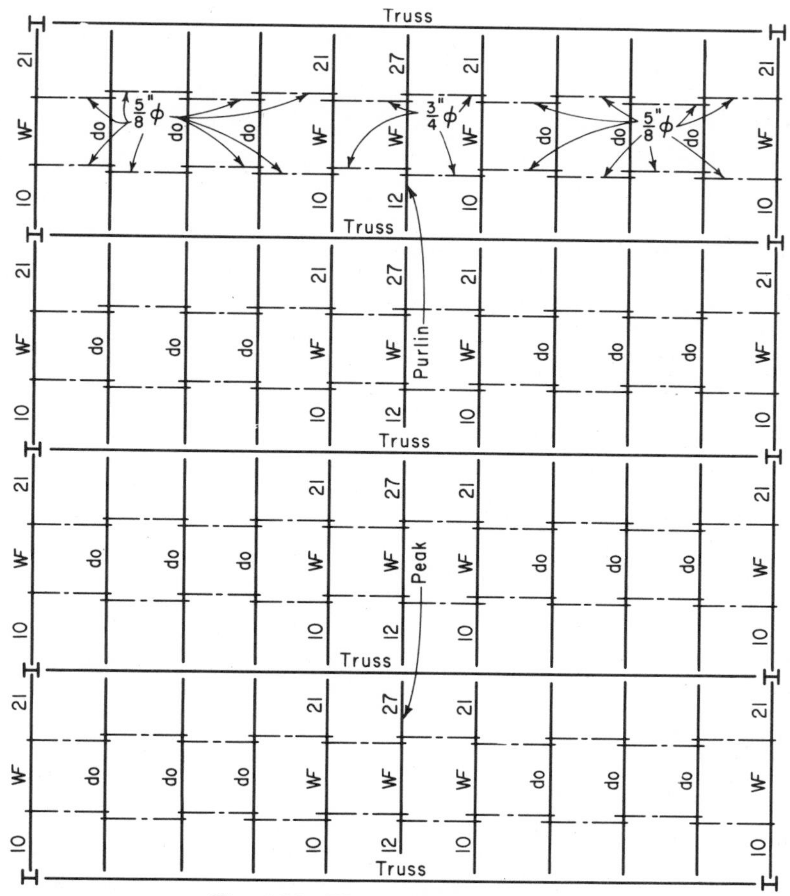

Fig. 4.13 Plan of Pitched Roof

The final design is indicated in Fig. 4.13 and detailed in Fig. 4.12.

4.15 Rigid and semirigid construction

As has been noted in Art. 4.10, the AISC Code recognizes Type I, Rigid Frame Construction, and Type III, Semirigid Construction as well as Type II, Conventional or Simple Framing. In general, the material in this chapter applies to Type II, Conventional Framing, which, in buildings of not over three or four stories, is normally the economical type to use.

It has been the author's experience that where a saving is indicated by any rigidity in framing, the saving in Type I, Rigid Frame Construction is greater than that in Type III, Semirigid, at little extra cost of fabrication and therefore, in the author's opinion, Type III may be ruled out in any analysis. With the increasing use of welding and high-strength bolting, together with the provision for wind bracing in any building the height of which exceeds $1\frac{1}{2}$ times the width of base, the use of Type I, Rigid Frame Construction appears to be desirable for all such frames across the short direction of the building, even though the filler beams and other framing in the longer direction may still use a conventional type of construction.

When Type I, Rigid Frame Construction, is used, the moment distribution method as described in Art. 3.21 may be used except that the introduction of columns above and below the floor to be analyzed will require the moment distribution to one additional member at each roof joint and to two additional members at each floor, as indicated in Fig. 4.14.

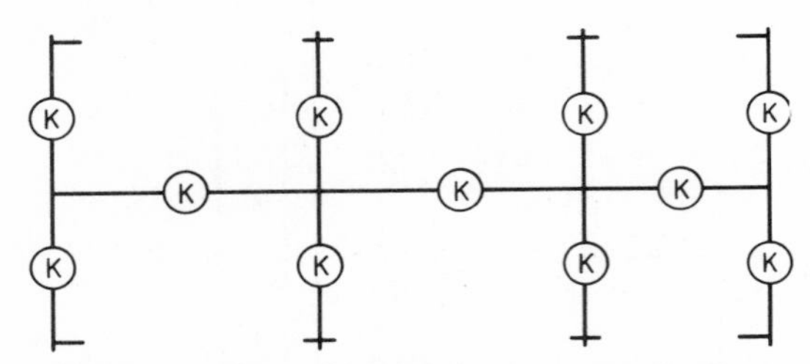

Fig. 4.14 Moment Distribution for Type 1 Rigid Frame Construction

In the application of this method, columns are assumed to be rigidly connected at the floors above and below the floor being analyzed. Preliminary size members are assumed for each beam and column, and the stiffness factor $K\left(=\dfrac{I}{L}\right)$ is put in each circle as designated. The moment distribution is carried through as in Art. 3.21; end moments are obtained for both beams and columns at the floor in question.

The United States General Services Administration gives a moment factor method of determining the minimum moments of beams which may be used for a preliminary design. These moment factors are given in Table 4.8. The moment factors are divided into three categories depending on the restraint at supports—negligible, moderate, and large.

1. Negligible restraint would mean web connections only or freely supported seat connections to columns.

2. Moderate restraint would mean a member rigidly connected to a column when the sum of the stiffness of the columns, above and below the connections, and the members opposite the member being analyzed is less than twice the stiffness of the beam under consideration.

3. Large restraint would mean a member where the combined stiffness of columns and beams as noted in paragraph 2 exceeds twice the stiffness of the beam under consideration.

Having made the preliminary computation for determination of member sizes, the stiffness factor K may be assigned and a complete moment distribution made. Regardless of what minimum beam sizes may be indicated by this analysis, the member should not be less than that determined from the factor coefficient of Table 4.8.

This method has been applied in Art. 10.22 on wind bracing.

Table 4.8 Continuous Frames Moment Factors C'

Restraint at supports		Moment location	Number of spans 1	2	3	Over 3
Negligible	End span	End support	1/24	1/24	1/24	1/24
		Mid span	1/8	1/10	1/10	1/10
		First interior support	...	1/8	1/9	1/9
	Interior span	Mid span	...	...	1/14	1/14
		Other supports	...	...	...	1/12
Moderate	End span	End support	1/16	1/16	1/16	1/16
		Mid span	1/10	1/12	1/14	1/14
		First interior support	...	1/10	1/10	1/10
	Interior span	Mid span	...	...	1/16	1/16
		Other supports	...	...	...	1/12
Large	End span	End support	1/12	1/12	1/12	1/12
		Mid span	1/12	1/16	1/16	1/16
		First interior support	...	1/11	1/11	1/11
	Interior span	Mid span	...	...	1/16	1/16
		Other supports	...	...	...	1/12

4.16 Lintels

a Lintels are ordinarily steel members used in masonry walls to carry the masonry over an opening. The type of lintel used depends on (1) the span, (2) the wall thickness, (3) the type of masonry, and (4) the type of window or door over which the lintel is used. The simplest type of lintel is a loose angle lintel, used in a brick wall when the span is not too great. The flat leg of the angle, not over $3\frac{1}{2}$ in. wide, provides a seat for one course of brick, and the upstanding leg provides the strength. The smallest angle used is usually $3 \times 3 \times \frac{1}{4}$ in. and the maximum span as governed by deflection is limited in feet to about $1\frac{2}{3}$ times the depth in inches. Thus a 3×3-in. angle under light load should span not over 5 ft. Other angles often used for lintels are 4×3 in. and $5 \times 3\frac{1}{2}$ in., with the longer leg vertical. It is seldom that section modulus controls the design of a lintel unless there is a concentrated load applied immediately over the lintel.

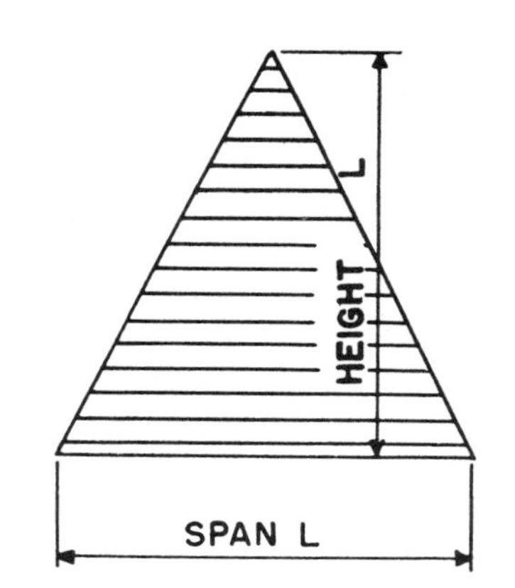

Fig. 4.15 Masonry Load to Be Carried on a Short Lintel

The masonry load ordinarily figured on a simple short lintel is the weight of a triangle of masonry of a height equal to the span, plus any concentrations which come immediately over the opening. Certainly a very safe assumption is a rectangle of height equal to the span. A good simple rule for bearing for angle lintels is 1 in. of bearing at each end for each foot of clear span, but not less than 4 in. at each end.

Problem Assume a 12-in. brick wall on a clear span of 4 ft 6 in. with no load except the wall itself. On the basis of a triangle of masonry carried, with sides equal to the span,

$$W = \frac{4.5 \times 120 \times 4.5}{2} = 1{,}215$$

$$S = 0.546\left(\frac{1.215 \times 4.5}{6}\right) = 0.50$$

It is advisable to use three angles to carry the three-brick thickness of walls. Using the minimum angle specified above, $3 \times 3 \times \frac{1}{4}$ in., the section modulus furnished is 1.74, and our final lintel is three angles, $3 \times 3 \times \frac{1}{4}$ in., 5 ft 4 in. long. This allows for 5 in. of bearing at each end.

Most masonry construction nowadays—even brick construction—uses brick for the facing only with concrete masonry backup. Frequently for this type of construction the brick facing is carried by a steel angle, and the backup is carried by a precast concrete lintel of the thickness of the concrete masonry it carries. Such a precast lintel should have sufficient end bearing on solid concrete masonry and should be reinforced top and bottom to protect against possible error in setting the lintel upside down.

b If span or section modulus requirements are beyond the limitations of $5 \times 3\frac{1}{2} \times \frac{3}{8}$-in. angles, it is ordinarily better to use a fabricated lintel—the simplest lintel to use being built up of two channels with separators between them and an angle riveted or welded to the outer channel to carry the face brick, as shown in Fig. 4.16.

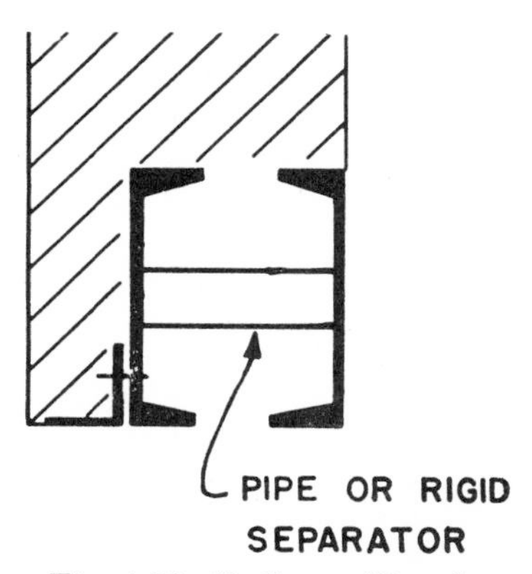

Fig. 4.16 Built-up Lintel

c It has been mentioned that the type of window to some extent controls the lintel. This is an architectural rather than a structural provision, but it is well to call attention to it. Figure 4.17 shows a very common condition—the outer angle is set about $\frac{3}{4}$ in. lower than the inner angles so as to form a weather break and a point for calking the frame.

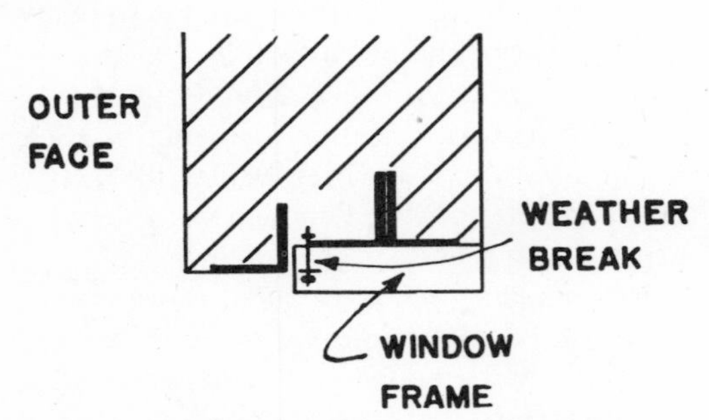

Fig. 4.17 Angle Lintel—Showing Weather Break

d Where factory-type steel sash is used, it is necessary to provide a surface against which the sash section can finish and a means of fastening the steel sash in place. Since factory-type sash units are usually longer than the economical span for loose angle lintels, and in fact are frequently continuous under an eave purlin, provision must usually be made to fabricate the lintel accordingly. The catalogs of the various steel-sash companies suggest methods of providing for the various details to support the sash, but two of the most common details are shown in Fig. 4.18.

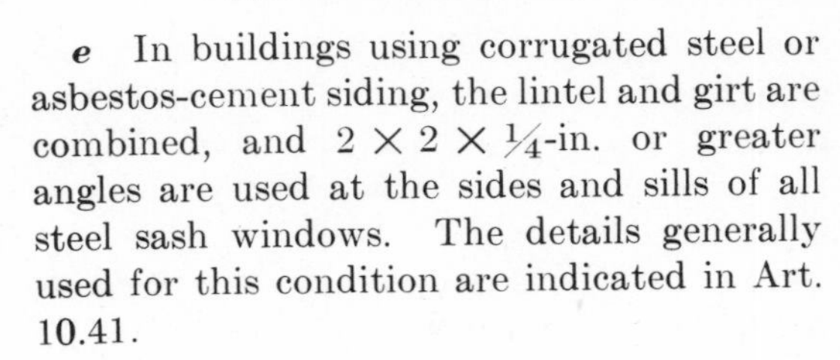

e In buildings using corrugated steel or asbestos-cement siding, the lintel and girt are combined, and $2 \times 2 \times \frac{1}{4}$-in. or greater angles are used at the sides and sills of all steel sash windows. The details generally used for this condition are indicated in Art. 10.41.

4.17 Torsion in lintels and spandrels

a For reasons of practicability, lintels or spandrel beams are frequently placed in a wall in such a manner that they are subject to torsion as well as to direct load stress, and in some instances horizontal wall cracks develop at or above the top flange of the spandrel as a result of such torsion. This condition is indicated in Fig. 4.19. In this construction, the only concentric load is a light-weight concrete masonry backup wall, frequently without the benefit of slab or joist bearing on the wall. The 4-in. face brick, which is usually heavier than the 8-in. backup, applies an eccentric load on an approximate 6-in. lever arm that, particularly

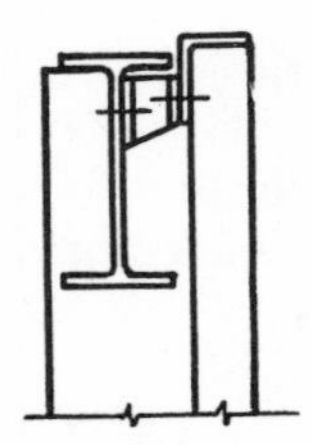

Fig. 4.19 Spandrel Subject to Torsion

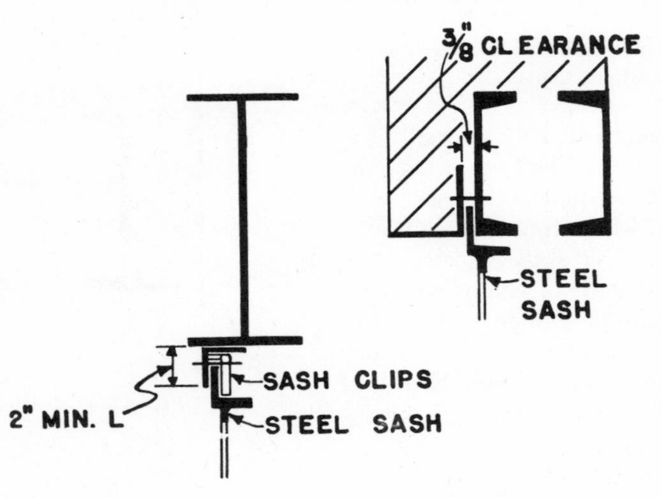

Fig. 4.18 Lintel Types for Steel Sash

in a long beam, causes the beam to roll at the center of the span, the bottom flange deflecting inward and the upper flange outward.

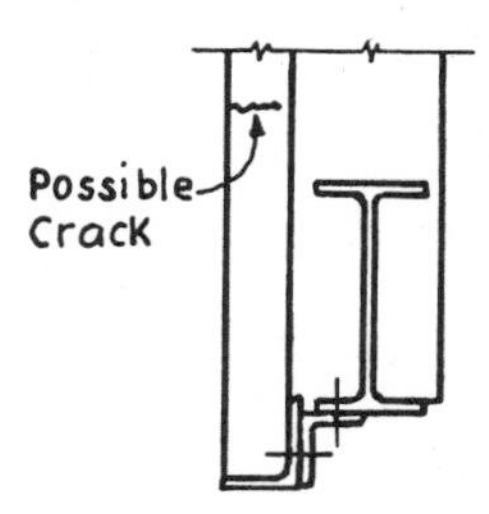

Fig. 4.20 Spandrel Subject to Torsion

The detail shown in Fig. 4.20 will usually aggravate the tendency to crack. The method of attachment of the shelf angle by means of clip angles spaced 2 to 3 ft will permit the roll of the shelf angle with relation to the web of the beam itself, a condition which is prevented by the method of attachment shown in Fig. 4.19. It should be noted that metal wall ties do not make a rigid wall, and the face brick can apply load independently of the backup.

b Inasmuch as an accurate computation of stresses is involved and factors to be used for built-up sections are not readily obtainable, it is preferable to design against the occurrence of such a condition. If beams can be framed into the spandrel or attached to the top flange, this tendency to roll can usually be eliminated. A poured-concrete slab resting on the top flange or the ends of bar joists tack-welded to the beam will accomplish the same purpose. Where this is impossible, a section should be used which is capable of resisting lateral as well as vertical stresses. Some engineers make a practice of tying the facing to the top flange so that the moment produced by eccentricity is restricted by a moment couple, and building up the top and bottom flanges to take the horizontal stresses produced by this moment couple. The section shown in Fig. 4.16 will resist such stresses, particularly if rigid separators are used. If the beam required is too long to permit the use of standard channels, the proper resistance may sometimes be developed by using a flatwise channel on the top flange and building up the bottom flange by rigidly connecting the shelf angle and putting a continuous angle at the inner edge of the bottom flange with the leg turned out and up.

c Another type of spandrel which is subject to torsion is a curved beam, as indicated in Fig. 4.21. This is usually carried by

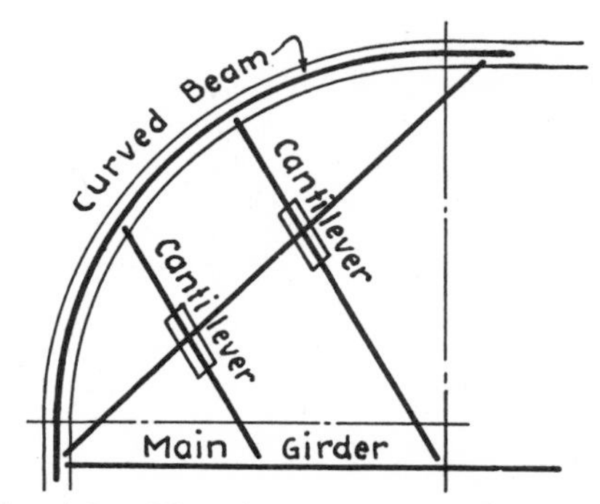

Fig. 4.21 Plan Showing Curved Beam

breaking the curve up into short lengths so that the beam between supports has only a small curvature, and carrying the support points by means of cantilevers from the main construction, using the cantilever detail shown in Figs. 4.8 or 4.9.

4.20 PLATE GIRDERS

a When the required section modulus is

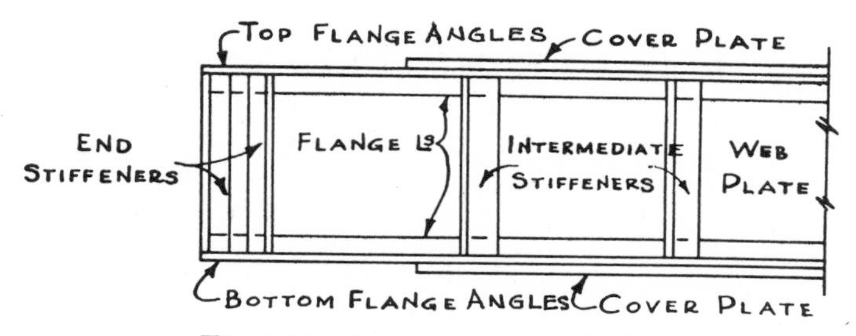

Fig. 4.22 Parts of a Plate Girder

greater than that available in rolled wide-flange sections or economical cover-plated wide-flange sections, it becomes necessary to build up a plate-girder section.

Until the 1963 AISC Code was adopted, a plate girder was normally built up of plates and angles as shown in Fig. 4.22. The changes in the Code, however, have liberalized welding to such an extent that it is probable that the days of the built-up plate and angle girder are numbered. However, it seems advisable to include in this text information on both types of plate girder.

b The following sections from the 1963 AISC Code deal primarily with the design of plate girders.

SECTION 1.10 PLATE GIRDERS

1.10.1 *Proportions*

Riveted and welded plate girders, cover-plated beams and rolled beams shall in general be proportioned by the moment of inertia of the gross section. No deduction shall be made for shop or field rivet or bolt holes in either flange, except that in cases where the reduction of the area of either flange by such holes, calculated in accordance with the provisions of Sect. 1.14.3, exceeds 15 percent of the gross flange area, the excess shall be deducted.

Section 1.14.3 referred to above is quoted in Art. 2.13.

1.10.2 *Web*

The clear distance between flanges in inches shall not exceed

$$\frac{14{,}000{,}000}{\sqrt{F_y(F_y + 16{,}500)}}$$

times the web thickness.

1.10.3 *Flanges*

The thickness of outstanding parts of flanges shall conform to the requirements of Sect. 1.9.

Each flange of welded plate girders shall in general consist of a single plate rather than two or more plates superimposed. The single plate may comprise a series of shorter plates, laid end-to-end and joined by complete penetration butt welds.

Unstiffened cover plates on riveted girders shall not extend more than $3{,}000/\sqrt{F_y}$ times the thickness of the thinnest outside plate beyond the outer row of rivets or bolts connecting them to the angles. The total cross-sectional area of cover plates of riveted girders shall not exceed 70 percent of the total flange area.

1.10.4 *Flange development*

Rivets, high strength bolts or welds connecting flange to web, or cover plate to flange, shall be proportioned to resist the total horizontal shear resulting from the bending forces on the girder. The longitudinal distribution of these rivets or intermittent welds shall be in proportion to the intensity of the shear. But the longitudinal spacing shall not exceed the maximum permitted, respectively, for compression or tension members in Sect. 1.18.2.3 or 1.18.3.1. Additionally, rivets or welds connecting flange to web shall be proportioned to transmit to the web any loads applied directly to the flange unless provision is made to transmit such loads by direct bearing.

Partial length cover plates shall be extended beyond the theoretical cut-off point and the extended portion shall be attached to the beam or girder by rivets, high strength bolts (friction-type joint), or fillet welds adequate, at stresses allowed in Sect. 1.5.2 or 1.5.3 or Sect. 1.7, to develop the cover plate's portion of the flexural stresses in the beam or girder at the theoretical cut-off point. In addition, for welded cover plates, the welds connecting the cover plate termination to the beam or girder in the length a', defined below, shall be adequate, at the allowed stresses, to develop the cover plate's portion of the flexural stresses in the beam or girder at the distance a' from the end of the cover plate.* The length a', measured from the end of the cover plate, shall be:

1. A distance equal to the width of the cover plate when there is a continuous weld equal to or larger than ¾ of the plate thickness across the end of the plate and continued welds along both edges of the cover plate in the length a'.
2. A distance equal to 1½ times the width of the cover plate when there is a continuous weld smaller than ¾ of the plate thickness across the end of the plate and continued welds along both edges of the cover plate in the length a'.
3. A distance equal to 2 times the width of the cover plate when there is no weld across the end of the plate but continuous welds along both edges of the cover plate in the length a'.

* This may require the cover plate termination to be placed at a point in the beam or girder that has lower bending stress than the stress at the theoretical cut-off point.

1.10.5 *Stiffeners*

1.10.5.1 Bearing stiffeners shall be placed in pairs at unframed ends on the webs of plate girders and, where required* at points of concentrated loads. Such stiffeners shall have a close bearing against the flange, or flanges, through which they receive their loads or reactions, and shall extend approximately to the edge of the flange plates or flange angles. They shall be designed as columns subject to the provisions of Sect. 1.5.1, assuming the column section to comprise the pair of stiffeners and a centrally located strip of the web whose width is equal to not more than 25 times its thickness at interior stiffeners or a width equal to not more than 12 times its thickness when the stiffeners are located at the end of the web. The effective length shall be taken as not less than ¾ of the length of the stiffeners in computing the ratio l/r. Only that portion of the stiffener outside of the angle fillet or the flange-to-web welds shall be considered effective in bearing.

* For provisions governing welded plate girders see Sect. 1.10.10.

1.10.5.2 The largest average web shear f_v in any panel between stiffeners (total shear force divided by web cross-sectional area), in pounds per square inch, computed for any condition of complete or partial loading, shall not exceed the value given by Formula (8) or (9),** as applicable.

$$F_v = \frac{F_y}{2.89}\left[C_v + \frac{1 - C_v}{1.15\sqrt{1 + (a/h)^2}}\right] \qquad (8)$$

when C_v is less than 1.0;

$$F_v = \frac{F_y}{2.89}(C_v) \qquad (9)$$

but not more than $0.4F_y$, when C_v is more than 1.0 or when intermediate stiffeners are omitted; where

a = clear distance between transverse stiffeners, in inches

h = clear distance between flanges, in inches

$C_v = \dfrac{45{,}000{,}000k}{F_y(h/t)^2}$, when C_v is less than 0.8

$= \dfrac{6{,}000}{h/t}\sqrt{\dfrac{k}{F_y}}$, when C_v is more than 0.8

t = thickness of web, in inches

$k = 4.00 + \dfrac{5.34}{(a/h)^2}$, when a/h is less than 1.0

$= 5.34 + \dfrac{4.00}{(a/h)^2}$, when a/h is more than 1.0

When a/h is more than 3 its value shall be taken as infinity. In this case Formula (8) reduces to Formula (9) and $k = 5.34$.

** For values of F_v corresponding to various stiffener spacing see Tables 4.9 and 4.10 of this book.

1.10.5.3 Intermediate stiffeners are not required when the ratio h/t is less than 260 and the maximum web shear stress f_v is less than that permitted by Formula (9).

The spacing of intermediate stiffeners, when stiffeners are required, shall be such that the web shear stress will not exceed the value for F_v given by Formula (8) or (9), as applicable, and the ratio a/h shall not exceed $\left(\dfrac{260}{h/t}\right)^2$ nor 3.0.

The spacing between stiffeners at end panels and panels containing large holes shall be such that the smaller panel dimension, a or h, shall not exceed $\dfrac{11{,}000t}{\sqrt{f_v}}$.

1.10.5.4 The gross area, in square inches, of intermediate stiffeners spaced in accordance with Formula (8) (*total* area, when stiffeners are furnished in pairs) shall be not less than that computed by Formula (10).

$$A_{st} = \frac{1 - C_v}{2}\left[\frac{a}{h} - \frac{(a/h)^2}{\sqrt{1 + (a/h)^2}}\right] YDht \qquad (10)$$

where

C_v, a, h and t are as defined in Sect. 1.10.5.2

$Y = \dfrac{\text{yield point of web steel}}{\text{yield point of stiffener steel}}$

D = 1.0 for stiffeners furnished in pairs
= 1.8 for single angle stiffeners
= 2.4 for single plate stiffeners

When the greatest shear stress f_v in a panel is less than that permitted by Formula (8) this gross area requirement may be reduced in like proportion.

The moment of inertia of a pair of stiffeners, or a single stiffener, with reference to an axis in the plane of the web, shall not be less than $(h/50)^4$.

Intermediate stiffeners may be stopped short of the tension flange a distance not to exceed 4 times the web thickness, provided bearing is not needed to transmit a concentrated load or reaction. When single stiffeners are used they shall be attached to the compression flange, if it consists of a rectangular plate, to resist any uplift tendency due to torsion in the plate. When lateral bracing is attached to a stiffener, or a pair of stiffeners, these, in turn, shall be connected to the compression flange to transmit 1 percent of the total flange stress, unless the flange is composed only of angles.

Intermediate stiffeners required by the provisions of Sect. 1.10.5.3 shall be connected for a total shear transfer, in pounds per linear inch of single stiffener or pair of stiffeners, not less than that computed by the formula

$$f_{vs} = h\sqrt{\left(\frac{F_y}{3{,}400}\right)^3}$$

where F_y = yield point of web steel.

This shear transfer may be reduced in the same proportion that the largest computed shear stress f_v in the adjacent panels is less than that permitted by Formula (8). However, rivets and welds in intermediate stiffeners which are required to transmit to the web an applied concentrated load or reaction shall be proportioned for not less than the applied load or reaction.

Rivets connecting stiffeners to the girder web shall be spaced not more than 12 inches on center. If intermittent fillet welds are used, the clear distance between welds shall not be more than 16 times the web thickness nor more than 10 inches.

1.10.6 *Reduction in flange stress*

When the web depth-to-thickness ratio exceeds $24{,}000/\sqrt{F_b}$, the maximum stress in the compression flange shall not exceed

$$F'_b \leq F_b \left[1.0 - 0.0005 \frac{A_w}{A_f} \left(\frac{h}{t} - \frac{24{,}000}{\sqrt{F_b}} \right) \right] \qquad (11)$$

where

F_b = applicable bending stress given in Sect. 1.5.1

A_w = area of the web

A_f = area of compression flange

1.10.7 *Combined shear and tension stress*

Plate girder webs, subject to a computed average shear stress in excess of that permitted by Formula (9), shall be so proportioned that bending tensile stress, due to moment in the plane of the girder web, shall not exceed $0.6F_y$ nor

$$\left(0.825 - 0.375 \frac{f_v}{F_v} \right) F_y \qquad (12)$$

where

f_v = computed average web shear stress (total shear divided by web area)

F_v = allowable web shear stress according to Formula (8) or (9)

1.10.8 *Splices*

Butt welded splices in plate girders and beams shall be complete penetration groove welds and shall develop the full strength of the smaller spliced section. Other types of splices in cross-sections of plate girders and in beams shall develop the strength required by the stresses, at the point of splice, but in no case less than 50 percent of the effective strength of the material spliced.

1.10.9 *Horizontal forces*

The flanges of plate girders supporting cranes or other moving loads shall be proportioned to resist the horizontal forces produced by such loads. (See Sect. 1.3.4.)

1.10.10 *Web crippling*

1.10.10.1 Webs of beams and welded plate girders shall be so proportioned that the compressive stress at the web toe of the fillets, resulting from concentrated loads not supported by bearing stiffeners, shall not exceed the value of $0.75F_y$ pounds per square inch allowed in Sect. 1.5.1; otherwise, bearing stiffeners shall be provided. The governing formulas shall be:

For interior loads,

$$\frac{R}{t(N + 2k)} = \text{not over } 0.75F_y \text{ pounds per square inch} \qquad (13)$$

For end-reactions,

$$\frac{R}{t(N + k)} = \text{not over } 0.75F_y \text{ pounds per square inch} \qquad (14)$$

where R = concentrated load or reaction, in pounds

t = thickness of web, in inches

N = length of bearing in inches (not less than k for end reactions)

k = distance from outer face of flange to web toe of fillet, in inches

1.10.10.2 Webs of plate girders shall also be so proportioned or stiffened that the sum of the compression stresses resulting from concentrated and distributed loads, bearing directly on or through a flange plate, upon the compression edge of the web plate, and not supported directly by bearing stiffeners, shall not exceed

$$\left[5.5 + \frac{4}{(a/h)^2} \right] \frac{10{,}000{,}000}{(h/t)^2} \text{ pounds per square inch} \qquad (15)$$

when the flange is restrained against rotation, nor

$$\left[2 + \frac{4}{(a/h)^2} \right] \frac{10{,}000{,}000}{(h/t)^2} \text{ pounds per square inch} \qquad (16)$$

when the flange is not so restrained.

These stresses shall be computed as follows:

> Concentrated loads and loads distributed over partial length of a panel shall be divided by the product of the web thickness and the girder depth or the length of panel in which the load is placed, whichever is the lesser panel dimension.
>
> Any other distributed loading, in pounds per linear inch of length, shall be divided by the web thickness.

In Section 1.10.3 reference is made to Section 1.9, the following parts of which are applicable.

SECTION 1.9 WIDTH-THICKNESS RATIOS

1.9.1 *Projecting elements under compression*

Projecting elements of members subjected to axial compression or compression due to bending shall have ratios of width-to-thickness not greater than the following:

Angles or plates projecting from girders, or other compression members; compression flanges of beams; stiffeners on plate girders. . . . . $3{,}000/\sqrt{F_y}$

The width of plates shall be taken from the free edge to the first row of rivets, bolts or welds; the width of legs of angles, channels and zees, and of the stems of tees, shall be taken as the full nominal dimension; the width of flanges of beams and tees shall be taken as one-half the full nominal width. The thickness of a sloping flange shall be measured halfway between a free edge and the corresponding face of the web.

When a projecting element exceeds the width-to-thickness ratio prescribed in the preceding paragraph, but would conform to same and would satisfy the stress requirements with a portion of its width considered as removed, the member will be acceptable.

1.9.2 *Compression elements supported along two edges*

In compression members the unsupported width of web, cover or diaphragm plates, between the nearest lines of fasteners or welds, or between the roots of the flanges in case of rolled sections, shall not exceed $8{,}000/\sqrt{F_y}$ times its thickness.

When the unsupported width exceeds this limit, but a portion of its width no greater than $8{,}000/\sqrt{F_y}$ times the thickness would satisfy the stress requirements, the member will be considered acceptable.

The unsupported width of cover plates perforated with a succession of access holes may exceed $8{,}000/\sqrt{F_y}$, but shall not exceed $10{,}000/\sqrt{F_y}$, times the thickness. The gross width of the plate less the width of the widest access hole shall be assumed available to resist compression.

The reference to Section 1.18.2.3 or 1.18.3.1 is as follows:

1.18.2.3 The longitudinal spacing for intermediate rivets, bolts or intermittent welds in built-up members shall be adequate to provide for the transfer of calculated stress. However, where a component of a built-up compression member consists of an outside plate, the maximum spacing shall not exceed the thickness of the thinner outside plate times $4{,}000/\sqrt{F_y}$ when rivets are provided on all gage lines at each section, or when intermittent welds are provided along the edges of the components, but this spacing shall not exceed 12 inches. When rivets or bolts are staggered, the maximum spacing on each gage line shall not exceed the thickness of the thinner outside plate times $6{,}000/\sqrt{F_y}$ nor 18 inches. The maximum longitudinal spacing of rivets, bolts or intermittent welds connecting two rolled shapes in contact with one another shall not exceed 24 inches.

1.18.3.1 The longitudinal spacing of rivets, bolts and intermittent fillet welds connecting a plate and a rolled shape in a built-up tension member, or two plate components in contact with one another, shall not exceed 24 times the thickness of the thinner plate nor 12 inches. The longitudinal spacing of rivets, bolts and intermittent welds connecting two or more shapes in contact with one another in a tension member shall not exceed 24 inches. Tension members composed of two or more shapes or plates separated from one another by intermittent fillers shall be connected to one another at these fillers at intervals such that the slenderness ratio of either component between the fasteners does not exceed 240.

The references in Sections 1.10.4, 1.10.5.1, and 1.10.10 refer to portions of Section 1.5, which are quoted in various articles in Chapter 9.

In accordance with the Code, the applicable formulae and stresses listed in Art. 4.10*a* have been set up by the AISC and in addition, the following which are particularly applicable to plate girders.

SECTION 1.10 PLATE GIRDERS AND ROLLED BEAMS A36 STEEL

1.10.2 *Web*

Maximum clear distance between flanges $h = 320t$

1.10.5 *Stiffeners*

1.10.5.3 For required stiffener spacing and gross area of intermediate stiffeners see Table 3.36. (In AISC Manual, page 5-70.)*

* Table 4.9 in this book.

1.10.5.4 Maximum shear between web and intermediate stiffeners in pounds per linear inch of stiffeners or pair of stiffeners $f_{vs} = 35h$

1.10.6 *Reduction in flange stress*

When h/t exceeds $24{,}000/\sqrt{F_b}$ the maximum compression flange stress shall not exceed

$$F_b\left[1.0 - 0.0005\frac{A_w}{A_f}\left(\frac{h}{t} - \frac{24{,}000}{\sqrt{F_b}}\right)\right] \quad (11)$$

1.10.7 *Combined shear and tension*

$$F_b = 29{,}500 - 13{,}500\left(\frac{f_v}{F_v}\right) \leq 22{,}000 \text{ psi} \quad (12)$$

1.10.10 *Web crippling*

1.10.10.1 Use stiffeners under concentrated interior loads when

$$\frac{R}{t(N + 2k)} \text{ would exceed 27,000 psi} \quad (13)$$

and under end reactions when

$$\frac{R}{t(N + k)} \text{ would exceed 27,000 psi} \quad (14)$$

1.10.10.2 The compression stress, in pounds per square inch, produced by loads applied to girder webs, except through stiffeners, shall not exceed

$$\left[5.5 + \frac{4}{(a/h)^2}\right]\frac{10{,}000{,}000}{(h/t)^2} \quad (15)$$

when flange is restrained against rotation; otherwise

$$\left[2 + \frac{4}{(a/h)^2}\right]\frac{10{,}000{,}000}{(h/t)^2} \quad (16)$$

The compression stresses to be limited by formulas (15) and (16) shall be computed as follows:

Concentrated loads and total distributed loads over partial length of a panel shall be divided by the product of the web thickness and the girder depth or the length of the panel in which the load is placed, whichever is the lesser panel dimension. Any other distributed loading, in pounds per linear inch of length, shall be divided by the web thickness.

SECTION 1.10 PLATE GIRDERS AND ROLLED BEAMS $F_y = 50{,}000$ psi.

1.10.2 *Web*

Maximum clear distance between flanges $h = 243t$

1.10.5 *Stiffeners*

1.10.5.3 For required stiffener spacing and gross area of intermediate stiffeners see Table 3.50. (In AISC Manual, page 5-94.)*

* Table 4.10 in this book.

1.10.5.4 Maximum shear between web and intermediate stiffeners, in pounds per linear inch of stiffeners or pair of stiffeners $f_{vs} = 56h$

1.10.6 *Reduction in flange stress*

When h/t exceeds $24{,}000/\sqrt{F_b}$ the maximum compression flange stress shall not exceed

$$F_b\left[1.0 - 0.0005\frac{A_w}{A_f}\left(\frac{h}{t} - \frac{24{,}000}{\sqrt{F_b}}\right)\right] \quad (11)$$

1.10.7 *Combined shear and tension*

$$F_b = 41{,}000 - 18{,}500\left(\frac{f_v}{F_v}\right) \leq 30{,}000 \text{ psi} \quad (12)$$

1.10.10 *Web crippling*

1.10.10.1 Use stiffeners under concentrated interior loads when

$$\frac{R}{t(N + 2k)} \text{ would exceed 37,500 psi} \quad (13)$$

and under end reactions when

$$\frac{R}{t(N + k)} \text{ would exceed 37,500 psi} \quad (14)$$

1.10.10.2 The compression stress, in pounds per square inch, produced by loads applied to girder webs, except through stiffeners, shall not exceed

$$\left[5.5 + \frac{4}{(a/h)^2}\right]\frac{10{,}000{,}000}{(h/t)^2} \quad (15)$$

when flange is restrained against rotation; otherwise

$$\left[2 + \frac{4}{(a/h)^2}\right]\frac{10{,}000{,}000}{(h/t)^2} \quad (16)$$

The compression stresses to be limited by formulas (15) and (16) shall be computed as follows:

Concentrated loads and total distributed loads over partial length of a panel shall be divided by the product of the web thickness and the girder depth or the length of the panel in which the load is placed, whichever is the lesser panel dimension.

Any other distributed loading, in pounds per linear inch of length, shall be divided by the web thickness.

In the application of the AISC Code, a plate girder is not considered a compact section and therefore $F_b = 0.6F_y$.

c The design of plate girders is well illustrated by the discussion and problems given in the AISC Manual, pages 2-55 through 2-65, and the accompanying tables on pages 2-66 through 2-79, for riveted and welded girders, and pages 2-80 through 2-83, and the accompanying tables on pages 2-84 through 2-85, for welded girders.

The moment design is by the same method as has been described in Chapter 3, and the design of the section is in accordance with Art. 1.31, the properties of compound sections, and the tables referred to above are primarily simplifications of and extensions of the material in Art. 1.31. The author has seen fit to duplicate here those tables he has found most useful in his own practice: Tables 4.9 through 4.14.

It will be noted that in the design of riveted plate girders, the distance back-to-back of angles—the base line from which properties are computed—is assumed as the web depth plus ½ to ¾ in. to allow for any irregularities in the plate depth in shop fabrication. Since plates are rolled in even inch widths, it is advisable to adhere to these sizes in selection of both web and cover plates. It is obvious that if the necessary bending and shear can be provided by the use of a rolled WF beam, it is not economical to use a plate girder even if, in order to obtain the required section, it may be necessary to go to a 50,000-psi yield point steel.

In the selection of the section of a plate girder to be used, the depth is usually dictated by maximum allowable architectural requirement and for this determination an appropriate selection of girder from Table 4.14 may be made, even though the girder to be used will be an equivalent plate and angle section. The same relationship of depth to span as dictated by deflection requirements for beams applies with equal force to plate girders, but deflection requirements usually will control only in the case of a long girder with comparatively light loads.

Table 4.13 values may be extended in a straight-line ratio or interpolated. In each column, there are four cross lines. The upper cross line indicates that, for a plate thickness below this line, the plate area is less than 50 percent of the flange. The lower cross line indicates that, for plate thickness below this line, the plate area is over 65 percent of the flange. These tables are based on net section, but may be adjusted for gross section.

d In the AISC Manual, page 2-55, the text suggests that for riveted girders it is generally expeditious to select a trial section by the "flange area method," then check this preliminary section by the "moment of inertia method." The flange area method was for many years the basic method of solution of a plate girder and requires the following steps:

1. With the required moment and shear known, assume an approximate overall depth of the girder in accordance with the factors listed in Art. 4.20*c*.

2. Assume the distance between centers of gravity of flanges jd to be 0.9 of the overall depth of the girder.

3. Assume a trial web plate of a depth to the nearest even inch of approximately jd and compute the required thickness to take the shear from the formula

$$t = \frac{V}{F_v d}$$

4. Calculate the required flange area $A = \frac{Md}{F_b(jd)^2}$. In the above M is given in inch-pounds. This formula is derived from Formula (3.1) in Chapter 3.

5. One-sixth of the gross area of the web is considered to be part of the flange area, so deduct this area from the A which has been computed in step 4.

6. From Table 4.13, select a combination of plates and angles to furnish the net area determined in step 5.

7. Determine the true value of jd of the designed section—the distance back-to-back of angles—less the figure $2y$ for the flange section selected from Table 4.13. Where

Table 4.9 Allowable Shear Stresses (F_v) in Plate Girders (ksi) for 36 ksi Specified Yield Point Steel, F_y = 36,999

Required gross area of intermediate stiffeners, as percent of web area, shown in *italics*.

Slenderness ratios h/t: web depth to web thickness	Aspect ratios a/h: stiffener spacing to web depth													
	0.5	0.6	0.7	0.8	0.9	1.0	1.2	1.4	1.6	1.8	2.0	2.5	3.0	over 3
70											14.3	14.0	13.7	13.1
80							14.2	13.5	13.1	12.8	12.6 *0.7*	12.3 *0.3*	12.1 *0.4*	11.5
90					14.4	13.9	12.6	12.3 *0.6*	12.1 *0.9*	11.9 *1.1*	11.7 *1.2*	11.4 *1.3*	11.2 *1.2*	10.2
100				14.0	13.0	12.4 *0.5*	12.0 *1.4*	11.7 *1.8*	11.4 *2.1*	11.2 *2.1*	11.0 *2.2*	10.4 *2.3*	10.1 *2.1*	8.4
110			14.0	12.7	12.3 *1.0*	12.0 *1.8*	11.6 *2.5*	11.1 *3.1*	10.6 *3.5*	10.3 *3.6*	9.9 *3.6*	9.3 *3.4*	8.9 *3.1*	6.9
120		14.4	12.8	12.3 *1.1*	12.0 *2.1*	11.6 *2.9*	10.9 *4.1*	10.4 *4.7*	9.9 *4.9*	9.5 *4.9*	9.1 *4.8*	8.5 *4.3*	8.0 *3.8*	5.8
130		13.3	12.4 *0.9*	12.0 *2.2*	11.6 *3.2*	11.1 *4.3*	10.4 *5.6*	9.8 *5.9*	9.3 *6.0*	8.9 *5.8*	8.5 *5.6*	7.8 *5.0*	7.4 *4.4*	5.0
140	14.3	12.5 *0.3*	12.1 *1.9*	11.7 *3.2*	11.1 *4.8*	10.6 *5.9*	9.9 *6.7*	9.3 *6.9*	8.8 *6.8*	8.4 *6.6*	8.0 *6.3*	7.3 *5.5*	6.8 *4.9*	4.3
150	13.4	12.3 *1.2*	11.9 *2.8*	11.3 *4.7*	10.8 *6.1*	10.3 *7.1*	9.5 *7.6*	8.9 *7.7*	8.4 *7.5*	8.0 *7.2*	7.6 *6.8*	6.9 *6.0*	6.4 *5.2*	3.7
160	12.6 *0.1*	12.1 *2.1*	11.6 *4.1*	11.0 *6.0*	10.4 *7.2*	10.0 *8.0*	9.2 *8.4*	8.6 *8.3*	8.1 *8.1*	7.7 *7.7*	7.3 *7.3*	6.6 *6.3*		
170	12.4 *0.9*	12.0 *2.8*	11.3 *5.3*	10.7 *7.0*	10.2 *8.1*	9.7 *8.7*	9.0 *9.0*	8.3 *8.9*	7.8 *8.5*	7.4 *8.1*	7.0 *7.7*			
180	12.3 *1.6*	11.7 *4.0*	11.0 *6.4*	10.5 *7.9*	10.0 *8.8*	9.5 *9.4*	8.8 *9.6*	8.1 *9.3*	7.6 *8.9*	7.2 *8.5*	6.8 *8.0*			
200	12.0 *2.9*	11.3 *6.0*	10.7 *8.0*	10.1 *9.2*	9.6 *10.0*	9.2 *10.4*	8.4 *10.4*	7.8 *10.0*	7.3 *9.5*					
220	11.6 *4.8*	10.9 *7.5*	10.4 *9.2*	9.8 *10.2*	9.4 *10.8*	8.9 *11.1*	8.2 *11.0*	7.5 *10.6*						
240	11.3 *6.2*	10.7 *8.6*	10.1 *10.1*	9.6 *11.0*	9.2 *11.5*	8.7 *11.7*	8.0 *11.5*							
260	11.1 *7.3*	10.5 *9.5*	10.0 *10.8*	9.5 *11.6*	9.0 *12.0*									
280	10.9 *8.2*	10.3 *10.2*	9.8 *11.4*	9.3 *12.1*	8.9 *12.4*									
300	10.8 *9.0*	10.2 *10.8*	9.7 *11.8*	9.2 *12.4*										
320	10.7 *9.5*	10.1 *11.2*	9.6 *12.2*	9.1 *12.8*										

Girders so proportioned that the computed shear is less than that given in right-hand column do not require intermediate stiffeners.

the figure is preceded in the tables by a plus sign, $2y$ should be added to the back-to-back depth of angles instead of subtracted.

8. On the basis of the true values as determined, recalculate the required flange area as described in step 4 above. If this area does not agree with the selected flange area, repeat steps 5 through 8 until results are satisfactory.

***e** Length of cover plates.* For practical purposes and for economy, in riveted plate girders any cover plate over ⅞ in. thick is usually made up of two or more plates, and the plates may be stopped off at varying lengths as required by moment considerations. Good shop practice requires that cover plates shall be equal in thickness or shall diminish in thickness from the flange

Table 4.10 Allowable Shear Stresses (F_v) in Plate Girders (ksi) for 50 ksi Specified Yield Point Steel, F_y = 50,000

Required gross area of intermediate stiffeners, as percent of web area, shown in *italics*.

Slenderness ratios h/t: web depth to web thickness	Aspect ratios a/h: stiffener spacing to web depth													
	0.5	0.6	0.7	0.8	0.9	1.0	1.2	1.4	1.6	1.8	2.0	2.5	3.0	over 3
60							20.0	20.0	20.0	20.0	19.7	19.1	18.8	18.1
70					20.0	20.0	19.1	18.2	17.6	17.3	17.1	16.7	16.5	15.5
										0.2	*0.4*	*0.6*	*0.6*	
80				20.0	19.1	17.9	17.1	16.7	16.4	16.1	15.8	15.4	15.0	13.1
							0.6	*1.2*	*1.4*	*1.6*	*1.6*	*1.6*	*1.5*	
90			20.0	18.3	17.3	16.9	16.3	15.8	15.3	14.8	14.3	13.5	13.0	10.4
					0.4	*1.3*	*2.1*	*2.5*	*2.8*	*3.1*	*3.1*	*3.0*	*2.8*	
100		20.0	18.1	17.2	16.7	16.3	15.4	14.6	13.9	13.4	12.9	12.0	11.4	8.4
				0.9	*1.9*	*2.6*	*3.8*	*4.4*	*4.6*	*4.6*	*4.5*	*4.1*	*3.7*	
110	20.0	18.5	17.2	16.7	16.2	15.4	14.4	13.6	12.9	12.3	11.8	10.9	10.3	6.9
			0.9	*2.2*	*3.2*	*4.5*	*5.5*	*5.9*	*5.9*	*5.8*	*5.6*	*5.0*	*4.4*	
120	19.7	17.4	16.8	16.2	15.4	14.7	13.7	12.8	12.1	11.5	11.0	10.1	9.4	5.8
		0.4	*2.0*	*3.4*	*5.0*	*6.1*	*6.9*	*7.0*	*6.9*	*6.7*	*6.4*	*5.8*	*4.9*	
130	18.2	17.0	16.4	15.6	14.8	14.1	13.1	12.2	11.5	10.9	10.4	9.4	8.7	5.0
		1.5	*3.1*	*5.1*	*6.5*	*7.4*	*7.9*	*7.9*	*7.7*	*7.4*	*7.0*	*6.1*	*5.3*	
140	17.3	16.7	15.9	15.1	14.3	13.6	12.6	11.8	11.0	10.4	9.9	8.9	8.2	4.3
	0.5	*2.5*	*4.7*	*6.5*	*7.7*	*8.4*	*8.7*	*8.6*	*8.3*	*7.9*	*7.5*	*6.5*	*5.7*	
150	17.1	16.4	15.5	14.6	13.9	13.3	12.2	11.4	10.7	10.0	9.5	8.5	7.7	3.7
	1.4	*3.6*	*6.0*	*7.6*	*8.6*	*9.2*	*9.4*	*9.2*	*8.8*	*8.4*	*7.9*	*6.8*	*5.9*	
160	16.9	16.0	15.1	14.3	13.6	13.0	11.9	11.1	10.3	9.7	9.2	8.1		
	2.2	*4.9*	*7.1*	*8.5*	*9.4*	*9.8*	*9.9*	*9.7*	*9.2*	*8.7*	*8.2*	*7.1*		
170	16.7	15.6	14.8	14.0	13.3	12.7	11.7	10.8	10.1	9.5	8.9			
	2.9	*6.0*	*8.0*	*9.2*	*10.0*	*10.4*	*10.4*	*10.0*	*9.6*	*9.0*	*8.5*			
180	16.3	15.4	14.5	13.8	13.1	12.5	11.5	10.6	9.9	9.2	8.7			
	4.1	*7.0*	*8.8*	*10.0*	*10.5*	*10.9*	*10.8*	*10.4*	*9.8*	*9.3*	*8.7*			
200	15.8	14.9	14.1	13.4	12.8	12.2	11.1	10.3	9.5					
	5.9	*8.4*	*9.9*	*10.8*	*11.4*	*11.6*	*11.4*	*10.9*	*10.3*					
220	15.4	14.6	13.8	13.2	12.5	11.9	10.9	10.0						
	7.3	*9.5*	*10.8*	*11.6*	*12.0*	*12.1*	*11.8*	*11.3*						
240	15.1	14.3	13.6	13.0	12.3	11.7	10.7							
	8.3	*10.3*	*11.5*	*12.1*	*12.4*	*12.5*	*12.1*							

Girders so proportioned that the computed shear is less than that given in right-hand column do not require intermediate stiffeners.

angles outward. No plate shall be thicker than the flange angles. The total cross-sectional area of cover plates shall not exceed 70 percent of total flange area.

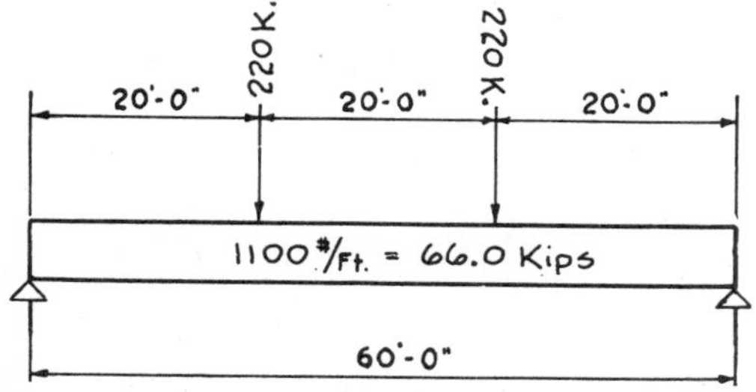

Fig. 4.23 Load Diagram for Plate Girder—Problem

The AISC Manual, page 2-58, gives a simple method of determining the length of cover plates when the moment induced is entirely through uniform load. Frequently, however, the moment results from a combination of uniform and concentrated load as shown in Fig. 4.23. A computation of this girder indicates that the following section would be satisfactory:

- 1 web 44 × 9/16 in.
- 4 flange angles 8 × 8 × 1 in.
- Total flange plates 20 × 2¼ in. (which will be divided into 3 plates, each 20 × ¾ in.).

Table 4.11 Moment of Inertia of Vertical Plate

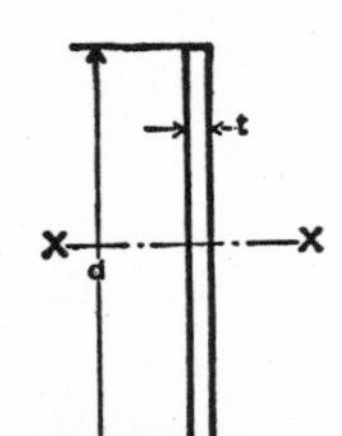

MOMENT OF INERTIA

OF ONE PLATE ABOUT AXIS X-X

To obtain the moment of inertia for any thickness of plate not listed below, multiply the value for a plate one inch thick by the desired thickness.

Depth d Inches	Thickness t, Inches							
	3/8	7/16	1/2	9/16	5/8	3/4	7/8	1
10	31.3	36.5	41.7	46.9	52.1	62.5	72.9	83.3
11	41.6	48.5	55.5	62.4	69.3	83.2	97.1	110.9
12	54.0	63.0	72.0	81.0	90.0	108.0	126.0	144.0
13	68.7	80.1	91.5	103.0	114.4	137.3	160.2	183.1
14	85.8	100.0	114.3	128.6	142.9	171.5	200.1	228.7
15	105.5	123.0	140.6	158.2	175.8	210.9	246.1	281.3
16	128.0	149.3	170.7	192.0	213.3	256.0	298.7	341.3
17	153.5	179.1	204.7	230.3	255.9	307.1	358.2	409.4
18	182.3	212.6	243.0	273.4	303.8	364.5	425.3	486.0
19	214.3	250.1	285.8	321.5	357.2	428.7	500.1	571.6
20	250.0	291.7	333.3	375.0	416.7	500.0	583.3	666.7
21	289.4	337.6	385.9	434.1	482.3	578.8	675.3	771.8
22	332.8	388.2	443.7	499.1	554.6	665.5	776.4	887.3
23	380.2	443.6	507.0	570.3	633.7	760.4	887.2	1013.9
24	432.0	504.0	576.0	648.0	720.0	864.0	1008.0	1152.0
25	488.3	569.7	651.0	732.4	813.8	976.6	1139.3	1302.1
26	549.3	640.8	732.3	823.9	915.4	1098.5	1281.6	1464.7
27	615.1	717.6	820.1	922.6	1025.2	1230.2	1435.2	1640.3
28	686.0	800.3	914.7	1029.0	1143.3	1372.0	1600.7	1829.3
29	762.2	889.2	1016.2	1143.2	1270.3	1524.3	1778.4	2032.4
30	843.8	984.4	1125.0	1265.6	1406.3	1687.5	1968.8	2250.0
31	931.0	1086.1	1241.3	1396.5	1551.6	1861.9	2172.3	2482.6
32	1024.0	1194.7	1365.3	1536.0	1706.7	2048.0	2389.3	2730.7
33	1123.0	1310.2	1497.4	1684.5	1871.7	2246.1	2620.4	2994.8
34	1228.3	1433.0	1637.7	1842.4	2047.1	2456.5	2865.9	3275.3
35	1339.8	1563.2	1786.5	2009.8	2233.1	2679.7	3126.3	3572.9
36	1458.0	1701.0	1944.0	2187.0	2430.0	2916.0	3402.0	3888.0
37	1582.9	1846.7	2110.5	2374.4	2638.2	3165.8	3693.4	4221.1
38	1714.8	2000.5	2286.3	2572.1	2857.9	3429.5	4001.1	4572.7
39	1853.7	2162.7	2471.6	2780.6	3089.5	3707.4	4325.3	4943.3
40	2000.0	2333.3	2666.7	3000.0	3333.3	4000.0	4666.7	5333.3
41	2153.8	2512.7	2871.7	3230.7	3589.6	4307.6	5025.5	5743.4
42	2315.3	2701.1	3087.0	3472.9	3858.8	4630.5	5402.3	6174.0
43	2484.6	2898.7	3312.8	3726.9	4141.0	4969.2	5797.4	6625.6
44	2662.0	3105.7	3549.3	3993.0	4436.7	5324.0	6211.3	7098.7
45	2847.7	3322.3	3796.9	4271.5	4746.1	5695.3	6644.5	7593.8
46	3041.8	3548.7	4055.7	4562.6	5069.6	6083.5	7097.4	8111.3
47	3244.5	3785.2	4326.0	4866.7	5407.4	6488.9	7570.4	8651.9
48	3456.0	4032.0	4608.0	5184.0	5760.0	6912.0	8064.0	9216.0
49	3676.5	4289.3	4902.0	5514.8	6127.6	7353.1	8578.6	9804.1
50	3906.3	4557.3	5208.3	5859.4	6510.4	7812.5	9114.6	10417
51	4145.3	4836.2	5527.1	6218.0	6908.9	8290.7	9672.5	11054
52	4394.0	5126.3	5858.7	6591.0	7323.3	8788.0	10253	11717
53	4652.4	5427.8	6203.2	6978.6	7754.0	9304.8	10856	12406
54	4920.8	5740.9	6561.0	7381.1	8201.3	9841.5	11482	13122

Table 4.11 Moment of Inertia of Vertical Plate (*Continued*)

MOMENT OF INERTIA

OF ONE PLATE ABOUT AXIS X-X

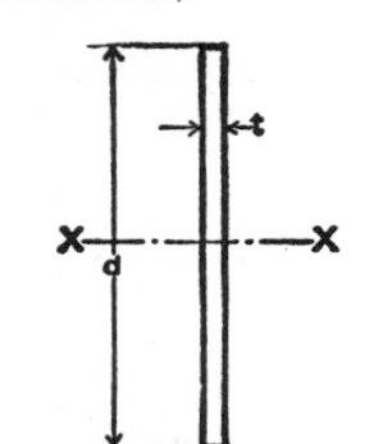

To obtain the moment of inertia for any thickness of plate not listed below, multiply the value for a plate one inch thick by the desired thickness.

Depth d Inches	Thickness t, Inches							
	3/8	7/16	1/2	9/16	5/8	3/4	7/8	1
55	5199.2	6065.8	6932.3	7798.8	8665.4	10398	12132	13865
56	5488.0	6402.7	7317.3	8232.0	9146.7	10976	12805	14635
57	5787.3	6751.8	7716.4	8680.9	9645.5	11575	13504	15433
58	6097.3	7113.5	8129.7	9145.9	10162	12195	14227	16259
59	6418.1	7487.8	8557.5	9627.1	10697	12836	14976	17115
60	6750.0	7875.0	9000.0	10125	11250	13500	15750	18000
61	7093.2	8275.3	9457.5	10640	11822	14186	16551	18915
62	7447.8	8689.0	9930.3	11172	12413	14896	17378	19861
63	7814.0	9116.3	10419	11721	13023	15628	18232	20837
64	8192.0	9557.3	10923	12288	13653	16384	19115	21845
65	8582.0	10012	11443	12873	14303	17164	20025	22885
66	8984.3	10482	11979	13476	14974	17969	20963	23958
67	9398.8	10965	12532	14098	15665	18798	21931	25064
68	9826.0	11464	13101	14739	16377	19652	22927	26203
69	10266	11977	13688	15399	17110	20532	23954	27376
70	10719	12505	14292	16078	17865	21438	25010	28583
72	11664	13608	15552	17496	19440	23328	27216	31104
74	12663	14774	16884	18995	21105	25327	29548	33769
76	13718	16004	18291	20577	22863	27436	32009	36581
78	14830	17301	19773	22245	24716	29660	34603	39546
80	16000	18667	21333	24000	26667	32000	37333	42667
82	17230	20102	22974	25845	28717	34461	40204	45947
84	18522	21609	24696	27783	30870	37044	43218	49392
86	19877	23190	26502	29815	33128	39754	46379	53005
88	21296	24845	28395	31944	35493	42592	49691	56789
90	22781	26578	30375	34172	37969	45563	53156	60750
92	24334	28390	32445	36501	40557	48668	56779	64891
94	25956	30282	34608	38934	43260	51912	60563	69215
96	27648	32256	36864	41472	46080	55296	64512	73728
98	29412	34314	39216	44118	49020	58825	68629	78433
100	31250	36458	41667	46875	52083	62500	72917	83333
102	33163	38690	44217	49744	55271	66326	77380	88434
104	35152	41011	46869	52728	58587	70304	82021	93739
106	37219	43422	49626	55829	62032	74439	86845	99251
108	39366	45927	52488	59049	65610	78732	91854	104976
110	41594	48526	55458	62391	69323	83188	97052	110917
112	43904	51221	58539	65856	73173	87808	102443	117077
114	46298	54015	61731	69447	77164	92597	108029	123462
116	48778	56908	65037	73167	81297	97556	113815	130075
118	51345	59902	68460	77017	85575	102690	119804	136919
120	54000	63000	72000	81000	90000	108000	126000	144000
122	56745	66203	75660	85118	94575	113491	132406	151321
124	59582	69512	79443	89373	99303	119164	139025	158885
126	62512	72930	83349	93768	104186	125024	145861	166698
128	65536	76459	87381	98304	109227	131072	152917	174763

Table 4.12

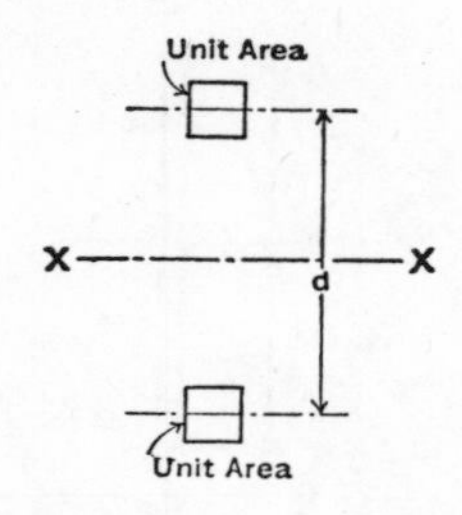

MOMENT OF INERTIA OF A PAIR OF UNIT AREAS ABOUT AXIS X-X

d	.0	.1	.2	.3	.4	.5	.6	.7	.8	.9
10	50	51	52	53	54	55	56	57	58	59
11	61	62	63	64	65	66	67	68	70	71
12	72	73	74	76	77	78	79	81	82	83
13	85	86	87	88	90	91	92	94	95	97
14	98	99	101	102	104	105	107	108	110	111
15	113	114	116	117	119	120	122	123	125	126
16	128	130	131	133	134	136	138	139	141	143
17	145	146	148	150	151	153	155	157	158	160
18	162	164	166	167	169	171	173	175	177	179
19	181	182	184	186	188	190	192	194	196	198
20	200	202	204	206	208	210	212	214	216	218
21	221	223	225	227	229	231	233	235	238	240
22	242	244	246	249	251	253	255	258	260	262
23	265	267	269	271	274	276	278	281	283	286
24	288	290	293	295	298	300	303	305	308	310
25	313	315	318	320	323	325	328	330	333	335
26	338	341	343	346	348	351	354	356	359	362
27	365	367	370	373	375	378	381	384	386	389
28	392	395	398	400	403	406	409	412	415	418
29	421	423	426	429	432	435	438	441	444	447
30	450	453	456	459	462	465	468	471	474	477
31	481	484	487	490	493	496	499	502	506	509
32	512	515	518	522	525	528	531	535	538	541
33	545	548	551	554	558	561	564	568	571	575
34	578	581	585	588	592	595	598	602	606	609
35	613	616	620	623	627	630	634	637	641	644
36	648	652	655	659	662	666	670	673	677	681
37	685	688	692	696	699	703	707	711	714	718
38	722	726	730	733	737	741	745	749	753	757
39	761	764	768	772	776	780	784	788	792	796
40	800	804	808	812	816	820	824	828	832	836
41	841	845	849	853	857	861	865	869	874	878
42	882	886	890	895	899	903	907	912	916	920
43	925	929	933	937	942	946	950	955	959	964
44	968	972	977	981	986	990	995	999	1004	1008
45	1013	1017	1022	1026	1031	1035	1040	1044	1049	1053
46	1058	1063	1067	1072	1076	1081	1086	1090	1095	1100
47	1105	1109	1114	1119	1123	1128	1133	1138	1142	1147
48	1152	1157	1162	1166	1171	1176	1181	1186	1191	1196
49	1201	1205	1210	1215	1220	1225	1230	1235	1240	1245

Table 4.12 (*Continued*)

MOMENT OF INERTIA OF A PAIR OF UNIT AREAS ABOUT AXIS X-X

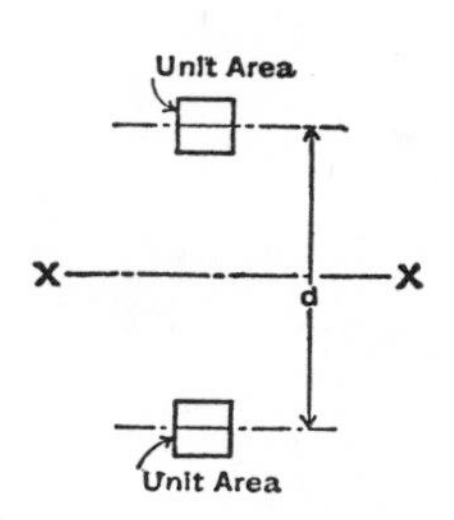

d	.0	.1	.2	.3	.4	.5	.6	.7	.8	.9
50	1250	1255	1260	1265	1270	1275	1280	1285	1290	1295
51	1301	1306	1311	1316	1321	1326	1331	1336	1342	1347
52	1352	1357	1362	1368	1373	1378	1383	1389	1394	1399
53	1405	1410	1415	1420	1426	1431	1436	1442	1447	1453
54	1458	1463	1469	1474	1480	1485	1491	1496	1502	1507
55	1513	1518	1524	1529	1535	1540	1546	1551	1557	1562
56	1568	1574	1579	1585	1590	1596	1602	1607	1613	1619
57	1625	1630	1636	1642	1647	1653	1659	1665	1670	1676
58	1682	1688	1694	1699	1705	1711	1717	1723	1729	1735
59	1741	1746	1752	1758	1764	1770	1776	1782	1788	1794
60	1800	1806	1812	1818	1824	1830	1836	1842	1848	1854
61	1861	1867	1873	1879	1885	1891	1897	1903	1910	1916
62	1922	1928	1934	1941	1947	1953	1959	1966	1972	1978
63	1985	1991	1997	2003	2010	2016	2022	2029	2035	2042
64	2048	2054	2061	2067	2074	2080	2087	2093	2100	2106
65	2113	2119	2126	2132	2139	2145	2152	2158	2165	2171
66	2178	2185	2191	2198	2204	2211	2218	2224	2231	2238
67	2245	2251	2258	2265	2271	2278	2285	2292	2298	2305
68	2312	2319	2326	2332	2339	2346	2353	2360	2367	2374
69	2381	2387	2394	2401	2408	2415	2422	2429	2436	2443
70	2450	2457	2464	2471	2478	2485	2492	2499	2506	2513
71	2521	2528	2535	2542	2549	2556	2563	2570	2578	2585
72	2592	2599	2606	2614	2621	2628	2635	2643	2650	2657
73	2665	2672	2679	2686	2694	2701	2708	2716	2723	2731
74	2738	2745	2753	2760	2768	2775	2783	2790	2798	2805
75	2813	2820	2828	2835	2843	2850	2858	2865	2873	2880
76	2888	2896	2903	2911	2918	2926	2934	2941	2949	2957
77	2965	2972	2980	2988	2995	3003	3011	3019	3026	3034
78	3042	3050	3058	3065	3073	3081	3089	3097	3105	3113
79	3121	3128	3136	3144	3152	3160	3168	3176	3184	3192
80	3200	3208	3216	3224	3232	3240	3248	3256	3264	3272
81	3281	3289	3297	3305	3313	3321	3329	3337	3346	3354
82	3362	3370	3378	3387	3395	3403	3411	3420	3428	3436
83	3445	3453	3461	3469	3478	3486	3494	3503	3511	3520
84	3528	3536	3545	3553	3562	3570	3579	3587	3596	3604
85	3613	3621	3630	3638	3647	3655	3664	3672	3681	3689
86	3698	3707	3715	3724	3732	3741	3750	3758	3767	3776
87	3785	3793	3802	3811	3819	3828	3837	3846	3854	3863
88	3872	3881	3890	3898	3907	3916	3925	3934	3943	3952
89	3961	3969	3978	3987	3996	4005	4014	4023	4032	4041

Table 4.13 Properties of Plate-Girder Flanges

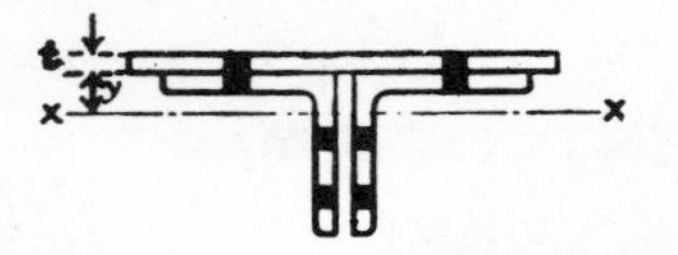

Angles 6 × 6 in., cover plates 14 in. wide, ⅞-in. rivets

Cover plate thickness	Thickness of angles																%
	⅜		7/16		½		9/16		⅝		¾		⅞		1		
	A	$2y$	A	$2y$	A	$2y$	A	$2y$	A	$2y$	A	$2y$	A	$2y$	A	$2y$	
0	7.22	3.28	8.37	3.32	9.50	3.36	10.61	3.42	11.72	3.46	13.88	3.56	15.96	3.64	18.00	3.72	
⅝	14.72	1.32															
¾	16.22	1.08	17.37	1.24													
⅞	17.72	0.85	18.87	1.02	20.00	1.17	21.11	1.32									
1	19.22	0.64	20.37	0.82	21.50	0.96	22.61	1.12	23.72	1.24							
1⅛	20.72	0.44	21.87	0.62	23.00	0.77	24.11	0.92	25.22	1.04							
1¼			23.27	0.42	24.50	0.58	25.61	0.72	26.72	0.86	28.88	1.11					
1⅜			24.87	0.24	26.00	0.39	27.11	0.54	28.22	0.68	30.38	0.93	32.46	1.14			
1½					27.50	0.22	28.61	0.36	29.72	0.50	31.88	0.75	33.96	0.97			50
1⅝							30.11	0.20	31.22	0.32	33.38	0.58	35.46	0.80	37.50	1.00	
1¾							31.61	0.03	32.72	0.16	34.88	0.42	36.96	0.64	39.00	0.84	
1⅞									34.22	0.00	36.38	0.25	38.46	0.48	40.50	0.68	45
2											37.88	0.09	39.96	0.32	42.00	0.52	
																	40
																	35

Angles 6 × 6 in., cover plates 16 in. wide, ⅞-in. rivets

Cover plate thickness	⅜ A	⅜ $2y$	7/16 A	7/16 $2y$	½ A	½ $2y$	9/16 A	9/16 $2y$	⅝ A	⅝ $2y$	¾ A	¾ $2y$	⅞ A	⅞ $2y$	1 A	1 $2y$	%
0	7.22	3.28	8.37	3.32	9.50	3.36	10.61	3.42	11.72	3.46	13.88	3.56	15.96	3.64	18.00	3.72	
⅝	15.97	1.20	17.12	1.36													
¾	17.72	0.94	18.87	1.12	20.00	1.26	21.11	1.40									
⅞	19.47	0.72	20.62	0.89	21.75	1.03	22.86	1.18	23.97	1.30							
1	21.22	0.52	22.37	0.68	23.50	0.82	24.61	0.98	25.72	1.10	27.88	1.34					
1½			24.12	0.48	25.25	0.62	26.36	0.78	27.47	0.90	29.63	1.14					
1¼					27.00	0.43	28.11	0.58	29.22	0.70	31.38	0.94	33.46	1.16			
1⅜					28.75	0.25	29.86	0.39	30.97	0.51	33.13	0.76	35.21	0.97	37.25	1.16	50
1½							31.61	0.22	32.72	0.34	34.88	0.58	36.96	0.80	39.00	1.00	
1⅝									34.47	0.18	36.63	0.40	38.71	0.63	40.75	0.82	45
1¾											38.38	0.24	40.46	0 46	42.50	0.66	
1⅞											40.13	0.08	42.21	0.30	44.25	0.49	
2											41.88	+0.08	43.96	0.14	46.00	0.34	40
																	35

Dimensions in inches and square inches. Values of A are net areas of one flange only and are based on deduction of three holes from each angle and two from each cover plate—all ⅛ in. larger than the nominal diameter of the rivet.

Heavy rules indicate relationship of gross area of angles to total gross area of flange, expressed in last column as percentage.

Courtesy Bethlehem Steel Co.

Table 4.13 Properties of Plate-Girder Flanges (*Continued*)

Angles 8 × 6 in. (long leg vertical), cover plates 14 in. wide, $\frac{7}{8}$-in. rivets

Cover plate thickness	Thickness of angles: $\frac{7}{16}$		$\frac{1}{2}$		$\frac{9}{16}$		$\frac{5}{8}$		$\frac{3}{4}$		$\frac{7}{8}$		1		%
	A	$2y$	A	$2y$	A	$2y$	A	$2y$	A	$2y$	A	$2y$	A	$2y$	
0	9.24	4.90	10.50	4.94	11.75	5.00	12.97	5.04	15.38	5.12	17.71	5.22	20.00	5.30	
$\frac{7}{8}$	19.74	1.97													
1	21.24	1.71	22.50	1.92											
$1\frac{1}{8}$	22.74	1.46	24.00	1.68	25.25	1.88									
$1\frac{1}{4}$	24.24	1.23	25.50	1.45	26.75	1.64	27.97	1.82							
$1\frac{3}{8}$	25.74	1.01	27.00	1.23	28.25	1.42	29.47	1.60	31.88	1.92					
$1\frac{1}{2}$	27.24	0.81	28.50	1.02	29.75	1.22	30.97	1.40	33.38	1.72					
$1\frac{5}{8}$	28.74	0.62	30.00	0.82	31.25	1.02	32.47	1.20	34.88	1.52	37.21	1.82			
$1\frac{3}{4}$			31.50	0.62	32.75	0.82	33.97	1.00	36.38	1.32	38.71	1.62			**50**
$1\frac{7}{8}$					34.25	0.63	35.47	0.82	37.88	1.14	40.21	1.44	42.50	1.70	
2					35.75	0.45	36.97	0.64	39.38	0.96	41.71	1.26	44.00	1.52	
$2\frac{1}{8}$							38.47	0.46	40.88	0.78	43.21	1.08	45.50	1.34	
$2\frac{1}{4}$							39.97	0.28	42.38	0.60	44.71	0.90	47.00	1.17	**45**
$2\frac{3}{8}$									43.88	0.42	46.21	0.72	48.50	0.99	
$2\frac{1}{2}$									45.38	0.26	47.71	0.56	50.00	0.82	**40**
															35

Angles 8 × 6 in. (long leg vertical), cover plates 16 in. wide, $\frac{7}{8}$-in. rivets

Cover plate thickness	Thickness of angles: $\frac{7}{16}$		$\frac{1}{2}$		$\frac{9}{16}$		$\frac{5}{8}$		$\frac{3}{4}$		$\frac{7}{8}$		1		%
	A	$2y$	A	$2y$	A	$2y$	A	$2y$	A	$2y$	A	$2y$	A	$2y$	
0	9.24	4.90	10.50	4.94	11.74	5.00	12.97	5.04	15.38	5.12	17.71	5.22	20.00	5.30	
$\frac{7}{8}$	21.48	1.77	22.75	1.98											
1	23.24	1.52	24.50	1.72	25.74	1.92	26.97	2.08							
$1\frac{1}{8}$	24.98	1.28	26.25	1.48	27.50	1.68	28.72	1.84							
$1\frac{1}{4}$	26.74	1.04	28.00	1.24	29.24	1.44	30.47	1.62	32.88	1.93					
$1\frac{3}{8}$	28.48	0.82	29.75	1.02	31.00	1.22	32.22	1.39	34.63	1.71	36.96	2.00			
$1\frac{1}{2}$			31.50	0.82	32.74	1.02	33.97	1.18	36.38	1.50	38.71	1.78			
$1\frac{5}{8}$			33.25	0.62	34.50	0.82	35.72	0.98	38.13	1.30	40.46	1.58	42.75	1.84	**50**
$1\frac{3}{4}$					36.24	0.62	37.47	0.78	39.88	1.10	42.21	1.40	44.50	1.64	
$1\frac{7}{8}$							39.22	0.59	41.63	0.91	43.96	1.20	46.25	1.45	**45**
2							40.97	0.42	43.38	0.73	45.71	1.01	48.00	1.27	
$2\frac{1}{8}$									45.13	0.55	47.46	0.84	49.75	1.09	
$2\frac{1}{4}$									46.88	0.37	49.21	0.66	51.50	0.91	
$2\frac{3}{8}$											50.96	0.48	53.25	0.74	**40**
$2\frac{1}{2}$											52.71	0.32	55.00	0.58	
															35

Dimensions in inches and square inches. Values of A are net areas of one flange only and are based on deduction of three holes from each angle and two from each cover plate—all $\frac{1}{8}$ in. larger than the nominal diameter of the rivet.

Heavy rules indicate relationship of gross area of angles to total gross area of flange, expressed in last column as percentage.

Courtesy Bethlehem Steel Co.

Table 4.13 Properties of Plate-Girder Flanges (*Continued*)

Angles 8 × 6 in. (short leg vertical), cover plates 18 in. wide, $\frac{7}{8}$-in. rivets

Cover plate thickness	Thickness of angles $\frac{7}{16}$		$\frac{1}{2}$		$\frac{9}{16}$		$\frac{5}{8}$		$\frac{3}{4}$		$\frac{7}{8}$		1		%
	A	$2y$	A	$2y$	A	$2y$	A	$2y$*	A	$2y$*	A	$2y$*	A	$2y$*	
0	10.11	2.90	11.50	2.94	12.87	3.00	14.22	3.04	16.88	3.12	19.46	3.22	22.00	3.30	
$\frac{7}{8}$	24.11	0.74	25.50	0.88	26.87	1.02	28.22	1.14							
1	26.11	0.54	27.50	0.68	28.87	0.82	30.22	0.94							
$1\frac{1}{8}$	28.11	0.36	29.50	0.50	30.87	0.64	32.22	0.76	34.88	0.98					
$1\frac{1}{4}$	30.11	0.18	31.50	0.32	32.87	0.46	34.22	0.58	36.88	0.80	39.46	1.00			
$1\frac{3}{8}$			33.50	0.14	34.87	0.28	36.22	0.40	38.88	0.62	41.46	0.84			50
$1\frac{1}{2}$					36.87	0.12	38.22	0.24	40.88	0.46	43.46	0.66	46.00	0.86	
$1\frac{5}{8}$							40.22	0.08	42.88	0.30	45.46	0.50	48.00	0.70	
$1\frac{3}{4}$							42.22	+0.09	44.88	0.14	47.46	0.34	50.00	0.54	45
$1\frac{7}{8}$									46.88	+0.02	49.46	0.18	52.00	0.38	
2									48.88	+0.18	51.46	0.04	54.00	0.22	
$2\frac{1}{8}$											53.46	+0.12	56.00	0.07	40
$2\frac{1}{4}$											55.46	+0.28	58.00	+0.08	
$2\frac{3}{8}$											57.46	+0.42	60.00	+0.23	
$2\frac{1}{2}$													62.00	+0.37	
															35

Dimensions in inches and square inches. Values of A are net areas of one flange only and are based on deduction of three holes from each angle and two from each cover plate—all $\frac{1}{8}$ in. larger than the nominal diameter of the rivet.

Heavy rules indicate relationship of gross area of angles to total gross area of flange, expressed in last column as percentage.

*Values marked + must be added to back-to-back of angles to obtain center-to-center of flanges.

Courtesy Bethlehem Steel Co.

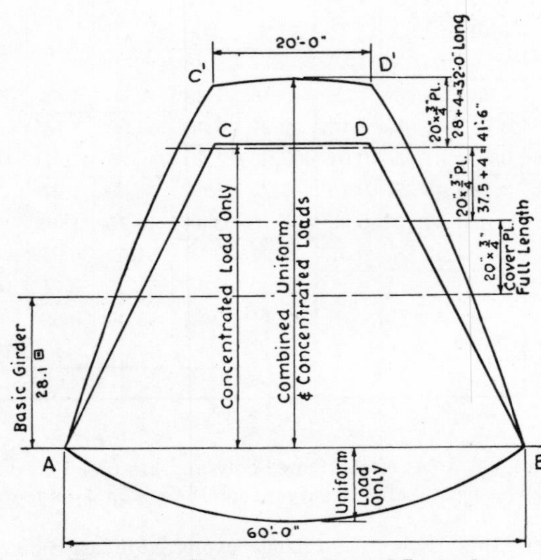

Fig. 4.24 Determination of Length of Cover Plates

The moment diagram due to these loads is shown in Fig. 4.24. On this diagram, lines have been drawn parallel with base A-B, the distance between these lines being proportional to the I of each plate and of the basic section. While this is not absolutely accurate since I and S do not vary uniformly, the variation is so slight that it does not result in any material difference.

4.21 Cover-plated beams

Sometimes it is advisable in practice to increase the section modulus of a rolled section by riveting or welding on top and bottom cover plates. This method does not, of course, increase the shear or buckling value so care should be taken to check against overstress in these. It is not necessary to

Table 4.13 Properties of Plate-Girder Flanges (*Continued*)

Angles 8 × 8 in., cover plates 18 in. wide, $\frac{7}{8}$-in. rivets

Cover plate thickness	Thickness of angles														%
	$\frac{1}{2}$		$\frac{9}{16}$		$\frac{5}{8}$		$\frac{3}{4}$		$\frac{7}{8}$		1		$1\frac{1}{8}$		
	A	$2y$	A	$2y$	A	$2y$	A	$2y$	A	$2y$	A	$2y$	A	$2y$	
0	12.50	4.38	13.99	4.42	15.47	4.46	18.38	4.56	21.21	4.64	24.00	4.74	26.71	4.82	
$\frac{7}{8}$	26.50	1.72													
1	28.50	1.48	29.99	1.66											
$1\frac{1}{8}$	30.50	1.26	31.99	1.44	33.47	1.59									
$1\frac{1}{4}$	32.50	1.04	33.99	1.22	35.47	1.38	38.38	1.68							
$1\frac{3}{8}$	34.50	0.84	35.99	1.02	37.47	1.18	40.38	1.48							
$1\frac{1}{2}$	36.50	0.64	37.99	0.82	39.47	0.98	42.38	1.28	45.21	1.54					
$1\frac{5}{8}$	38.50	0.46	39.99	0.63	41.47	0.80	44.38	1.09	47.21	1.35	50.00	1.60			
$1\frac{3}{4}$			41.99	0.44	43.47	0.60	46.38	0.90	49.21	1.17	52.00	1.41			**50**
$1\frac{7}{8}$					45.47	0.42	48.38	0.72	51.21	0.97	54.00	1.23	56.71	1.45	
2					47.47	0.24	50.38	0.55	53.21	0.81	56.00	1.06	58.71	1.28	
$2\frac{1}{8}$							52.38	0.38	55.21	0.64	58.00	0.90	60.71	1.12	
$2\frac{1}{4}$							54.38	0.21	57.21	0.47	60.00	0.72	62.71	0.95	**45**
$2\frac{3}{8}$							56.38	0.04	59.21	0.30	62.00	0.56	64.71	0.78	
$2\frac{1}{2}$									61.21	0.14	64.00	0.40	66.71	0.62	**40**
															35

Angles 8 × 8 in., cover plates 20 in. wide, $\frac{7}{8}$-in. rivets

Cover plate thickness	Thickness of angles														%
	$\frac{1}{2}$		$\frac{9}{16}$		$\frac{5}{8}$		$\frac{3}{4}$		$\frac{7}{8}$		1		$1\frac{1}{8}$		
	A	$2y$	A	$2y$	A	$2y$	A	$2y$	A	$2y$	A	$2y$	A	$2y$	
0	12.50	4.38	13.99	4.42	15.47	4.46	18.38	4.56	21.21	4.64	24.00	4.74	26.71	4.82	
$\frac{7}{8}$	28.25	1.60	29.74	1.77											
1	30.50	1.35	31.99	1.52	33.47	1.68									
$1\frac{1}{8}$	32.75	1.12	34.24	1.29	35.72	1.45	38.63	1.74							
$1\frac{1}{4}$	35.00	0.90	36.49	1.07	37.97	1.23	40.88	1.53							
$1\frac{3}{8}$	37.25	0.70	38.74	0.86	40.22	1.02	43.13	1.32	45.96	1.58					
$1\frac{1}{2}$	39.50	0.50	40.99	0.66	42.47	0.82	45.38	1.12	48.21	1.38	51.00	1.62			
$1\frac{5}{8}$			43.24	0.48	44.72	0.64	47.63	0.93	50.46	1.18	53.25	1.42	55.96	1.64	**50**
$1\frac{3}{4}$					46.97	0.46	49.88	0.75	52.71	1.00	55.50	1.24	58.21	1.46	
$1\frac{7}{8}$							52.13	0.57	54.96	0.82	57.75	1.06	60.46	1.28	
2							54.38	0.39	57.21	0.64	60.00	0.88	62.71	1.10	**45**
$2\frac{1}{8}$							56.63	0.22	59.46	0.48	62.25	0.72	64.96	0.93	
$2\frac{1}{4}$									61.71	0.30	64.50	0.54	67.21	0.76	
$2\frac{3}{8}$									63.96	0.14	66.75	0.38	69.46	0.60	
$2\frac{1}{2}$									66.21	+0.02	69.00	0.22	71.71	0.44	**40**
															35

Dimensions in inches and square inches. Values of A are net areas of one flange only and are based on deduction of three holes from each angle and two from each cover plate—all $\frac{1}{8}$ in. larger than the nominal diameter of the rivet.

Heavy rules indicate relationship of gross area of angles to total gross area of flange, expressed in last column as percentage.

*Values marked + must be added to back-to-back of angles to obtain center-to-center of flanges.

Courtesy Bethlehem Steel Co.

Table 4.14

86-61

WELDED PLATE GIRDERS
Dimensions and properties

Nominal Size / h/t Ratio	Wt. per Foot	Area	Depth d	Flange Width b_f	Flange Thick t_f	Web Depth h	Web Thick t	Axis X-X I	Axis X-X S	Axis X-X $^aS'$	^br	^cR	$\frac{d}{A_f}$
in.	lb.	in.^{2}	in.	in.	in.	in.	in.	in.^{4}	in.^{3}	in.^{3}	in.	kips	in.$^{-1}$
86 × 28	749.7	220.50	90.00	28	3	84	5/8	348894.0	7753.2	68.6	7.69	241.6	1.07
$h/t = 134$	654.5	192.50	89.00	28	2½	84	5/8	292821.7	6580.3	69.4	7.62	241.6	1.27
	559.3	164.50	88.00	28	2	84	5/8	237995.3	5409.0	70.2	7.52	241.6	1.57
	511.7	150.50	87.50	28	1¾	84	5/8	211045.0	4823.9	70.6	7.45	241.6	1.79
	464.1	136.50	87.00	28	1½	84	5/8	184401.0	4239.1	71.0	7.35	241.6	2.07
	416.5	122.50	86.50	28	1¼	84	5/8	158061.5	3654.6	71.4	7.23	241.6	2.47
	368.9	108.50	86.00	28	1	84	5/8	132024.7	3070.3	71.8	7.06	241.6	3.07
	345.1	101.50	85.75	28	7/8	84	5/8	119119.3	2778.3	72.0	6.94	241.6	3.50
80 × 26	696.2	204.75	84.00	26	3	78	5/8	280712.3	6683.6	58.8	7.14	260.2	1.08
$h/t = 125$	607.8	178.75	83.00	26	2½	78	5/8	235392.1	5672.1	59.6	7.08	260.2	1.28
	519.4	152.75	82.00	26	2	78	5/8	191150.9	4662.2	60.3	6.98	260.2	1.58
	475.2	139.75	81.50	26	1¾	78	5/8	169430.9	4157.8	60.7	6.91	260.2	1.79
	431.0	126.75	81.00	26	1½	78	5/8	147975.8	3653.7	61.0	6.83	260.2	2.08
	386.8	113.75	80.50	26	1¼	78	5/8	126783.9	3149.9	61.4	6.71	260.2	2.48
	342.6	100.75	80.00	26	1	78	5/8	105853.6	2646.3	61.8	6.55	260.2	3.08
	320.5	94.25	79.75	26	7/8	78	5/8	95486.0	2394.6	62.0	6.44	260.2	3.51
74 × 24	627.3	184.50	78.00	24	3	72	9/16	220104.0	5643.7	49.8	6.62	205.5	1.08
$h/t = 128$	545.7	160.50	77.00	24	2½	72	9/16	184066.0	4780.9	50.5	6.57	205.5	1.28
	464.1	136.50	76.00	24	2	72	9/16	148952.0	3919.8	51.2	6.49	205.5	1.58
	423.3	124.50	75.50	24	1¾	72	9/16	131737.8	3489.7	51.5	6.43	205.5	1.80
	382.5	112.50	75.00	24	1½	72	9/16	114750.0	3060.0	51.8	6.36	205.5	2.08
	341.7	100.50	74.50	24	1¼	72	9/16	97987.2	2630.5	52.2	6.26	205.5	2.48
	300.9	88.50	74.00	24	1	72	9/16	81448.0	2201.3	52.5	6.12	205.5	3.08
	280.5	82.50	73.75	24	7/8	72	9/16	73261.7	1986.8	52.7	6.03	205.5	3.51
68 × 22	561.0	165.00	72.00	22	3	66	½	169191.0	4699.8	41.6	6.10	157.5	1.09
$h/t = 132$	486.2	143.00	71.00	22	2½	66	½	141073.2	3973.9	42.2	6.06	157.5	1.29
	411.4	121.00	70.00	22	2	66	½	113736.3	3249.6	42.8	5.99	157.5	1.59
	374.0	110.00	69.50	22	1¾	66	½	100357.4	2888.0	43.1	5.94	157.5	1.81
	336.6	99.00	69.00	22	1½	66	½	87169.5	2526.7	43.4	5.88	157.5	2.09
	299.2	88.00	68.50	22	1¼	66	½	74171.4	2165.6	43.7	5.80	157.5	2.49
	261.8	77.00	68.00	22	1	66	½	61361.7	1804.8	44.0	5.68	157.5	3.09
	243.1	71.50	67.75	22	7/8	66	½	55027.0	1624.4	44.2	5.60	157.5	3.52
	224.4	66.00	67.50	22	¾	66	½	48738.9	1444.1	44.4	5.50	157.5	4.09
61 × 20	429.3	126.25	65.00	20	2½	60	7/16	105583.3	3248.7	34.6	5.54	116.0	1.30
$h/t = 137$	361.3	106.25	64.00	20	2	60	7/16	84781.7	2649.4	35.2	5.48	116.0	1.60
	327.3	96.25	63.50	20	1¾	60	7/16	74621.5	2350.3	35.4	5.44	116.0	1.81
	293.3	86.25	63.00	20	1½	60	7/16	64620.0	2051.4	35.7	5.39	116.0	2.10
	259.3	76.25	62.50	20	1¼	60	7/16	54776.0	1752.8	36.0	5.33	116.0	2.50
	225.3	66.25	62.00	20	1	60	7/16	45088.3	1454.5	36.3	5.23	116.0	3.10
	208.3	61.25	61.75	20	7/8	60	7/16	40302.7	1305.4	36.4	5.16	116.0	3.53
	191.3	56.25	61.50	20	¾	60	7/16	35555.6	1156.3	36.6	5.08	116.0	4.10
	174.3	51.25	61.25	20	5/8	60	7/16	30847.0	1007.2	36.7	4.97	116.0	4.90

[a] S' = Additional section modulus corresponding to 1/16″ increase in web thickness.

[b] r = Radius of gyration of the "T" section comprising the compression flange plus 1/6 the web area, about an axis in the plane of the web.

[c] R = Maximum end reaction permissible without intermediate stiffeners for tabulated web plate.

The width-thickness ratios for girders in this table comply with AISC Specification Section 1.9 for ASTM A36 steel. For steels of higher yield strengths, check flanges for compliance with this section.

See Section 1.10.5 for design of stiffeners.

Welds not included in tabulated weight per foot.

AMERICAN INSTITUTE OF STEEL CONSTRUCTION

Table 4.14 *(Continued)*

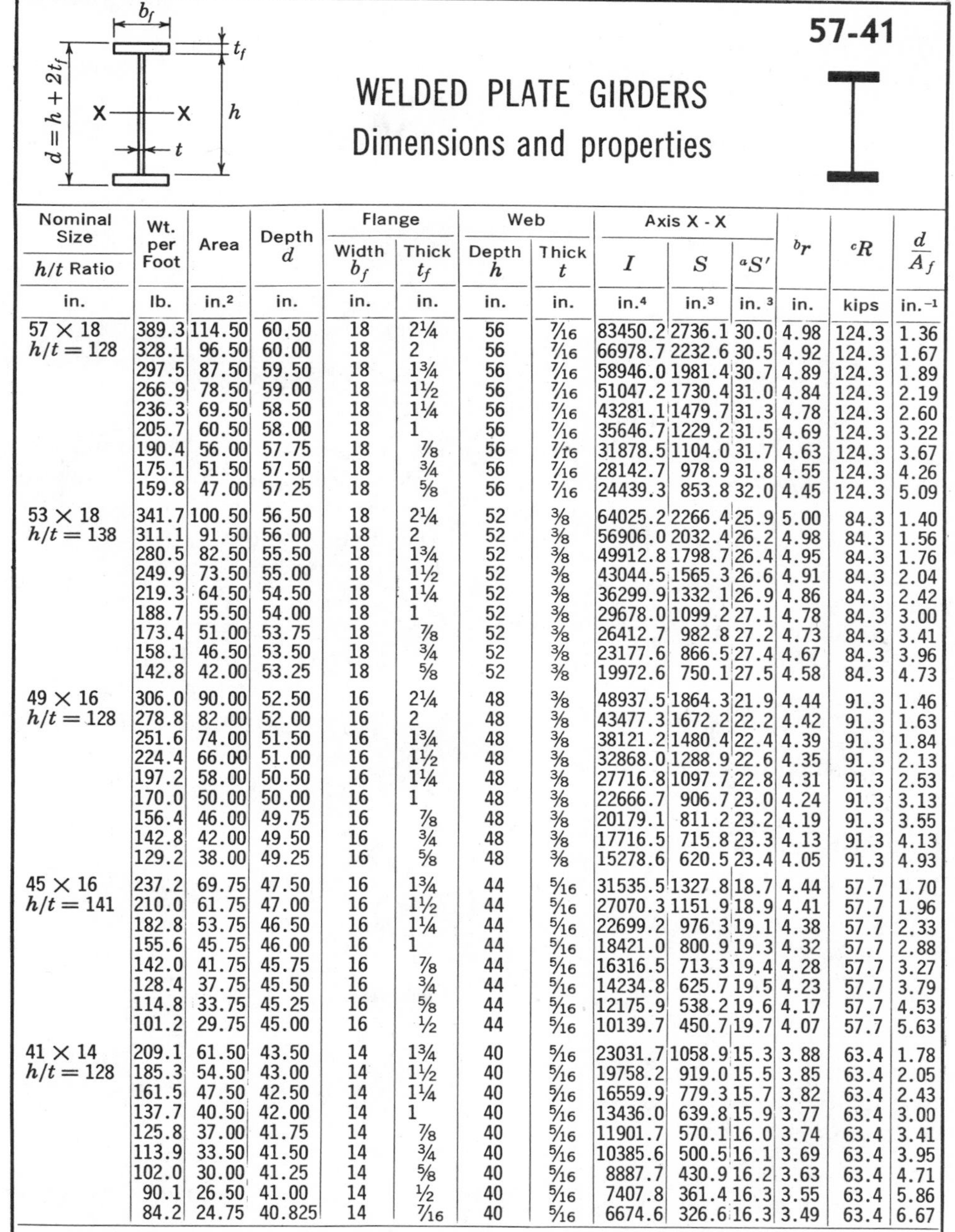

57-41

WELDED PLATE GIRDERS
Dimensions and properties

Nominal Size / h/t Ratio	Wt. per Foot	Area	Depth d	Flange Width b_f	Flange Thick t_f	Web Depth h	Web Thick t	Axis X-X I	Axis X-X S	Axis X-X $^aS'$	^br	^cR	$\frac{d}{A_f}$
in.	lb.	in.^{2}	in.	in.	in.	in.	in.	in.^{4}	in.^{3}	in.^{3}	in.	kips	in.$^{-1}$
57 × 18	389.3	114.50	60.50	18	2¼	56	7/16	83450.2	2736.1	30.0	4.98	124.3	1.36
$h/t = 128$	328.1	96.50	60.00	18	2	56	7/16	66978.7	2232.6	30.5	4.92	124.3	1.67
	297.5	87.50	59.50	18	1¾	56	7/16	58946.0	1981.4	30.7	4.89	124.3	1.89
	266.9	78.50	59.00	18	1½	56	7/16	51047.2	1730.4	31.0	4.84	124.3	2.19
	236.3	69.50	58.50	18	1¼	56	7/16	43281.1	1479.7	31.3	4.78	124.3	2.60
	205.7	60.50	58.00	18	1	56	7/16	35646.7	1229.2	31.5	4.69	124.3	3.22
	190.4	56.00	57.75	18	⅞	56	7/16	31878.5	1104.0	31.7	4.63	124.3	3.67
	175.1	51.50	57.50	18	¾	56	7/16	28142.7	978.9	31.8	4.55	124.3	4.26
	159.8	47.00	57.25	18	⅝	56	7/16	24439.3	853.8	32.0	4.45	124.3	5.09
53 × 18	341.7	100.50	56.50	18	2¼	52	⅜	64025.2	2266.4	25.9	5.00	84.3	1.40
$h/t = 138$	311.1	91.50	56.00	18	2	52	⅜	56906.0	2032.4	26.2	4.98	84.3	1.56
	280.5	82.50	55.50	18	1¾	52	⅜	49912.8	1798.7	26.4	4.95	84.3	1.76
	249.9	73.50	55.00	18	1½	52	⅜	43044.5	1565.3	26.6	4.91	84.3	2.04
	219.3	64.50	54.50	18	1¼	52	⅜	36299.9	1332.1	26.9	4.86	84.3	2.42
	188.7	55.50	54.00	18	1	52	⅜	29678.0	1099.2	27.1	4.78	84.3	3.00
	173.4	51.00	53.75	18	⅞	52	⅜	26412.7	982.8	27.2	4.73	84.3	3.41
	158.1	46.50	53.50	18	¾	52	⅜	23177.6	866.5	27.4	4.67	84.3	3.96
	142.8	42.00	53.25	18	⅝	52	⅜	19972.6	750.1	27.5	4.58	84.3	4.73
49 × 16	306.0	90.00	52.50	16	2¼	48	⅜	48937.5	1864.3	21.9	4.44	91.3	1.46
$h/t = 128$	278.8	82.00	52.00	16	2	48	⅜	43477.3	1672.2	22.2	4.42	91.3	1.63
	251.6	74.00	51.50	16	1¾	48	⅜	38121.2	1480.4	22.4	4.39	91.3	1.84
	224.4	66.00	51.00	16	1½	48	⅜	32868.0	1288.9	22.6	4.35	91.3	2.13
	197.2	58.00	50.50	16	1¼	48	⅜	27716.8	1097.7	22.8	4.31	91.3	2.53
	170.0	50.00	50.00	16	1	48	⅜	22666.7	906.7	23.0	4.24	91.3	3.13
	156.4	46.00	49.75	16	⅞	48	⅜	20179.1	811.2	23.2	4.19	91.3	3.55
	142.8	42.00	49.50	16	¾	48	⅜	17716.5	715.8	23.3	4.13	91.3	4.13
	129.2	38.00	49.25	16	⅝	48	⅜	15278.6	620.5	23.4	4.05	91.3	4.93
45 × 16	237.2	69.75	47.50	16	1¾	44	5/16	31535.5	1327.8	18.7	4.44	57.7	1.70
$h/t = 141$	210.0	61.75	47.00	16	1½	44	5/16	27070.3	1151.9	18.9	4.41	57.7	1.96
	182.8	53.75	46.50	16	1¼	44	5/16	22699.2	976.3	19.1	4.38	57.7	2.33
	155.6	45.75	46.00	16	1	44	5/16	18421.0	800.9	19.3	4.32	57.7	2.88
	142.0	41.75	45.75	16	⅞	44	5/16	16316.5	713.3	19.4	4.28	57.7	3.27
	128.4	37.75	45.50	16	¾	44	5/16	14234.8	625.7	19.5	4.23	57.7	3.79
	114.8	33.75	45.25	16	⅝	44	5/16	12175.9	538.2	19.6	4.17	57.7	4.53
	101.2	29.75	45.00	16	½	44	5/16	10139.7	450.7	19.7	4.07	57.7	5.63
41 × 14	209.1	61.50	43.50	14	1¾	40	5/16	23031.7	1058.9	15.3	3.88	63.4	1.78
$h/t = 128$	185.3	54.50	43.00	14	1½	40	5/16	19758.2	919.0	15.5	3.85	63.4	2.05
	161.5	47.50	42.50	14	1¼	40	5/16	16559.9	779.3	15.7	3.82	63.4	2.43
	137.7	40.50	42.00	14	1	40	5/16	13436.0	639.8	15.9	3.77	63.4	3.00
	125.8	37.00	41.75	14	⅞	40	5/16	11901.7	570.1	16.0	3.74	63.4	3.41
	113.9	33.50	41.50	14	¾	40	5/16	10385.6	500.5	16.1	3.69	63.4	3.95
	102.0	30.00	41.25	14	⅝	40	5/16	8887.7	430.9	16.2	3.63	63.4	4.71
	90.1	26.50	41.00	14	½	40	5/16	7407.8	361.4	16.3	3.55	63.4	5.86
	84.2	24.75	40.825	14	7/16	40	5/16	6674.6	326.6	16.3	3.49	63.4	6.67

$^a S'$ = Additional section modulus corresponding to 1/16″ increase in web thickness.

$^b r$ = Radius of gyration of the "T" section comprising the compression flange plus ⅙ the web area, about an axis in the plane of the web.

$^c R$ = Maximum end reaction permissible without intermediate stiffeners for tabulated web plate.

The width-thickness ratios for girders in this table comply with AISC Specification Section 1.9 for ASTM A36 steel. For steels of higher yield strengths, check flanges for compliance with this section.

See Section 1.10.5 for design of stiffeners.

Welds not included in tabulated weight per foot.

run cover plates the full length of the beam, but they may be stopped off as described in Art. 4.20*e* for plate girders.

Problem Find the section modulus of a 36 WF 300 with top and bottom cover plates, $20 \times 3/4$ in., welded on.

$$I_1 \text{ (36 WF)} = 20{,}290.2$$

$$I_2 \text{ (cover plate about axis)} = 2 \times \frac{20 \times 0.75^3}{12} = 1.4$$

$$A_2 x_2^2 = 2 \times 15 \times 18.74^2 = \underline{10{,}533.0}$$

$$\text{Total } I = 30{,}824.6$$

$$c = 18.36 + 0.75 = 19.11$$

$$S = \frac{30{,}824.6}{19.11} = 1{,}614$$

Problem In a second type of problem frequently encountered, architectural limitations may prevent the use of any beam greater than a given depth. Let us assume that we must carry a total uniform load of 380 kips on a span of 36 ft with a maximum depth of 30 in., using A36 steel and disregarding deflection.

$$S \text{ required} = \frac{380 \times 36}{16} = 864$$

From the definition of section modulus,

$$I = 864 \times 15 = 12{,}960.0$$

$$I \text{ furnished by 27 WF 145} = \underline{5{,}414.3}$$

$$I \text{ required in plates} = 7{,}545.7$$

For a 30-in. maximum depth, the permissible plate thickness is

$$\frac{30 - 26.88}{2} = 1.56 \text{ in. or } 1\tfrac{9}{16} \text{ in.}$$

From Table 4.12, the unit I for a depth of $(26.88 + 1.56 =)$ 28.44 is 404. Therefore the plate required must be

$$\frac{7{,}545.7}{1.56 \times 404} = 11.97 \text{ in.}$$

Use $1\tfrac{9}{16}$ in. (Total) $\times$ 12-in. cover plates, cut to length as may be determined by the method of Art. 4.20*e*.

4.22 Compound steel beams

a Occasionally it may prove simpler to use a compound steel beam—or, as it is sometimes called, a "tandem beam"— than to use a plate girder or cover-plated beam. This type of beam is formed by riveting or welding one beam on top of the other instead of placing them side by side. Although the beams used are usually of the same size, the use of existing material may dictate the use of different-sized beams. Two beams joined thus will furnish from 20 to 30 percent more strength than if used side by side, as well as providing a much greater stiffness (Fig. 4.25).

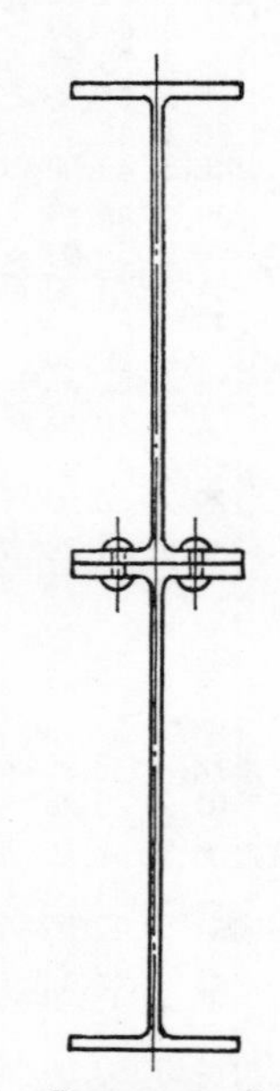

Fig. 4.25 Compound Steel Beam

b In the design of a compound beam, compute the moment and shear, and lay out a shear diagram. Compute the required section modulus, and select the beams to make up this required section modulus, estimating the strength of the combined section from the assumption made in paragraph *a*. Check the section modulus of the assumed section. Since the only holes taken out are at or near the neutral axis, it is perfectly sound to base the analysis on gross section. If the original assumption was high or low, revise the assumed beams and recheck.

c It is necessary to get the longitudinal or horizontal shear per lineal inch along the two contact flanges to compute the riveting required. The longitudinal shear is found from the formula

$$v_h = \frac{VM_s}{I} \tag{4.4}$$

where V is the vertical shear at the section and M_s the static moment of the section. This static moment is found by a tabulation of the parts as shown in the following problem. Theoretically, the horizontal shear increases directly as the vertical shear increases, and the total horizontal shear in any section is the average shear per inch from Formula (4.4), multiplied by the length in inches. This total horizontal shear divided by the value of a rivet in single shear gives the number of rivets required to develop the horizontal shear and the consequent spacing of rivets. The beam is usually divided into sections and the riveting computed for each section.

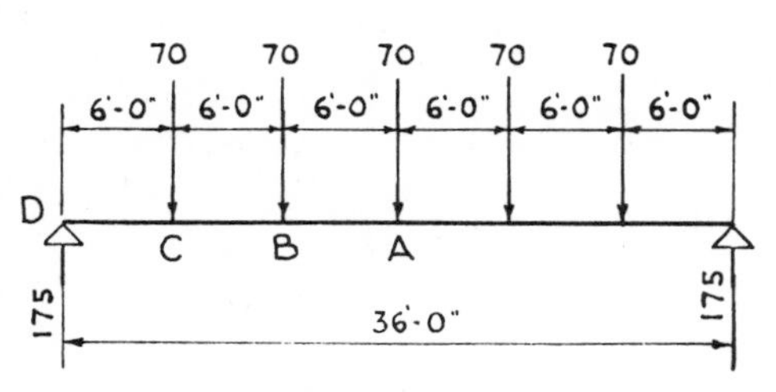

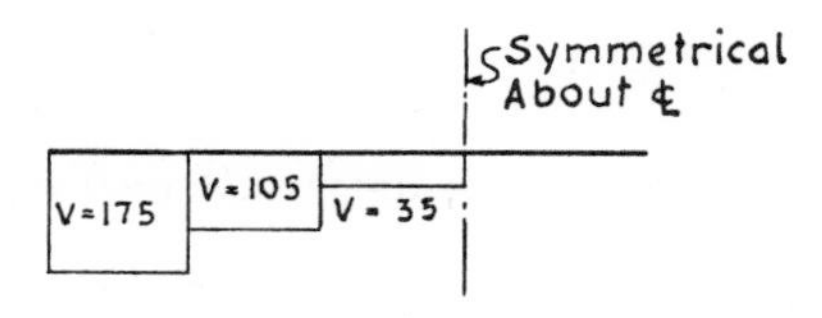

Fig. 4.26 Problem Diagrams—Compound Steel Beams

Problem (See load diagram, Fig. 4.26.) Using 22,000 psi, S required (see Art. 4.10*e*) $= 0.409 \times 70 \times 36 = 1{,}031$. If we assume that the saving by using a tandem beam is 20 percent, the S required of each of the beams would be

$$\frac{0.8 \times 1{,}031}{2} = 412.4$$

Try two 36 WF 135. The properties of the section are

$$A = 2 \times 39.70 = 79.4$$
$$I = 7{,}796.1 \times 2 = 15{,}592$$
$$79.4 \times 17.78^2 = \underline{25{,}100}$$
$$40{,}692$$
$$S = \frac{I}{c} = \frac{40{,}692}{35.55} = 1{,}144$$

The static moment is calculated as follows:

Area of one beam $= 39.70$
$\bar{y} = 17.78$
Static moment about line of horizontal shear $= A\bar{y} = 705.9$.

$$\text{Shear } a \text{ to } b = \frac{35 \times 705.9 \times 72}{40{,}692} = 43.7$$

¾ in. A141 rivets required $= \dfrac{43.7}{6.67} = 6\frac{1}{2}$ rivets; use 8.

$$\text{Shear } b \text{ to } c = \frac{105 \times 705.9 \times 72}{40{,}692} = 131.2$$

$$\text{Rivets required} = \frac{131.2}{6.67} = 20 \text{ rivets.}$$

$$\text{Shear } c \text{ to } d = \frac{175 \times 705.9 \times 72}{40{,}692} = 218.6$$

Rivets required $= \dfrac{218.6}{6.67} = 32\frac{1}{2}$ rivets; use 34.

The method described may be used to provide additional moment capacity for an existing beam as long as vertical shear, web buckling, and end connections or supports are provided for in the original member. If the two beams are of different sizes, the method of approach is as follows:

1. Compute the moment of inertia, the section modulus, and the position of the neutral axis by the methods of Art. 1.31*f* and 1.31*g*.
2. Compute the static moment about the

splice line between the two beams. This static moment is the product of the area of the smaller of the two beams times half the depth of the beam.

3. The horizontal shear and rivets required are computed in the manner already described for two equal beams.

4.23 Crane-runway girders

a One of the problems encountered in the design of steel-mill buildings is the design of crane runways. It involves not only the design of the crane girder, but also the column and bracing. Shop superintendents frequently condemn a crane for its excessive power consumption or high maintenance costs when the fault lies chiefly with the design of the crane runways. The stiffness of the whole structure is of the utmost importance. Lack of stiffness brings about bumpy movement of the crane, throwing rails out of alignment and causing excessive strains on the crane and its supporting structure and excessive wear on moving parts.

Regardless of minimum requirements as set forth in the Code, the design of a crane runway girder is a poor place to economize and the author has frequently been called upon to design proper stiffness into such an existing girder where the addition of one hundred dollars in the original structure would have saved thousands of dollars in maintenance through the years. For example, even though the lateral thrust of the crane may be provided for by the section modulus of the top flange of the girder, it is good economy to lay on top of the girder a channel of sufficient strength to take the entire moment required, as shown in Fig. 4.28. Moreover, it is good policy to brace the channel laterally at each column to provide against the possible "bump" as the wheels pass from one span to the next, either by bracketing against the column or by providing an angle splice across the joint.

The design of the column is a problem in eccentricity and is given in Art. 7.10. A crane runway in a mill building also depends for its stiffness and rigidity largely on the bottom chord bracing of the roof trusses. Some suggestions on this bracing are given in Art. 10.40, Mill Buildings.

b In addition to those portions of the AISC Code which apply to any steel girder, the following sections also apply to the design of crane-runway girders.

1.3.3 *Impact*

For structures carrying live loads which induce impact, the assumed live load shall be increased sufficiently to provide for same.

If not otherwise specified, the increase shall be:

For supports of elevators......	100 percent
For traveling crane support girders and their connections.	25 percent
For supports of light machinery, shaft or motor driven, not less than......................	20 percent
For supports of reciprocating machinery or power driven units, not less than	50 percent
For hangers supporting floors and balconies...............	33 percent

1.3.4 *Crane runway horizontal forces*

The lateral force on crane runways to provide for the effect of moving crane trolleys shall, if not otherwise specified, be 20 percent of the sum of the weights of the lifted load and of the crane trolley (but exclusive of other parts of the crane), applied at the top of rail, one-half on each side of the runway; and shall be considered as acting in either direction normal to the runway rail.

The longitudinal force shall, if not otherwise specified, be taken as 10 percent of the maximum wheel loads of the crane applied at the top of rail.

SECTION 1.7 MEMBERS AND CONNECTIONS SUBJECT TO REPEATED VARIATION OF STRESS

1.7.1 *Up to 10,000 complete stress reversals*

The stress carrying area of members, connection material and fasteners* need not be increased because of repeated variation or reversal of stress unless the maximum stress allowed by Sect. 1.5 and 1.6 is expected to occur over 10,000[a] times in the life of the structure.

* As used in this Section, "fasteners" comprise welds, rivets and bolts.

[a] Approximately equivalent to one application per day for 25 years.

1.7.2 *10,000 to 100,000 cycles of maximum load*

Members, connection material and fasteners (except high strength bolts in friction-type joints) subject to more than 10,000 but not over 100,000[b] applications of maximum design loading shall be proportioned, at unit stresses allowed in Sect. 1.5 and 1.6 for the kind of steel and fasteners used, to support the algebraic difference** of the maximum computed stress and two-thirds of the minimum computed stress, but the stress-carrying area shall not be less than that required in proportioning the member, connection material and fasteners to support either the maximum or minimum computed stress at the values allowed in Sect. 1.5 and 1.6 for the kind of steel and fasteners used.

[b] Approximately equivalent to ten applications per day for 25 years.

** In determining the algebraic difference, tensile stress is designated as positive and compression stress as negative.

1.7.3 *100,000 to 2,000,000 cycles of maximum load*

Members, connection material and fasteners (except high strength bolts in friction-type joints) subject to more than 100,000 but not more than 2,000,000[c] applications of maximum design loading shall be proportioned at unit stresses allowed in Sect. 1.5 and 1.6 for A7 steel, A141 rivet steel and E60XX and submerged arc Grade SAW-1 welds to support the algebraic difference of the maximum computed stress and ⅔ of the minimum computed stress, but the stress-carrying area shall not be less than that required in proportioning the member, connection material and fasteners to support either the maximum or minimum computed stress at the values allowed in Sect. 1.5 and 1.6 for the kind of steel and fasteners used.

[c] Approximately equivalent to 200 applications per day for 25 years.

1.10.9 *Horizontal forces*

The flanges of plate girders supporting cranes or other moving loads shall be proportioned to resist the horizontal forces produced by such loads.

Wheel loads and wheel spacing are given in the catalog of the various manufacturers. No table of standards which provides for general use can be given.

c With the wheel load and wheel base known, the design of the crane girder is a problem in moving-load moment and shear as described in Art. 3.10*q*. Usually the channel to be used will be approximately half the depth of the girder, so the question of lateral support of top flange does not ordinarily enter the problem.

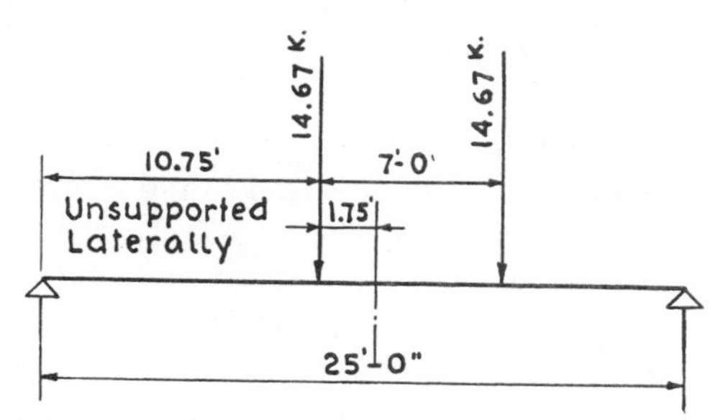

Fig. 4.27 Position of Crane for Maximum Moment

Problem Design the crane-runway girder for a 10-ton crane on a 50-ft crane bridge span—the runway-girder span to be 25 ft. The wheel base is 7 ft and the wheel load is 14,670 lb (see Fig. 4.27)

$$M = 14.67 \frac{(25 - 3.5)^2}{50} = 135.5 \text{ ft-kips}$$

At 22,000 psi, $S = 0.545 \times 135.5 = 73.8$. Use 18 WF 45. From Art. 4.23*b* the lateral moment on the top flange, assuming (*a*) that the maximum lateral moving load is half the total load, (*b*) that the lateral moment is divided between the two girders, and (*c*) that the number of reversals of stress is between 10 per day and 200 per day, the S required for the flatwise channel is 73.8 ×

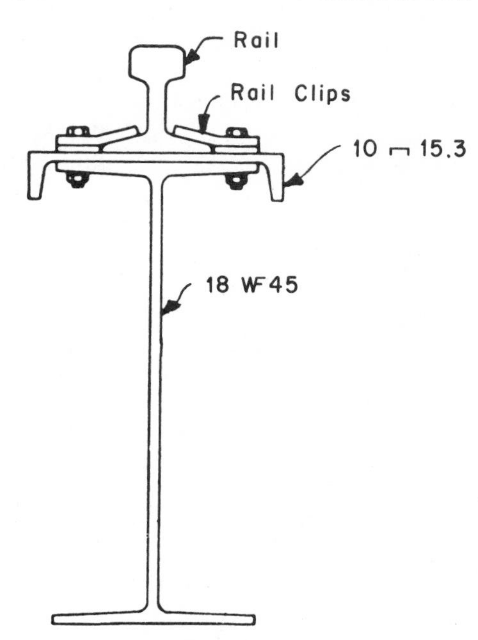

Fig. 4.28 Typical Crane Girder

$\frac{0.20}{4} \times 1.67 = 6.15$. The flange width of an 18 WF 45 is 7½ in. and from the table of detail dimensions of channels in the AISC Manual, page 1-26, the smallest channel which will provide this T distance is a 10 in. ⊓ 15.3 with an S of 13.4—more than double the computed S required. The section to be used is shown in Fig. 4.28.

4.30 OPEN-WEB STEEL JOISTS

a Open-web steel joists are standardized light-weight steel trusses, either welded or expanded, and capable of carrying the loads given in Tables 4.15, 4.16, 4.17, or 4.18, depending on the type and quality of material specified in the design. The standard type joists listed in Tables 4.15 and 4.16 have an end depth of 2½ in., while the long-span joists LA and LH are built to meet job conditions as shown in Fig. 4.34 and 4.35 except that for those joists shown as "underslung" the standard end depth may vary with the different manufacturers. The J and LA series are fabricated from A7 or A36 steel while the H and LH are of high-strength steel. Tables 4.15 and 4.17 are based on the use of A36 steel.

The following extracts from the Standard Specification of the AISC and the Steel Joist Institute represent good construction practice: for J and H joists, Section 204, Application, and Section 205, Handling and Erection.

SECTION 204 APPLICATION

204.1 *Usage*

These specifications for use of joists shall apply in any type of structure where floors and roof decks are to be supported directly by steel joists installed as hereinafter specified. Where joists are used otherwise than on simple spans under uniformly distributed loading, as prescribed in Section 203.1, or where their compression flange is not stayed sufficiently to prevent lateral buckling, their stresses shall be analyzed and the joists modified if necessary to make their design conform to the specifications listed in Section 203.1.

204.2 *Span*

The clear span of a joist shall not exceed 24 times its depth. In floor construction, the span shall not be greater than 20 times the nominal depth of the joist.

204.3 *End supports*

The ends of joists shall extend a distance of not less than 4 inches over masonry or poured concrete supports and not less than 2½ inches over steel supports except where opposite joists butt over a narrow steel support and positive attachment to the support is made by welding or bolting. In such cases a shorter end bearing length may be used when proper design provisions are made.

204.4 *Bridging*

(a) *Type*

Bridging shall consist of either (a) pairs of horizontal steel members acting as tension ties between top chords and between bottom chords of adjacent joists, or (b) steel cross-bracing acting as struts from top chord of each joist to the bottom chord of adjacent joists.

(b) *Sizing*

The maximum slenderness ratios (l/r) of bridging members shall not exceed 200 for struts nor 300 for tension members, except that when round rod horizontal bridging is used the rod shall be not less than ½ inch in diameter. The l shall be taken as the clear distance between attachments including the connection at the point of intersection in the case of cross-brace bridging.

(c) *Spacing*

Lines of bridging shall be approximately equally spaced throughout the joist span. On joist spans up to 21′-0 the spacing center-to-center of lines of bridging shall not exceed 7′-0. On joist spans over 21′-0 to 48′-0 the spacing center-to-center of lines of bridging shall not exceed 8′-0.

204.5 *Installation of bridging*

All bridging shall be completely installed before any construction loads are placed on the joists, except the weight of the workmen necessary to install the bridging.

All bridging members shall be secured to joist chords by welding, bolting or other mechanical means capable of resisting a horizontal force of 500 pounds.

Bridging shall support the top chords against lateral movement during the construction period and shall hold the steel joists in an approximately vertical plane.

The ends of all bridging lines terminating at walls or beams shall be anchored thereto at top and bottom chords.

204.6 *End anchorage*

(a) *Masonry supports*

Joists resting on masonry supports shall be bedded in mortar and attached to the support with an anchor equivalent to a 3/8 inch round steel bar not less than 8 inches long. Every third joist in floors and every joist in roofs shall be anchored. In roofs where masonry parapet walls are not present, two 1/2 inch anchor bolts or other equal means shall be used in lieu of the steel bar.

(b) *Steel supports*

Ends of joists resting on steel supports shall be connected thereto with not less than two 1/8 inch fillet welds 1 inch long, a 1/2 inch bolt, or a 3/16 inch round steel anchor fastened over the beam flange. The 3/16 inch round steel anchor shall not be used where wind or similar forces must be considered.

(c) *Uplift*

Where uplift forces are a design consideration, roof joists shall be anchored to resist such forces.

204.7 *Joist spacing*

Joists shall be spaced so that the loading on each joist does not exceed the allowable load for the particular joist design. The spacing shall not exceed the safe span of the deck or slab.

204.8 *Floors and roof decks*

(a) *Material*

Floors and roof decks may consist of poured or precast concrete or gypsum, formed steel, wood, or other suitable material capable of supporting the required load at the specified joist spacing.

(b) *Thickness*

Poured slabs shall have a minimum thickness of 2 inches.

(c) *Centering*

Centering for poured slabs may be ribbed metal lath, corrugated steel sheets, paper-backed welded wire fabric, removable centering or any other suitable material capable of supporting the slab at the designated joist spacing.

Centering shall not cause lateral displacement of the top chord of joists during installation of the centering or placing of the concrete.

(d) *Bearing*

Slabs or decks shall bear uniformly along the top chords of the joists.*

Attachments of slab or deck to top chords of joists, where required, shall be capable of staying the top chords laterally. The spacing of such attachments along the top chords of joists shall not exceed 36 inches.

Where wood nailers are used, such nailers in conjunction with the deck or slab shall be firmly attached to the top chords of the joists.

*** Tests on steel joists have demonstrated that concentrated loads applied to the top chord (such as loads developed in bulb-tee roof construction) may be treated the same as uniform loads when their total does not exceed the uniform load allowed on the joist by the design and their spacing does not exceed 33 inches.**

204.9 *Deflection*

The deflection due to the design live load shall not exceed the following:

Floors: 1/360 of span

Roots: 1/360 of span where a plaster ceiling is attached or suspended

1/240 of span for all other cases

204.10 *Camber*

Any vertical camber shall be convex upward but shall not exceed the following:

Top Chord Length	Maximum Camber
Up to 20′	3/8″
20′ to 30′	1/2″
30′ to 40′	5/8″
Over 40′	1″

204.11 *Inspection*

Before shipment, all joists shall be thoroughly inspected by the manufacturer. If the purchaser wishes an inspection of the joists by someone other than the manufacturer's own inspectors, he may reserve the right to do so in his "invitation to bid" or the accompanying job specifications. Arrangements shall be made with the manufacturer for such inspection of the joists at the manufacturing shop by the purchaser's inspectors at the purchaser's expense.

SECTION 205 HANDLING AND ERECTION

Care shall be exercised at all times to avoid damage through careless handling during unloading, storing and erecting. Dumping onto the ground shall not be permitted.

As soon as joists are erected, all bridging shall be completely installed and the joists permanently fastened into place before the application of any loads except the weight of the erectors.

During the construction period, the contractor shall provide means for the adequate distribution of concentrated loads so that the carrying capacity of any joist is not exceeded.

Field welding shall not damage the joists. The total length of weld at any one point on cold-formed members whose yield point has been attained by cold working and whose as-formed strength is used in the design shall not exceed 50 percent of the overall developed width of the cold-formed section.

b Some of the following comments and suggestions may be deemed by some manufacturers and dealers as being ultraconservative, but they are based on the author's experience over a period extending back to the first few months of manufacture of these joists.

Steel joist construction is not suited to tall buildings because it lacks the stiffening effect of heavier, more rigidly connected construction. This is particularly true where wind bracing is required and where the plate action of a heavy concrete slab acts to distribute wind pressure. Neither is it a desirable construction for heavy concentrated or vibratory floor loads. Because the construction is built more or less to form a drum, it is advisable in many types of buildings to use acoustical ceilings. Vibration, which is probably the greatest inherent fault of the system, can be minimized by proper details of installation. Details other than those recommended by or shown in some of the steel joist catalogs have been found preferable, some of which are indicated in the following paragraph.

In the case of long-span joists of the LA and LH series, there are a few variations from the foregoing regulations as described in the AISC Manual, pages 5-195 through 5-197 and pages 5-210 through 5-213.

Standard Load Table for Open Web Steel Joists, J-Series

Based on allowable stress of 22,000 psi

Adopted by the American Institute of Steel Construction, Inc., June 19, 1963

The following table gives the TOTAL safe uniformly-distributed load-carrying capacities, in pounds per linear foot, of J-Series Open Web Steel Joists. The weight of DEAD loads, including the joists, must in all cases be deducted to determine the LIVE load-carrying capacities of the joists.

Loads above heavy stepped lines are governed by shear.

Loads below dashed lines are to be used for roof construction only.

This table is in accordance with Simplified Practice Recommendation filed with the Commodity Standards Division, Office of Technical Services, U. S. Department of Commerce.

Footnotes—See tables

* Indicates Nominal Depth of Steel Joists only.

** Approximate Weights per Linear Foot of Steel Joists only. Accessories and nailer strip not included.

See manufacturers' catalogs for detailed information on specific joist types.

Table reproduced by courtesy of Steel Joist Institute. (Dashed lines for roof construction limitation added.)

Table 4.15

STANDARD LOAD TABLE FOR OPEN WEB STEEL JOISTS, J-SERIES

Allowable total safe loads in pounds per linear foot based on allowable stress of 22,000 psi

Joist Designation	8J2	10J2	10J3	10J4	12J2	12J3	12J4	12J5	12J6
*Depth in Inches	8	10	10	10	12	12	12	12	12
Resisting Moment in Inch Kips	56	70	89	111	85	108	135	161	196
Max. End Reaction in Pounds	1900	2000	2200	2400	2200	2300	2500	2700	3000
**Approx. Joist Wgt. Pounds per Foot	4.2	4.2	4.8	6.0	4.5	5.1	6.0	7.0	8.1
Span in Feet									
8	475								
9	422								
10	373	400	440	480					
11	309	364	400	436					
12	259	324	367	400	367	383	417	450	500
13	221	276	338	369	335	354	385	415	462
14	190	238	303	343	289	329	357	386	429
15	166	207	264	320	252	307	333	360	400
16	146	182	232	289	221	281	313	338	375
17		161	205	256	196	249	294	318	353
18		144	183	228	175	222	278	300	333
19		129	164	205	157	199	249	284	316
20		117	148	185	142	180	225	268	300
21					128	163	204	243	286
22					117	149	186	222	270
23					107	136	170	203	247
24					98	125	156	186	227

Table 4.15 (*Continued*)

STANDARD LOAD TABLE FOR OPEN WEB STEEL JOISTS, J-SERIES

Allowable total safe loads in pounds per linear foot based on allowable stress of 22,000 psi

Joist Designation	14J3	14J4	14J5	14J6	14J7	16J4	16J5	16J6	16J7	16J8
*Depth in Inches	14	14	14	14	14	16	16	16	16	16
Resisting Moment in Inch Kips	127	159	190	230	276	173	216	258	310	359
Max. End Reaction in Pounds	2400	2800	3100	3400	3700	3000	3300	3600	4000	4300
**Approx. Joist Wgt. Pounds per Foot	5.2	6.4	7.3	8.4	9.7	6.6	7.6	8.5	10.1	11.3
Span in Feet										
14	343	400	443	486	529					
15	320	373	413	453	493					
16	300	350	388	425	463	375	413	450	500	538
17	282	329	365	400	435	353	388	424	471	506
18	261	311	344	378	411	333	367	400	444	478
19	235	294	326	358	389	316	347	379	421	453
20	212	265	310	340	370	288	330	360	400	430
21	192	240	287	324	352	262	314	343	381	410
22	175	219	262	309	336	238	298	327	364	391
23	160	200	239	290	322	218	272	313	348	374
24	147	184	220	266	308	200	250	299	333	358
25	135	170	203	245	294	185	230	275	320	344
26	125	157	187	227	272	171	213	254	306	331
27	116	145	174	210	252	158	198	236	283	319
28	108	135	162	196	235	147	184	219	264	305
29						137	171	205	246	285
30						128	160	191	230	266
31						120	150	179	215	249
32						113	141	168	202	234

Table 4.15 (*Continued*)

STANDARD LOAD TABLE FOR OPEN WEB STEEL JOISTS, J-SERIES

Allowable total safe loads in pounds per linear foot based on allowable stress of 22,000 psi

Joist Designation	18J5	18J6	18J7	18J8	20J5	20J6	20J7	20J8
*Depth in Inches	18	18	18	18	20	20	20	20
Resisting Moment in Inch Kips	243	293	352	406	265	316	382	455
Max. End Reaction in Pounds	3500	3900	4200	4500	3800	4100	4300	4600
**Approx. Joist Wgt. Pounds per Foot	7.9	9.0	10.2	11.3	8.1	9.2	10.6	11.9
Span in Feet								
18	389	433	467	500				
19	368	411	442	474				
20	350	390	420	450	380	410	430	460
21	333	371	400	429	362	390	410	438
22	318	355	382	409	345	373	391	418
23	304	339	365	391	330	357	374	400
24	281	325	350	375	307	342	358	383
25	259	312	336	360	283	328	344	368
26	240	289	323	346	261	312	331	354
27	222	268	311	333	242	289	319	341
28	207	249	299	321	225	269	307	329
29	193	232	279	310	210	250	297	317
30	180	217	261	300	196	234	283	307
31	169	203	244	282	184	219	265	297
32	158	191	229	264	173	206	249	288
33	149	179	215	249	162	193	234	279
34	140	169	203	234	153	182	220	262
35	132	159	192	221	144	172	208	248
36	125	151	181	209	136	163	197	234
37					129	154	186	222
38					122	146	176	210
39					116	139	167	199
40					110	132	159	190

Table 4.15 (*Continued*)

STANDARD LOAD TABLE FOR OPEN WEB STEEL JOISTS, J-SERIES

Allowable total safe loads in pounds per linear foot based on allowable stress of 22,000 psi

Joist Designation	22J6	22J7	22J8	24J6	24J7	24J8
*Depth in Inches	22	22	22	24	24	24
Resisting Moment in Inch Kips	335	420	493	367	460	540
Max. End Reaction in Pounds	4200	4500	4800	4400	4700	5000
**Approx. Joist Wgt. Pounds per Foot	9.6	10.5	11.9	9.9	11.1	12.4
Span in Feet						
22	382	409	436			
23	365	391	417			
24	350	375	400	367	392	417
25	336	360	384	352	376	400
26	323	346	369	338	362	385
27	306	333	356	326	348	370
28	285	321	343	312	336	357
29	266	310	331	291	324	345
30	248	300	320	272	313	333
31	232	290	310	255	303	323
32	218	273	300	239	294	313
33	205	257	291	225	282	303
34	193	242	282	212	265	294
35	182	229	268	200	250	286
36	172	216	254	189	237	278
37	163	205	240	179	224	263
38	155	194	228	169	212	249
39	147	184	216	161	202	237
40	140	175	205	153	192	225
41	133	167	196	146	182	214
42	127	159	186	139	174	204
43	121	151	178	132	166	195
44	115	145	170	126	158	186
45				121	151	178
46				116	145	170
47				111	139	163
48				106	133	156

Standard Load Table for Open Web Steel Joists, H-Series

Based on allowable stress of 30,000 psi

Adopted by the American Institute of Steel Construction, Inc., June 19, 1963

The **bold face** figures in the following table give the TOTAL safe uniformly distributed load-carrying capacities in pounds per linear foot, of H-Series High Strength Steel Joists. The weight of DEAD loads, including the joists, must in all cases be deducted to determine the LIVE load-carrying capacities of the joists.

The light face figures in this load table are the LIVE loads per linear foot of joist which will produce an approximate deflection of 1/360 of the span. LIVE loads which will produce a deflection of 1/240 of the span may be obtained by multiplying the figures in light face by 1.5. **In no case shall the total load capacity of the joist be exceeded.*****

Loads above heavy stepped lines are governed by shear.

Loads below dashed lines are to be used for roof construction only.

This table is in accordance with Simplified Practice Recommendation filed with the Commodity Standards Division, Office of Technical Services, U. S. Department of Commerce.

Footnotes—See tables and text above

* Indicates Nominal Depth of Steel Joists only.

** Approximate Weights per Linear Foot of Steel Joists only. Accessories and nailer strip not included.

See manufacturers' catalogs for detailed information on specific joist types.

*** Section 204.9 of the J- and H-Series Specifications limits the design LIVE load deflection as follows: Floors, $\frac{1}{360}$ of span. Roofs, $\frac{1}{360}$ of span where a plaster ceiling is attached or suspended; $\frac{1}{240}$ of span for all other cases.

Table reproduced by courtesy of Steel Joist Institute. (Dashed lines for roof construction limitation added.)

Table 4.16

STANDARD LOAD TABLE OPEN WEB STEEL JOISTS, H-SERIES

Allowable total safe loads in pounds per linear foot based on allowable stress of 30,000 psi

Joist Designation	8H2	10H2	10H3	10H4	12H2	12H3	12H4	12H5	12H6
*Depth in Inches	8	10	10	10	12	12	12	12	12
Resisting Moment in Inch Kips	73	91	116	148	111	140	180	222	260
Maximum End Reaction in Pounds	2000	2200	2500	2800	2400	2800	3200	3600	3900
**Approximate Weight in Pounds per Foot	4.2	4.2	5.0	6.1	4.5	5.2	6.2	7.1	8.2
Span in Feet									
8	**500**								
9	**444**								
10	**400**	**440**	**500**	**560**					
11	**364** 319	**400**	**455**	**509**					
12	**333** 246	**367**	**417**	**467**	**400**	**467**	**533**	**600**	**650**
13	**288** 193	**338** 300	**385** 382	**431**	**369**	**431**	**492**	**554**	**600**
14	**248** 155	**310** 240	**357** 306	**400** 368	**343**	**400**	**457**	**514**	**557**
15	**216** 126	**270** 195	**333** 249	**373** 299	**320** 286	**373** 365	**427**	**480**	**520**
16	**190** 103	**237** 161	**302** 205	**350** 247	**289** 236	**350** 300	**400** 364	**450** 440	**488**
17		**210** 134	**268** 171	**329** 206	**256** 196	**323** 250	**376** 304	**424** 367	**459** 437
18		**187** 113	**239** 144	**305** 173	**228** 165	**288** 211	**356** 256	**400** 309	**433** 368
19		**168** 96	**214** 122	**273** 147	**205** 141	**259** 179	**332** 217	**379** 263	**411** 313
20		**152** 82	**193** 105	**247** 126	**185** 120	**233** 154	**300** 186	**360** 225	**390** 268
21					**168** 104	**212** 133	**272** 161	**336** 194	**371** 232
22					**153** 90	**193** 115	**248** 140	**306** 169	**355** 202
23					**140** 79	**176** 101	**227** 122	**280** 148	**328** 176
24					**128** 69	**162** 89	**208** 108	**257** 130	**301** 155

Table 4.16 (*Continued*)

STANDARD LOAD TABLE OPEN WEB STEEL JOISTS, H-SERIES

Allowable total safe loads in pounds per linear foot based on allowable stress of 30,000 psi

Joist Designation	14H3	14H4	14H5	14H6	14H7	16H4	16H5	16H6	16H7	16H8
*Depth in Inches	14	14	14	14	14	16	16	16	16	16
Resisting Moment in Inch Kips	165	212	259	307	369	221	289	344	413	478
Maximum End Reaction in Pounds	3200	3500	3800	4200	4600	3800	4300	4600	4900	5200
**Approximate Weight in Pounds per Foot	5.5	6.5	7.4	8.6	10.0	6.6	7.8	8.6	10.3	11.4
Span in Feet										
14	**457**	**500**	**543**	**600**	**657**					
15	**427**	**467**	**507**	**560**	**613**					
16	**400**	**438**	**475**	**525**	**575**	**475**	**538**	**575**	**613**	**650**
17	**376** 346	**412**	**447**	**494**	**541**	**447**	**506**	**541**	**570**	**612**
18	**340** 292	**389** 354	**422**	**467**	**511**	**422**	**478**	**511**	**544**	**578**
19	**305** 248	**368** 301	**400** 364	**442** 437	**484**	**400** 399	**453**	**484**	**516**	**547**
20	**275** 212	**350** 258	**380** 312	**420** 374	**460** 435	**368** 342	**430** 413	**460**	**490**	**520**
21	**249** 183	**320** 223	**362** 270	**400** 323	**438** 375	**334** 295	**410** 357	**438** 429	**467**	**495**
22	**227** 159	**292** 194	**345** 235	**382** 281	**418** 326	**304** 257	**391** 310	**418** 373	**445** 434	**473**
23	**208** 140	**267** 169	**326** 205	**365** 246	**400** 286	**279** 225	**364** 272	**400** 326	**426** 379	**452** 434
24	**191** 123	**245** 149	**300** 181	**350** 217	**383** 251	**256** 198	**334** 239	**383** 287	**408** 334	**433** 382
25	**176** 109	**226** 132	**276** 160	**327** 191	**368** 222	**236** 175	**308** 211	**367** 254	**392** 295	**416** 338
26	**163** 96	**209** 117	**255** 142	**303** 170	**354** 198	**218** 155	**285** 188	**339** 226	**377** 262	**400** 300
27	**151** 86	**194** 104	**237** 127	**281** 152	**337** 176	**202** 139	**264** 168	**315** 201	**363** 234	**385** 268
28	**140** 77	**180** 94	**220** 114	**261** 136	**314** 158	**188** 124	**246** 150	**293** 181	**350** 210	**371** 240
29						**175** 112	**229** 135	**273** 162	**327** 189	**359** 216
30						**164** 101	**214** 122	**255** 147	**306** 171	**347** 195
31						**153** 91	**200** 111	**239** 133	**287** 155	**332** 177
32						**144** 83	**188** 101	**224** 121	**269** 141	**311** 161

Table 4.16 (*Continued*)

STANDARD LOAD TABLE OPEN WEB STEEL JOISTS, H-SERIES

Allowable total safe loads in pounds per linear foot based on allowable stress of 30,000 psi

Joist Designation	18H5	18H6	18H7	18H8	20H5	20H6	20H7	20H8
*Depth in Inches	18	18	18	18	20	20	20	20
Resisting Moment in Inch Kips	325	383	466	540	365	406	499	602
Maximum End Reaction in Pounds	4500	4800	5200	5400	4800	5100	5400	5600
**Approximate Weight in Pounds per Foot	8.0	9.2	10.4	11.6	8.4	9.6	10.7	12.2
Span in Feet								
18	**500**	**533**	**578**	**600**				
19	**474**	**505**	**547**	**568**				
20	**450**	**480**	**520**	**540**	**480**	**510**	**540**	**560**
21	**429**	**457**	**495**	**514**	**457**	**486**	**514**	**533**
22	**409** 398	**436**	**473**	**491**	**436**	**464**	**491**	**509**
23	**391** 348	**417** 417	**452**	**470**	**417**	**443**	**470**	**487**
24	**375** 306	**400** 367	**433** 427	**450**	**400** 380	**425**	**450**	**467**
25	**347** 271	**384** 325	**416** 378	**432**	**384** 336	**408** 405	**432**	**448**
26	**321** 241	**369** 289	**400** 336	**415** 385	**360** 299	**392** 360	**415**	**431**
27	**297** 215	**350** 258	**385** 300	**400** 344	**334** 267	**371** 321	**400** 375	**415**
28	**276** 193	**326** 231	**371** 269	**386** 308	**310** 239	**345** 288	**386** 336	**400** 385
29	**258** 173	**304** 208	**359** 242	**372** 278	**289** 215	**322** 259	**372** 302	**386** 346
30	**241** 157	**284** 188	**345** 219	**360** 251	**270** 194	**301** 234	**360** 273	**373** 313
31	**225** 142	**266** 170	**323** 198	**348** 227	**253** 176	**282** 212	**346** 247	**361** 284
32	**212** 129	**249** 155	**303** 180	**338** 206	**238** 160	**264** 193	**325** 225	**350** 258
33	**199** 117	**234** 141	**285** 164	**327** 188	**223** 146	**249** 176	**305** 205	**339** 235
34	**187** 107	**221** 129	**269** 150	**311** 172	**210** 133	**234** 161	**288** 187	**329** 215
35	**177** 98	**208** 118	**254** 137	**294** 158	**199** 122	**221** 147	**272** 172	**320** 197
36	**167** 90	**197** 108	**240** 126	**278** 145	**188** 112	**209** 135	**257** 158	**310** 181
37					**178** 103	**198** 125	**243** 145	**293** 167
38					**169** 95	**187** 115	**230** 134	**278** 154
39					**160** 88	**178** 106	**219** 124	**264** 142
40					**152** 82	**169** 98	**208** 115	**251** 132

Table 4.16 (*Continued*)

STANDARD LOAD TABLE OPEN WEB STEEL JOISTS, H-SERIES

Allowable total safe loads in pounds per linear foot based on allowable stress of 30,000 psi

Joist Designation	22H6	22H7	22H8	24H6	24H7	24H8
*Depth in Inches	22	22	22	24	24	24
Resisting Moment in Inch Kips	422	526	653	462	576	716
Maximum End Reaction in Pounds	5400	5600	5800	5600	5800	6000
**Approximate Weight in Pounds per Foot	9.7	10.7	12.0	10.3	11.5	12.7
Span in Feet						
22	**491**	**509**	**527**			
23	**470**	**487**	**504**			
24	**450**	**467**	**483**	**467**	**483**	**500**
25	**432**	**448**	**464**	**448**	**464**	**480**
26	**415**	**431**	**446**	**431**	**446**	**462**
27	**386**	**415**	**430**	**415**	**430**	**444**
28	**359** 351	**400**	**414**	**393**	**414**	**429**
29	**335** 316	**386** 368	**400**	**366**	**400**	**414**
30	**313** 286	**373** 332	**387** 382	**342** 342	**387**	**400**
31	**293** 259	**361** 301	**374** 346	**320** 310	**374** 361	**387**
32	**275** 235	**342** 274	**363** 315	**301** 282	**363** 329	**375**
33	**258** 214	**322** 250	**352** 287	**283** 257	**352** 300	**364** 344
34	**243** 196	**303** 228	**341** 263	**266** 235	**332** 274	**353** 315
35	**230** 180	**286** 209	**331** 241	**251** 215	**313** 251	**343** 288
36	**217** 165	**271** 192	**322** 221	**238** 198	**296** 231	**333** 265
37	**206** 152	**256** 177	**314** 204	**225** 182	**280** 212	**324** 244
38	**195** 140	**243** 163	**301** 188	**213** 168	**266** 196	**316** 225
39	**185** 130	**231** 151	**286** 174	**202** 155	**252** 181	**308** 208
40	**176** 120	**219** 140	**272** 161	**193** 144	**240** 168	**298** 193
41	**167** 112	**209** 130	**259** 149	**183** 134	**228** 156	**284** 179
42	**159** 104	**199** 121	**247** 139	**175** 124	**218** 145	**271** 167
43	**152** 97	**190** 113	**235** 130	**167** 116	**208** 135	**258** 155
44	**145** 90	**181** 105	**225** 121	**159** 108	**198** 126	**247** 145
45				**152** 101	**190** 118	**236** 135
46				**146** 94	**181** 110	**226** 127
47				**139** 89	**174** 103	**216** 119
48				**134** 83	**167** 97	**207** 111

Standard Load Table for Longspan LA-Series Joists

Based on allowable stress of 22,000 psi

Adopted by the American Institute of Steel Construction, Inc., July 1, 1961
Adopted by the Steel Joist Institute, July 1, 1961

This table is based on 22,000 psi allowable stress for A36 steel. When A7, A245 or A303 steel is used, all load carrying capacities shall be reduced by 10 percent. Joists designed of steel other than A36 shall be designated by L rather than LA as shown in the table.

The following table gives the TOTAL safe uniformly distributed load-carrying capacities in pounds per linear foot of span.

This load table applies to joists with either parallel chords or standard pitched top chores. When top chords are pitched, the carrying capacities are determined by the nominal depth of the joists at center of the span.

Standard pitch is 1/8″ per foot. If pitch exceeds this standard, the load table does not apply.

Loads to the right of the dashed vertical line to be used for roof construction only.

Loads below heavy stepped line are governed by maximum end reaction.

The weight of dead loads, including the weight of joists, must in all cases be deducted to determine the live load-carrying capacities which must be reduced for concentrated loads. Approximate weights per linear foot of joists include accessories.

When holes are required in top or bottom chords the above carrying capacities must be reduced in proportion to reduction of chord areas. The top chords are considered as being stayed laterally by floor slab or roof deck.

Table 4.17

STANDARD LOAD TABLE FOR LONGSPAN LA-SERIES JOISTS

Pounds per Linear Foot Based on Allowable Stress of 22,000 psi

Joist Designation	Approx. Wt. Lbs. per Lin. Ft.	Nominal Depth in Inches	Maximum End Reaction Lbs.	Clear Opening or Net Span in Feet											
				25	26	27	28	29	30	31	32	33	34	35	36
18LA02	13	18	4,031	314	295	278	263	248	235	222	211	200	190	181	172
18LA03	14	18	4,549	354	334	315	297	281	266	252	239	227	216	205	196
18LA04	16	18	5,493	428	402	378	355	335	316	299	283	268	255	241	228
18LA05	17	18	5,970	465	438	413	390	369	349	331	314	298	283	270	257
18LA06	19	18	7,135	556	522	490	462	435	411	388	368	348	330	312	295
18LA07	21	18	7,648	596	574	540	508	479	453	428	406	385	364	344	326
18LA08	23	18	8,334	649	625	602	581	547	516	487	461	436	414	393	373
18LA09	25	18	8,611	671	646	623	601	581	562	530	502	475	450	428	406
18LA10	27	18	9,217	718	691	666	643	621	601	582	550	520	493	468	444
18LA11	29	18	9,539	743	715	690	665	643	622	602	584	567	550	523	497
18LA12	31	18	10,216	796	766	739	713	689	666	645	625	607	589	573	544

Joist Designation	Approx. Wt. Lbs. per Lin. Ft.	Nominal Depth in Inches	Maximum End Reaction Lbs.	25	26	27	28	29	30	31	32	33	34	35	36	37	38	39	40
20LA03	14	20	4,703	366	347	328	311	295	280	266	253	241	230	219	209	200	191	183	175
20LA04	16	20	5,776	450	424	400	377	357	338	320	304	288	274	261	249	237	226	216	207
20LA05	17	20	6,177	481	455	431	408	387	368	349	332	316	301	287	274	262	251	240	230
20LA06	19	20	7,349	573	551	520	491	464	439	416	395	375	356	339	323	308	294	281	269
20LA07	21	20	7,887	615	591	570	539	510	483	458	435	413	393	374	357	340	325	311	297
20LA08	23	20	8,525	664	639	616	595	575	556	526	498	473	449	427	406	387	369	352	337
20LA09	25	20	9,063	706	680	655	632	611	591	572	542	515	489	465	442	422	402	384	367
20LA10	27	20	9,552	744	716	690	666	644	623	603	585	567	538	511	486	463	441	421	402
20LA11	29	20	10,129	789	760	732	707	683	661	640	620	602	584	568	541	515	492	470	449
20LA12	31	20	10,692	833	802	773	746	721	697	675	655	635	617	600	583	568	541	516	493
20LA13	36	20	11,587	903	869	838	808	781	756	732	709	688	669	650	632	615	599	584	570

Joist Designation	Approx. Wt. Lbs. per Lin. Ft.	Nominal Depth in Inches	Maximum End Reaction Lbs.	33	34	35	36	37	38	39	40	41	42	43	44	45	46	47	48
24LA04	16	24	5,333	317	303	289	277	265	254	243	233	224	215	207	199	191	184	177	171
24LA05	17	24	5,693	338	324	311	298	286	274	264	253	244	235	226	218	210	202	195	188
24LA06	19	24	6,942	412	394	376	360	344	330	316	303	291	280	269	259	249	240	231	223
24LA07	21	24	7,606	452	432	413	395	379	363	348	334	321	308	296	285	275	265	255	246
24LA08	23	24	8,892	528	504	481	459	439	420	402	385	369	354	340	327	314	302	291	280
24LA09	25	24	9,668	574	548	523	499	477	457	437	419	402	386	370	356	342	329	317	306
24LA10	27	24	10,199	606	588	572	556	531	508	485	465	445	427	409	393	378	363	350	337
24LA11	29	24	10,791	641	623	605	589	573	558	535	512	491	472	453	435	419	403	388	374
24LA12	31	24	11,595	689	669	650	632	616	600	585	570	546	524	503	483	464	446	429	413
24LA13	36	24	12,753	758	736	715	696	677	660	643	627	612	598	584	571	549	528	507	486
24LA14	38	24	13,290	790	767	745	725	706	687	670	654	638	623	609	595	582	570	546	524

Table 4.17 (*Continued*)

STANDARD LOAD TABLE FOR LONGSPAN LA-SERIES JOISTS

Pounds per Linear Foot Based on Allowable Stress of 22,000 psi

Joist Designation	Approx. Wt. Lbs. per Lin. Ft.	Nominal Depth in Inches	Maximum End Reaction Lbs.	Clear Opening or Net Span in Feet															
				41	42	43	44	45	46	47	48	49	50	51	52	53	54	55	56
28LA06	19	28	6,530	313	302	291	281	271	262	253	244	236	228	221	214	207	201	194	188
28LA07	21	28	7,156	343	331	319	308	298	288	278	269	260	251	243	236	228	221	215	208
28LA08	23	28	8,390	403	388	373	359	346	334	322	311	300	290	281	272	263	254	246	239
28LA09	25	28	9,126	438	422	406	391	377	364	351	339	327	316	306	296	286	277	269	260
28LA10	27	28	10,239	491	472	454	437	421	406	391	377	364	351	340	328	317	307	297	288
28LA11	29	28	11,147	535	515	496	478	461	445	429	414	400	387	374	362	350	339	329	319
28LA12	31	28	12,180	585	571	558	537	517	498	480	463	447	432	417	403	390	377	365	354
29LA13	36	28	13,539	650	635	620	606	593	580	568	548	529	511	494	477	462	447	432	419
28LA14	38	28	14,302	687	670	655	640	626	613	600	588	576	565	545	526	509	491	473	457
28LA15	43	28	14,776	709	693	677	662	647	633	620	607	595	583	572	561	551	531	512	494
				49	50	51	52	53	54	55	56	57	58	59	60	61	62	63	64
32LA07	21	32	6,800	274	266	258	250	243	236	229	223	217	211	205	199	194	189	184	179
32LA08	23	32	8,003	322	312	302	293	284	275	267	259	252	245	238	231	225	218	212	207
32LA09	25	32	8,705	351	340	329	319	309	300	291	282	274	266	259	252	245	238	232	225
32LA10	27	32	9,796	394	382	369	358	346	336	326	316	306	297	289	280	272	265	258	250
32LA11	29	32	10,637	428	415	402	390	378	367	356	346	336	326	317	308	300	291	284	276
32LA12	31	32	12,034	485	469	454	439	426	413	400	388	376	365	355	345	335	326	317	308
32LA13	36	32	14,043	565	554	537	520	504	488	473	459	445	432	420	408	396	385	375	365
32LA14	38	32	15,061	606	595	583	572	561	543	526	510	495	480	465	452	439	426	414	403
32LA15	43	32	15,878	639	627	615	603	592	581	570	560	551	532	514	498	482	466	452	438
32LA16	48	32	17,314	697	683	670	657	645	633	622	611	600	590	580	571	562	553	535	519
				57	58	59	60	61	62	63	64	65	66	67	68	69	70	71	72
36LA08	23	36	7,685	267	259	252	246	239	233	227	221	216	210	205	200	195	190	186	182
36LA09	25	36	8,359	290	282	275	267	260	254	247	241	235	229	223	218	213	207	203	198
36LA10	27	36	9,437	327	318	309	301	293	285	277	270	263	256	250	244	238	232	226	221
36LA11	29	36	10,215	354	345	336	327	318	310	302	295	287	280	273	267	260	254	248	242
36LA12	31	36	11,596	402	391	380	370	360	350	341	332	324	315	307	300	292	285	278	271
36LA13	36	36	13,709	475	462	450	437	426	414	403	393	383	373	364	355	346	337	329	321
36LA14	38	36	15,398	534	519	504	490	476	463	450	438	427	415	405	394	384	375	365	356
36LA15	43	36	16,744	581	571	561	552	537	522	509	495	482	467	453	440	428	416	404	393
36LA16	48	36	18,396	638	627	617	606	597	587	578	569	560	546	532	518	505	493	479	466
36LA17	54	36	19,598	680	668	657	646	636	625	616	606	597	588	579	571	563	555	541	528

Table 4.17 (*Continued*)

STANDARD LOAD TABLE FOR LONGSPAN LA-SERIES JOISTS

Pounds per Linear Foot Based on Allowable Stress of 22,000 psi

Joist Designation	Approx. Wt. Lbs. per Lin. Ft.	Nominal Depth in Inches	Maximum End Reaction Lbs.	Clear Opening or Net Span in Feet															
				65	66	67	68	69	70	71	72	73	74	75	76	77	78	79	80
40LA09	25	40	8,061	246	240	234	229	224	218	214	209	204	200	195	191	187	183	179	175
40LA10	27	40	9,131	278	271	265	258	252	246	241	235	230	225	220	215	210	205	201	197
40LA11	29	40	9,851	300	293	286	280	273	267	261	255	250	244	239	234	229	224	219	214
40LA12	31	40	11,224	342	334	325	318	310	303	296	289	283	276	270	264	258	253	247	242
40LA13	36	40	13,269	404	394	385	376	367	358	350	342	334	327	319	312	306	299	293	286
40LA14	38	40	14,948	455	444	433	422	412	402	392	383	374	365	357	349	341	333	326	319
40LA15	43	40	16,801	512	499	487	475	464	453	442	432	422	412	403	394	385	375	366	357
40LA16	48	40	18,977	578	569	561	553	539	527	514	502	491	479	468	458	448	438	428	419
40LA17	54	40	20,338	619	610	601	592	584	576	568	560	552	540	527	516	504	493	482	472
40LA18	61	40	22,010	670	660	651	641	632	623	614	606	598	590	582	574	567	560	553	540
				73	74	75	76	77	78	79	80	81	82	83	84	85	86	87	88
44LA10	27	44	8,864	241	235	230	226	221	216	212	207	203	199	195	191	187	184	180	177
44LA11	29	44	9,528	259	253	248	243	238	233	229	224	220	215	210	205	201	197	192	188
44LA12	31	44	10,896	296	289	283	277	271	266	260	255	250	245	240	235	230	226	222	217
44LA13	36	44	12,880	350	342	335	328	321	314	308	302	295	290	284	278	273	267	262	257
44LA14	38	44	14,555	395	386	378	370	361	354	346	339	332	325	318	312	305	299	293	287
44LA15	43	44	16,342	444	434	425	416	407	398	390	382	374	366	359	352	345	338	331	325
44LA16	48	44	19,024	516	505	494	484	473	463	453	444	435	426	417	409	401	393	385	378
44LA17	54	44	21,040	571	564	556	544	533	521	510	500	489	479	470	460	451	442	434	425
44LA18	61	44	22,872	621	613	605	597	589	582	574	567	560	553	542	531	520	509	499	489
44LA19	68	44	24,621	668	659	651	642	634	626	618	610	603	596	589	582	575	568	562	555
				81	82	83	84	85	86	87	88	89	90	91	92	93	94	95	96
48LA11	29	48	9,234	226	222	217	212	208	203	199	195	191	187	183	179	176	172	169	166
48LA12	31	48	10,601	260	255	250	245	240	236	231	227	223	219	215	211	207	203	199	195
48LA13	36	48	12,530	307	301	295	290	284	279	274	269	264	259	254	250	245	240	236	231
48LA14	38	48	14,204	348	341	334	328	321	315	309	303	297	292	286	281	276	271	266	261
48LA15	43	48	15,930	390	383	375	368	361	354	347	341	335	328	322	317	311	305	300	295
48LA16	48	48	18,550	454	445	437	428	420	412	404	397	389	382	375	368	362	355	349	343
48LA17	54	48	20,875	511	501	491	482	473	464	455	446	438	430	422	415	407	400	393	386
48LA18	61	48	23,735	581	574	567	561	550	539	528	518	508	498	489	480	471	462	454	445
48LA19	68	48	25,460	624	616	609	601	594	588	581	574	568	562	555	545	535	525	515	506

Standard Load Table for Longspan LH-Series Joists

Based on Allowable Stress of 30,000 psi

Adopted by the American Institute of Steel Construction, Inc., June 21, 1962
Adopted by the Steel Joist Institute, June 21, 1962

The bold face figures in the following table give the TOTAL safe uniformly-distributed load-carrying capacities, in pounds per linear foot, of LH-Series joists. The weight of DEAD loads, including the joists,* must in all cases be deducted to determine the LIVE load-carrying capacities of the joists.

The light face figures in this load table are the LIVE loads per linear foot of joist which will produce an approximate deflection of $1/360$ of the span. LIVE loads which will produce a deflection of $1/240$ of the span may be obtained by multiplying the light face figures by 1.5. (Note: The tabulated loads corresponding to these deflection limitations have been computed on the basis of 30,000 psi allowable stress provisions. For joists designed to a lower working stress these loads may be increased in the ratio of 30,000 psi to the design stress used, in order to meet the same deflection limitations.) **In no case shall the total load capacity of the joist be exceeded.****

This load table applies to joists with either parallel chords or standard pitched top chords. When top chords are pitched, the carrying capacities are determined by the nominal depth of the joists at center of the span.

When holes are required in top or bottom chords, the above carrying capacities must be reduced in proportion to reduction of chord areas.

The top chords are considered as being stayed laterally by floor slab or roof deck.

Loads to the right of the heavy dashed vertical line are to be used for roof construction only.

Loads below heavy stepped line are governed by maximum end reaction.

Standard pitch is $1/8''$ per foot. If pitch exceeds this standard, the load table does not apply.

---

* The weight of joists per linear foot will vary with the design but will not exceed that given in the standard specification for LA-Series longspan joists of corresponding designation.

** Section 204.10 of this LH-Series specification limits the design LIVE load deflection as follows: Floors—$1/360$ of span. Roofs—$1/360$ of span where a plaster ceiling is attached or suspended; $1/240$ of span for all other cases.

Table 4.18

STANDARD LOAD TABLE FOR LONGSPAN LH-SERIES JOISTS

Pounds per Linear Foot Based on Allowable Stress of 30,000 psi

Joist Designation	Nominal Depth in Inches	Maximum End Reaction Lbs.	Clear Opening or Net Span in Feet											
			25	26	27	28	29	30	31	32	33	34	35	36
18LH02	18	6,006	**468** 363	**442** 323	**418** 289	**391** 260	**367** 235	**345** 212	**324** 193	**306** 176	**289** 160	**273** 147	**259** 135	**245** 124
18LH03	18	6,686	**521** 388	**493** 346	**467** 310	**438** 279	**409** 251	**382** 227	**359** 207	**337** 188	**317** 172	**299** 157	**283** 144	**267** 133
18LH04	18	7,751	**604** 435	**571** 388	**535** 347	**500** 312	**469** 281	**440** 255	**413** 231	**388** 211	**365** 192	**344** 176	**325** 162	**308** 149
18LH05	18	8,778	**684** 509	**648** 454	**614** 406	**581** 365	**543** 330	**508** 298	**476** 271	**448** 247	**421** 225	**397** 206	**375** 189	**355** 174
18LH06	18	10,382	**809** 554	**749** 494	**696** 442	**648** 397	**605** 358	**566** 324	**531** 295	**499** 268	**470** 245	**443** 224	**418** 206	**396** 190
18LH07	18	10,790	**840** 628	**809** 560	**780** 502	**726** 451	**678** 407	**635** 368	**595** 334	**559** 304	**526** 278	**496** 255	**469** 234	**444** 215
18LH08	18	11,243	**876** 696	**843** 620	**812** 555	**784** 499	**758** 450	**717** 408	**680** 370	**641** 337	**604** 308	**571** 282	**540** 259	**512** 238
18LH09	18	12,017	**936** 753	**901** 671	**868** 601	**838** 540	**810** 488	**783** 441	**759** 401	**713** 365	**671** 333	**633** 305	**598** 280	**566** 258

Joist Designation	Nominal Depth in Inches	Maximum End Reaction Lbs.	25	26	27	28	29	30	31	32	33	34	35	36	37	38	39	40
20LH02	20	5,672	**442**	**437** 404	**431** 362	**410** 325	**388** 293	**365** 265	**344** 241	**325** 220	**307** 200	**291** 184	**275** 169	**262** 155	**249** 143	**237** 132	**225** 122	**215** 114
20LH03	20	6,018	**469**	**463** 432	**458** 387	**452** 348	**434** 314	**414** 284	**395** 258	**372** 235	**352** 215	**333** 196	**316** 180	**299** 166	**283** 153	**269** 141	**255** 131	**243** 122
20LH04	20	7,366	**574** 545	**566** 486	**558** 435	**528** 391	**496** 353	**467** 319	**440** 290	**416** 264	**393** 241	**372** 221	**353** 203	**335** 187	**318** 172	**303** 159	**289** 147	**275** 137
20LH05	20	7,905	**616**	**609** 568	**602** 508	**595** 457	**571** 412	**544** 373	**513** 339	**484** 309	**458** 282	**434** 258	**411** 237	**390** 218	**371** 201	**353** 186	**336** 172	**321** 160
20LH06	20	10,554	**822** 695	**791** 620	**763** 555	**723** 499	**679** 450	**635** 407	**596** 370	**560** 337	**527** 308	**497** 282	**469** 259	**444** 238	**421** 220	**399** 203	**379** 188	**361** 174
20LH07	20	11,273	**878** 788	**845** 702	**814** 629	**786** 565	**760** 510	**711** 462	**667** 419	**627** 382	**590** 349	**556** 319	**526** 293	**497** 270	**471** 249	**447** 230	**425** 213	**404** 198
20LH08	20	11,653	**908** 875	**873** 780	**842** 698	**813** 628	**785** 566	**760** 513	**722** 466	**687** 424	**654** 387	**621** 355	**588** 326	**558** 300	**530** 276	**503** 255	**479** 237	**457** 220
20LH09	20	12,709	**990** 951	**953** 848	**918** 759	**886** 682	**856** 615	**828** 557	**802** 506	**778** 461	**755** 421	**712** 386	**673** 354	**636** 326	**603** 300	**572** 278	**544** 257	**517** 239
20LH10	20	13,710	**1068** 1067	**1028** 951	**991** 851	**956** 765	**924** 690	**894** 625	**865** 568	**839** 517	**814** 472	**791** 433	**748** 397	**707** 365	**670** 337	**636** 312	**604** 289	**575** 268

Table 4.18 (*Continued*)

STANDARD LOAD TABLE FOR LONGSPAN LH-SERIES JOISTS

Pounds per Linear Foot Based on Allowable Stress of 30,000 psi

Joist Designation	Nominal Depth in Inches	Maximum End Reaction Lbs.	Clear Opening or Net Span In Feet															
			33	34	35	36	37	38	39	40	41	42	43	44	45	46	47	48
24LH03	24	5,757	**342** 314	**339** 288	**336** 264	**323** 243	**307** 224	**293** 207	**279** 192	**267** 178	**255** 166	**244** 154	**234** 144	**224** 134	**215** 126	**207** 118	**199** 110	**191** 104
24LH04	24	7,053	**419** 355	**398** 325	**379** 298	**360** 275	**343** 253	**327** 234	**312** 217	**298** 201	**285** 187	**273** 174	**262** 162	**251** 152	**241** 142	**231** 133	**222** 125	**214** 117
24LH05	24	7,558	**449** 414	**446** 379	**440** 348	**419** 320	**399** 295	**380** 273	**363** 253	**347** 235	**331** 218	**317** 203	**304** 189	**291** 177	**280** 166	**269** 155	**258** 145	**248** 137
24LH06	24	10,167	**604** 455	**579** 417	**555** 382	**530** 352	**504** 325	**480** 300	**457** 278	**437** 258	**417** 240	**399** 223	**381** 208	**364** 194	**348** 182	**334** 170	**320** 160	**307** 150
24LH07	24	11,194	**665** 514	**638** 471	**613** 432	**588** 398	**565** 367	**541** 339	**516** 314	**491** 291	**468** 271	**446** 252	**426** 235	**407** 220	**389** 206	**373** 193	**357** 181	**343** 170
24LH08	24	11,901	**707** 574	**677** 525	**649** 482	**622** 444	**597** 409	**572** 378	**545** 351	**520** 325	**497** 302	**475** 282	**455** 263	**435** 245	**417** 230	**400** 215	**384** 202	**369** 190
24LH09	24	14,006	**832** 627	**808** 574	**785** 527	**764** 485	**731** 447	**696** 414	**663** 383	**632** 355	**602** 330	**574** 308	**548** 287	**524** 268	**501** 251	**480** 235	**460** 221	**441** 207
24LH10	24	14,848	**882** 701	**856** 642	**832** 590	**809** 543	**788** 500	**768** 463	**737** 428	**702** 398	**668** 370	**637** 344	**608** 321	**582** 300	**556** 281	**533** 263	**511** 247	**490** 232
24LH11	24	15,616	**927** 776	**900** 710	**875** 652	**851** 600	**829** 554	**807** 512	**787** 474	**768** 440	**734** 409	**701** 381	**671** 355	**642** 332	**616** 310	**590** 291	**567** 273	**544** 256
			41	42	43	44	45	46	47	48	49	50	51	52	53	54	55	56
28LH05	28	7,020	**337** 277	**323** 258	**310** 240	**297** 224	**286** 210	**275** 197	**265** 185	**255** 173	**245** 163	**237** 154	**228** 145	**220** 137	**213** 129	**206** 122	**199** 116	**193** 110
28LH06	28	9,333	**448** 332	**429** 310	**412** 289	**395** 270	**379** 252	**364** 237	**350** 222	**337** 208	**324** 196	**313** 185	**301** 174	**291** 164	**281** 155	**271** 147	**262** 139	**253** 132
28LH07	28	10,520	**505** 375	**484** 349	**464** 326	**445** 304	**427** 285	**410** 267	**394** 250	**379** 235	**365** 221	**352** 208	**339** 196	**327** 185	**316** 175	**305** 166	**295** 157	**285** 149
28LH08	28	11,250	**540** 385	**517** 359	**496** 335	**475** 313	**456** 293	**438** 274	**420** 257	**403** 242	**387** 227	**371** 214	**357** 202	**344** 191	**331** 180	**319** 170	**308** 161	**297** 153
28LH09	28	13,895	**667** 461	**639** 429	**612** 400	**586** 374	**563** 350	**540** 328	**519** 308	**499** 289	**481** 272	**463** 256	**446** 241	**430** 228	**415** 215	**401** 204	**387** 193	**374** 183
28LH10	28	15,187	**729** 514	**704** 479	**679** 447	**651** 417	**625** 390	**600** 366	**576** 343	**554** 322	**533** 303	**513** 286	**495** 269	**477** 254	**460** 240	**444** 227	**429** 215	**415** 204
28LH11	28	16,256	**780** 569	**762** 530	**736** 495	**711** 462	**682** 432	**655** 405	**629** 380	**605** 357	**582** 336	**561** 316	**540** 298	**521** 282	**502** 266	**485** 252	**468** 238	**453** 226
28LH12	28	17,873	**857** 645	**837** 600	**818** 560	**800** 523	**782** 490	**766** 459	**737** 430	**709** 404	**682** 380	**656** 358	**632** 338	**609** 319	**587** 301	**566** 285	**546** 270	**527** 256
28LH13	28	18,649	**895** 707	**874** 659	**854** 614	**835** 574	**816** 537	**799** 503	**782** 472	**766** 444	**751** 417	**722** 393	**694** 371	**668** 350	**643** 331	**620** 313	**598** 296	**577** 281

Table 4.18 (*Continued*)

STANDARD LOAD TABLE FOR LONGSPAN LH-SERIES JOISTS

Pounds per Linear Foot Based on Allowable Stress of 30,000 psi

Joist Designation	Nominal Depth in Inches	Maximum End Reaction Lbs.	Clear Opening or Net Span in Feet															
			49	50	51	52	53	54	55	56	57	58	59	60	61	62	63	64
32LH06	32	8,393	**338**	**326**	**315**	**304**	**294**	**284**	**275**	**266**	**257**	**249**	**242**	**234**	**227**	**220**	**214**	**208**
			244	229	216	204	193	182	173	164	155	148	140	133	127	121	115	110
32LH07	32	9,411	**379**	**366**	**353**	**341**	**329**	**318**	**308**	**298**	**288**	**279**	**271**	**262**	**254**	**247**	**240**	**233**
			274	258	243	230	217	205	194	184	175	166	158	150	143	136	130	124
32LH08	32	10,206	**411**	**397**	**383**	**369**	**357**	**345**	**333**	**322**	**312**	**302**	**293**	**284**	**275**	**267**	**259**	**252**
			301	284	268	253	239	226	214	203	192	183	174	165	157	150	143	136
32LH09	32	12,814	**516**	**498**	**480**	**463**	**447**	**432**	**418**	**404**	**391**	**379**	**367**	**356**	**345**	**335**	**325**	**315**
			362	341	321	303	287	271	257	243	231	219	208	198	189	180	171	164
32LH10	32	14,179	**571**	**550**	**531**	**512**	**495**	**478**	**462**	**445**	**430**	**416**	**402**	**389**	**376**	**364**	**353**	**342**
			380	358	337	318	301	285	269	255	242	230	219	208	198	189	180	172
32LH11	32	15,520	**625**	**602**	**580**	**560**	**541**	**522**	**505**	**488**	**473**	**458**	**443**	**429**	**416**	**403**	**390**	**378**
			421	397	374	353	334	316	299	284	269	256	243	231	220	210	200	191
32LH12	32	18,227	**734**	**712**	**688**	**664**	**641**	**619**	**598**	**578**	**559**	**541**	**524**	**508**	**492**	**477**	**463**	**449**
			505	476	449	424	401	379	359	340	323	307	291	277	264	251	240	229
32LH13	32	20,303	**817**	**801**	**785**	**771**	**742**	**715**	**690**	**666**	**643**	**621**	**600**	**581**	**562**	**544**	**527**	**511**
			555	522	493	465	440	416	394	373	354	336	320	304	290	276	263	251
32LH14	32	20,937	**843**	**826**	**810**	**795**	**780**	**766**	**738**	**713**	**688**	**665**	**643**	**622**	**602**	**583**	**564**	**547**
			586	552	520	491	464	439	416	394	374	355	338	321	306	291	278	265
32LH15	32	21,625	**870**	**853**	**837**	**821**	**805**	**791**	**776**	**763**	**750**	**725**	**701**	**678**	**656**	**635**	**616**	**597**
			661	622	587	554	524	495	469	445	422	401	381	362	345	329	313	299
			57	58	59	60	61	62	63	64	65	66	67	68	69	70	71	72
36LH07	36	8,419	**292**	**283**	**274**	**266**	**258**	**251**	**244**	**237**	**230**	**224**	**218**	**212**	**207**	**201**	**196**	**191**
			224	212	202	192	183	174	166	159	151	145	138	132	127	121	116	112
36LH08	36	9,255	**321**	**311**	**302**	**293**	**284**	**276**	**268**	**260**	**253**	**246**	**239**	**233**	**227**	**221**	**215**	**209**
			230	218	208	197	188	179	171	163	156	149	142	136	130	125	120	115
36LH09	36	11,850	**411**	**398**	**386**	**374**	**363**	**352**	**342**	**333**	**323**	**314**	**306**	**297**	**289**	**282**	**275**	**267**
			281	267	253	241	230	219	209	199	190	182	174	166	159	152	146	140
36LH10	36	13,090	**454**	**440**	**426**	**413**	**401**	**389**	**378**	**367**	**357**	**347**	**338**	**328**	**320**	**311**	**303**	**295**
			311	295	281	267	254	242	231	220	210	201	192	184	176	169	162	155
36LH11	36	14,272	**495**	**480**	**465**	**451**	**438**	**425**	**412**	**401**	**389**	**378**	**368**	**358**	**348**	**339**	**330**	**322**
			345	327	311	296	282	269	256	244	233	223	213	204	195	187	179	172
36LH12	36	17,098	**593**	**575**	**557**	**540**	**523**	**508**	**493**	**478**	**464**	**450**	**437**	**424**	**412**	**400**	**389**	**378**
			389	370	352	334	318	303	289	276	264	252	241	230	221	211	203	194
36LH13	36	20,096	**697**	**675**	**654**	**634**	**615**	**596**	**579**	**562**	**546**	**531**	**516**	**502**	**488**	**475**	**463**	**451**
			454	432	410	390	372	354	338	322	308	294	281	269	258	247	237	227
36LH14	36	22,146	**768**	**755**	**729**	**706**	**683**	**661**	**641**	**621**	**602**	**584**	**567**	**551**	**535**	**520**	**505**	**492**
			481	457	434	413	393	375	357	341	326	311	298	285	273	261	250	240
36LH15	36	23,326	**809**	**795**	**781**	**769**	**744**	**721**	**698**	**677**	**656**	**637**	**618**	**600**	**583**	**567**	**551**	**536**
			542	515	489	465	443	422	403	384	367	351	335	321	307	294	282	271

Table 4.18 (*Continued*)

STANDARD LOAD TABLE FOR LONGSPAN LH-SERIES JOISTS

Pounds per Linear Foot Based on Allowable Stress of 30,000 psi

Joist Designation	Nominal Depth in Inches	Maximum End Reaction Lbs.	Clear Opening or Net Span in Feet															
			65	66	67	68	69	70	71	72	73	74	75	76	77	78	79	80
40LH08	40	8,339	**254** 194	**247** 185	**241** 177	**234** 170	**228** 162	**222** 156	**217** 149	**211** 143	**206** 137	**201** 132	**196** 127	**192** 122	**187** 117	**183** 113	**178** 108	**174** 104
40LH09	40	10,900	**332** 237	**323** 227	**315** 217	**306** 207	**298** 198	**291** 190	**283** 182	**276** 175	**269** 168	**263** 161	**256** 155	**250** 149	**244** 143	**239** 138	**233** 133	**228** 128
40LH10	40	12,049	**367** 248	**357** 237	**347** 226	**338** 217	**329** 207	**321** 199	**313** 191	**305** 183	**297** 175	**290** 168	**283** 162	**276** 156	**269** 150	**262** 144	**255** 139	**249** 133
40LH11	40	13,100	**399** 274	**388** 262	**378** 250	**368** 240	**358** 229	**349** 220	**340** 211	**332** 202	**323** 194	**315** 186	**308** 179	**300** 172	**293** 165	**286** 159	**279** 153	**273** 148
40LH12	40	15,957	**486** 329	**472** 315	**459** 301	**447** 288	**435** 276	**424** 264	**413** 253	**402** 243	**392** 233	**382** 224	**373** 215	**364** 207	**355** 199	**346** 191	**338** 184	**330** 177
40LH13	40	18,813	**573** 384	**557** 367	**542** 351	**528** 336	**514** 321	**500** 308	**487** 295	**475** 283	**463** 272	**451** 261	**440** 251	**429** 241	**419** 232	**409** 223	**399** 215	**390** 207
40LH14	40	21,538	**656** 407	**638** 389	**620** 372	**603** 356	**587** 341	**571** 327	**556** 313	**542** 300	**528** 288	**515** 277	**502** 266	**490** 256	**478** 246	**466** 237	**455** 228	**444** 220
40LH15	40	24,099	**734** 458	**712** 438	**691** 419	**671** 401	**652** 384	**633** 368	**616** 352	**599** 338	**583** 324	**567** 312	**552** 299	**538** 288	**524** 277	**511** 266	**498** 256	**486** 247
40LH16	40	26,535	**808** 537	**796** 513	**784** 491	**772** 470	**761** 450	**751** 431	**730** 413	**710** 396	**691** 380	**673** 365	**655** 351	**638** 337	**622** 324	**606** 312	**591** 300	**576** 289
			73	74	75	76	77	78	79	80	81	82	83	84	85	86	87	88
44LH09	44	10,018	**272** 186	**265** 179	**259** 172	**253** 165	**247** 159	**242** 153	**236** 147	**231** 142	**226** 137	**221** 132	**216** 127	**211** 123	**207** 118	**202** 114	**198** 110	**194** 107
44LH10	44	11,050	**300** 214	**293** 205	**286** 197	**279** 190	**272** 183	**266** 176	**260** 169	**254** 163	**249** 157	**243** 151	**238** 146	**233** 141	**228** 136	**223** 131	**218** 127	**214** 122
44LH11	44	11,970	**325** 237	**317** 227	**310** 218	**302** 210	**295** 202	**289** 194	**282** 187	**276** 180	**269** 174	**264** 167	**258** 161	**252** 156	**247** 150	**242** 145	**236** 140	**232** 136
44LH12	44	14,807	**402** 267	**393** 256	**383** 246	**374** 237	**365** 228	**356** 219	**347** 211	**339** 203	**331** 196	**323** 189	**315** 182	**308** 176	**300** 170	**293** 164	**287** 158	**280** 153
44LH13	44	17,569	**477** 313	**466** 301	**454** 289	**444** 278	**433** 267	**423** 257	**413** 248	**404** 239	**395** 230	**386** 222	**377** 214	**369** 206	**361** 199	**353** 192	**346** 186	**338** 180
44LH14	44	20,221	**549** 340	**534** 326	**520** 313	**506** 301	**493** 290	**481** 279	**469** 269	**457** 259	**446** 249	**436** 240	**425** 232	**415** 224	**406** 216	**396** 208	**387** 201	**379** 195
44LH15	44	23,536	**639** 396	**623** 381	**608** 366	**593** 352	**579** 338	**565** 325	**551** 313	**537** 302	**524** 291	**512** 280	**500** 270	**488** 261	**476** 252	**466** 243	**455** 235	**445** 227
44LH16	44	27,146	**737** 465	**719** 446	**701** 429	**684** 412	**668** 396	**652** 381	**637** 367	**622** 354	**608** 341	**594** 329	**580** 317	**568** 306	**555** 295	**543** 285	**531** 275	**520** 266
44LH17	44	29,125	**790** 528	**780** 507	**769** 487	**759** 469	**750** 451	**732** 434	**715** 417	**699** 402	**683** 388	**667** 374	**652** 360	**638** 348	**624** 336	**610** 324	**597** 313	**584** 303

Table 4.18 (*Continued*)

STANDARD LOAD TABLE FOR LONGSPAN LH-SERIES JOISTS

Pounds per Linear Foot Based on Allowable Stress of 30,000 psi

Joist Designation	Nominal Depth in Inches	Maximum End Reaction Lbs.	Clear Opening or Net Span in Feet															
			81	82	83	84	85	86	87	88	89	90	91	92	93	94	95	96
48LH10	48	10,045	**246**	**241**	**236**	**231**	**226**	**221**	**217**	**212**	**208**	**204**	**200**	**196**	**192**	**188**	**185**	**181**
			188	181	175	169	163	157	152	147	142	137	133	129	124	121	117	113
48LH11	48	10,861	**266**	**260**	**255**	**249**	**244**	**239**	**234**	**229**	**225**	**220**	**216**	**212**	**208**	**204**	**200**	**196**
			208	201	194	187	180	174	168	163	157	152	147	142	138	133	129	125
48LH12	48	13,720	**336**	**329**	**322**	**315**	**308**	**301**	**295**	**289**	**283**	**277**	**272**	**266**	**261**	**256**	**251**	**246**
			235	226	218	211	203	196	190	183	177	171	166	161	155	151	146	141
48LH13	48	16,415	**402**	**393**	**384**	**376**	**368**	**360**	**353**	**345**	**338**	**332**	**325**	**318**	**312**	**306**	**300**	**294**
			276	266	257	248	239	231	223	215	208	201	195	189	183	177	171	166
48LH14	48	19,395	**475**	**464**	**454**	**444**	**434**	**425**	**416**	**407**	**399**	**390**	**383**	**375**	**367**	**360**	**353**	**346**
			299	288	278	268	259	250	242	234	226	218	211	205	198	192	186	180
48LH15	48	22,252	**545**	**533**	**521**	**510**	**499**	**488**	**478**	**468**	**458**	**448**	**439**	**430**	**422**	**413**	**405**	**397**
			349	336	325	313	302	292	282	273	264	255	247	239	231	224	217	210
48LH16	48	25,684	**629**	**615**	**601**	**588**	**576**	**563**	**551**	**540**	**528**	**518**	**507**	**497**	**487**	**477**	**468**	**459**
			409	394	380	367	354	342	331	320	309	299	289	280	271	262	254	246
48LH17	48	28,828	**706**	**690**	**675**	**660**	**646**	**632**	**619**	**606**	**593**	**581**	**569**	**558**	**547**	**536**	**525**	**515**
			465	449	433	418	403	389	376	363	351	340	329	318	308	299	289	280

c With welding becoming more common and economical, the best bridging may be cross bridging of 1 × 1 × ⅛-in. angle with flattened ends, welded to top and bottom chords. The ends of joists bearing on steel beams should be tack welded rather than clipped to the beam. For wall-bearing ends,

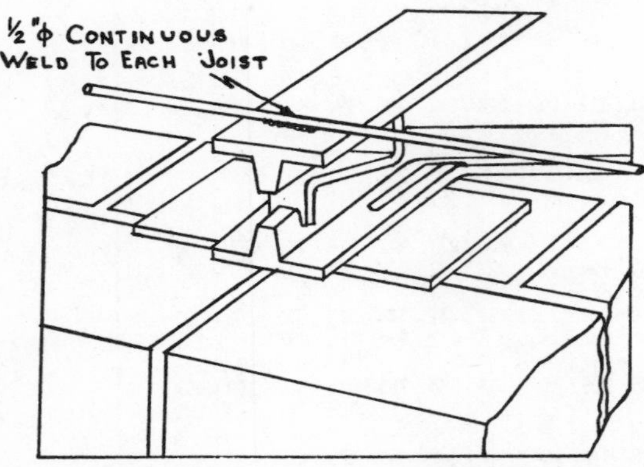

Fig. 4.29 Steel Joist Bearing on Wall

the most satisfactory wall anchor is a continuous ½-in. rod laid on top of the joists and tack welded to each joist. Where bracing beams are used parallel to a line of joists to stiffen columns, they should not be used in place of a bar joist, but should be dropped at least ½ in. lower than the plane of the top

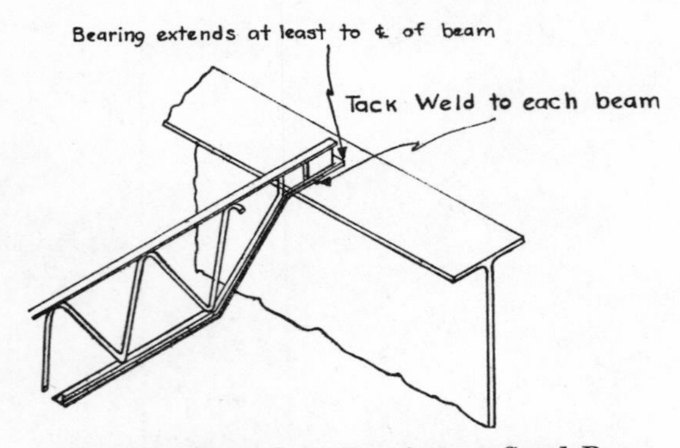

Fig. 4.30 Steel Joist Bearing on Steel Beam

chord to avoid cracking the slab. Where joist spans change directions, the joist parallel with the supporting beam should be properly anchored to the beam, and a 3-ft wide strip of wire mesh should be laid near the top of the slab, centered on the beam line. It is also good practice to lay a strip of metal lath over any beam carrying joists the span of which is over 15 times the joist depth.

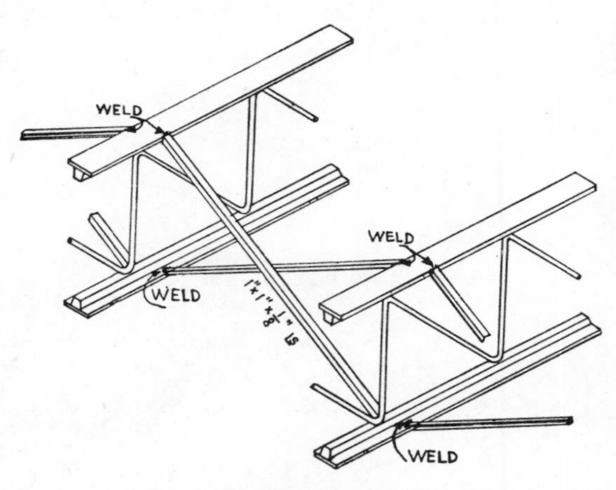

Fig. 4.31 Typical Welded Bridging for Steel Joists

If a wood floor is to be laid on steel joists, the most satisfactory construction is to bed the sleepers in the concrete topping at right angles to the direction of the joists, with at least ½ in. of concrete between. Wood-top nailer joists and wood nailers wired to the top flange of ordinary joists almost invariably result in squeaky floors.

The AISC and Steel Joist Institute specification permits the use of either of two types of bridging:

(*a*) Horizontal parallel steel bridging attached to top and bottom chord.
(*b*) X bracing acting as struts.

It has been the author's experience that if type (*a*) above is used, an X bracing should be used in every third or fourth space to insure that the rectangles formed by the joists and the bridging are not deformed into parallelograms, thus placing the joists in a leaning position and reducing their efficiency. The author has had better results with welded X bracing throughout.

d The joist tables published by the various manufacturers give allowable loads on spacings varying by 1-in. increments, implying that any desired spacing can be used. This is true with metal deck and tongue-and-groove concrete or gypsum plank, up to a

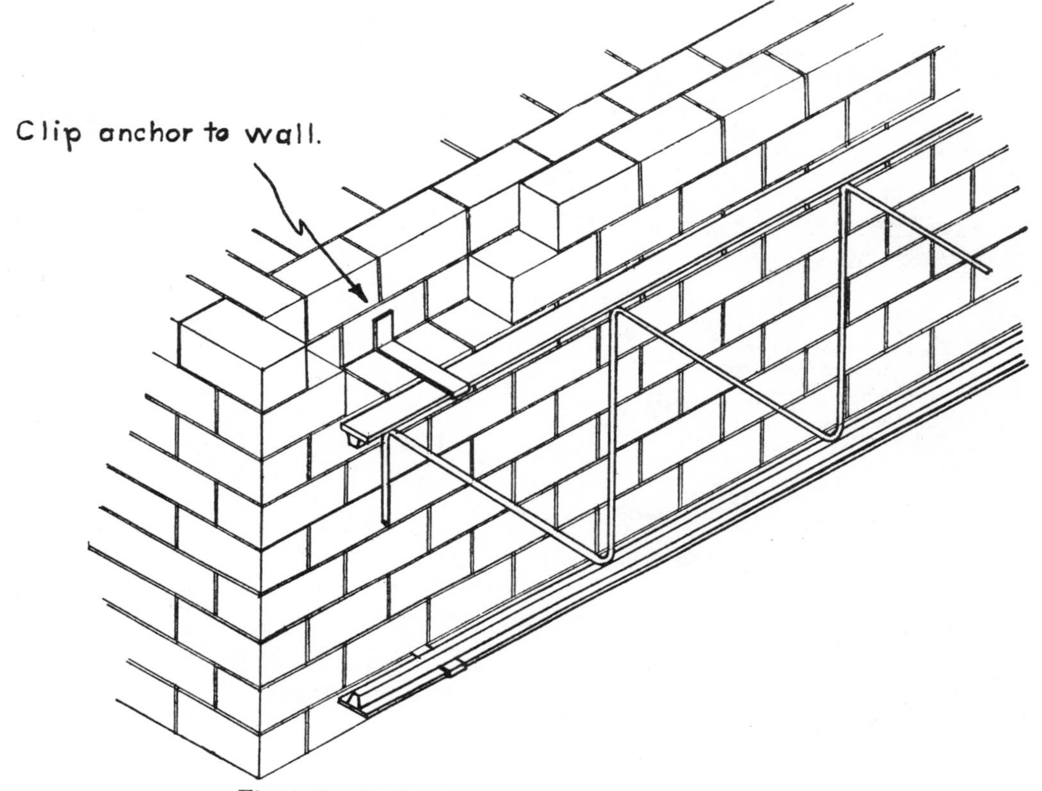

Fig. 4.32 Anchorage of Steel Joist Parallel with Wall

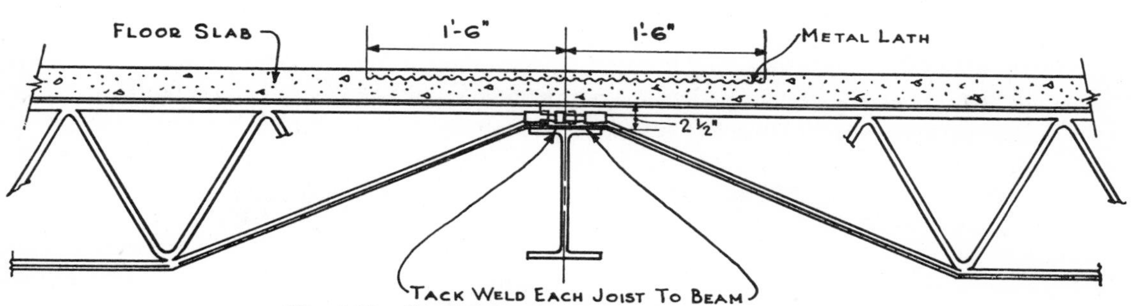

Fig. 4.33 Steel Joist Construction at Steel Beam

1. Parallel chords, underslung

4. Top chord pitched one way, square ends

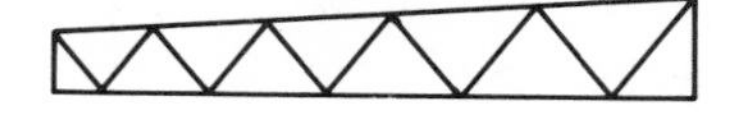

2. Parallel chords, square ends

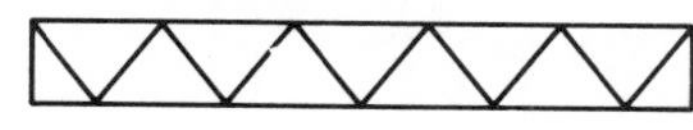

5. Top chord pitched two ways, underslung

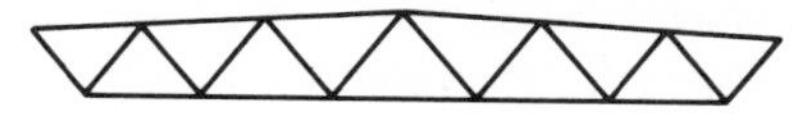

3. Top chord pitched one way, underslung

6. Top chord pitched two ways, square ends

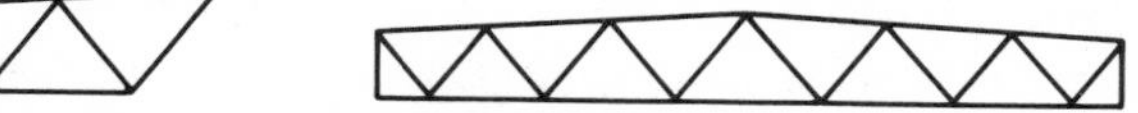

Fig. 4.34 Long Span Joists—Types

WELDED OR BOLTED CONNECTIONS

Where **LA**-Series Joists are supported on structural steel members, they are generally field welded. The number and length of welds should be specified. Where bolted connections are desired, round or slotted holes are provided in the bearing plates for this purpose. Holes are generally 13⁄16″ diameter to receive a ¾-inch bolt.

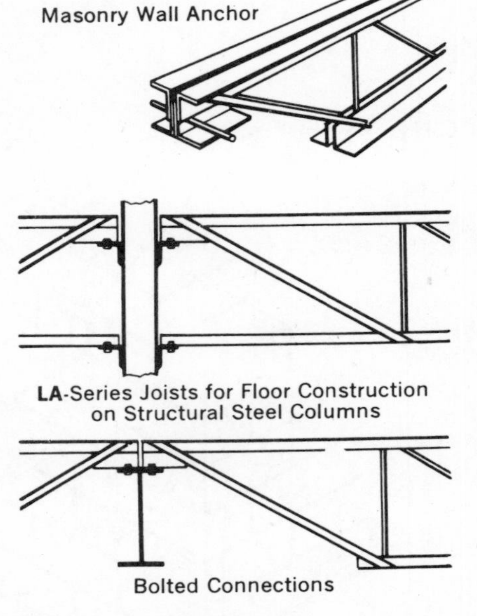

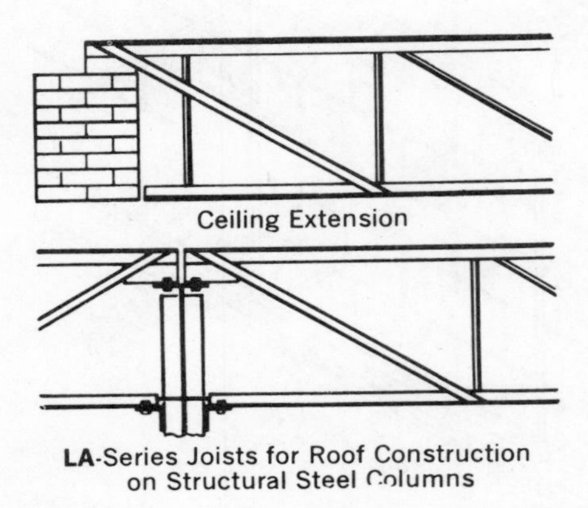

Fig. 4.35 Long Span Joists—Accessories and Details

maximum roof spacing of 30 in., but for poured concrete it is well to use spacings which will assure a lap of the floor lath and ceiling lath at a joist. If a span is picked at random, the lap point will come between joists more often than over joists, and there will almost surely be trouble when concrete is being poured. Assuming an 8-ft-long sheet of lath, with about 2- to 2½-in. lap, the following spaces will work out satisfactorily: 13½, 15¾, 18¾, and 23½ in. It is always well to double joists under partitions parallel with the joists, and to allow for the added effect of cross partitions within the middle third of a joist.

The AISC and Steel Joist Institute specification does not place any limitation on the spacing of the joists except the allowable span of the deck which is used. However, it is inadvisable to use a spacing much in excess of twice the depth of the joist used because of the reduced efficiency of the bridging as it becomes flatter. It is always well to remember what has been said earlier in this article, that the chief objection to the use of steel joists is the tendency to vibration with its resultant psychological effect on the public.

4.40 CORRUGATED SIDING AND ROOFING

a Because of certain properties of strength obtained through corrugating the sheets, corrugated siding and roofing is a popular low-priced covering used primarily for industrial buildings. The commonest material used is corrugated steel, either black, galvanized, or coated with asphalt and asbestos, but corrugated aluminum and asbestos-cement are also used. If the roof pitch is held to a minimum of 1 on 2½, or 1 on 3 if the edges are cemented, no applied roofing is required over the corrugated material. The corrugations in aluminum and asbestos-cement have not yet been standardized, but Figs. 4.36 and 4.37 and Table 4.19 give the standards adopted for corrugated steel sheets.

Standard sheets are of all single-foot lengths from 4 to 12 ft. The minimum end lap for roofing with a pitch of less than 1 on 3 is 8 in., and for roofing with a pitch of over 1 on 3 it is 6 in. The minimum end lap for siding is 4 in.

b Because failure of corrugated roofing or siding is usually caused by deflection, by wind, or—regardless of how carefully it may

Table 4.19 Corrugated Steel Construction

Dimensions and properties of sheets

U. S. Mfr's Gage	Thickness Inches	*Weight Lb. per square foot			Properties (per foot of corrugated width)					
		Flat	Corrugated		2⅔ × ½			3 × ¾		
			2⅔ × ½	3 × ¾	A in.[2]	I in.[4]	S in.[3]	A in.[2]	I in.[4]	S in.[3]
UNCOATED (BLACK) CORRUGATED STEEL SHEETS										
12	.1046	4.38	4.77	5.05	1.356	.0410	.136	1.444	.104	.243
14	.0747	3.13	3.41	3.61	.968	.0288	.100	1.031	.0736	.179
16	.0598	2.50	2.73	2.89	.775	.0229	.0818	.825	.0588	.145
18	.0478	2.00	2.18	2.31	.620	.0182	.0665	.660	.0469	.118
20	.0359	1.50	1.64	1.73	.465	.0136	.0509	.495	.0352	.0895
22	.0299	1.25	1.36	1.44	.388	.0113	.0428	.413	.0293	.0751
24	.0239	1.00	1.09	1.15	.310	.00906	.0346	.330	.0234	.0605
26	.0179	.75	.82	.87	.232	.00678	.0262	.247	.0175	.0456
28	.0149	.63	.68	.72	.193	.00564	.0219	.206	.0146	.0381
29	.0135	.56	.60	.65	.175	.00511	.0199	.186	.0132	.0346
GALVANIZED CORRUGATED STEEL SHEETS										
12	.1084	4.53	4.94	5.23	1.379	.0417	.138	1.468	.1058	.247
14	.0785	3.28	3.58	3.79	.991	.0295	.102	1.056	.0755	.183
16	.0635	2.66	2.90	3.07	.797	.0236	.0839	.849	.0605	.149
18	.0516	2.16	2.35	2.49	.643	.0189	.0688	.684	.0487	.122
20	.0396	1.66	1.81	1.91	.487	.0143	.0532	.519	.0369	.0936
22	.0336	1.41	1.53	1.62	.410	.0120	.0451	.436	.0310	.0792
24	.0276	1.16	1.26	1.33	.332	.00971	.0369	.353	.0251	.0646
26	.0217	.91	.99	1.05	.255	.00746	.0287	.272	.0193	.0501
28	.0187	.78	.85	.90	.216	.00632	.0245	.230	.0163	.0426
29	.0172	.72	.78	.83	.197	.00575	.0223	.210	.0149	.0389

* No allowance for side or end laps.

Allowable working stresses for corrugated sheets may be taken as $F_b = .60\ F_y$. Laps at sides and ends of sheets are ordinarily ignored in strength calculations. For wind loads, design stress may be increased by ⅓.

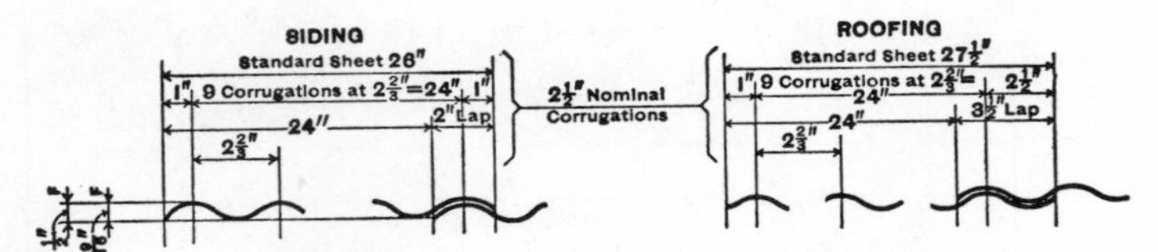

Fig. 4.36 Standard Corrugated Steel Sheets

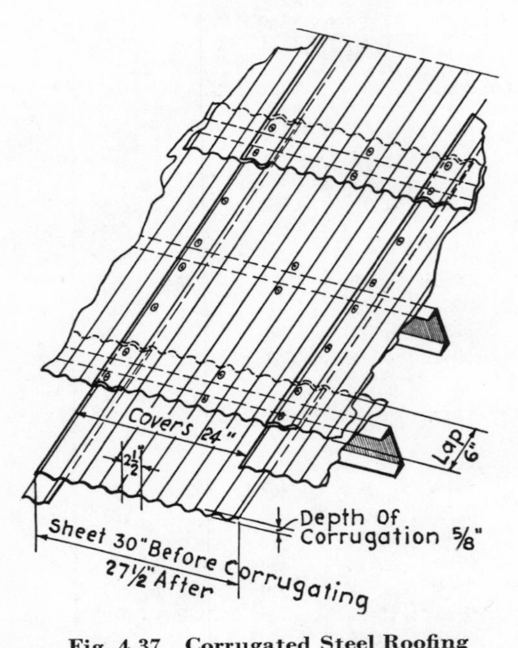

Fig. 4.37 Corrugated Steel Roofing

be maintained—by rusting around the bolt or the rivet, it is good economy to reduce the allowable stress well below the $0.60F_y$ given in the footnote to Table 4.19. If we assume alternation of stresses resulting from flutter due to wind, the limitations of Section 1.7.3 would reduce the allowable stress to $\frac{22,000}{1.67}$ or 13,200 psi and it is common practice to limit this stress to 12,000 psi. On this basis, using a total load of 35 psf, the allowable span for $2\frac{1}{2} \times \frac{1}{2}$-in. corrugations would be

$$l = 274\sqrt{t}$$

The maximum span for this loading would be

For 12 gage, 89.5 in.—7 ft 5½ in.
For 14 gage, 76.2 in.—6 ft 4 in.
For 16 gage, 67.5 in.—5 ft 7½ in.
For 18 gage, 60.5 in.—5 ft ½ in.
For 20 gage, 52.5 in.—4 ft 4½ in.
For 22 gage, 48.2 in.—4 ft 0 in.

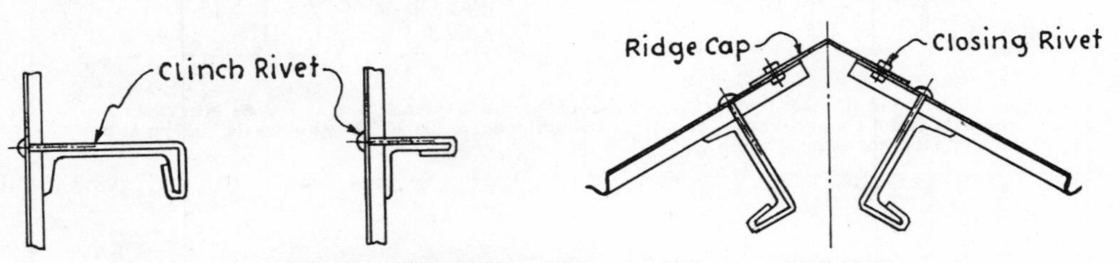

Fig. 4.38 Standard Connection by Clinch Rivets

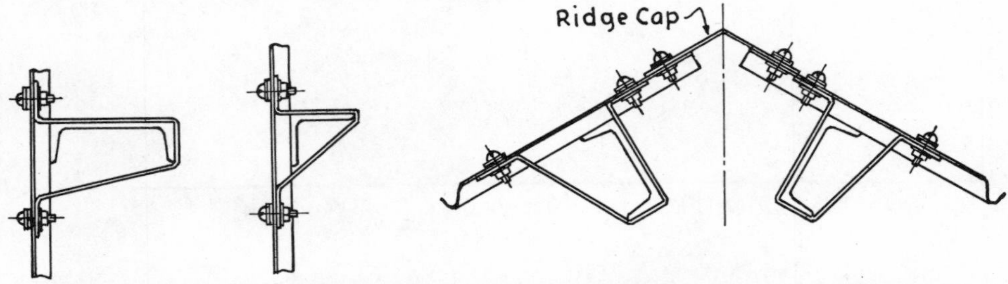

¼-in. galvanized bolts, galvanized washers with lead or asphalt-saturated felt washers underneath 18-ga galvanized metal straps, 1 in. wide

Fig. 4.39 Standard Connection by Straps and Bolts

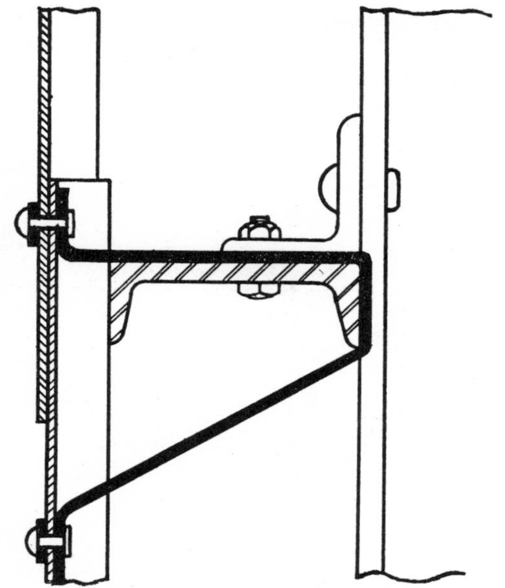

Fig. 4.40 **Strap Fastening to Girts**

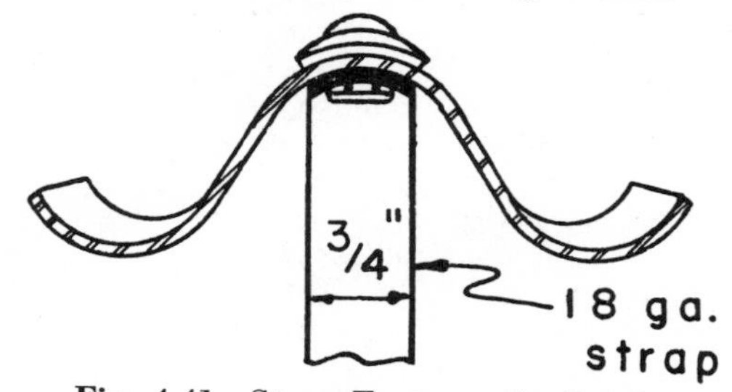

Fig. 4.41 **Strap Fastener for Roofing**

In climates where snow load need not be considered but only uplift due to wind pressure, the formula may be reduced to

$$l = 363\sqrt{t}$$

and the maximum spans would be

For 16 gage, 89.5 in.—7 ft 5½ in.
For 18 gage, 80 in. —6 ft 8 in.
For 20 gage, 69.6 in.—5 ft 9½ in.
For 22 gage, 64 in. —5 ft 0 in.

It is inadvisable to use less than 22-gage sheets for roofing. For siding, with a 15-psf wind load, the spacing of girts may be

$$l = 420\sqrt{t}$$

and the maximum spans would be

For 20 gage, 80.5 in.—6 ft 8½ in.
For 22 gage, 74 in. —6 ft 2 in.
For 24 gage, 66.4 in.—5 ft 6 in.

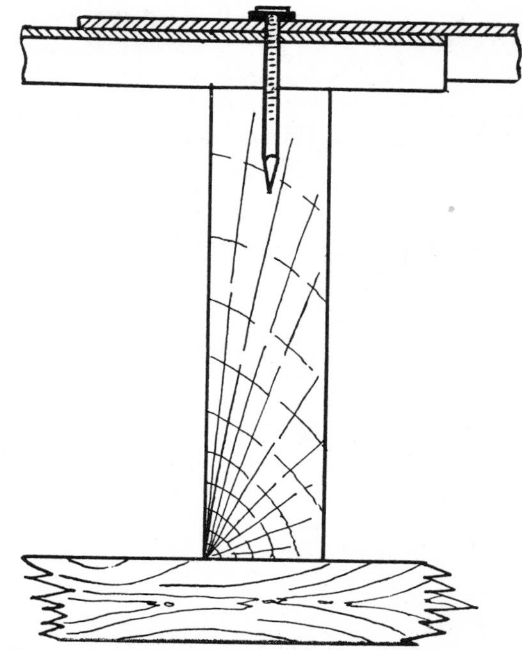

Fig. 4.42 **Fastening of Corrugated Steel to Wood**

Fig. 4.43 **Ridge Roll**

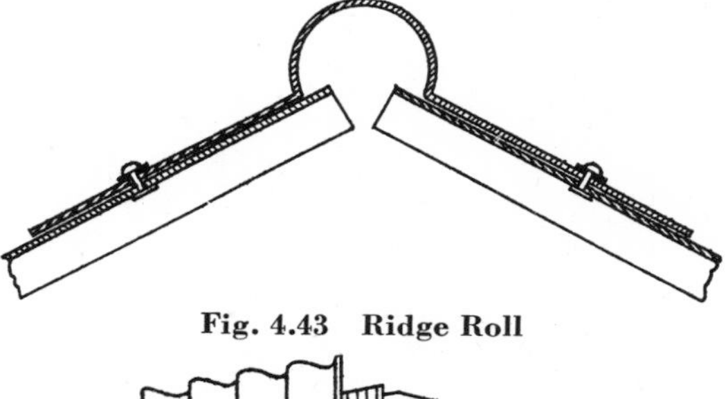

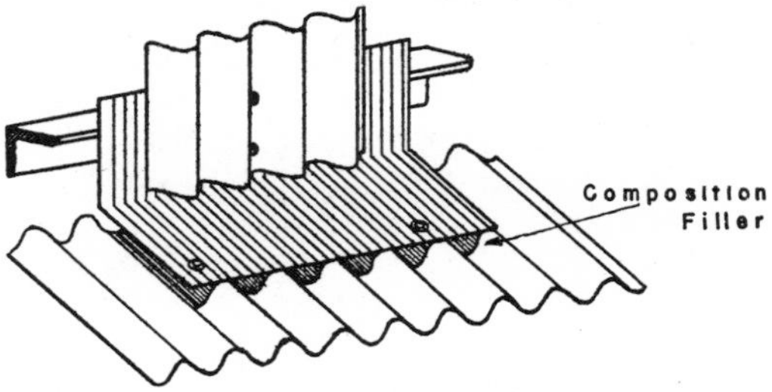

Fig. 4.44 **Flashing from Monitor Siding to Roofing**

c Details for the fastening of corrugated steel are standardized also. The method of fastening to girts and purlins shown in Fig. 4.38 may be used, although where the structure is subject to high winds, the details shown in Figs. 4.39 and 4.40 are preferable. Other details are shown in Art. 10.41, Girts.

Table 4.20 Standard Subway Gratings

U = Safe uniform load, psf
C = Safe concentrated load, lb per ft of width
D = Deflection, in., based on fiber stress of 16,000 psi

Loads to right of heavy lines cause excessive deflection.

Bearing bar, in. (all types)	Reticulated		Rectangular		Load and deflection	Span, ft								
	Crimp bar, in.	Weight, psf	Cross bar, in.	Weight, psf		2	$2\frac{1}{2}$	3	$3\frac{1}{2}$	4	$4\frac{1}{2}$	5	$5\frac{1}{2}$	6
$\frac{3}{4} \times \frac{1}{8}$	$\frac{3}{4} \times \frac{1}{8}$	6.5	$\frac{5}{8} \times \frac{1}{8}$	4.2	U	330	222	143						
					D	.085	.134	.192						
					C	330	265	215						
					D	.068	.108	.154						
$\frac{3}{4} \times \frac{3}{16}$	$\frac{3}{4} \times \frac{1}{8}$	8.0	$\frac{5}{8} \times \frac{1}{8}$	5.9	U	500	320	217						
					D	.085	.134	.192						
					C	500	400	325						
					D	.068	.108	.154						
$1 \times \frac{1}{8}$	$\frac{3}{4} \times \frac{1}{8}$	7.5	$\frac{3}{4} \times \frac{1}{8}$	5.5	U	600	384	267	188	150				
					D	.064	.099	.143	.195	.256				
					C	600	480	400	330	300				
					D	.051	.080	.115	.156	.205				
$1\frac{1}{4} \times \frac{1}{8}$	$\frac{3}{4} \times \frac{1}{8}$	8.5	$\frac{3}{4} \times \frac{1}{8}$	6.9	U	950	600	420	303	232	184	146	120	
					D	.051	.081	.115	.157	.205	.259	.321	.389	
					C	950	750	630	530	465	415	365	330	
					D	.041	.064	.092	.125	.163	.207	.256	.310	
$1\frac{1}{4} \times \frac{3}{16}$	$\frac{3}{4} \times \frac{1}{8}$	10.5	$\frac{3}{4} \times \frac{1}{8}$	9.4	U	1,425	900	633	457	350	278	220	182	
					D	.051	.081	.115	.157	.205	.259	.321	.389	
					C	1,425	1,125	950	800	700	625	550	500	
					D	.041	.064	.092	.125	.163	.207	.256	.310	
$1\frac{1}{2} \times \frac{1}{8}$	$\frac{3}{4} \times \frac{1}{8}$	9.5	$1 \times \frac{1}{8}$	8.1	U	1,365	880	610	445	340	266	220	182	150
					D	.043	.067	.094	.131	.166	.216	.267	.324	.385
					C	1,365	1,100	915	785	680	600	550	500	450
					D	.034	.053	.077	.104	.137	.173	.214	.259	.308
$1\frac{1}{2} \times \frac{3}{16}$	$\frac{3}{4} \times \frac{1}{8}$	12.0	$1 \times \frac{1}{8}$	11.6	U	2,050	1,320	917	672	512	400	330	272	225
					D	.043	.067	.094	.131	.166	.216	.267	.324	.385
					C	2,050	1,650	1,375	1,175	1,025	900	825	750	675
					D	.034	.053	.077	104	.137	.173	.214	.259	.308
$1\frac{3}{4} \times \frac{3}{16}$	$\frac{3}{4} \times \frac{1}{8}$	13.5	$1 \times \frac{1}{8}$	13.3	U	2,800	1,780	1,230	915	700	544	440	364	308
					D	.038	.057	.082	.112	.147	.185	.229	.276	.330
					C	2,800	2,225	1,860	1,600	1,400	1,225	1,100	1,000	925
					D	.029	.046	.066	.090	117	.148	.183	.221	.264
$2 \times \frac{3}{16}$	$1 \times \frac{1}{8}$	16.0	$1 \times \frac{1}{8}$	15.1	U	3,650	2,340	1,618	1,200	912	723	580	482	400
					D	.032	.050	.072	.099	.128	.163	.201	.243	.289
					C	3,650	2,925	2,425	2,100	1,825	1,625	1,450	1,325	1,200
					D	.026	.040	.057	.078	.102	.129	.160	.193	.230
$2\frac{1}{4} \times \frac{3}{16}$	$1 \times \frac{1}{8}$	17.5	$1 \times \frac{1}{8}$	16.8	U	4,650	2,960	2,065	1,515	1,150	912	740	608	516
					D	.027	.044	.064	.087	.113	.148	.177	.214	.255
					C	4.650	3,700	3,100	2,650	2,300	2,050	1,850	1,675	1,550
					D	.023	.035	.051	.070	.091	.115	.142	.172	.204

4.50 METAL DECK, SUBWAY GRATING

a Metal deck is a type of formed sheet steel manufactured by a number of companies for use as a flat roof deck (see Fig. 4.45). It is used in conjunction with tar and gravel or other standard membrane-type roofing, and sometimes carries a layer of insulation on which the roofing is laid. It may be attached to the steel beams by means of strap anchors or clips, or it may be tack welded to the beams. If this latter

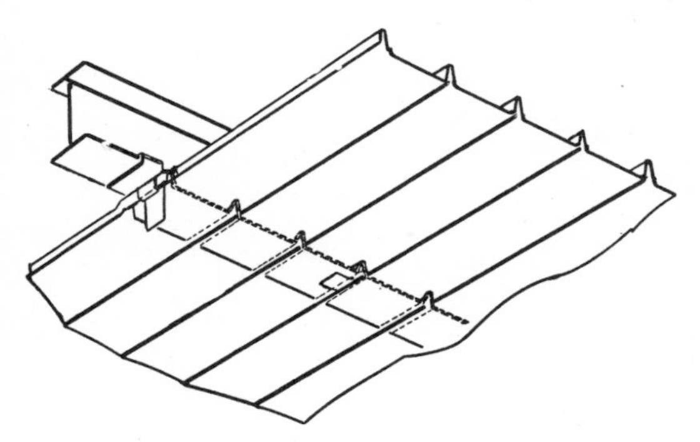

Fig. 4.45 Metal Deck

method is used, the specifications should require that all welds shall be field painted, since the heat of the weld burns off the asphaltic paint with which the deck is usually covered, and unless it is retouched, the deck will rust out at this point. It is probably inadvisable to use a metal deck of less than 20 gage regardless of what the strength tables of the manufacturer may indicate. Moreover, because of the springiness and the

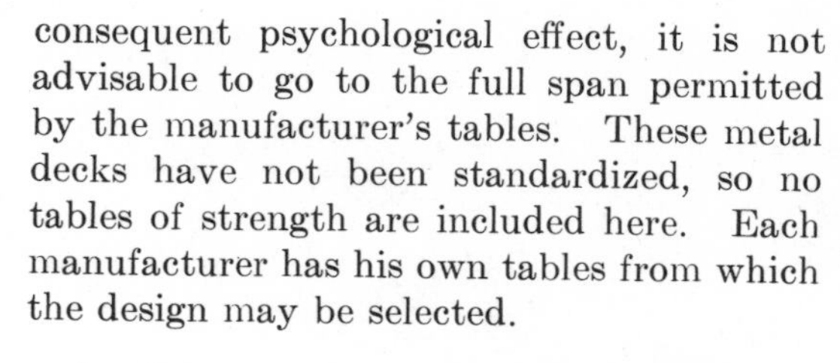

consequent psychological effect, it is not advisable to go to the full span permitted by the manufacturer's tables. These metal decks have not been standardized, so no tables of strength are included here. Each manufacturer has his own tables from which the design may be selected.

b Subway grating is a type of flooring material used for factory mezzanine floors and platforms. Because of its open construction, it is frowned upon by state labor departments except for the above-named uses. Moreover, it is too costly to compete with standard reinforced concrete slabs except for such special uses.

Subway grating is made in two general types—rectangular, and reticulated or crimped, as shown in plan in Fig. 4.46. The safe loads for standard types manufactured by the different manufacturers in accordance with the U.S. Government Federal Specification of Jan. 30, 1936, are listed in Table 4.20.

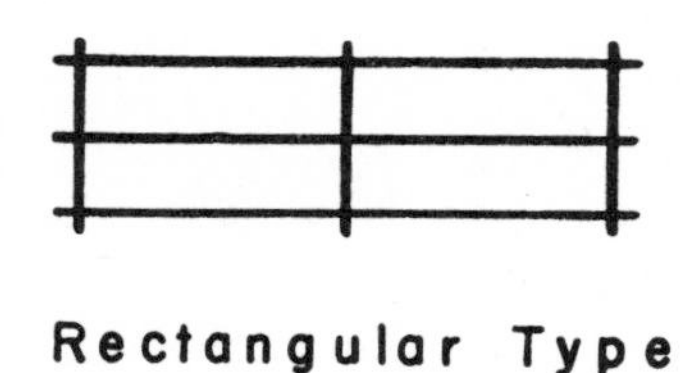

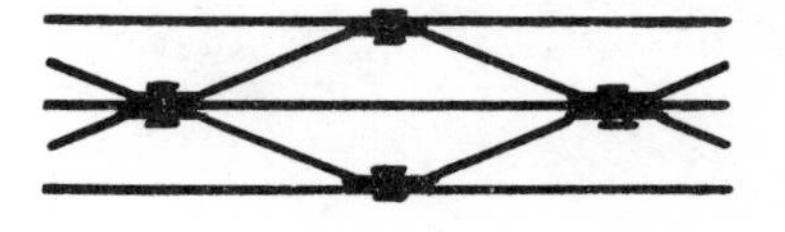

Fig. 4.46 Subway Grating

5

Reinforced Concrete in Bending

5.10 REINFORCED CONCRETE

a Steel can be used to resist a given tensile stress much more economically than an equal compressive stress. A bar of any shape will offer equal resistance, per square inch, to tension, whereas for compression the steel must be made into special shapes to provide sufficient lateral rigidity. Even then, the strength in compression is less than the strength in tension. Concrete, on the other hand, can only be used in tension to a very limited extent, but its compressive strength is comparatively high. In strength per dollar, the two materials are working advantageously when compression is taken by concrete and tension by steel, as is done in reinforced concrete beams. In compression members of appreciable length, such as columns, a combination of the two materials is also quite advantageous. The two materials bond satisfactorily, so that they work in unison—that is, their coefficient of thermal expansion is almost identical.

b It is well for the designer to know the economical limits of reinforced concrete construction. It is naturally a heavier-dead-load material than structural steel and is much more economical when the live loads are heavy as in warehouses. For light floor loads the advantage is with structural steel. For high buildings the area of concrete columns in the lower stories becomes so great that they occupy too much rentable area to be economical. This limitation sometimes governs the choice of design even though live loads are heavy. To cut down the size of concrete columns in high buildings, it is common to put in a structural-steel core for the lower stories, but this practice is not economical.

Another objection sometimes offered to the use of reinforced concrete for factory buildings is that it is not so easy to make alterations or hang such equipment as trolley beams—in short, that the material is not so adaptable as steel. On the other hand, reinforced concrete is fireproof, while a steel skeleton must be encased to achieve the same result.

c The strength of reinforced concrete was formerly rated on the basis of the mix, as 1:2:4 or 1:1½:3 or 1:1:2, indicating proportions respectively of cement, sand, and coarse aggregate (gravel or crushed stone). In 1918, however, the water-cement ratio law was put forth as the basis of determination of strength of concrete mixtures. The fundamental law may be stated as follows: For plastic mixtures, using sound and clean aggregates, the strength and other desirable properties of concrete under given job conditions are governed by the net quantity of mixing water used per sack of cement.

The water-cement ratio law was first established with respect to compressive strength.

Subsequent studies, however, have shown that it applies equally well to flexural and tensile strength, to the resistance of concrete to wear, and to the bond between concrete and steel. Still more recently, investigations have established that the properties of watertightness and resistance to weathering are controlled in the same way by the proportion of water to cement.

This principle may be more readily understood if the cement and water are thought of as forming a paste, which, on hardening, binds the aggregate particles together. The strength of this paste is determined by the proportions of cement and water. Increasing the water content dilutes the paste and reduces the strength. Similarly, the watertightness of the concrete is determined by the watertightness of the paste—a thin, watery paste, on hardening, will not form the dense, impermeable binding medium essential to watertight concrete.

It should be noted that in the statement of the water-cement ratio law above, its application is limited to plastic mixtures and to given job conditions.

As a result of the change in thinking brought about by the discovery and use of the water-cement ratio law, the old method of designating strength of concrete on the basis of mix has given place to stating allowable stresses as a proportion of the 28-day compression test strength. Tables commonly recognize 2,500-, 3,000-, and 4,000-psi strength.

It has become common practice in many offices to base the specifications and the design on 3,000-psi concrete, and it has seemed advisable to the author, therefore, where tables are dependent on the strength of concrete, as for foundations, etc., to base these designs on the use of 3,000-psi concrete unless otherwise noted.

5.11 Reinforced concrete design

a The tables and formulae given here are to be used for all reinforced concrete design, whether beams, slabs, footings, retaining walls, or other structures. Other special formulae and tables are given in specific articles pertaining to special aspects of design, such as T beams, doubly reinforced beams, and so on.

b *Formulae for rectangular-beam and slab design*

$$k = \sqrt{2pn + (pn)^2} - pn = \frac{1}{1 + \dfrac{f_s}{nf_c}}$$

$$j = 1 - \frac{k}{3}$$

$$p = \frac{A_s}{bd} = \frac{\frac{1}{2}}{\dfrac{f_s}{f_c}\left(\dfrac{f_s}{nf_c} + 1\right)} = \frac{f_c k}{2f_s}$$

$$R = \tfrac{1}{2}f_c kj \quad \text{or} \quad pf_s j \text{ (in beam design)}$$

$$M = \tfrac{1}{2}f_c kjbd^2 = Rbd^2 \quad \text{or} \quad bd^2 = \frac{2M}{f_c kj} = \frac{M}{R}$$

$$M = pf_s jbd^2 \quad \text{or} \quad bd^2 = \frac{M}{pf_s j}$$

$$A_s = \frac{M}{f_s jd} \quad \text{or} \quad f_s = \frac{M}{A_s jd}$$

$$f_c = \frac{2f_s p}{k} \quad \text{or} \quad \frac{f_s k}{n(1 - k)} \quad \text{or} \quad \frac{2M}{kjbd^2}$$

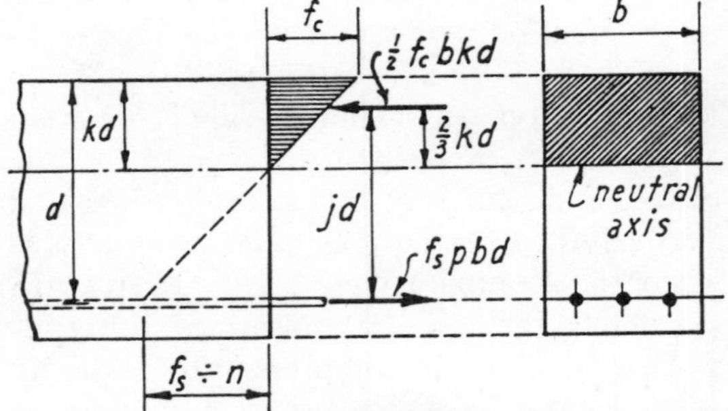

Courtesy Portland Cement Association

Fig. 5.1 Formulae for Rectangular Beam and Slab Design

5.12 Plain concrete

a The ACI Code does not recognize plain unreinforced concrete as a material to resist bending except in footings and walls. In accordance with Table 1002 (a) (Table 1.2 of this book), the extreme fiber stress in tension in footings and walls is established at

$$f_c = 1.6\sqrt{f'_c}$$

For the various grades of concrete normally used, therefore, the following may be used:

For 2,500-psi concrete, $f_c = 80$
For 3,000-psi concrete, $f_c = 88$
For 4,000-psi concrete, $f_c = 102$
For 5,000-psi concrete, $f_c = 113$

b Although it is not advisable to consider the use of unreinforced concrete for ordinary beam and slab design, there are certain parts of buildings or structures which may be of plain concrete, designed to resist bending; for example, the various parts of a retaining wall (as described in Art. 8.30), the offsets of a footing (as in Art. 8.11), and basement retaining walls (as in Art. 8.31). A problem in this third category will serve as an example of the use of these rules of design.

Problem Assume the basement retaining wall shown in Fig. 5.2. The resultant beam diagram is shown in Fig. 5.3. From similar triangles, the distance x may be determined from the equation

$$15x_1^2 = 617$$
$$x_1 = 6.41$$
$$x = 8.41$$

Table 5.1 Reinforced Rectangular Concrete-Beam Constants

	n	10	9	8	7
	f'_c	2,500	3,000	4,000	5,000
	f_c	1,125	1,350	1,800	2,250
$f_s = 20{,}000$ psi	R	174.2	222.6	324.0	418.7
	k	.362	.379	.419	.440
	j	.879	.871	.860	.853
	p	.0099	.0128	.0188	.0248
	f_sj	17.6	17.4	17.2	17.1
$f_s = 24{,}000$ psi	R	160.5	201.5	295.3	387.2
	k	.319	.336	.375	.396
	j	.894	.888	.875	.868
	p	.0075	.0094	.0141	.0186
	f_sj	21.5	21.3	21.0	20.8
$f_s = 30{,}000$ psi	R	139.4	175.9	260.3	342.7
	k	.273	.288	.324	.344
	j	.909	.904	.892	.885
	p	.0051	.0065	.0097	.0129
	f_sj	27.3	27.1	26.8	26.6

Table 5.2 Sectional Area of Various Numbers of Bars

Square inches

Bar no.	3	4	5	6	7	8	9	10	11	14S	18S
Size, in.	⅜φ	½φ	⅝φ	¾φ	⅞φ	1φ	1□	1⅛□	1¼□	1½□	2□
Weight per ft, lb	0.376	0.668	1.043	1.502	2.044	2.670	3.400	4.303	5.313	7.65	13.60
Perimeter, in.	1.18	1.57	1.96	2.36	2.75	3.14	3.54	3.99	4.43	5.32	7.09
1 bar	0.11	0.20	0.31	0.44	0.60	0.79	1.00	1.27	1.56	2.25	4.00
2 bars	0.22	0.40	0.62	0.88	1.20	1.58	2.00	2.54	3.12	4.50	8.00
3 bars	0.33	0.60	0.93	1.32	1.80	2.37	3.00	3.81	4.68	6.75	12.00
4 bars	0.44	0.80	1.24	1.76	2.40	3.16	4.00	5.08	6.24	9.00	16.00
5 bars	0.55	1.00	1.55	2.20	3.00	3.95	5.00	6.35	7.80	11.25	20.00
6 bars	0.66	1.20	1.86	2.64	3.60	4.74	6.00	7.62	9.36	13.50	24.00
7 bars	0.77	1.40	2.17	3.08	4.20	5.53	7.00	8.89	10.92	15.75	28.00
8 bars	0.88	1.60	2.48	3.52	4.80	6.32	8.00	10.16	12.48	18.00	32.00
9 bars	0.99	1.80	2.79	3.96	5.40	7.11	9.00	11.43	14.04	20.25	36.00
10 bars	1.10	2.00	3.10	4.40	6.00	7.90	10.00	12.70	15.60	22.50	40.00

$$M = 617 \times 8.41$$
$$\underline{-617 \times 2.14}$$
$$617 \times 6.27 = 3{,}870 \text{ ft-lb}$$
$$= 46{,}430 \text{ in.-lb}$$

Therefore, assuming a 3,000-psi concrete,

$$\text{required } S = \frac{46{,}430}{88} = 527.6 = \frac{bd^2}{6}$$

or since $b = 12$ in., $S = 2d^2$.

$$d = \sqrt{\frac{527.6}{2}} = \sqrt{263.8} = 16.25 \text{ in.}$$

Use 16-in. thickness.

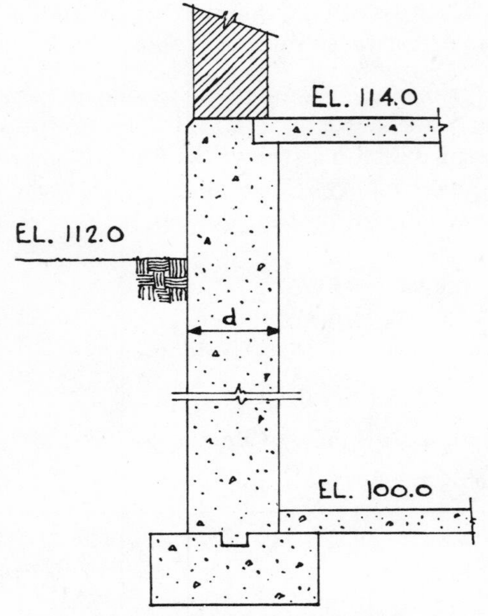

Fig. 5.2 Plain Concrete Basement Retaining Wall

5.20 REINFORCED CONCRETE BEAM DESIGN

a Reinforced concrete beam design involves more computations than the design of steel beams. Because of the monolithic nature of concrete construction, beams are designed for continuity, and the sign of the moment at the end of the beam usually is reversed from the moment at the center of the beam.

In the design of a reinforced concrete beam the following steps are necessary:

1. Compute bending moments and shears in accordance with the methods of Art. 3.10 for simple beams, or Art. 3.20 or 3.21 for continuous beams.

2. Determine the concrete-beam size by the method of Art. 5.20*b* or 5.21. (Frequently it is advisable for the economy of form reuse or for architectural requirements to determine one of several beam sizes for an entire area, and make all the beams this size. This may waste concrete but save more in form work.)

3. Check the assumed size of beam for shear by the method of Art. 5.30.

4. Compute the steel area required by means of Art. 5.20*b* and select the bars to furnish this steel, both for positive and negative moment. Positive moment through the center of the span is provided for by furnishing sufficient straight and bent bars for the total positive area. Negative moment at the supports is provided for by bending up enough bars from each of the two adjacent beams so that the sum of bar areas at this point is sufficient for the negative area or by providing sufficient straight top bars (see Fig. 5.4).

5. Compute the stirrups required, if any, and determine spacing by the method of Art. 5.30.

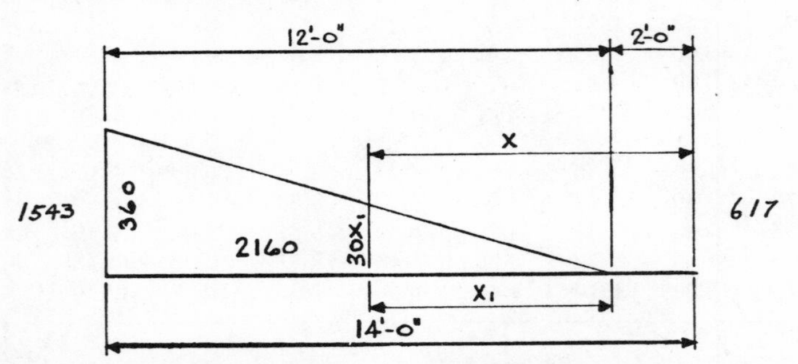

Fig. 5.3 Beam Load Diagram—Basement Retaining Wall

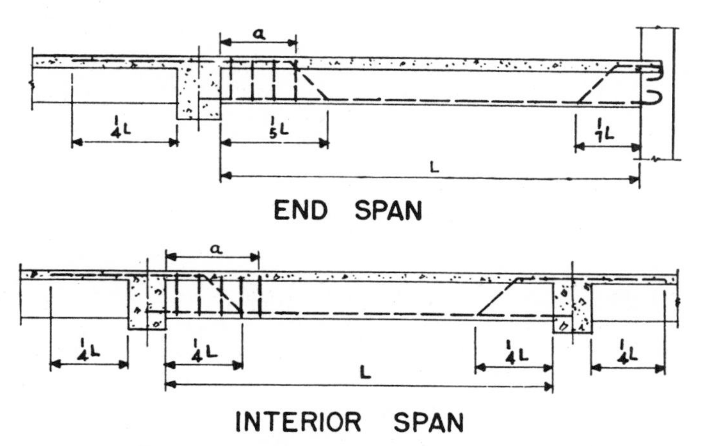

Fig. 5.4 Typical Beam Reinforcement

b In addition to those portions of the ACI Code quoted in Art. 1.22 the following sections of the Code will apply to beam design.

904 Frame Analysis—General

(a) All members of frames or continuous construction shall be designed to resist at all sections the maximum effects of the prescribed loads as determined by the theory of elastic frames in which the simplifying assumptions of Section 905 may be used.

(b) Except in the case of prestressed concrete, approximate methods of frame analysis may be used for buildings of usual types of construction, spans, and story heights.

(c) Except for prestressed concrete, in the case of two or more approximately equal spans (the larger of two adjacent spans not exceeding the shorter by more than 20 percent) with loads uniformly distributed, where the unit live load does not exceed three times the unit dead load, the following moments and shears may be used in design in lieu of more accurate analyses.

Positive moment

End spans

If discontinuous end is unrestrained .... $\frac{1}{11} wl'^2$

If discontinuous end is integral with the support .... $\frac{1}{14} wl'^2$

Interior spans .... $\frac{1}{16} wl'^2$

Negative moment at exterior face of first interior support

Two spans .... $\frac{1}{9} wl'^2$

More than two spans .... $\frac{1}{10} wl'^2$

Negative moment at other faces of interior supports .... $\frac{1}{11} wl'^2$

Negative moment at face of all supports for, (a) slabs with spans not exceeding 10 ft, and (b) beams and girders where ratio of sum of column stiffnesses to beam stiffness exceeds eight at each end of the span .... $\frac{1}{12} wl'^2$

Negative moment at interior faces of exterior supports for members built integrally with their supports

Where the support is a spandrel beam or girder .... $\frac{1}{24} wl'^2$

Where the support is a column .... $\frac{1}{16} wl'^2$

Shear in end members at first interior support .... $1.15 \frac{wl'}{2}$

Shear at all other supports .... $\frac{wl'}{2}$

905 Frame Analysis—Details

(a) *Arrangement of live load*

1. The live load may be considered to be applied only to the floor or roof under consideration, and the far ends of the columns may be assumed as fixed.

2. Consideration may be limited to combinations of dead load on all spans with full live load on two adjacent spans and with full live load on alternate spans.

(b) *Span length*

1. The span length, l, of members that are not built integrally with their supports shall be considered the clear span plus the depth of the slab or beam but shall not exceed the distance between centers of supports.

2. In analysis of continuous frames, center to center distances shall be used in the determination of moments. Moments at faces of supports may be used for design of beams and girders.

3. Solid or ribbed slabs with clear spans of not more than 10 ft that are built integrally with their supports may be designed as continuous slabs on knife edge supports with spans equal to the clear spans of the slab and the width of beams otherwise neglected.

(c) *Stiffness*

1. Any reasonable assumptions may be adopted for computing the relative flexural stiffness of columns, of walls, and of floor and roof systems. The assumptions made shall be consistent throughout the analysis.

2. In computing the value of I for the relative flexural stiffness of slabs, beams, girders, and columns, the reinforcement may be neglected. In T-shaped sections allowance shall be made for the effect of flange.

3. If the total torsional stiffness in the plane of a continuous system at a joint does not exceed 20 percent of the flexural stiffness at the joint, the torsional stiffness need not be taken into account in the analysis.

(d) *Haunched members*

1. The effect of haunches shall be considered both in determining bending moments and in design of members.

907 Effective Depth of Beam or Slab

(a) The effective depth, d, of a beam or slab shall be taken as the distance from the centroid of its tensile reinforcement to its compression face.

(b) Any floor finish not placed monolithically with the floor slab shall not be included as a part of the structural member. When the top of a monolithic slab is the wearing surface and unusual wear is expected as in buildings of the warehouse or industrial class, there shall be placed an additional depth of ½ in. over that required by the design of the member.

908 Distance between Lateral Supports

(a) The effects of lateral eccentricity of load shall be taken into account in determining the spacing of lateral supports for a beam, which shall never exceed 50 times the least width, b, of compression flange or face.

909 Control of Deflections*

(a) Reinforced concrete members subject to bending shall be designed to have adequate stiffness to prevent deflections or other deformations which may adversely affect the strength or serviceability of the structure.

(b) The minimum thicknesses, t, stipulated in Table 909(b) shall apply to flexural members of normal weight concrete, except when calculations of deflections prove that lesser thicknesses may be used without adverse effects.

(c) Where deflections are to be computed, those which occur immediately upon application of service load shall be computed by the usual methods and formulas for elastic deflections, using the modulus of elasticity for concrete specified in Section 1102. The moment of inertia shall be based on the gross section when pf_y is equal to or less than 500 and on the transformed cracked

TABLE 909(b)—MINIMUM THICKNESS OR DEPTH OF FLEXURAL MEMBERS UNLESS DEFLECTIONS ARE COMPUTED

Member	Minimum thickness or depth t			
	Simply supported	One end continuous	Both ends continuous	Cantilever
One-way slabs	$l/25$	$l/30$	$l/35$	$l/12$
Beams	$l/20$	$l/23$	$l/26$	$l/10$

section when pf_y is greater. In continuous spans, the moment of inertia may be taken as the average of the values obtained for the positive and negative moment regions.

(d) The additional long-time deflections may be obtained by multiplying the immediate deflection caused by the sustained part of the load by 2.0 when $A_s' = 0$; 1.2 when $A_s' = 0.5A_s$; and 0.8 when $A_s' = A_s$.

(e) Maximum limits for immediate deflection due to live load computed as above are:

1. For roofs which do not support plastered ceilings.................. $l/180$
2. For roofs which support plastered ceilings or for floors which do not support partitions.............. $l/360$

(f) For a floor or roof construction intended to support or to be attached to partitions or other construction likely to be damaged by large deflections of the floor, the allowable limit for the sum of the immediate deflection due to live load and the additional deflection due to shrinkage and creep under all sustained loads computed as above shall not exceed $l/360$.

* See Art. 2.30*f*.

910 Deep Beams

(a) Beams with depth/span ratios greater than 2/5 for continuous spans, or 4/5 for simple spans shall be designed as deep beams taking account of nonlinear distribution of stress, lateral buckling, and other pertinent effects. The minimum horizontal and vertical reinforcement in the faces shall be as in Section 2202(f); the minimum tensile reinforcement as in Section 911.

911 Minimum Reinforcement of Flexural Members

(a) Wherever at any section of a flexural member (except slabs of uniform thickness) positive reinforcement is required by analysis, the ratio, p, supplied shall not be less than $200/f_y$, unless the area of reinforcement provided at every section, positive or negative, is at least one-third greater than that required by analysis.

(b) In structural slabs of uniform thickness, the minimum amount of reinforcement in the direction of the span shall not be less than that required for shrinkage and temperature reinforcement (see Section 807).

The spacing of bars, required beam width protection, and anchorage of reinforcement are covered by the following sections.

804 Spacing of Bars

(a) The clear distance between parallel bars (except in columns and between multiple layers of bars in beams) shall be not less than the nominal diameter of the bars, 1⅓ times the maximum size of the coarse aggregate, nor 1 in.

(b) Where reinforcement in beams or girders is placed in two or more layers, the clear distance between layers shall be not less than 1 in., and the bars in the upper layers shall be placed directly above those in the bottom layer.

808 Concrete Protection for Reinforcement

(a) The reinforcement of footings and other principal structural members in which the concrete is deposited against the ground shall have not less than 3 in. of concrete between it and the ground contact surface. If concrete surfaces after removal of the forms are to be exposed to the weather or be in contact with the ground, the reinforcement shall be protected with not less than 2 in. of concrete for bars larger than #5 and 1½ in. for #5 bars or smaller.

(b) The concrete protective covering for any reinforcement at surfaces not exposed directly to the ground or weather shall be not less than ¾ in. for slabs and walls, and not less than 1½ in. for beams and girders. In concrete joist floors in which the clear distance between joists is not more than 30 in., the protection of reinforcement shall be at least ¾ in.

(d) Concrete protection for reinforcement shall in all cases be at least equal to the diameter of bars, except for concrete slabs and joists as in (b).

918 Anchorage Requirements—General

(a) The calculated tension or compression in any bar at any section must be developed on each side of that section by proper embedment length, end anchorage, or hooks. A tension bar may be anchored by bending it across the web at an angle of not less than 15 deg with the longitudinal portion of the bar and making it continuous with the reinforcement on the opposite face of the member.

(b) Except at supports, every reinforcing bar shall be extended beyond the point at which it is no longer needed to resist flexural stress, for a distance equal to the effective depth of the member or 12 bar diameters, whichever is greater.

(c) No flexural bar shall be terminated in a tension zone unless *one* of the following conditions is satisfied:

1. The shear is not over half that normally

permitted, including allowance for shear reinforcement, if any.

2. Stirrups in excess of those normally required are provided each way from the cut off a distance equal to three-fourths of the depth of the beam. The excess stirrups shall be at least the minimum specified in Section 1206(b) or 1706(b). The stirrup spacing shall not exceed $d/8r_b$ where r_b is the ratio of the area of bars cut off to the total area of bars at the section.

3. The continuing bars provide double the area required for flexure at that point or double the perimeter required for flexural bond.

(d) Tensile negative reinforcement in any span of a continuous, restrained or cantilever beam, or in any member of a rigid frame shall be adequately anchored by bond, hooks, or mechanical anchors in or through the supporting member.

(e) At least one-third of the total reinforcement provided for negative moment at the support shall be extended beyond the extreme position of the point of inflection a distance not less than $\frac{1}{16}$ of the clear span, or the effective depth of the member, whichever is greater.

(f) At least one-third the positive moment reinforcement in simple beams and one-fourth the positive moment reinforcement in continuous beams shall extend along the same face of the beam into the support at least 6 in.

(g) Plain bars (as defined in Section 301) in tension, except bars for shrinkage and temperature reinforcement, shall terminate in standard hooks except that hooks shall not be required on the positive reinforcement at interior supports of continuous members.

(h) Standard hooks (Section 801) in tension may be considered as developing 10,000 psi in Part IV-A or 19,000 psi in Part IV-B in the bars or may be considered as extensions of the bars at appropriate bond stresses.

(i) Hooks shall not be considered effective in adding to the compressive resistance of bars.

(j) Any mechanical device capable of developing the strength of the bar without damage to the concrete may be used in lieu of a hook. Test results showing the adequacy of such devices must be presented.

The width determined by these clearances may be the governing width of the beam required (see Table 5.3).

Table 5.3 shows minimum beam width when stirrups are used. If no stirrups are required, deduct $\frac{3}{4}$ in. from the figures shown.

c Since reinforced concrete is not a homogeneous material, the location of the neutral axis will vary, depending on the compression stress in the concrete and the tensile stress in the steel. The location of the neutral axis may be found from either of the following formulae

$$k = \sqrt{2pn + (pn)^2} - pn \qquad (5.1)$$

$$k = \frac{n}{m + n} \qquad (5.2)$$

The effective depth between centroids of compression and tension jd is easily found after k has been found.

$$j = 1 - \frac{k}{3} \qquad (5.3)$$

With reference to Art. 3.10, it will be seen that the external moment M is resisted by an internal moment couple of force P times the lever arm jd. This force in tension may be replaced by the product of a steel stress and a steel area, f_sA_s, and the resisting moment in tension is

$$M_s = f_sA_sjd \qquad (5.4)$$

The force P in compression may be replaced by the area of the triangle $\frac{1}{2}f_ckd$ multiplied by width b, or

Table 5.3 Minimum Beam Widths, in. (ACI Code)

Bar no.	Number of bars in single layer					Add for each added bar
	2	3	4	5	6	
4	$5\frac{3}{4}$	$7\frac{1}{4}$	$8\frac{3}{4}$	$10\frac{1}{4}$	$11\frac{3}{4}$	$1\frac{1}{2}$
5	6	$7\frac{3}{4}$	$9\frac{1}{4}$	11	$12\frac{1}{2}$	$1\frac{5}{8}$
6	$6\frac{1}{4}$	8	$9\frac{3}{4}$	$11\frac{1}{2}$	$13\frac{1}{4}$	$1\frac{3}{4}$
7	$6\frac{1}{2}$	$8\frac{1}{2}$	$10\frac{1}{4}$	$12\frac{1}{4}$	14	$1\frac{7}{8}$
8	$6\frac{3}{4}$	$8\frac{3}{4}$	$10\frac{3}{4}$	$12\frac{3}{4}$	$14\frac{3}{4}$	2
9	$7\frac{1}{4}$	$9\frac{1}{2}$	$11\frac{3}{4}$	14	$16\frac{1}{4}$	$2\frac{1}{4}$
10	$7\frac{3}{4}$	$10\frac{1}{4}$	$12\frac{3}{4}$	$15\frac{1}{4}$	$17\frac{3}{4}$	$2\frac{5}{8}$
11	8	11	$13\frac{3}{4}$	$16\frac{1}{2}$	$19\frac{1}{2}$	$2\frac{7}{8}$

$$M = \tfrac{1}{2}f_c kjbd^2 \qquad (5.5)$$

Since for any given combination of steel and concrete stresses, $\tfrac{1}{2}f_c kj$ is replaced by a constant R, Formula (5.5) becomes

$$M_c = Rbd^2 \qquad (5.6)$$

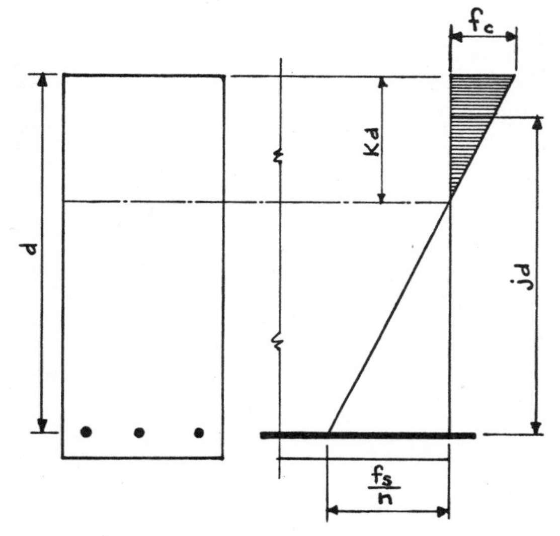

Fig. 5.5 Graphical Indication of Stresses—Concrete Beam

The majority of concrete-beam problems can be solved by the use of Formulae (5.4) and (5.6) above. For practical design purposes, j may be taken as 0.875. Constants for the various combinations of steel and concrete stresses ordinarily encountered in various codes are listed in Table 5.1. The constants given in this table are based on "balanced reinforcement"—that is, both steel and concrete are working to their maximum allowable working capacity. Within the limits of these constants, any change in steel or concrete will be only toward the side of safety and, therefore, design according to the methods here stated is sufficiently accurate.

Since A432 high-strength billet steel with an allowable tensile stress of 24,000 psi costs the same as A15 intermediate grade steel with an allowable tensile stress of 20,000 psi, it is of course advisable to use the higher strength steel where it is feasible. However, the A432 steel is more difficult to bend in the field than A15, so if field bending is required, the A15 steel is preferable. There is an increasing use of straight bars, and this method of construction eliminates the necessity and the cost of bending.

Problem Design for moment only a rectangular concrete beam for a simple span of 20 ft to carry a uniform load of 1,100 lb per ft, using a 3,000-psi concrete and 1963 ACI stresses ($f_c = 1{,}350$, $f_s = 24{,}000$).

$$W = 20 \times 1{,}100 = 22{,}000 \text{ lb}$$

$$M = \frac{22{,}000 \times 20 \times 12}{8} = 660{,}000 \text{ in.-lb}$$

$$M_c = 660{,}000 = Rbd^2$$

From Table 5.1, $R = 201.5$

$$bd^2 = \frac{660{,}000}{201.5} = 3{,}275$$

Use 12 × 16-in. net.

$$M_s = 660{,}000 = f_s A_s jd$$

From Table 5.1, $f_s j = 21{,}300$

$$A_s = \frac{660{,}000}{21{,}300 \times 16} = 1.94 \text{ sq in.}$$

Use two No. 7 straight and one No. 8 bent.

The computed depth d is from the top of the beam to the center line of the steel. Ordinary code requirement for cover of steel reinforcement in beams and girders is $1\tfrac{1}{2}$ in. The gross depth of the beam, therefore, is 16 in. (computed above) + $\tfrac{1}{2}$ in. (half the depth of the bars) + $1\tfrac{1}{2}$-in. cover = 18 in. total depth.

d Occasionally it is necessary to solve a problem using special materials or stresses for which the constants given in Table 5.2 will not work. For example, suppose it is required to design a beam using a specially prepared aggregate with the characteristics $f_c' = 3{,}500$ and $E_c = 2{,}800{,}000$, and with standard steel stresses $f_s = 20{,}000$ and $E_s = 29{,}000{,}000$. Using $f_c = 0.45f_c' = 1{,}575$,

$$m = \frac{20{,}000}{1{,}575} = 12.7 \quad \text{and} \quad n = \frac{29}{2.8} = 10.4$$

From Formula (5.2)

$$k = \frac{n}{m+n} = \frac{10.4}{12.7 + 10.4} = \frac{10.4}{23.1} = 0.45$$

$$j = 1 - 0.15 = 0.85$$

The same result is obtained graphically in Fig. 5.6.

$$R = \tfrac{1}{2}f_ckj = \tfrac{1}{2} \times 1{,}575 \times 0.45 \times 0.85 = 301.2$$

Then for the problem given in paragraph *c*

$$bd^2 = \frac{660{,}000}{301.2} = 2{,}191$$

Assume 12 in. width; then

$$d = \sqrt{\frac{2{,}191}{12}} = \sqrt{182.6} = 13.5$$

Use 12 × 14-in. net.

$$A_s = \frac{660{,}000}{20{,}000 \times 0.85 \times 14} = 2.77$$

Use two No. 9 bent and 1 No. 8 straight. Total depth = 14 in. + ½ in. (half bar thickness) + 1½-in. cover = 16 in. Therefore the beam is 12 × 16 in.

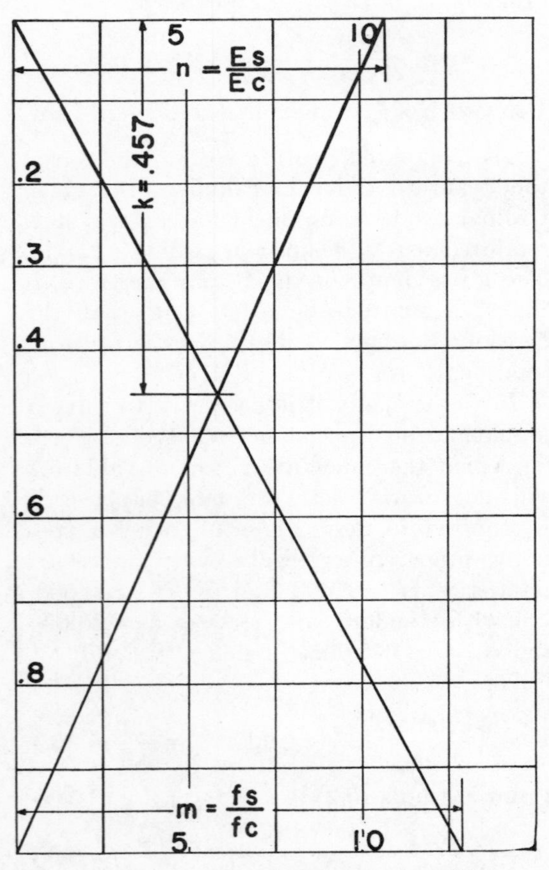

Fig. 5.6 Graphical Computation of *k*

A simple method of obtaining *k* graphically, based on Formula (5.2), is shown in Fig. 5.6. The use of the diagram is self-explanatory.

e Another form of problem is the determination of actual stresses in steel and concrete in a given beam under given load conditions.

Assume a 720,000-in.-lb moment in a 12 × 23-in. beam, 3,000-psi concrete, 1963 ACI Code, three 1-in. square bars. The net depth of beam is 21 in. and the steel percentage

$$p = \frac{3}{12 \times 21} = 0.012, \quad n = 9, \quad pn = 0.108$$

From Formula (5.1)

$$\begin{aligned} k &= \sqrt{0.216 + 0.01166} - 0.108 \\ &= \sqrt{0.2277} - 0.108 \\ &= 0.477 - 0.108 = 0.369 \\ j &= 1 - 0.123 = 0.877 \end{aligned}$$

From Formula (5.4)

$$f_s = \frac{720{,}000}{3 \times 0.877 \times 21} = 13{,}030 \text{ psi}$$

From Formula (5.5)

$$f_c = \frac{720{,}000}{0.5 \times 0.369 \times 0.877 \times 12 \times 441} = 856 \text{ psi}$$

f Before any steel bar can develop its tensile strength it must develop sufficient bond strength to the concrete to equal the tensile value. Otherwise, the tensile value of that bar must be reduced in proportion to the reduced bond value. Ordinarily bond plays very little part in the design of any beam since sufficient anchorage is usually provided.

The 1963 ACI Code includes the following on "Bond and Anchorage."

1300 NOTATION

- d = distance from extreme compression fiber to centroid of tension reinforcement
- D = nominal diameter of bar, inches
- f_c' = compressive strength of concrete (see Section 301)
- j = ratio of distance between centroid of compression and centroid of tension to the depth, d

Σo = sum of perimeters of all effective bars crossing the section on the tension side if of uniform size; for mixed sizes, substitute $4A_s/D$, where A_s is the total steel area and D is the largest bar diameter. For bundled bars use the sum of the exposed portions of the perimeters

u = bond stress

V = total shear

1301 COMPUTATION OF BOND STRESS IN FLEXURAL MEMBERS

(a) In flexural members in which the tension reinforcement is parallel to the compression face, the flexural bond stress at any cross section shall be computed by

$$u = \frac{V}{\Sigma ojd} \qquad (13\text{-}1)$$

Bent-up bars that are not more than $d/3$ from the level of the main longitudinal reinforcement may be included. Critical sections occur at the face of the support, at each point where tension bars terminate within a span, and at the point of inflection.

(b) To prevent bond failure or splitting, the calculated tension or compression in any bar at any section must be developed on each side of that section by proper embedment length, end anchorage, or, for tension only, hooks. Anchorage or development bond stress, u, shall be computed as the bar forces divided by the product of Σo times the embedment length.

(c) The bond stress, u, computed as in (a) or (b) shall not exceed the limits given below, except that flexural bond stress need not be considered in compression, nor in those cases of tension where anchorage bond is less than 0.8 of the permissible.

(1) For tension bars with sizes and deformations conforming to ASTM A 305:

Top bars* $\quad \dfrac{3.4\sqrt{f_c'}}{D}$ nor 350 psi

Bars other than top bars $\quad \dfrac{4.8\sqrt{f_c'}}{D}$ nor 500 psi

(2) For tension bars with sizes and deformations conforming to ASTM A 408:

Top bars* $\quad 2.1\sqrt{f_c'}$

Bars other than top bars $\quad 3\sqrt{f_c'}$

(3) For all deformed compression bars:

$6.5\sqrt{f_c'}$ nor 400 psi

(4) For plain bars the allowable bond stresses shall be one-half of those permitted for bars conforming to ASTM A 305 but not more than 160 psi

(d) Adequate anchorage shall be provided for the tension reinforcement in all flexural members to which Eq. (13-1) does not apply, such as sloped, stepped or tapered footings, brackets, or beams in which the tension reinforcement is not parallel to the compression face.

Let us check the unit bond stress in a continuous beam with a total uniform load of 24,000 lb, a net depth of 16 in., and with two No. 7 and one No. 9 bars, assuming the point of inflection ⅕ of the span from the support.

At this point $V = 0.3 \times 24{,}000 = 7{,}200$ lb. The perimeter of bars = 5.5 (two No. 7) + 3.54 (one No. 9) = 9.04 in.

$$u = \frac{8 \times 7{,}200}{7 \times 9.04 \times 16} = 56.9 \text{ psi}$$

* Top bars, in reference to bond, are horizontal bars so placed that more than 12 in. of concrete is cast in the member below the bar.

5.21 Concrete T Beams

a Because floor slabs are ordinarily poured monolithic with the supporting beams, the more common condition is to treat the construction as a T-beam construction instead of rectangular beams. The only point of saving in the use of T-beam construction is in the fact that the compression may be developed in the combination of web and flange instead of the web alone. The constants change slightly but the amount of steel is the same as for rectangular beams, and the shear must be taken by the web only. Obviously, if the neutral axis comes within the flange, that is, if t/d is 0.375 or over, the beam is treated as a rectangular beam, but the width b used in computing the beam in compression may be the full flange width determined by the limitations of the ACI Code.

b The following sections quoted from the 1963 ACI Code may be taken as good practice in the design of T beams.

906 REQUIREMENTS FOR T-BEAMS

(a) In T-beam construction the slab and beam shall be built integrally or otherwise effectively bonded together.

(b) The effective flange width to be used in the design of symmetrical T-beams shall not exceed one-fourth of the span length of the beam, and its overhanging width on either side of the web shall not exceed eight times the thickness of the slab nor one-half the clear distance to the next beam.

(c) Isolated beams in which the T-form is used only for the purpose of providing additional compression area, shall have a flange thickness not less than one-half the width of the web and a total flange width not more than four times the width of the web.

(d) For beams having a flange on one side only, the effective overhanging flange width shall not exceed 1/12 of the span length of the beam, nor six times the thickness of the slab, nor one-half the clear distance to the next beam.

(e) Where the principal reinforcement in a slab which is considered as the flange of a T-beam (not a joist in concrete joist floors) is parallel to the beam, transverse reinforcement shall be provided in the top of the slab. This reinforcement shall be designed to carry the load on the portion of the slab required for the flange of the T-beam. The flange shall be assumed to act as a cantilever. The spacing of the bars shall not exceed five times the thickness of the flange, nor in any case 18 in.

(f) The overhanging portion of the flange of the beam shall not be considered as effective in computing the shear and diagonal tension resistance of T-beams.

(g) Provision shall be made for the compressive stress at the support in continuous T-beam construction, care being taken that the provisions of Section 804 relating to the spacing of bars, and of Section 604 relating to the placing of concrete shall be fully met.

The following formulae may be used in the solution of T beams.

$$k = \frac{1}{1 + \dfrac{fs}{nf_c}} = \frac{pn + \frac{1}{2}\left(\dfrac{t}{d}\right)^2}{pn + \dfrac{t}{d}} \qquad (5.7)$$

$$kd = \frac{2ndA_s + bt^2}{2nA_s + 2bt}$$

$$d - jd = \frac{3kd - 2t}{2kd - t} \times \frac{t}{3}$$

$$j = \frac{6 - 6\left(\dfrac{t}{d}\right) + 2\left(\dfrac{t}{d}\right)^2 + \left(\dfrac{t}{d}\right)^3\left(\dfrac{1}{2pn}\right)}{6 - 3\left(\dfrac{t}{d}\right)}$$

$$= 1 - \frac{t}{d}\left(\frac{3k - 2t/d}{6k - 3t/d}\right) \qquad (5.8)$$

$$f_s = \frac{M}{A_s jd} = \frac{M}{pjbd^2}$$

$$f_c = \frac{f_s k}{n\,(1 - k)}$$

$$M_s = f_s A_s jd$$

$$M_c = f_c\left(1 - \frac{t}{2kd}\right) bt \times jd = Rbd^2 \qquad (5.9)$$

c When the neutral axis lies in the web, the amount of compression in the web below the flange is commonly small compared with that in the flange and is neglected in the formulae.

The value of R of the formula may be selected from Table 5.4. Note that in using this chart b is the flange width conforming with the limitations of Art. 5.21*b*.

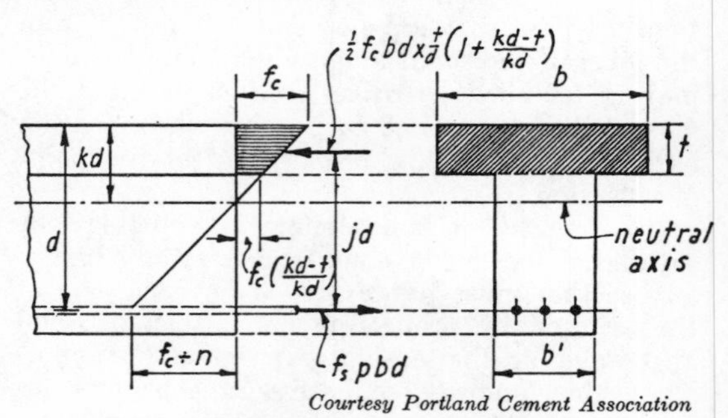

Courtesy Portland Cement Association

Fig. 5.7 Formulae for T-Beam Design

Problem Design for moment only a reinforced concrete T beam for a moment of 1,600,000 in.-lb, using a beam with a total depth of 18 in. (16 in. net) and an adjacent slab 4 in. thick, and assuming that all the limitations of Art. 5.21*b* will be met. Use 1963 ACI Code specifications for 3,000-psi concrete, with f_s = 20,000 psi.

$$\frac{t}{d} = \frac{4}{16} = 0.25$$

From Table 5.4,

$$R = 202.5$$

$$b = \frac{1{,}600{,}000}{202.5 \times 256} = 30.9 \text{ in.}$$

$$A_s = \frac{1{,}600{,}000}{20{,}000 \times 0.871 \times 16} = 5.74 \text{ sq. in.}$$

Use two No. 11 bars, A = 3.16, and two No. 10 bars, A = 2.54. A_s furnished = 5.70 sq in.

From Table 5.3 the minimum width of stem would be 13½ in. Use 14-in. stem.

compressive value, but not over 2 percent of the area bd.

Table 5.4 Value of *R* for T Beams

t/d		$f_s = 20{,}000$				$f_s = 24{,}000$			
	f'_c	2,500	3,000	4,000	5,000	2,500	3,000	4,000	5,000
	f_c	1,125	1,350	1,800	2,250	1,125	1,350	1,800	2,250
0.06		60	72	95	122	59	72	96	121
0.08		77	93	125	157	76	91	124	156
0.10		93	112	151	190	90	109	149	187
0.12		107	129	175	220	104	126	171	216
0.14		119	145	196	247	115	140	192	242
0.16		130	158	216	272	125	153	210	266
0.18		140	171	234	295	134	164	226	288
0.20		149	180	250	316	141	173	241	306
0.22		157	191	265	334	147	181	253	323
0.24		163	199	275	350	153	188	264	338
0.26		168	206	287	365	155	193	273	350
0.28		172	213	297	377	159	197	281	360
0.30		173	216	304	387	160	200	286	369
0.32		173	220	310	396	160	201	290	376
0.34		174	222	316	404	160	201	294	381
0.36		174	222	320	410	160	201	295	385
0.38		174	222	323	414	160	201	295	386
0.40		174	222	324	416	160	201	295	387

5.22 Concrete beams reinforced for compression

a Although it is more economical to use rectangular or T beams without compression reinforcement, sometimes isolated cases require that the beam be limited to a certain width and depth so that it is impracticable to get the b and d required. In such cases steel reinforcement is used to obtain additional

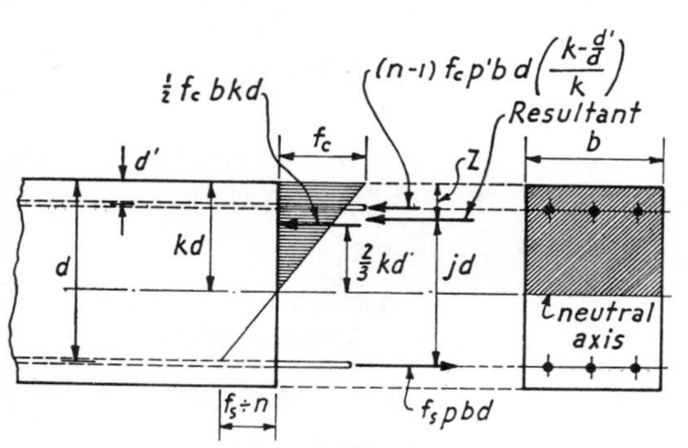

Courtesy Portland Cement Association

Fig. 5.8 Formulae for Double Reinforced Beams

Accurate methods of computation for this problem are quite lengthy as may be noted from the formulae below, to be used in conjunction with Fig. 5.8. The ACI Code permits the compression reinforcement in a double reinforced beam to be taken at twice the tabular value, but not more than the allowable stress in tension.

$$k = \sqrt{2n\left(p + p'\frac{d'}{d}\right) + n^2(p + p')^2} - n(p + p')$$

$$z = \frac{\frac{1}{3}k^3d + 2p'nd'\left(k - \frac{d'}{d}\right)}{k^2 + 2p'n\left(k - \frac{d'}{d}\right)}$$

$$jd = d - z$$

$$f_c = \frac{6M}{bd^2\left[3k - k^2 + \frac{6p'n}{k}\left(k - \frac{d'}{d}\right)\left(1 - \frac{d'}{d}\right)\right]}$$

$$f_s = \frac{M}{pjbd^2} = nf_c\frac{1 - k}{k} \qquad f'_s = nf_c\frac{k - \frac{d'}{d}}{k}$$

$$A_s = \frac{M}{f_s jd} \text{ or } pbd \qquad A_s' = p'bd$$

$$M = Kbd^2 \qquad d = \sqrt{\frac{M}{Kb}}$$

$$p = \frac{f_c}{f_s}\left[\frac{k}{2} + \frac{n-1}{k}\left(k - \frac{d'}{d}\right)p'\right]$$

$$K = f_c\left(\frac{k}{2} - \frac{k^2}{6} + \frac{n-1}{k}\,p'\left[(1-k)\left(k - \frac{d'}{d}\right) + \left(k - \frac{d'}{d}\right)^2\right]\right)$$

b Without going through the long complicated solution by the foregoing formulae, a close approximation of the area of top steel may be arrived at as follows:

$$A_s' = \frac{M - Rbd^2}{2cd}$$

In the above equation

$$c = 12{,}500 - 37{,}500\,\frac{d'}{d}$$

Problem Assume the same beam as in the preceding problem except that it must be a rectangular beam with no T. The rectangular beam is 14 in. wide, 18 in. total depth (16 in. net).

From these dimensions

$$Rbd^2 = 222.6 \times 14 \times 256 = 798{,}250 \text{ in.-lb}$$

$$\frac{d'}{d} = \frac{2}{16} = 0.125$$

$$c = 12{,}500 - (37{,}500 \times 0.125) = 12{,}500 - 4{,}688 = 7{,}812$$

$$A_s' = \frac{1{,}600{,}000 - 798{,}250}{2 \times 7{,}812 \times 16} = 3.21 \text{ sq in.}$$

Use five No. 8.

$$p' = \frac{3.9}{16 \times 14} = 0.0174 \text{ or } 1.74 \text{ percent}$$

c The change in center of compression area location which may result from the combination of two moment couples, one steel and concrete, the other of steel only, may change the amount of compression steel required although in any beam whose net depth is over 16 in., the error will be on the safe side. In the above problem, the jd distance for the moment carried by the concrete is $0.871 \times 16 = 13.93$ in. From steel to steel, the distance is $16 - 2 = 14$ in. Thus using the moment of 1,600,000 with $j = 7/8$, the area of compression steel would be identical with that found by using the area arrived at by adding

$$A_s = \frac{798{,}250}{13.93 \times 20{,}000} \text{ and } \frac{801{,}750}{14 \times 20{,}000}$$

However, if the depth becomes great enough to appreciably increase the net depth steel to steel, a saving may be made by adding the A_s required to balance the concrete to the A_s required to balance the steel.

5.30 SHEAR IN CONCRETE

a Because reinforced concrete is nonhomogeneous and because the relations between vertical and horizontal shear are complicated, the ordinary conception of simple shear as considered in Art. 2.11 is not applicable to reinforced concrete. Instead, the subject is treated under the joint subject of shear and diagonal tension. Much of the design data, particularly with reference to diagonal tension and the use of stirrups, has been obtained from tests rather than pure theory.

As a result of several rather unfortunate failures, the subject of shear in concrete has been restudied and the 1963 ACI Code has been radically revised. The new code is much more conservative than were the former codes, and regardless of whether the 1963 Code has or has not been adopted by the community in which the work is to be done, it is advisable that the applicable portion of the 1963 ACI Code be used in shear analysis.

b The following portions of the 1963 ACI Code apply to shear and diagonal tension under working stress design.

SHEAR AND DIAGONAL TENSION—WORKING STRESS DESIGN*

1200 NOTATION

A_g = gross area of section
A_s = area of tension reinforcement
A_v = total area of web reinforcement in ten-

sion within a distance, s, measured in a direction parallel to the longitudinal reinforcement

α = angle between inclined web bars and longitudinal axis of member

b = width of compression face of flexural member

b' = width of web in I- and T-sections

b_o = periphery of critical section for slabs and footings

d = distance from extreme compression fiber to centroid of tension reinforcement

f_c' = compressive strength of concrete (see Section 301)

f_v = tensile stress in web reinforcement

F_{sp} = ratio of splitting tensile strength to the square root of compressive strength (see Section 505)

M = bending moment

M' = modified bending moment

N = load normal to the cross section, to be taken as positive for compression, negative for tension, and to include the effects of tension due to shrinkage and creep

p_w = $A_s/b'd$

s = spacing of stirrups or bent bars in a direction parallel to the longitudinal reinforcement

t = total depth of section

v = shear stress

v_c = shear stress carried by concrete

V = total shear

V' = shear carried by web reinforcement

* The provisions for shear are based on recommendations of ACI-ASCE Committee 326. The term j is omitted in determination of normal shear stress.

1201 Shear Stress

(a) The nominal shear stress, as a measure of diagonal tension, in reinforced concrete members shall be computed by:

$$v = V/bd \qquad (12\text{-}1)$$

For design, the maximum shear shall be considered as that at the section a distance, d, from the face of the support. Wherever applicable, effects of torsion shall be added and effects of inclined flexural compression in variable-depth members shall be included.

(b) For beams of I- or T-section, b' shall be substituted for b in Eq. (12-1).

(c) The shear stress, v_c, permitted on an unreinforced web shall not exceed $1.1\sqrt{f_c'}$ at a distance d from the face of the support unless a more detailed analysis is made in accordance with (d) or (e). The shear stresses at sections between the face of the support and the section a distance d therefrom shall not be considered critical. For members with axial tension, v_c shall not exceed the value given in (e).

(d) The shear stress permitted on an unreinforced web shall not exceed that given by:

$$v_c = \sqrt{f_c'} + 1300\,\frac{p_w V d}{M} \qquad (12\text{-}2)$$

but not to exceed $1.75\sqrt{f_c'}$. The shear stresses at sections between the face of the support and the section a distance d therefrom shall not be considered critical. V and M are the shear and bending moment at the section considered, but M shall be not less than Vd.

(e) For members subjected to axial load in addition to shear and flexure, Eq. (12-2) shall apply except that M' shall be substituted for M where

$$M' = M - N\,\frac{(4t - d)}{8} \qquad (12\text{-}3)$$

and v_c shall not exceed

$$v_c = 1.75\sqrt{f_c'(1 + 0.004N/A_g)} \qquad (12\text{-}4)$$

When all longitudinal reinforcement at a section acts in compression use Eq. (12-4).

1202 Web Reinforcement

(a) Wherever the value of the shear stress, v, computed by Eq. (12-1), plus effects of torsion, exceeds the shear stress, v_c, permitted for the concrete of an unreinforced web by Sections 1201(c), (d), or (e), web reinforcement shall be provided to carry the excess. Such web reinforcement shall also be provided for a distance equal to the depth, d, of the member beyond the point theoretically required. Web reinforcement between the face of the support and the section at a distance d therefrom shall be the same as required at that section.

(b) Web reinforcement may consist of:

1. Stirrups perpendicular to the longitudinal reinforcement
2. Stirrups making an angle of 45 deg or more with the longitudinal tension reinforcement
3. Longitudinal bars bent so that the axis of the bent bar makes an angle of 30 deg or more with the axis of the longitudinal portion of the bar
4. Combinations of 1 or 2 with 3

(c) Stirrups or other bars to be considered effective as web reinforcement shall be anchored at both ends according to the provisions of Section 919.

1203 Stirrups

(a) The area of steel required in stirrups placed perpendicular to the longitudinal reinforcement shall be computed by:

$$A_v = V's/f_v d \qquad (12\text{-}5)$$

(b) The area of inclined stirrups shall be computed by Eq. (12-7).

1204 Bent Bars

(a) Only the center three-fourths of the inclined portion of any longitudinal bar that is bent up for web reinforcement shall be considered effective for that purpose.

(b) When the web reinforcement consists of a single bent bar or of a single group of parallel bars all bent up at the same distance from the support, the required area shall be computed by:

$$A_v = \frac{V'}{f_v \sin \alpha} \qquad (12\text{-}6)$$

in which V' shall not exceed $1.5bd\sqrt{f_c'}$.

(c) Where there is a series of parallel bars or groups of bars bent up at different distances from the support, the required area shall be computed by:

$$A_v = \frac{V's}{f_v d(\sin \alpha + \cos \alpha)} \qquad (12\text{-}7)$$

(d) Bent bars used alone as web reinforcement shall be so spaced that the effective inclined portion defined in (a) meets the requirements of Section 1206(a).

(e) Where more than one type of web reinforcement is used to reinforce the same portion of the web, the total shear resistance shall be computed as the sum of the resistances computed for the various types separately. In such computations, the resistance of the concrete, v_c, shall be included only once, and no one type of reinforcement shall be assumed to resist more than $2V'/3$.

1205 Stress Restrictions

(a) The tensile stress in web reinforcement, f_v, shall not exceed the values given in Section 1003.

(b) The shear stress, v, shall not exceed $5\sqrt{f_c'}$ in sections with web reinforcement.

1206 Web Reinforcement Restrictions

(a) Where web reinforcement is required, it shall be so spaced that every 45-deg line, representing a potential diagonal crack and extending from middepth, $d/2$, of the member to the longitudinal tension bars, shall be crossed by at least one line of web reinforcement. When the shear stress exceeds $3\sqrt{f_c'}$, every such 45-deg line shall be crossed by at least two lines of web reinforcement.

(b) Where web reinforcement is required, its area shall not be less than 0.15 percent of the area, bs, computed as the product of the width of the web and the spacing of the web reinforcement along the longitudinal axis of the member.

1207 Shear Stress in Slabs and Footings*

(a) The shear capacity of slabs and footings in the vicinity of concentrated loads or concentrated reactions shall be governed by the more severe of two conditions:

1. The slab or footing acting essentially as a wide beam, with a potential diagonal crack extending in a plane across the entire width. This case shall be considered in accordance with Section 1201.
2. Two-way action existing for the slab or footing, with potential diagonal cracking along the surface of a truncated cone or pyramid around the concentrated load or reaction. The slab or footing in this case shall be designed as required in the remainder of this section.

(b) The critical section for shear to be used as a measure of diagonal tension shall be perpendicular to the plane of the slab and located at a distance $d/2$ out from the periphery of the concentrated load or reaction area.

(c) The nominal shear stress shall be computed by:

$$v = V/b_o d \qquad (12\text{-}8)$$

in which V and b_o are taken at the critical section specified in (b). The shear stress, v, so computed shall not exceed $2\sqrt{f_c'}$, unless shear reinforcement is provided in accordance with (d), in which case v shall not exceed $3\sqrt{f_c'}$.

(d) When v exceeds $2\sqrt{f_c'}$, shear reinforcement shall be provided in accordance with Sections 1202 to 1206, except that the allowable stress in shear reinforcement shall be 50 percent of that prescribed in Section 1003. Shear reinforcement consisting of bars, rods or wire shall not be considered effective in members with a total thickness of less than 10 in.

* For transfer of moments and effects of openings see Section 920.

1208 Lightweight Aggregate Concretes

(a) When structural lightweight aggregate concretes are used, the provisions of this chapter shall apply with the following modifications:

1. The shear stress, v_c, permitted on an unreinforced web in Section 1201 (c) shall be

$$0.17F_{sp}\sqrt{f_c'} \qquad (12\text{-}9)$$

2. Eq. (12-2) shall be replaced by:

$$v_c = 0.15F_{sp}\sqrt{f_c'} + 1300\frac{p_w V d}{M} \qquad (12\text{-}10)$$

3. The limiting value for shear stress in slabs

and footings, v_c, in Section 1207(c) and (d) shall be:

$$0.3F_{sp}\sqrt{f_c'} \qquad (12\text{-}11)$$

(b) The value of F_{sp} shall be 4.0 unless determined in accordance with Section 505 for the particular aggregate to be used.

919 Anchorage of Web Reinforcement

(a) The ends of bars forming simple U- or multiple U-stirrups shall be anchored by one of the following methods:

1. By a standard hook, considered as developing 50 percent of the allowable stress in the bar, plus embedment sufficient to develop by bond the remaining stress in the bar, in conformance with Chapters 13 and 18. The effective embedment of a stirrup leg shall be taken as the distance between the middepth of the member, $d/2$, and the center of radius of bend of the hook.
2. Welding to longitudinal reinforcement.
3. Bending tightly around the longitudinal reinforcement through at least 180 deg.
4. Embedment above or below the middepth, $d/2$, of the beam on the compression side, a distance sufficient to develop by bond the stress to which the bar will be subjected, at the bond stresses permitted by Sections 1301 and 1801 but, in any case, a minimum of 24 bar diameters.

(b) Between the anchored ends, each bend in the continuous portion of a simple U- or multiple U-stirrup shall be made around a longitudinal bar.

(c) Hooking or bending stirrups around the longitudinal reinforcement shall be considered effective only when these bars are perpendicular to the longitudinal reinforcement or make an angle of at least 45 deg with deformed longitudinal bars.

(d) Longitudinal bars bent to act as web reinforcement shall, in a region of tension, be continuous with the longitudinal reinforcement and in a compression zone shall be anchored as in (a)1 or (a)4.

(e) In all cases web reinforcement shall be carried as close to the compression surface of the beam as fireproofing regulations and the proximity of other steel will permit.

It will be noted that although the permissible stresses without stirrups have been materially decreased, the factor j which appeared in former editions of the Code has been eliminated. Moreover, the allowable stress on a higher strength concrete member with stirrups has been decreased, although not so radically. The allowable shear on 3,000-psi concrete beams without stirrups in accordance with the limitations of the ACI Code is given in Table 5.5.

Theoretically in accordance with the ACI Code, bent-up bars may be figured to resist diagonal tension but most engineers disregard this allowance and assume that this merely gives an added bonus to the safety factor. In order to take advantage of bent-up bars, the bend points and angle of slope must be computed and detailed on the drawings. Moreover, there is an increasing use of straight bars, both top and bottom, to provide the necessary tensile reinforcement, thus reducing the cost of total reinforcement since bending of the bars adds to the cost of the steel, especially in areas where union regulations require steel to be bent on the job.

Problem To check the beam designed in Art. 5.20*b* for shear.

The critical point for shear is 16 in. from the support, and at this point, the shear is

$$1{,}100 \times (10 - 1.33) = 9.53 \text{ kips}$$

From Table 5.5, the allowable shear without stirrups is 11.52. Therefore stirrups are not required.

c Ordinarily two-leg or U stirrups are used, as shown in Fig. 5.9, but where shear values are high, it may be advisable to use four-leg or W stirrups.

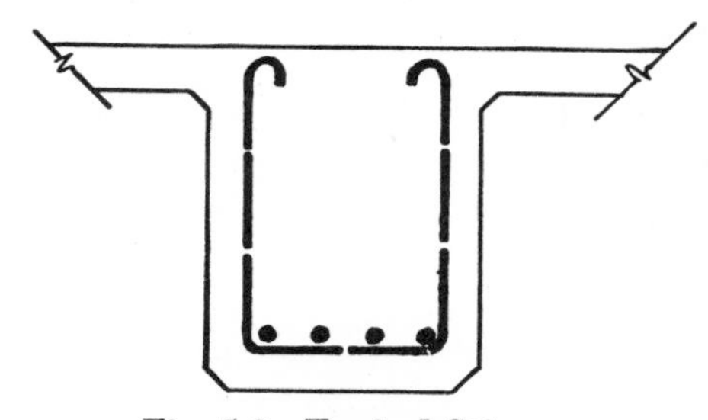

Fig. 5.9 Typical Stirrups

The allowable fiber stress in the stirrup is limited to 20,000 psi when the size of bar is such that bond and anchorage requirements on the stirrup are within proper limits, as set by the ACI Code, Art. 919, as previously quoted, and Art. 1301 as quoted in Art. 5.20*f* of this book.

Table 5.5 Allowable Shear in Kips, Beams Without Stirrups

Based on 3,000-psi concrete, V_c = 60 psi

Net d, in.	b = 8 in.	b = 10 in.	b = 12 in.	b = 14 in.	b = 16 in.
8	3.84	4.80			
10	4.80	6.00			
12	5.76	7.20	8.64		
14	6.72	8.40	10.08	11.76	
16	7.68	9.60	11.52	13.44	15.36
18	8.64	10.80	12.96	15.12	17.28
20	9.60	12.00	14.40	16.80	19.20
22	10.56	13.20	15.84	18.48	21.12
24	11.52	14.40	17.28	20.16	23.04
26		15.60	18.72	21.84	24.96
28		16.80	20.16	23.52	26.88
30		18.00	21.60	25.20	28.80
32		19.20	23.04	26.88	30.72

d The method of computing stirrups recommended by the American Concrete Institute assumes that the concrete takes its full shear value, as given in Art. 1.22, and the steel stirrups take the remainder. The spacing of stirrups at any given point as obtained by this method is

$$s = \frac{A_v f_v d}{V'} \qquad (5.10)$$

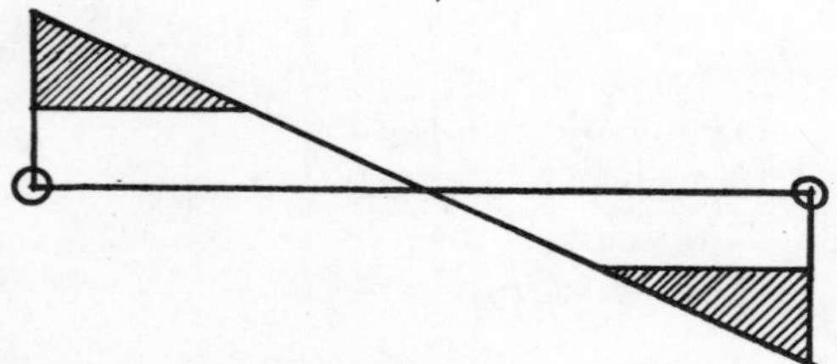

Fig. 5.10 Shear Diagram for Stirrups

The hatched area in Fig. 5.10 represents graphically the shear resisted by stirrups computed by the above method.

Since for any size stirrup selected A_v represents the area of the two legs of the stirrup, f_v is taken as 20,000, and we may derive the following from Formula (5.11):

For No. 3 U Stirrups, $s_1 = \dfrac{4{,}400}{V'}$

For No. 4 U Stirrups, $s_1 = \dfrac{8{,}000}{V'}$

For No. 5 U Stirrups, $s_1 = \dfrac{12{,}000}{V'}$

For No. 6 U Stirrups, $s_1 = \dfrac{17{,}600}{V'}$

Maximum spacing of stirrups should not exceed $\frac{1}{2}d$.

There are several graphical methods of determining stirrup requirements from shear diagrams (see Art. 3.12), but many engineers prefer to use the above method, using the Stirrup Dividend Table, Table 5.6. This table does not take into account reduction of values where required in accordance with the ACI Code.

Table 5.6 Stirrup Dividends

Net depth d, in.	No. 3 stirrup	No. 4 stirrup	No. 5 stirrup
8	35,200	64,000	96,000
10	44,000	80,000	120,000
12	52,800	96,000	144,000
14	61,600	112,000	168,000
16	70,400	128,000	192,000
18	79,200	144,000	216,000
20	88,000	160,000	240,000
22	96,800	176,000	264,000
24	105,600	192,000	288,000
26	114,400	208,000	312,000
28	123,200	224,000	336,000
30	132,000	240,000	360,000
32	140,800	256,000	384,000
34	149,600	272,000	408,000
36	158,400	288,000	432,000

e If the unit shearing stress for a 3,000-psi concrete beam (computed from ACI Code Sec. 1201(a) as quoted in Art. 5.30*b*) exceeds 60 psi, stirrups are required. Table 5.5

gives the allowable shear for a 3,000-psi concrete beam without stirrups. Values for beams not listed may be interpolated. To use Table 5.6 for computation of stirrups, subtract the shear value of the plain concrete beam from the shear at distance d from the end, to obtain V' the shear to be taken by stirrups. From Table 5.6, for d and the size stirrup to be used, select the proper stirrup dividend. The spacing at the end of the beam is this dividend divided by V'. The first stirrup should be placed at the center of this spacing, and the second one a full spacing from it, but not over $\frac{1}{2}d$. The reduction in shear up to this point is subtracted from V' and a new spacing computed based on the revised V', repeating this operation until V' is zero. Stirrups will be used to a distance d beyond this zero point.

Frequently it is accurate enough to compute a spacing, and use this spacing until the shear is reduced enough to use a greater predetermined spacing. Thus, assume that the spacing computed for end reaction is 6 in. Then compute the allowable shear for a 9-in. spacing, and carry the 6-in. spacing until this point is reached. Next, check the allowable shear for 12-in. spacing and carry 9-in. spacing until the new point is reached, and so on. If the load on the beam is concentrated, since the shear does not reduce except at points of concentrated load, the spacing from the reaction to the point of concentration will be the same with no change. Thus, if 6 in. is the computed spacing and the concentrated load is 4 ft out on the beam, the spacing will be one at 3 in. and eight at 6 in., making 51 in. to the last stirrup, plus the beam depth beyond this point.

Problem Assume a T beam carrying 3,000 lb per ft (250 lb per in.), 15 ft long, and a central concentrated load of 20,000 lb, with a stem 16 in. wide, 16 in. net depth, 18 in. total depth, ACI Code for 3,000-psi concrete. Compute the required stirrups (Fig. 5.11).

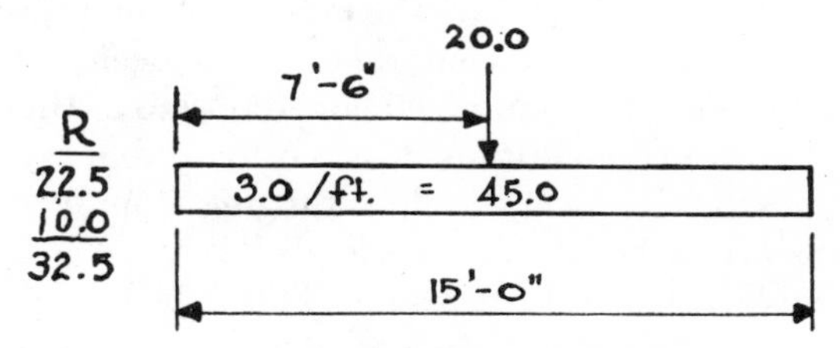

Fig. 5.11 Stirrup Design Problem

The critical V is the end shear 32.5 kips less the load on a length of 16 in. (net depth), or $32.5 - 4.0 = 28.5$.

Allowable V' without stirrups from Table 5.5 = 15.36.

$$\begin{aligned} \text{Critical } V &= 28.5 \\ -V_c &= 15.36 \\ \hline V' &= 13.14 \end{aligned}$$

Since the shear reduces at the rate of 3 kips per ft, stirrups will be required over a distance of

$$\frac{13.14}{3} = 4.38 \text{ ft} + 1.33 = 5.71 \text{ ft} = 68\tfrac{1}{2} \text{ in.}$$

With No. 3 U stirrups, from Table 5.6

$$s_1 = \frac{70.4}{13.14} = 5.36 \text{ in.} \quad \text{Use } 5\tfrac{1}{2} \text{ in.}$$

This is the spacing at 16 in. from the support. Use one at 3 in. from the support plus three at $5\frac{1}{2}$ in. = 3 in. + $16\frac{1}{2}$ in. = $19\frac{1}{2}$ in. total.

The maximum spacing of stirrups is $d/2$ or 8 in. For this spacing, the allowable V' is $\frac{70{,}400}{8} = 8.8$ kips. This point is $\frac{13.14 - 8.8}{3.0} = 1.45$ ft beyond the critical shear point, or $33\frac{1}{2}$ in. from the support. The spacing therefore would be

1 at 3 in.		3 in. from support
3 at $5\frac{1}{2}$ in.	Add $16\frac{1}{2}$	= $19\frac{1}{2}$ in. from support
1 at 6 in.	Add 6	= $25\frac{1}{2}$ in. from support
1 at 7 in.	Add 7	= $32\frac{1}{2}$ in. from support
4 at 8 in.	Add 32	= $64\frac{1}{2}$ in. from support

$10 \times 2 = 20$ No. 3 U Stirrups

5.40 SOLID ONE-WAY REINFORCED CONCRETE SLABS

a For many years the most common type of fireproof floor construction has been the solid one-way reinforced concrete slab. Under normal conditions it is most suitable and most economical for slabs ranging from 6 to 8½ ft, although for light loads it may be economical up to 12 ft span. This type of slab is designed as a rectangular beam in accordance with the method of Art. 5.20, with the main reinforcement the short direction of the slab. At right angles to the main bars, shrinkage and temperature reinforcement is provided for to the extent of $0.002bd$ for floors and $0.0025bd$ for roofs. If the main steel area required is not over 0.2 sq in. and the supporting members are steel beams, it is usually more economical to use welded wire-mesh reinforcement at an allowable stress of 30,000 psi, which may be unrolled and placed with much less labor cost than bars. Table 5.7 will assist in selecting bars to furnish any given area of steel.

b In addition to those portions of the ACI 1963 Code applicable to flexure in concrete, most of which has been quoted elsewhere in this book, there are several sections of this code which apply to slabs only.

803 PLACING REINFORCEMENT

(a) *Supports*—Reinforcement shall be accurately placed and adequately supported by concrete, metal, or other approved chairs; spacers; or ties and secured against displacement within tolerances permitted.

(c) *Draped fabric*—When wire or other reinforcement, not exceeding ¼ in. in diameter, is used as reinforcement for slabs not exceeding 10 ft in span, the reinforcement may be curved from a point near the top of the slab over the support to a point near the bottom of the slab at midspan, provided such reinforcement is either continuous over, or securely anchored to, the support.

804 SPACING OF BARS

(c) In walls and slabs other than concrete joist construction, the principal reinforcement shall be centered not farther apart than three times the wall or slab thickness nor more than 18 in.

807 SHRINKAGE AND TEMPERATURE REINFORCEMENT

(a) Reinforcement for shrinkage and temperature stresses normal to the principal reinforcement shall be provided in structural floor and roof slabs where the principal reinforcement extends in one direction only. Such reinforcement shall provide at least the following ratios of reinforcement area to gross concrete area, but in no case shall such reinforcing bars be placed farther apart than five times the slab thickness or more than 18 in.

Table 5.7 Area of Steel for Rods Spaced at Various Intervals

Square inches per linear foot of slab

Spacing o.c., in.	Rod No.								
	3	4	5	6	7	8	9	10	11
3	0.44	0.80	1.24	1.76	2.40	3.16	4.00	5.08	
3½	0.38	0.69	1.06	1.51	2.06	2.71	3.43	4.35	5.35
4	0.33	0.60	0.93	1.32	1.80	2.37	3.00	3.81	4.68
4½	0.29	0.53	0.83	1.17	1.60	2.11	2.67	3.39	4.16
5	0.26	0.48	0.74	1.06	1.44	1.90	2.40	3.05	3.74
5½	0.24	0.44	0.68	0.96	1.31	1.72	2.18	2.77	3.40
6	0.22	0.40	0.62	0.88	1.20	1.58	2.00	2.54	3.12
6½	0.20	0.37	0.57	0.81	1.11	1.46	1.85	2.34	2.88
7	0.19	0.34	0.53	0.75	1.03	1.35	1.71	2.18	2.67
7½	0.18	0.32	0.50	0.70	0.96	1.26	1.60	2.03	2.50
8	0.16	0.30	0.46	0.66	0.90	1.18	1.50	1.90	2.34
8½	0.15	0.28	0.44	0.62	0.85	1.12	1.41	1.79	2.20
9		0.27	0.41	0.59	0.80	1.05	1.33	1.69	2.08
9½		0.25	0.39	0.56	0.76	1.00	1.26	1.60	1.97
10		0.24	0.37	0.53	0.72	0.95	1.20	1.52	1.87

Slabs where plain bars are used....... 0.0025
Slabs where deformed bars with specified yield strengths less than 60,000 psi are used..................... 0.0020
Slabs where deformed bars with 60,000 psi specified yield strength or welded wire fabric having welded intersections not farther apart in the direction of stress than 12 in. are used... 0.0018

See also Sections 808(b), 904(c), 905(b) item 3, 907(b), 909(b) and Table 909(b), and 911(b) as quoted in Art. 5.20.

920 Transfer of Moments and Effect of Openings in Slabs and Footings

(b) When openings in slabs are located at a distance less than ten times the thickness of the slab from a concentrated load or reaction or when openings in flat slabs are located within the column strips as defined in Section 2101(d), that part of the periphery of the critical section for shear which is covered by radial projections of the openings to the centroid of the loaded area shall be considered ineffective.

1003 Allowable Stresses in Reinforcement

Unless otherwise provided in this code, steel for concrete reinforcement shall not be stressed in excess of the following limits:

(a) *In tension*

For main reinforcement, ⅜ in. or less in diameter, in one-way slabs of not more than 12-ft span, 50 percent of the minimum yield strength specified by the American Society for Testing and Materials for the reinforcement used, but not to exceed...................... 30,000 psi

c If adjacent slab spans vary by not over 20 percent, the bending moment under uniform load conditions may be computed by means of coefficients in accordance with Table 5.8.

Problem Design a slab to carry a total load of 200 psf on a continuous span of 10 ft 6 in., using A432 steel (f_s = 24,000 psi) and 3,000-psi concrete and the ACI Code.

$$W = 10.5 \times 200 = 2{,}100$$

$$M = \frac{2{,}100 \times 10.5 \times 12}{11} = 24{,}054$$

$$d = \sqrt{\frac{24{,}054}{201.5 \times 12}} = \sqrt{10} = 3.16$$

Allow for ¾-in. cover plus half thickness of bar = 3.16 + 1 = 4.16. Use 4½-in. slab.

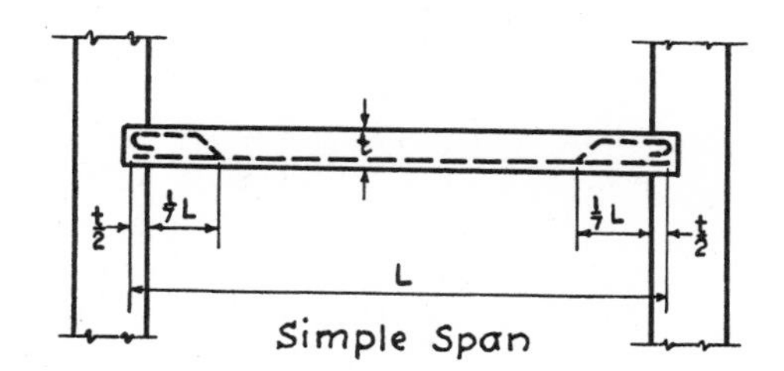

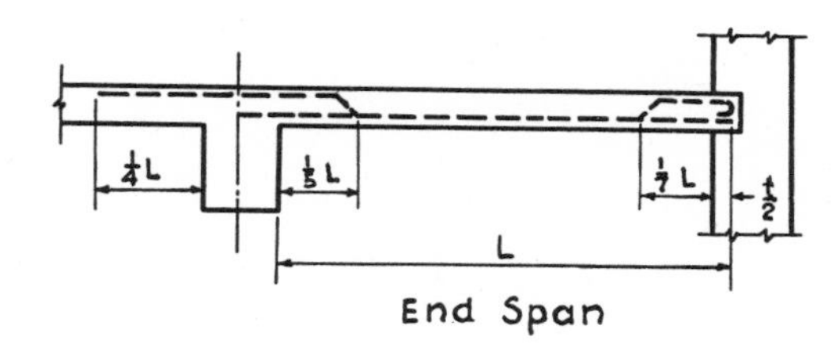

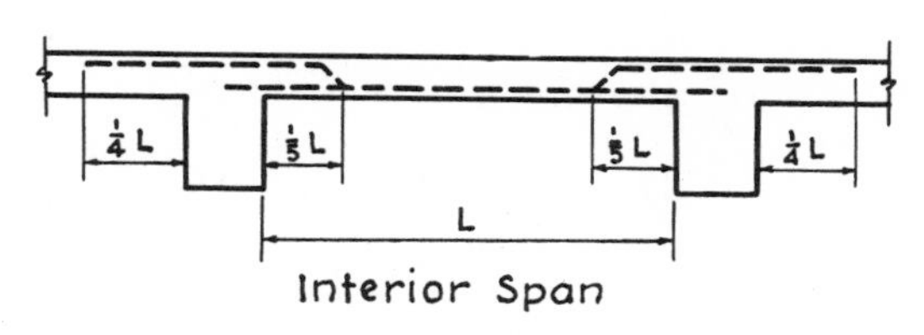

Fig. 5.12 Typical Concrete Slabs

Table 5.8 Bending Moment in Slabs for Uniform Loads

Coefficients of wl^2

Length of spans	Number of spans	End spans		Intermediate spans		Length of spans	Number of spans	End spans		Intermediate spans	
		$+M$ at center	$-M$ at face of 2d support	$+M$ at center	$-M$ at face of support			$+M$ at center	$-M$ at face of 2d support	$+M$ at center	$-M$ at face of support
Equal to or less than 10 ft	Two	$\frac{1}{14}$	$\frac{1}{9}$			Greater than 10 ft	Two	$\frac{1}{14}$	$\frac{1}{12}$		
	More than two	$\frac{1}{14}$	$\frac{1}{10}$	$\frac{1}{16}$	$\frac{1}{12}$		More than two	$\frac{1}{14}$	$\frac{1}{10}$	$\frac{1}{16}$	$\frac{1}{11}$

This is within the limitation of $\frac{1}{35}$ of the span.

$$A_s = \frac{24{,}054}{21{,}300 \times 3.5} = 0.324 \text{ sq in.}$$

From Table 5.7 use No. 4 bars at 7 in. o.c.; $A_s = 0.34$.

Table 5.9 is useful for the selection of moment direct. It gives moment for simple, end, and continuous spans varying by 3-in. increments from 4 to 13 ft, and for 100 psf. For other loads the moments given in the series of tables for 100 psf may be multiplied by the correct multiplier. For example, for 165 psf, use 1.65 times the moment given.

d If a slab is required to carry a concentrated load, the simplest method of design is a simplification of the Hardy Cross method of moment distribution, as described in Art. 3.21 and as given below. (Many codes, however, permit the weight of partitions to be spread as described in Art. 1.50*b*.)

The variation consists of determining the fixed end moment at the heavy end due to partition load, in accordance with the coeffi-

Table 5.9 Required Moment, Inch-Pounds per 100 psf

Span	W	$WL/8$	$WL/9$	$WL/10$	$WL/11$	$WL/12$	$WL/14$	$WL/16$
4′0″	400	2,400	2,133	1,920	1,745	1,600	1,371	1,200
4′3″	425	2,710	2,408	2,160	1,970	1,810	1,548	1,355
4′6″	450	3,040	2,700	2,430	2,209	2,020	1,736	1,520
4′9″	475	3,380	3,008	2,710	2,461	2,250	1,934	1,690
5′0″	500	3,750	3,333	3,000	2,727	2,500	2,143	1,875
5′3″	525	4,130	3,675	3,310	3,007	2,760	2,363	2,065
5′6″	550	4,540	4,033	3,630	3,300	3,020	2,593	2,270
5′9″	575	4,960	4,408	3,960	3,607	3,300	2,834	2,480
6′0″	600	5,400	4,800	4,320	3,927	3,600	3,086	2,700
6′3″	625	5,860	5,208	4,680	4,261	3,910	3,348	2,930
6′6″	650	6,340	5,633	5,070	4,609	4,230	3,621	3,170
6′9″	675	6,840	6,075	5,470	4,970	4,560	3,905	3,420
7′0″	700	7,350	6,533	5,880	5,345	4,900	4,200	3,675
7′3″	725	7,880	7,008	6,310	5,734	5,260	4,505	3,940
7′6″	750	8,440	7,500	6,750	6,136	5,630	4,821	4,220
7′9″	775	9,020	8,008	7,210	6,552	6,010	5,148	4,510
8′0″	800	9,600	8,533	7,680	6,982	6,400	5,486	4,800
8′3″	825	10,210	9,075	8,170	7,426	6,810	5,834	5,105
8′6″	850	10,830	9,633	8,670	7,883	7,220	6,193	5,415
8′9″	875	11,480	10,188	9,180	8,354	7,650	6,563	5,740
9′0″	900	12,150	10,800	9,710	8,839	8,100	6,943	6,075
9′3″	925	12,820	11,408	10,250	9,338	8,560	7,334	6,410
9′6″	950	13,530	12,033	10,820	9,850	9,020	7,736	6,765
9′9″	975	14,250	12,675	11,410	10,376	9,500	8,148	7,125
10′0″	1,000	15,000	13,333	12,000	10,910	10,000	8,571	7,500
10′3″	1,025	15,750	14,008	12,610	11,470	10,500	9,005	7,875
10′6″	1,050	16,530	14,700	13,230	12,030	11,020	9,450	8,265
10′9″	1,075	17,350	15,408	13,860	12,603	11,560	9,905	8,675
11′0″	1,100	18,150	16,133	14,530	13,200	12,100	10,371	9,075
11′3″	1,125	19,000	16,875	15,200	13,806	12,660	10,848	9,500
11′6″	1,150	19,820	17,633	15,870	14,427	13,230	11,336	9,910
11′9″	1,175	20,710	18,408	16,580	15,061	13,810	11,834	10,355
12′0″	1,200	21,600	19,200	17,290	15,709	14,400	12,343	10,800
12′3″	1,225	22,500	20,008	18,010	16,368	15,020	12,863	11,250
12′6″	1,250	23,420	20,833	18,750	17,045	15,630	13,393	11,710
12′9″	1,275	24,400	21,675	19,520	17,734	16,250	13,934	12,200
13′0″	1,300	25,340	22,533	20,290	18,437	16,900	14,486	12,670

cient from Table 3.1, and using half of this moment added to the normal uniform load moment found by the coefficient method described in Art. 5.40*c*.

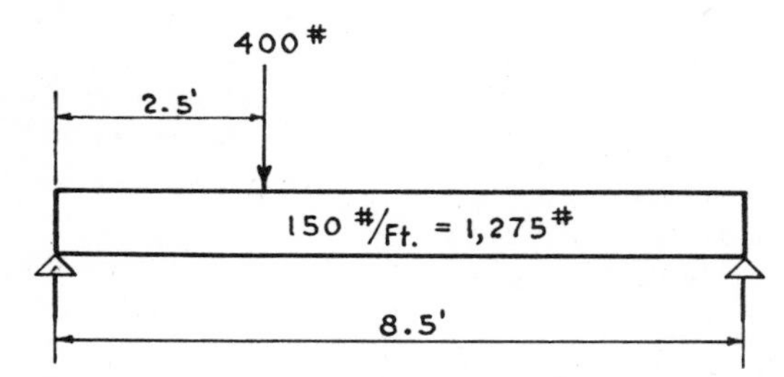

Fig. 5.13 Slab Carrying Concentrated Load

Problem Design an intermediate slab on a span of 8 ft 6 in., carrying a total uniform load of 150 psf and a partition load of 400 lb per ft at right angles to the main reinforcement, at a point 2 ft 6 in. from the support (see Fig. 5.13).

For the uniform load,

$$W = 150 \times 8.5 = 1.275$$

$$M_1 = \frac{1{,}275 \times 8.5 \times 12}{12} = 10{,}837$$

For the concentrated load, the position of the load is 2.5/8.5 = 0.29 of span. From Table 3.1

$$M_2 = \frac{0.146 \times 400 \times 8.5 \times 12}{2} = 2{,}980$$

The divisor 2 is an approximate moment distribution factor. The total moment to design for is 10,840 + 2,980 = 13,820.

$$d = \sqrt{\frac{13{,}820}{201.5 \times 12}} = \sqrt{5.69} = 2.38$$

Add ¾ in. for cover, and for half the thickness of the bar,

$$\text{Total } t = 2.38 + 1 = 3.38 \text{ in.}$$

Use 3½ in. slab.

$$A_s = \frac{13{,}820}{21{,}300 \times 2.5} = 0.26 \text{ sq in.}$$

Use No. 4 bars, 8½ in. o.c.; $A_s = 0.28$, or No. 3 bars, 5 in. o.c.; $A_s = 0.26$.

e If there are a large number of slabs to design, it is profitable for the designer as well as the contractor to standardize them, using slabs such as those listed in Table 5.10. The slab is designated by means of a number and a letter—the number representing the gross thickness of the structural slab, the letter indicating the reinforcement. Thus a 4G slab is 4 in. thick reinforced with 4 × 16, 4/9 welded wire mesh and is good for a moment of 6,120 in.-lb. The area of steel required is also given in case it is desired to use bars instead of mesh for reinforcement. It is advisable that the designer understand the designation used for standard mesh rein-

Table 5.10 Reinforced Concrete Slabs—Allowable Moments

Inch-pounds per foot width
Using standard welded wire mesh, 3,000-psi concrete, $f_s = 30{,}000$ psi

Welded wire mesh		A_s	3 in.	3½ in.	4 in.	4½ in.	5 in.	5½ in.	6 in.
A	4 × 12, $\frac{10}{12}$	.043	2,310						
B	4 × 12, $\frac{9}{12}$	.052	2,800	3,500					
C	4 × 12, $\frac{8}{12}$	.062	3,340	4,170					
D	4 × 16, $\frac{7}{11}$	.074	3,980	4,980	5,970				
E	4 × 16, $\frac{6}{10}$	.087	4,680	5,850	7,020	8,190	9,360		
F	4 × 16, $\frac{5}{10}$	.101	5,430	6,790	8,150	9,510	10,870	12,230	
G	4 × 16, $\frac{4}{9}$	.120	6,460	8,070	9,684	11,300	12,900	14,530	16,140
H	4 × 16, $\frac{3}{8}$	.140	7,530	9,410	11,300	13,180	14,060	15,950	17,830
J	4 × 16, $\frac{2}{8}$	.162	8,720	10,990	13,170	15,350	17,530	19,710	21,890
K	2 × 16, $\frac{6}{10}$	.174	9,360	11,700	14,040	16,380	18,720	21,060	23,400
L	2 × 16, $\frac{5}{10}$	.202	10,830	13,580	16,300	19,020	21,740	24,460	27,180
M	2 × 16, $\frac{4}{9}$	.239	12,860	16,070	19,290	22,500	25,720	28,930	32,150
N	2 × 16, $\frac{3}{8}$	.280		18,820	22,600	26,360	28,120	33,900	35,660
P	2 × 16, $\frac{2}{8}$	.325			26,230	30,600	34,970	39,340	43,710
Q	2 × 16, $\frac{1}{7}$	.377				35,500	40,580	45,650	50,720

forcement. The first two figures (4 × 16) indicate the spacing in inches of the main and cross reinforcement respectively, and the fractional number (4/9) indicates the gage of the main and cross wires in the mesh. Thus the reinforcement called for above, 4 × 16, 4/9, calls for 4-gage wires spaced 4 in. o.c. for the main reinforcement, and 9-gage wires spaced 16 in. o.c. for the cross wires.

f Various methods are used to simplify the construction of slabs so that the same mesh may be used over as many spans as possible. If there is unlimited choice of beam spacings, it will be found that if the end spans are made equal to 0.91 of the intermediate span, the moments for $WL/10$ in the end span and $WL/12$ in the intermediate span will be approximately equal. For instance, assume a roof of 60-ft length and allow for 4-in. bearing on the wall at each end. Select spans of between 7 ft 6 in. and 8 ft to use the same size mesh throughout. This will give a total of eight spans, or two end spans at $0.91L$ and six intermediate spans, or

$$7.82L = 60.67 \text{ ft}$$
$$L = 7.76$$

Use 7 ft 9 in. The end spans will be

$$\frac{60 \text{ ft } 8 \text{ in.} - 46 \text{ ft } 6 \text{ in.}}{2} = 7 \text{ ft } 1 \text{ in.}$$

Using 100 psf total load, the intermediate moment from Table 5.9 is 6,010 in.-lb. The end-span load is 7.08 × 100 = 708, and the end span moment is

$$\frac{708 \times 7.08 \times 12}{10} = 6{,}015 \text{ in.-lb}$$

Thus the same slab may be used for both end span and intermediate.

Sometimes the same mesh is used throughout, but the end-span forms are dropped ½ in. to increase the capacity of the steel. Thus if we assume 150 psf on 7 ft 9 in. spans throughout, both end span and intermediate, we will have an intermediate moment of 9,000 in.-lb and an end-span moment of 10,800 in.-lb. Referring to Table 5.10, we may use a $3K$ intermediate span and $3\frac{1}{2}K$ end span, thus using the same mesh throughout.

Another method of providing for the increased moment is to use the same mesh throughout and increase the steel area in the same end span by adding ¼-in. round bars spaced as required to provide the extra area. Note that additional steel is required for negative moment only.

g Long time deflection as caused by plastic flow can become quite serious without any way affecting the safety of the slab. A long slab for instance may deflect sufficiently to permit a masonry partition to arch itself away from the floor. The ACI Code, Section 909(d), specifies protective measures against this deflection. Since the deflection increases proportionately to the square of the span and decreases proportionately to the depth, it is advisable to maintain the greatest depth of slab feasible with economy, especially in long spans.

Since long spans with flat exposed ceilings are being used more and more, one of the methods of obtaining economy is the use of sonovoids, hollow tubes, as shown in Fig. 5.14. The properties of slabs of this type are listed in Table 5.11.

As floor slab spans become longer, there is an increasing probability of obtaining "set" in the slab; that is, a deflection which results from plastic flow and does not return to original level when the live load is re-

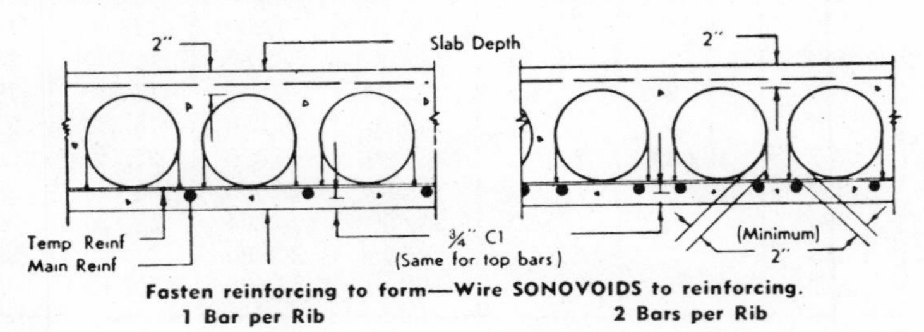

Fig. 5.14 Sonovoid Slab Construction

Table 5.11 Sonovoid Concrete Slabs—3,000-lb Concrete

	M/ft width	Wt./sq ft, lb
8″ slab with 4″ tubes @ 6″ o.c.	8.7 ft-kips	75
10″ slab with 6″ tubes @ 8″ o.c.	14.2 ft-kips	84
12″ slab with 8″ tubes @ 10″ o.c.	20.1 ft-kips	89
14″ slab with 10″ tubes @ 13″ o.c.	27.6 ft-kips	103
16″ slab with 12″ tubes @ 16″ o.c.	35.9 ft-kips	116

moved. A practical method of guarding against this difficulty is to use half as much steel in the top of the slab as has been used in the bottom. The necessity for this expense does not usually occur until the span in feet is more than double the slab depth in inches.

5.41 One-way concrete-joist floor construction

a One-way concrete-joist floors are frequently used in hospital, school, or apartment-house construction. Precast concrete joists, of which there are several types on the market, may be discussed more properly under Art. 5.42. This discussion and the methods proposed are best fitted to use in metal-pan ribbed floors—sometimes known as tin-pan construction—and one-way floors with block fillers. Depth of metal pans runs from 8 to 14 in., and a top slab 2½ or 3 in. thick is used to form a T beam. Block fillers are run from 6 to 12 in. thick plus the top slab.

One of the lightest types of concrete floors results from the use of metal-pan fillers between concrete joists. Numerous kinds of such pans are available. Some are fabricated of light-gage material and are intended to be left in place. Other pans are made of heavier metal and are removed when the forms are stripped, to be used again or returned to the owner, if leased. Both styles are furnished with either straight or tapered closed ends. Tapered pans are highly desirable on long spans, because with their use the width of the joists is increased at points where such width is needed, namely, near the supports where the shearing stresses are the highest. The usual width of metal pans is 20 in., although 10-, 15-, and 30-in. pans are generally available. Figures 5.16 and 5.17 indicate this type of construction.

Metal-pan ribbed floors reach their highest economy on long spans with light loads. The weight is very light and the depth of the floor produces great rigidity. These floors are not so well suited to support concentrated loads, as the topping between joists is comparatively thin (2½ to 3 in.). Care must be taken to reinforce the topping across construction joints to prevent them from opening.

Hollow filler blocks of light-weight concrete or clay tile laid in rows in the bottom

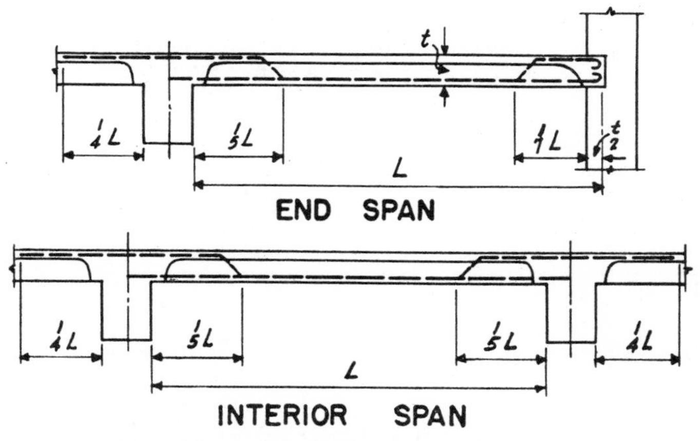

Fig. 5.15 Typical Concrete Joist Slabs

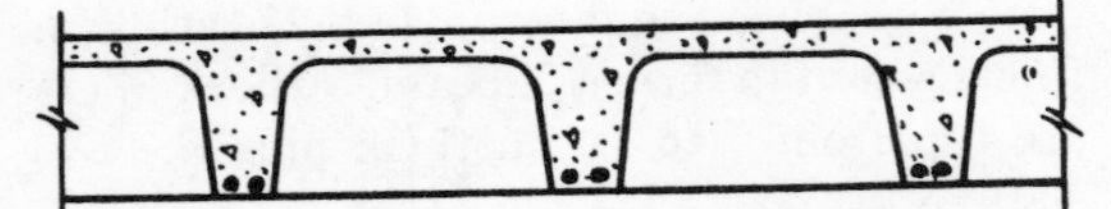

Fig. 5.16 Section Through Metal Pan Ribbed Floor

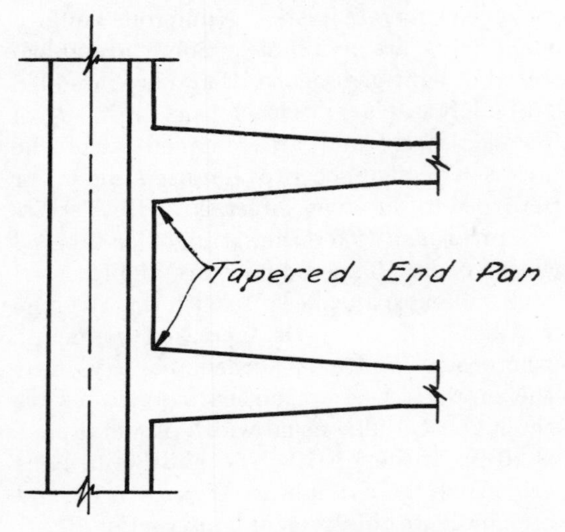

Fig. 5.17 Use of Tapered End Pans

of concrete slabs constitute a form of ribbed slab. The dead load is considerably reduced in comparison with a solid slab of equal load-carrying capacity, although the total depth of slab is increased. The width of the concrete joists, separating the rows of filler and encasing the reinforcement, may be made any desired size to meet the strength requirements. It is customary to include a solid concrete top of 2 in. or more in depth over the blocks. This serves the double purpose of providing a space for concealing small pipes and conduits, and also forms a T section with the joists, thus adding considerable strength to them. Introduction of filler blocks improves sound and heat insulation of a slab.

Although plaster may be applied directly to the slab, it has been noted that at times a slight discoloration takes place when clay-tile filler is used. This is caused by the use of materials of different densities and absorption qualities next to one another, and may be eliminated by placing the soffit pieces in the bottoms of the joists to form an all-tile ceiling.

To insure straight joists, care should be taken that the filler blocks are kept accurately in line. As the blocks are porous, they should be thoroughly sprinkled to prevent absorption of the water in the concrete, particularly in warm weather.

It is advisable to maintain a minimum joist width of 5 in. using metal pans and 4 in. using filler blocks, to facilitate the placing of bars.

b The following sections quoted from the 1963 ACI Code may be taken as representing good practice in the design of concrete-joist floor construction.

2001 CONCRETE JOIST FLOOR CONSTRUCTION

(a) In concrete joist floor construction consisting of concrete joists and slabs placed monolithically with or without burned clay or concrete tile fillers, the joists shall not be farther apart than 30 in. face to face. The ribs shall be straight, not less than 4 in. wide, and of a depth not more than three times the width.

(b) When burned clay or concrete tile fillers of material having a unit compressive strength at least equal to that of the specified strength of the concrete in the joists are used, the vertical shells of the fillers in contact with the joists may be included in the calculations involving shear or negative bending moment. No other portion of the fillers may be included in the design calculations.

(c) The concrete slab over the fillers shall be not less than 1½ in. in thickness, nor less in thickness than $\frac{1}{12}$ of the clear distance between joists. Shrinkage reinforcement shall be provided in the slab at right angles to the joists equal to that required in Section 807.

(d) Where removable forms or fillers not com-

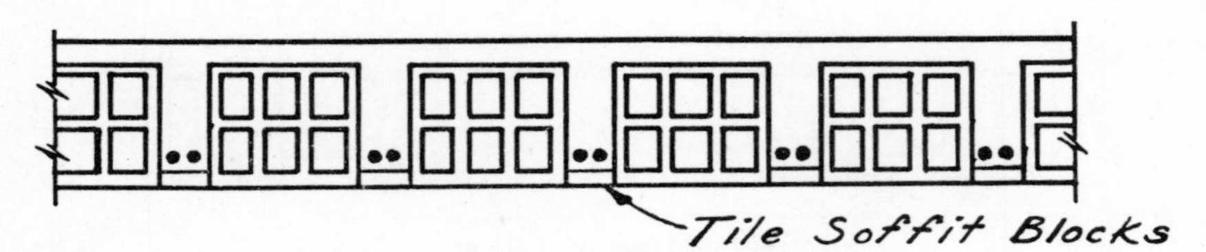

Fig. 5.18 Tile and Joist Ribbed Floor

plying with (b) are used, the thickness of the concrete shall not be less than 1/12 of the clear distance between joists and in no case less than 2 in. Such slab shall be reinforced at right angles to the joists with at least the amount of reinforcement required for flexure, giving due consideration to concentrations, if any, but in no case shall the reinforcement be less than that required by Section 807.

(e) When the finish used as a wearing surface is placed monolithically with the structural slab in buildings of the warehouse or industrial class, the thickness of the concrete over the fillers shall be ½ in. greater than the thickness used for design purposes.

(f) Where the slab contains conduits or pipes as allowed in Section 703, the thickness shall not be less than 1 in. plus the total over-all depth of such conduits or pipes at any point. Such conduits or pipes shall be so located as not to impair the strength of the construction.

(g) Shrinkage reinforcement is not required in the slab parallel to the joists.

(h) The shear stress, v_c, may be increased 10 percent over those prescribed in sections 1201 or 1701.

c Concrete-joist construction in general is subject to the design methods for T beams as described in Art. 5.21. In many instances, however, the layout is such that the coefficient method of design (Table 5.8) is not applicable, as, for example, in schools, where the classroom spans are approximately 23 ft, while the corridor span is only about 10 ft or less. Under such conditions it is necessary to compute the moment by methods of moment distribution as described in Art. 3.21. Since the maximum negative moment occurs with two adjacent spans loaded—whereas the maximum positive moment occurs with one span loaded and the adjacent spans unloaded—it is necessary to run through design calculations under the several possible load conditions and determine the combination of dead and live loads which gives maximum moment conditions. A simplification of this computation is given in Art. 5.41*d*.

After the maximum positive and negative moments in each span have been determined, select the joist and top slab required for the design, and the reinforcement in the joists. Table 5.12 is taken from the standard practice of the author's office. It is based on the use of 20-in.-wide metal forms.

In Table 5.12 the column $+M_c$ is the maximum positive moment in the joists as a T beam; $-M_c$ is the maximum negative moment, unless compression steel is used. The moment of inertia is given in case it is desired to compute stiffness factors. The weight given is the average weight per square foot for joist and top slab using 5-in.-wide joists and 20-in.-wide metal forms. If block fillers are used the weight must be increased accordingly. In Tables 5.12 and 5.13 concrete joists are shown thus: 12 + 2½cd. This indicates 12-in.-deep pans with 2½ in. top slab. The letters indicate reinforcement as tabulated. The first letter is the straight bar, the second, bent. An indication such as 12 + 2½c/a indicates a (c) top bar across the entire span, and an (a) straight bottom bar. Except as otherwise noted, joists are

Table 5.12 Allowable Moment in Concrete-joist Spans

Inch-kips per foot of width of slab

Depth, in.	(a) No. 3	(b) No. 4	(c) No. 5	(d) No. 6	(e) No. 7	(f) No. 8	(g) No. 9	(h) No. 10	(j) No. 11	$+M_c$	$-M_c$	Weight, psf
8 + 2	7.9	14.25	21.4	31.4	42.8	56.0	71.25	91.6	111.3	151.5	41.0	48
8 + 2½	8.35	15.1	22.65	33.2	45.3	59.1	75.5	97.0	117.0	181.0	46.0	54
10 + 2	9.75	17.6	26.4	38.9	52.8	69.0	88.0	113.0	138.0	207.0	63.5	53
10 + 2½	10.2	18.5	27.75	40.7	55.5	72.3	92.5	118.6	144.7	249.0	68.5	59
10 + 3	10.6	19.3	28.8	42.4	57.8	74.3	96.4	123.6	151.0	288.0	75.0	65
12 + 2	11.56	21.0	31.5	46.3	63.0	82.3	105.0	134.6	164.5	263.0	88.3	59
12 + 2½	12.0	21.8	32.7	48.2	65.4	85.6	109.0	139.8	170.8	317.0	95.5	65
12 + 3	12.5	22.6	33.9	50.0	67.8	89.1	113.0	145.0	177.0	370.0	103.0	71
14 + 2½	13.87	25.1	37.65	55.5	75.3	99.0	125.5	161.2	197 0	389.0	127.5	70
14 + 3	14.35	26.0	39.0	57.4	78.0	102.2	130.0	167.0	204.0	450.0	136.0	76

Table 5.13 Allowable Shear in Concrete-joist Spans

Pounds per foot of width of slab

Depth in.	Width of joist								
	4 in.	4½ in.	5 in.	5½ in.	6 in.	6½ in.	7 in.	7½ in.	8 in.
8 + 2	1,182	1,299	1,413	1,523	1,630	1,734	1,834	1,931	2,025
8 + 2½	1,247	1,372	1,493	1,610	1,723	1,832	1,938	2,040	2,138
10 + 2	1,444	1,588	1,732	1,867	2,000	2,128	2,252	2,372	2,488
10 + 2½	1,510	1,663	1,812	1,953	2,088	2,220	2,348	2,472	2,591
10 + 3	1,575	1,733	1,886	2,034	2,177	2,316	2,448	2,576	2,699
12 + 2	1,707	1,877	2,040	2,199	2,358	2,500	2,644	2,784	2,920
12 + 2½	1,772	1,949	2,121	2,287	2,449	2,604	2,754	2,894	3,038
12 + 3	1,838	2,024	2,204	2,376	2,543	2,705	2,861	3,010	3,153
14 + 2½	2,034	2,238	2,436	2,628	2,813	2,992	3,164	3,332	3,488
14 + 3	2,100	2,307	2,509	2,706	2,897	3,082	3,261	3,434	3,601

5 in. wide. A figure in parentheses, e.g., (8), indicates a joist width which varies from the standard. The indication (t) after the size means tapered end pans to increase shear or negative-moment area. Tables 5.12 and 5.13 are for $f_s = 20{,}000$ psi and $f'_c = 3{,}000$ psi. For $f_s = 24{,}000$ psi, a conservative approximation in Table 5.12 would be to add 20 percent to the allowable moment.

d As has been noted in paragraph *c*, the computation of maximum positive and negative moments for three-span symmetrical joists such as may be used in schools, hospitals, etc., may be simplified by the use of curves. These curves are shown in Fig. 5.19 and are used in accordance with the following procedure.

1. Compute ratio r of center span to side span.
2. Compute factor $K = (\text{side span}/10)^2$.
3. Enter the chart (Fig. 5.19) with the ratio r computed in step 1, and read the value of $-M_1$ fully loaded, $-M_2$ maximum, $-M_3$ side spans loaded, and $-M_4$ center span loaded.
4. Set down dead load per square foot on side span and center span. This load is on the span regardless of any variation of live-load condition. Set down separately the live load per square foot on side span and center span. This may be on classrooms or corridor separately or on both.
5. The maximum negative moment over the support occurs with two adjacent spans loaded, and dead load only on the third span. The maximum positive load occurs in the side spans with both side spans loaded, and dead load only on the corridor, and this condition of loading will also cause the maximum negative moment in the center span. The maximum positive moment in the center span (if the spans are such that positive moment can be obtained in the center span) will occur with the center span fully loaded and with dead load only on side spans.

Compute $-M$ for each of the above cases by multiplying $-M$ as selected in step 3 by $w/100$ for each of the cases noted, and by the factor K. The maximum net $+M$ in the side spans is $(WL/8) + (-M/2)$ and in the center span $(WL/8) + (-M)$.

Problem A school has classroom joists 23 ft long with 9-ft corridor spans. The construction is 12 + 2½ slabs, using 5-in. joists, 25 in. on centers. The classroom finish weighs 15 psf, with a 10-psf ceiling. The corridor finish weighs 30 psf with a 10-psf ceiling. The classroom live load is 50 psf, and the corridor live load 80 psf. Design the joists.

$$\text{Ratio } r = \frac{9}{23} = 0.391$$

$$\text{Factor } K = 2.3^2 = 5.29$$

From Fig. 5.19

$$-M_1 = 10.03$$

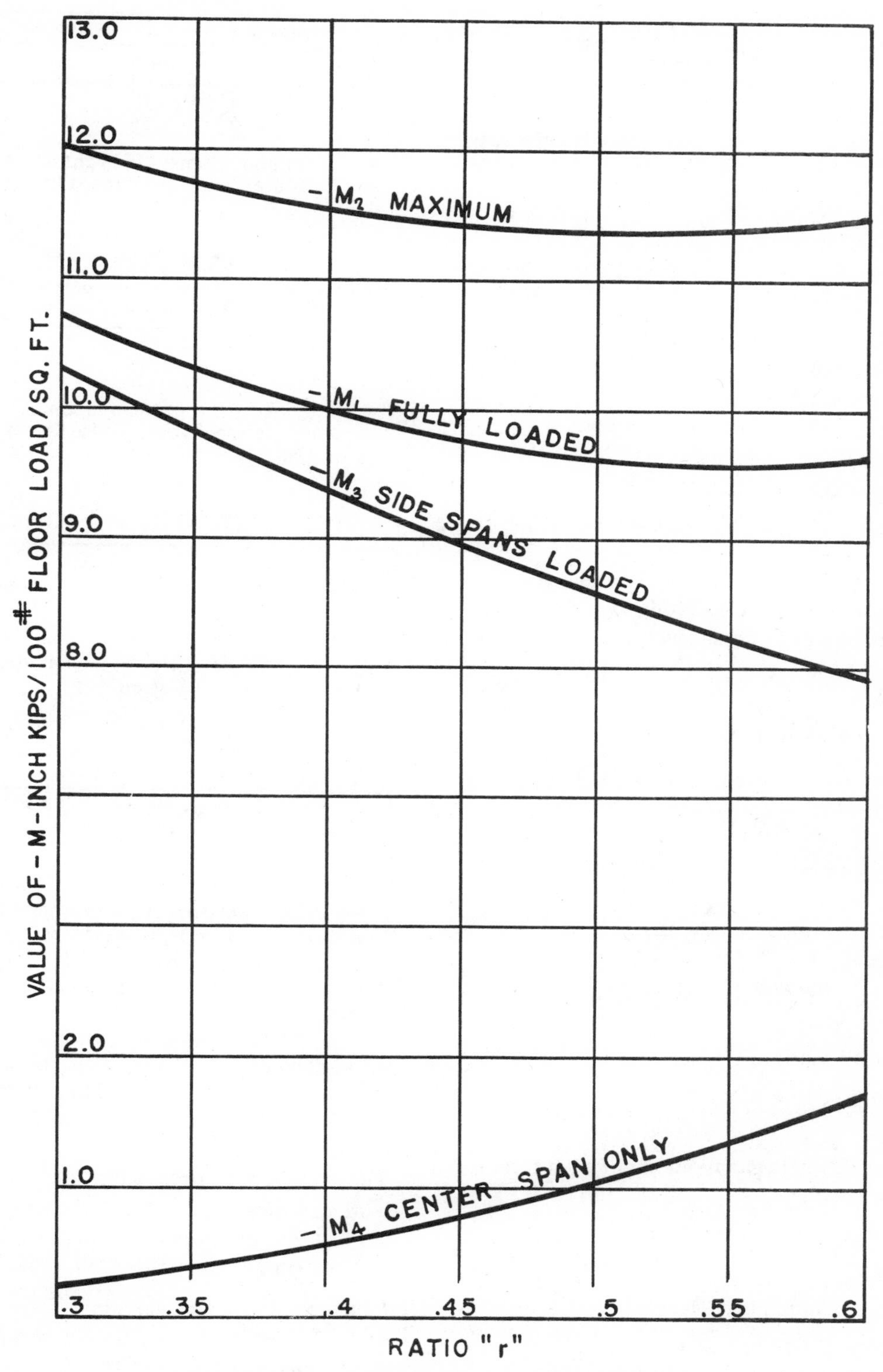

Fig. 5.19 Moments—Three-Span Symmetrical Slabs

$$-M_2 = 11.6$$
$$-M_3 = 9.4$$
$$-M_4 = 0.58$$

	Side spans	Center span
Dead loads:		
12 + 2½ slab	65	65
Finish	15	30
Ceiling	10	10
Total dead load	90	105
Live load	50	80
Total load	140	185

The maximum negative moment is

$$\begin{aligned} 1.4 \times 11.6 &= 16.24 \\ +0.45 \times 0.58 &= 0.26 \\ 1.85 \quad & \quad 16.50 \times 5.29 = \\ & -87.3 \text{ in.-kips} \end{aligned}$$

The maximum positive moment in side spans is obtained as follows. Maximum positive moment on simple span is

$$\frac{WL}{8} = \frac{140 \times 23 \times 23 \times 12}{8} = 111.1 \text{ in.-kips}$$

This is reduced by half the following:

$$\begin{aligned} 1.05 \times 10.03 &= 10.53 \\ 0.35 \times \ 9.4 &= 3.29 \\ 1.40 \quad & \quad 13.82 \times 5.29 = -73.1 \end{aligned}$$

$+M$ in side span $= 111.1 - 36.55 = +74.55$

The maximum positive moment on the center span as a simple span, with live plus dead load, is

$$\frac{WL}{8} = \frac{185 \times 9 \times 9 \times 12}{8} = 22.5 \text{ in.-kips}$$

For dead load only it is

$$\frac{105 \times 9 \times 9 \times 12}{8} = 12.76 \text{ in.-kips}$$

To check for negative moment at the center of the span,

$$M = -73.1 + 12.76 = -60.34$$

To check for possible positive moment, the negative moment at the ends of the span is

$$\begin{aligned} 0.90 \times (-10.03) &= -9.03 \\ +0.95 \times (-0.58) &= -0.57 \\ & -9.60 \times 5.29 = \\ & -50.8 \end{aligned}$$

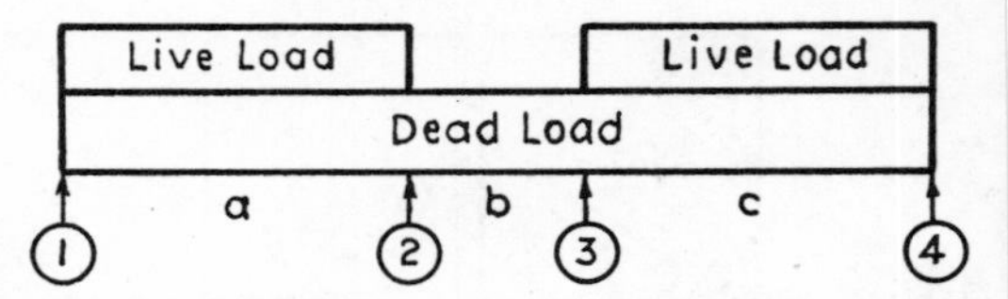

(a) For Maximum Moment In Spans "a" and "c". For Possible Negative Moment In Slab "b"

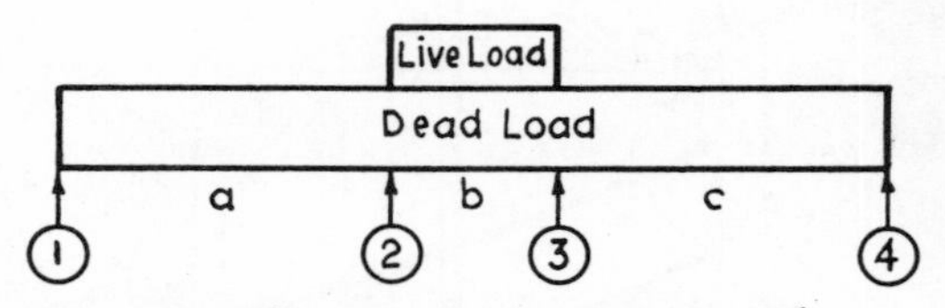

(b) For Maximum Moment In Span "b". For Possible Negative Moment In Spans "a" and "c"

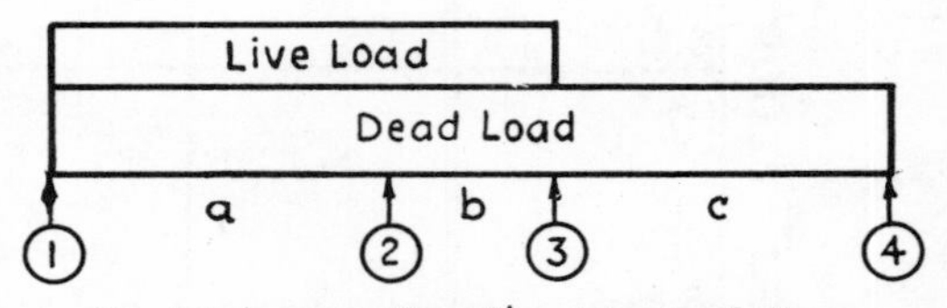

(c) For Maximum Negative Moment Shear And Reactions At Support "2"

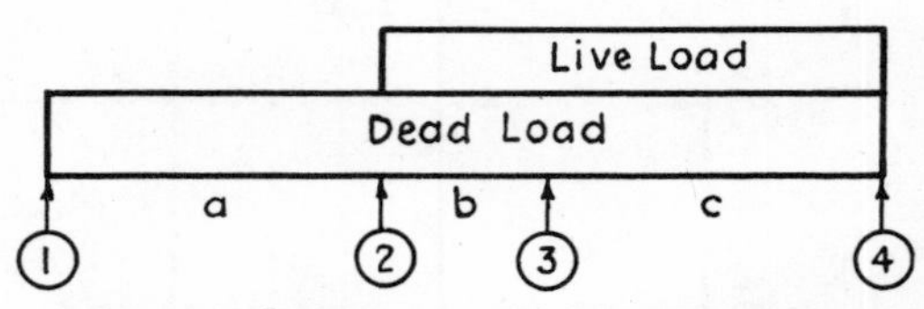

(d) For Maximum Negative Moment Shear and Reactions At Support "3"

Fig. 5.20 Possible Maximum Moment Conditions—Three-Span Slab

Therefore, with a minimum negative moment at the ends of 50.8 in.-kips and a maximum positive moment of 22.5, there can never be any net positive moment in the center of the span.

The moments to design for are,

$+M$ in side spans, $+74.55$ in.-kips
$-M$ at center supports, -87.3 in.-kips
$-M$ at center of corridor, 60.34 in.-kips

To furnish M in the corridor (referring to Table 5.12) will require one No. 7 top bar, which furnishes 65.4 in.-kips.

At the supports we need in addition

$$87.4 - 65.4 = 22.0 \text{ in.-kips}$$

This will require one No. 5 bar, which will be bent up from the side spans. This furnishes 32.7 in.-kips.

The side span will require in addition

$$74.55 - 32.7 = 41.85 \text{ in.-kips}$$

This will require one No. 6 bar, straight.

In accordance with the system of designation in Table 5.12, the side span will be designated 12 + 2½dc, and the center span 12 + 2½ e/a.

Although the calculations indicate no positive moment in the center span, it is good policy to put in a minimum-sized bar (No. 3) for possible erection stresses.

The computation of reactions takes into account internal moment also, as described in Art. 3.20*d*. If there were no such internal moment, the reactions would be half the load in each span, or for the side span,

$$R = \frac{140 \times 23}{2} = 1{,}610 \text{ lb per ft}$$

and for the center span,

$$R = \frac{185 \times 9}{2} = 833 \text{ lb per ft}$$

The correction for internal moment at the outer end is

$$\frac{73{,}100}{23 \times 12} = 265 \text{ lb}$$

and the corrected reaction is

$$1{,}610 - 265 = 1{,}345 \text{ lb}$$

The correction for internal moment at the inner end of the side span is

$$\frac{87{,}400}{23 \times 12} = 317$$

and the corrected reaction is

$$1{,}610 + 317 = 1{,}927 \text{ lb}$$

Although it is not absolutely accurate, it is safe to assume that the reaction from the corridors is the uncorrected reaction of 833 lb, since it is hardly likely that there can be enough variation in these moments to affect materially a variation in load. The total load for which to design the corridor beam would therefore be

$$1{,}927 + 833 = 2{,}760 \text{ lb per ft}$$

5.42 Precast concrete floors and roofs

a There are various patented precast floor and roof systems available, their economy and usability being determined largely by freight rates from the point of manufacture. In this category we may list five classes as most common.

1. Precast roof tile, which requires no roof finish of any kind (Fig. 5.21)
2. Short-span flat tongue-and-groove plank or channel-type slabs (Fig. 5.22)
3. Long-span cored plank (Fig. 5.23)
4. Leap slab (Fig. 5.24)
5. Precast concrete joists (Fig. 5.25)

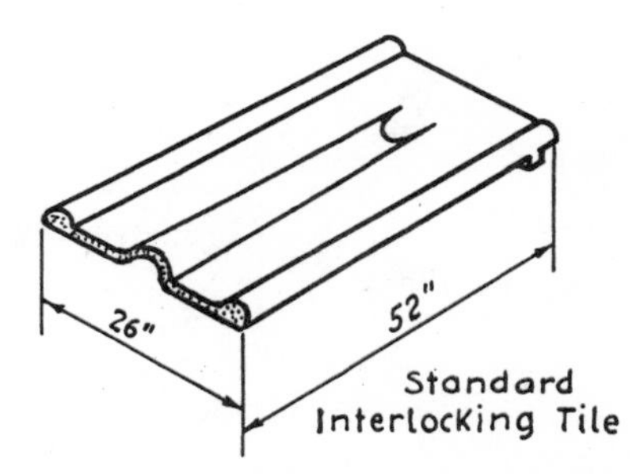

Fig. 5.21 Precast Concrete Roof Tile

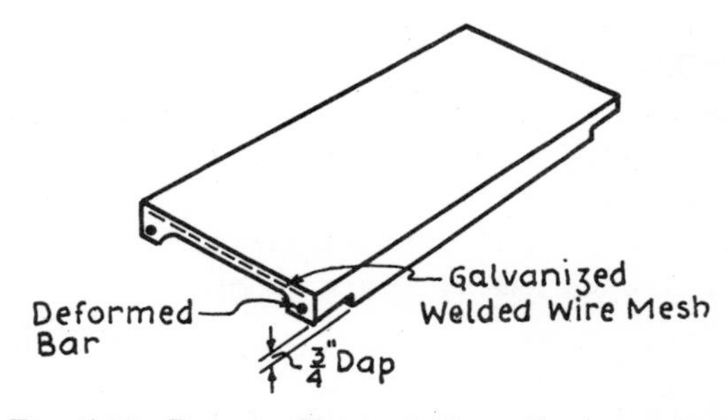

Fig. 5.22 Precast Channel-Type Concrete Slab

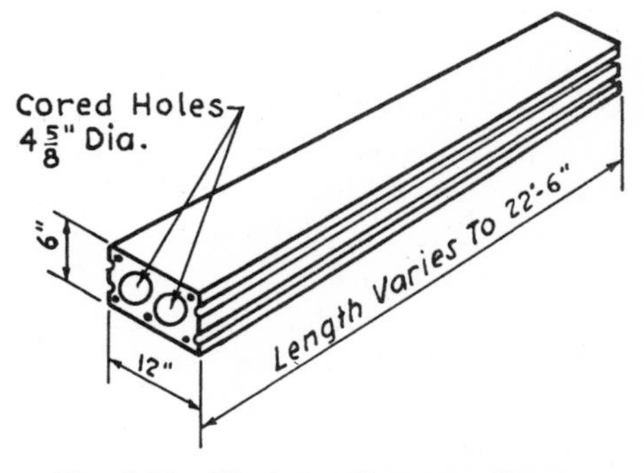

Fig. 5.23 Flexicore Concrete Plank

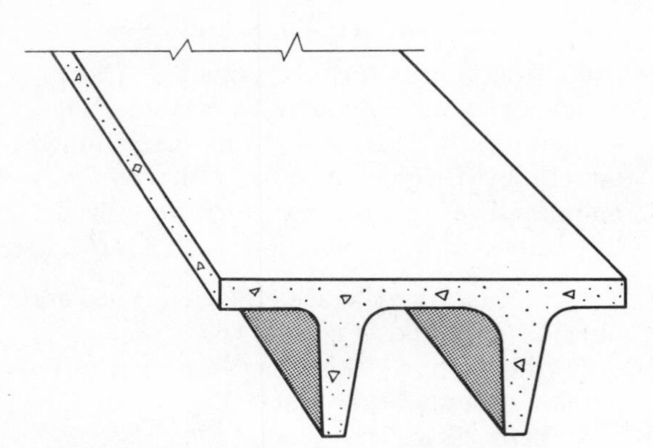

Fig. 5.24 Section Through Leap Slab

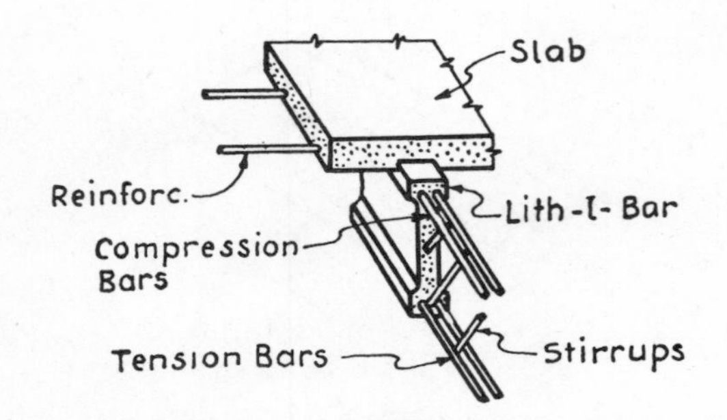

Fig. 5.25 Lith-I-Bar Concrete Joist Construction

The various manufacturers publish load tables and specifications for application, and some of the manufacturers will bid on the material erected in place. Only an actual bid from a contractor will give a comparison of cost with cast-in-place systems.

b Any precast system lacks stiffness in itself and requires more stiffness to be built into the frame than does a poured-in-place slab. Most precast systems do not provide lateral bracing for the top flange of steel beams, and frequently rod bracing must be installed with them. Each system has its economical limitations of span, and some systems are only economical if spans are kept uniform. The designer should investigate the requirements of the system before proceeding with the design.

To offset the disadvantages, precast systems save forming costs, are cleaner in installation, are faster to install provided delivery is made in time, are usually lighter in deal load, and can be installed regardless of freezing weather.

5.43 Kneed reinforced concrete slabs

a Concrete slabs with a knee, similar to a stair slab and landing poured monolithically without support at the knee, are actually rigid frames, and under all conditions of loading they will have negative moment at the knee. The accurate computation of such a slab is a complex problem involving a number of variables, but a satisfactory solution which errs on the side of safety may be obtained by simple approximate methods. The method is largely graphical and involves the use of a moment diagram.

Although the method given here is applied to a slab with an upward knee as shown in Fig. 5.26, step 1, it applies equally to a slab with a downward knee such as the opposite run in a flight of stairs.

b The steps in the computation of this slab are as follows.

1. In order to determine the position of the theoretical knee, draw the horizontal center line of the landing slab *bc* and the sloping center line of the slab supporting the stairs *ab*. The intersection of these two lines (*b*) determines the position of the knee (Fig. 5.26, step 1).

2. Compute the simple beam moment on the slab, allowing for full live and dead load on all parts of the slab, and lay out the moment diagram (Fig. 5.26, step 2). At the point where the line of the knee *b* cuts the moment diagram, lay off a negative moment *ef* beyond the curve of the moment diagram. This negative moment is 0.4 of the maximum positive moment *gh* on the simple span. Draw lines from the extreme outer point of the negative moment *f* to the two ends of the moment diagram. These lines, *af* and *fc*, become the new base lines of the moment diagram; the area intercepted on the original moment diagram represents positive moment on the span, the area outside the original moment diagram represents negative moment. These are indicated by the shaded area in Fig. 5.26, step 2. This gives the negative moment for which the slab must be

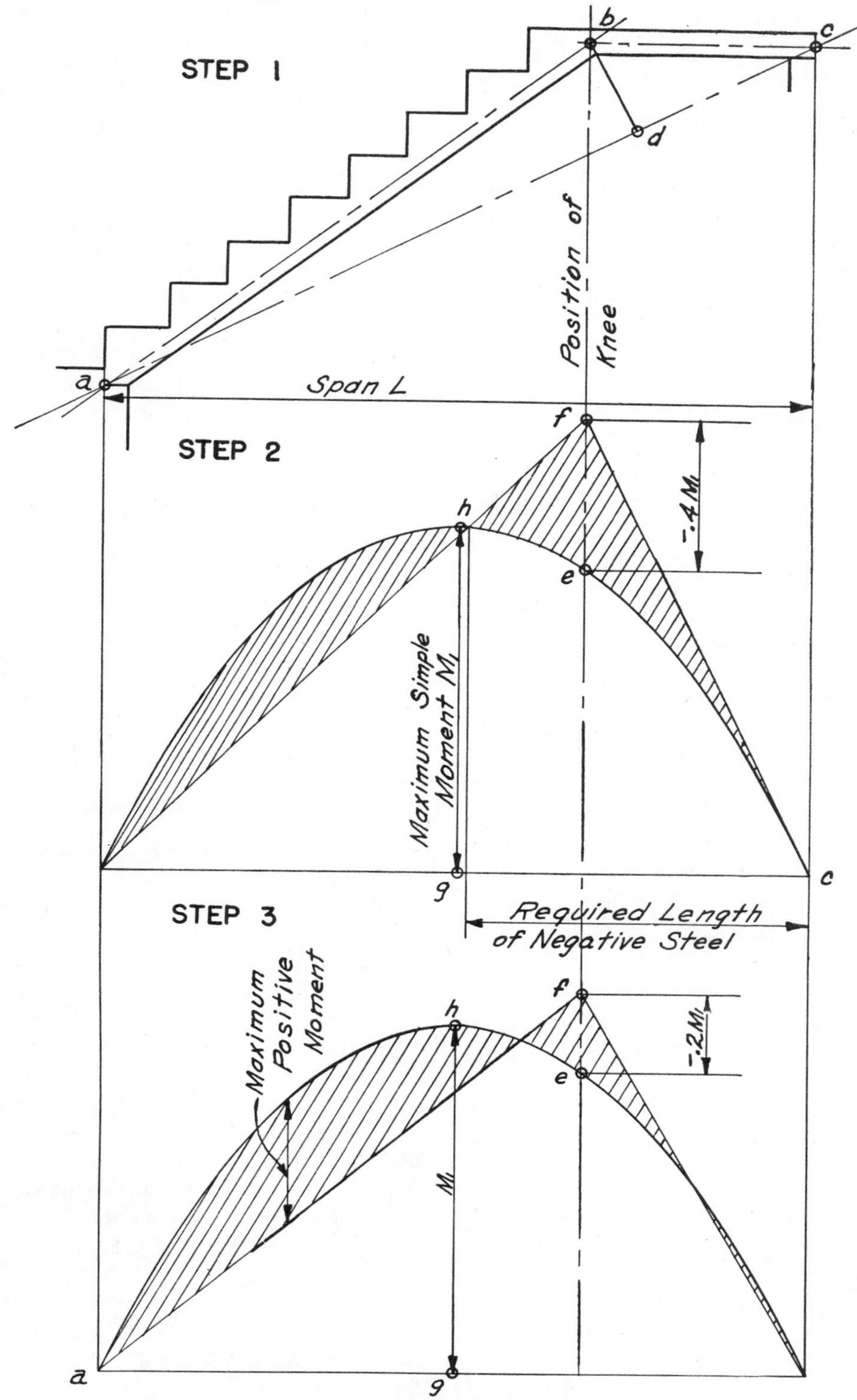

Fig. 5.26 Computation of Moments—Kneed Concrete Slab

designed, and the length through which negative steel must be provided. Negative bars should be developed in tension beyond this point.

3. Under some conditions of span or loading, the negative moment will not be as great as designed for in step 2; and since the maximum positive moment may result when the negative moment is a minimum, the same procedure as that given in step 2 should be followed through, except that the negative moment should be reduced to 0.2 of the maximum positive moment. The intercepts thus obtained will give the maximum positive moment, and the length through which positive moment occurs.

4. If a straight line be drawn between the two points of support (*ac* in Fig. 5.26, step 1), it indicates the direction of thrust on the supports of an upward knee, or pull on the supports of a downward knee. The amount of this stress is obtained by dividing the maximum negative moment found in step 2 by the height *bd* perpendicular to the line *ac*. By means of a vector analysis the stress may be divided into vertical and horizontal components, or it may be taken out at the end of the supporting span by means of ties or struts.

Problem Assume the stair slab shown in Fig. 5.26, step 1, with eight risers at 7½ in., using 10-in. runs and a 4-ft landing. Design for a live load of 125 psf.

The span L becomes

Seven treads at 10 in. = 70 in.	= 5 ft 10 in.
Landing	= 4 ft
Span	= 9 ft 10 in.

Assume a 4-in. slab having a constant dead load of 50 psf. In addition, spread the load of the steps through a length of 6 ft 8 in. This load is made up of eight triangular prisms of concrete, each 7½ in. high by 10 in. long, or 2.08 cu ft of concrete, which would weigh 312 lb, or 48 psf. Add to these the live load of 125 psf.

$$M = \frac{4.76 \times 12 \times 1{,}063}{2} = 30{,}360 \text{ on simple span}$$

$$-M = 0.4 \times 30{,}360 = 12{,}150 \text{ in.-lb}$$

Although the moment diagram is made up of segments of several parabolas, it will be accurate enough to plot a single parabola with a maximum height of 30,360 in.-lb and a span of 9.83 ft in order to obtain maximum positive and negative moments.

The moment diagrams in Fig. 5.26 steps 2 and 3 have been drawn to such a scale that M_1 is 30,360 in.-lb, $-0.4M_1$ is 12,150 in.-lb, and $-0.2M_1$ is 6,075 in.-lb. The maximum positive moment scaled from the diagram in step 3 of 11,250 in.-lb.

On the basis of a positive moment of 11,250 in.-lb and a negative moment of 12,150 in.-lb, using 3,000-psi concrete with A15 steel, R is 222.6, and the minimum slab thickness may be computed thus:

$$d = \sqrt{\frac{12{,}150}{222.6 \times 12}} = \sqrt{4.55} = 2.15 \text{ in. net}$$

Therefore, allowing for proper coverage, a 3½-in. slab could be used. However, there are the practical considerations of getting two layers of steel—positive and negative moment resistance—in the slab, which would make it more practical to use a 4-in. minimum slab.

On the basis of a 4-in. slab of 3 in. net depth, the steel required would be

$$-A_s = \frac{12{,}150}{0.875 \times 20{,}000 \times 3} = 0.231 \text{ sq in.} = \text{No. 3 at } 5\tfrac{1}{2} \text{ in. o.c.}$$

$$+A_s = \frac{11{,}250}{0.875 \times 20{,}000 \times 3} = 0.214 \text{ sq in.} = \text{No. 3 at 6 in. o.c.}$$

The moment diagram in step 3 indicates that a small amount of positive steel may be required throughout almost the entire length of the span, and it is advisable to extend half the positive steel through this area.

The thrust or direct pull may be computed next. The scaled distance *bd* on the diagram is 1 ft 5 in., or 17 in. Therefore

$$H = \frac{12{,}150}{17} = 711 \text{ lb per ft width}$$

Assume that the stairs are 4 ft wide for each flight, as shown in plan in Fig. 5.28. The resultant loads and reactions are shown in

Fig. 5.29. To carry the thrust for the run from the landing down, a tensile area of 1,422/20,000 = 0.07 sq in. will be required. Use one No. 3 rod buried in the stair partition.

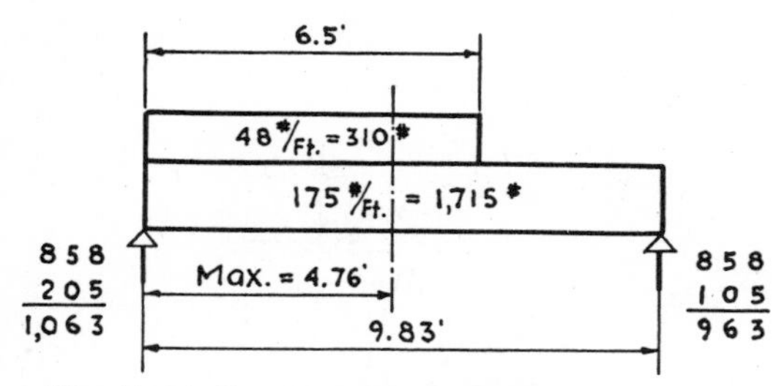

Fig. 5.27 Load Diagram—Kneed Concrete Slab Problem

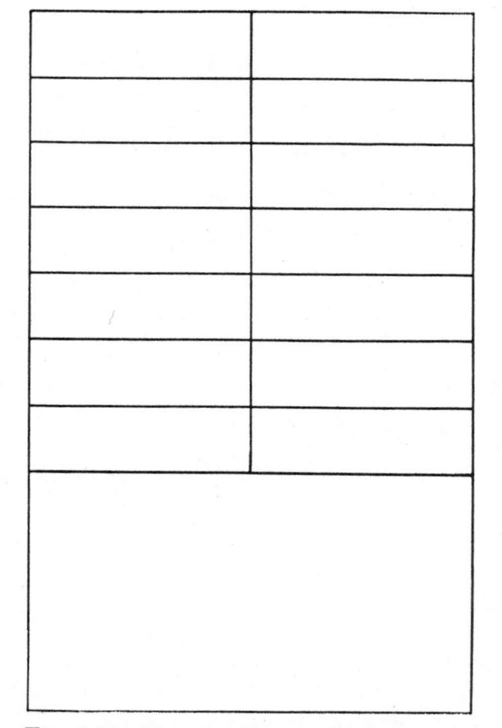

Fig. 5.28 Plan for Kneed Concrete Slab Problem

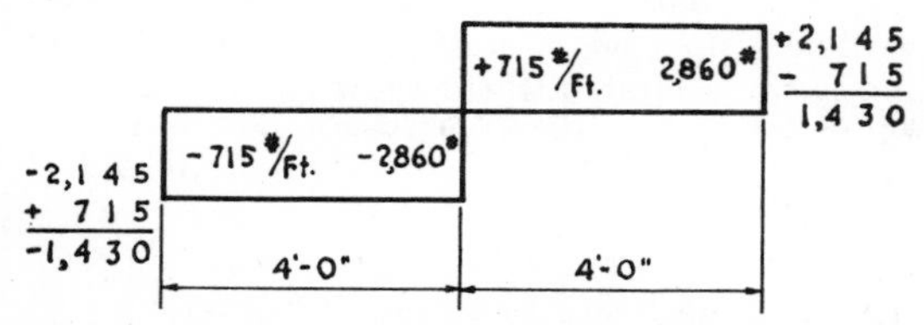

Fig. 5.29 Thrust Diagram—Kneed Concrete Slab Problem

To carry the inward thrust on the run from the landing up, a compression member capable of carrying 1,422 lb on an unsupported length of 11 ft will be required. For this, a member with $l/r = 1.1$ will be required if we use 120 as a minimum or 0.61 if we use 200 as the minimum. Even on this latter basis, it would require a standard 1½-in. pipe, which would be stressed to only 1,422/0.799 = 1,780 psi in compression. It might sometimes be better to pour a concrete curb to take this thrust and use this curb as the base of the stair wall.

The final detail of the stairs is shown in Fig. 5.30. This figure indicates several points relative to the reinforcement which should be noted for stair slabs. Reinforcement required to pass a reentrant angle should not be merely bent around the corner, since under tension it has a tendency to straighten out and spall off the concrete. It is, therefore, carried through and hooked beyond the reentrant angle as is indicated in Fig. 5.30. Also, it is advisable for the protection of the nosing to put a single nosing bar in the nosing of each tread.

5.50 TWO-WAY CONCRETE SLABS

a One of the most popular of the newer types of floor systems is the so-called "two-way slab." This is a floor slab reinforced in two directions, with supporting beams or walls on all four sides, and is not to be confused with the "flat-slab system," which is a system of girderless slabs. The slabs may be either solid concrete or concrete ribs with tile or concrete-block fillers, or without fillers, the so-called waffle slab. The system using square tile fillers, either with or without topping, is known as the Shuster system and is marketed by the Whitacre Fireproofing Co. The system using specially shaped concrete-block fillers is known as the Republic system and is sold by the Republic Fireproofing Co. In these systems using fillers, the reinforcement is concentrated in the ribs, approximately 2 ft apart. The shear is taken care of by widening the ribs at the ends, by using narrower fillers. Usu-

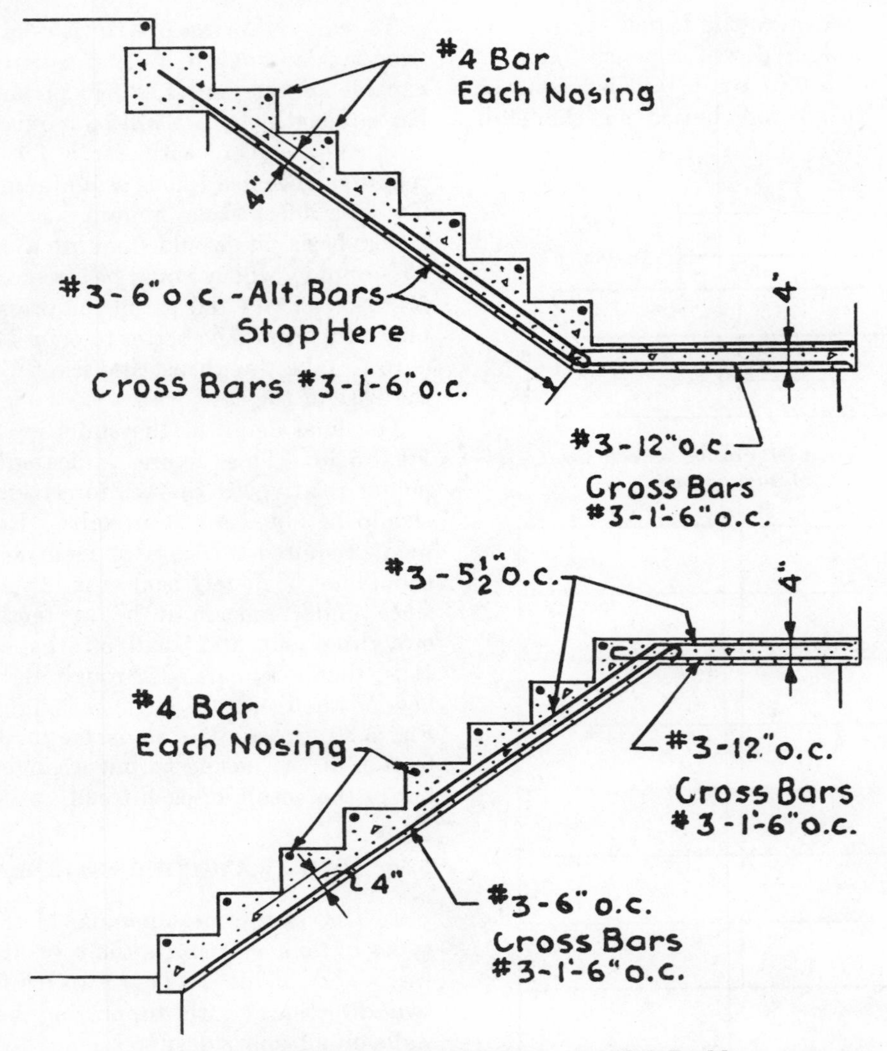

Fig. 5.30 Final Design—Kneed Concrete Slab Problem

ally, only straight bars are used—straight bottom bars from beam to beam and straight top bars from quarter point to quarter point. Only bars at the end of the span are hooked at the outer end.

This type of construction becomes more efficient as it approaches a square. As the ratio of sides approaches 2:1, the slab approximates a one-way slab. The slab is well suited to carry heavy concentrations through plate action.

The design of two-way slabs is specified in the 1963 ACI Code as follows:

2002 TWO-WAY FLOOR SYSTEMS WITH SUPPORTS ON FOUR SIDES

(a) This construction, reinforced in two directions, includes solid reinforced concrete slabs concrete joists with fillers of hollow concrete units or clay tile, with or without concrete top slabs; and concrete joists with top slabs placed monolithically with the joists. The slab shall be supported by walls or beams on all sides and if not securely attached to supports, shall be reinforced in each direction as specified in (b).

(b) Special reinforcement shall be provided at exterior corners in both the bottom and top of the slab. This reinforcement shall be provided for a distance in each direction from the corner equal to one-fifth the longest span. The reinforcement in the top of the slab shall be parallel to the diagonal from the corner. The reinforcement in the bottom of the slab shall be at right angles to the diagonal or may be of bars in two directions parallel to the sides of the slab. The reinforcement in each band shall be of equivalent size and spacing to that required for the maximum positive moment in the slab.

(c) The slab shall be designed by approved methods which shall take into account the effect of continuity and fixity at supports, the ratio of length to width of slab and the effect of two-way action.*

(d) The supports of two-way slabs shall be designed by accepted methods taking into account the effect of continuity. The loading on the supports may be computed from the coefficients of the approved methods.

(e) In no case shall the slab thickness be less than 3½ in. nor less than the perimeter of the slab divided by 180. The center to center spacing of reinforcement shall be not more than three times the slab thickness and the ratio of reinforcement in each direction shall be not less than required by Section 807.

* The requirements of this section are satisfied by any of the methods of design shown in Appendix A.

In the Appendix to the 1963 ACI Code, the method of design is spelled out fully in the following sections. A choice of three methods of approach is given, any one of which is acceptable.

DESIGN OF TWO-WAY SLABS—APPENDIX A—ACI CODE

There are several satisfactory methods for designing two-way slabs. Although they may give somewhat different results in details, the resulting floors give reasonable over-all safety factors. Three methods which have been used extensively with satisfactory results are given in this appendix. These methods of design are to implement the provisions of Section 2002.

A2001 Method 1

Notation

L = length of clear span

L_1 = length of clear span in the direction normal to L

g = ratio of span between lines of inflection to L in the direction of span L, when span L only is loaded

g_1 = ratio of span between lines of inflection to L_1 in the direction of span L_1, when span L_1, only is loaded

r = gL/g_1L_1

w = total uniform load per sq ft

W = total uniform load between opposite supports on slab strip of any width or total slab load on beam when considered as one-way construction

x = ratio of distance from support to any section of slab or beam, to span L or L_1

B = bending moment coefficient for one-way construction

C = factor modifying bending moments prescribed for one-way construction for use in proportioning the slabs and beams in the direction of L of slabs supported on four sides

C_s = ratio of the shear at any section of a slab strip distant xL from the support to the total load W on the strip in direction of L

C_b = ratio of the shear at any section of a beam distant xL from the support to the total load W on the beam in the direction of L

W_1, C_1, C_{s1}, C_{b1}, are corresponding values of W, C, C_s, C_b, for slab strip or beam in direction of L_1.

(a) *Lines of inflection for determination of r*—The lines of inflection shall be determined by elastic analysis of the continuous structure in each direction, when only the span under consideration is loaded.

When the span L or L_1 is at least ⅔ and at most 3⁄2 of the adjacent continuous span or spans, the values of g or g_1 may be taken as 0.87 for exterior spans and 0.76 for interior spans (see Fig. 5.31).

For spans discontinuous at both ends, g or g_1 shall be taken as unity.

(b) *Bending moments and shear*—Bending moments shall be determined in each direction with the coefficients prescribed for one-way construction in Chapter 9 and modified by factor C or C_1 from Tables 5.14 or 5.15.

When the coefficients prescribed in Section 904(c) are used, the average value of Cw or C_1w for the two spans adjacent to a support shall be used in determining the negative bending moment at the face of the support.

The shear at any section distant xL or xL_1 from supports shall be determined by modifying the total load on the slab strip or beam by the factors C_s, C_{s1}, C_b, or C_{b1} taken from Tables 5.14 or 5.15.

(c) *Arrangement of reinforcement*

1. In any panel, the area of reinforcement per unit width in the long direction shall be at least one-third that provided in the short direction.

		In L direction	In L_1 direction
Bending moment	Slab strip	$M = CBWL$	$M_1 = C_1BW_1L_1$
	Beam	$M = (1-C)BWL$	$M_1 = (1-C_1)BW_1L_1$

Method 1—Table 5.14 Design of Two-way Concrete Slabs

Upper figure	Lower figure	C_s					C
		C_{s1}					C_1
		Values of x					
r	$r_1 = \frac{1}{r}$	0.0	0.1	0.2	0.3	0.4	
0.00		0.50	0.40	0.30	0.20	0.10	1.00
	∞	0.00	0.00	0.00	0.00	0.00	0.00
0.50		0.44	0.36	0.27	0.18	0.09	0.89
	2.00	0.06	0.03	0.02	0.00	0.00	0.06
0.55		0.43	0.33	0.23	0.15	0.07	0.79
	1.82	0.07	0.04	0.02	0.01	0.00	0.08
0.60		0.41	0.30	0.20	0.12	0.05	0.70
	1.67	0.09	0.05	0.03	0.01	0.00	0.10
0.65		0.39	0.28	0.18	0.10	0.04	0.64
	1.54	0.11	0.06	0.03	0.01	0.00	0.13
0.70		0.37	0.26	0.16	0.09	0.03	0.58
	1.43	0.13	0.08	0.04	0.01	0.00	0.15
0.80		0.33	0.22	0.13	0.07	0.02	0.48
	1.25	0.17	0.10	0.06	0.02	0.00	0.21
0.90		0.29	0.19	0.11	0.05	0.01	0.40
	1.11	0.21	0.13	0.07	0.03	0.01	0.27
1.00		0.25	0.16	0.09	0.04	0.01	0.33
	1.00	0.25	0.16	0.09	0.04	0.01	0.33
1.10		0.21	0.13	0.07	0.03	0.01	0.28
	0.91	0.29	0.19	0.11	0.05	0.01	0.39
1.20		0.18	0.11	0.06	0.02	0.00	0.23
	0.83	0.32	0.21	0.13	0.06	0.02	0.45
1.30		0.16	0.10	0.05	0.02	0.00	0.19
	0.77	0.34	0.23	0.14	0.07	0.03	0.51
1.40		0.13	0.08	0.04	0.02	0.00	0.16
	0.71	0.37	0.25	0.16	0.09	0.03	0.57
1.50		0.11	0.07	0.04	0.01	0.00	0.14
	0.67	0.39	0.27	0.17	0.10	0.04	0.61
1.60		0.10	0.06	0.03	0.01	0.00	0.12
	0.63	0.40	0.29	0.19	0.11	0.05	0.66
1.80		0.07	0.04	0.02	0.01	0.00	0.08
	0.55	0.43	0.33	0.23	0.15	0.07	0.79
2.00		0.06	0.03	0.02	0.00	0.00	0.06
	0.50	0.44	0.36	0.27	0.18	0.09	0.89
∞		0.00	0.00	0.00	0.00	0.00	0.00
	0.00	0.50	0.40	0.30	0.20	0.10	1.00

2. The area of positive moment reinforcement adjacent to a continuous edge only and for a width not exceeding one-fourth of the shorter dimension of the panel may be reduced 25 percent.

3. At a noncontinuous edge the area of negative moment reinforcement per unit width shall be at least one-half of that required for maximum positive moment.

A2002 METHOD 2

Notation

C = moment coefficient for two-way slabs as given in Table 5.14

m = ratio of short span to long span for two-way slabs

S = length of short span for two-way slabs. The span shall be considered as the center-to-center distance between supports or the clear span plus twice the thickness of slab, whichever value is the smaller.

w = total uniform load per sq ft

(a) *Limitations*—These recommendations are intended to apply to slabs (solid or ribbed), isolated or continuous, supported on all four sides by walls or beams, in either case built monolithically with the slabs.

		In L direction	In L_1 direction
Shear	Slab strip	$V = C_s W$	$V_1 = C_{s1} W_1$
	Beam	$V = C_b W$	$V_1 = C_{b1} W_1$

Method 1—Table 5.15 Design of Beams for Two-way Concrete Slabs

Upper figure C_b Lower figure C_{b1}		Values of x					$1-C$ $1-C_1$
r	$r_1 = \frac{1}{r}$	0.0	0.1	0.2	0.3	0.4	
0.00	∞	0.00 0.50	0.00 0.40	0.00 0.30	0.00 0.20	0.00 0.10	0.00 1.00
0.50	2.00	0.06 0.44	0.04 0.37	0.03 0.28	0.02 0.20	0.01 0.10	0.11 0.94
0.55	1.82	0.07 0.43	0.07 0.36	0.07 0.28	0.05 0.19	0.03 0.10	0.21 0.92
0.60	1.67	0.09 0.41	0.10 0.35	0.10 0.27	0.08 0.19	0.05 0.10	0.30 0.90
0.65	1.54	0.11 0.39	0.12 0.34	0.12 0.27	0.10 0.19	0.06 0.10	0.36 0.87
0.70	1.43	0.13 0.37	0.14 0.32	0.14 0.26	0.11 0.19	0.07 0.10	0.42 0.85
0.80	1.25	0.17 0.33	0.18 0.30	0.17 0.24	0.13 0.18	0.08 0.10	0.52 0.79
0.90	1.11	0.21 0.29	0.21 0.27	0.19 0.23	0.15 0.17	0.09 0.09	0.60 0.73
1.00	1.00	0.25 0.25	0.24 0.24	0.21 0.21	0.16 0.16	0.09 0.09	0.67 0.67
1.10	0.91	0.29 0.21	0.27 0.21	0.23 0.19	0.17 0.15	0.09 0.09	0.72 0.61
1.20	0.83	0.32 0.18	0.29 0.19	0.24 0.17	0.18 0.14	0.10 0.08	0.77 0.55
1.30	0.77	0.34 0.16	0.30 0.17	0.25 0.16	0.18 0.13	0.10 0.07	0.81 0.49
1.40	0.71	0.37 0.13	0.32 0.15	0.26 0.14	0.18 0.11	0.10 0.07	0.84 0.43
1.50	0.67	0.39 0.11	0.33 0.13	0.26 0.13	0.19 0.10	0.10 0.06	0.86 0.39
1.60	0.63	0.40 0.10	0.34 0.11	0.27 0.11	0.19 0.09	0.10 0.05	0.88 0.34
1.80	0.55	0.43 0.07	0.36 0.07	0.28 0.07	0.19 0.05	0.10 0.03	0.92 0.21
2.00	0.50	0.44 0.06	0.37 0.04	0.28 0.03	0.20 0.02	0.10 0.01	0.94 0.11
∞	0.00	0.50 0.00	0.40 0.00	0.30 0.00	0.20 0.00	0.10 0.00	1.00 0.00

A two-way slab shall be considered as consisting of strips in each direction as follows:

A middle strip one-half panel in width, symmetrical about panel center line and extending through the panel in the direction in which moments are considered.

A column strip one-half panel in width, occupying the two quarter-panel areas outside the middle strip.

Where the ratio of short to long span is less than 0.5, the middle strip in the short direction shall be considered as having a width equal to the difference between the long and short span, the remaining area representing the two column strips.

The critical sections for moment calculations are referred to as principal design sections and are located as follows:

For negative moment, along the edges of the panel at the faces of the supporting beams.

For positive moment, along the center lines of the panels.

(b) *Bending moments*—The bending moments for the middle strips shall be computed from the formula

$$M = CwS^2$$

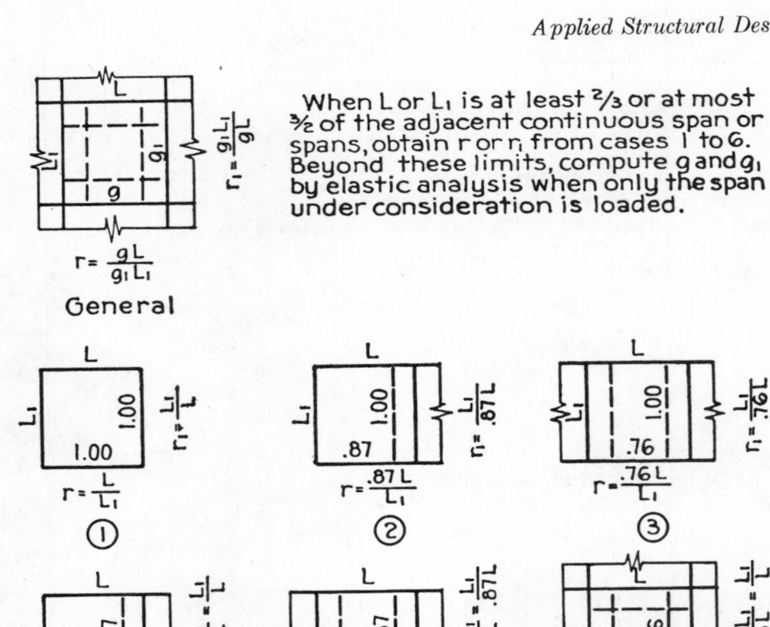

Fig. 5.31 Two-Way Concrete Slabs—Values of "g"

The average moments per foot of width in the column strip shall be two-thirds of the corresponding moments in the middle strip. In determining the spacing of the reinforcement in the column strip, the moment may be assumed to vary from a maximum at the edge of the middle strip to a minimum at the edge of the panel.

Where the negative moment on one side of a support is less than 80 percent of that on the other side, two-thirds of the difference shall be distributed in proportion to the relative stiffnesses of the slabs.

(c) *Shear*—The shear stresses in the slab may be computed on the assumption that the load is distributed to the supports in accordance with (d).

(d) *Supporting beams*—The loads on the supporting beams for a two-way rectangular panel may be assumed as the load within the tributary areas of the panel bounded by the intersection of 45-deg lines from the corners with the median line of the panel parallel to the long side.

The bending moments may be determined approximately by using an equivalent uniform load per lineal foot of beam for each panel supported as follows:

For the short span: $\dfrac{wS}{3}$

For the long span: $\dfrac{wS}{3}\dfrac{(3 - m^2)}{2}$

A2003 Method 3

Notation

A = length of clear span in short direction

B = length of clear span in long direction

C = moment coefficients for two-way slabs as given in Tables 5.17, 5.18, and 5.19. Coefficients have identifying indexes, such as $C_{A\,\text{neg}}$, $C_{B\,\text{neg}}$, $C_{A\,\text{DL}}$, $C_{B\,\text{DL}}$, $C_{A\,\text{LL}}$, $C_{B\,\text{LL}}$.

m = ratio of short span to long span for two-way slabs

w = uniform load per sq ft. For negative moments and shears, w is the total dead load plus live load for use in Table 5.17. For positive moments, w is to be separated into dead and live loads for use in Tables 5.18 and 5.19.

w_A, w_B = percentages of load w in A and B directions according to Table 5.20. These shall be used for computations of shear and for loadings on supports.

(a) *Limitations*—A two-way slab shall be considered as consisting of strips in each direction as follows:

A middle strip one-half panel in width,

Method 2—Table 5.16 Two-way Concrete Slabs—Moment Coefficients

Moments	Short span, Values of m: 1.0	0.9	0.8	0.7	0.6	0.5 and less	Long span, all values of m
Case 1—Interior panels							
Negative moment at—							
Continuous edge	0.033	0.040	0.048	0.055	0.063	0.083	0.033
Discontinuous edge	—	—	—	—	—	—	—
Positive moment at midspan	0.025	0.030	0.036	0.041	0.047	0.062	0.025
Case 2—One edge discontinuous							
Negative moment at—							
Continuous edge	0.041	0.048	0.055	0.062	0.069	0.085	0.041
Discontinuous edge	0.021	0.024	0.027	0.031	0.035	0.042	0.021
Positive moment at midspan	0.031	0.036	0.041	0.047	0.052	0.064	0.031
Case 3—Two edges discontinuous							
Negative moment at—							
Continuous edge	0.049	0.057	0.064	0.071	0.078	0.090	0.049
Discontinuous edge	0.025	0.028	0.032	0.036	0.039	0.045	0.025
Positive moment at midspan	0.037	0.043	0.048	0.054	0.059	0.068	0.037
Case 4—Three edges discontinuous							
Negative moment at—							
Continuous edge	0.058	0.066	0.074	0.082	0.090	0.098	0.058
Discontinuous edge	0.029	0.033	0.037	0.041	0.045	0.049	0.029
Positive moment at midspan	0.044	0.050	0.056	0.062	0.068	0.074	0.044
Case 5—Four edges discontinuous							
Negative moment at—							
Continuous edge	—	—	—	—	—	—	—
Discontinuous edge	0.033	0.038	0.043	0.047	0.053	0.055	0.033
Positive moment at midspan	0.050	0.057	0.064	0.072	0.080	0.083	0.050

symmetrical about panel center line and extending through the panel in the direction in which moments are considered.

A column strip one-half panel in width, occupying the two quarter-panel areas outside the middle strip.

Where the ratio of short to long span is less than 0.5, the slab shall be considered as a one-way slab and is to be designed in accordance with Chapter 9 except that negative reinforcement, as required for a ratio of 0.5, shall be provided along the short edge.

At discontinuous edges, a negative moment one-third (1/3) of the positive moment is to be used.

Critical sections for moment calculations are located as follows:

For negative moment along the edges of the panel at the faces of the supports.

For positive moment, along the center lines of the panels.

(b) *Bending moments*—The bending moments for the middle strips shall be computed by the use of Tables 5.17, 5.18, and 5.19.

$$M_A = CwA^2 \quad \text{and} \quad M_B = CwB^2$$

The bending moments in the column strips shall be gradually reduced from the full value M_A and M_B from the edge of the middle strip to one-third (1/3) of these values at the edge of the panel.

Where the negative moment on one side of a support is less than 80 percent of that on the other side, the difference shall be distributed in proportion to the relative stiffnesses of the slabs.

(c) *Shear*—The shear stresses in the slab may be computed on the assumption that the load is distributed to the supports in accordance with Table 5.20.

(d) *Supporting beams*—The loads on the supporting beams for a two-way rectangular panel shall be computed using Table 5.20 for the percentages of loads in "A" and "B" directions. In no case shall the load on the beam along the short edge be less than that of an area bounded by the intersection of 45-deg lines from the corners. The equivalent uniformly distributed load per linear foot on this short beam is

$$\frac{wA}{3}$$

Method 3—Table 5.17 Two-way Concrete Slabs—Coefficients for Negative Moments*

$$\left.\begin{array}{l} M_{A\ neg} = C_{A\ neg} \times w \times A^2 \\ M_{B\ neg} = C_{B\ neg} \times w \times B^2 \end{array}\right\} \text{where } w = \text{ total uniform dead plus live load}$$

Ratio		Case 1 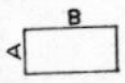	Case 2 	Case 3 	Case 4 	Case 5 	Case 6 	Case 7 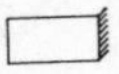	Case 8 	Case 9 
$m = \frac{A}{B}$										
1.00	$C_{A\ neg}$		0.045		0.050	0.075	0.071		0.033	0.061
	$C_{B\ neg}$		0.045	0.076	0.050			0.071	0.061	0.033
0.95	$C_{A\ neg}$		0.050		0.055	0.079	0.075		0.038	0.065
	$C_{B\ neg}$		0.041	0.072	0.045			0.067	0.056	0.029
0.90	$C_{A\ neg}$		0.055		0.060	0.080	0.079		0.043	0.068
	$C_{B\ neg}$		0.037	0.070	0.040			0.062	0.052	0.025
0.85	$C_{A\ neg}$		0.060		0.066	0.082	0.083		0.049	0.072
	$C_{B\ neg}$		0.031	0.065	0.034			0.057	0.046	0.021
0.80	$C_{A\ neg}$		0.065		0.071	0.083	0.086		0.055	0.075
	$C_{B\ neg}$		0.027	0.061	0.029			0.051	0.041	0.017
0.75	$C_{A\ neg}$		0.069		0.076	0.085	0.088		0.061	0.078
	$C_{B\ neg}$		0.022	0.056	0.024			0.044	0.036	0.014
0.70	$C_{A\ neg}$		0.074		0.081	0.086	0.091		0.068	0.081
	$C_{B\ neg}$		0.017	0.050	0.019			0.038	0.029	0.011
0.65	$C_{A\ neg}$		0.077		0.085	0.087	0.093		0.074	0.083
	$C_{B\ neg}$		0.014	0.043	0.015			0.031	0.024	0.008
0.60	$C_{A\ neg}$		0.081		0.089	0.088	0.095		0.080	0.085
	$C_{B\ neg}$		0.010	0.035	0.011			0.024	0.018	0.006
0.55	$C_{A\ neg}$		0.084		0.092	0.089	0.096		0.085	0.086
	$C_{B\ neg}$		0.007	0.028	0.008			0.019	0.014	0.005
0.50	$C_{A\ neg}$		0.086		0.094	0.090	0.097		0.089	0.088
	$C_{B\ neg}$		0.006	0.022	0.006			0.014	0.010	0.003

*A cross-hatched edge indicates that the slab continues across or is fixed at the support; an unmarked edge indicates a support at which torsional resistance is negligible.

Method 3—Table 5.18 Two-way Concrete Slabs—Coefficients for Dead Load Positive Moments*

$$\left.\begin{array}{l} M_{A\,pos\,DL} = C_{A\,DL} \times w \times A^2 \\ M_{B\,pos\,DL} = C_{B\,DL} \times w \times B^2 \end{array}\right\} \text{where } w = \text{total uniform dead load}$$

Ratio $m = \frac{A}{B}$		Case 1	Case 2	Case 3	Case 4	Case 5	Case 6	Case 7	Case 8	Case 9
1.00	$C_{A\,DL}$	0.036	0.018	0.018	0.027	0.027	0.033	0.027	0.020	0.023
	$C_{B\,DL}$	0.036	0.018	0.027	0.027	0.018	0.027	0.033	0.023	0.020
0.95	$C_{A\,DL}$	0.040	0.020	0.021	0.030	0.028	0.036	0.031	0.022	0.024
	$C_{B\,DL}$	0.033	0.016	0.025	0.024	0.015	0.024	0.031	0.021	0.017
0.90	$C_{A\,DL}$	0.045	0.022	0.025	0.033	0.029	0.039	0.035	0.025	0.026
	$C_{B\,DL}$	0.029	0.014	0.024	0.022	0.013	0.021	0.028	0.019	0.015
0.85	$C_{A\,DL}$	0.050	0.024	0.029	0.036	0.031	0.042	0.040	0.029	0.028
	$C_{B\,DL}$	0.026	0.012	0.022	0.019	0.011	0.017	0.025	0.017	0.013
0.80	$C_{A\,DL}$	0.056	0.026	0.034	0.039	0.032	0.045	0.045	0.032	0.029
	$C_{B\,DL}$	0.023	0.011	0.020	0.016	0.009	0.015	0.022	0.015	0.010
0.75	$C_{A\,DL}$	0.061	0.028	0.040	0.043	0.033	0.048	0.051	0.036	0.031
	$C_{B\,DL}$	0.019	0.009	0.018	0.013	0.007	0.012	0.020	0.013	0.007
0.70	$C_{A\,DL}$	0.068	0.030	0.046	0.046	0.035	0.051	0.058	0.040	0.033
	$C_{B\,DL}$	0.016	0.007	0.016	0.011	0.005	0.009	0.017	0.011	0.006
0.65	$C_{A\,DL}$	0.074	0.032	0.054	0.050	0.036	0.054	0.065	0.044	0.034
	$C_{B\,DL}$	0.013	0.006	0.014	0.009	0.004	0.007	0.014	0.009	0.005
0.60	$C_{A\,DL}$	0.081	0.034	0.062	0.053	0.037	0.056	0.073	0.048	0.036
	$C_{B\,DL}$	0.010	0.004	0.011	0.007	0.003	0.006	0.012	0.007	0.004
0.55	$C_{A\,DL}$	0.088	0.035	0.071	0.056	0.038	0.058	0.081	0.052	0.037
	$C_{B\,DL}$	0.008	0.003	0.009	0.005	0.002	0.004	0.009	0.005	0.003
0.50	$C_{A\,DL}$	0.095	0.037	0.080	0.059	0.039	0.061	0.089	0.056	0.038
	$C_{B\,DL}$	0.006	0.002	0.007	0.004	0.001	0.003	0.007	0.004	0.002

*A cross-hatched edge indicates that the slab continues across or is fixed at the support; an unmarked edge indicates a support at which torsional resistance is negligible.

Method 3—Table 5.19 Two-way Concrete Slabs—Coefficients for Live Load Positive Moments*

$$\left.\begin{array}{l} M_{A\ pos\ LL} = C_{A\ LL} \times w \times A^2 \\ M_{B\ pos\ LL} = C_{B\ LL} \times w \times B^2 \end{array}\right\} \text{where } w = \text{total uniform live load}$$

Ratio $m = \frac{A}{B}$		Case 1 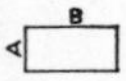	Case 2 	Case 3 	Case 4 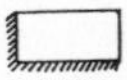	Case 5 	Case 6 	Case 7 	Case 8 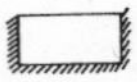	Case 9 
1.00	$C_{A\ LL}$	0.036	0.027	0.027	0.032	0.032	0.035	0.032	0.028	0.030
	$C_{B\ LL}$	0.036	0.027	0.032	0.032	0.027	0.032	0.035	0.030	0.028
0.95	$C_{A\ LL}$	0.040	0.030	0.031	0.035	0.034	0.038	0.036	0.031	0.032
	$C_{B\ LL}$	0.033	0.025	0.029	0.029	0.024	0.029	0.032	0.027	0.025
0.90	$C_{A\ LL}$	0.045	0.034	0.035	0.039	0.037	0.042	0.040	0.035	0.036
	$C_{B\ LL}$	0.029	0.022	0.027	0.026	0.021	0.025	0.029	0.024	0.022
0.85	$C_{A\ LL}$	0.050	0.037	0.040	0.043	0.041	0.046	0.045	0.040	0.039
	$C_{B\ LL}$	0.026	0.019	0.024	0.023	0.019	0.022	0.026	0.022	0.020
0.80	$C_{A\ LL}$	0.056	0.041	0.045	0.048	0.044	0.051	0.051	0.044	0.042
	$C_{B\ LL}$	0.023	0.017	0.022	0.020	0.016	0.019	0.023	0.019	0.017
0.75	$C_{A\ LL}$	0.061	0.045	0.051	0.052	0.047	0.055	0.056	0.049	0.046
	$C_{B\ LL}$	0.019	0.014	0.019	0.016	0.013	0.016	0.020	0.016	0.013
0.70	$C_{A\ LL}$	0.068	0.049	0.057	0.057	0.051	0.060	0.063	0.054	0.050
	$C_{B\ LL}$	0.016	0.012	0.016	0.014	0.011	0.013	0.017	0.014	0.011
0.65	$C_{A\ LL}$	0.074	0.053	0.064	0.062	0.055	0.064	0.070	0.059	0.054
	$C_{B\ LL}$	0.013	0.010	0.014	0.011	0.009	0.010	0.014	0.011	0.009
0.60	$C_{A\ LL}$	0.081	0.058	0.071	0.067	0.059	0.068	0.077	0.065	0.059
	$C_{B\ LL}$	0.010	0.007	0.011	0.009	0.007	0.008	0.011	0.009	0.007
0.55	$C_{A\ LL}$	0.088	0.062	0.080	0.072	0.063	0.073	0.085	0.070	0.063
	$C_{B\ LL}$	0.008	0.006	0.009	0.007	0.005	0.006	0.009	0.007	0.006
0.50	$C_{A\ LL}$	0.095	0.066	0.088	0.077	0.067	0.078	0.092	0.076	0.067
	$C_{B\ LL}$	0.006	0.004	0.007	0.005	0.004	0.005	0.007	0.005	0.004

*A cross-hatched edge indicates that the slab continues across or is fixed at the support; an unmarked edge indicates a support at which torsional resistance is negligible.

Method 3—Table 5.20 Ratio of Load w in A and B Directions for Shear in Slab and Load on Supports*

Ratio $m = \frac{A}{B}$		Case 1	Case 2	Case 3	Case 4	Case 5	Case 6	Case 7	Case 8	Case 9
1.00	W_A	0.50	0.50	0.17	0.50	0.83	0.71	0.29	0.33	0.67
	W_B	0.50	0.50	0.83	0.50	0.17	0.29	0.71	0.67	0.33
0.95	W_A	0.55	0.55	0.20	0.55	0.86	0.75	0.33	0.38	0.71
	W_B	0.45	0.45	0.80	0.45	0.14	0.25	0.67	0.62	0.29
0.90	W_A	0.60	0.60	0.23	0.60	0.88	0.79	0.38	0.43	0.75
	W_B	0.40	0.40	0.77	0.40	0.12	0.21	0.62	0.57	0.25
0.85	W_A	0.66	0.66	0.28	0.66	0.90	0.83	0.43	0.49	0.79
	W_B	0.34	0.34	0.72	0.34	0.10	0.17	0.57	0.51	0.21
0.80	W_A	0.71	0.71	0.33	0.71	0.92	0.86	0.49	0.55	0.83
	W_B	0.29	0.29	0.67	0.29	0.08	0.14	0.51	0.45	0.17
0.75	W_A	0.76	0.76	0.39	0.76	0.94	0.88	0.56	0.61	0.86
	W_B	0.24	0.24	0.61	0.24	0.06	0.12	0.44	0.39	0.14
0.70	W_A	0.81	0.81	0.45	0.81	0.95	0.91	0.62	0.68	0.89
	W_B	0.19	0.19	0.55	0.19	0.05	0.09	0.38	0.32	0.11
0.65	W_A	0.85	0.85	0.53	0.85	0.96	0.93	0.69	0.74	0.92
	W_B	0.15	0.15	0.47	0.15	0.04	0.07	0.31	0.26	0.08
0.60	W_A	0.89	0.89	0.61	0.89	0.97	0.95	0.76	0.80	0.94
	W_B	0.11	0.11	0.39	0.11	0.03	0.05	0.24	0.20	0.06
0.55	W_A	0.92	0.92	0.69	0.92	0.98	0.96	0.81	0.85	0.95
	W_B	0.08	0.08	0.31	0.08	0.02	0.04	0.19	0.15	0.05
0.50	W_A	0.94	0.94	0.76	0.94	0.99	0.97	0.86	0.89	0.97
	W_B	0.06	0.06	0.24	0.06	0.01	0.03	0.14	0.11	0.03

*A cross-hatched edge indicates that the slab continues across or is fixed at the support; an unmarked edge indicates a support at which torsional resistance is negligible.

5.51 Flat slab

a A flat slab is a special type of two-way concrete slab which is an economical type of construction, particularly for heavy loads and multistory construction—for example, a warehouse type of building. Figure 5.32 shows a typical cross section of such a slab indicating figures which have proven to be good economic proportions. The arrangement of the bands for design purposes is indicated in Fig. 5.33.

b The method of design of flat slabs is specified in detail in the 1963 ACI Code as follows.

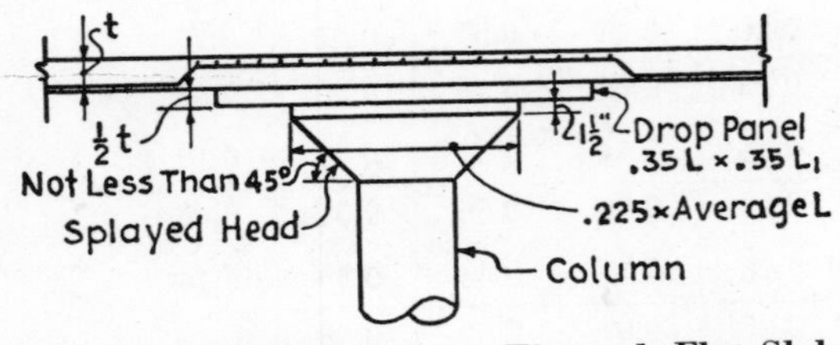

Fig. 5.32 Typical Section Through Flat Slab

FLAT SLABS WITH SQUARE OR RECTANGULAR PANELS

2100 NOTATION

A = distance in the direction of span from center of support to the intersection of the center line of the slab thickness with the extreme 45-deg diagonal line lying wholly within the concrete section of slab and column or other support, including drop panel, capital and bracket

b_o = periphery of critical section for shear

c = effective support size [see Section 2104(c)]

d = distance from extreme compression fiber to centroid of tension requirement

f_c' = compressive strength of concrete (see Section 301)

h = distance from top of slab to bottom of capital

H = story height in feet of the column or support of a flat slab center to center of slabs

K = ratio of moment of inertia of column

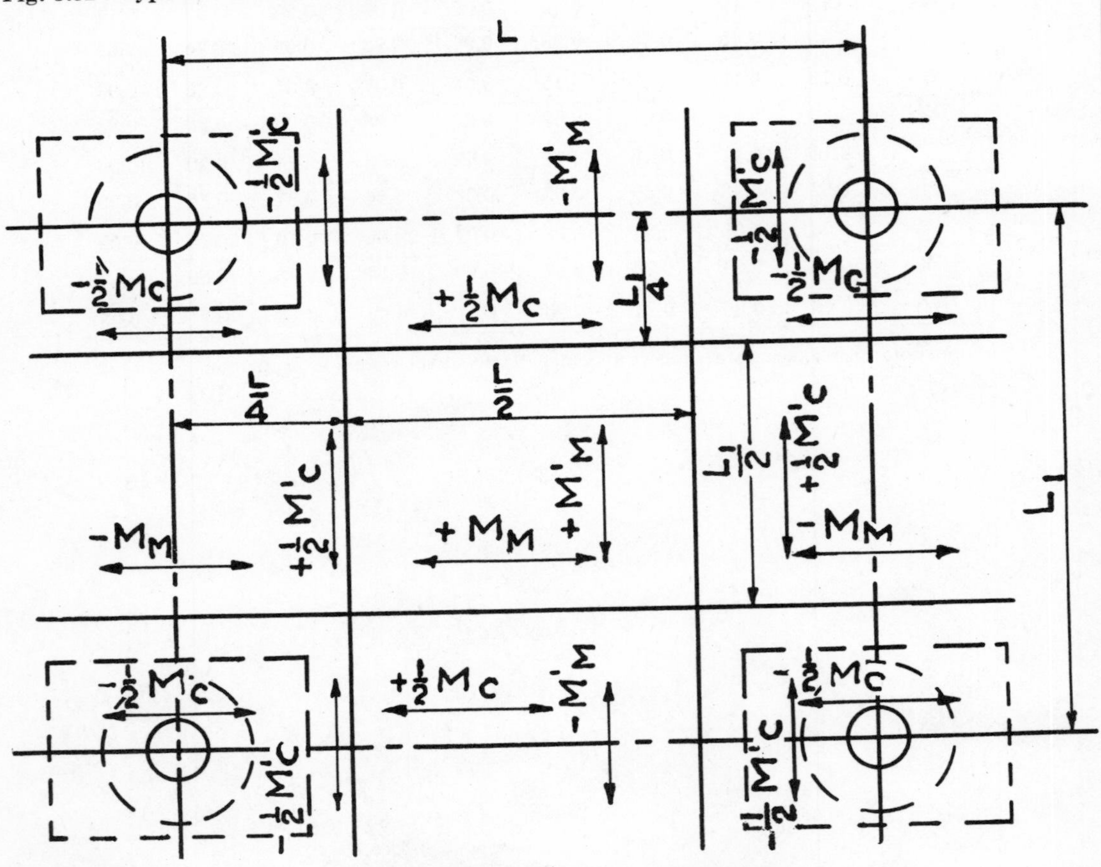

Fig. 5.33 Typical Plan of Flat Slab Bay

provided to I_c required by Eq. (21-1)

L = span length of a flat slab panel center to center of supports

M_o = numerical sum of assumed positive and average negative moments at the critical design sections of a flat slab panel [see Section 2104(f)1]

R_n = factor for increasing negative moment [Section 2104, Eq. (21-2)]

R_p = factor for increasing positive moment [Section 2104, Eq. (21-3)]

t = thickness in inches of slab at center of panel

t_1 = thickness in inches of slab without drop panels, or through drop panel, if any

t_2 = thickness in inches of slab with drop panels at points beyond the drop panel

w' = uniformly distributed unit dead and live load

W = total dead and live load on panel

W_D = total dead load on panel

W_L = total live load on panel, uniformly distributed

2101 Definitions and Scope

(a) *Flat slab*—A concrete slab reinforced in two or more directions, generally without beams or girders to transfer the loads to supporting members. Slabs with recesses or pockets made by permanent or removable fillers between reinforcing bars may be considered flat slabs. Slabs with paneled ceilings may be considered as flat slabs provided the panel of reduced thickness lies entirely within the area of intersecting middle strips, and is at least two-thirds the thickness of the remainder of the slab, exclusive of the drop panel, and is not less than 4 in. thick.

(b) *Column capital*—An enlargement of the end of a column designed and built to act as an integral unit with the column and flat slab. No portion of the column capital shall be considered for structural purposes which lies outside of the largest right circular cone with 90-deg vertex angle that can be included within the outlines of the column capital. Where no capital is used, the face of the column shall be considered as the edge of the capital.

(c) *Drop panel*—The structural portion of a flat slab which is thickened throughout an area surrounding the column, column capital, or bracket.

(d) *Panel strips*—A flat slab shall be considered as consisting of strips in each direction as follows:

A middle strip one-half panel in width, symmetrical about panel center line.

A column strip consisting of the two adjacent quarter-panels, one each side of the column center line.

(e) *Ultimate strength design*—Flat slabs shall be proportioned by Part IV-A only, except that Part IV-B may be used if the following modifications are made in the design:

1. For either empirical or elastic analysis the numerical sum of the positive and negative bending moments in the direction of either side of a rectangular panel shall be assumed as not less than

$$M_o = 0.10\, WLF \left(1 - \frac{2c}{3L}\right)^2$$

in which $F = 1.15 - c/L$ but not less than 1.

2. The thickness of slab shall not be less than shown in Table 5.21.

Table 5.21 Minimum Slab Thickness

f_y	With drop panels*	Without drop panels
40,000	$L/40$ or 4 in.	$L/36$ or 5 in.
50,000	$L/36$ or 4 in.	$L/33$ or 5 in.
60,000	$L/33$ or 4 in.	$L/30$ or 5 in.

* To be considered effective, the drop panel shall have a length of at least one-third the parallel span length and a projection below the slab of at least one-fourth the slab thickness.

2102 Design Procedures

(a) *Methods of analysis*—All flat slab structures shall be designed in accordance with a recognized elastic analysis subject to the limitations of Sections 2102 and 2103, except that the empirical method of design given in Section 2104 may be used for the design of flat slabs conforming with the limitations given therein. Flat slabs within the limitations of Section 2104, when designed by elastic analysis, may have resulting analytical moments reduced in such proportion that the numerical sum of the positive and average negative bending moments used in design procedure need not exceed the sum of the corresponding values as determined from Table 5.23 in Section 2104.

(b) *Critical sections*—The slab shall be proportioned for the bending moments prevailing at every section except that the slab need not be proportioned for a greater negative moment than that prevailing at a distance A from the support center line.

(c) *Size and thickness of slabs and drop panels*

1. Subject to limitations of Section 2102(c)4, the thickness of a flat slab and the size and thickness of the drop panel, where used, shall be such that the compression due to bending at any section, and the shear about the column, column capital, and drop panel shall not exceed those permitted in Part IV-A or Part IV-B. When designed under Section 2104, three-fourths of the width of the strip shall be used as the width of the section in computing compression due to bending, except that on a

section through a drop panel, three-fourths of the width of the drop panel shall be used. Account shall be taken of any recesses which reduce the compressive area.

2. The shear on vertical sections which follow a periphery, b_o, at distance, $d/2$, beyond the edges of the column, column capital, or drop panel, and concentric with them, shall be computed as required and limited in Chapters 12 or 17.

3. If shear reinforcement is used, the first line shall be not further than $d/2$ from the face of the support.

4. Slabs with drop panels whose length is at least one-third the parallel span length and whose projection below the slab is at least one-fourth the slab thickness shall be not less than $L/40$ nor 4 in. in thickness.

Slabs without drop panels as described above shall be not less than $L/36$ nor 5 in. in thickness.

5. For determining reinforcement, the thickness of the drop panel below the slab shall not be assumed to be more than one-fourth of the distance from the edge of the drop panel to the edge of the column capital.

(d) *Arrangement of slab reinforcement*

1. The spacing of the bars at critical sections shall not exceed two times the slab thickness, except for those portions of the slab area which may be of cellular or ribbed construction. In the slab over the cellular spaces, reinforcement shall be provided as required by Section 807.

2. In exterior panels, except for bottom bars adequately anchored in the drop panel, all positive reinforcement perpendicular to the discontinuous edge shall extend to the edge of the slab and have embedment, straight or hooked, of at least 6 in. in spandrel beams, walls, or columns where provided. All negative reinforcement perpendicular to the discontinuous edge shall be bent, hooked, or otherwise anchored in spandrel beams, walls, or columns.

3. The area of reinforcement shall be determined from the bending moments at the critical sections but shall be not less than required by Section 807.

4. Required splices in bars may be made wherever convenient, but preferably away from points of maximum stress. The length of any such splice shall conform to Section 805.

5. Bars shall be spaced approximately uniformly across each panel strip, except:

a. At least 25 percent of required negative reinforcement in the column strip shall cross the periphery located at a distance of d from the column or column capital.

b. At least 50 percent of the required negative reinforcement in the column strip shall cross the drop panel, if any.

c. The spacing for the remainder of the column strip may vary uniformly from that required for a or b to that required for the middle strip.

(e) *Openings in flat slabs*

1. Openings of any size may be provided in flat slabs if provision is made for the total positive and negative moments and for shear without exceeding the allowable stresses except that when design is based on Section 2104, the limitations given therein shall not be exceeded.

2. When openings are provided within the area common to two column strips, that part of the critical section shall be considered ineffective which either passes through an opening, or is covered by a radial projection of any opening to the centroid of the support.

(f) *Design of columns*

1. All columns supporting flat slabs shall be designed as provided in Chapter 14 or 19 with the additional requirements of this chapter.

(g) *Transfer of bending moment between column and slab*—When unbalanced gravity load, wind or earthquake causes transfer of bending moment between column and slab, the stresses on the critical section shall be investigated by a rational analysis, and the section proportioned accordingly by the requirements of Part IV-A or IV-B. Concentration of reinforcement over the column head by additional reinforcement or closer spacing may be used to resist the moment of the section. A slab width between lines that are $1.5t$ each side of the column may be considered effective.

2103—Design by Elastic analysis

(a) *Assumptions*—In design by elastic analysis the following assumptions may be used and all sections shall be proportioned for the moments and shears thus obtained.

1. The structure may be considered divided into a number of bents, each consisting of a row of columns or supports and strips of supported slabs, each strip bounded laterally by the center line of the panel on either side of the center line of columns or supports. The bents shall be taken longitudinally and transversely of the building.

2. Each such bent may be analyzed in its entirety or each floor thereof and the roof may be analyzed separately with its adjacent columns as they occur above and below, the columns being assumed fixed at their remote ends. Where slabs are thus analyzed separately, it may be assumed in determining the bending at a given support that the slab is fixed at any support two panels distant therefrom provided the slab continues beyond that point.

3. The joints between columns and slabs may be considered rigid, and this rigidity (infinite moment of inertia) may be assumed to extend in the slabs from the center of the column to the edge of the capital, and in the column from the top of slab to the bottom of the capital. The change in length of columns and slabs due to direct stress, and deflections due to shear, may be neglected.

4. Where metal column capitals are used, account may be taken of their contributions to stiffness and resistance to bending and shear.

5. The moment of inertia of the slab or column at any cross section may be assumed to be that of the cross section of the concrete. Variation in the moments of inertia of the slabs and columns along their axes shall be taken into account.

6. Where the load to be supported is definitely known, the structure shall be analyzed for that load. Where the live load is variable but does not exceed three-quarters of the dead load, or the nature of the live load is such that all panels will be loaded simultaneously, the maximum bending may be assumed to occur at all sections under full live load. For other conditions, maximum positive bending near midspan of a panel may be assumed to occur under three-quarters of the full live load in the panel and in alternate panels; and maximum negative bending in the slab at a support may be assumed to occur under three-quarters of the full live load in the adjacent panels only. In no case, shall the design moments be taken as less than those occurring with full live load on all panels.

(b) *Critical sections*—The critical section for negative bending, in both the column strip and middle strip, may be assumed as not more than the distance A from the center of the column or support and the critical negative moment shall be considered as extending over this distance.

(c) *Distribution of panel moments*—Bending at critical sections across the slabs of each bent may be apportioned between the column strip and middle strip, as given in Table 5.22. For design purposes, any of these percentages may be varied by not more than 10 percent of its value, but their sum for the full panel width shall not be reduced.

2104 Empirical Method

(a) *General limitations*—Flat slab construction may be designed by the empirical provisions of this section when they conform to all of the limitations on continuity and dimensions given herein.

1. The construction shall consist of at least three continuous panels in each direction.

2. The ratio of length to width of panels shall not exceed 1.33.

3. The grid pattern shall consist of approximately rectangular panels. The successive span lengths in each direction shall differ by not more than 20 percent of the longer span. Within these limitation, columns may be offset a maximum of 10 percent of the span, in direction of the offset, from either axis between center lines of successive columns.

4. The calculated lateral force moments from wind or earthquake may be combined with the critical moments as determined by the empirical method, and the lateral force moments shall be distributed between the column and middle strips in the same proportions as specified for the negative moments in the strips for structures not exceeding 125 ft high with maximum story height not exceeding 12 ft 6 in.

(b) *Columns*

1. The minimum dimension of any column shall be as determined by a and b below, but in no case less than 10 in.

a. For columns or other supports of a flat slab, the required minimum average moment of inertia, I_c, of the gross concrete section of the columns above and below the slab shall be determined from Eq. (21-1) and shall be not less than 1000 in.^{4} If there is no column above the slab, the I_c of the column below shall be $(2 - 2.3h/H)$ times that given by the formula with a minimum of 1000 in.^{4}

$$I_c = \frac{t^3H}{0.5 + \dfrac{W_D}{W_L}} \qquad (21\text{-}1)$$

where t need not be taken greater than t_1 or t_2 as determined in (d), H is the average story height of the columns above and below the slab, and W_L is the greater value of any two adjacent spans under consideration.

b. Columns smaller than required by Eq. (21-1) may be used provided the bending moment coefficients given in Table 5.23 are increased in the following ratios:

For negative moments

$$R_n = 1 + \frac{(1 - K)^2}{2.2(1 + 1.4W_D/W_L)} \qquad (21\text{-}2)$$

For positive moments

$$R_p = 1 + \frac{(1 - K)^2}{1.2(1 + 0.10W_D/W_L)} \qquad (21\text{-}3)$$

The required slab thickness shall be modified by multiplying w' by R_n in Eq. (21-4) and (21-5).

2. Columns supporting flat slabs designed by the empirical method shall be proportioned for the bending moments developed by unequally loaded panels, or uneven spacing of columns. Such bending moment shall be the

PANEL		INTERIOR		EXTERIOR						
MOMENT		SUPPORT	CENTER OF SPAN	1ST. INTERIOR SUPPORT	CENTER OF SPAN	EXTERIOR SUPPORT		1ST. INTERIOR SUPPORT	CENTER OF SPAN	EXTERIOR SUPPORT
END SUPPORT						B	A			C
MARGINAL HALF COLUMN STRIP — SIDE SUPPORT	3	-12	+6	-13	+7	-8	-10	-17	+10	-3
	2	-18	+9	-19	+11	-12	-15	-25	+15	-3
	1	-23	+11	-25	+14	-16	-20	-33	+20	-3
MIDDLE STRIP		-16*	+16	-18*	+20	-20	-10	-24*	+28	-6
COLUMN STRIP		-46	+22	-50	+28	-32	-40	-66	+40	-6

← DIRECTION OF ALL MOMENTS →

* Increase negative moments 30 percent when middle strip is continuous across a support of Type B or C; no other values need be increased.

Fig. 5.34 Moments in Flat Slab Panels in Percentage of M_0—Without Drops

(See Table 5.23 for notes and classification of conditions of end supports and side supports.)

PANEL		INTERIOR		EXTERIOR						
MOMENT		SUPPORT	CENTER OF SPAN	I ST. INTERIOR SUPPORT	CENTER OF SPAN	EXTERIOR SUPPORT		IST. INTERIOR SUPPORT	CENTER OF SPAN	EXTERIOR SUPPORT
END SUPPORT						B	A			C
MARGINAL HALF COLUMN STRIP — SIDE SUPPORT	3	-13	+5	-14	+6	-9	-11	-18	+9	-3
	2	-19	+8	-21	+9	-14	-17	-27	+14	-3
	1	-25	+10	-28	+12	-18	-22	-36	+18	-3
MIDDLE STRIP		-15*	+15	-17*	+20	-20	-10	-22*	+26	-6
COLUMN STRIP		-50	+20	-56	+24	-36	-44	-72	+36	-6

DIRECTION OF ALL MOMENTS

* Increase negative moments 30 percent when middle strip is continuous across a support of Type B or C; no other values need be increased.

Fig. 5.35 Moments in Flat Slab Panels in Percentage of M_0—With Drops
(See Table 5.23 for notes and classification of conditions of end supports and side supports.)

maximum value derived from

$$\frac{WL_1 - W_DL_2}{f}$$

L_1 and L_2 being lengths of the adjacent spans ($L_2 = 0$ when considering an exterior column) and f is 30 for exterior and 40 for interior columns.

This moment shall be divided between the columns immediately above and below the floor or roof line under consideration in direct proportion to their stiffness and shall be applied without further reduction to the critical sections of the columns.

(c) *Determination of "c" (effective support size)*

1. Where column capitals are used, the value of c shall be taken as the diameter of the cone described in Section 2101(b) measured at the bottom of the slab or drop panel.

2. Where a column is without a concrete capital, the dimension c shall be taken as that of the column in the direction considered.

3. Brackets capable of transmitting the negative bending and the shear in the column strips to the columns without excessive unit stress may be substituted for column capitals at exterior columns. The value of c for the span where a bracket is used shall be taken as twice the distance from the center of the column to a point where the bracket is 1½ in. thick, but not more than the thickness of the column plus twice the depth of the bracket.

4. Where a reinforced concrete beam frames into a column without capital or bracket on the same side with the beam, for computing bending for strips parallel to the beam, the value of c for the span considered may be taken as the width of the column plus twice the projection of the beam above or below the slab or drop panel.

5. The average of the values of c at the two supports at the ends of a column strip shall be used to evaluate the slab thickness t_1 or t_2 as prescribed in (d).

(d) *Slab thickness*

1. The slab thickness, span L being the longest side of the panel, shall be at least:

$L/36$ for slab without drop panels conforming with (e), or where a drop panel is omitted at any corner of the panel, but not less than 5 in. nor t_1 as given in Eq. (21-4).

$L/40$ for slabs with drop panels conforming to (e) at all supports, but not less than 4 in. nor t_2 as given in Eq. (21-5).

2. The total thickness, t_1, in inches, of slabs without drop panels, or through the drop panel if any, shall be at least

$$t_1 = 0.028L\left(1 - \frac{2c}{3L}\right)\sqrt{\frac{w'}{f_c'/2000}} + 1\tfrac{1}{2} \qquad \text{(21-4)*}$$

3. The total thickness, t_2, in inches, of slabs with drop panels, at points beyond the drop panel shall be at least

$$t_2 = 0.024L\left(1 - \frac{2c}{3L}\right)\sqrt{\frac{w'}{f_c'/2000}} + 1 \qquad \text{(21-5)*}$$

4. Where the exterior supports provide only negligible restraint to the slab, the values of t_1 and t_2 for the exterior panel shall be increased by at least 15 percent.

* In these formulas t_1 and t_2 are in inches, L and c are in feet, and w' is in pounds per square foot.

(e) *Drop panels*

1. The maximum total thickness at the drop panel used in computing the negative steel area for the column strip shall be $1.5t_2$.

2. The side or diameter of the drop panel shall be at least 0.33 times the span in the parallel direction.

3. The minimum thickness of slabs where drop panels at wall columns are omitted shall equal $(t_1 + t_2)/2$ provided the value of c used in the computations complies with (c).

(f) *Bending moment coefficients*

1. The numerical sum of the positive and negative bending moments in the direction of either side of a rectangular panel shall be assumed as not less than

$$M_o = 0.09\,WLF\left(1 - \frac{2c}{3L}\right)^2 \qquad \text{(21-6)}$$

in which $F = 1.15 - c/L$ but not less than 1.

2. Unless otherwise provided, the bending moments at the critical sections of the column and middle strips shall be at least those given in Table 5.23.

3. The average of the values of c at the two supports at the ends of a column strip shall be used to evaluate M_o in determining bending in the strip. The average of the values of M_o, as determined for the two parallel half column strips in a panel, shall be used in determining bending in the middle strip.

4. Bending in the middle strips parallel to a discontinuous edge shall be assumed the same as in an interior panel.

5. For design purposes, any of the moments determined from Table 5.23 may be varied by not more than 10 percent, but the numerical sum of the positive and negative moments in a panel shall be not less than the amount specified.

(g) *Length of reinforcement*—In addition to the requirements of Section 2102(d), reinforcement shall have the minimum lengths given in Tables 5.24 and 5.25. Where adjacent spans are unequal, the extension of negative reinforcement

Table 5.22 Distribution Between Column Strips and Middle Strips in Percent of Total Moments at Critical Sections of a Panel

Strip		Moment section: Negative moment at interior support	Moment section: Positive moment	Negative moment at exterior support: Slab supported on columns and on beams of total depth equal to the slab thickness*	Negative moment at exterior support: Slab supported on reinforced concrete bearing wall or columns with beams of total depth equal or greater than 3 times the slab thickness*
Column strip		76	60	80	60
Middle strip		24	40	20	40
Half column strip adjacent and parallel to marginal beam or wall	Total depth of beam equal to slab thickness*	38	30	40	30
	Total depth of beam or wall equal to or greater than 3 times slab thickness*	19	15	20	15

*Interpolate for intermediate ratios of beam depth to slab thickness.

Note: The total dead and live reaction of a panel adjacent to a marginal beam or wall may be divided between the beam or wall and the parallel half column strip in proportion to their stiffness, but the moment provided in the slab shall not be less than that given in Table 5.22.

on each side of the column center line as prescribed in Table 5.24 shall be based on the requirements of the longer span.

(h) *Openings in flat slabs*

1. Openings of any size may be provided in a flat slab in the area common to two intersecting middle strips provided the total positive and negative steel areas required in (f) are maintained.

2. In the area common to two column strips, not more than one-eighth of the width of strip in any span shall be interrupted by openings. The equivalent of all bars interrupted shall be provided by extra steel on all sides of the openings. The shear stresses given in Section 2102(c)2 shall not be exceeded following the procedure of Section 920(b).

3. In any area common to one column strip and one middle strip, openings may interrupt one-quarter of the bars in either strip. The equivalent of the bars so interrupted shall be provided by extra steel on all sides of the opening.

4. Any opening larger than described above shall be analyzed by accepted engineering principles and shall be completely framed as required to carry the loads to the columns.

c Under certain conditions there is an application of the flat-slab system to short spans which may be found particularly useful and economical. This application (which actually is a variation of the flat-slab requirements of the ACI Code) quoted from a local building code:

In flat slabs without drop panels where t is

Table 5.23 Moments in Flat Slab Panels in Percentages of M_o

Strip	Column head	Side support type	End support type	Exterior panel: Exterior negative moment	Exterior panel: Positive moment	Exterior panel: Interior negative moment	Interior panel: Positive moment	Interior panel: Negative moment
Column strip	With drop		A	44				
			B	36	24	56	20	50
			C	6	36	72		
	Without drop		A	40				
			B	32	28	50	22	46
			C	6	40	66		
Middle strip	With drop		A	10				
			B	20	20	17*	15	15*
			C	6	26	22*		
	Without drop		A	10				
			B	20	20	18*	16	16*
			C	6	28	24*		
Half column strip adjacent to marginal beam or wall	With drop	1	A	22				
			B	18	12	28	10	25
			C	3	18	36		
		2	A	17				
			B	14	9	21	8	19
			C	3	14	27		
		3	A	11				
			B	9	6	14	5	13
			C	3	9	18		
	Without drop	1	A	20				
			B	16	14	25	11	23
			C	3	20	33		
		2	A	15				
			B	12	11	19	9	18
			C	3	15	25		
		3	A	10				
			B	8	7	13	6	12
			C	3	10	17		

Percentage of panel load to be carried by marginal beam or wall in addition to loads directly superimposed thereon	Type of support listed in Table 5.23: Side support parallel to strip	Side or end edge condition of slabs of depth t	End support at right angles to strip
0	1	Columns with no beams	
20	2	Columns with beams of total depth $1\frac{1}{4}t$	A
40	3	Columns with beams of total depth $3t$ or more	B
		Reinforced concrete bearing walls integral with slab	B
		Masonry or other walls providing negligible restraint	C

*Increase negative moments 30 percent of tabulated values when middle strip is continuous across support of Type B or C. No other values need be increased.

Note: For intermediate proportions of total beam depth to slab thicknesses, values for loads and moments may be obtained by interpolation. See also Fig. 5.34 and 5.35.

greater than $.05L$, a continuous mat of reinforcement sufficient for the total positive moments of the panel may be used in the bottom of the slab and a mat of reinforcement sufficient for the total negative moments of the panel may be used over each supporting column in the top of the slab. The reinforcement in the mats for positive moment shall be spliced only on column center

Table 5.24 Flat Slab Minimum Length of Negative Reinforcement

Strip	Percentage of required reinforcing steel area to be extended at least as indicated	Minimum distance beyond center line of support to end of straight bar or to bend point of bent bar*			
		Flat slabs without drop panels		Flat slabs with drop panels	
		Straight	Bend point where bars bend down and continue as positive reinforcement	Straight	Bend point where bars bend down and continue as positive reinforcement
Column strip reinforcement	Not less than 33 percent	0.30L†		0.33L‡	
	Not less than an additional 34 percent	0.27L†		0.30L‡	
	Remainder†	0.25L or	0.20L	0.25L or	To edge of drop but at least 0.20L
Middle strip reinforcement	Not less than 50 percent	0.25L		0.25L	
	Remainder§	0.25L or	0.15L	0.25L or	0.15L

*At exterior supports where masonry walls or other construction provide only negligible restraint to the slab, the negative reinforcement need not be carried further than 0.20L beyond the center line of such support.

†Where no bent bars are used, the 0.27L bars may be omitted, provided the 0.30L bars are at least 50 percent of total required.

‡Where no bent bars are used, the 0.30L bars may be omitted provided the 0.33L bars provide at least 50 percent of the total required.

§Bars may be straight, bent, or any combination of straight and bent bars. All bars are to be considered straight bars for the end under consideration unless bent at that end and continued as positive reinforcement.

Note: See also Fig. 5.36.

lines, and shall lap not less than 40 diameters of the bars. The reinforcement in the mats for negative moments shall extend not less than .3L in each direction from the column center lines.

Obviously this use of the flat slab is intended for short spans. It has been used over the crawl space under the first floor of school buildings, where column spacing of 10 to 12 ft in each direction does not interfere with the use of this space for pipes or plenum chambers.

5.52 Reinforced concrete slabs on the ground

a Reinforcement in slabs on compacted or undisturbed earth of uniform bearing value does not contribute enough to load-carrying capacity to influence the design. However, it is good practice to put reinforcement in the slab for the purpose of preventing unsightly shrinkage cracks, and the area of reinforcement per foot width in each direction should be $A_s = wL/25{,}000$.

b Under certain unusual conditions, reinforced concrete slabs on the ground may be computed as flat slabs or two-way slabs, in accordance with Art. 5.50 or 5.51.

The first of these conditions is a slab supported on filled ground, particularly a floor designed to carry heavy live loads, such as heavy-duty machine-shop floors or heavy warehouses.

A second condition where a slab of this type must be used is to resist the uplift of water pressure. In the design for this condition, the dead weight of the concrete slab

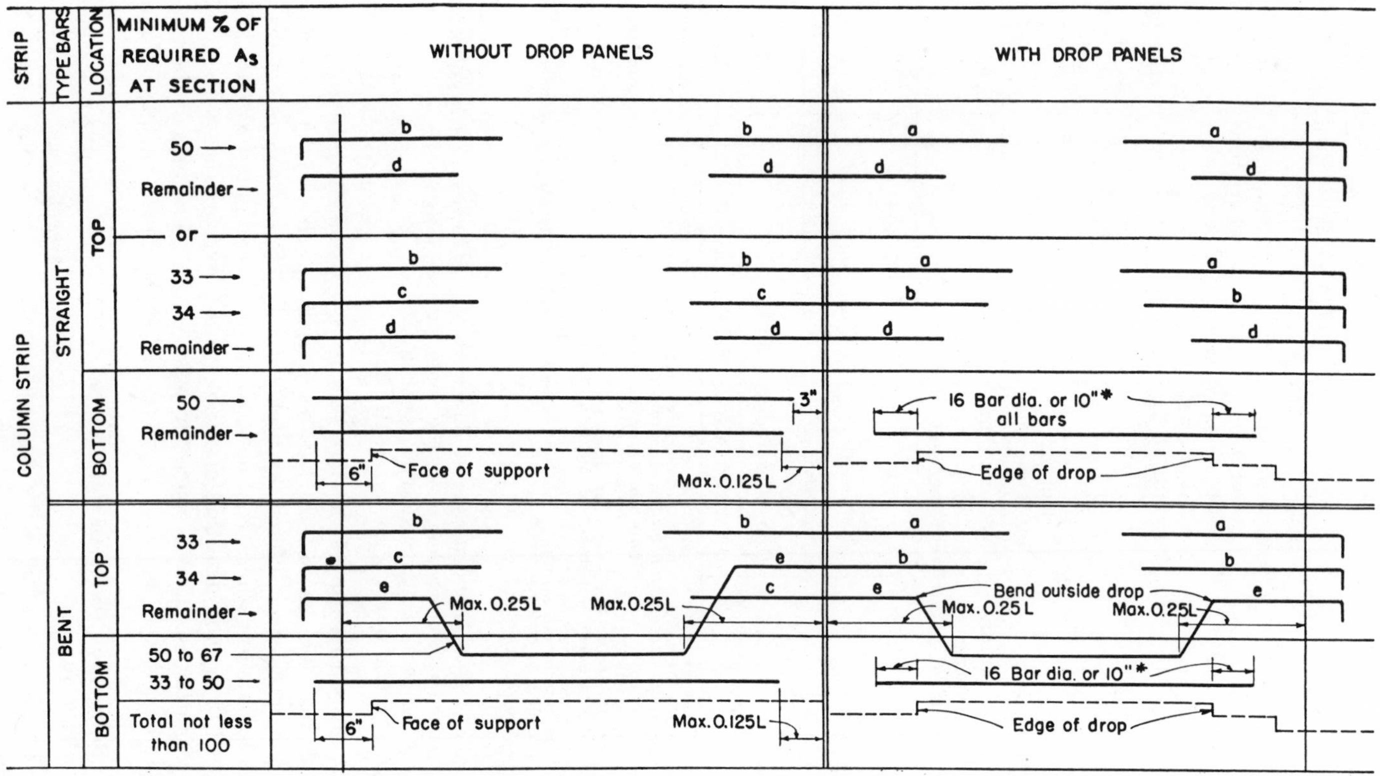

STRIP	TYPE BARS	LOCATION	MINIMUM % OF REQUIRED A_s AT SECTION	WITHOUT DROP PANELS	WITH DROP PANELS
COLUMN STRIP	STRAIGHT	TOP	50 → Remainder → or 33 → 34 → Remainder →	b, d, b, d b, c, d, b, c, d	a, d, a, d a, b, d, a, b, d
		BOTTOM	50 → Remainder →	3" 6" Face of support Max. 0.125L	16 Bar dia. or 10"* all bars Edge of drop
	BENT	TOP	33 → 34 → Remainder →	b, c, e, b, e, c Max. 0.25L Max. 0.25L	a, b, e, a, b, e Bend outside drop Max. 0.25L Max. 0.25L
		BOTTOM	50 to 67 → 33 to 50 → Total not less than 100	6" Face of support Max. 0.125L	16 Bar dia. or 10"* Edge of drop

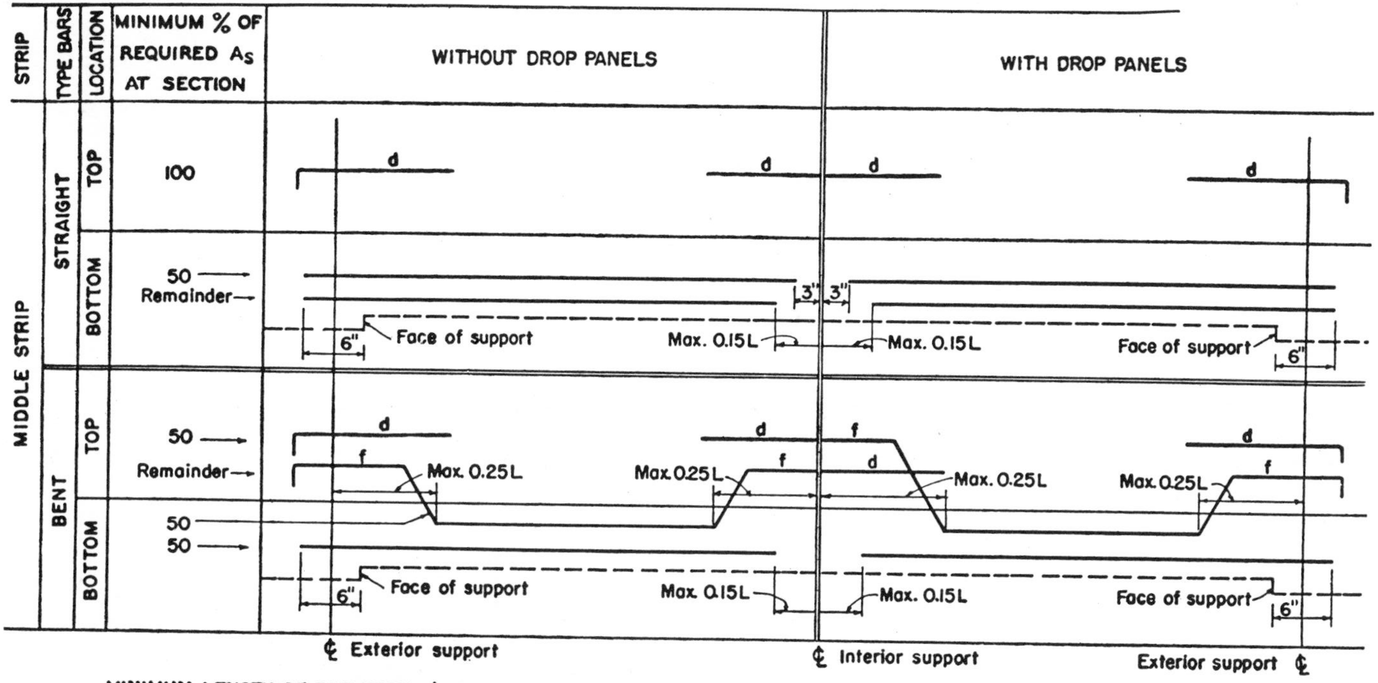

STRIP	TYPE BARS	LOCATION	MINIMUM % OF REQUIRED A_S AT SECTION	WITHOUT DROP PANELS	WITH DROP PANELS
MIDDLE STRIP	STRAIGHT	TOP	100	d; d	d; d
		BOTTOM	50 → Remainder →	6"; Face of support; Max. 0.15L; 3"	3"; Max. 0.15L; Face of support; 6"
	BENT	TOP	50 → Remainder →	d; f; Max. 0.25L; Max. 0.25L; d; f	f; d; Max. 0.25L; Max. 0.25L; d; f
		BOTTOM	50 50 →	6"; Face of support; Max. 0.15L	Max. 0.15L; Face of support; 6"

℄ Exterior support ℄ Interior support Exterior support ℄

MINIMUM LENGTH OF BAR FROM ℄ SUPPORT

MARK	a	b	c	d	e	f
LENGTH	0.33L	0.30L	0.27L	0.25L	0.20L	0.15L

At interior supports, L is longer of adjacent spans.

*For bars not terminating in drop panel use lengths shown for panels without drops.

Fig. 5.36 Minimum Length of Flat Slab Reinforcement
(At exterior supports, where masonry walls or other construction provide only negligible restraint to the slab, the negative reinforcement need not be carried further than $0.20L$ beyond the center line of such support; any combination of straight and bent bars may be used provided minimum requirements are met.)

Table 5.25 Flat Slab Minimum Length of Positive Reinforcement

Strip	Percentage of required reinforcing steel area to be extended at least as indicated	Maximum distance from center line of support to end of straight bar or bend point of bent bar: Flat slabs without drop panels — Straight	Flat slabs without drop panels — Bend point where bars bend up and continue as negative reinforcement	Flat slabs with drop panels — Straight	Flat slabs with drop panels — Bend point where bars bend up and continue as negative reinforcement
Column strip reinforcement	Not less than 33 percent	$0.125L$		Minimum embedment in drop panel of 16 bar diameters but at least 10 in.	
	Not less than 50 percent*	3 in.	or $0.25L$		
	Remainder*	$0.125L$	or $0.25L$	Minimum embedment in drop panel of 16 bar diameters but at least 10 in.	or $0.25L$
Middle strip reinforcement	50 percent	$0.15L$		$0.15L$	
	50 percent*	3 in.	or $0.25L$	3 in.	or $0.25L$

*Bars may be straight, bent, or any combination of straight and bent bars. All bars are to be considered straight bars for the end under consideration unless bent at that end and continued as negative reinforcement.

Note: See also Fig. 5.36.

is subtracted from the pressure head per square foot, and the slab is designed to resist the difference. If the head to be designed against is great, it may be necessary to provide additional dead load in the slab to assist the dead load of the structure above in resisting the uplift.

Obviously, since slabs and beams on the ground are formed by the earth, it is advisable to form beams and footings using sloping sides. For waterproofing, a subslab is often used, on which waterproofing is laid, and the supporting slab laid over the waterproofing. Remember that the load on a waterproofing slab is a load upward and the location of the steel reinforcement is reversed —the positive-moment steel being located in the top of the slab and the negative moment in the bottom.

A typical problem will illustrate the use of this type of construction to resist the uplift of water pressure.

Problem Assume a column spacing of 20 ft in each direction (an ideal situation for flat-slab construction). Design a slab to resist a water pressure of 6 ft of head. The columns carry a load of 200 kips on soil with a carrying capacity of 6,000 psf. From Tables 8.2 and 8.3, we find that the foundation required is 6 ft square and 1 ft 7 in. thick. The slab is designed in accordance with the flat-slab formulae given in Art. 5.51. The design load per square foot is arrived at as follows:

Upward load = 6-ft head × 62.5 = 375

Dead load:	3 in. subfloor	
	$1\frac{1}{2}$ in. finish	
	6 in. slab (assumed)	
	$10\frac{1}{2}$ in. total	= 125
Net design load		= 250 psf

The foundation in this instance must serve both as a foundation and as an inverted splayed head, and its depth must be either the depth required for the footing or for the splayed head adjacent to the column, whichever is the greater. Using the splayed-head dimension ratio shown on Fig. 5.32, the desirable width of splayed head for the span is $0.225 \times 20 = 4.5$ ft. Since the splayed head must be below the slab, this is the side of the square at the bottom of the footing.

Assuming an 18-in. column, the depth of splay at 45 deg is $\dfrac{4.5 - 1.5}{2} = 1.5$ ft, or 18 in. To this should be added the depth of the shoulder to arrive at the total depth—2 in. plus the depth of slab, assumed at 6 in., thus giving a total depth of footing design of 26 in. This will undoubtedly satisfy conditions either for footing design under full live and dead load without uplift or for flat slab designed for uplift and dead load only. Assuming a 45-deg slope for the excavation, the side of square for punching shear in the slab is $4.5 + 1.5 + 1.5 = 7.5$.

The total uplift load per bay = $250 \times 20 \times 20 = 100{,}000$ lb.

The shear is taken at a distance d outside the splay, and with a 6-in. slab we may use $4\frac{1}{2}$ in. net for d; or the area to be deducted for shear is (7 ft 6 in. + 9 in.) square, and the load on this area to be deducted for shear is $8.25 \times 8.25 \times 250 = 17{,}015$. The net shear is $100{,}000 - 17{,}015 = 82{,}985$ lb.

The unit shear on a perimeter 99 in. square is

$$v = \frac{82{,}985}{99 \times 4 \times 4.5 \times 0.875} = 53.2 \text{ lb}$$

which is within the allowable.

The design of reinforcement is in accordance with the method of Art. 5.51.

The depth of the splayed head for a column under these conditions would be

Depth of slab	$= 7\frac{1}{2}$ in.
Depth of shoulder	= 2 in.
Depth of splay, assuming an 18-in. column $= \dfrac{4 \text{ ft } 6 \text{ in.} - 1 \text{ ft } 6 \text{ in.}}{2} = 1$ ft 6 in.	= 18 in.
	$27\frac{1}{2}$ in.
	$= 2$ ft $3\frac{1}{2}$ in.

Therefore, since this is the greater of the two depths, it should be used. Figure 5.37 shows a section through the construction.

c A slab on filled ground is frequently combined with pile foundations because of the nature of the subsoil. Although the column spacing is frequently satisfactory for flat-slab construction (a ratio of not over $1\frac{1}{3}$ to 1), many installations are not suited to flat-slab column spacing—for example, machine shops, which are often equipped with crane spans and other irregular conditions. Beams laid in trenches, either as one-way or two-way slab systems, are quite satisfactory for this condition, as indicated in the accompanying plan and beam sections (Fig. 5.38).

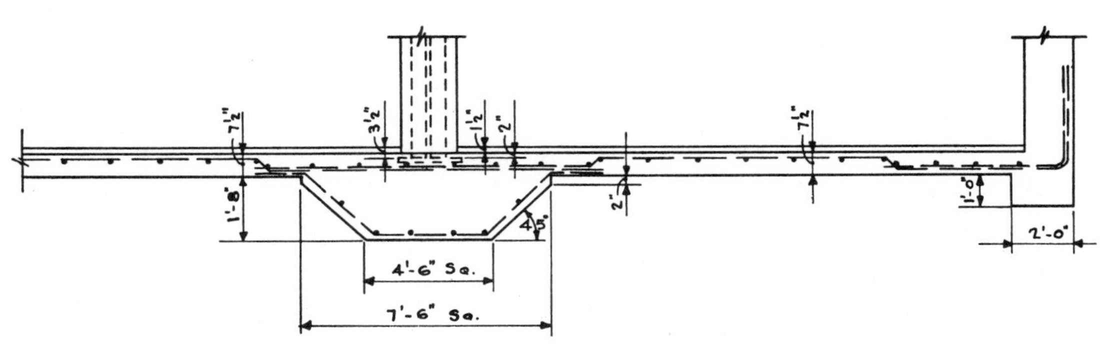

Fig. 5.37 Flat Slab Reinforced Against Water Pressure

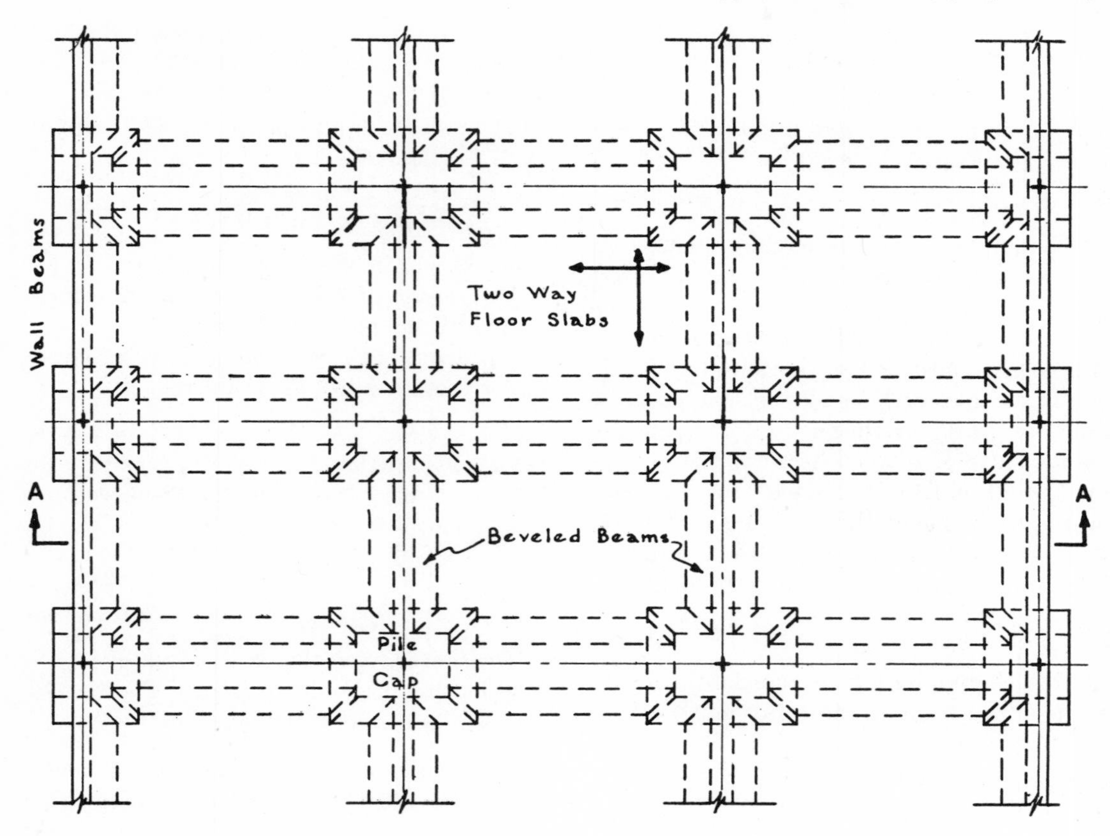

PLAN

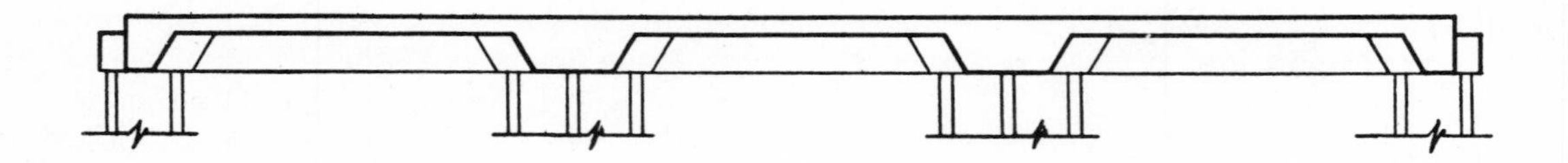

SEC. A-A

Fig. 5.38 Supported First-Floor Slabs

5.60 GYPSUM SYSTEMS

a Because in some areas they fill certain needs at a low cost, and because the sales-promotion work has been well handled, gypsum floor and roof slabs have come into fairly common use in sections where freight rates do not work against them. Their chief advantages may be summed up as follows: (1) light-weight; (2) low cost; (3) good fire resistance; (4) if precast, can be applied in winter; (5) if precast, can be used satisfactorily for sloping roofs; and (6) if precast, can be nailed, and nails will hold reasonably well. The chief arguments against their use seem to be: (1) lack of strength renders them undesirable for heavy or concentrated loads; (2) if not properly protected during construction they will soften and disintegrate in wet weather; (3) all precast systems lack the ability to stiffen the structure.

b All gypsum systems are sold and in-

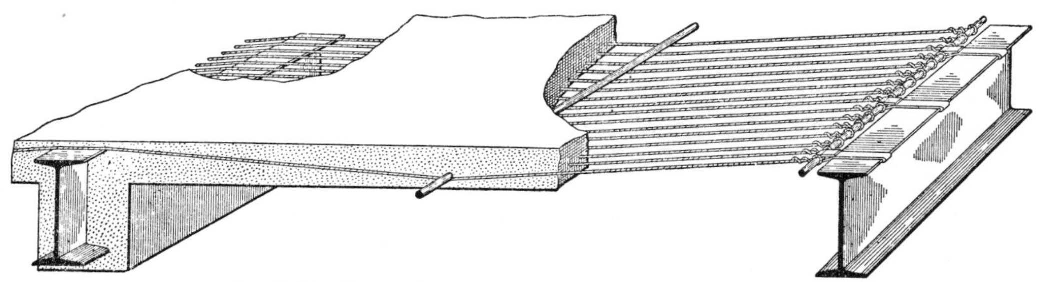

Fig. 5.39 Long-Span Poured-In-Place Gypsum Slab

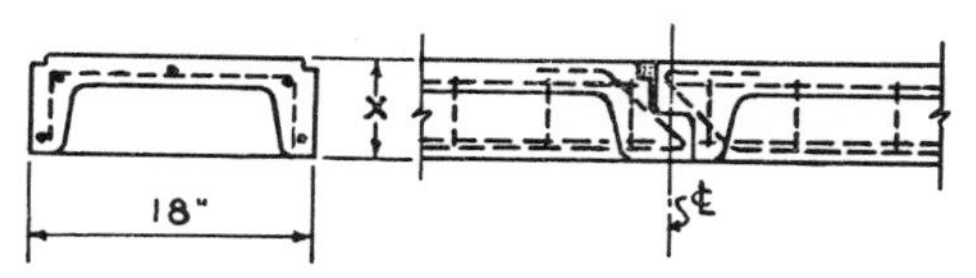

Fig. 5.40 Channel-Type Gypsum Slab

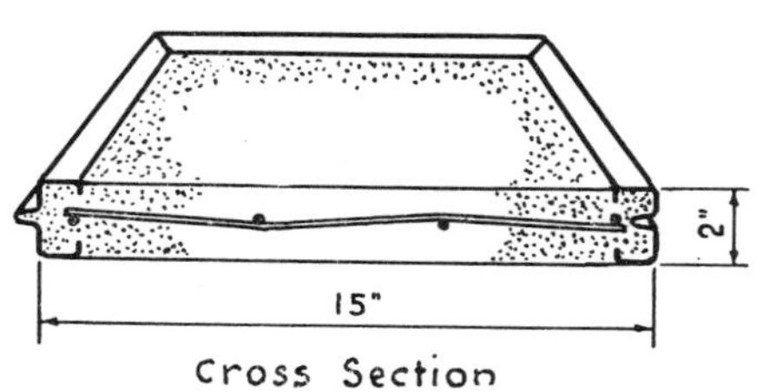

Fig. 5.41 Gypsteel Plank

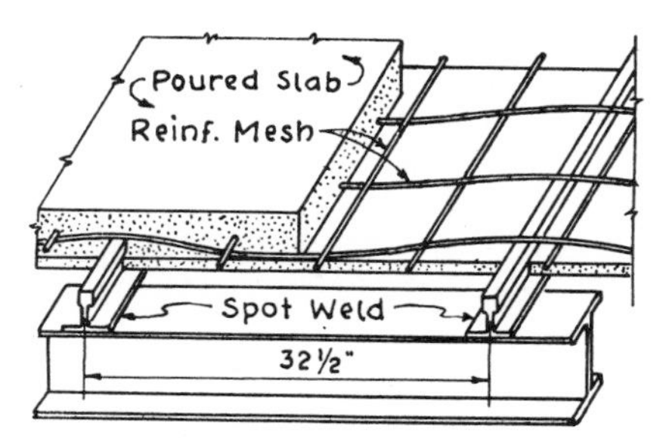

Fig. 5.42 Short-Span Poured-In-Place Gypsum Slab

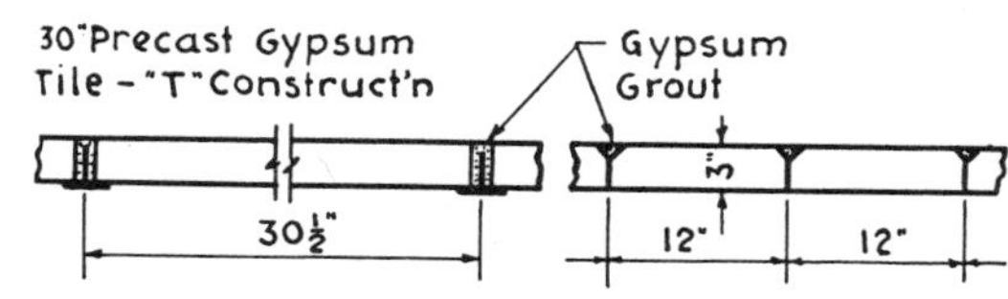

Spacing: 30½ in. for tees, 31⅛ in. for 12-lb rail, 31⅜ in. for 16-lb rail, and 31½ in. for 20-lb rail

Fig. 5.43 30-in. Gypsum-Tile T Construction

stalled by specialists, and the details, load tables, and other data are available from their catalogs and from their salesmen. The long-span poured-in-place systems (spans up to about 7 ft 6 in.) are designed on the basis of the suspension-bridge principle, and the end-bay steel must be braced to provide anchorage (see Fig. 5.39). This system is seldom used nowadays. The long-span precast may be designed on the same principle, or it may be of rigid-channel type (Fig. 5.40) or tongue-and-groove plank sections. With either type of precast system, the steel beams must have the top flange braced to reduce the l/b of the beam, since the slab does not furnish lateral support for the beam. Gypsum planks, either 2 or 2½ in. thick, usually with tongue-and-groove metal edges (Fig. 5.41), are one of the most common forms of long-span gypsum slabs; and although the metal clips by means of which the slab is clamped to the beam do furnish considerable lateral support to the top flange of the beam, there are many building departments which still require the additional lateral bracing of tie rods. The short-span roof systems, usually 30- or 32-in. span (Figs. 5.42 and 5.43), either precast or cast in place on plasterboard forms, are supported on steel T sections or rail sections, which are usually furnished by the gypsum contractor and welded to steel beams about 7 or 8 ft apart. These welded T sections in themselves will furnish adequate lateral bracing to the steel beams.

c Most of the design data for gypsum systems is found in the manufacturer's catalogs. The load capacity of gypsum slabs or planks is tabulated for various spans. Similarly the maximum span of Ts or rails for

Table 5.26 Capacity of Ts for Short-Span Gypsum Systems

Weight, pounds per foot

Span	4.1 lb	4.9 lb	5.5 lb	6.4 lb	6.7 lb	7.5 lb	8.9 lb	10.8 lb
5′0″	66.25	85.0	103.75	122.5				
5′6″	55.0	70.0	85.0	101.5	127.5			
6′0″		58.75	71.25	83.75	106.25	121.25		
6′6″			60.0	71.25	90.0	102.50	118.75	133.75
7′0″				61.25	77.5	87.5	102.50	115.0
7′6″					67.5	77.5	88.75	98.75
8′0″					58.75	66.25	76.25	86.25
8′6″						58.75	76.5	76.25
9′0″							60.0	67.5

the short-span systems is usually given in the manufacturer's catalogs, although it may sometimes be necessary to use standard Ts as listed below instead of the rail Ts or other sections sold by the manufacturer.

Table 5.27 Capacity of Rails for Short-Span Gypsum Systems

Weight, pounds per yard

Span	12 lb	16 lb	20 lb
5′0″	132.5		
5′6″	110.0		
6′0″	92.5		
6′6″	77.5	125	
7′0″	67.5	107.5	
7′6″	58.75	93.75	132.5
8′0″		80.0	116.25
8′6″		72.5	102.5
9′0″		63.75	90.0
9′6″		56.25	81.25
10′0″			72.5
10′6″			65.0
11′0″			60.0

Since some form of short-span system is most commonly used, a problem and tables of capacities of the T and rail sections ordinarily used are included in Tables 5.26 and 5.27. Note that in Table 5.27 the weight of supporting rails is given in pounds per yard (as is common practice for rails) instead of pounds per foot as for Ts.

Problem Select the required T or rail to carry a 2-in. cast-in-place gypsum slab with 6-psf roof covering and 30-psf roof load plus suspended ceiling, spacing of beams to be 7 ft 6 in.

Roof load	=	30
2-in. gypsum slab	=	8
Roofing	=	6
Ceiling	=	10
Steel	=	3
		57 psf

From Table 5.27 select 12-lb rail.

6

Timber and Other Materials in Bending

6.10 TIMBER BEAMS AND JOISTS

a Timber beams are widely used in building construction as joists or girders carrying floor or roof loads. Strength limitations prevent their use for very heavy loads or very long spans. Art. 1.23, with Table 1.3, gives standard practice and permissible stresses for various kinds of lumber, but, since availability in various parts of the country is limited by the economical source of supply, the designer should check locally as to the lumber to be used in the area in which he is working. Lumber is listed in even inch sizes but the actual size is less than the nominal size and properties must be computed on the basis of actual sizes as listed in Table 6.4.

b In addition to the normal requirements for bending as contained in Chapter 3 of this book and those sections of the NLMA Code given in Art. 1.23, the following sections of the above named code apply to the design of beams and girders. Since the nomenclature of this code does not necessarily follow that of the other codes, it is necessary to include the applicable nomenclature here.

400-A. *Notation.*

400-A-1. Except where otherwise noted, the following symbols are used in the formulas for beams:

b = breadth of section, inches.
Δ = deflection, inches.
E = modulus of elasticity.
f = unit stress in extreme fiber in bending.
h = depth of section, inches.
H = unit stress in horizontal shear.
I = moment of inertia of the section.
M = bending or resisting moment, maximum.
n = distance from neutral axis to extreme fiber, inches.
P = total concentrated load, pounds.
L = span, feet.
l = span, inches.
S = section modulus.
V = total vertical end shear or end reaction, pounds.
W = total uniformly distributed load, pounds.

400-B. *Beam Span.*

400-B-1. For simple beams, the span shall be taken as the distance from face to face of supports, plus one-half the required length of bearing at each end; for continuous beams, the span is the distance between centers of bearings on supports over which the beam is continuous.

400-C-7. Lateral moment distribution of a concentrated load from a critically loaded beam to adjacent parallel beams should be calculated. (See Appendix A.)

LATERAL DISTRIBUTION OF A CONCENTRATED LOAD

A. *Lateral Distribution of a Concentrated Load for Moment*

The lateral distribution of a concentrated load for computing bending moment may be determined by the following method (see Reference 12, Appendix I):

When a concentrated load at the center of the beam span is distributed to adjacent parallel beams by a wood (plank or laminated strip) or concrete-slab floor, the load on the beam nearest the point of application shall be determined by multiplying the load by the following factors:

Kind of floor:	Load on Critical beam**
2-inch wood	S/4.0
4-inch wood	S/4.5
6-inch wood	S/5.0
Concrete, structurally designed	S/6.0

*S = average spacing of beams in feet.

* In case S exceeds the denominator of the factor, the load on the two adjacent beams shall be the reactions of the load, with the assumption that the floor slab between the beams acts as a simple beam.

** For more than one traffic lane, see additional data in reference 12, Appendix I.

B. *Lateral Distribution of a Concentrated Load for Shear*

When the load distribution for moment at the center of a beam is known or assumed to correspond to specific values in the first two columns of Table A-1, the distribution to adjacent parallel beams when loaded at or near the quarter point, i.e., the approximate point of maximum shear, shall be assumed to be the corresponding values in the last two columns of Table A-1.

TABLE A-1 *Distribution in terms of proportion of total load.*

Load applied at center of span		Load applied at ¼ point of span	
Center beam	Distribution to side beams	Center beam	Distribution to side beams
1.00	0	1.00	0
.90	.10	.94	.06
.80	.20	.87	.13
.70	.30	.79	.21
.60	.40	.69	.31
.50	.50	.58	.42
.40	.60	.44	.56
.33	.67	.33	.67

205-ALLOWABLE UNIT STRESSES FOR FLEXURE

205-A. *Joist and Plank Grades.*

205-A-1. Allowable unit stresses in flexure for Joist and Plank grades apply to material with the load applied to either the narrow or wide face.

205-B. *Beam and Stringer Grades.*

205-B-1. Allowable unit stresses in flexure for Beam and Stringer grades apply only to material with the load applied to the narrow face.

400-D. *Vertical Shear.*

400-D-1. Ordinarily it is unnecessary to compute or check the strength of beams in cross-grain (vertical) shear.

400-I. *Deflection.*

400-I-1. If vertical deflection is a factor in design, it shall be calculated by standard methods of engineering mechanics. Under full design load permanently applied, timber acquires a permanent set about equal to the initial deflection, but the strength is not reduced.

400-I-2. Floor joists, having a depth-to-thickness ratio of 6 or more, should be supported laterally by bridging installed at intervals not exceeding 8 feet. Bridging may be omitted at the ends of joists which are nailed or otherwise fastened to framing members.

400-I-3. For rectangular beams and roof joists, the designer should apply the following approximate rules, based on nominal dimensions, in providing lateral restraint (Reference 2, Appendix I):

(a) If the ratio of depth to thickness is 2 to 1, no lateral support is needed.

(b) If the ratio is 3 to 1, the ends shall be held in position.

(c) If the ratio is 4 to 1, the piece shall be held in line as in a well-bolted chord member in a truss.

(d) If the ratio is 5 to 1, one edge shall be held in line.

(e) If the ratio is 6 to 1, the provisions of section 400-I-2 may be applied.

(f) If the ratio is 7 to 1, both edges shall be held in line.

(g) If a beam is subject to both flexure and compression parallel to grain, the ratio may be as much as 5 to 1, if one edge is held firmly in line; e.g., by rafters (or by roof joists) and diagonal sheathing. If the dead load is sufficient to induce tension on the under side of the rafters, the ratio for the beam may be 6 to 1.

209-B. *Support on Ledger.*

209-B-1. In joists supported on a ribbon or ledger board and spiked to the studs, the allowable unit stresses in Tables 1, 20 and 21, for compression perpendicular to grain, may be increased 50 per cent.

209-C. *Length of Bearing.*

209-C-1. Allowable unit stresses in compression perpendicular to grain may be increased for bearings less than 6 inches in length and located 3 inches or more from the end of a timber in accordance with section 400-H.

400-H. *Compression Perpendicular to Grain.*

400-H-1. The allowable unit stresses for compression perpendicular to grain, in Tables 1, 20 and 21, apply to bearings of any length at the ends of the beam, and to all bearings 6 inches or more in length at any other location. When calculating the bearing area at the ends of beams, no allowance shall be made for the fact that, as the beam bends, the pressure upon the inner edge of the bearing is greater than at the end of the beam.

For bearings of less than 6 inches in length and not nearer than 3 inches to the end of a member, the maximum allowable load per square inch is obtained by multiplying the allowable unit stresses in compression perpendicular to grain by the following factor:

$$\frac{l + 3/8}{l}$$

in which l is the length of bearing in inches measured along the grain of the wood.

The multiplying factors for indicated lengths of bearing on such small areas as plates and washers become:

Length of bearing in inches	½	1	1½	2	3	4	6 or more
Factor.....	1.75	1.38	1.25	1.19	1.13	1.10	1.00

In using the preceding formula and table for round washers or bearing areas, use a length equal to the diameter.

400-E. *Horizontal Shear.*

400-E-1. The maximum horizontal shear stress in a solid-sawn or glued-laminated wood beam shall be calculated by means of the formula:

$$H = \frac{VQ}{It}$$

in which

Q = the statical moment of the area above or below the neutral axis.
t = the width of the beam at the neutral axis.

For a rectangular beam b inches wide and h inches deep, this becomes

$$H = \frac{3V}{2bh}$$

Except as provided in section 400-E-2, the unit shear stress, H, shall not exceed the allowable for the species and grade, as given in Table 1 for solid-sawn lumber, and in Tables 20 and 21 for glued laminated lumber, adjusted for duration of loading, as provided in sections 203 and 204.

When calculating the reaction, V:

(a) Take into account any relief to the beam resulting from distribution of load to adjacent parallel beams by flooring or other members. (See Appendix A.)

(b) Neglect all loads within a distance from either support equal to the depth of the beam.

(c) With a single moving load, or one moving load that is considerably larger than any of the others, place the load at a distance from the support equal to the depth of the beam, keeping others in their normal relation. With two or more moving loads of about equal weight and in proximity, place loads in the position that produces the highest reaction V, neglecting any load within a distance from the support equal to the depth of the beam.

400-E-2. In the stress grades of solid-sawn beams, allowances for checks, end splits, and shakes have been made in the assigned unit stresses. For such members, the formula in section 400-E-1 does not indicate the actual horizontal shear resistance because of the redistribution of shear stress that occurs in checked beams. (See Appendix E.)

Thus, for a solid-sawn beam which does not qualify under section 400-E-1, the end reaction or vertical end shear, V, may be determined on the following basis:

(a) Calculate the reaction as in section 400-E-1, except that if there is a single moving load or one moving load that is considerably larger than any of the others, place that load at a distance from the support equal to three times the depth of the beam, or at the quarter point, whichever is closer. Consider all other loads in the usual manner.

(b) In lieu of the procedure in paragraph (a), the reaction may be determined by the following formulas, except that the formula for concentrated loads shall be used when there are two or more moving loads of about equal weight and in proximity:

For concentrated load

$$V' = \frac{10P(l - x)(x/h)^2}{9l[2 + (x/h)^2]}$$

For uniformly distributed load

$$V' = \frac{W}{2}\left(1 - \frac{2h}{l}\right)$$

in which:

h = depth of beam in inches.

l = span in inches.
P = concentrated load in pounds.
V' = modified end reaction, or end shear in pounds. (For a combination of concentrated loads, V' is the sum of $V'_1 + V'_2 + \cdots V'_n$ resulting from each such load at its distance $x_1, x_2 \ldots$ or x_n from the reaction. The load combination should be placed in that position which provides the maximum value for V'.)
W = total uniformly distributed load in pounds.
x = distance in inches from reaction to load.

(c) The methods for determining end reaction, set forth in paragraphs (a) and (b), shall not result in horizontal shear stress, H, when calculated as in section 400-E-1, that exceeds the following values (in pounds per square inch), adjusted for duration of loading as provided in sections 203 and 204:

Species	psi
Douglas Fir, Coast Region	145
Douglas Fir	145
Hemlock, Eastern	110
Hemlock, West Coast	120
Larch	145
Pine, Southern	175
Pine, Southern (KD Grades)	200
Pine, Norway	130
Redwood	110
Spruce, Eastern	130

400-C-6. Insofar as possible, notching of beams should be avoided. For a beam notched at or near the middle of the span, the net depth shall be taken when determining the flexural strength. Notches at the ends do not affect flexural strength directly. For effect of notch on shear strength, see section 400-F. The stiffness of a beam, as determined from its cross section, is practically unaffected by notches of reasonable depth and length, i.e., a notch depth equal to one-sixth the beam depth and a notch equal to one-third the beam depth at any point along the beam.

A gradual change in cross section compared with a square notch increases the shearing strength nearly to that computed for the actual depth above the notch.

400-F. *Horizontal Shear in Notched Beams.*

400-F-1. For notched beams, the shearing strength of a short, relatively deep beam notched on the lower face at the end is decreased by an amount depending on the relation of the depth of the notch to the depth of the beam. When designing a beam having a notch at its end, the desired bending load shall be checked against the load obtained by the formula:

$$V = \left(\frac{2Hbd}{3}\right)\left(\frac{d}{h}\right)$$

in which d is the actual end depth above the notch.

c If a fiber stress of 1,200 psi is used, $S = 1.25WL$ where W is the total load and L is the length in feet. If 1,600 psi is used, $S = 0.94WL$.

Problem Select floor joists for a residence span of 60-psf total load on a span of 17 ft using a 1,200-psi stress grade number.

$W = 17 \times 60 = 1020$

$S = 1.25 \times 1.02 \times 17 = 21.7$

Use 2×10, 12 in. o.c. ($S = 24.4$)
or 2×12, 16 in. o.c.

$$\left(S = \frac{35.8}{1.33} = 26.9 \text{ per ft width}\right)$$

The 2×10 would require

$$\frac{2 \times 10}{12} = 1.67 \text{ bd ft per sq ft}$$

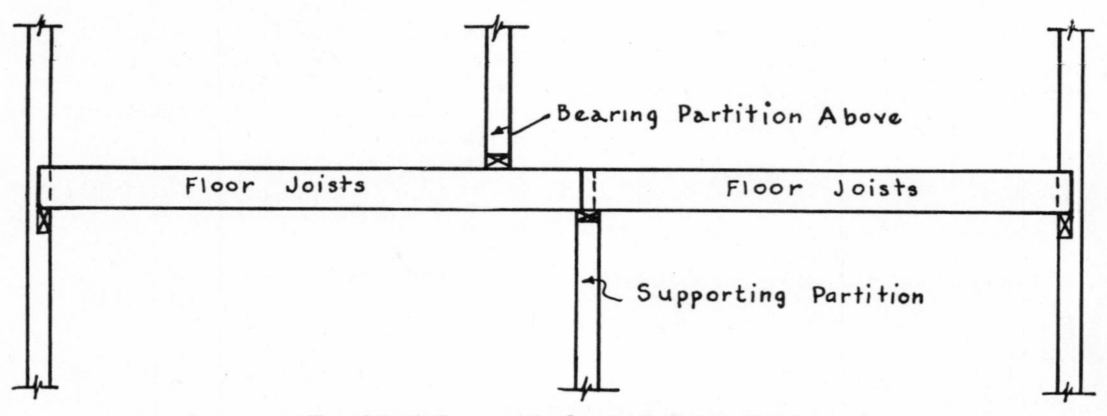

Fig. 6.1 A Frequently Overlooked Joist Load

The 2 × 12 would require

$$\frac{2 \times 12}{12 \times 1.33} = 1.5 \text{ bd ft per sq ft}$$

Therefore, other factors being equal, it would be preferable to use the 2 × 12; certainly the construction would be more rigid.

Table 6.1 gives the joist sizes required for various spans and loads using an allowable fiber stress of 1,200 psi. This table does not take into account deflection and, if the span in feet is in excess of 1.5 times the depth in inches, and if a plaster ceiling is supported by the joists, the size should be modified as indicated in Table 6.2. Both of the above tables are taken from the Southern Pine Manual of Standard Wood Construction. These comments are based on normal use of floor joists carrying uniform load only. Figure 6.1 shows a type of joist loading frequently encountered and seldom provided for. In the design of joists carrying load from above as shown here, the computation of moment is by the method of Chapter 3 and the selection of the required joist is to be made from Table 6.4.

d As has been noted, deflection may control the selection of a timber joist or beam and should be checked. Deflection may be computed by the regular deflection formula, but for simpler application the following formula, based on the standard formula for deflection of a uniformly distributed loaded beam, may be derived, using deflection in inches and span in feet,

$$\Delta = \frac{15fL^2}{Ec}$$

In case of steel or concrete under a given load on a span, the beam will take a deflection and, within the elastic limit, when the load is removed the beam will assume its original position. This is not strictly true with wood. If a timber is loaded excessively with a constant load, it will deflect and the deflection will continue to increase under the same load, even within the elastic limit of the material. Then, when the load has been removed, it will be found that the material has taken a permanent set or deflection.

Problem Check the deflection on a joist figured in paragraph *c* using the 12-in. joist.

$$f = \frac{21.7}{26.9} \times 1{,}200 = 968 \text{ psi}$$

$$\Delta = \frac{15 \times 968 \times 17^2}{1{,}600{,}000 \times 5.75} = 0.456 \text{ in.}$$

The permissible $\Delta = 17/30 = 0.567$ in. (see Table 6.2).

e The NLMA Code quoted previously warns against the use of notched beams and notes provisions to be taken in their design. Notches may occur in joists or in beams as indicated in Fig. 6.2, or the cut may be for the purpose of permitting the passage of a pipe or conduit within the ceiling space. If the notch or cut is made at any place other than the end of the beam, the allowable moment will be reduced and the section

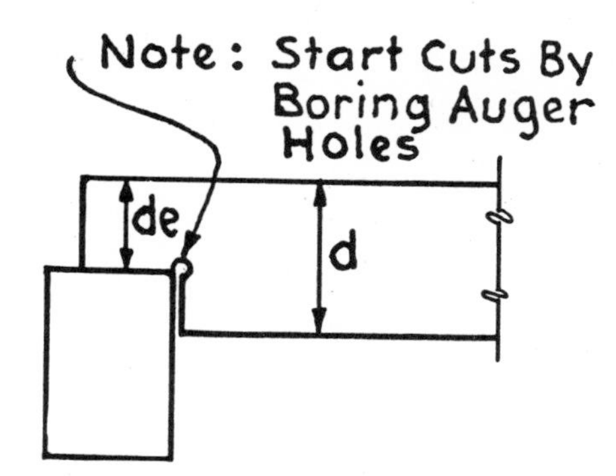

Fig. 6.2 Notched End of Joist

modulus computed on the basis of net depth will be the controlling factor.

f In many instances, careless or insufficient nailing is responsible for shoddy construction. Table 6.3, Recommended Nailing Schedule, from the Code Manual of the New York State Building Code, represents good practice in this detail.

6.20 FLITCH-PLATE OR COMBINATION BEAMS

a Formula (2.3) may be written $\delta = f/E$. Therefore if two materials work to resist the same type of stress in unison,

$$\delta = \frac{f_1}{E_1} = \frac{f_2}{E_2} \quad \text{or} \quad f_1 = \left(\frac{E_1}{E_2}\right) f_2 \quad \text{or}$$

$$f_1 = nf_2$$

Table 6.1 Floor Joists for 12-in. and 16-in. Spacing

Designed for live load and dead load.
Fiber stress of 1,200 psi, shear of 125 psi.
Where several sizes of joists are possible, they are given so that the most economical size may be selected. Narrower-faced joists result in a saving of wall height.
D.L. (dead load) includes weight of joist, 1-in. subfloor, and 1-in. finish floor.
A 10-psf plastered ceiling can be provided for by selecting the size for a 10-psf heavier live load.

Span, feet	30 psf + D.L.		40 psf + D.L.		50 psf + D.L.		60 psf + D.L.		70 psf + D.L.	
	12 in.	16 in.	12 in.	16 in.	12 in.	16 in.	12 in.	16 in.	12 in.	16 in.
9	2 × 6	2 × 6	2 × 6	2 × 6	2 × 6	2 × 6	2 × 6	2 × 8	2 × 6	2 × 8
10	2 × 6	2 × 6	2 × 6	2 × 6	2 × 6	2 × 8	2 × 8	2 × 8	2 × 8	2 × 8
11	2 × 6	2 × 6	2 × 6	2 × 8	2 × 8	2 × 8	2 × 8	2 × 8	2 × 8	2 × 10 3 × 8
12	2 × 6	2 × 8	2 × 8	2 × 8	2 × 8	2 × 8	2 × 8	2 × 10	2 × 8	2 × 10 3 × 8
13	2 × 6	2 × 8	2 × 8	2 × 8	2 × 8	2 × 10 3 × 8	2 × 8	2 × 10 3 × 8	2 × 10 3 × 8	2 × 10 3 × 8
14	2 × 8	2 × 8	2 × 8	2 × 10 3 × 8	2 × 8	2 × 10 3 × 8	2 × 10 3 × 8	2 × 10 3 × 8	2 × 10 3 × 8	2 × 12 3 × 10
15	2 × 8	2 × 8	2 × 8	2 × 10 3 × 8	2 × 10 3 × 8	2 × 10 3 × 8	2 × 10 3 × 8	2 × 12 3 × 10	2 × 10 3 × 8	2 × 12 3 × 10
16	2 × 8	2 × 10 3 × 8	2 × 10 3 × 8	2 × 10 3 × 8	2 × 10 3 × 8	2 × 12 3 × 10	2 × 10 3 × 8	2 × 12 3 × 10	2 × 12 3 × 10	2 × 12 3 × 10
17	2 × 8	2 × 10 3 × 8	2 × 10 3 × 8	2 × 10 3 × 8	2 × 10 3 × 8	2 × 12 3 × 10	2 × 12 3 × 10	2 × 12 3 × 10	2 × 12 3 × 10	2 × 14 3 × 10
18	2 × 10 3 × 8	2 × 10 3 × 8	2 × 10 3 × 8	2 × 12 3 × 10	2 × 10 3 × 8	2 × 12 3 × 10	2 × 12 3 × 10	2 × 14 3 × 10	2 × 12 3 × 10	2 × 14 3 × 12 4 × 10
19	2 × 10 3 × 8	2 × 10 3 × 8	2 × 10 3 × 8	2 × 12 3 × 10	2 × 12 3 × 10	2 × 12 3 × 10	2 × 12 3 × 10	2 × 14 3 × 12 4 × 10	2 × 14 3 × 10	2 × 14 3 × 12 4 × 10
20	2 × 10 3 × 8	2 × 12 3 × 10	2 × 12 3 × 10	2 × 12 3 × 10	2 × 12 3 × 10	2 × 14 3 × 12 4 × 10	2 × 12 3 × 10	2 × 14 3 × 12 4 × 10	2 × 14 3 × 12 4 × 10	2 × 16 3 × 12
21	2 × 10 3 × 8	2 × 12 3 × 10	2 × 12 3 × 10	2 × 14 3 × 10	2 × 12 3 × 10	2 × 14 3 × 12 4 × 10	2 × 14 3 × 12 4 × 10	2 × 16 3 × 12	2 × 14 3 × 12 4 × 10	2 × 16 3 × 14 4 × 12
22	2 × 10 3 × 8	2 × 12 3 × 10	2 × 12 3 × 10	2 × 14 3 × 12 4 × 10	2 × 14 3 × 10	2 × 14 3 × 12 4 × 10	2 × 14 3 × 12 4 × 10	2 × 16 3 × 12	2 × 14 3 × 12 4 × 10	2 × 16 3 × 14 4 × 12
23	2 × 12 3 × 10	2 × 12 3 × 10	2 × 12 3 × 10	2 × 14 3 × 12 4 × 10	2 × 14 3 × 12 4 × 10	2 × 16 3 × 12	2 × 14 3 × 12 4 × 10	3 × 14 4 × 12	2 × 16 3 × 12	3 × 14 4 × 12
24	2 × 12 3 × 10	2 × 14 3 × 10	2 × 14 3 × 10	2 × 14 3 × 12 4 × 10	2 × 14 3 × 12 4 × 10	2 × 16 3 × 14 4 × 12	2 × 16 3 × 12 4 × 10	3 × 14 4 × 12	2 × 16 3 × 14 4 × 12	3 × 14 3 × 16
25	2 × 12 3 × 10	2 × 14 3 × 12 4 × 10	2 × 14 3 × 12 4 × 10	2 × 16 3 × 12	2 × 14 3 × 12 4 × 10	2 × 16 3 × 14 4 × 12	2 × 16 3 × 12	3 × 14 4 × 12	2 × 16 3 × 14 4 × 12	3 × 16
26	2 × 12 3 × 10	2 × 14 3 × 12 4 × 10	2 × 14 3 × 12 4 × 10	2 × 16 3 × 14 4 × 12	2 × 16 3 × 12	3 × 14 4 × 12	2 × 16 3 × 14 4 × 12	3 × 16	3 × 14 4 × 12	3 × 16

Table 6.2 Floor Joists Based on Deflection

Designed for live load and dead load.
Spans limited to deflection of $\frac{1}{360}$ of span.
D.L. (dead load) includes weight of joist, 1-in. subfloor, and 1-in. finish floor.
A 10-psf plastered ceiling can be provided for by selecting the size for a 10-psf heavier live load.

	30 psf + D.L.		40 psf + D.L.		50 psf + D.L.	
	12 in. o.c.	16 in. o.c.	12 in. o.c.	16 in. o.c.	12 in. o.c.	16 in. o.c.
2 × 6	11′6″	10′7″	10′7″	9′8″	10′0″	9′1″
2 × 8	15′2″	14′1″	14′1″	12′11″	13′2″	12′1″
3 × 8	17′6″	16′3″	16′3″	14′11″	15′4″	14′0″
2 × 10	19′1″	17′9″	17′9″	16′3″	16′8″	15′3″
3 × 10	21′11″	20′5″	20′5″	18′9″	19′3″	17′8″
2 × 12	23′0″	21′4″	21′4″	19′6″	20′1″	18′4″
3 × 12	23′11″	24′6″	24′6″	22′7″	23′2″	21′3″
2 × 14	26′9″	24′11″	24′11″	22′11″	23′5″	21′6″

In other words, if two materials having different moduli of elasticity are used together, they do not both work to the limit of their stress, but each carries stress in proportion to the above formula.

Problem Let us assume that an 8 × 12 long-leaf yellow pine beam is reinforced by the addition of two 12-in. 20.7-lb channels properly bolted together. What percentage is added to the strength of the wood beam?

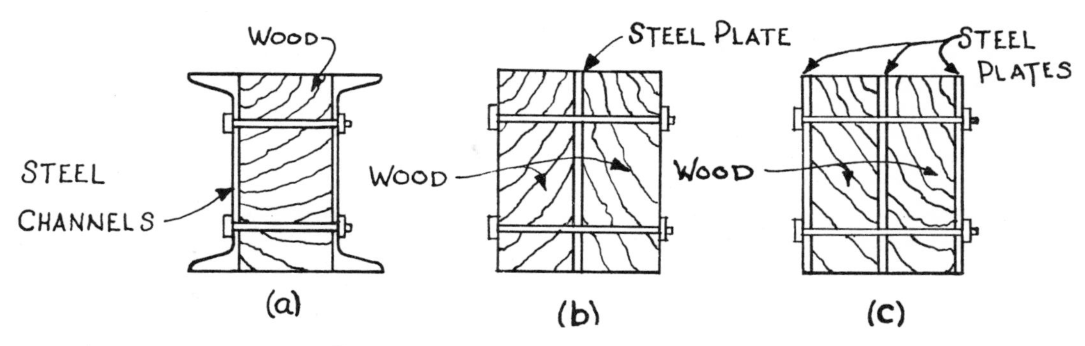

Fig. 6.3 Types of Flitch-Plate Girders

Assume $E_{yp} = 1{,}600{,}000$, $E_s = 29{,}000{,}000$, and $n = 29/1.6 = 18.1$.

If $f_{yp} = 1{,}600$, $f_s = 28{,}960$ which would overstress A36 steel. Therefore, if f_s is kept to 22,000 psi, $f_{yp} = 1.215$ psi.

On the basis of long-leaf yellow pine only,

$$M = 1{,}600 \times 165 = 264{,}000 \text{ in.-lb}$$

The combined girder will carry

$$\begin{aligned} 1{,}215 \times 165 &= 207{,}900 \\ 22{,}000 \times 42.8 &= 941{,}600 \\ & 1{,}149{,}500 \\ & -\ 264{,}000 \\ \text{Increase} &= 885{,}500 \text{ in.-lb} \end{aligned}$$

$$\text{Percent increase} = \frac{885.5}{264} = 336 \text{ percent}$$

b The above type of beam, using two wood beams on either side of a vertical steel or wrought-iron plate, properly bolted, was used quite extensively until steel beams were developed, and was known as a flitched beam or flitch-plate girder. Occasionally nowadays this method may be found economical to reinforce a steel beam by the addition of two wood beams, or to reinforce a wood beam by the addition of two steel channels or plates. It should be noted also that this

Table 6.3 Recommended Nailing Schedule

Building Element	Nail Type	Number and Distribution
Stud to sole plate	Common-toe-nail	3—16d
Stud to cap plate	Common-end-nail	2—16d
Double studs	Common-direct	10d 12″ o.c. or 16d 30″ o.c.
Corner studs	Common-direct	16d 30″ o.c.
Sole plate to joist or blocking	Common	20d 16″ o.c.
Double cap plate	Common-direct	16d 24″ o.c.
Cap plate laps	Common-direct	3—16d
Ribbon strip, 6″ or less	Common-direct	2—10d each bearing
Ribbon strip, over 6″	Common-direct	3—10d each bearing
Roof rafter to plate	Common-toe-nail	3—16d
Roof rafter to ridge	Common-toe-nail	2—16d
Jack rafter to hip	Common-toe-nail	3—10d
Floor joists to studs (no ceiling joists)	Common-direct	5—10d or 3—16d
Floor joists to studs (with ceiling joists)	Common-direct	2—10d
Floor joists to sill or girder	Common-toe-nail	2—16d
Ledger strip	Common-direct	3—20d at each joist
Ceiling joists to plate	Common-toe-nail	2—16d
Ceiling joists to every rafter	Common-direct	(See table below)
Ceiling joists (laps over partitions)	Common-direct	3—16d
Collar beam	Common-direct	4—10d
Bridging to joists	Common-direct	2—8d each end
Diagonal brace (to stud & plate)	Common-direct	2—8d each bearing
Tail beams to headers (when nailing permitted)	Common-end	1—20d each 4 sq. ft. floor area
Header beams to trimmers (when nailing permitted)	Common-end	1—20d each 8 sq. ft. floor area
1″ Subflooring (6″ or less in width)	Common-direct	2—8d each joist
1″ Subflooring (8″ or more in width)	Common-direct	3—8d each joist
2″ Subflooring	Common-direct	2—20d each joist
1″ Wall sheathing (8″ or less in width)	Common-direct	2—8d each stud
1″ Wall sheathing (over 8″ in width)	Common-direct	3—8d each stud
Plywood sheathing	Common-direct	6d 6″ o.c. exterior edges 6d 12″ o.c. intermediate
1″ Roof sheathing (6″ or less in width)	Common-direct	2—8d each rafter
1″ Roof sheathing (over 6″ in width)	Common-direct	3—8d each rafter
Shingles, wood	Corrosion-resistive	2—No. 14 B&S each bearing
Weather boarding	Corrosion-resistive	2—8d each bearing

Shingle nails should penetrate not less than ¾ inches into nailing strips, sheathing or supporting construction, unless approved fastenings are used.

CEILING JOIST NAILING TO EVERY RAFTER
(Number of 16-penny nails)

Slope of roof		4/12		5/12		6/12		7/12		9/12		12/12	
Rafter spacing, o.c.		16″	24″	16″	24″	16″	24″	16″	24″	16″	24″	16″	24″
Width of building	Up to 24′	5	8	4	7	3	5	3	4	3	3	3	3
	24′ to 30′	7	11	6	9	4	7	3	6	3	4	3	3

Table 6.4 Properties of Wood Sections

Nominal size	Actual size	A	S	Weight per ft	fbm per lin ft
2 × 4	$1\frac{5}{8} \times 3\frac{5}{8}$	5.89	3.56	1.64	.67
2 × 6	$1\frac{5}{8} \times 5\frac{5}{8}$	9.14	8.57	2.54	1.0
2 × 8	$1\frac{5}{8} \times 7\frac{1}{2}$	12.2	15.3	3.39	1.33
2 × 10	$1\frac{5}{8} \times 9\frac{1}{2}$	15.4	24.4	4.29	1.67
2 × 12	$1\frac{5}{8} \times 11\frac{1}{2}$	18.7	35.8	5.19	2.0
2 × 14	$1\frac{5}{8} \times 13\frac{1}{2}$	21.9	49.4	6.09	2.33
3 × 8	$2\frac{5}{8} \times 7\frac{1}{2}$	19.7	24.6	5.47	2.0
3 × 10	$2\frac{5}{8} \times 9\frac{1}{2}$	24.9	39.5	6.93	2.5
3 × 12	$2\frac{5}{8} \times 11\frac{1}{2}$	30.2	57.9	8.39	3.0
3 × 14	$2\frac{5}{8} \times 13\frac{1}{2}$	35.4	79.7	9.84	3.5
4 × 6	$3\frac{5}{8} \times 5\frac{5}{8}$	20.4	19.1	5.66	2.0
4 × 8	$3\frac{5}{8} \times 7\frac{1}{2}$	27.2	34.0	7.55	2.67
4 × 10	$3\frac{5}{8} \times 9\frac{1}{2}$	34.4	54.5	9.57	3.33
4 × 12	$3\frac{5}{8} \times 11\frac{1}{2}$	41.7	79.5	11.6	4.0
4 × 14	$3\frac{5}{8} \times 13\frac{1}{2}$	48.9	110.0	13.6	4.67
4 × 16	$3\frac{5}{8} \times 15\frac{1}{2}$	56.2	145.0	15.6	5.33
6 × 6	$5\frac{1}{2} \times 5\frac{1}{2}$	30.3	27.7	8.4	3.0
6 × 8	$5\frac{1}{2} \times 7\frac{1}{2}$	41.3	51.6	11.4	4.0
6 × 10	$5\frac{1}{2} \times 9\frac{1}{2}$	52.3	82.7	14.5	5.0
6 × 12	$5\frac{1}{2} \times 11\frac{1}{2}$	63.3	121.0	17.5	6.0
6 × 14	$5\frac{1}{2} \times 13\frac{1}{2}$	74.3	167.0	20.6	7.0
6 × 16	$5\frac{1}{2} \times 15\frac{1}{2}$	85.3	220.0	23.6	8.0
8 × 6	$7\frac{1}{2} \times 5\frac{1}{2}$	41	51.56	11.4	4.0
8 × 8	$7\frac{1}{2} \times 7\frac{1}{2}$	56	70.31	15.7	5.33
8 × 10	$7\frac{1}{2} \times 9\frac{1}{2}$	71	112.8	19.8	6.67
8 × 12	$7\frac{1}{2} \times 11\frac{1}{2}$	86	165.3	24.0	8.0
8 × 14	$7\frac{1}{2} \times 13\frac{1}{2}$	101	227.8	28.0	9.33
8 × 16	$7\frac{1}{2} \times 15\frac{1}{2}$	116	300.3	32.5	10.67
10 × 10	$9\frac{1}{2} \times 9\frac{1}{2}$	90.3	143	25.0	8.33
10 × 12	$9\frac{1}{2} \times 11\frac{1}{2}$	109	209	30.3	10.0
10 × 14	$9\frac{1}{2} \times 13\frac{1}{2}$	128	289	35.6	11.67
10 × 16	$9\frac{1}{2} \times 15\frac{1}{2}$	147	380	40.9	13.33
10 × 18	$9\frac{1}{2} \times 17\frac{1}{2}$	166	485	46.1	15.0
12 × 1	$11\frac{1}{2} \times \frac{25}{32}$	9.04	1.17	2.5	1
12 × $1\frac{1}{4}$	$11\frac{1}{2} \times 1\frac{1}{16}$	12.4	2.23	3.44	1.25
12 × $1\frac{1}{2}$	$11\frac{1}{2} \times 1\frac{5}{16}$	15.0	3.3	4.16	1.5
12 × 2	$11\frac{1}{2} \times 1\frac{5}{8}$	18.6	5.07	5.17	2
12 × $2\frac{1}{2}$	$11\frac{1}{2} \times 2\frac{1}{8}$	24.4	8.61	6.78	2.5
12 × 3	$11\frac{1}{2} \times 2\frac{5}{8}$	30.2	13.24	8.38	3
12 × 4	$11\frac{1}{2} \times 3\frac{5}{8}$	41.6	25.1	11.57	4
12 × 6	$11\frac{1}{2} \times 5\frac{5}{8}$	64.6	60.6	18.0	6
12 × 8	$11\frac{1}{2} \times 7\frac{1}{2}$	86	108	23.9	8
12 × 10	$11\frac{1}{2} \times 9\frac{1}{2}$	109	172.5	30.3	10
12 × 12	$11\frac{1}{2} \times 11\frac{1}{2}$	132	253	36.7	12
12 × 14	$11\frac{1}{2} \times 13\frac{1}{2}$	155	349	43.1	14
12 × 16	$11\frac{1}{2} \times 15\frac{1}{2}$	178	460	49.5	16
14 × 14	$13\frac{1}{2} \times 13\frac{1}{2}$	182	410	50.6	16.33
14 × 16	$13\frac{1}{2} \times 15\frac{1}{2}$	209	541	58.1	18.66
14 × 18	$13\frac{1}{2} \times 17\frac{1}{2}$	236	680	65.6	21
16 × 16	$15\frac{1}{2} \times 15\frac{1}{2}$	240	621	66.7	21.33
16 × 18	$15\frac{1}{2} \times 17\frac{1}{2}$	270	791	75.3	24

method may only be used if the component parts are of equal depth.

6.30 COMPOSITE BEAMS

a Composite construction as defined by the AISC Code differs from the definition in the ACI Code in that the AISC Code recognizes the composite action of a steel beam and concrete slab whereas the ACI Code refers to the interaction of a precast or prestressed concrete beam with a cast-in-place concrete slab. Composite beams as referred to in this article are in accordance with the Section 1.11 of the AISC Code, as follows.

SECTION 1.11 COMPOSITE CONSTRUCTION

1.11.1 *Definition*

Composite construction shall consist of steel beams or girders supporting a reinforced concrete slab, so inter-connected that the beam and slab act together to resist bending. When the slab extends on both sides of the beam, the effective width of the concrete flange shall be taken as not more than one-fourth of the span of the beam, and its effective projection beyond the edge of the beam shall not be taken as more than one-half the clear distance to the adjacent beam, nor more than eight times the slab thickness. When the slab is present on only one side of the beam, the effective width of the concrete flange (projection beyond the beam) shall be taken as not more than one-twelfth of the beam span, nor six times its thickness nor one-half the clear distance to the adjacent beam.

Beams totally encased 2 inches or more on their sides and soffit in concrete poured integrally with the slab may be assumed to be inter-connected to the concrete by natural bond, without additional anchorage, provided the top of the beam is at least $1\frac{1}{2}$ inches below the top and 2 inches above the bottom of the slab, and provided that the encasement has adequate mesh or other reinforcing steel throughout the whole depth and across the soffit of the beam. When shear connectors are provided in accordance with Sect. 1.11.4, encasement of the beam to achieve composite action is not required.

1.11.2 *Design Assumptions*

1.11.2.1 Encased beams shall be proportioned to support unassisted all dead loads applied prior to the hardening of the concrete (unless these loads are supported temporarily on shoring) and, acting in conjunction with the slab, to support all dead and live loads applied after hardening of the concrete, without exceeding a computed bending stress of $0.66F_y$, where F_y is the yield point of the steel beam. The bending stress produced by loads after the concrete has hardened shall be computed on the basis of the moment of inertia of the composite section. Concrete tension stresses below the neutral axis of the composite section shall be neglected. Alternatively, the steel beam alone may be proportioned to resist unassisted the moment produced by all loads, live and dead, using a bending stress equal to $0.76F_y$, in which case temporary shoring is not required.

1.11.2.2 When shear connectors are used in accordance with Sect. 1.11.4 the composite section shall be proportioned to support all of the loads without exceeding the allowable stress prescribed in Sect. 1.5.1.4.1 or 1.5.1.4.4 as applicable. The moment of inertia I_{tr} of the composite section shall be computed in accordance with the elastic theory. Concrete tension stresses below the neutral axis of the composite section shall be neglected. The compression area of the concrete above the neutral axis shall be treated as an equivalent area of steel by dividing it by the modular ratio n.

For construction without temporary shoring the value of the section modulus of the transformed composite section used in stress calculations (referred to the tension flange) shall not exceed

$$S_{tr} = \left(1.35 + 0.35\frac{M_L}{M_D}\right) S_s \tag{17}$$

where M_L and M_D are, respectively, the live load and dead load moments and S_s is the section modulus of the steel beam (referred to its tension flange) and provided that the steel beam alone, supporting the loads before the concrete has hardened, is not stressed to more than the applicable bending stress given in Sect. 1.5.1.

1.11.3 *End Shear*

The web and the end connections of the steel beam shall be designed to carry the total dead and live load.

1.11.4 *Shear Connectors*

Except in the case of encased beams as defined in Sect. 1.11.1, the entire horizontal shear at the junction of the steel beam and the concrete slab shall be assumed to be transferred by shear connectors welded to the top flange of the beam and embedded in the concrete. The total horizontal shear to be thus resisted between the point of maximum positive moment and each end of the steel beam (or between the point of maximum positive moment and a point of contraflexure in continuous beams) shall be taken as the smaller value using the formulas

$$V_h = \frac{0.85f_c'A_c}{2} \tag{18}$$

and

$$V_h = \frac{A_sF_y}{2} \tag{19}$$

where f_c' = specified compression strength of concrete at 28 days

A_c = actual area of effective concrete flange defined in Sect. 1.11.1

A_s = area of steel beam

The number of connectors resisting this shear, each side of the point of maximum moment, shall not be less than that determined by the relationship V_h/q, where q, the allowable shear load for one connector, or one pitch of a spiral bar, is as given in Table 1.11.4.

Table 1.11.4

Connector	**Allowable Horizontal Shear Load (q) (kips) (Applicable only to concrete made with ASTM C33 aggregates)**		
	f'_c = 3,000	f'_c = 3,500	f'_c = 4,000
½″ diam. × 2″ hooked or headed stud	5.1	5.5	5.9
⅝″ diam. × 2½″ hooked or headed stud	8.0	8.6	9.2
¾″ diam. × 3″ hooked or headed stud	11.5	12.5	13.3
⅞″ diam. × 3½″ hooked or headed stud	15.6	16.8	18.0
3″ channel, 4.1 lb.	4.3w	4.7w	5.0w
4″ channel, 5.4 lb.	4.6w	5.0w	5.3w
5″ channel, 6.7 lb.	4.9w	5.3w	5.6w
½″ diam. spiral bar	11.9	12.4	12.8
⅝″ diam. spiral bar	14.8	15.4	15.9
¾″ diam. spiral bar	17.8	18.5	19.1

w = length of channel in inches.

The required number of shear connectors may be spaced uniformly between the sections of maximum and zero moment.

Shear connectors shall have at least 1 inch of concrete cover in all directions.

The proper use of composite construction can be used to reduce construction costs because the concrete slab is normally used as part of the compression area, where concrete works to greater efficiency, and thus permits a larger section of the steel beam to take the tension. A suspended ceiling of proper specification may be used instead of complete concrete haunching of a steel beam to obtain the necessary fireproofing, and the suspended ceiling is normally considerably cheaper than concrete haunching.

b If the composite construction is obtained by means of haunching as shown in Fig. 6.4, the revised S may be derived as shown in the following problem.

Problem Assume an 18 WF 50, as shown in Fig. 6.4, with a 4-in. thick concrete slab. What is the comparative strength of this combination figured as a composite beam, with the beam treated as a simple steel beam laterally supported?

Assume $n = 9$. The concrete may be converted into equivalent steel and the method of Art. 1.31 (f and g) used.

The concrete to be considered is $16 \times 4 + 7.5$ in. $= 71.5$ in. wide and 4 in. thick. The equivalent steel area is

$$\frac{\text{Concrete area}}{n} = \frac{71.5 \times 4}{9} = 31.8 \text{ sq in.}$$

The equivalent steel I is

$$\frac{I_c}{n} = \frac{71.5 \times 4^3}{12 \times 9} = 42.2$$

Using the center line of the steel beam as a reference axis, we may tabulate the results:

	A	y	Ay	Ay^2	I
Concrete	31.8	8.5	270.3	2,297.6	42.2
Steel	14.71	—	0	0	800.6
	46.51		270.3	2,297.6	842.8
					2,297.6
					3,140.4

$\bar{y} = \dfrac{270.3}{46.51} = 5.81$ in. above center of 18-in. beam

$c = 5.81 + 9 = 14.81$

$A\bar{y}^2 = A\bar{y} \times \bar{y} = 270.3 \times 5.81 = 1,570.4$

I about axis of beam $= 3,140.4$
$\quad -1,570.4$

I about combined neutral axis $= 1,570.0$

$S = \dfrac{1,570.0}{14.81} = 106.0$

S of original beam $= 89.0$

Increase in strength $= 19$ percent

Section 1.11.2 above recognizes two methods of connecting the slab to the beam in order to obtain composite action, either (1) a fully encased beam as shown in Fig. 6.4 or (2) a construction in which the composite action is obtained by the attachment of shear connectors to the top flange of the steel section. Three types of shear connectors are recognized as shown in Fig. 6.5.

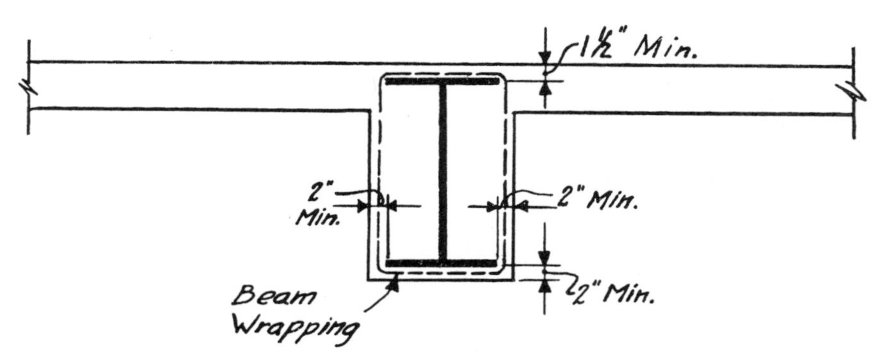

Fig. 6.4 Composite Beam

c Composite action obtained through the use of connectors is more economical than by

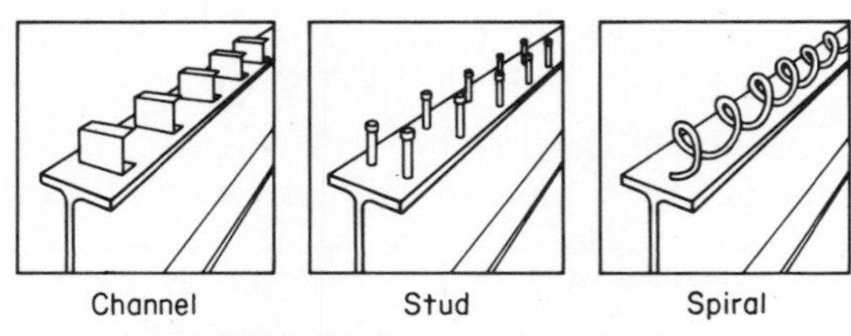

Fig. 6.5 Types of Connectors—Composite Beams

encasement of the beam if fire resistance is obtained by the use of a hung ceiling, a condition which normally exists in office building, hotel, or hospital construction, or where fire requirements do not require protection of the steel. The design may be in accordance with the AISC Manual, pages 2-86 through 2-117.

Unfortunately, the design properties listed in the tables in the AISC Manual assume the bottom of the slab to be flush with the top of the beam, a condition which exists in comparatively few jobs. In order to achieve this, either steel forms must be welded to the beam flange, or plywood forms must be cut and wedged to the side of the flange in such a way as to prevent seepage from the concrete as it is poured. The normal method of slab construction as shown in Fig. 4.1 permits the form work to be set under the beam flange so that the same form panel may be used for variations in span with comparatively little seepage.

Despite the above criticism, the tables referred to in the AISC Manual, pages 2–86 through 2–117, may be used by entering the tables with a slab thickness ½ in. greater than the thickness of the slab used. In other words, if the slab used, as shown in Fig. 4.1, is a 4-in. thick slab, make the selection from the AISC table for beams with a 4½-in. slab. The variation will not be great enough to warrant any further investigation, and it will be on the side of safety.

6.40 TRUSSED BEAMS

Although a trussed beam—or as it is commonly called by carpenters, a belly-rod truss—might more properly be treated as a truss, the common use to which it is put nowadays makes it more logical to classify it as a beam. Formerly it was a form of combination wood and wrought-iron truss, using a wood beam for the top chord and bent rods to form an inverted king-post or queen-post truss. It was used for spans or loads which were too great for the economic use of a single wooden beam.

Nowadays, however, the use of trussed beams is largely limited to strengthening a wood beam which is overloaded, cracked, or sagging, or where it is desirable to provide for additional beam capacity. They are seldom used in the design of a new building, since the cost of labor involved is greater than that of a structural-steel beam of equal or greater capacity. For this reason, the discussion of the problem presented here will be limited to the installation of belly-rods for strengthening an existing beam, which may, because of practical considerations, be somewhat different from the design for a new building.

Problem Assume a 12 × 12 beam in place in a building on a span of 12 ft center to center of support on timber bolsters, as shown in Fig. 6.6. This beam is of 1,200-psi-stress grade southern pine and is carrying 16 ft of floor at a total load of 150 psf.

$$W = 12 \times 16 \times 150 = 28{,}800 \text{ lb}$$

$$M = \frac{28{,}800 \times 12 \times 12}{8} = 518.4 \text{ in.-kips}$$

The section modulus of a 12 × 12 from Table 6.4 is 253. Therefore the stress in the beam is 518,400/253 = 2,049 psi, which exceeds the allowable limit.

Whatever method of correction is used must permit installation in a practical manner. The method indicated here has been used satisfactorily. Undoubtedly a number of other satisfactory details could be worked out.

In order to turn up the nut on the tension rods, they are connected into a 15-in. channel under the 12 × 12-in. strut block. The truss depth is, therefore,

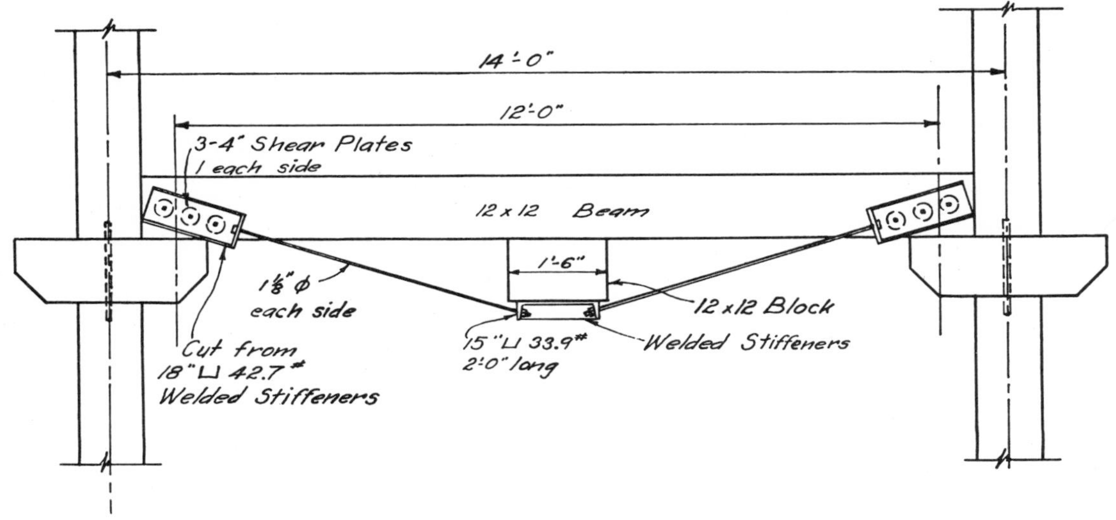

Fig. 6.6 Trussed Beams

Half the beam depth	=	5.75
12 × 12-in. block	=	11.5
To gage line of 15-in. channel	=	2
		19.25 in.

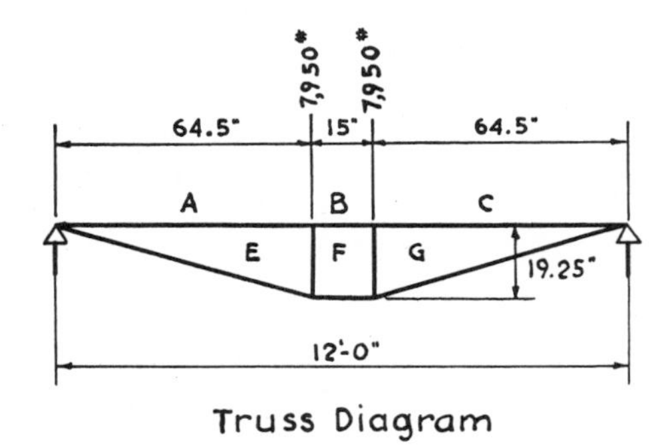

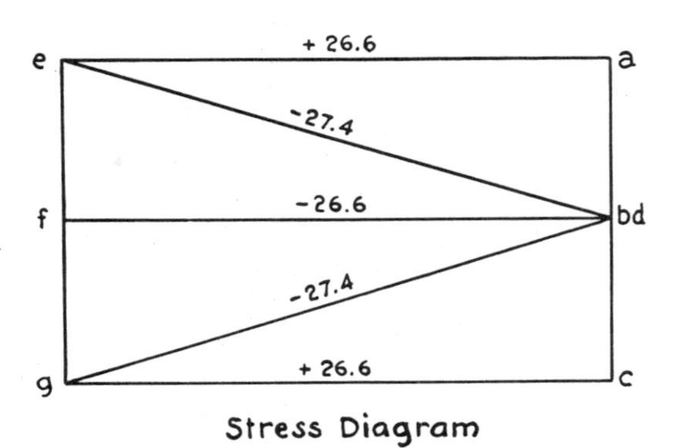

Fig. 6.7 Stress Diagram—Trussed Beams

The truss and stress diagrams are as shown in Fig. 6.7. Because of the width of the strut, it is indicated as two members, and the truss becomes an inverted queen-post truss.

The panel-point load is

$$\frac{79.5}{2 \times 12} \times 16 \times 150 = 7{,}950 \text{ lb}$$

On this basis, the two truss rods carry 27.4 kips, or 13.7 kips each (see Art. 10.10*b*). This requires an area of $13.7/22 = 0.622$ sq in. each at the root of the thread, or 1⅛-in. rods (see Table 2.4).

To check the strength of the beam,

$$\text{Direct stress} = \frac{26{,}600}{132.25} = 201$$

$$\text{Bending stress (roughly)} = \frac{2{,}049}{4} = \underline{512}$$

(original stress ÷ 2^2)

$$\text{Total} = 713$$

This stress is within the safe limit.

The channel into which the rods frame at the bottom should be strengthened by welding in stiffeners to prevent the thin web from deflecting away from the strut block and permitting the flange to bend.

The greatest problem in detail is the anchorage of the upper end of the rods to the beam. This is accomplished by means of shear plates. Shear plates are a variation of split-ring connectors, described in Art. 9.30*b*. The information relative to their design is obtained from "Teco Design Manual" of the Timber Engineering Co., Washington, D.C. From the load diagram, the angle with the grain is approximately 18 deg, and for group B lumber the approximate value of one ring would be 5,000 lb. It would, therefore, require three shear plates to develop the stress in each rod, and they would be carrying 91 percent of their capacity. On this basis, the minimum edge distance is 2¾ in. and spacing is 7 in. The connecting member may be cut from 18-in. channel and strengthened with welded stiffeners as described for the bottom channel.

7

Columns

7.10 STEEL COLUMNS

a In Art. 2.12 attention was called to the fact that in all steel columns direct compression is combined with some buckling, and therefore we cannot adopt a single unit stress that applies to all columns. The ratio l/r—the length between lateral supports divided by the radius of gyration of the section—is the factor on the basis of which the allowable stress is reduced. Moreover, a K factor—a multiplier for the actual length—is introduced to reflect the relative end conditions and structural rigidity of the complete frame, in accordance with the 1963 AISC specifications.

Figure 7.1 indicates graphically six types of buckling curvature, three of which (c), (e), and (f) involve sidesway. The "Commentary on the AISC Specification" contains this statement:

While ordinarily the existence of masonry walls provides enough lateral support for tier building frames to prevent sidesway, the increas-

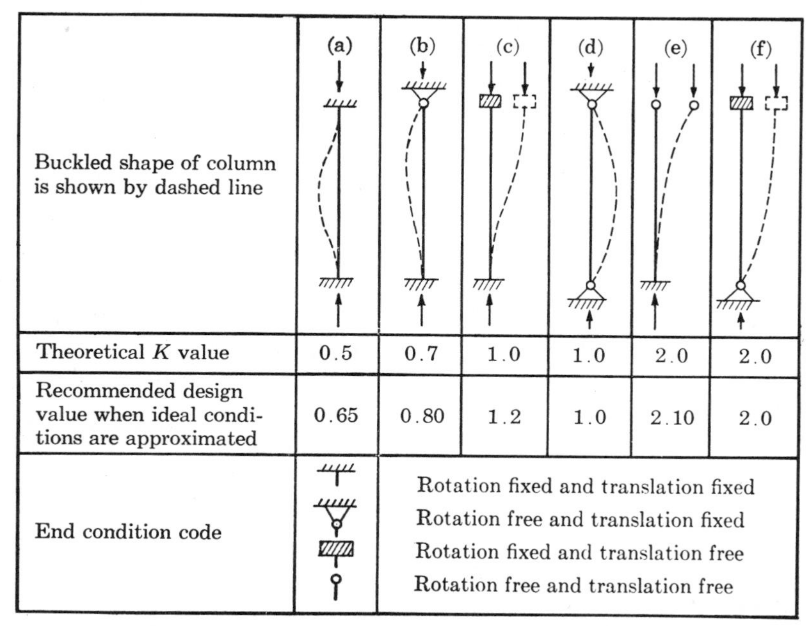

	(a)	(b)	(c)	(d)	(e)	(f)
Buckled shape of column is shown by dashed line						
Theoretical K value	0.5	0.7	1.0	1.0	2.0	2.0
Recommended design value when ideal conditions are approximated	0.65	0.80	1.2	1.0	2.10	2.0
End condition code	Rotation fixed and translation fixed Rotation free and translation fixed Rotation fixed and translation free Rotation free and translation free					

Fig. 7.1 Selection of *K* Values—Steel Columns

ing use of light curtain wall construction and wide column spacing for high-rise structures not provided with a positive system of diagonal bracing can create a situation where only the bending stiffness of the frame itself provides this support.

It has been the author's experience that even in the case of light prefabricated curtain wall construction of aluminum, concrete, or stainless steel, the panel is more or less rigidly attached to the steel framework in such a manner as to preclude sidesway. Moreover, we are not designing a single column in a structure. Any building with a concrete floor slab requires all columns to move in unison, and rigidity at any point in the floor plan will provide sufficient rigidity to prevent sidesway in any column of the building. Therefore, in probably 90 percent of the buildings we design, sidesway may be neglected.

Furthermore, the type of curvature indicated in Fig. 7.1 in (b) and (d) assumes pin connections—a condition which does not occur in buildings where end connection angles on the beams or cap and seat connections ensure some transfer of moment. Normally, the type of curvature indicated in Fig. 7.1(a) is the only type to be considered in most of the buildings ordinarily encountered. However, if the designer is doubtful of his structure and wishes to concede that there may be a tendency toward pin action in light connections or at the column base, he may use a K factor of 1 instead of 0.65 as recommended in Fig. 7.1. Usually the variation in this factor will not change the column size or, if it does change it, the variation will be one size upward for a more conservative design.

b The following sections from the AISC Code govern the design of columns for axial load only. The section governing combined bending and direct stress is given in Art. 7.11.

1.5.1.3 *Compression*

1.5.1.3.1 On the gross section of axially loaded compression members when Kl/r, the largest effective slenderness ratio of any unbraced segment as defined in Sect. 1.8, is less than C_c

$$F_a = \frac{\left[1 - \dfrac{(Kl/r)^2}{2C_c^2}\right] F_y}{\text{F.S.}} \tag{1}$$

where

$$\text{F.S.} = \text{factor of safety} = \frac{5}{3} + \frac{3(Kl/r)}{8C_c} - \frac{(Kl/r)^3}{8C_c^3}$$

and

$$C_c = \sqrt{\frac{2\pi^2 E}{F_y}}$$

1.5.1.3.2 On the gross section of axially loaded columns when l/r exceeds C_c

$$F_a = \frac{149{,}000{,}000}{(Kl/r)^2} \tag{2}$$

1.5.1.3.3 On the gross section of axially loaded bracing and secondary members, when l/r exceeds 120*

$$F_{as} = \frac{F_a \text{ (by Formula 1 or 2)}}{1.6 - \dfrac{l}{200r}} \tag{3}$$

* For this case, K is taken as unity.

1.5.6 *Wind and Seismic Stresses*

Allowable stresses may be increased one-third above the values provided in Sect. 1.5.1, 1.5.2, 1.5.3, 1.5.4 and 1.5.5 when produced by wind or seismic loading, acting alone or in combination with the design dead and live loads, provided the required section computed on this basis is not less than that required for the design dead and live load and impact (if any), computed without the one-third stress increase, nor less than that required by Sect. 1.7, if it is applicable.

1.8.1 *Definition—Slenderness Ratios*

In determining the slenderness ratio of an axially loaded compression member, except as provided in Sect. 1.5.1.3.3, the length shall be taken as its effective length Kl and r as the corresponding radius of gyration.

1.8.2 *Sidesway Prevented*

In frames where lateral stability is provided by diagonal bracing, shear walls, attachment to an adjacent structure having adequate lateral stability, or by floor slabs or roof decks secured horizontally by walls or bracing systems parallel to the plane of the frame, and in trusses the effective length factor, K, for the compression members shall be taken as unity, unless analysis shows that a smaller value may be used.

1.8.3 *Sidesway Not Prevented*

The effective length Kl of compression members in a frame which depends upon its own bending stiffness for lateral stability, shall be determined by a rational method and shall not be less than the actual unbraced length.

1.15.8 *Compression Members with Bearing Joints*

Where compression members bear on bearing plates, and where tier-building columns are finished to bear, there shall be sufficient rivets, bolts or welds to hold all parts securely in place.

Where other compression members are finished to bear, the splice material and its riveting, bolting or welding shall be arranged to hold all parts in line and shall be proportioned for 50 percent of the computed stress.

All of the foregoing joints shall be proportioned to resist any tension that would be developed by specified lateral forces acting in conjunction with 75 percent of the calculated dead load stress and no live load.

1.18.2 *Compression Members—Built up*

1.18.2.1 All parts of built-up compression members and the transverse spacing of their lines of fasteners shall meet the requirements of Sect. 1.8 and 1.9.

1.18.2.2 At the ends of built-up compression members bearing on base plates or milled surfaces, all components in contact with one another shall be connected by rivets or bolts spaced longitudinally not more than 4 diameters apart for a distance equal to $1\frac{1}{2}$ times the maximum width of the member, or by continuous welds having a length not less than the maximum width of the member.

1.18.2.3 The longitudinal spacing for intermediate rivets, bolts or intermittent welds in built-up members shall be adequate to provide for the transfer of calculated stress. However, where a component of a built-up compression member consists of an outside plate, the maximum spacing shall not exceed the thickness of the thinner outside plate times $4{,}000/\sqrt{F_y}$ when rivets are provided on all gage lines at each section, or when intermittent welds are provided along the edges of the components, but this spacing shall not exceed 12 inches. When rivets or bolts are staggered, the maximum spacing on each gage line shall not exceed the thickness of the thinner outside plate times $6{,}000/\sqrt{F_y}$ nor 18 inches. The maximum longitudinal spacing of rivets, bolts or intermittent welds connecting two rolled shapes in contact with one another shall not exceed 24 inches.

1.18.2.4 Compression members composed of two or more rolled shapes separated from one another by intermittent fillers shall be connected to one another at these fillers at intervals such that the slenderness ratio l/r of either shape, between the fasteners, does not exceed the governing slenderness ratio of the built-up member. The least radius of gyration r shall be used in computing the slenderness ratio of each component part.

1.18.2.5 Open sides of compression members built up from plates or shapes shall be provided with lacing having tie plates at each end, and at intermediate points if the lacing is interrupted. Tie plates shall be as near the ends as practicable. In main members carrying calculated stress the end tie plates shall have a length of not less than the distance between the lines of rivets, bolts or welds connecting them to the components of the member. Intermediate tie plates shall have a length not less than one-half of this distance. The thickness of tie plates shall be not less than $\frac{1}{50}$ of the distance between the lines of rivets, bolts or welds connecting them to the segments of the members. In riveted and bolted construction the pitch in tie plates shall be not more than 6 diameters and the tie plates shall be connected to each segment by at least three fasteners. In welded construction, the welding on each line connecting a tie plate shall aggregate not less than one-third the length of the plate.

1.18.2.6 Lacing, including flat bars, angles, channels or other shapes employed as lacing, shall be so spaced that the ratio l/r of the flange included between their connections shall not exceed the governing ratio for the member as a whole. Lacing shall be proportioned to resist a shearing stress normal to the axis of the member equal to 2 percent of the total compressive stress in the member. The ratio l/r for lacing bars arranged in single systems shall not exceed 140. For double lacing this ratio shall not exceed 200. Double lacing bars shall be joined at their intersections. In determining the required section for lacing bars, Formula (1) or (3) shall be used, l being taken as the unsupported length of the lacing bar between rivets or welds connecting it to the components of the built-up member for single lacing and 70 percent of that distance for double lacing. The inclination of lacing bars to the axis of the member shall preferably be not less than 60 degrees for single lacing and 45 degrees for double lacing. When the distance between the lines of rivets or welds in the flanges is more than 15 inches, the lacing shall preferably be double or be made of angles.

1.18.2.7 The function of tie plates and lacing may be performed by continuous cover plates perforated with a succession of access holes. The width of such plates at access holes, as defined in Sect. 1.9.2, is assumed available to resist axial stress, provided that: the width-to-thickness ratio conforms to the limitations of

Table 7.1 Allowable Stress (ksi) for Compression Members of 36 ksi Specified Yield Point Steel, F_y = 36 ksi

Main and Secondary Members Kl/r not over 120						Main Members Kl/r 121 to 200				Secondary Members* l/r 121 to 200			
$\frac{Kl}{r}$	F_a (ksi)	$\frac{Kl}{r}$	F_a (ksi)	$\frac{Kl}{r}$	F_a (ksi)	$\frac{Kl}{r}$	F_a (ksi)	$\frac{Kl}{r}$	F_a (ksi)	$\frac{l}{r}$	F_{as} (ksi)	$\frac{l}{r}$	F_{as} (ksi)
1	21.56	41	19.11	81	15.24	121	10.14	161	5.76	121	10.19	161	7.25
2	21.52	42	19.03	82	15.13	122	9.99	162	5.69	122	10.09	162	7.20
3	21.48	43	18.95	83	15.02	123	9.85	163	5.62	123	10.00	163	7.16
4	21.44	44	18.86	84	14.90	124	9.70	164	5.55	124	9.90	164	7.12
5	21.39	45	18.78	85	14.79	125	9.55	165	5.49	125	9.80	165	7.08
6	21.35	46	18.70	86	14.67	126	9.41	166	5.42	126	9.70	166	7.04
7	21.30	47	18.61	87	14.56	127	9.26	167	5.35	127	9.59	167	7.00
8	21.25	48	18.53	88	14.44	128	9.11	168	5.29	128	9.49	168	6.96
9	21.21	49	18.44	89	14.32	129	8.97	169	5.23	129	9.40	169	6.93
10	21.16	50	18.35	90	14.20	130	8.84	170	5.17	130	9.30	170	6.89
11	21.10	51	18.26	91	14.09	131	8.70	171	5.11	131	9.21	171	6.85
12	21.05	52	18.17	92	13.97	132	8.57	172	5.05	132	9.12	172	6.82
13	21.00	53	18.08	93	13.84	133	8.44	173	4.99	133	9.03	173	6.79
14	20.95	54	17.99	94	13.72	134	8.32	174	4.93	134	8.94	174	6.76
15	20.89	55	17.90	95	13.60	135	8.19	175	4.88	135	8.86	175	6.73
16	20.83	56	17.81	96	13.48	136	8.07	176	4.82	136	8.78	176	6.70
17	20.78	57	17.71	97	13.35	137	7.96	177	4.77	137	8.70	177	6.67
18	20.72	58	17.62	98	13.23	138	7.84	178	4.71	138	8.62	178	6.64
19	20.66	59	17.53	99	13.10	139	7.73	179	4.66	139	8.54	179	6.61
20	20.60	60	17.43	100	12.98	140	7.62	180	4.61	140	8.47	180	6.58
21	20.54	61	17.33	101	12.85	141	7.51	181	4.56	141	8.39	181	6.56
22	20.48	62	17.24	102	12.72	142	7.41	182	4.51	142	8.32	182	6.53
23	20.41	63	17.14	103	12.59	143	7.30	183	4.46	143	8.25	183	6.51
24	20.35	64	17.04	104	12.47	144	7.20	184	4.41	144	8.18	184	6.49
25	20.28	65	16.94	105	12.33	145	7.10	185	4.36	145	8.12	185	6.46
26	20.22	66	16.84	106	12.20	146	7.01	186	4.32	146	8.05	186	6.44
27	20.15	67	16.74	107	12.07	147	6.91	187	4.27	147	7.99	187	6.42
28	20.08	68	16.64	108	11.94	148	6.82	188	4.23	148	7.93	188	6.40
29	20.01	69	16.53	109	11.81	149	6.73	189	4.18	149	7.87	189	6.38
30	19.94	70	16.43	110	11.67	150	6.64	190	4.14	150	7.81	190	6.36
31	19.87	71	16.33	111	11.54	151	6.55	191	4.09	151	7.75	191	6.35
32	19.80	72	16.22	112	11.40	152	6.46	192	4.05	152	7.69	192	6.33
33	19.73	73	16.12	113	11.26	153	6.38	193	4.01	153	7.64	193	6.31
34	19.65	74	16.01	114	11.13	154	6.30	194	3.97	154	7.59	194	6.30
35	19.58	75	15.90	115	10.99	155	6.22	195	3.93	155	7.53	195	6.28
36	19.50	76	15.79	116	10.85	156	6.14	196	3.89	156	7.48	196	6.27
37	19.42	77	15.69	117	10.71	157	6.06	197	3.85	157	7.43	197	6.26
38	19.35	78	15.58	118	10.57	158	5.98	198	3.81	158	7.39	198	6.24
39	19.27	79	15.47	119	10.43	159	5.91	199	3.77	159	7.34	199	6.23
40	19.19	80	15.36	120	10.28	160	5.83	200	3.73	160	7.29	200	6.22

*** *K* taken as 1.0 for secondary members.**

Sect. 1.9.2; the ratio of length (in direction of stress) to width of hole shall not exceed 2; the clear distance between holes in the direction of stress shall be not less than the transverse distance between nearest lines of connecting rivets, bolts or welds; and the periphery of the holes at all points shall have a minimum radius of 1½ inches.

c In recognition of the fact that increased working of the metal through the rolling mill produces a steel of higher yield strength, the 1963 AISC Code sets up three specified yield points for high-strength steel: ASTM A242, A440, and A441. The yield points are based on the average thickness of metal

Table 7.2 Allowable Stress (ksi) for Compression Members of 42 ksi Specified Yield Point Steel, F_y = 42 ksi

Main and Secondary Members Kl/r not over 120						Main Members Kl/r 121 to 200				Secondary Members* l/r 121 to 200			
$\frac{Kl}{r}$	F_a (ksi)	$\frac{Kl}{r}$	F_a (ksi)	$\frac{Kl}{r}$	F_a (ksi)	$\frac{Kl}{r}$	F_a (ksi)	$\frac{Kl}{r}$	F_a (ksi)	$\frac{l}{r}$	F_{as} (ksi)	$\frac{l}{r}$	F_{as} (ksi)
1	25.15	41	21.98	81	16.92	121	10.20	161	5.76	121	10.25	161	7.25
2	25.10	42	21.87	82	16.77	122	10.03	162	5.69	122	10.13	162	7.20
3	25.05	43	21.77	83	16.62	123	9.87	163	5.62	123	10.02	163	7.16
4	24.99	44	21.66	84	16.47	124	9.71	164	5.55	124	9.91	164	7.12
5	24.94	45	21.55	85	16.32	125	9.56	165	5.49	125	9.80	165	7.08
6	24.88	46	21.44	86	16.17	126	9.41	166	5.42	126	9.70	166	7.04
7	24.82	47	21.33	87	16.01	127	9.26	167	5.35	127	9.59	167	7.00
8	24.76	48	21.22	88	15.86	128	9.11	168	5.29	128	9.49	168	6.96
9	24.70	49	21.10	89	15.71	129	8.97	169	5.23	129	9.40	169	6.93
10	24.63	50	20.99	90	15.55	130	8.84	170	5.17	130	9.30	170	6.89
11	24.57	51	20.87	91	15.39	131	8.70	171	5.11	131	9.21	171	6.85
12	24.50	52	20.76	92	15.23	132	8.57	172	5.05	132	9.12	172	6.82
13	24.43	53	20.64	93	15.07	133	8.44	173	4.99	133	9.03	173	6.79
14	24.36	54	20.52	94	14.91	134	8.32	174	4.93	134	8.94	174	6.76
15	24.29	55	20.40	95	14.75	135	8.19	175	4.88	135	8.86	175	6.73
16	24.22	56	20.28	96	14.59	136	8.07	176	4.82	136	8.78	176	6.70
17	24.15	57	20.16	97	14.43	137	7.96	177	4.77	137	8.70	177	6.67
18	24.07	58	20.03	98	14.26	138	7.84	178	4.71	138	8.62	178	6.64
19	24.00	59	19.91	99	14.09	139	7.73	179	4.66	139	8.54	179	6.61
20	23.92	60	19.79	100	13.93	140	7.62	180	4.61	140	8.47	180	6.58
21	23.84	61	19.66	101	13.76	141	7.51	181	4.56	141	8.39	181	6.56
22	23.76	62	19.53	102	13.59	142	7.41	182	4.51	142	8.32	182	6.53
23	23.68	63	19.40	103	13.42	143	7.30	183	4.46	143	8.25	183	6.51
24	23.59	64	19.27	104	13.25	144	7.20	184	4.41	144	8.18	184	6.49
25	23.51	65	19.14	105	13.08	145	7.10	185	4.36	145	8.12	185	6.46
26	23.42	66	19.01	106	12.90	146	7.01	186	4.32	146	8.05	186	6.44
27	23.33	67	18.88	107	12.73	147	6.91	187	4.27	147	7.99	187	6.42
28	23.24	68	18.75	108	12.55	148	6.82	188	4.23	148	7.93	188	6.40
29	23.15	69	18.61	109	12.37	149	6.73	189	4.18	149	7.87	189	6.38
30	23.06	70	18.48	110	12.19	150	6.64	190	4.14	150	7.81	190	6.36
31	22.97	71	18.34	111	12.01	151	6.55	191	4.09	151	7.75	191	6.35
32	22.88	72	18.20	112	11.83	152	6.46	192	4.05	152	7.69	192	6.33
33	22.78	73	18.06	113	11.65	153	6.38	193	4.01	153	7.64	193	6.31
34	22.69	74	17.92	114	11.47	154	6.30	194	3.97	154	7.59	194	6.30
35	22.59	75	17.78	115	11.28	155	6.22	195	3.93	155	7.53	195	6.28
36	22.49	76	17.64	116	11.10	156	6.14	196	3.89	156	7.48	196	6.27
37	22.39	77	17.50	117	10.91	157	6.06	197	3.85	157	7.43	197	6.26
38	22.29	78	17.35	118	10.72	158	5.98	198	3.81	158	7.39	198	6.24
39	22.19	79	17.21	119	10.55	159	5.91	199	3.77	159	7.34	199	6.23
40	22.08	80	17.06	120	10.37	160	5.83	200	3.73	160	7.29	200	6.22

*** K taken as 1.0 for secondary members.**

as follows:

For the above named steels over 1½ in. in thickness, up to and including 4 in., F_y = 42,000 psi

Over ¾ in. up to and including 1½ in. in thickness, F_y = 46,000 psi

For thicknesses up to and including ¾ in.,

F_y = 50,000 psi

On the basis of the specified yield point for A36 steel and the yield points listed above and in accordance with Section 1.5.1.3 of the AISC Code, Tables 7.1, 7.2, 7.3, and 7.4 give the allowable stress for each of the above yield points for a complete range of

Table 7.3 Allowable Stress (ksi) for Compression Members of 46 ksi Specified Yield Point Steel, f_y = 46 ksi

Main and Secondary Members Kl/r not over 120						Main Members Kl/r 121 to 200				Secondary Members* l/r 121 to 200			
$\frac{Kl}{r}$	F_a (ksi)	$\frac{Kl}{r}$	F_a (ksi)	$\frac{Kl}{r}$	F_a (ksi)	$\frac{Kl}{r}$	F_a (ksi)	$\frac{Kl}{r}$	F_a (ksi)	$\frac{l}{r}$	F_{as} (ksi)	$\frac{l}{r}$	F_{as} (ksi)
1	27.54	41	23.85	81	17.91	121	10.20	161	5.76	121	10.25	161	7.25
2	27.48	42	23.73	82	17.74	122	10.03	162	5.69	122	10.13	162	7.20
3	27.42	43	23.60	83	17.56	123	9.87	163	5.62	123	10.02	163	7.16
4	27.36	44	23.48	84	17.39	124	9.71	164	5.55	124	9.91	164	7.12
5	27.30	45	23.35	85	17.21	125	9.56	165	5.49	125	9.80	165	7.08
6	27.23	46	23.22	86	17.03	126	9.41	166	5.42	126	9.70	166	7.04
7	27.16	47	23.09	87	16.85	127	9.26	167	5.35	127	9.59	167	7.00
8	27.09	48	22.96	88	16.67	128	9.11	168	5.29	128	9.49	168	6.96
9	27.02	49	22.83	89	16.48	129	8.97	169	5.23	129	9.40	169	6.93
10	26.95	50	22.69	90	16.30	130	8.84	170	5.17	130	9.30	170	6.89
11	26.87	51	22.56	91	16.12	131	8.70	171	5.11	131	9.21	171	6.85
12	26.79	52	22.42	92	15.93	132	8.57	172	5.05	132	9.12	172	6.82
13	26.72	53	22.28	93	15.74	133	8.44	173	4.99	133	9.03	173	6.79
14	26.63	54	22.14	94	15.55	134	8.32	174	4.93	134	8.94	174	6.76
15	26.55	55	22.00	95	15.36	135	8.19	175	4.88	135	8.86	175	6.73
16	26.47	56	21.86	96	15.17	136	8.07	176	4.82	136	8.78	176	6.70
17	26.38	57	21.72	97	14.97	137	7.96	177	4.77	137	8.70	177	6.67
18	26.29	58	21.57	98	14.78	138	7.84	178	4.71	138	8.62	178	6.64
19	26.21	59	21.43	99	14.58	139	7.73	179	4.66	139	8.54	179	6.61
20	26.11	60	21.28	100	14.39	140	7.62	180	4.61	140	8.47	180	6.58
21	26.02	61	21.13	101	14.19	141	7.51	181	4.56	141	8.39	181	6.56
22	25.93	62	20.98	102	13.99	142	7.41	182	4.51	142	8.32	182	6.53
23	25.83	63	20.83	103	13.79	143	7.30	183	4.46	143	8.25	183	6.51
24	25.73	64	20.68	104	13.58	144	7.20	184	4.41	144	8.18	184	6.49
25	25.64	65	20.53	105	13.38	145	7.10	185	4.36	145	8.12	185	6.46
26	25.54	66	20.37	106	13.17	146	7.01	186	4.32	146	8.05	186	6.44
27	25.43	67	20.22	107	12.96	147	6.91	187	4.27	147	7.99	187	6.42
28	25.33	68	20.06	108	12.75	148	6.82	188	4.23	148	7.93	188	6.40
29	25.23	69	19.90	109	12.54	149	6.73	189	4.18	149	7.87	189	6.38
30	25.12	70	19.74	110	12.33	150	6.64	190	4.14	150	7.81	190	6.36
31	25.01	71	19.58	111	12.12	151	6.55	191	4.09	151	7.75	191	6.35
32	24.90	72	19.42	112	11.90	152	6.46	192	4.05	152	7.69	192	6.33
33	24.79	73	19.26	113	11.69	153	6.38	193	4.01	153	7.64	193	6.31
34	24.68	74	19.10	114	11.49	154	6.30	194	3.97	154	7.59	194	6.30
35	24.56	75	18.93	115	11.29	155	6.22	195	3.93	155	7.53	195	6.28
36	24.45	76	18.76	116	11.10	156	6.14	196	3.89	156	7.48	196	6.27
37	24.33	77	18.60	117	10.91	157	6.06	197	3.85	157	7.43	197	6.26
38	24.21	78	18.43	118	10.72	158	5.98	198	3.81	158	7.39	198	6.24
39	24.10	79	18.26	119	10.55	159	5.91	199	3.77	159	7.34	199	6.23
40	23.97	80	18.08	120	10.37	160	5.83	200	3.73	160	7.29	200	6.22

*** K taken as 1.0 for secondary members.**

slenderness ratios up to 120. Although these tables give allowable stresses for main members, or bracing members up to l/r = 200, it is inadvisable to go beyond 120. Slenderness ratios in excess of this allow appreciable vibration due to wind or to heavy traffic and, although no actual danger may be indicated, the psychological effect may be damaging to the reputation of the designer.

Table 7.4 Allowable Stress (ksi) for Compression Members of 50 ksi Specified Yield Point Steel, F_y = 50 ksi

Main and Secondary Members Kl/r not over 120						Main Members Kl/r 121 to 200				Secondary Members* l/r 121 to 200			
$\frac{Kl}{r}$	F_a (ksi)	$\frac{Kl}{r}$	F_a (ksi)	$\frac{Kl}{r}$	F_a (ksi)	$\frac{Kl}{r}$	F_a (ksi)	$\frac{Kl}{r}$	F_a (ksi)	$\frac{l}{r}$	F_{as} (ksi)	$\frac{l}{r}$	F_{as} (ksi)
1	29.94	41	25.69	81	18.81	121	10.20	161	5.76	121	10.25	161	7.25
2	29.87	42	25.55	82	18.61	122	10.03	162	5.69	122	10.13	162	7.20
3	29.80	43	25.40	83	18.41	123	9.87	163	5.62	123	10.02	163	7.16
4	29.73	44	25.26	84	18.20	124	9.71	164	5.55	124	9.91	164	7.12
5	29.66	45	25.11	85	17.99	125	9.56	165	5.49	125	9.80	165	7.08
6	29.58	46	24.96	86	17.79	126	9.41	166	5.42	126	9.70	166	7.04
7	29.50	47	24.81	87	17.58	127	9.26	167	5.35	127	9.59	167	7.00
8	29.42	48	24.66	88	17.37	128	9.11	168	5.29	128	9.49	168	6.96
9	29.34	49	24.51	89	17.15	129	8.97	169	5.23	129	9.40	169	6.93
10	29.26	50	24.35	90	16.94	130	8.84	170	5.17	130	9.30	170	6.89
11	29.17	51	24.19	91	16.72	131	8.70	171	5.11	131	9.21	171	6.85
12	29.08	52	24.04	92	16.50	132	8.57	172	5.05	132	9.12	172	6.82
13	28.99	53	23.88	93	16.29	133	8.44	173	4.99	133	9.03	173	6.79
14	28.90	54	23.72	94	16.06	134	8.32	174	4.93	134	8.94	174	6.76
15	28.80	55	23.55	95	15.84	135	8.19	175	4.88	135	8.86	175	6.73
16	28.71	56	23.39	96	15.62	136	8.07	176	4.82	136	8.78	176	6.70
17	28.61	57	23.22	97	15.39	137	7.96	177	4.77	137	8.70	177	6.67
18	28.51	58	23.06	98	15.17	138	7.84	178	4.71	138	8.62	178	6.64
19	28.40	59	22.89	99	14.94	139	7.73	179	4.66	139	8.54	179	6.61
20	28.30	60	22.72	100	14.71	140	7.62	180	4.61	140	8.47	180	6.58
21	28.19	61	22.55	101	14.47	141	7.51	181	4.56	141	8.39	181	6.56
22	28.08	62	22.37	102	14.24	142	7.41	182	4.51	142	8.32	182	6.53
23	27.97	63	22.20	103	14.00	143	7.30	183	4.46	143	8.25	183	6.51
24	27.86	64	22.02	104	13.77	144	7.20	184	4.41	144	8.18	184	6.49
25	27.75	65	21.85	105	13.53	145	7.10	185	4.36	145	8.12	185	6.46
26	27.63	66	21.67	106	13.29	146	7.01	186	4.32	146	8.05	186	6.44
27	27.52	67	21.49	107	13.04	147	6.91	187	4.27	147	7.99	187	6.42
28	27.40	68	21.31	108	12.80	148	6.82	188	4.23	148	7.93	188	6.40
29	27.28	69	21.12	109	12.57	149	6.73	189	4.18	149	7.87	189	6.38
30	27.15	70	20.94	110	12.34	150	6.64	190	4.14	150	7.81	190	6.36
31	27.03	71	20.75	111	12.12	151	6.55	191	4.09	151	7.75	191	6.35
32	26.90	72	20.56	112	11.90	152	6.46	192	4.05	152	7.69	192	6.33
33	26.77	73	20.38	113	11.69	153	6.38	193	4.01	153	7.64	193	6.31
34	26.64	74	20.19	114	11.49	154	6.30	194	3.97	154	7.59	194	6.30
35	26.51	75	19.99	115	11.29	155	6.22	195	3.93	155	7.53	195	6.28
36	26.38	76	19.80	116	11.10	156	6.14	196	3.89	156	7.48	196	6.27
37	26.25	77	19.61	117	10.91	157	6.06	197	3.85	157	7.43	197	6.26
38	26.11	78	19.41	118	10.72	158	5.98	198	3.81	158	7.39	198	6.24
39	25.97	79	19.21	119	10.55	159	5.91	199	3.77	159	7.34	199	6.23
40	25.83	80	19.01	120	10.37	160	5.83	200	3.73	160	7.29	200	6.22

*** K taken as 1.0 for secondary members.**

d For axial load only, it is not necessary to compute the column area since for ordinary WF column sections, the size may be selected directly from Table 7.5. In accordance with the comment in paragraph *c* above, this table is laid out only up to an effective slenderness ratio of 120. It has been prepared in accordance with the allowable stresses given in Tables 7.1 through 7.4. For each column section, there are two series

Table 7.5 Allowable Column Loads

	14 WF 426		14 WF 398		14 WF 370		14 WF 342	
r_x/r_y	1.67		1.66		1.66		1.66	
B_x	.177		.178		.179		.180	
B_y	.443		.445		.451		.456	
KL	P_c	P_h*	P_c	P_h*	P_c	P_h*	P_c	P_h*
7	2,585	3,002	2,413	2,802	2,243	2,604	2,073	2,407
8	2,564	2,974	2,393	2,776	2,224	2,580	2,056	2,384
9	2,541	2,945	2,372	2,749	2,204	2,554	2,037	2,360
10	2,518	2,915	2,350	2,721	2,184	2,527	2,018	2,336
11	2,494	2,884	2,328	2,691	2,162	2,500	1,998	2,309
12	2,469	2,851	2,304	2,660	2,140	2,471	1,977	2,282
13	2,443	2,817	2,280	2,629	2,117	2,441	1,956	2,254
14	2,416	2,783	2,254	2,596	2,093	2,410	1,933	2,225
15	2,389	2,747	2,228	2,562	2,069	2,378	1,910	2,195
16	2,360	2,709	2,201	2,527	2,043	2,345	1,887	2,165
17	2,331	2,671	2,174	2,491	2,017	2,311	1,862	2,139
18	2,301	2,632	2,145	2,454	1,991	2,276	1,837	2,100
19	2,270	2,592	2,116	2,416	1,963	2,240	1,812	2,067
20	2,239	2,550	2,087	2,377	1,935	2,203	1,786	2,032
21	2,206	2,508	2,056	2,337	1,906	2,165	1,759	1,997
22	2,173	2,465	2,025	2,296	1,877	2,127	1,731	1,961
23	2,139	2,420	1,993	2,254	1,847	2,087	1,703	1,924
24	2,105	2,375	1,960	2,212	1,816	2,047	1,674	1,886
P_{a+b}	2,805	3,156	2,574	2,948	2,393	2,741	2,213	2,535

	14 WF 314		14 WF 287		14 WF 264		14 WF 246	
r_x/r_y	1.64		1.63		1.63		1.62	
B_x	.180		.181		.182		.182	
B_y	.459		.464		.467		.469	
KL	P_c	P_h*	P_c	P_h*	P_c	P_h*	P_c	P_h*
7	1,901	2,208	1,737	2,017	1,598	1,855	1,488	1,728
8	1,885	2,186	1,722	1,997	1,584	1,837	1,475	1,711
9	1,868	2,164	1,706	1,977	1,569	1,818	1,461	1,693
10	1,850	2,141	1,690	1,955	1,554	1,798	1,447	1,674
11	1,831	2,116	1,672	1,933	1,538	1,777	1,432	1,654
12	1,812	2,091	1,655	1,910	1,521	1,755	1,416	1,634
13	1,792	2,065	1,636	1,886	1,504	1,733	1,400	1,613
14	1,771	2,038	1,617	1,861	1,486	1,710	1,383	1,591
15	1,750	2,010	1,597	1,835	1,468	1,686	1,366	1,569
16	1,728	1,982	1,577	1,808	1,449	1,661	1,349	1,546
17	1,705	1,952	1,556	1,781	1,429	1,636	1,330	1,529
18	1,682	1,922	1,535	1,753	1,410	1,610	1,312	1,498
19	1,658	1,891	1,513	1,724	1,389	1,583	1,292	1,472
20	1,634	1,859	1,490	1,695	1,368	1,555	1,273	1,447
21	1,609	1,826	1,467	1,664	1,347	1,527	1,253	1,420
22	1,583	1,792	1,443	1,633	1,325	1,498	1,232	1,393
23	1,557	1,758	1,419	1,602	1,302	1,469	1,211	1,365
24	1,530	1,723	1,394	1,569	1,279	1,439	1,189	1,337
P_{a+b}	2,031	2,326	1,856	2,126	1,708	1,956	1,591	1,823

Table 7.5 Allowable Column Loads (*Continued*)

	14 WF 237		14 WF 228		14 WF 219		14 WF 211	
r_x/r_y	1.62		1.61		1.62		1.61	
B_x	.182		.182		.183		.183	
B_y	.472		.473		.475		.477	
KL	P_c	P_h*	P_c	P_h*	P_c	P_h*	P_c	P_h*
7	1,434	1,664	1,379	1,601	1,323	1,536	1,276	1,617
8	1,421	1,647	1,367	1,585	1,312	1,521	1,265	1,600
9	1,408	1,630	1,354	1,569	1,299	1,504	1,253	1,582
10	1,394	1,612	1,341	1,550	1,286	1,487	1,240	1,563
11	1,379	1,594	1,327	1,533	1,273	1,470	1,227	1,543
12	1,364	1,574	1,312	1,514	1,259	1,452	1,213	1,522
13	1,349	1,553	1,297	1,494	1,244	1,433	1,199	1,501
14	1,332	1,532	1,282	1,474	1,229	1,413	1,185	1,479
15	1,316	1,512	1,266	1,453	1,213	1,393	1,170	1,456
16	1,299	1,489	1,249	1,432	1,197	1,372	1,154	1,432
17	1,281	1,466	1,232	1,410	1,181	1,351	1,138	1,408
18	1,263	1,442	1,215	1,387	1,164	1,329	1,122	1,383
19	1,244	1,417	1,197	1,363	1,147	1,306	1,105	1,357
20	1,225	1,392	1,178	1,339	1,129	1,283	1,088	1,331
21	1,206	1,367	1,159	1,314	1,111	1,259	1,070	1,303
22	1,186	1,341	1,140	1,289	1,092	1,234	1,052	1,276
23	1,165	1,314	1,120	1,263	1,073	1,210	1,034	1,247
24	1,145	1,287	1,100	1,237	1,054	1,184	1,015	1,218
P_{a+b}	1,533	1,756	1,475	1,690	1,416	1,622	1,366	1,564

	14 WF 202		14 WF 193		14 WF 184		14 WF 176	
r_x/r_y	1.60		1.61		1.61		1.60	
B_x	.183		.183		.183		.184	
B_y	.477		.479		.480		.483	
KL	P_c	P_h	P_c	P_h	P_c	P_h	P_c	P_h
7	1,221	1,547	1,166	1,478	1,111	1,408	1,063	1,347
8	1,210	1,531	1,155	1,462	1,101	1,393	1,053	1,332
9	1,198	1,513	1,144	1,445	1,090	1,377	1,043	1,316
10	1,186	1,495	1,133	1,427	1,079	1,360	1,032	1,300
11	1,174	1,476	1,121	1,409	1,068	1,342	1,021	1,283
12	1,161	1,456	1,108	1,390	1,056	1,324	1,009	1,266
13	1,147	1,435	1,095	1,370	1,044	1,305	996	1,248
14	1,133	1,414	1,082	1,350	1,031	1,286	985	1,229
15	1,119	1,392	1,068	1,329	1,017	1,266	972	1,210
16	1,104	1,369	1,054	1,307	1,004	1,245	959	1,189
17	1,089	1,346	1,039	1,285	990	1,224	946	1,169
18	1,073	1,322	1,024	1,262	975	1,201	932	1,148
19	1,057	1,297	1,009	1,238	961	1,179	918	1,126
20	1,040	1,272	993	1,214	946	1,156	903	1,103
21	1,023	1,246	977	1,189	930	1,132	888	1,080
22	1,006	1,219	960	1,163	914	1,107	873	1,057
23	988	1,192	943	1,137	898	1,082	857	1,033
24	970	1,164	926	1,110	881	1,057	841	1,008
P_{a+b}	1,307	1,497	1,248	1,546	1,190	1,472	1,138	1,408

Table 7.5 Allowable Column Loads (*Continued*)

	14 WF 167		14 WF 158		14 WF 150		14 WF 142	
r_x/r_y	1.60		1.60		1.60		1.59	
B_x	.184		.183		.184		.185	
B_y	.485		.485		.487		.491	
KL	P_c	P_h	P_c	P_h	P_c	P_h	P_c	P_h
7	1,008	1,278	954	1,209	905	1,147	859	1,088
8	999	1,264	946	1,196	897	1,134	851	1,076
9	989	1,249	936	1,182	888	1,121	843	1,064
10	979	1,234	927	1,167	879	1,107	834	1,050
11	969	1,217	917	1,152	869	1,092	825	1,036
12	969	1,217	917	1,152	869	1,092	825	1,036
12	958	1,201	906	1,136	859	1,077	815	1,022
13	946	1,183	895	1,120	849	1,062	805	1,007
14	935	1,166	884	1,103	838	1,045	795	991
15	922	1,147	873	1,085	827	1,029	785	975
16	910	1,128	861	1,067	816	1,011	774	959
17	897	1,108	849	1,048	804	994	763	942
18	884	1,088	836	1,029	793	975	751	924
19	870	1,067	823	1,009	780	957	740	906
20	856	1,046	810	989	768	937	728	888
21	842	1,024	796	968	756	917	715	869
22	827	1,002	783	947	742	897	703	850
23	813	979	768	925	728	876	690	830
24	797	955	754	903	714	855	676	809
P_{a+b}	1,080	1,355	1,022	1,283	970	1,217	921	1,155

	14 WF 136		14 WF 127		14 WF 119		14 WF 111	
r_x/r_y	1.67		1.67		1.67		1.67	
B_x	.185		.185		.185		.185	
KL	P_c	P_h	P_c	P_h	P_c	P_h	P_c	P_h
7	818	1,122	764	1,047	716	981	667	915
8	810	1,107	756	1,034	708	969	661	903
9	801	1,092	748	1,019	701	955	653	891
10	792	1,076	739	1,005	693	941	646	878
11	783	1,060	731	989	684	927	638	864
12	773	1,043	721	973	676	912	630	850
13	763	1,025	712	956	667	896	622	835
14	752	1,007	702	939	658	880	613	820
15	742	987	692	921	648	863	604	804
16	730	968	682	903	638	845	595	787
17	719	947	671	884	628	828	585	771
18	707	926	660	864	618	809	576	753
19	695	905	648	844	607	790	566	735
20	683	883	637	823	596	771	555	717
21	670	860	625	802	585	751	545	698
22	657	837	613	780	573	730	534	679
23	643	813	600	758	562	709	523	659
24	630	789	587	735	550	687	511	639
P_{a+b}	876	1,103	821	1,030	770	966	718	901

Table 7.5 Allowable Column Loads (*Continued*)

	14 WF 103		14 WF 95		14 WF 87		14 WF 84	
r_x/r_y	1.67		1.66		1.66		2.03	
B_x	.185		.186		.185		.189	
B_y	.525		.529		.530		.659	
KL	P_c	P_h	P_c	P_h	P_c	P_h	P_c	P_h
7	618	848	571	783	522	716	497	678
8	612	837	565	772	517	706	490	665
9	605	825	559	762	511	697	482	653
10	599	813	552	750	505	686	475	639
11	591	800	546	739	499	675	467	625
12	584	787	539	726	493	664	458	611
13	576	773	532	714	486	652	450	595
14	568	759	524	700	479	640	441	579
15	560	744	516	687	472	628	432	563
16	551	729	508	673	465	615	422	546
17	542	714	500	658	457	601	412	528
18	533	697	492	643	449	588	402	510
19	524	681	483	628	441	574	392	492
20	514	664	474	612	433	559	381	472
21	504	646	465	596	425	544	370	453
22	494	628	456	579	416	529	359	432
23	484	610	446	562	408	513	347	411
24	473	591	436	545	399	497	335	389
P_{a+b}	666	835	615	838	562	767	544	682

	14 WF 78		14 WF 74		14 WF 68		14 WF 61	
r_x/r_y	2.09		2.44		2.45		2.44	
B_x	.189		.194		.194		.195	
B_y	.665		.821		.830		.834	
KL	P_c	P_h	P_c	P_h	P_c	P_h	P_c	P_h
7	461	628	428	580	393	533	352	477
8	454	617	420	566	385	519	345	465
9	447	605	411	551	377	505	338	453
10	440	593	402	535	369	491	331	440
11	433	579	393	519	360	475	323	426
12	425	566	383	501	351	459	315	411
13	417	551	373	483	342	443	306	396
14	409	537	363	465	332	425	297	380
15	400	521	352	445	322	407	288	364
16	391	505	340	425	311	388	279	347
17	382	489	329	404	301	368	269	329
18	372	472	317	382	289	348	259	311
19	362	454	304	359	278	327	248	292
20	352	436	291	336	266	305	237	272
21	342	418	278	312	253	283	226	252
22	331	398	264	287	241	259	215	231
23	320	379	250	262	227	237	203	211
24	309	258	236	241	214	218	191	194
P_{a+b}	505	688	478	601	440	600	395	538

Table 7.5 Allowable Column Loads (*Continued*)

	14 WF 53		14 WF 48		14 WF 43	
r_x/r_y	3.07		3.07		3.08	
B_x	.200		.201		.202	
B_y	1.090		1.102		1.119	
KL	P_c	P_h	P_c	P_h	P_c	P_h
7	294	394	266	356	238	316
8	286	380	259	343	231	304
9	277	364	251	329	224	291
10	268	347	242	314	216	278
11	258	330	233	298	208	263
12	248	312	224	281	200	248
13	237	292	214	264	191	232
14	226	272	204	245	181	216
15	214	251	193	226	172	198
16	202	229	182	206	162	180
17	190	206	171	185	151	162
18	177	184	159	165	140	144
19	163	165	146	148	129	130
P_{a+b}	343	468	310	423	278	377

	12 WF 190		12 WF 161		12 WF 133		12 WF 120	
r_x/r_y	1.79		1.77		1.77		1.76	
B_x	.212		.213		.214		.214	
B_y	.600		.610		.620		.620	
KL	P_c	P_h	P_c	P_h	P_c	P_h	P_c	P_h
7	1,130	1,427	957	1,209	789	996	712	899
8	1,116	1,406	945	1,190	779	981	702	884
9	1,101	1,383	932	1,171	768	964	693	869
10	1,085	1,360	919	1,150	757	947	682	854
11	1,069	1,335	904	1,129	745	929	671	837
12	1,052	1,309	890	1,106	733	910	660	820
13	1,035	1,283	875	1,083	720	891	648	802
14	1,017	1,255	859	1,059	707	871	636	784
15	998	1,226	843	1,034	693	850	624	764
16	979	1,197	826	1,008	679	828	611	745
17	959	1,166	808	982	664	806	598	724
18	938	1,134	791	954	649	783	584	703
19	917	1,102	772	926	634	759	570	681
20	895	1,068	753	897	618	734	555	659
21	873	1,034	734	867	602	709	540	636
22	850	999	714	837	585	683	525	612
23	827	962	694	805	568	657	509	588
24	827	962	694	805	568	657	509	588
P_{a+b}	1229	1308	1042	1308	860	1,079	777	976

Table 7.5 Allowable Column Loads (*Continued*)

	12 WF 106		12 WF 99		12 WF 92		12 WF 75	
r_x/r_y	1.76		1.76		1.75		1.75	
B_x	.216		.216		.216		.216	
B_y	.634		.637		.641		.642	
KL	P_c	P_h	P_c	P_h	P_c	P_h	P_c	P_h
7	628	959	586	800	545	744	503	686
8	620	844	578	786	537	731	496	674
9	611	828	570	771	530	717	489	662
10	602	812	561	756	521	703	481	648
11	592	795	552	740	513	688	473	635
12	582	777	542	724	504	672	465	620
13	572	759	533	706	495	656	457	605
14	561	740	522	688	486	639	448	589
15	550	720	512	669	476	622	439	573
16	538	699	501	650	466	604	429	557
17	526	678	490	630	455	585	420	539
18	514	656	478	610	444	566	410	521
19	502	634	466	588	433	546	399	503
20	489	610	454	567	422	526	389	484
21	475	587	442	544	410	505	378	464
22	462	562	429	521	398	483	367	444
23	448	537	416	497	386	461	355	424
24	433	511	402	472	373	438	343	402
P_{a+b}	686	861	640	803	595	747	550	689

	12 WF 79		12 WF 72		12 WF 65		12 WF 58	
r_x/r_y	1.75		1.75		1.75		2.10	
B_x	.217		.217		.217		.218	
B_y	.649		.653		.657		.797	
KL	P_c	P_h	P_c	P_h	P_c	P_h	P_c	P_h
7	467	637	425	581	384	524	336	456
8	461	626	420	570	379	514	330	445
9	454	614	413	560	373	505	323	433
10	447	602	407	548	367	494	316	421
11	439	589	400	536	361	484	309	408
12	432	575	393	524	345	472	302	395
13	424	561	386	511	348	460	294	381
14	415	547	378	497	341	448	286	367
15	407	531	370	484	334	435	277	352
16	398	516	362	469	326	422	269	336
17	389	499	354	454	319	409	260	320
18	380	483	345	439	311	395	250	303
19	370	465	336	423	303	380	241	286
20	360	448	327	407	295	365	231	268
21	350	429	318	390	286	350	221	249
22	339	410	308	373	277	334	210	230
23	328	391	299	355	268	318	199	211
24	317	371	288	336	259	301	188	194
P_{a+b}	511	697	466	635	420	573	375	512

Table 7.5 Allowable Column Loads (*Continued*)

	12 WF 53		12 WF 50		12 WF 45		12 WF 40	
r_x/r_y	3.11		2.64		2.65		2.64	
B_x	.221		.227		.227		.227	
B_y	.812		1.051		1.068		1.070	
KL	P_c	P_h	P_c	P_h	P_c	P_h	P_c	P_h
7	307	416	279	374	251	336	223	298
8	301	406	271	361	244	323	217	288
9	295	395	263	346	236	310	210	276
10	288	384	255	331	228	297	203	264
11	282	372	246	315	220	282	196	251
12	275	359	236	298	212	267	188	237
13	267	346	227	281	203	251	180	223
14	260	333	216	263	193	234	172	208
15	252	319	206	243	184	216	163	192
16	244	304	195	223	174	198	154	176
17	235	289	183	202	163	179	145	159
18	227	274	171	181	152	159	135	142
19	218	258	159	162	141	143	125	127
20	209	241						
21	199	223						
22	189	205						
23	179	188						
24	169	173						
P_{a+b}	343	468	324	441	291	397	259	353

	10 WF 112		10 WF 100		10 WF 89		10 WF 77	
r_x/r_y	1.75		1.74		1.73		1.73	
B_x	.261		.262		.263		.263	
B_y	.728		.738		.744		.753	
KL	P_c	P_h	P_c	P_h	P_c	P_h	P_c	P_h
7	653	888	583	793	519	705	448	609
8	642	869	573	775	510	689	440	595
9	631	848	563	757	500	673	432	581
10	618	827	552	738	490	655	424	565
11	606	805	541	718	480	637	414	549
12	592	781	529	696	469	618	405	533
13	579	757	516	674	458	598	395	515
14	564	731	503	651	447	577	385	497
15	550	705	490	627	435	556	374	478
16	534	678	476	603	422	534	363	458
17	519	649	462	577	409	511	352	438
18	502	620	447	551	396	487	340	417
19	485	590	432	523	382	462	328	395
20	468	558	416	495	368	437	316	373
21	450	526	400	466	354	410	303	349
22	432	493	383	435	339	383	290	325
23	413	458	366	404	323	355	276	300
24	394	423	349	372	307	326	262	276
P_{a+b}	724	909	647	812	576	723	499	626

Table 7.5 Allowable Column Loads (*Continued*)

	10 WF 72		10 WF 66		10 WF 60		10 WF 54	
r_x/r_y	1.72		1.72		1.72		1.72	
B_x	.264		.263		.263		.263	
B_y	.759		.761		.765		.767	
KL	P_c	P_h	P_c	P_h	P_c	P_h	P_c	P_h
7	419	569	383	521	349	474	313	426
8	411	556	377	509	343	463	308	416
9	404	542	370	496	336	451	302	405
10	395	528	362	483	329	439	296	394
11	387	513	354	469	322	426	289	383
12	378	497	346	455	314	413	282	371
13	369	480	337	439	307	399	275	358
14	359	463	329	424	299	385	268	345
15	349	445	319	407	290	370	260	332
16	339	427	310	390	281	354	253	317
17	328	408	300	372	272	338	244	303
18	317	388	290	354	263	321	236	288
19	306	368	279	336	254	304	227	272
20	294	346	269	316	244	286	218	256
21	282	325	258	296	234	268	209	239
22	270	302	246	275	223	248	200	222
23	257	279	234	253	212	229	190	204
24	244	256	222	233	201	210	180	187
P_{a+b}	466	585	427	582	339	530	349	476

	10 WF 49		10 WF 45		10 WF 39		10 WF 33	
r_x/r_y	1.71		2.17		2.16		2.16	
B_x	.264		.270		.275		.272	
B_y	.774		.995		1.084		1.055	
KL	P_c	P_h	P_c	P_h	P_c	P_h	P_c	P_h
7	284	385	252	338	218	293	184	246
8	279	376	245	326	212	282	179	237
9	273	367	238	314	206	271	173	228
10	268	357	231	301	199	260	167	217
11	262	346	223	287	193	247	161	207
12	256	335	215	272	185	235	155	196
13	249	324	206	257	178	221	149	184
14	242	312	197	241	170	207	142	171
15	235	299	188	224	162	192	135	159
16	228	286	178	207	153	177	127	145
17	221	273	168	189	145	161	120	131
18	213	259	158	170	135	144	112	117
19	205	245	147	152	126	129	103	105
20	197	230	136	137				
21	188	214						
22	180	198						
23	171	182						
24	161	167						
P_{a+b}	317	432	291	397	253	344	214	291

Table 7.5 Allowable Column Loads (*Continued*)

	8 WF 67		8 WF 58		8 WF 48		8 WF 40	
r_x/r_y	1.75		1.74		1.74		1.73	
B_x	.326		.328		.327		.331	
B_y	.921		.937		.941		.972	
KL	P_c	P_h	P_c	P_h	P_c	P_h	P_c	P_h
7	379	510	327	441	270	364	225	302
8	370	494	319	427	264	352	219	292
9	360	477	311	412	257	339	213	281
10	350	459	302	396	249	326	206	270
11	339	440	293	379	241	312	200	258
12	328	420	283	362	233	297	193	245
13	316	399	273	343	224	282	185	232
14	304	378	262	324	215	266	177	218
15	292	355	251	304	206	249	170	204
16	279	331	239	284	197	232	161	189
17	265	307	228	262	187	214	153	173
18	251	281	215	239	176	195	144	157
19	236	254	202	216	165	175	134	141
20	221	230	189	195	154	158	125	127
21	206	208	175	177				
P_{a+b}	433	544	375	471	310	423	259	353

	8 WF 35		8 WF 31		8 WF 28		8 WF 24	
r_x/r_y	1.72		1.73		2.13		2.12	
B_x	.331		.333		.337		.339	
B_y	.972		.991		1.244		1.261	
KL	P_c	P_h	P_c	P_h	P_c	P_h	P_c	P_h
7	197	264	174	235	150	198	128	169
8	191	255	169	225	144	188	123	161
9	186	246	164	217	138	177	118	152
10	180	236	159	208	132	166	113	142
11	174	225	154	198	125	154	107	131
12	168	214	148	188	118	141	101	120
13	162	202	142	178	111	128	94	109
14	155	190	136	167	103	114	88	97
15	148	177	130	155	95	100	81	84
16	141	164	123	143	86	87	73	74
17	133	150	117	131				
18	125	136	110	118				
19	117	122	102	106				
20	109	110	95	96				
P_{a+b}	227	309	201	274	181	247	155	211

of load capacities, one marked P_c for A36 steel and the other marked P_h for high-strength steel in accordance with Table 7.2, 7.3, or 7.4, whichever is applicable. For loads beyond the slenderness ratio of 120, the designer is referred to the AISC Manual.

Table 7.5 Allowable Column Loads (*Continued*)

	6 WF 25		6 WF 20		6 WF 15.5	
r_x/r_y	1.77		1.77		1.77	
B_x	.439		.440		.457	
B_y	1.316		1.341		1.444	
KL	P_c	P_h	P_c	P_h	P_c	P_h
7	132	173	105	138	81	107
8	126	163	101	130	78	100
9	120	153	96	121	74	93
10	114	142	91	112	70	85
11	107	130	85	102	65	77
12	100	117	80	92	60	69
13	93	104	74	81	55	60
14	85	90	67	70	50	51
15	77	78	61	61		
P_{a+b}	165	221	130	177	102	139

	5 WF 18.5		5 WF 16	
r_x/r_y	1.69		1.69	
B_x	.548		.551	
B_y	1.540		1.567	
KL	P_c	P_h	P_c	P_h
7	92	118	79	101
8	87	109	74	93
9	81	99	69	84
10	75	88	64	74
11	69	76	58	64
12	62	64	52	54
P_{a+b}	120	164	103	141

	4 WF 13	
r_x/r_y	1.74	
B_x	.701	
B_y	2.065	
KL	P_c	P_h
7	57	69
8	51	59
9	45	48
P_{a+b}	84	115

Table 7.5 gives the allowable axial load for effective lengths from 7 to 24 ft, the normal range for most buildings. At the bottom of the table for each column is a line marked P_{a+b}, the allowable stress at 22 ksi for A36 steel, and 25.2, 27.6, or 30 ksi for A440 and A441 steel, the allowable strength for high-strength steel depending on the thickness of flange.

There are three types of lines across Table

7.5, a dash line, a solid line, and a double solid line. The dash line indicates that above this line, the flange width is over 1/13 of the unbraced height and an allowable bending stress of $0.66F_y$ is permissible. The single solid line indicates that below this line the L_u value of the section is exceeded so that the allowable bending stress must be reduced. The above two sets of lines are used only for a combination of axial stress and moment. The double solid line applies only to the high-strength steel and indicates that below this line, the additional cost of steel is greater than the saving in poundage, so there is no apparent economy in the use of a high-strength steel beyond this. In addition to those sections with the double lines, there is no economy in the use of any of the sections where P_h is marked with an asterisk. For other sections where there is no "economy" line shown, the high-strength steel is economical throughout the entire length shown in Table 7.5.

The above statements relative to economy are based on the following cost ratios, which may change from time to time.

For sections with a web thickness of over 0.75 in., the high-strength steel costs 30 percent more than the basic A36 steel and the maximum increase in capacity is only about 26 percent for columns with an effective length of 7 ft, decreasing as this length increases. Therefore, there is no apparent economy in the use of high-strength steel for any of these heavier column sections. This applies to all 14-in. columns above 167 lb, all 12-in. columns over 133 lb, and the 10 WF 112. For web thicknesses from 3/8 to 3/4 in., the material costs 22½ percent more than A36 steel. At this rate, for the following sections, the high-strength steel shows economy up to l/r of 57. Above this, there is no apparent economy. This includes 14 WF 74, 14 WF 84, and all 14-in. columns from 103 to 158 lb; all 12-in. columns from 85 through 120 lb, and all 10-in. columns from 72 through 100 lb, and 8-in. columns from 58 through 67 lb.

The following sections, with material cost increase of 22½ percent but with greater carrying capacity because of flange thickness of less than 3/4 in., will show a price advantage in high-strength steel up to l/r of 83: 14 WF 61, 68, 78, 87, 95, 12 WF 65, 72, 79, 10 WF 60 and 66, and 8 WF 48. For all other sections in the WF column lists, the price premium is 17 percent and for columns below the weights listed above, the high-strength steel shows economy up to l/r of 94. Above this there is no economy. It may be that other factors besides apparent economy may have bearing on the selection of the quality of steel but, in many cases, the above facts will influence your choice.

In most instances, a column is braced at the floor line in both directions. It is not necessary that it be braced by a steel beam. It has been found that any support which will furnish 2 percent of the capacity of the column is sufficient to prevent movement, and any concrete slab will furnish more than this. Table 7.5 is laid out on the basis of a slenderness ratio based on the least l/r; that is, the l/r about the y axis of the section. There may be occasional cases where some additional support in the weak direction has been applied as, for example, a column built into a masonry wall. In this case, we are interested in the capacity of the column based on a slenderness ratio of l/r. The selection from the table may be made on the basis of the effective length in the stronger direction divided by the ratio of r_x/r_y given in Table 7.5 for the column, using the new effective length as the length to select for. For example, let us assume a load of 450 kips on an effective length in the stronger direction of 12 ft, but built into a wall the other direction. From Table 7.5, for a 12-in. WF column, the ratio r_x/r_y is 1.75. The length to use in Table 7.5 is $12/1.75 = 6.83$. Entering the table for A36 steel, for a 7 ft effective length, we find that a 12 WF 79 will carry 467 kips and is therefore the economic column to use.

e Square and rectangular tubing in a wide range of sizes is now available in A36 steel, and these selections lend themselves to use as columns under several conditions. Rectangular sections, used in panel curtain walls or in window walls, receive enough lateral support from the connection of the panels or windows so that the greater l/r only need be considered, and connections of beams may be made to welded seat angles, or to vertical fins for clip angle connections.

There are certain conditions and some locations where ordinary round pipe columns may be either more satisfactory or more readily available than either structural shapes or square or rectangular tubing. Under such conditions, the designer should specify the grade of pipe noted in the AISC Manual, page 3-44. The corresponding tables for steel pipe columns will be found in the AISC Manual, pages 3-45 through 3-47.

The AISC Manual, pages 3-48 through 3-51, gives the properties of and allowable loads on rectangular structural tubing columns. In this latter group, the unsupported length is assumed to be the narrow direction or weak way of the column. In many instances, such as book stack or window wall columns, the rectangular tubing is braced in the weaker direction and, in order to select the proper column, the KL dimension is to be divided by the ratio r_x/r_y at the bottom of this table to determine the equivalent KL dimension and the allowable load selected for the new length.

7.11 Steel columns—axial plus bending stress

a When bending stress in addition to axial stress occurs in a column section as hereinafter described in Art. 7.11*b*, the following section of the AISC Code is applicable in addition to those sections quoted in Art. 7.10.

1.6.1 *Axial Compression and Bending*

Members subjected to both axial compression and bending stresses shall be proportioned to satisfy the following requirements:

When $f_a/F_a \leq 0.15$

$$\frac{f_a}{F_a} + \frac{f_b}{F_b} \leq 1.0 \qquad (6)$$

When $f_a/F_a > 0.15$

$$\frac{f_a}{F_a} + \frac{C_m f_b}{\left(1 - \frac{f_a}{F_e'}\right) F_b} \leq 1.0 \qquad (7a)$$

and, in addition, at points braced in the plane of bending,

$$\frac{f_a}{0.6F_y} + \frac{f_b}{F_b} \leq 1.0 \qquad (7b)$$

where F_a = axial stress that would be permitted if axial force alone existed

F_b = compressive bending stress that would be permitted if bending moment alone existed

$F_e' = \dfrac{149{,}000{,}000}{(Kl_b/r_b)^2}$ (In the expression for F_e', l_b is the actual unbraced length *in the plane of bending* and r_b is the corresponding radius of gyration. K is the effective length factor *in the plane of bending*. As in the case of F_a, F_b and 0.6 F_y, F_e' may be increased one-third in accordance with Sect. 1.5.6.)

f_a = computed axial stress

f_b = computed compressive bending stress at the point under consideration

C_m = a coefficient whose value shall be taken as follows:

1. For compression members in frames subject to joint translation (sidesway), $C_m = 0.85$.
2. For restrained compression members in frames braced against joint translation and not subject to transverse loading between their supports in the plane of bending,

$$C_m = 0.6 + 0.4\frac{M_1}{M_2}, \text{ but not less than 0.4,}$$

where M_1/M_2 is the ratio of the smaller to larger moments at the ends of that portion of the member, unbraced in the plane of bending under consideration. M_1/M_2 is positive when the member is bent in single curvature and negative when it is bent in reverse curvature.

3. For compression members in frames braced against joint translation in the plane of loading and subjected to transverse loading between their supports, the value of C_m may be determined by rational analysis. However, in lieu of such analysis, the following values may be used: (a) for members whose ends are restrained, $C_m = 0.85$, (b) for members whose ends are unrestrained, $C_m = 1.0$.

There are several conditions under which it becomes necessary to design a steel column for eccentricity or a combination of bending and direct stress. Figure 7.2 indicates

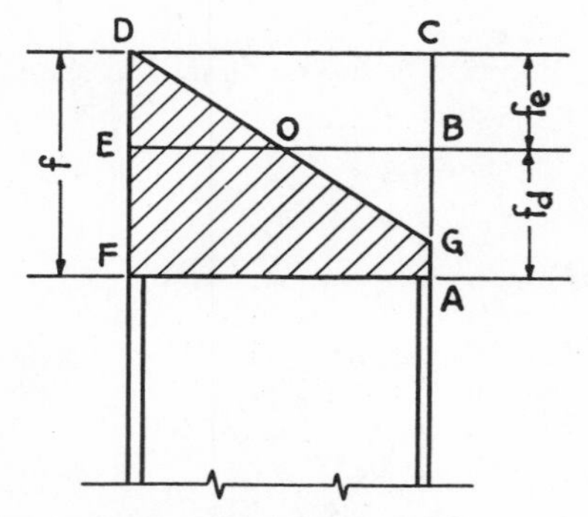

Fig. 7.2 Representation of Stresses on Column Section

graphically the stress on a column due to combined stresses. The rectangle *ABEF* indicates the total direct stress load on the column, equal to Af_d. The compression area *ODE* and the corresponding tensile area *OBG* indicate the effect of bending. The simplest way to design for the combined stress is to convert the moment into an equivalent direct stress, *BCDE* equal to Af_e, which is added to the total direct stress. This conversion is performed by multiplying the moment in inch-pounds by a bending factor *B* in the column-load tables (Table 7.5).

This bending factor *B* for any section is equal to $\dfrac{A}{S}$ and since there are two values of *S* for each section, there are likewise B_x and B_y for each column. If there is moment about both axes the correct equivalent direct load to use is the sum of the loads. In accordance with the AISC Code, the stress obtained by the method here described is conservative but usually the variation is so slight that it may be disregarded.

b The various conditions under which the computations of a column for bending and direct stress is required are as follows.

1. In steel frames which are designed in accordance with Type I, Rigid Frame construction in accordance with the AISC Specifications, Section 1.2. This would, of course, include the columns of a wind-braced building where the moment results from wind stresses. The moment used in this instance is taken directly from the rigid frame or wind-bracing computations.

2. Offset beams, as shown in Fig. 7.3(a) where the moment is the product of the load on the two beams framed parallel with the column flange, times the distance from the column center line to the center of application of the load, less the counterbalancing effect of the load framed to the opposite flange of the column.

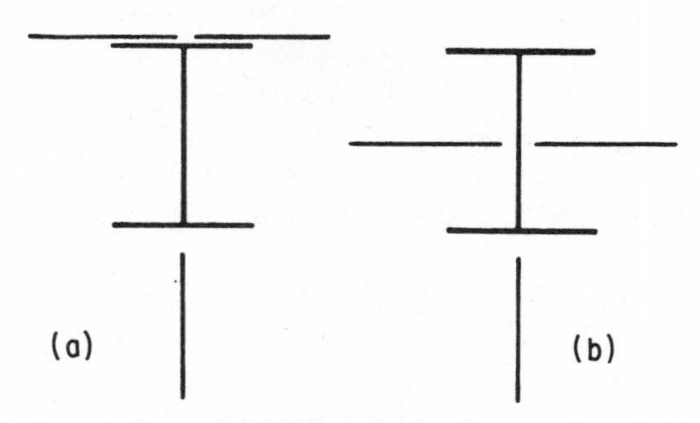

Fig. 7.3 Application of Eccentric Loads to Columns

3. Unbalanced load resulting from beams framing on the center line of the column (on either axis) but with one beam carrying a heavier load than that opposite. The moment resulting from this type of eccentricity is dependent on the type of connection used. For example, if a beam is carried to the flange of a column by means of a framed connection (AISC Manual, pages 4-12 through 4-16) it may be assumed that the lever arm to be used is the distance to the face of the column. If it is a seated connection (as shown in the Manual, pages 4-32 through 4-41) the lever arm must be taken to the approximate center of the seat, 2 in. off the face of the column. Thus it is possible for a heavier load to be balanced by a longer lever arm by varying the type of connection, and thus eliminating or reducing the resulting moment. A 30-kip load carried to the flange of an 8-in. WF column ($M = 30 \times 4 = 120$) may be balanced by a 20-kip load carried on a seat angle to the opposite flange ($20 \times 6 = 120$). On the other hand, if the opposite type of connection is used for each load, the moment could be $(30 \times 6) - (20 \times 4) = 180 - 80$ or 100 in.-kips.

4. Crane columns as described in Art.

7.11*d* following and as shown in Figs. 7.4 through 7.10.

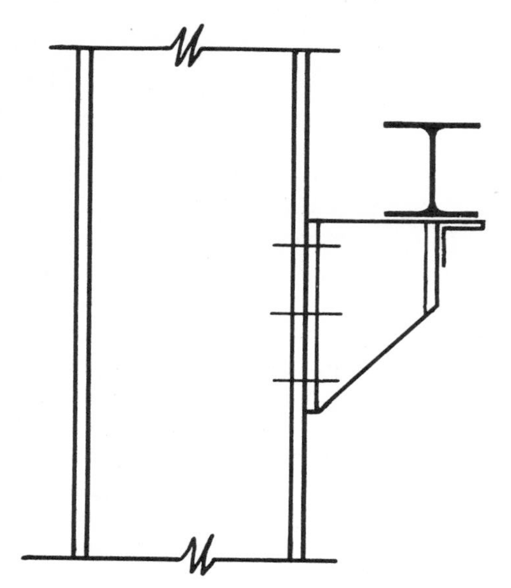

Fig. 7.4 Detail of Crane Support on Column

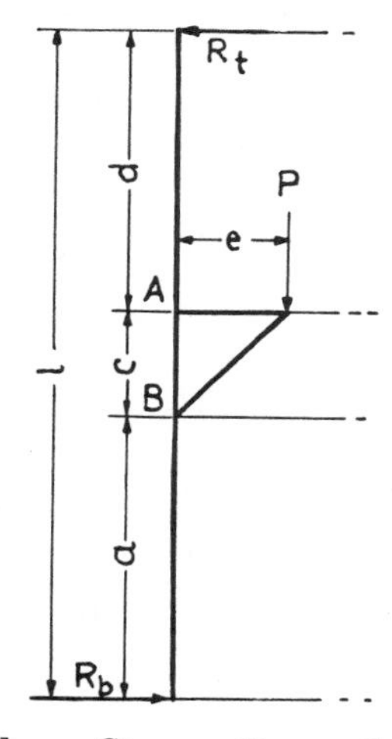

Fig. 7.5 Column Stresses Due to Crane Support

c As has been noted in Art. 7.11*a*, there are three possible formulae which may govern the design of a steel column for bending and direct stress—Formulae (6), (7*a*), and (7*b*). Formulae (6) and (7*b*) are relatively simple while (7*a*) is quite complex—so complex in fact that few engineers use it for the type of buildings ordinarily encountered. Under most conditions, the moment induced as described in paragraphs 1, 2, and 3 of Art. 7.11*b* above do not vary greatly from

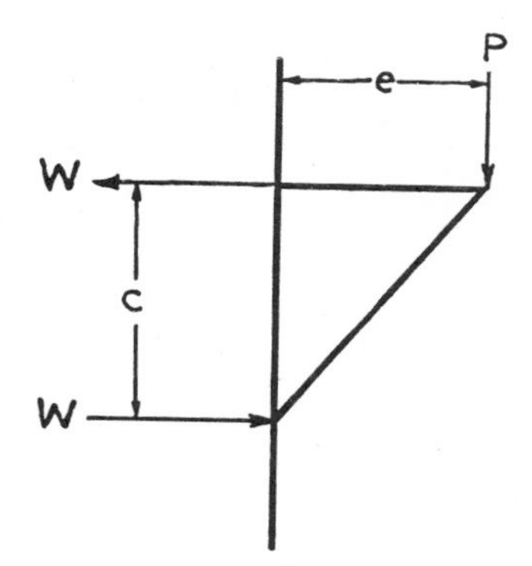

Fig. 7.6 Line Diagram of Crane Load on Column

floor to floor. In multistory structures, the loading conditions on one floor approximate those on the next floor above or below, even as to wind moment.

In Art. 7.10*a*, the possibility of sidesway was discussed and the author suggested the use of curve "a" of Fig. 7.1. Therefore, in accordance with the "Commentary on the AISC Specification," page 5-112 of the AISC Manual, the columns would come in category B of Table C 1.6.1.1. At the bottom of page 5-112, attention is called to the fact that "when M_1/M_2 is less than -0.5, formula (7*b*) would govern." Computation of a number of columns indicates that this is usually the case. A variation of the problem in the Manual of AISC, pages 3-10 and 3-11, will indicate this fact. This problem assumes full sidesway, in which case C_m is

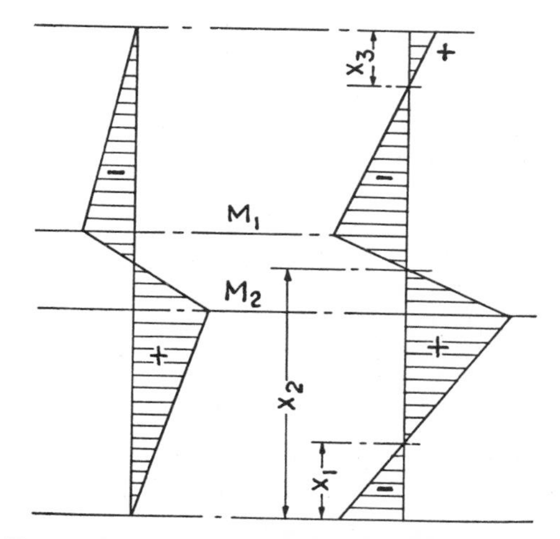

Figs. 7.7–7.8 Moment Diagram Due to Crane Load on Column

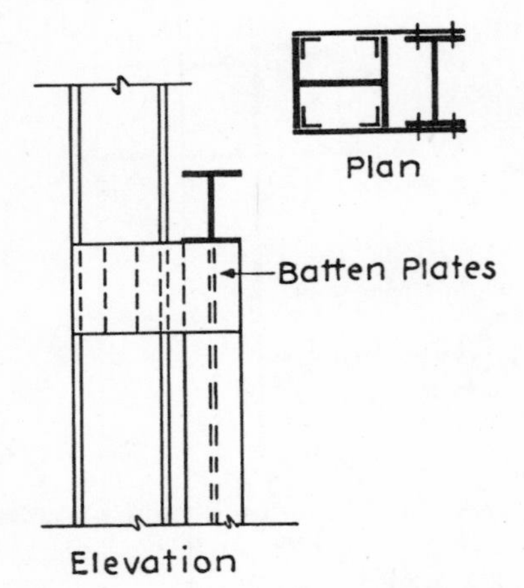

Fig. 7.9 Dual Column to Support Crane

assumed as 0.85. If, in accordance with category B of Table C.1.6.1.1, C_m is assumed as 0.4, it will be found that Formula (7*b*) rather than (7*a*) will be applicable.

The moment may be considered to be resisted jointly by the section above the floor and below the floor so the moment applied to any column section under this condition may be divided by two in the use of the bending factor. The column, therefore, must be checked for two conditions. (1) For axial strength only, select from Table 7.5 the column required for the effective length shown. (2) For bending stress, convert M in inch-kips to equivalent direct load by multiplying by the bending factor B_x or B_y whichever applies and, for any tier except the top, divide by two. Add this equivalent direct load and apply the allowable P_{a+b}, regardless of length, except as hereinafter noted. If the effective length is above the dash line, the bending factor may be reduced by dividing by 1.1. If the effective length is below the solid line, switch to another section in which the solid line comes below the effective length you are using because the allowable stress in bending drops very rapidly when L_u has been passed. If the section required to satisfy P_{a+b} is greater than that required for axial stress only, it must be used. Otherwise, the section required for axial load only is the section to use.

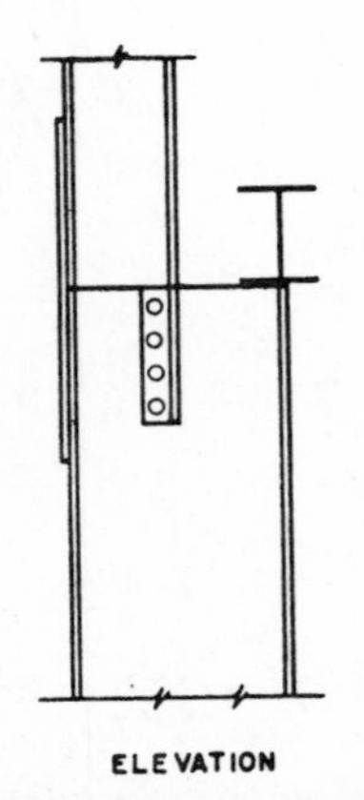

Fig. 7.10 Spliced Column to Support Crane

It is therefore common office practice to convert the moment in inch-pounds to equivalent load in kips, by applying the bending factor B, divide by two for any story except the roof tier, then add the axial load obtained by converting the moment and select the column directly from Table 7.5.

d The type of eccentricity noted in Art. 7-11*b* 4 above resolves itself into a rather complex series of computations. An analysis made by R. Fleming many years ago is in the opinion of the author the most direct approach he has found.

Referring to Figs. 7.5 and 7.6, we may consider that the moment Pe is resisted by a moment couple Wc, or $W = Pe/c$. Subscript b may be used for the base of the column and t for the top. If the column is considered to have a free end at top and bottom (Fig. 7.7) as it would be with a triangular truss and no knee brace at the top, and anchor bolts on the center lines only, the following formulae may be used:

$$M_t = M_b = 0, \quad R_t = R_b = -\frac{Pe}{l}$$

$$M_1 = \frac{Ped}{l}$$

$$M_2 = -\frac{Pea}{l}$$

If the ends are considered fixed, the following formulae may be used:

$$M_b = \frac{W}{l^3}[-d^2(l-d) + a(l-a)^2]$$

$$R_b = \frac{W}{l^3}[d^2(3l-2d) - (2a+l)(l-a)^2]$$

$$M_t = \frac{W}{l^3}[-(l-d)^2 d + a^2(l-a)]$$

$$R_t = \frac{W}{l^3}[(l-d)^2(l+2d) - a^2(3l-2a)]$$

$$M_1 = \frac{W}{l^3}\left[-d(l-d)^2 + a^2(l-a) + \frac{d}{l}(l-d)^2(l+2d) - \frac{d}{l}a^2(3l-2a)\right]$$

$$M_2 = \frac{W}{l^3}\left[-d^2(l-d) + a(l-a)^2 + \frac{a}{l}d^2(3l-2d) - \frac{a}{l}(l-a)^2(2a+1)\right]$$

In developing these end moments, the moment at the base must be provided for by anchorage—the moment couple between the center of one flange and the opposite anchor bolt, multiplied by the value of the anchor bolt at the root of the thread in tension. The moment at the top must be provided for by a knee brace to the column, or by a square-ended truss.

e The use of a traveling crane in a building usually combines another form of column bending with that discussed in paragraph *d* above. As the crane load is moved across the building, it puts a tractive effort into the crane bridge which is transmitted as a kick against the column. This is usually taken as 20 percent of the weight of the traveling crab plus load, and is applied to the column as a direct load at the line of the top of the bracket, or if the top of the crane girder is braced to the column, at this line.

If a simple column with a bracket is used, the moment and reactions are found as for an ordinary beam. If the column is of different size above and below the crane girder as in Fig. 7.10, the best method of approach is the indirect Hardy Cross method, as described in Art. 3.21*g*.

f The problem of designing columns for wind bracing is treated under the heading of wind bracing for mill buildings in Art. 10.21 and multistory buildings in Art. 10.22.

g For multistory columns it is advisable for a designer to standardize on a form for loads and calculations. This statement applies to all types of calculations, but it is most important in column design. Figure 7.11 shows the form used in the author's office. A typical six-story and roof office building is set up on this form. For each story the load is set down on the form as applied, in order, from the top (as the floor plan is laid out), right side, bottom, and left side. The length to floor below is set down and the moment in each direction.

It is common practice to splice columns at alternate floors and, if there are an odd number of stories, as in this problem, it is better to make the splice above the first floor. Ordinarily, all columns should be spliced at the same floors, instead of staggering splices. The splice point should be located about 1 ft 6 in. above the floor line. In this particular problem, the splices would be located above the odd numbered floors. These conditions may be varied by wind bracing.

We set down first the beam reactions for the roof and the sixth floor taken from the calculations for beams. The sum of these loads is 104.5 kips. The approximate weight of column to add in for this load from the formula in Art. 1.50*c* is $60 + (0.4 \times 103.5) = 101.4$ lb per foot. We will therefore add in 1.5 kips for the roof column, and 1.2 kips for the sixth floor. The top tier is therefore designed for a total load of 107.2 kips plus an unbalance of 16.5, or 123.7 kips on a height of $(0.65 \times 12 =)$ 7.8 ft. This would require an 8 WF 31.

The fifth and fourth floor loads from beam reactions are the same as the sixth, and the approximate total load at the fourth floor is found to be $107.2 + 59.7 + 59.7$ or 226.5. The weight to allow for columns will therefore be $60 + (0.4 \times 226.5) = 150.6$ lb per foot, or 1.8 kips per story. This is 0.6 kips greater than the load from the sixth floor, so we may add this to 59.7, or take down a total load per floor of $59.7 + 0.6 = 60.3$. However, in accordance with paragraph *a* above, we may start reducing the live load. The live load in this design is 50 psf on an area of 400 sq ft, or a total live load of 20 kips. The reduction on the fifth floor is 5 percent, or 1 kip. On the fourth floor, the reduction is 10 percent, or 2 kips. Adding in for unbalanced load, the equivalent total load is

COLUMN 6 - 400 SQ. FT.			JOB 1ST NATIONAL BANK BUILDING		
R R-11 R-21 R-11 R-22	T- 9.0 M1 R- 10.0 M2 B- 9.0 L- 18.0 COL 1.5 47.5 47.5 L= 15'		2ND 2-16 2-18 2-21 27.5 21.0 10.0 6"	T- 21.0 M1 22.9 R- 10.0 M2 6.5 B- 27.5 L- COL 3.5 62.0-4 = 58.0 341.8 L= 14'	12 WF 65
6TH T-12 T-16 T-12 T-19 10.0 27.5 11.0 10.0	T- 10.0 M1 16.5 R- 11.0 M2 B- 10.0 L- 27.0 COL 1.2 59.7 107.2 L= 12'	8 WF 28	1ST 1-11 1-12 1-11 1-12	T- 20.0 M1 R- 25.0 M2 B- 20.0 L- 25.0 COL 93.3-10 = 83.3 425.1 L= 11'	12 WF 79
5TH	T- M1 R- M2 B- L- COL 60.3-1 = 59.3 166.5 L= 12'		BASE AND FOUNDATION (FROM TABLE 8.2)	T- BILLET M1 R- 20×20×1¾" M2 B- L- FON. @ 8000#/SQ.FT. COL 7'-6"F L= (SEE TABLE 8.3)	
4TH	T- M1 16.5 R- M2 B- L- COL 60.3-2 = 58.3 224.8 L= 12'	10 WF 49		T- M1 R- M2 B- L- COL L=	
3RD	T- M1 R- M2 B- L- COL 62.0-3 = 59.0 283.8 L= 13'			T- M1 R- M2 B- L- COL L=	

Fig. 7.11 Column Calculation Sheet

224.8 + 16.5 = 241.3 kips. The economical column for this is 10 WF 45.

We may assume about the same load increase from the third and second floors as from the fifth and fourth floors, or an approximate total load at the second floor of 342 kips. The weight per foot of column would be $60 + (0.4 \times 342) = 265.2$ lb per foot. Thirteen feet of this load is 3.5 kips, or 1.7 heavier than the fourth floor, or

62.0 kips, but the live load reduction is 15 percent or 3 kips, and for the second 20 percent or 4 kips.

The load on the second floor is applied in a slightly different manner from those on the typical floor. Although the reactions are the same, the moment differs because of the point of application of load. For a load of 341.8 kips on $(0.65 \times 14 =)$ 9.1 ft height, the economical column would be 12 WF 65. The 10.0-kip load would be applied on a lever arm of $6 + 2 = 8$ in. The moment about axis 1-1 would therefore be

$$\begin{aligned} 10.0 \times 8 &= \quad 80 \\ -48.5 \times 6 &= -291 \\ \text{Net} &= -211 \end{aligned}$$

The equivalent direct load would be

$$\frac{211 \times 0.217}{2} = 22.9$$

The unbalance in the other direction is 6.5, so the total direct load would be $341.8 + 29.4 = 371.2$ kips. This would require a 12 WF 72.

The first floor is designed for 100 psf live load and there is no eccentricity. The approximate total load is $341.8 + 90 = 431.8$ kips. The weight of column to add would be 3.3 kips. The reduction would be 25 percent of 100 psf, or 25 psf $\times$ 400 sq ft $=$ 10.0 kips. The total load of 425.1 kips would require a 12 WF 79.

Certain liberties have been taken in the foregoing calculation, but it will be found that if every possible refinement were made, probably not a single column section in the entire height would be changed.

It will be found that time will be saved and fewer errors encountered if the calculations are made as follows.

1. From the work sheet—the plan layout on which beam calculation numbers have been noted—write down before each *T*, *R*, *B*, and *L* (top, right, bottom, left) for each floor the calculation number, and sketch odd moment conditions, as for the second floor in the problem we have considered. At the top of each sheet set down the floor area carried by the column.

2. Go through all calculations and set down the reactions and the live load for each floor.

3. Run the loads down and set down the moments, computing column allowances as the loads are run down.

4. Select the sizes.

It is wise to follow through step 1 with all columns, then step 2 with all, and so on. In this manner discrepancies will show up and errors will be avoided.

7.20 REINFORCED CONCRETE COLUMNS

a As has been previously stated, the 1963 ACI Code has not as yet been adopted into all building codes and the engineer should determine before proceeding with the design whether the authority under which he is working will approve the 1963 Code or is still working under the 1956 Code. This warning is particularly applicable to column design because the newer code is much more liberal in permissible column loads than was the former code. For example, the maximum load on a 10 $\times$ 12 tied column of 5,000-lb concrete under the 1956 Code was 185.3 kips, while under the 1963 Code, allowing for higher stresses and double the percentage of steel, this same small column may be loaded to 372.3 kips—an increase of 100 percent. This, of course, is an extreme instance of the liberalization of the new code, but it calls attention to the necessity for checking the acceptability of the code.

b The applicable portions of the 1963 ACI Code for the design of columns are as follows.

804 SPACING OF BARS

(d) In spirally reinforced and in tied columns, the clear distance between longitudinal bars shall be not less than 1½ times the bar diameter, 1½ times the maximum size of the coarse aggregate, nor 1½ in.

(e) The clear distance between bars shall also apply to the clear distance between a contact splice and adjacent splices or bars.

805 Splices in Reinforcement

(c) *Splices in reinforcement in which the critical design stress is compressive—*

1. Where lapped splices are used, the minimum amount of lap shall be:

With concrete having a strength of 3000 psi or more, the length of lap for deformed bars shall be 20, 24, and 30 bar diameters for specified yield strengths of 50,000 and under, 60,000, and 75,000 psi, respectively, nor less than 12 in. When the specified concrete strengths are less than 3000 psi, the amount of lap shall be one-third greater than the values given above.

For plain bars, the minimum amount of lap shall be twice that specified for deformed bars.

2. Welded splices or other positive connections may be used instead of lapped splices. Where the bar size exceeds #11, welded splices or other positive connections shall preferably be used. In bars required for compression only, the compressive stress may be transmitted by bearing of square-cut ends held in concentric contact by a suitably welded sleeve or mechanical device.

3. Where longitudinal bars are offset at a splice, the slope of the inclined portion of the bar with the axis of the column shall not exceed 1 in 6, and the portions of the bar above and below the offset shall be parallel to the axis of the column. Adequate horizontal support at the offset bends shall be treated as a matter of design, and shall be provided by metal ties, spirals, or parts of the floor construction. Metal ties or spirals so designed shall be placed near (not more than eight bar diameters from) the point of bend. The horizontal thrust to be resisted shall be assumed as 1½ times the horizontal component of the nominal stress in the inclined portion of the bar.

Offset bars shall be bent before they are placed in the forms. See Section 801(d).

4. Where column faces are offset 3 in. or more, splices of vertical bars adjacent to the offset face shall be made by separate dowels overlapped as specified above.

5. In tied columns the amount of reinforcement spliced by lapping shall not exceed a steel ratio of 0.04 in any 3 ft length of column.

(d) An approved welded splice is one in which the bars are butted and welded so that it will develop in tension at least 125 percent of the specified yield strength of the reinforcing bar. Approved positive connections for bars designed to carry critical tension or compression shall be equivalent in strength to an approved welded splice.

(e) Metal cores in composite columns shall be accurately milled at splices and positive provision shall be made for alignment of one core above another. At the column base, provision shall be made to transfer the load to the footing at safe unit stresses in accordance with Section 1002(a). The base of the metal section shall be designed to transfer the load from the entire composite column to the footing, or it may be designed to transfer the load from the metal section only, provided it is so placed in the pier or pedestal as to leave ample section of concrete above the base for the transfer of load from the reinforced concrete section of the column by means of bond on the vertical reinforcement and by direct compression on the concrete.

806 Lateral Reinforcement

(a) Spiral column reinforcement shall consist of evenly spaced continuous spirals held firmly in place and true to line by vertical spacers. At least two spacers shall be used for spirals 20 in. or less in diameter, three for spirals 20 to 30 in. in diameter, and four for spirals more than 30 in. in diameter. When spiral rods are ⅝ in. or larger, three spacers shall be used for spirals 24 in. or less in diameter and four for spirals more than 24 in. in diameter. The spirals shall be of such size and so assembled as to permit handling and placing without being distorted from the designed dimensions. The material used in spirals shall have a minimum diameter of ¼ in. for rolled bars or No. 4 AS&W gage for drawn wire. Anchorage of spiral reinforcement shall be provided by 1½ extra turns of spiral rod or wire at each end of the spiral unit. Splices when necessary in spiral rods or wires shall be made by welding or by a lap of 1½ turns. The center to center spacing of the spirals shall not exceed one-sixth of the core diameter. The clear spacing between spirals shall not exceed 3 in. nor be less than 1⅜ in. or 1½ times the maximum size of coarse aggregate used. The reinforcing spiral shall extend from the floor level in any story or from the top of the footing to the level of the lowest horizontal reinforcement in the slab, drop panel, or beam above. In a column with a capital, the spiral shall extend to a plane at which the diameter or width of the capital is twice that of the column.

(b) All bars for tied columns shall be enclosed by lateral ties at least ¼ in. in diameter spaced apart not over 16 bar diameters, 48 tie diameters, or the least dimension of the column. The ties shall be so arranged that every corner and alternate longitudinal bar shall have lateral support provided by the corner of a tie having an included angle of not more than 135 deg and no bar shall be farther than 6 in. from such a laterally supported bar. Where the bars are located around the periphery of a circle, a complete circular tie may be used.

808 Concrete Protection for Reinforcement

(a) The reinforcement of footings and other principal structural members in which the concrete is deposited against the ground shall have not less than 3 in. of concrete between it and the ground contact surface. If concrete surfaces after removal of the forms are to be exposed to the weather or be in contact with the ground, the reinforcement shall be protected with not less than 2 in. of concrete for bars larger than #5 and 1½ in. for #5 bars or smaller.

(c) Column spirals or ties shall be protected everywhere by a covering of concrete cast monolithically with the core, for which the thickness shall be not less than 1½ in. nor less than 1½ times the maximum size of the coarse aggregate.

(d) Concrete protection for reinforcement shall in all cases be at least equal to the diameter of bars.

912 Limiting Dimensions of Columns

(a) *Minimum size*—Columns constituting the principal supports of a floor or roof shall have a diameter of at least 10 in., or in the case of rectangular columns, a thickness of at least 8 in., and a gross area not less than 96 sq in. Auxiliary supports placed at intermediate locations and not continuous from story to story may be smaller but not less than 6 in. thick.

(b) *Isolated column with multiple spirals*—If two or more interlocking spirals are used in a column, the outer boundary of the column shall be taken at a distance outside the extreme limits of the spiral equal to the requirements of Section 808(c).

(c) *Limits of section of column built monolithically with wall*—For a spiral column built monolithically with a concrete wall or pier, the outer boundary of the column section shall be taken either as a circle at least 1½ in. outside the column spiral or as a square or rectangle, the sides of which are at least 1½ in. outside the spiral or spirals.

(d) *Equivalent circular columns*—As an exception to the general procedure of utilizing the full gross area of the column section, it shall be permissible to design a circular column and to build it with a square, octagonal, or other shaped section of the same least lateral dimension. In such case, the allowable load, the gross area considered, and the required percentages of reinforcement shall be taken as those of the circular column.

(e) *Limits of column section*—In a tied column which has a larger cross section than required by considerations of loading, a reduced effective area, A_g, not less than one-half of the total area may be used for determining minimum steel area and load capacity.

913 Limits for Reinforcement of Columns

(a) The vertical reinforcement for columns shall be not less than 0.01 nor more than 0.08 times the gross cross-sectional area. The minimum size of bar shall be #5. The minimum number of bars shall be six for spiral columns and four for tied columns.

(b) The ratio of spiral reinforcement, p_s, shall be not less than the value given by

$$p_s = 0.45(A_g/A_c - 1)f_c'/f_y \qquad (9\text{-}1)$$

wherein f_y is the yield strength of spiral reinforcement but not more than 60,000 psi.

914 Bending Moments in Columns

(a) Columns shall be designed to resist the axial forces from loads on all floors, plus the maximum bending due to loads on a single adjacent span of the floor under consideration. Account shall also be taken of the loading condition giving the maximum ratio of bending moment to axial load. In building frames, particular attention shall be given to the effect of unbalanced floor loads on both exterior and interior columns and of eccentric loading due to other causes. In computing moments in columns due to gravity loading, the far ends of columns which are monolithic with the structure may be considered fixed.

Resistance to bending moments at any floor level shall be provided by distributing the moment between the columns immediately above and below the given floor in proportion to their relative stiffnesses and conditions of restraint.

915 Length of Columns

(a) For purposes of determining the limiting dimensions of columns, the unsupported length of reinforced concrete columns shall be taken as the clear distance between floor slabs, except that

1. In flat slab construction, it shall be the clear distance between the floor and the lower extremity of the capital, the drop panel or the slab, whichever is least.

2. In beam and slab construction, it shall be the clear distance between the floor and the underside of the deeper beam framing into the column in each direction at the next higher floor level.

3. In columns restrained laterally by struts, it shall be the clear distance between consecutive struts in each vertical plane; provided

that to be an adequate support, two such struts shall meet the column at approximately the same level, and the angle between vertical planes through the struts shall not vary more than 15 deg from a right angle. Such struts shall be of adequate dimensions and anchorage to restrain the column against lateral deflection.

4. In columns restrained laterally by struts or beams, with brackets used at the junction, it shall be the clear distance between the floor and the lower edge of the bracket, provided that the bracket width equals that of the beam or strut and is at least half that of the column.

(b) For rectangular columns, that length shall be considered which produces the greatest ratio of length to radius of gyration of section.

(c) The effective length, h', of columns in structures where lateral stability or resistance to lateral forces is provided by shear walls or rigid bracing, by fastening to an adjoining structure of sufficient lateral stability, or by any other means that affords adequate lateral support, shall be taken as the unbraced length, h.

(d) Larger effective lengths, h', shall be used for all columns in structures which depend upon the column stiffness for lateral stability:

1. The end of a column shall be considered hinged in a plane if in that plane r' (see Section 900) exceeds 25.

2. For columns restrained against rotation at one end and hinged at the other end the effective length shall be taken as $h' = 2h(0.78 + 0.22r') \geq 2h$, where r' is the value at the restrained end.

3. For columns restrained against rotation at both ends the effective length h' shall be taken as $h' = h(0.78 + 0.22r') \geq h$, where r' is the average of the values at the two ends of the column.

4. For cantilever columns, that is, those fixed at one end and free at the other, the effective length h' shall be taken as twice the over-all length.

916 Strength Reductions for Length of Compression Members

(a) When compression governs the design of the section, the axial load and moment computed from the analysis shall be divided by the appropriate factor R as given in 1, 2, or 3 below, and the design shall be made using the appropriate formulas for short members in Chapters 14 and 19.

1. If relative lateral displacement of the ends of the member is prevented and the ends of the member are fixed or definitely restrained such that a point of contraflexure occurs between the ends, no correction for length shall be made unless h'/r exceeds 60. For h'/r between 60 and 100, the design shall be based on an analysis according to (d) or the following factor shall be used

$$R = 1.32 - 0.006h/r \leqq 1.0 \qquad (9\text{-}2)$$

If h/r exceeds 100, an analysis according to (d) shall be made.

2. If relative lateral displacement of the ends of the members is prevented and the member is bent in single curvature, the following factor shall be used.

$$R = 1.07 - 0.008h/r \leqq 1.0 \qquad (9\text{-}3)$$

3. The design of restrained members for which relative lateral displacement of the ends is not prevented shall be made using the factor given in Eq. (9-4); that is, with the effective length h' from Section 915 substituted for h.

$$R = 1.07 - 0.008h'/r \leqq 1.0 \qquad (9\text{-}4)$$

When the design is governed by lateral loads of short duration, such as wind or earthquake loading, the factor R may be increased by 10 percent, which is equivalent to using

$$R = 1.18 - 0.009h'/r \leqq 1.0 \qquad (9\text{-}5)$$

4. The radius of gyration, r, may be taken equal to 0.30 times the over-all depth in the direction of bending for a rectangular column and 0.25 times the diameter of circular columns. For other shapes r may be computed for the gross concrete section.

(b) When tension governs the design of the section, the axial load and moment computed from the analysis shall be increased as required in (a) except that the factor R shall be considered to vary linearly with axial load from the values given by Eq. (9-2) and (9-3) at the balanced condition [as defined in Section 1407 or 1900(b)] to a value of 1.0 when the axial load is zero.

(c) When a column design is governed by the minimum eccentricities specified for ultimate strength design in Section 1901(a) the effect of length on column strength shall be determined in one of the following ways:

1. Where the actual computed eccentricities at both ends are less than the specified minimum eccentricity, the strength reduction for length shall correspond to the actual conditions of curvature and end restraint.

2. If column moments have not been considered in the design of the column or if computations show that there is no eccentricity at one or both ends of the column, the factor in (a)2 shall be used.

(d) In lieu of other requirements of this section, an analysis may be made taking into account the effect of additional deflections on moments in columns.

In such an analysis a reduced modulus of elasticity, not greater than one-third the value specified in Section 1102, shall be used in calculations of deflections caused by sustained loads.

917 Transmission of Column Load through Floor System

(a) When the specified strength of concrete in columns exceeds that specified for the floor system by more than 40 percent, proper transmission of load through the weaker concrete shall be provided by one of the following:

1. Concrete of the strength specified for the column shall be placed in the floor for an area four times A_g, about the column, well integrated into floor concrete, and placed in accordance with Section 704(b).
2. The capacity of the column through the floor system shall be computed using the weaker concrete strength and adding vertical dowels and spirals as required.
3. For columns laterally supported on four sides by beams of approximately equal depth or by slabs, the capacity may be computed by using an assumed concrete strength in the column formulas equal to 75 percent of the column concrete strength plus 35 percent of the floor concrete strength.

1003 Allowable Stresses in Reinforcement

(*b*) *In compression, vertical column reinforcement*
Spiral columns, 40 percent of the minimum yield strength, but not to exceed . . . . 30,000 psi
Tied columns, 85 percent of the value for spiral columns, but not to exceed . . . . 25,500 psi
Composite and combination columns:
Structural steel sections
For ASTM A 36 Steel . . . . 18,000 psi
For ASTM A 7 Steel . . . . 16,000 psi
Cast iron sections . . . . 10,000 psi
Steel pipe . . . . see limitations of Section 1406

(*c*) *In compression, flexural members*
For compression reinforcement in flexural members see Section 1102

(*d*) *Spirals* [*yield strength for use in Eq.* (*9-1*)]
Hot rolled rods, intermediate grade 40,000 psi
Hot rolled rods, hard grade . . . . 50,000 psi
Hot rolled rods, ASTM A 432 grade and cold-drawn wire . . . . 60,000 psi

1004 Allowable Stresses—Wind and Earthquake Forces

(a) Members subject to stresses produced by wind or earthquake forces combined with other loads may be proportioned for stresses 33-1/3 percent greater than those specified in Sections 1002 and 1003, provided that the section thus required is not less than that required for the combination of dead and live load.

REINFORCED CONCRETE COLUMNS—WORKING STRESS DESIGN

1400 Notation

A_c = area of concrete within a pipe column
A_g = gross area of spirally reinforced or tied column
= the total area of the concrete encasement of combination column
= area of concrete of a composite column
A_r = area of steel or cast-iron core of a composite, combination, or pipe column
A_s = area of tension reinforcement
A_{st} = total area of longitudinal reinforcement
b = width of compression face of flexural member
d = distance from extreme compression fiber to centroid of tension reinforcement
d' = distance from extreme compression fiber to centroid of compression reinforcement
D_s = diameter of circle through centers of the longitudinal reinforcement in spiral columns
e = eccentricity of the resultant load on a column, measured from the gravity axis
e_b = maximum permissible eccentricity of N_b
F_b = allowable bending stress that would be permitted for bending alone
f_a = axial load divided by area of member, A_g
f_c' = compressive strength of concrete (see Section 301)
f_r = allowable stress in the metal core of a composite column
f_r' = allowable stress on unencased metal columns and pipe columns
f_s = allowable stress in column vertical reinforcement
f_y = yield strength of reinforcement (see Section 301)
h = unsupported length of column
j = ratio of distance between centroid of compression and centroid of tension to the depth, d
K_c = radius of gyration of concrete in pipe columns
K_s = radius of gyration of metal pipe in pipe columns
m = $f_y/0.85f_c'$
n = ratio of modulus of elasticity of steel to that of concrete
N = eccentric load normal to the cross section of a column
N_b = the value of N below which the allowable eccentricity is controlled by tension, and above which by compression
p = ratio of area of tension reinforcement to effective area of concrete
p' = ratio of area of compression reinforcement to effective area of concrete
p_g = ratio of area of vertical reinforcement to the gross area, A_g

P = allowable axial load on a reinforced concrete column without reduction for length or eccentricity
= allowable axial load on combination, composite, or pipe column without reduction for eccentricity

t = over-all depth of rectangular column or the diameter of a round column

1401 LIMITING DIMENSIONS

(a) The loads determined by the provisions of this chapter apply only when unsupported length reductions are not required by the provision of Sections 915 and 916. (See Section 912 for minimum size)

1402 SPIRALLY REINFORCED COLUMNS

(a) The maximum allowable axial load, P, on columns with closely spaced spirals (see Section 913) enclosing a circular core reinforced with vertical bars shall be that given by

$$P = A_g(0.25f_c' + f_s p_g) \tag{14-1}$$

where f_s = allowable stress in vertical column reinforcement, to be taken at 40 percent of the minimum specification value of the yield strength, but not to exceed 30,000 psi.

1403 TIED COLUMNS

(a) The maximum allowable axial load on columns reinforced with longitudinal bars and separate lateral ties shall be 85 percent of that given by Eq. (14-1).

1404 COMPOSITE COLUMNS

(a) The allowable load on a composite column, consisting of a structural steel or cast-iron column thoroughly encased in concrete reinforced with both longitudinal and spiral reinforcement, shall not exceed that given by

$$P = 0.225A_g f_c' + f_s A_{st} + f_r A_r \tag{14-2}$$

where f_r = allowable unit stress in metal core, not to exceed 18,000 psi for steel conforming to ASTM A36, 16,000 psi for steel conforming to ASTM A7, or 10,000 psi for a cast-iron core.

The column as a whole shall satisfy the requirements of Eq. (14-2) at any point. The reinforced concrete portion shall be designed to carry all loads imposed between metal core brackets or connections at a stress of not more than $0.35f_c'$ based on an area of A_g.

(b) *Metal core and reinforcement*—The cross-sectional area of the metal core shall not exceed 20 percent of the gross area of the column. If a hollow metal core is used it shall be filled with concrete. The amounts of longitudinal reinforcement and the requirements as to spacing of bars, details of splices and thickness of protective shell outside the spiral shall conform to the limiting values specified for a spiral column of the same over-all dimensions. Spiral reinforcement shall conform to Eq. (9-1). A clearance of at least 3 in. shall be maintained between the spiral and the metal core at all points except that when the core consists of a structural steel H-column, the minimum clearance may be reduced to 2 in.

Transfer of loads to the metal core shall be provided for by the use of bearing members such as billets, brackets, or other positive connections; these shall be provided at the top of the metal core and at intermediate floor levels where required.

The metal cores shall be designed to carry safely any construction or other loads to be placed upon them prior to their encasement in concrete.

1405 COMBINATION COLUMNS

(a) *Steel columns encased in concrete*—The allowable load on a structural steel column which is encased in concrete at least 2½ in. thick over all metal (except rivet heads), reinforced as hereinafter specified, shall be computed by

$$P = A_r f_r' \left[1 + \frac{A_g}{100A_r}\right] \tag{14-3}$$

The concrete used shall develop a compressive strength, f_c', of at least 2500 psi at 28 days. The concrete shall be reinforced by the equivalent of welded wire fabric having wires of No. 10 AS&W gage, the wires encircling the column being spaced not more than 4 in. apart and those parallel to the column axis not more than 8 in. apart. This fabric shall extend entirely around the column at a distance of 1 in. inside the outer concrete surface and shall be lap-spliced at least 40 wire diameters and wired at the splice. Special brackets shall be used to receive the entire floor load at each floor level. The steel column shall be designed to carry safely any construction or other loads to be placed upon it prior to its encasement in concrete.

1406 CONCRETE-FILLED PIPE COLUMNS

(a) The allowable load on columns consisting of steel pipe filled with concrete shall be determined by

$$P = 0.25f_c'\left(1 - 0.000025\frac{h^2}{K_c^2}\right)A_c + f_r'A_r \tag{14-4}$$

The value of f_r' shall be given by Eq. (14-5) when the pipe has a yield strength of at least 33,000 psi, and an h/K_s ratio equal to or less than 120

$$f_r' = 17{,}000 - 0.485\frac{h^2}{K_s^2} \tag{14-5}$$

1407 Columns Subjected to Axial Load and Bending

(a) The strength of the column is controlled by compression if the load, N, has an eccentricity, e, in each principal direction, no greater than that given by Eq. (14-6), (14-7), or (14-8) and by tension if e exceeds these values in either principal direction.

For symmetrical spiral columns:

$$e_b = 0.43\ p_g m D_s + 0.14t \qquad (14\text{-}6)$$

For symmetrical tied columns:

$$e_b = (0.67\ p_g m + 0.17)d \qquad (14\text{-}7)$$

For unsymmetrical tied columns:

$$e_b = \frac{p'm(d - d') + 0.1d}{(p' - p)m + 0.6} \qquad (14\text{-}8)$$

(b) Columns controlled by compression shall be proportioned by Eq. (14-9) except that the allowable load N shall not exceed the load, P, permitted when the column supports axial load only.

$$f_a/F_a + f_{bx}/F_b + f_{by}/F_b \text{ not greater than unity} \qquad (14\text{-}9)$$

where f_{bx} and f_{by} are the bending moment components about the x and y principal axes divided by the section modulus of the respective transformed uncracked section, $2n$ being assumed as the modular ratio for all vertical reinforcement, and

$$F_a = 0.34(1 + p_g m)f_c' \qquad (14\text{-}10)$$

(c) The allowable bending moment M on columns controlled by tension shall be considered to vary linearly with axial load, from M_o when the section is in pure flexure, to M_b when the axial load is equal to N_b; M_b and N_b shall be determined from e_b and Eq. (14-9); M_o from Eq. (14-11), (14-12), or (14-13).

For spiral columns:

$$M_o = 0.12A_{st}f_y D_s \qquad (14\text{-}11)$$

For symmetrical tied columns:

$$M_o = 0.40A_s f_y(d - d') \qquad (14\text{-}12)$$

For unsymmetrical tied columns:

$$M_o = 0.40A_s f_y jd \qquad (14\text{-}13)$$

(d) For bending about two axes

$$\frac{M_x}{M_{ox}} + \frac{M_y}{M_{oy}} \text{ not greater than unity} \qquad (14\text{-}14)$$

where M_x and M_y are bending moments about the X and Y principal axes, and M_{ox} and M_{oy} are the values of M_o for bending about these axes.

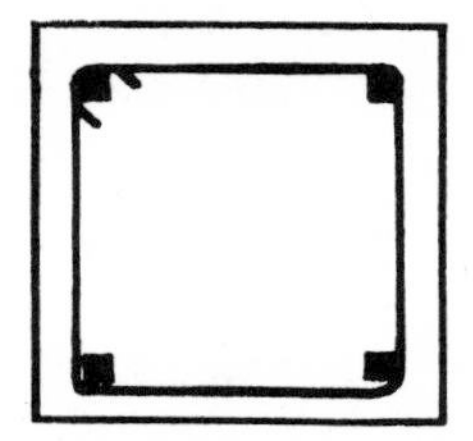

Fig. 7.12 Section of Concrete Column—4 Bars

The design of flat slab columns is covered by the code for flat slab construction in Art. 5.51.

The method of placing column bars and ties is shown in Figs. 7.12 through 7.17.

c The 1963 ACI Code has been changed to take into account the possible effect of bending even under concentric loading, as has the 1963 AISC Code for steel columns. Two factors are applied in the determination of applicability of this bending: (1) the possible increase in the effective length similar to the K factor in a steel column—this decision to be made on the basis of the ACI Code, Section 915; and (2) when h'/r (the effective length divided by the radius of gyration of the column) exceeds 60, the determination of a magnification factor which increases the effective load, as described in Section 916 of the ACI Code.

In accordance with Section 915(c) in structures where sidesway is inhibited by walls, partitions, or other normal bracing elements, the effective length h' may be determined as specified in Section 916(d) but in no case does it exceed $2h$. Under these conditions, it seems reasonable to assume that for most reinforced concrete structures, the effective height is the actual unsupported length of column as specified in Section 915(a) and (b) of the ACI Code.

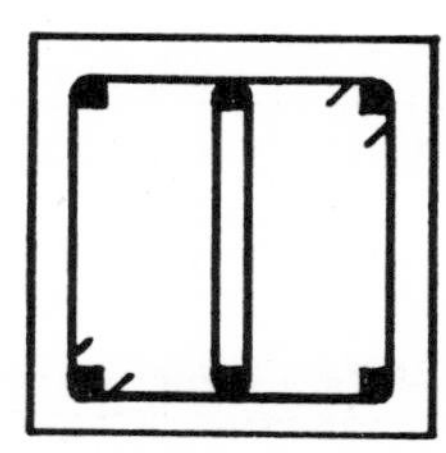

Fig. 7.13 Section of Concrete Column—6 Bars

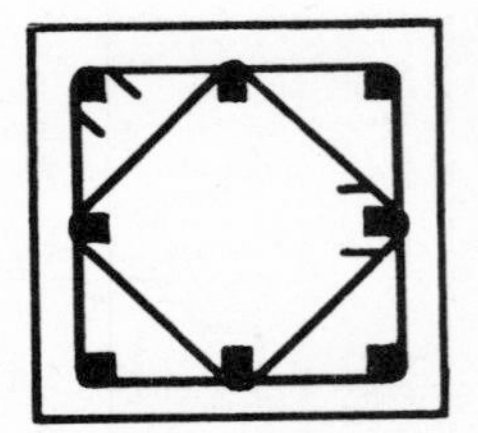

Fig. 7.14 Section of Concrete Column—8 Bars

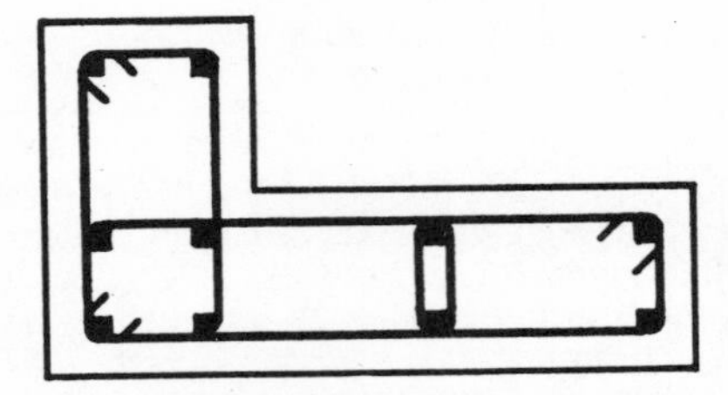

Fig. 7.16 Typical Arrangement of Corner Columns

In structures of this type, the magnification factor may be neglected if h' does not exceed 60. In columns of square or rectangular sections, $r = 0.3d$, so the above limitation would convert to $h' \leqq 18d$, and for round columns, $r = 0.25d$ or $h' \leqq 15d$. Therefore for most buildings use the actual design load plus moment when the height of the column does not exceed the above limitations. This simplification reduces an otherwise rather complicated portion of the ACI Code to usable office practice.

d There are a number of points relative to concrete column design which, although not included within the limits of the ACI Code, are nevertheless dictated by good practice and by economics. Among these are the following.

1. All columns within any one story height should be of the same quality concrete, i.e., 2,500-, 3,000-, 4,000-, or 5,000-psi concrete. This is to prevent possible error in pouring.
2. In carrying any column section down from top to bottom, no reduction should be made in any one of the following:

   (a) Size of column
   (b) Quality of concrete in column
   (c) Number of bars in section
   (d) Size of bars in section
   (e) Grade of steel in bars

3. Since forms are one of the most expensive factors in reinforced concrete construction, and since many architectural details are dependent on the column size, it is advisable to use as few variations as possible in column size throughout the height of a column, even to the possible extent of wasting some concrete in the upper tier columns.
4. Insofar as possible, use only four bars in a square or rectangular column section, thus cutting the cost of ties. This applies even though it is necessary to use 14S or 18S bars.
5. In the use of larger size bars, give consideration to the use of welded splices instead of lapped splices. Since the cost relationship of welding and lapping may vary in various areas, the best answer to determination of the economical relationship may be checked with a local contractor for reinforcement.

e In accordance with the preceding, when bending is not involved, allowable column loads may be arrived at, and columns designed, by the use of Tables 7.6 through 7.9.

A series of graphs, similar to those shown in Fig. 7.18, the basis of which is tabulated in Table 7.10, has been found to be useful in

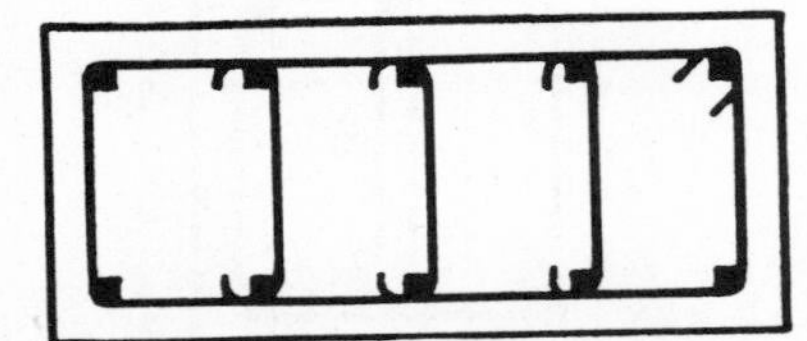

Fig. 7.15 Section of Concrete Column—Alternate Method of Tie Arrangement

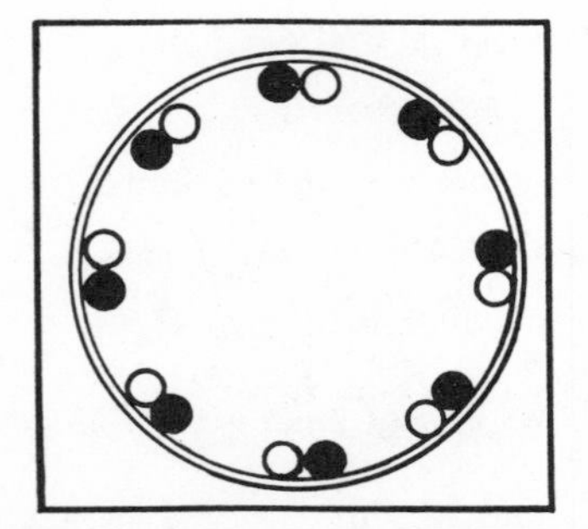

Fig. 7.17 Arrangement of Reinforcement—Spirally Reinforced Columns

Table 7.6 Allowable Loads—Tied Concrete Columns

Maximum load in kips on concrete and limiting load in kips on longitudinal reinforcement

Column size, in.	Gross area, A_g	Load on longitudinal reinforcement				Load on concrete			
		Intermediate grade		High strength $f_s = 60$ ksi		Concrete strength: f_c'			
		Min.	Max.	Min.	Max.	2,500	3,000	4,000	5,000
10 × 12	120	16.3	130.6	30.6	244.8	63.8	76.5	102.0	127.5
10 × 14	140	19.0	152.3	35.7	285.6	74.4	89.3	119.0	148.8
10 × 16	160	21.8	174.1	40.8	326.4	85.0	102.0	136.0	170.0
10 × 18	180	24.5	195.8	45.9	367.2	95.6	114.8	153.0	191.3
12 × 12	144	19.6	156.7	36.7	293.8	76.5	91.8	122.4	153.0
12 × 14	168	22.8	182.8	42.8	342.7	89.3	107.1	142.8	178.5
12 × 16	192	26.1	208.9	49.0	391.7	102.0	122.4	163.2	204.0
12 × 18	216	29.4	235.0	55.1	440.6	114.8	137.7	183.6	229.5
12 × 20	240	32.6	261.1	61.2	489.6	127.5	153.0	204.0	255.0
14 × 14	196	26.7	213.2	50.0	399.8	104.1	125.0	166.6	208.3
14 × 16	224	30.5	243.7	57.1	457.0	119.0	142.8	190.4	238.0
14 × 18	252	34.3	274.2	64.3	514.1	133.9	160.7	214.2	267.8
14 × 20	280	38.1	304.6	71.4	571.2	148.8	178.5	238.0	297.5
14 × 22	308	41.9	335.1	78.5	628.3	163.6	196.4	261.8	327.3
16 × 16	256	34.8	278.5	65.3	522.2	136.0	163.2	217.6	272.0
16 × 18	288	39.2	313.3	73.4	587.5	153.0	183.6	244.8	306.0
16 × 20	320	43.5	348.2	81.6	652.8	170.0	204.0	272.0	340.0
16 × 22	352	47.9	383.0	89.8	718.1	187.0	224.4	299.2	374.0
16 × 24	384	52.2	417.8	97.9	783.4	204.0	244.8	326.4	408.0
18 × 18	324	44.1	352.5	82.6	661.0	172.1	206.6	275.4	344.3
18 × 20	360	49.0	391.7	91.8	734.4	191.3	229.5	306.0	382.5
18 × 22	396	53.9	430.8	101.0	807.8	210.4	252.5	336.6	420.8
18 × 24	432	58.8	470.0	110.2	881.3	229.5	275.4	367.2	459.0
18 × 26	468	63.6	509.2	119.3	954.7	248.6	298.4	397.8	497.3
20 × 20	400	54.4	435.2	102.0	816.0	212.5	255.0	340.0	425.0
20 × 22	440	59.8	478.7	112.2	897.6	233.8	280.5	374.0	267.5
20 × 24	480	65.3	522.2	122.4	979.2	255.0	306.0	408.0	510.0
20 × 26	520	70.7	565.8	132.6	1,060.8	276.3	331.5	442.0	552.5
20 × 28	560	71.2	609.2	142.8	1,142,4	297.5	357.0	476.0	595.0
22 × 22	484	65.8	526.6	123.4	987.4	257.1	308.6	411.4	514.3
22 × 24	528	71.8	574.5	134.6	1,077.1	280.5	336.6	448.8	561.0
22 × 26	572	77.8	622.3	145.9	1,166.9	303.9	364.7	486.2	607.8
22 × 28	616	83.8	670.2	157.1	1,256.6	327.3	392.7	523.6	654.5
24 × 24	576	78.3	626.7	146.9	1,175.0	306.0	367.2	489.6	612.0
24 × 26	624	84.9	678.9	159.1	1,273.0	331.5	397.8	530.4	663.0
24 × 28	672	91.4	721.1	171.4	1,370.9	357.0	428.4	571.2	714.0
24 × 30	720	97.9	783.4	183.6	1,468.8	382.5	459.0	612.0	765.0
12 in. ϕ	113.1	15.4	123.1	28.8	230.7	60.1	72.1	96.1	120.2
14 in. ϕ	153.9	20.9	167.6	39.3	314.2	81.8	98.1	130.8	163.5
16 in. ϕ	201.1	27.3	218.7	51.3	410.2	106.9	128.2	170.9	213.8
18 in. ϕ	254.5	34.5	276.4	64.7	519.2	135.0	162.2	216.3	270.1
20 in. ϕ	314.2	43.0	339.7	80.1	640.6	166.9	200.3	267.1	338.8
22 in. ϕ	380.1	51.7	413.4	96.9	775.2	201.9	242.3	323.1	403.9
24 in. ϕ	452.4	61.5	491.8	115.3	922.1	240.3	288.4	384.5	480.7

Table 7.7 Spirally Reinforced Concrete Columns

Maximum load in kips on concrete and limiting load in kips on longitudinal reinforcement

Square		Load on longitudinal reinforcement				Load on concrete			
		Intermediate grade		High strength $f_s = 60$ ksi		Concrete strength: f_c'			
Column size, in.	Gross area, A_g	Min.	Max.	Min.	Max.	2,500	3,000	4,000	5,000
12	144	23.0	184.3	43.2	345.6	90.0	108.0	144.0	180.0
14	196	31.4	250.9	58.8	470.4	122.5	147.0	196.0	245.0
16	256	41.0	327.7	76.8	614.4	160.0	192.0	256.0	320.0
18	324	51.8	414.7	97.2	777.6	202.5	243.0	324.0	405.0
20	400	64.0	512.0	120.0	960.0	250.0	300.0	400.0	500.0
22	484	77.4	619.5	145.2	1,161.6	302.5	363.0	484.0	605.0
24	576	92.2	737.1	172.8	1,382.4	360.0	432.0	576.0	720.0
26	676	108.2	865.3	202.8	1,622.4	422.5	507.0	676.0	845.0
28	784	125.4	1,003.5	235.2	1,881.6	490.0	588.0	784.0	980.0
30	900	144.0	1,152.0	270.0	2,160.0	562.5	675.0	900.0	1,125.0

Round		Load on longitudinal reinforcement				Load on concrete			
		Intermediate grade		High strength $f_s = 60$ ksi		Concrete strength: f_c'			
Column size, in.	Gross area, A_g	Min.	Max.	Min.	Max.	2,500	3,000	4,000	5,000
12	113.1	18.1	144.8	33.9	271.4	70.7	84.8	113.1	141.4
14	153.9	24.6	197.0	46.2	369.4	96.2	115.4	153.9	192.4
16	201.1	32.2	257.4	60.3	482.6	125.7	150.8	201.1	251.4
18	254.5	40.7	325.8	76.4	610.8	154.1	190.9	254.5	318.1
20	314.2	50.3	402.2	94.3	754.1	196.4	235.7	314.2	392.7
22	380.1	60.8	486.5	114.0	912.2	237.6	285.1	380.1	475.1
24	452.4	72.4	579.1	135.7	1,085.8	282.8	339.3	452.4	565.5
26	530.9	84.9	679.6	159.3	1,274.2	331.8	398.2	530.9	663.6
28	615.8	98.8	788.2	184.7	1,477.9	384.9	461.9	615.8	769.8
30	706.9	113.0	904.4	212.1	1,696.6	441.8	530.2	706.9	883.6

column design. Figure 7.18, itself, is too crowded to be useful except to indicate the use of the graph. The designer may lay out his own series of graphs on several larger sheets. The design method, using this chart, is explained in the following problem.

Problem Let us assume a series of similar columns loaded as follows.

Below the roof—a total load of 50 kips

Below the third floor—a total load of 150 kips

Below the second floor—a total load of 250 kips

Below the first floor—a total load of 350 kips

We will assume that for economy in formwork, it is desired to use the same size forms throughout the entire height. Assuming a column 12 in. square (144 sq in.), the unit load, beginning with the bottom section, will be

$$\text{Bottom tier } \frac{350{,}000}{144} = 2{,}431 \text{ psi}$$

$$\text{2nd tier } \frac{250{,}000}{144} = 1{,}736 \text{ psi}$$

$$\text{3rd tier } \frac{150{,}000}{144} = 1{,}042 \text{ psi}$$

$$\text{Top tier } \frac{50{,}000}{144} = 347 \text{ psi}$$

From Fig. 7.18, reading vertically on the line 2431, we find that the following may be used:

Line	Percent	Steel	Concrete	
5	7.4	H.S.	2,500	Tied
12	7.3	Inter	5,000	Spiral
6	7.0	H.S.	3,000	Tied
7	6.2	H.S.	4,000	Tied
13	6.0	H.S.	2,500	Spiral
14	5.6	Inter	5,000	Spiral
8	5.3	H.S.	5,000	Tied
15	4.7	H.S.	4,000	Spiral
16	3.9	H.S.	5,000	Spiral

Any of the above designs will comply with the Code.

In the second tier up, following the vertical line for 1736, we find that the following may be used:

Line	Percent	Steel	Concrete	
9	7.0	H.S.	4,000	Tied
3	6.0	Inter	4,000	Tied
10	6.2	Inter	3,000	Spiral
4	4.9	Inter	5,000	Tied
5	4.7	H.S.	2,500	Tied
11	4.6	Inter	4,000	Spiral
6	4.3	H.S.	3,000	Tied
13	3.8	H.S.	2,500	Spiral
7	3.5	H.S.	4,000	Tied
14	3.3	H.S.	3,000	Spiral
12	3.0	H.S.	2,500	Spiral
8	2.6	H.S.	5,000	Tied
15	2.4	H.S.	4,000	Spiral
16	1.6	H.S.	5,000	Spiral

In the third tier, following the vertical line for 1042, we find that the following may be used:

Line	Percent	Steel	Concrete	
1	3.7	Inter	2,500	Tied
2	2.9	Inter	3,000	Tied
9	2.6	Inter	2,500	Spiral
5	2.0	H.S.	2,500	Tied
10	1.8	Inter	3,000	Spiral
6	1.6	H.S.	3,000	Tied
3	1.4	Inter	4,000	Tied
13	1.4	H.S.	2,500	Spiral

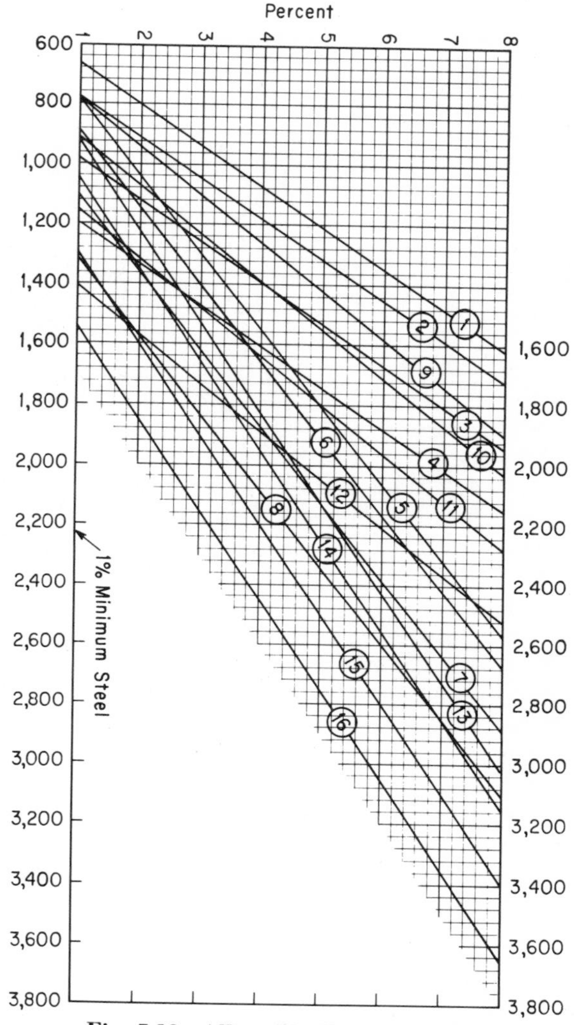

Fig. 7.18 Allowable Unit Stresses on Reinforced Concrete Columns (see Table 7.10)

For the top tier, the unit stress of 347 is less than the minimum for the 2,500-lb concrete with 1 percent intermediate grade steel.

On the basis of the above selections, and the limitations of good practice set by Art. 7.20*d*, the preferred design would be:

Bottom tier 5,000-psi concrete, 5.6 percent high-strength steel (= 8.06 sq in.) 4 bars, #14 S, A432.

Second tier 5,000-psi concrete, 4.9 percent intermediate steel (= 7.06 sq in.) 4 bars, #14 S, A15 intermediate.

Third tier 4,000-psi concrete, 1.4 percent intermediate steel (= 2.02 sq in.) 4 bars #7, A15 intermediate.

Table 7.8 Allowable Capacity of Steel Bars in Columns

Load in kips on longitudinal reinforcement (f_sA_s)

Tied columns

Inter. indicates A15 intermediate grade
Hi. St. indicates A432 high-strength steel

No. bars	Bar size	5	6	7	8	9	10	11	14S	18S
4	A_s	1.24	1.76	2.40	3.16	4.00	5.08	6.24	9.00	16.0
	Inter.	16.8	23.9	32.6	43.0	54.4	69.1	84.9	122.4	217.6
	Hi. St.	31.6	44.9	61.2	80.6	102.0	129.5	159.1	229.5	408.0
6	A_s	1.80	2.64	3.60	4.74	6.00	7.62	9.36	13.50	24.0
	Inter.	25.3	35.9	49.0	64.5	81.6	103.6	127.3	183.6	336.4
	Hi. St.	47.4	67.3	91.8	120.9	153.0	194.3	238.7	344.3	612.0
8	A_s	2.48	3.52	4.80	6.32	8.00	10.16	12.48	18.00	32.0
	Inter.	33.7	47.9	65.3	86.0	108.8	138.2	169.7	244.8	435.2
	Hi. St.	63.2	89.8	122.4	161.2	204.0	259.1	318.2	459.0	816.0
10	A_s	3.10	4.40	6.00	7.90	10.00	12.70	15.60	22.50	40.0
	Inter.	42.2	59.8	81.6	107.4	136.0	172.7	212.2	306.0	554.0
	Hi. St.	79.0	112.2	153.0	201.5	255.0	324.0	397.8	688.5	1,020.0

Spirally reinforced columns

Inter. indicates A15 intermediate grade
Hi. St. indicates A432 high-strength steel

No. bars	Bar size	5	6	7	8	9	10	11	14S	18S
6	A_s	1.80	2.64	3.60	4.74	6.00	7.62	9.36	13.50	24.0
	Inter.	29.8	42.2	58.2	75.8	96.0	121.9	149.8	316.0	384.0
	Hi. Sr.	58.8	79.2	128.0	142.2	180.0	228.6	280.8	405.0	720.0
8	A_s	2.48	3.52	4.80	6.32	8.00	10.16	12.48	18.00	32.0
	Inter.	39.7	56.3	76.8	101.1	128.0	162.6	199.7	288.0	512.0
	Hi. St.	74.4	105.6	144.0	189.6	240.0	304.8	374.4	540.0	960.0
10	A_s	3.10	4.40	6.00	7.90	10.00	12.70	15.60	22.50	40.0
	Inter.	49.6	70.4	95.4	126.4	160.0	203.2	249.6	360.0	640.0
	Hi. St.	93.0	132.0	180.0	237.0	300.0	381.0	468.0	675.0	1,200.0

Roof tier 2,500-psi concrete, 1 percent intermediate steel (= 1.44 sq in.) 4 bars #6, A15 intermediate.

7.21 Bending and direct stress

a Bending moment is usually brought into concrete columns through frame action, only occasionally through offset beams.

b Section 1407 of the 1963 ACI Code, as quoted in Art. 7.20*b*, covers the design of reinforced concrete columns for bending and direct stress.

Table 7.11 gives the value of "*m*" in the various equations of this section of the Code for the several combinations of steel and concrete stresses.

c The eccentricity in the above formulae may be obtained by dividing the moment in inch-pounds by the total load applied to the column below this floor line. Thus if the total load on the column below any floor is 250 kips, and the eccentricity induced by frame action to the column is 30 ft-kips (= 360 inch-kips), $e = 360/250 = 1.44$ in.

Table 7.9 Maximum Numbers of Bars in Spiral Columns

At lapped splices in outer ring (*O*) and in inner ring (*I*) if area required is greater than that obtainable in outer ring only.

Column size, in., $1\frac{1}{2}$ in. protection	Bar size														Column size, in., 2 in. protection
	No. 5		No. 6		No. 7		No. 8		No. 9		No. 10		No. 11		
	O	*I*	*O*	*I*	*O*	*I*	*O*	*I*	*O*	*I*	*O*	*I*	*O*	*I*	
14	11		10		9		8		7		6				15
16	13	8	11	7	10	6	9	5	8	4	7		6		17
18	15	10	13	9	12	8	11	7	10	5	8	4	8		19
20	17	12	15	11	14	9	13	8	11	6	10	5	9	4	21
22	19	14	17	13	16	11	15	10	13	8	11	6	10	5	23
24	21	16	19	15	18	13	16	12	14	9	12	8	11	6	25
26	23	19	21	17	20	15	18	13	16	11	14	9	12	8	27
28	26	21	24	19	22	17	20	15	17	13	15	11	14	9	29
30	28	23	26	21	24	19	22	17	19	14	17	12	15	10	31
32	30	26	28	23	26	21	24	19	21	16	18	13	16	11	33
34	33	28	30	25	27	23	25	21	22	17	19	15	17	13	35

Courtesy Portland Cement Association

Table 7.10 Maximum and Minimum Unit Stresses in Reinforced Concrete Columns

Line	Steel	Concrete		1%	8%
1	Inter.	2,500	Tied	667.25	1,619.25
2	Inter.	3,000	Tied	773.5	1,725.5
3	Inter.	4,000	Tied	986	1,938
4	Inter.	5,000	Tied	1,198.5	2,150.5
5	Hi. St.	2,500	Tied	786.25	2,571.25
6	Hi. St.	3,000	Tied	892.5	2,677.75
7	Hi. At.	4,000	Tied	1,105	2,890
8	Hi. St.	5,000	Tied	1,317.5	3,102.5
9	Inter.	2,500	Spiral	785	1,905
10	Inter.	3,000	Spiral	910	2,030
11	Inter.	4,000	Spiral	1,160	2,280
12	Inter.	5,000	Spiral	1,410	2,530
13	Hi. St.	2,500	Spiral	925	3,025
14	Hi. St.	3,000	Spiral	1,050	3,150
15	Hi. St.	4,000	Spiral	1,300	3,400
16	Hi. St.	5,000	Spiral	1,550	3,650

This is to be compared with the computed eccentricity derived for the particular column from Formula (14-6), (14-7), or (14-8), whichever is applicable.

7.30 WOOD COLUMNS

a Wood columns are designed in accordance with the requirements of the National Lumber Manufacturers Association (NLMA) as quoted in Art. 1.23 and in the following sections and tables from this Code.

Table 7.11 Value of "m"—Bending and Direct Stress—Reinforced Concrete Columns

$$m = \frac{f_y}{0.85 f_c'}$$

f_c'	A15 Intermediate grade (f_y 40,000 psi)	A432 High strength (f_y 60,000 psi)
2,500	18.8	28.2
3,000	13.7	23.5
4,000	11.8	17.7
5,000	9.4	14.1

210-D. *Adjustments for Service Conditions.*

210-D-1. The values in Table 7.12 apply under conditions continuously dry, such as in most covered structures. When used under other conditions, the values in Table 7.12 for glued-laminated timber and sawn lumber, 4 inches and less in thickness, shall be reduced 27 percent, and for sawn lumber, more than 4 inches in thickness, shall be reduced 9 percent.

206-A. *Based on l/d Ratio.*

206-A-1. The allowable unit stresses for compression parallel to grain, in Table 1.3 subject to the adjustments provided in sections 202, 203 and 204, shall not be exceeded in any column nor in the individual members of a spaced column.

Within this limitation, the allowable unit stresses for various l/d ratios shall be determined by the column formulas given in section 401.

Table 7.12 Allowable Unit Stresses for End Grain in Bearing

Species	Sawn Lumber 4″ and less in thickness, and Glued Laminated Lumber	Sawn Lumber more than 4″ in thickness
Ash, white	2,200	1,750
Beech and Birch	2,400	1,950
Cedar, Alaska	1,600	1,250
Cedar, Port Orford	1,800	1,450
Cedar, Western red	1,450	1,150
Cypress, Southern	2,200	1,750
Douglas Fir, coast region, dense	2,600	2,050
Douglas Fir, coast region, close grained	2,350	1,900
Douglas Fir, coast region	2,200	1,750
Douglas Fir, dense	2,600	2,050
Douglas Fir, close grained	2,350	1,900
Douglas Fir	2,200	1,750
Fir, white	1,450	1,150
Gum	1,600	1,250
Hemlock, Eastern	1,450	1,150
Hemlock, West Coast	1,800	1,450
Hickory and Pecan	3,000	2,400
Larch, dense	2,600	2,050
Larch, close grained	2,350	1,900
Larch	2,200	1,750
Maple, hard	2,400	1,950
Oak, red and white	2,050	1,650
Pine, Northern (Eastern) white	1,550	1,200
Pine, Ponderosa, sugar, Idaho white	1,550	1,200
Pine, Lodgepole	1,450	1,150
Pine, Norway	1,600	1,250
Pine, Southern, dense	2,600	2,050
Pine, Southern	2,200	1,750
Redwood, close grained	2,200	1,750
Redwood	2,050	1,650
Spruce, Engelmann	1,200	950
Spruce, Eastern and Sitka	1,600	1,250

401-WOOD COLUMNS

401-A. *Notation.*

401-A-1. Except where otherwise noted, the symbols used in the column formulae are as follows:

A = area of column cross section, square inches

c = allowable unit stress in compression parallel to grain, adjusted in accordance with sections 202, 203 and 204.

d = dimension of least side of simple solid column and of least side of individual members of spaced columns, inches.

E = modulus of elasticity.

l = unsupported overall length, in inches, between points of lateral support of simple columns or from center to center of lateral supports of continuous or spaced columns.

l_2 = distance in spaced columns from center of connector in end blocks to center of spacer block, inches.

P = total load, pounds.

P/A = load per unit of cross-sectional area.

r = least radius of gyration of section.

401-B. *Column Classifications.*

401-B-1. Simple Solid Wood Columns. Simple columns consist of a single piece or of pieces properly glued together to form a single member.

401-B-2. Spaced Columns, Connector Joined. Spaced columns are formed of two or more individual members with their longitudinal axes parallel, separated at the ends and middle points of their length by blocking and joined at the ends

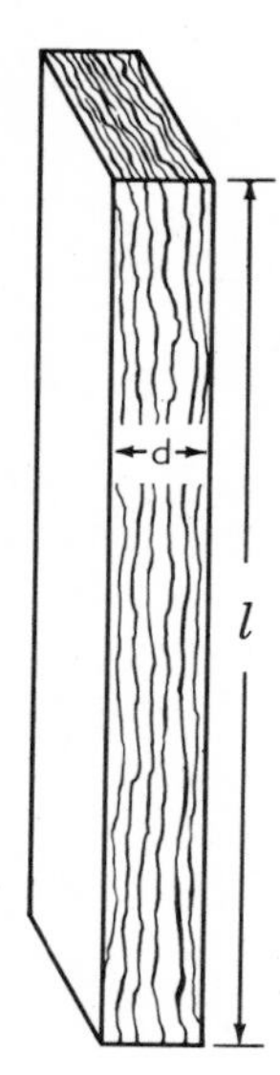

Fig. 7.19 Simple Solid Wood Column

by timber connectors capable of developing the required shear resistance. To obtain spaced-column action, end blocks with connectors and spacer blocks are required when the individual members of a spaced-column assembly have an l/d ratio greater than

$$\sqrt{\frac{0.30E}{c}}$$

For an assembly of members having a lesser l/d ratio, the individual members are designed as simple solid columns. Spaced columns are classified as to fixity, i.e., condition "a" or condition "b", which introduces a multiplying factor applicable in the design of its individual members. (See Fig. 7.20.)

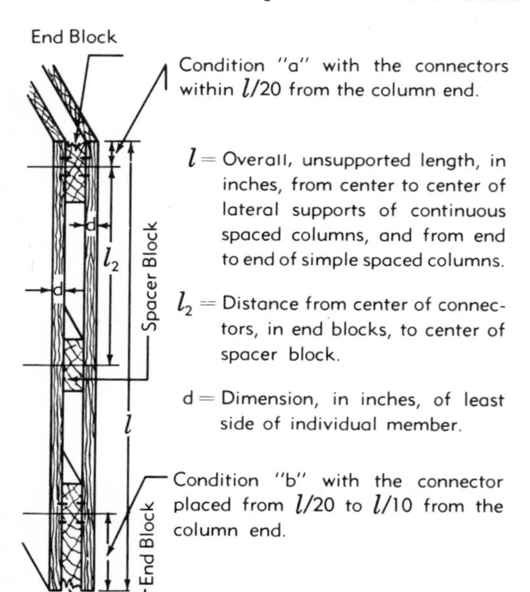

Fig. 7.20 Spaced Column—Connector Joined

401-B-3. Built-up Columns. Built-up columns, other than connector-joined spaced columns and glued-laminated columns, shall not be designed as solid columns but shall be designed in accordance with Reference 2, Appendix I, except that the allowable unit stresses provided herein shall apply.

401-C. *Limitations on l/d Ratios.*

401-C-1. For simple solid columns, l/d shall not exceed 50.

401-C-2. For individual members of a spaced column, l/d shall not exceed 80, nor shall l_2/d exceed 40.

401-D. *Compression Members Bearing End to End.*

401-D-1. For end-grain bearing of wood on wood and on metal plates or straps the allowable stresses at the bearing shall be as provided in section 210.

401-E. *Simple Solid-Column Design.*

401-E-1. End Conditions. These formulae for simple solid columns are based on pin-end conditions but shall be applied also to square-end conditions.

401-E-2. Allowable Unit Loads for Simple Solid Columns. Allowable unit loads in pounds per square inch of cross-sectional area of simple solid columns shall be determined by the following formula but the maximum unit load shall not exceed the values for compression parallel to grain c in Table 1.3 adjusted in accordance with the provisions of sections 202, 203 and 204.

$$P/A = \frac{\pi^2 E}{2.727(l/r)^2} = \frac{3.619E}{(l/r)^2}$$

For columns of square or rectangular cross section this formula becomes:

$$P/A = \frac{0.30E}{(l/d)^2}$$

401-E-3. Adjustments for Loading Durations. The values for P/A, as obtained from the formula in 401-E-2, are subject to duration of load adjustments as provided in sections 203 and 204.

401-F. *Spaced-Column Design.*

401-F-1. The individual members in a spaced column are considered to act together to carry the total column load. Each member is designed separately on the basis of its l/d ratio.

A greater l/d ratio than allowed for simple solid columns is permitted because of the end fixity developed by the connectors and end blocks. This fixity is effective only in the thickness direction. The l/d ratio in the direction of width is subject to the provisions for simple solid columns.

A higher unit stress than would be permitted for simple solid columns, under section 401-E-2, is allowed for spaced columns, due to end fixity.

401-F-2. Location of Spacer and End Blocks. When a single spacer block is located within the middle tenth of the column length (l), connectors are not required for this block. If there are two or more spacer blocks, connectors are required and the distance between two adjacent blocks shall not exceed one-half the distance between centers of connectors in the end blocks.

For spaced columns used as compression members of a truss, a panel point which is stayed laterally shall be considered as the end of the spaced column, and the portion of the web members, between the individual pieces making up a spaced column, may be considered as the end blocks.

If there are two or more connectors in a contact face, the center of gravity of the group shall be used in measuring the distance from connectors in the end block to the end of the column for determining fixity condition "a" or "b" (Fig. 7.20).

401-F-3. Dimensions of Spacer and End Blocks. Thickness of spacer and end blocks shall not be less than that of individual members of the spaced column nor shall thickness, width, and length of spacer and end blocks be less than required for connectors of a size and number capable of carrying the load computed from section 401-F-4. Blocks thicker than a side member do not appreciably increase load capacity.

401-F-4. Load Capacity of Connectors in End Spacer Blocks. To obtain spaced-column action the connectors in mutually contacting surfaces of end blocks and individual members at each end of a spaced column shall be of a size and number to provide a load capacity in pounds equal to the required cross-sectional area in square inches of one of the individual members times the appropriate end-spacer block constant in Table 7.13.

If spaced columns are a part of a truss system or other similar framing, the connectors required by joint design may be sufficient for end block connectors but should be checked against Table 7.13.

401-F-5. Allowable Loads for Spaced Columns. The total allowable load for a spaced column is the sum of the allowable loads for each of its individual members. Allowable unit stresses shall be determined as follows, but the maximum unit stress shall not exceed the values for compression parallel to grain "c" in Table 1.3 as adjusted in sections 202, 203 and 204, nor shall the load exceed that permitted by section 502.

For condition "a." The allowable unit stress for individual members of a spaced column, in which the connectors in end blocks are placed at a distance not exceeding $l/20$ from the ends, shall be determined by the formula:

$$P/A = \frac{0.75E}{(l/d)^2}$$

For condition "b." The allowable load for the individual members of a spaced column in which the connectors in end blocks are placed a distance of $l/20$ to $l/10$ from the ends and the blocks extend to the ends of the column shall be determined by the formula:

$$P/A = \frac{0.90E}{(l/d)^2}$$

The total load capacity determined by the foregoing procedure should be checked against the sum of the load capacities of the individual members taken as simple solid columns without regard to fixity, using their greater d and the l between the lateral supports which provide restraint in a direction parallel to the greater d.

401-F-6. Adjustments for Loading Durations. The values for P/A, as obtained from the provisions of 401-F-5, are subject to the duration of load adjustments as provided in sections 203 and 204.

Table 7.13 End Spacer Block Constants for Connector Joined Spaced Columns

l/d ratio of individual member in the spaced column[1]	End spacer block constant			
	Group A connector loads	Group B connector loads	Group C connector loads	Group D connector loads
0 to 11	0	0	0	0
15	38	33	27	21
20	86	73	61	48
25	134	114	94	75
30	181	155	128	101
35	229	195	162	128
40	277	236	195	154
45	325	277	229	181
50	372	318	263	208
55	420	358	296	234
60 to 80	468	399	330	261

[1] Constants for intermediate l/d ratios may be obtained by straight line interpolation.

401-G. *Round Columns.*

401-G-1. The allowable load for a column of round cross section shall not exceed that for a square column of the same cross-sectional area, or as determined by the formula:

$$P/A = \frac{3.619E}{(l/r)^2}$$

provided, however, that P/A shall not exceed c adjusted as provided herein.

401-G-2. The values for P/A, as obtained from the formula in 401-G-1, are subject to duration of load adjustments as provided in sections 203 and 204.

401-H. *Tapered Columns.*

401-H-1. In determining the d for tapered column design, the diameter of a round column or the least dimension of a column of rectangular section, tapered at one or both ends, shall be taken as the sum of the minimum diameter or least dimension and one-third the difference between the minimum and maximum diameters or lesser dimensions.

401-I. *Wood Column Bracing.*

401-I-1. Column bracing shall be installed where necessary to resist wind or other lateral forces. (See Appendix F.)

FORMULAE FOR WOOD COLUMNS WITH SIDE LOADS AND ECCENTRICITY

A. *General*

The following information is provided for use when more accurate calculation of the maximum direct compression load that can be put upon an eccentrically loaded column or one with a side load is desired.

The most accurate of all the column formulae for the critical load on a long, originally straight, centrally loaded column of uniform cross section is the Euler formula, and for long columns with an eccentric application of load is the secant formula.

The inaccuracies of these formulae are too small to be detected by the ordinary methods of testing such columns and are much smaller than the inaccuracy of calculating the stiffness of wood beams neglecting the distortions due to shear.

The formulae here presented can be made to have an accuracy fairly comparable to that for the Euler column, but in their simplified form they do not pretend to such accuracy; however, they do have an accuracy well within an acceptable range for most engineering purposes. (Reference 11, Appendix 1.)

B. *Assumptions*

The following assumptions have been made in arriving at the following simplified equations; normally, they change the results but little:

1. The stresses which cause a given deflection as a sinusoidal curve are the same as those for a beam, with a uniform side load.
2. For a single concentrated side load the stress under the load can be used, regardless of the position of the load with reference to the length of the column.
3. The stress to use with a system of side loads is the maximum stress due to the system. (With large side loads near each end some slight error on the side of overload will occur.)
4. The equations can be used for solving P/A for rectangular wood columns with l/d ratio greater than

$$\sqrt{\frac{0.30E}{c}}$$

5. For columns with an l/d ratio of

$$\sqrt{\frac{0.30E}{c}}$$

or less, stress due to deflection of the column may be neglected.

C. *Notation*

P/A = direct compressive stress in pounds per square inch induced by axial load.

M/S = flexural stress in pounds per square inch induced by side loads.

c = the allowable unit stress in compression parallel to the grain in pounds per square inch that would be permitted for the column if axial compressive stress only existed; that is, the allowable unit stress for the l/d of the column under consideration.

f = the allowable unit stress in flexure in pounds per square inch that would be permitted if flexural stress only existed.

e = eccentricity in inches.

l = length of column in inches.

d = side in inches of a rectangular column, measured in the direction of the side loads.

D. *Formulae*

The following formulae, which are for pin-end columns of rectangular cross sections, apply when a member is subjected to both axial compression and moment (from eccentricity or side loads). The formulae given are for determining the maximum permissible unit load or stress, P/A or M/S, which is permissible under the conditions of combined loading, i.e., the P/A and M/S induced by loading shall not exceed the values therefor obtained from the formulae.

1. Columns having an l/d ratio of

$$\sqrt{\frac{0.30E}{c}}$$

or less:

(a) Concentric End and Side Loads:

$$\frac{M/S}{f} + \frac{P/A}{c} = 1$$

(b) Eccentric Load:

$$\frac{P/A\left(\frac{6e}{d}\right)}{f} + \frac{P/A}{c} = 1$$

(c) Combined End Loads, Side Loads and Eccentricity:

$$\frac{M/S + P/A\left(\frac{6e}{d}\right)}{f} + \frac{P/A}{c} = 1$$

2. Columns having an l/d ratio greater than $\sqrt{\frac{0.30E}{c}}$:

(a) Concentric End and Side Loads:

$$\frac{M/S}{f - P/A} + \frac{P/A}{c} = 1$$

(b) Eccentric Load:

$$\frac{P/A\left(\frac{15e}{2d}\right)}{f - P/A} + \frac{P/A}{c} = 1$$

(c) Combined End Load, Side Loads and Eccentricity:

$$\frac{M/S + P/A\left(\frac{15e}{2d}\right)}{f - P/A} + \frac{P/A}{c} = 1$$

E. *Columns with Side Brackets*

An exact solution of the stresses in a column with end and bracket loads is difficult. The following simple rule for determining loads to be used in the combined loading formulae is safe, and for brackets in the upper quarter of the height of a column is sufficiently accurate.

Instead of using the bracket load P in its actual position, add to the end load on the column an axial end load P equal to the load on the bracket and calculate the bending stress from an assumed side load P', concentrated at the center of the length of the column and determined as follows (see Fig. 7.21):

$$P' = 3 \times \frac{l'}{l} \times \frac{a}{l} P = \frac{3al'P}{l^2}$$

in which

P = actual load in pounds on bracket.
P' = assumed horizontal side load in pounds assumed to be placed at center of height of column.
a = horizontal distance in inches from load on bracket to center of column.
l = total length of column in inches.
l' = distance in inches measured vertically from point of application of load on bracket to farther end of column.

Use P' to determine the induced unit flexural stress M/S, as if the column were a beam with a concentrated load P' at midlength. Then combine these stresses with those from other loads and apply the appropriate combined stress formulae given under the preceding section D.

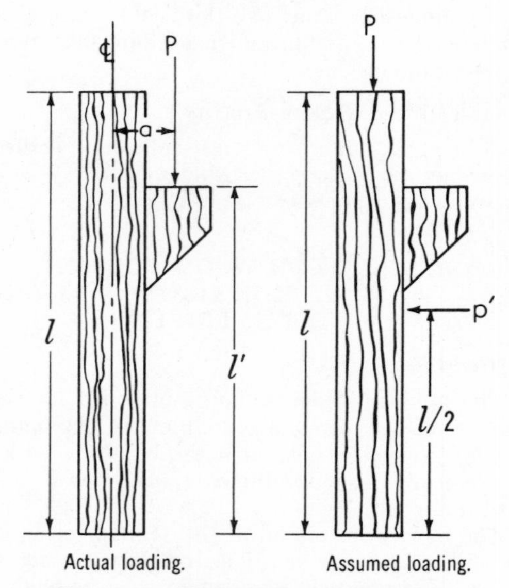

Fig. 7.21 Wood Column Brackets

b It will be noted from the NLMA Code that up to a certain length, the allowable unit stress may be taken from column 6 in Table 1.3 for the particular combination of unit stress and modulus of elasticity listed for the lumber used. The various combinations of unit stress and moduli of elasticity are listed in Table 7.14 and the l/d for which these stresses may be used. Beyond this break point, a reduction formula must be used as given in Section 401-E-2 of the Code.

The maximum length which may be used is $l = 50d$ and the allowable unit stress (which is dependent on E only) is given at the bottom of Table 7.14. There are too many variables to permit the use of any simple table of capacities, but the allowable length of column for each l/d within the limits of Table 7.14 are given in Table 7.15.

Table 7.14 Maximum l/d for Combinations of "c" and "E"

c	1,210,000	1,320,000	1,540,000	1,760,000
600	24.6	25.7		
700	22.8	23.8		
800	21.3	22.3		
900	20.1	21.0	22.7	24.2
1,000		19.9	21.5	23.0
1,100		19.0	20.6	21.9
1,200		18.2	19.6	21.0
1,300		17.4	18.9	20.2
1,400		16.8		19.4
1,500		16.2		18.8
1,600		15.7		18.2
1,700				17.6
1,800				17.1
Max. f at 50 l/d	145 psi	158 psi	185 psi	211 psi

Table 7.15 Length of Columns for Various l/d

l/d	4″ sq (3⅝″)	6″ sq (5½″)	8″ sq (7½″)	10″ sq (9½″)	12″ sq (11½″)
$A =$	13.14	30.25	56.25	90.25	132.25
15.7	4′ 9″	7′ 2″	9′ 10″	12′ 6″	15′ 1″
16	4′ 10″	7′ 4″	10′ 0″	12′ 8″	15′ 4″
17	5′ 2″	7′ 10″	10′ 7″	13′ 5″	16′ 3″
18	5′ 5″	8′ 3″	11′ 3″	14′ 3″	17′ 3″
19	5′ 9″	8′ 9″	11′ 10″	15′ 0″	18′ 2″
20	6′ 1″	9′ 2″	12′ 6″	15′ 10″	19′ 2″
21	6′ 4″	9′ 8″	13′ 1″	16′ 7″	20′ 1″
22	6′ 8″	10′ 1″	13′ 9″	17′ 5″	21′ 1″
23	6′ 11″	10′ 7″	14′ 4″	18′ 2″	22′ 0″
24	7′ 3″	11′ 0″	15′ 0″	19′ 0″	23′ 0″
25	7′ 7″	11′ 6″	15′ 7″	19′ 9″	23′ 11″
26	7′ 10″	11′ 11″	16′ 3″	20′ 7″	24′ 11″
50	15′ 1″	22′ 11″	31′ 3″	39′ 7″	47′ 11″

For example, assume a load of 100 kips on a wood column of Douglas Fir, Coast Region P. and T. (see Table 1.3) on a height of 12 ft. From column 6 in Table 1.3, $c = 1{,}500$ and from column 7 in the same table, $E = 1{,}760{,}000$. From Table 7.14, the maximum l/d is 18.8 unless the reduction formula is used. From Table 7.15, an 8-in. square column (or larger) may be used up to a height of about 11 ft 8 in. and the capacity of this column would be $56.25 \times 1{,}500 = 84{,}375$, which is not enough. A 10-in. square column will carry $90.25 \times 1{,}500 = 135{,}375$ on a column up to 14 ft 9 in. and is therefore the column to use.

7.31 Wood-stud partitions

Wood-stud partitions, bridged at the center of the height and otherwise braced by lath and plaster, are used to carry vertical loads in ordinary light-load structures. Most buildings in which wood-stud partitions are used as load-carrying walls have loads so light that it is not necessary to check the capacity of the studs. However, for the occasion when it is necessary to check this strength Table 7.16 is included. This table is taken from the "Manual of the Southern Pine Association."

7.40 COMPOSITE COLUMNS

Columns in which structural steel shapes and concrete are both used to carry load are called composite columns. Design methods as set forth in the 1963 ACI Code, Section 1404, are quoted in Art. 7.20*b*.

The most commonly used of the three systems is the combination column described in Section 1405. In some parts of the country the concrete-filled pipe column similar to the familiar Lally column—as described in Section 1406—is used.

7.50 COLUMN BASE PLATES

a Steel base plates or billets are generally used under columns for distributing the column loads over a sufficient area of the concrete foundation. The AISC Code, Section 1.5.5, specifies a bearing value of $0.25f_c'$ on the full area of concrete or $0.375f_c'$ on one third of the area. The value of $0.25f_c'$

Table 7.16 Strength of Wood Stud Partitions

Nominal size	Actual size, in.	Distance on centers, in.	Height, ft	Per lineal foot of partitions		
				Safe load, lb	Weight, lb	Amount, fbm
2 × 4	$1\frac{5}{8} \times 3\frac{5}{8}$	12	8	3,675	16.30	6.66
2 × 4	$1\frac{5}{8} \times 3\frac{5}{8}$	12	10	2,374	19.56	8.00
2 × 4	$1\frac{5}{8} \times 3\frac{5}{8}$	12	12	1,637	22.82	9.33
2 × 4	$1\frac{5}{8} \times 3\frac{5}{8}$	16	8	2,756	13.04	5.33
2 × 4	$1\frac{5}{8} \times 3\frac{5}{8}$	16	10	1,780	15.50	6.33
2 × 4	$1\frac{5}{8} \times 3\frac{5}{8}$	16	12	1,228	18.00	7.33
2 × 6	$1\frac{5}{8} \times 5\frac{5}{8}$	12	8	8,546	25.30	10.00
2 × 6	$1\frac{5}{8} \times 5\frac{5}{8}$	12	10	7,677	30.56	12.00
2 × 6	$1\frac{5}{8} \times 5\frac{5}{8}$	12	12	6,096	35.42	14.00
2 × 6	$1\frac{5}{8} \times 5\frac{5}{8}$	16	8	6,409	20.24	8.00
2 × 6	$1\frac{5}{8} \times 5\frac{5}{8}$	16	10	5,758	24.03	9.50
2 × 6	$1\frac{5}{8} \times 5\frac{5}{8}$	16	12	4,572	27.83	11.00
3 × 6	$2\frac{5}{8} \times 5\frac{5}{8}$	12	8	13,810	41.00	15.00
3 × 6	$2\frac{5}{8} \times 5\frac{5}{8}$	12	10	12,406	49.20	18.00
3 × 6	$2\frac{5}{8} \times 5\frac{5}{8}$	12	12	9,851	57.20	21.00
3 × 6	$2\frac{5}{8} \times 5\frac{5}{8}$	16	8	10,357	32.80	12.00
3 × 6	$2\frac{5}{8} \times 5\frac{5}{8}$	16	10	9,304	38.95	14.25
3 × 6	$2\frac{5}{8} \times 5\frac{5}{8}$	16	12	7,388	45.10	16.50

Safe load based on studs being bridged at center
Weight and strength based on actual size.
Board measure based on nominal size.
Add weight of plaster or ceiling.
Single plate top and bottom included, same size as studs

seems to be rather useless since even on a concrete pedestal, an edge distance of concrete of $2\frac{1}{2}$ to 4 in. is allowed. Normally however, we will use the higher value since the area on which the bearing plate rests will usually be much in excess of three times the area of the base plate. The allowable bearing stresses therefore are 937.5 psi for 2,500-psi concrete, 1,125 for 3,000-psi concrete, 1,500 for 4,000-psi concrete, and 1,875 for 5,000-psi concrete.

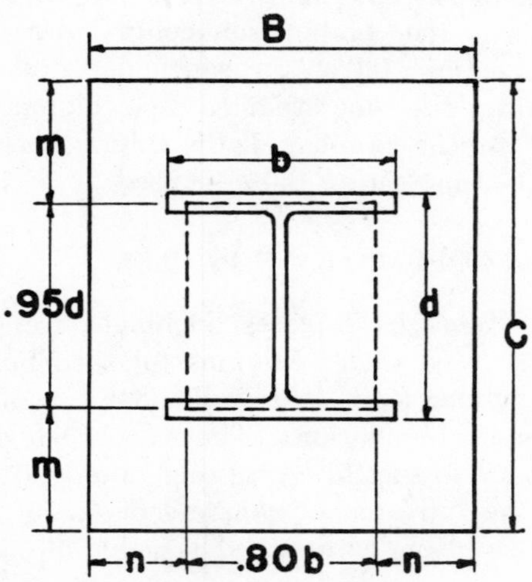

Fig. 7.22 Column Base Plate

In accordance with the AISC Code, Section 1.5.1.4.8, the bending stress $F_b = 0.75F_y$ or 27,000 psi for A36 steel which is normally used. It is common practice to use $f_c' =$ 3,000 psi, so $F_p = 1{,}125$. The method of design recommended by the AISC and described in full, together with the acceptable tables, is given in the AISC Manual of Steel Construction on pages 3-75 through 3-83 from which the following is taken.

P = Total column load, kips
$A = B \times C$ = Area of plate, square inches
F_b = Allowable bending stress in base plate, ksi
F_p = Allowable bearing pressure on support, ksi
f_c' = Compressive strength of concrete, ksi
t = Thickness of plate, inches

1. Establish bearing value of concrete, F_p, ksi
2. Determine the required area, $A = P/F_p$
3. Establish B and C, preferably rounded to full inches, so that m and n are approximately equal, and $B \times C \geq A$

4. Determine

$$m = \frac{C - .95d}{2} \quad \text{and} \quad n = \frac{B - .80b}{2}$$

5. Determine actual bearing pressure on concrete, $F_p = P/(B \times C)$
6. Use the larger of the values, m or n, to solve for t by whichever is the applicable formula:

$$t = \sqrt{\frac{3F_p m^2}{F_b}} \quad \text{or} \quad t = \sqrt{\frac{3F_p n^2}{F_b}}$$

Using the values quoted above, these formulae become $t = \sqrt{0.125m^2}$ or $t = \sqrt{0.125n^2}$.

Standard rolled sizes for bearing plates are as follows and should be used wherever possible:

14 × 1¼ in., 1½ in.
16 × 1½ in., 2 in.
20 × 2 in., 2½ in., 3 in.
24 × 2 in., 2½ in., 3 in.
28 × 3 in., 3½ in.
32 × 3½ in., 4 in.
36 × 4 in., 4½ in.
40 × 4½ in., 5 in.
44 × 5 in., 5½ in.
48 × 5½ in., 6 in., 6½ in.
52 × 6 in., 6½ in.
56 × 6½ in., 7 in., 8 in.

The base plate or billet should be cut or burned from one of the above sizes. Thickness less than the above may also be used.

Problem A 14 WF 95 column carries a total load of 450 kips to a footing of 3,000-psi concrete. Design a steel base plate for this column.

$$A = \frac{450}{1.125} = 400 \text{ sq in.} = 20\text{-in. square}$$

From observation, n is greater than m.

$$n = \frac{20 - (0.8 \times 14.12)}{2} = \frac{8.7}{2} = 4.35$$

$$t = \sqrt{0.125 \times 4.35^2} = \sqrt{2.37} = 1.54$$

Use 20 × 20 in. × 1¾ in. billet.

c Steel billets are normally bedded in grout on top of the concrete footings, and allowance for this should be made in setting the elevation of foundations. The minimum allowance for grout should be 1 in. and a greater depth should be allowed for bases or billets in excess of 12 in. square. Because of practical difficulties in placing grout under larger billets, and because of curl of the billet from shearing or burred edges from burning, it is well to allow approximately 1 in. depth of grout for each foot of side of billet, with a maximum of 3 in.

7.51 Grillages

a For bearing of a column on a wall or along a footing at the lot line or for spreading a truss or girder load along a wall, the most economical method is the use of a single-tier grillage. It is preferable to use a pair of channels back to back instead of an I beam in order to obtain a greater web thickness and to reduce the width of necessary bearing. The general method of figuring a grillage is to compute the area required under the grillage in bearing, then cut and try until the economical beam or pair of channels is found to furnish the bearing area required without exceeding the allowable fiber stress of the beam. The following formula will give the length of beam which is safe to use for any beam size selected.

$$l = \sqrt{\frac{176{,}000S}{F_p b}}$$

In this case F_p is the allowable bearing pressure, b the flange width, and S and l have their usual meanings. The length times the unit pressure times the width gives the capacity of this beam.

b It will be noted that the above formula may be broken down into two parts, one which we shall call K, a function of allowable bearing pressure, the other which we shall call C, a function of the steel section used. Their values are

$$K = \sqrt{\frac{176{,}000}{F_p}}$$

$$C = \sqrt{S/b}$$

$$l = CK$$

Table 7.17 gives the constants for bearing conditions usually specified and for channel sections used, and the length and bearing for a single channel. The use of this table is shown by the following problem.

Problem A column carrying a load of

Table 7.17 Channel Grillages—A36 Steel

	C	250 psi, good brick, $K = 28.5$	625 psi, 2,500-lb concrete wall, $K = 16.8$	750 psi, 3,000-lb concrete wall, $K = 15.3$	1,125 psi, 3,000-lb concrete footing, $K = 12.2$
6-in. channel 8.2	1.5	42.75 in. 20,600 lb	25.2 in. 30,400 lb	22.95 in. 33,000 lb	18.3 in. 39,500 lb
7-in. channel 9.8	1.7	48.45 in. 25,400 lb	28.56 in. 37,300 lb	26.01 in. 41,000 lb	20.74 in. 48,800 lb
8-in. channel 11.5	1.9	54.15 in. 30,600 lb	31.92 in. 45,300 lb	29.07 in. 49,500 lb	23.18 in. 59,000 lb
9-in. channel 13.4	2.08	59.28 in. 36,200 lb	34.94 in. 53,400 lb	31.82 in. 57,500 lb	25.38 in. 69,700 lb
10-in. channel 15.3	2.27	64.7 in. 42,000 lb	38.14 in. 62,800 lb	34.73 in. 68,000 lb	27.69 in. 81,400 lb
12-in. channel 20.7	2.7	76.95 in. 56,800 lb	45.36 in. 83,500 lb	41.31 in. 91,000 lb	32.94 in. 109,000 lb
12-in. channel 25.0	2.81	80.1 in. 61,200 lb	47.21 in. 90,000 lb	43.00 in. 99,000 lb	34.65 in. 119,000 lb
12-in. channel 30.0	2.91	82.94 in. 65,600 lb	48.89 in. 96,800 lb	44.52 in. 106,000 lb	35.87 in. 128,500 lb
15-in. channel 33.9	3.5	99.75 in. 85,000 lb	58.80 in. 125,000 lb	53.55 in. 136,600 lb	42.70 in. 164,000 lb
15-in. channel 40.0	3.62	103.17 in. 92,000 lb	60.82 in. 134,000 lb	55.39 in. 146,000 lb	44.16 in. 175,000 lb
15-in. channel 50.0	3.78	107.73 in. 100,000 lb	62.66 in. 145,000 lb	57.83 in. 161,000 lb	46.12 in. 194,000 lb
18-in. channel 42.7	3.94	112.29 in. 111,000 lb	66.19 in. 163,500 lb	60.28 in. 179,000 lb	48.07 in. 214,000 lb
18-in. channel 45.8	3.99	113.71 in. 113,710 lb	67.03 in. 167,600 lb	61.05 in. 183,200 lb	48.68 in. 219,000 lb
18-in. channel 51.9	4.11	117.14 in. 120,100 lb	69.05 in. 177,000 lb	62.88 in. 193,600 lb	50.14 in. 232,000 lb
18-in. channel 58.0	4.21	120.00 in. 126,000 lb	70.73 in. 185,300 lb	64.41 in. 219,000 lb	51.36 in. 243,000 lb

500 kips bears on a four-channel grillage on top of a concrete basement wall. The code under which we are working calls for a maximum allowable unit pressure of 750 psi. Design the grillage.

$$P = \frac{500}{4} = 125{,}000 \text{ lb on one channel}$$

One 15-in. channel, 33.9 lb, 53.55 in. long, will carry 136,600 lb. Therefore, 125,000 will require a length of

$$\frac{125{,}000}{136{,}000} \times 53.55 = 49 \text{ in.}$$

The required grillage, therefore, is four 15-in. channels, 33.9 lb, 4 ft 1 in. long.

c It may be necessary to spread a beam or truss load into the top of a tile or concrete masonry wall where the allowable bearing is much less than that listed for poor brick. For example, let us design a beam grillage to carry a beam load of 30,000 lb on a tile wall with an allowable bearing stress of 80 psf. From the formulae in paragraph *b*,

$$K = \sqrt{\frac{176{,}000}{80}} = \sqrt{2{,}200} = 47.0$$

Try two 8 WF 17 grillage.

$$C = \sqrt{\frac{14.1}{5.25}} = \sqrt{2.68} = 1.64$$

The allowable length = 47.0 × 1.64 = 77.08 in. The allowable load = 77.08 × 5.25 × 2 × 80 = 64,750 lb, which is greater than required. Try 6 I 12.5.

$$C = \sqrt{\frac{7.3}{3.33}} = \sqrt{2.2} = 1.47$$

The allowable length = 47.0 × 1.47 = 69.1 in. The allowable load = 69.1 × 3.33 × 2 × 80 = 36,848 lb. Therefore, use two 6 I 12.5, length $= \frac{30}{36.8} \times 69.1 = 56.4$. Use 4 ft 9 in.

As a check for strength, a single cantilever carries 30,000/4 = 7,500 lb. The moment on the cantilever is 7,500 × 2.38/2 = 8,920 ft-lb.

$$f_s = \frac{8{,}920 \times 12}{7.3} = 14{,}660 \text{ psi}$$

Despite the fact that this is well under the allowable 24,000-psi stress, it will be found that the use of a 5 I will cause an overstress.

8

Foundations and Walls

8.10 FOUNDATIONS—GENERAL

a The type of foundation to be used depends on:

1. The type of soil and its bearing capacity
2. The amount of superimposed load
3. The location with regard to lot line, etc.
4. The depth to rock

b We may divide column foundations as to type as follows:

1. Spread footings, centrally loaded
   - (*a*) Plain concrete
   - (*b*) Reinforced concrete
   - (*c*) Grillage
2. Combined footings
   - (*a*) Simple combined
   - (*b*) Cantilever
   - (*c*) Mat
3. Piles
   - (*a*) Wood
   - (*b*) Concrete
   - (*c*) Steel pipes or H sections
4. Caissons
   - (*a*) Open
   - (*b*) Pneumatic

c Although improper decisions as to foundation design may not bring about the complete failure of any portion of the building, unequal settlement may bring about deterioration of the building to an extent far in excess of the cost which might be involved in obtaining proper advice from a specialist, and making a conservative design to provide against such unequal settlement. The foundation of a building is certainly not an advisable place to economize unless there is ample reason for all decisions which may be made. It is well to remember that the more rigid the footing or the combination of footings, the more uniform will be the pressure on the base, and the less trouble may be expected from unequal settlement.

At the present time, building foundations are designed very largely by rule-of-thumb methods to spread the load on the subsoil within limits set by the code under which the engineer is working. In recent years, however, the science of soil mechanics has advanced rapidly, and some codes are modifying the old methods of design in accordance with our newly found knowledge. The proper application of the principles of soil mechanics depends on laboratory investigations and is a job for a specialist.

The determination of soil conditions, including water conditions and allowable pressures, should be made by the methods described in Art. 11.30 before proceeding with the design. Determination should also be made of all other conditions affecting the foundation layout, as described in Art. 11.20. This involves the determination of legal restrictions if footings are placed immediately adjacent to the lot line or the street line —both as to your own rights and the rights of support of the foundation on the adjacent lot.

If the class of subsoil is known and there is no information or code requirement that modifies the determination of bearing capacity, the values shown in Table 8.1 may be assumed for bearing values of the various soils.

In computing the vertical pressure of supporting materials, the weight of the excavated material permanently removed may be deducted from the weight of the structure provided the material does not heave or swell when the excavation is made. This may be compensated for by adding the weight of a column of the excavated material to the bearing value of the soil. Thus, if a loose coarse sand (classification 8 in Table 8.1), good for 3 tons per sq ft, weighing 100 lb per cu ft, is assumed, footings may be designed to carry 7,000 psf at a depth 10 ft below the surface.

Table 8.1 Allowable Bearing Value for Footings

Class and material	Tons per sq ft
1. Massive crystalline bedrocks, like granite, gneiss, or trap rock, in sound condition	100
2. Foliated rocks—schist, slate—in sound condition	40
3. Sedimentary rocks—sandstones, limestones, hard shales, conglomerates	15
4. Soft or broken bedrock except shale	10
5. Hardpan, partially cemented sands, and gravels	10
6. Compact gravel, or sand and gravel mixture	6
7. Loose gravel, compact coarse sand	4
8. Loose coarse sand, sand and gravel mixture, or compact fine sand	3
9. Loose fine sand	2
10. Stiff clay and soft shale	4
11. Medium stiff clay	$2\frac{1}{2}$
12. Medium soft clay and soft broken shale	$1\frac{1}{2}$

When the bearing materials at a lower level are of lesser allowable bearing value than the bearing value selected, the weight on the footing should be recomputed at the lower level on the basis that the load spreads uniformly at an angle of 60 deg with the horizontal. Thus, if we assume a footing carrying 200 kips on a stiff clay (classification 10) good for 4 tons, or 8,000 lb, per sq ft, but resting 4 ft above a medium soft clay (classification 12) good for $1\frac{1}{2}$ tons, or 3,000 lb, per sq ft, both planes must be investigated for bearing (Fig. 8.1).

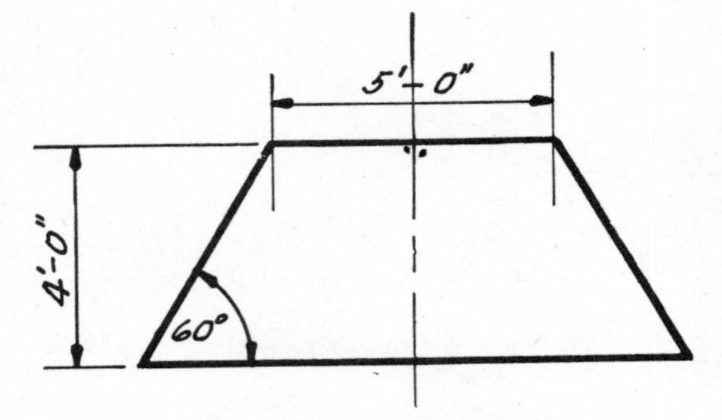

Fig. 8.1 Spread of Load under Footings

At the upper plane, the footing required is $^{200}/_{8}$, or 25 sq ft, which is 5 ft square. At the lower level, this load is spread over the area a^2, which is $[5 + 2(4 \cot 60°)]^2$ or $9.618^2 = 92.5$ sq ft. On this basis, the lower layer is loaded to $200/92.5 = 2{,}160$ psf, which is within the allowable limit.

Footings should be provided under all columns, piers, and walls to distribute the loads as uniformly as possible on level beds. Since dead load is on the footings continuously and live load may never be fully applied, it is becoming common practice to design footings on the basis of dead load rather than total load. This requires that loads in the footings be divided between live and dead loads. Then the footing having the largest percentage of live load is designed on the basis of the allowable soil pressure. From this the unit soil pressure due to dead load only is determined, and the area of each footing computed on this basis. The design of footing as to thickness and reinforcement, however, is made on the basis of total load.

As an example of the application of this method, let us assume five columns, each with a total load of 150 kips, as follows:

	Live load, kips	Dead load, kips	Dead load, percent	Design soil pressure, psf
Column 1	30	120	80	2,500
Column 2	45	105	70	2,857
Column 3	60	90	60	3,333
Column 4	75	75	50	4,000
Column 5	90	60	40	5,000

Assume 5,000-psf maximum soil pressure for total load for the footing with the lowest percentage dead load. Then column 5 footing will be 30 sq ft. The soil pressure for column 1 will be $^{40}/_{80}$ of 5,000, or 2,500 psf. The other footings are proportioned in the same manner, as tabulated above. Selection of footings should be made for 150 kips load from the proper table in Art. 8.11, or interpolated.

d Obviously, the amount of the superimposed load influences the type of foundation, particularly if the load is quite heavy. There is a point in the curve of costs beyond which it is cheaper to go even 50 or 100 ft to rock rather than attempt to support the load on spread footings. Most buildings higher than 15 or 20 stories are carried on rock, as are many buildings of lesser height, if rock is shallow.

e The location with regard to lot lines influences the type of footing required. On a footing carrying a single load, the load must be concentric on the footing. When the column is an interior column there is nothing difficult in this requirement, but when the column is a wall column adjacent to the lot line you cannot extend your footing under the adjacent property. The solution is to balance two or more columns on one footing so that the center of gravity of the computed loads coincides with the center of gravity of the footing area. This is accomplished by means of simple combined footings, cantilever footings, or mats.

f As stated in paragraph *d*, even with light loads it is frequently advisable to run the footing to rock if rock is not far below the level to which the footing would normally be carried.

If a structure rests partially on rock and partially on yielding soil, some codes require that the bearing capacity of the yielding soil be taken at not more than half the tabulated capacity. Experience indicates, however, that it is impossible to avoid settlement cracks if the building rests partially on rock and partially on yielding soil, or partially on piles and partially on spread footings, and such conditions should always be avoided. Even though the variations over the site may not be so marked as those indicated, any appreciable variations in soil conditions should be referred to a competent soils engineer for his advice before proceeding with the design.

g In all construction in a northern climate exterior walls or footings must run at least 4 ft below grade to get below the frost line. Otherwise upheaval and cracks due to frost pressure must be expected. In a building with a basement, of course, this requirement does not usually determine the depth. Ordinarily, interior footing depths are determined by placing the top of the footing under the basement floor, but if any part of the building is dropped to a lower grade, the line from the edge of one footing to the nearest bottom edge of any other footing cannot make an angle with the horizontal of over 1 vertical on 2 horizontal. If an angle exceeds this value, the adjacent footing must be lowered until the 1:2 condition is satisfied, unless sheet piling is introduced to support the soil. This condition is indicated in Fig. 8.2.

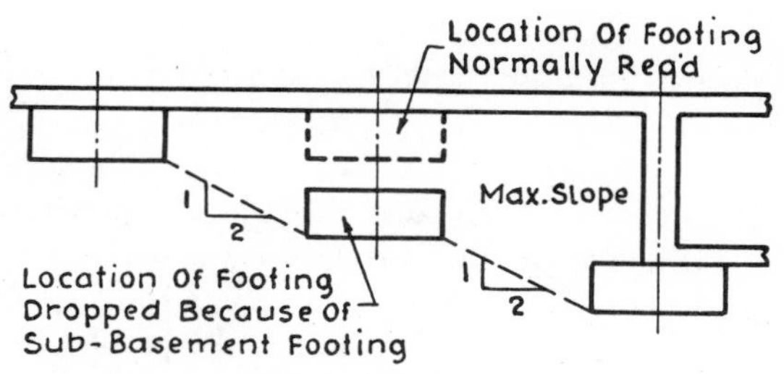

Fig. 8.2 Elevations of Footings as Controlled by Conditions

All footings must be carried below the depth of vegetation and organic matter in the soil. This usually requires depths of from

1 ft 6 in. to 2 ft into the undisturbed earth, although greater depths are required in fields which have been cultivated for a number of years.

h *Machinery foundations.* It is frequently desirable to isolate machinery foundations from the building structure. If the foundation of a machine is attached to the floor or walls of a building, the vibration of the machine may be transmitted to other machines and to other parts of the building and cause settlement. If necessary, floors adjacent to a foundation superstructure, such as a second-floor support for a motor-generator set, may be simply supported on a shelf on a foundation, provided that it is insulated in such a manner that no movement is transmitted.

The General Electric Company's recommendations for the design of foundations for motor-generator sets may be taken as good practice in design of foundations for most moving or vibrating machines. The company recommends that all parts be designed for the maximum stresses due to the worst possible combination of vertical loads, torque, longitudinal and transverse forces, stresses due to temperature variation, and foundation dead load. Where the foundation layout is partially built up of beam and column sections, the following unit stresses are recommended:

Concrete, compression	=	350 psi
Concrete, bending	=	400 psi
Steel, tension	=	10,000 psi
Steel, compression (n = 10)	=	4,000 psi

No reliance should be placed on concrete in shear. Suitably sized closed stirrups should be provided.

Use straight bars in both flanges of beams, and avoid bent-up bars. The subfoundation—that part of the foundation on which the superstructure rests—is usually a reinforced concrete slab or mat. It should be of substantial thickness and reinforced top and bottom.

In computing soil loading, the center of gravity of the machine and foundation loads should coincide with the center of gravity of the footing. Although sometimes impossible to attain, it is the ideal condition.

Occasionally, if the subsoil is soft clay or other material that may flow, it is advisable to drive a row of steel sheet pilings around the entire perimeter to prevent the loss of subsoil and a resultant settlement of the foundation. In order to protect foundations immediately adjacent, it may be necessary also to enclose a machine footing in steel sheet piling.

Because machinery foundations depend largely on mass to dampen vibration, it is desirable to pour the foundation as nearly monolithic as possible. If this cannot be done, suitable keys and dowels should be provided between pours.

The General Electric Company's practice requires that anchor bolts be placed to an accuracy of $\frac{1}{16}$ in. without pipe sleeves, using built-up steel angle templates, well cross-braced. Some other manufacturers are satisfied to have anchor bolts placed with the upper third of the bolt in a pipe sleeve, then poured in after the machine has been placed over the anchor bolts.

i If any load applied to a footing is eccentric, the pressure is not uniform under the footing, but varies uniformly from a maximum at the edge nearest the point of load, to a minimum at the opposite edge. Graphically, we may represent the resultant pressures by a trapezoid as shown in Fig. 1.14. In the figure a unit length is represented by area A for the sake of simplifying the method. The pressure f increases and f_1 decreases very rapidly as the eccentricity of the load increases, until the eccentricity is $\frac{1}{6}A$, at which point f_1 becomes double the average pressure.

The unit pressure at the edge is obtained from the formula

$$f = \frac{P}{A}\left(1 \pm \frac{6e}{b}\right)$$

In the above formula, + is used to compute the f value and − to compute f_1. Although a slight eccentricity, while economical, is permissible, an eccentricity in excess of $\frac{1}{10}$ may cause a dangerous unequal settlement. The best practice is to keep the load concentric.

8.11 Square column footings centrally loaded

a For consideration of design methods, it is well to consider all centrally loaded spread footings either square or rectangular, under one heading. The most economical and most common type of footing is square, concentrically loaded.

Square footings may be divided into three classifications for design purposes:

1. Plain concrete—no reinforcement
2. Reinforced concrete
3. Steel grillages

Normally, plain concrete footings are economical up to about 4 ft square, although the optimum size may vary depending on considerations of loads, available depths, or the like.

Table 8.2 is a table of footings used in the author's office. These tables are arranged so that when the load in kips and the allowable soil pressure are known, a steel billet size and footing designation (such as 5 ft 6 in. E) may be selected. Then from Table 8.3 the design of a footing using a 3,000-psi concrete may be selected and depth of footing and reinforcement determined. Table 8.2 advances by 10-kip intervals to 1,000 kips for soil pressures from 3,000 to 10,000 psf. In Table 8.2 note that the letters *A* through *G* refer to the same soil pressure throughout. The footings are arranged by 3-in. variations in side of square for simplicity of forming and reinforcement.

b Square reinforced concrete footings. The following sections quoted from the 1963 ACI Code may be taken as good practice in reinforced concrete footing design:

2302 LOADS AND REACTIONS

(a) Isolated footings shall be proportioned to sustain the applied loads and induced reactions without exceeding the stresses or strengths prescribed in Parts IV-A and IV-B, and as further provided in this chapter.

(b) In cases where the footing is concentrically loaded and the member being supported does not transmit any moment to the footing, computations for moments and shears shall be based on an upward reaction assumed to be uniformly distributed per unit area or per pile and a downward applied load assumed to be uniformly distributed over the area of the footing covered by the column, pedestal, wall, or metallic column base.

(c) In cases where the footing is eccentrically loaded and/or the member being supported transmits a moment to the footing, proper allowance shall be made for any variation that may exist in the intensities of reaction and applied load consistent with the magnitude of the applied load and the amount of its actual or virtual eccentricity.

(d) In the case of footings on piles, computations for moments and shears may be based on the assumption that the reaction from any pile is concentrated at the center of the pile.

2303 SLOPED OR STEPPED FOOTINGS

(a) In sloped or stepped footings, the angle of slope or depth and location of steps shall be such that the allowable stresses are not exceeded at any section.

(b) In sloped or stepped footings, the effective cross section in compression shall be limited by the area above the neutral plane.

(c) Sloped or stepped footings that are designed as a unit shall be cast as a unit.

2304 BENDING MOMENT

(a) The external moment on any section shall be determined by passing through the section a vertical plane which extends completely across the footing, and computing the moment of the forces acting over the entire area of the footing on one side of said plane.

(b) The greatest bending moment to be used in the design of an isolated footing shall be the moment computed in the manner prescribed in (a) at sections located as follows:

1. At the face of the column, pedestal or wall, for footings supporting a concrete column, pedestal or wall
2. Halfway between the middle and the edge of the wall, for footings under masonry walls
3. Halfway between the face of the column or pedestal and the edge of the metallic base, for footings under metallic bases

(c) The width resisting compression at any section shall be assumed as the entire width of the top of the footing at the section under consideration.

(d) In one-way reinforced footings, the total tensile reinforcement at any section shall provide a moment of resistance at least equal to the moment computed as prescribed in (a); and the reinforcement thus determined shall be distrib-

Table 8.2 Footing Table

Load, kips	Billets, in.	Footings (refer to Table 8.3)						
		3,000, *A*	4,000, *B*	5,000, *C*	6,000, *D*	7,000, *E*	8,000, *F*	10,000, *G*
10	8 × 9 × ½	2′0″	1′9″	1′6″	1′6″	1′6″	1′6″	1′6″
20	8 × 9 × ½	2′9″	2′6″	2′3″	2′0″	1′9″	1′9″	1′6″
30	8 × 9 × ½	3′3″	3′0″	2′6″	2′6″	2′3″	2′0″	1′9″
40	8 × 9 × ½	3′9″	3′3″	3′0″	2′9″	2′6″	2′6″	2′3″
50	8 × 9 × ⅝	4′3″	3′9″	3′3″	3′0″	2′9″	2′9″	2′6″
60	8 × 9 × ⅝	4′9″	4′0″	3′9″	3′3″	3′0″	3′0″	2′6″
70	8 × 9 × ⅝	5′0″	4′6″	4′0″	3′6″	3′3″	3′0″	2′9″
80	8 × 9 × ⅝	5′6″	4′9″	4′3″	3′9″	3′6″	3′3″	3′0″
90	10 × 9 × ¾	5′9″	5′0″	4′6″	4′0″	3′9″	3′6″	3′3″
100	10 × 9 × ¾	6′0″	5′3″	4′9″	4′3″	4′0″	3′9″	3′3″
110	10 × 10 × 1	6′6″	5′6″	5′0″	4′6″	4′3″	3′9″	3′6″
120	10 × 11 × 1	6′9″	5′9″	5′0″	4′9″	4′3″	4 0″	3′6″
130	10 × 12 × 1	7′0″	6′0″	5′3″	4′9″	4′6″	4′3″	3′9″
140	12 × 12 × 1⅛	7′3″	6′3″	5′6″	5′0″	4′9″	4′3″	4′0″
150	12 × 12 × 1⅛	7′6″	6′6″	5′9″	5′3″	4′9″	4′6″	4′0″
160	12 × 12 × 1⅛	7′9″	6′9″	6′0″	5′3″	5′0″	4′9″	4′3″
170	12 × 13 × 1⅛	8′0″	6′9″	6′0″	5′6″	5′0″	4′9″	4′3″
180	12 × 14 × 1⅛	8′3″	7′0″	6′3″	5′9″	5′3″	5′0″	4′6″
190	14 × 13 × 1¼	8′6″	7′3″	6′6″	5′9″	5′6″	5′0″	4′6″
200	14 × 13 × 1¼	8′6″	7′6″	6′6″	6′0″	5′6″	5′3″	4′9″
210	14 × 14 × 1⅜	8′9″	7′6″	6′9″	6 3″	5′9″	5′3″	4′9″
220	14 × 15 × 1½	9′0″	7′9″	7′0″	6′3″	5′9″	5′6″	4′9″
230	14 × 15 × 1½	9′3″	8′0″	7′0″	6′6″	6′0″	5′6″	5′0″
240	14 × 15 × 1⅝	9′6″	8′0″	7′3″	6′6″	6′0″	5′9″	5′0″
250	16 × 15 × 1⅝	9′9″	8′3″	7′3″	6′9″	6′3″	5′9″	5′3″
260	16 × 15 × 1⅝	9′9″	8′6″	7′6″	6′9″	6′3″	6′0″	5′3″
270	16 × 16 × 1⅝	10′0″	8′6″	7′9″	7′0″	6′6″	6′0″	5′3″
280	16 × 16 × 1⅝	10′3″	8′9″	7′9″	7′0″	6′6″	6′0″	5′6″
290	16 × 17 × 1⅝	10′6″	9′0″	8′0″	7′3″	6′9″	6′3″	5′6″
300	16 × 18 × 1⅝	10′6″	9′0″	8′0″	7′3″	6′9″	6′3″	5′9″
310	18 × 17 × 1⅝	10′9″	9′3″	8′3″	7′6″	7′0″	6′6″	5′9″
320	18 × 17 × 1⅝	11′0″	9′3″	8′3″	7′6″	7′0″	6′6″	5′9″
330	18 × 18 × 1⅝	11′3″	9′6″	8′6″	7′9″	7′0″	6′9″	6′0″
340	18 × 18 × 1⅝	11′3″	9′9″	8′6″	7′9″	7′3″	6′9″	6′0″
350	18 × 18 × 1⅝	11′6″	9′9″	8′9″	8′0″	7′3″	6′9″	6′0″
360	18 × 18 × 1¾	11′9″	10′0″	8′9″	8′0″	7′6″	7′0″	6′3″
370	18 × 19 × 1¾	11′9″	10′0″	9′0″	8′3″	7′6″	7′0″	6′3″
380	18 × 19 × 1¾	12′0″	10′3″	9′0″	8′3″	7′9″	7′0″	6′3″
390	20 × 18 × 1¾	12′3″	10′3″	9′3″	8′3″	7′9″	7′3″	6′6″
400	20 × 18 × 1¾	12′3″	10′6″	9′3″	8′6″	7′9″	7′3″	6′6″
410	20 × 19 × 1¾	12′6″	10′9″	9′6″	8′6″	8′0″	7′6″	6′6″
420	20 × 19 × 1¾	12′6″	10′9″	9′6″	8′9″	8′0″	7′6″	6′9″
430	20 × 20 × 1¾	12′9″	11′0″	9′9″	8′9″	8′0″	7′6″	6′9″
440	20 × 20 × 1⅞	13′0″	11′0″	9′9″	9′0″	8′3″	7′9″	6′9″
450	20 × 21 × 1⅞	13′0″	11′3″	10′0″	9′0″	8′3″	7′9″	7′0″
460	20 × 21 × 1⅞	13′3″	11′3″	10′0″	9′0″	8′6″	7′9″	7′0″
470	22 × 20 × 1⅞	13′3″	11′6″	10′0″	9′3″	8′6″	8′0″	7′0″
480	22 × 20 × 1⅞	13′6″	11′6″	10′3″	9′3″	8′6″	8′0″	7′3″
490	22 × 20 × 1⅞	13′9″	11′9″	10′3″	9′6″	8′9″	8′0″	7′3″
500	22 × 21 × 1⅞	14′0″	11′9″	10′6″	9′6″	8′9″	8′3″	7′3″
510	22 × 21 × 1⅞	14′0″	12′0″	10′6″	9′6″	8′9″	8′3″	7′3″
520	22 × 22 × 1⅞	14′0″	12′0″	10′9″	9′9″	9′0′	8′3″	7′6″

Table 8.2 Footing Table (*Continued*)

Load, kips	Billets, in.	Footings (refer to Table 8.3)						
		3,000, *A*	4,000, *B*	5,000, *C*	6,000, *D*	7,000, *E*	8,000, *G*	10,000, *G*
530	22 × 22 × 1⅞	14′3″	12′3″	10′9″	9′9″	9′0″	8′6″	7′6″
540	22 × 23 × 1⅞	14′3″	12′3″	11′0″	9′9″	9′0″	8′6″	7′6″
550	22 × 23 × 1⅞	14′6″	12′3″	11′0″	10′0″	9′3″	8′6″	7′9″
560	22 × 23 × 1⅞	14′6″	12′6″	11′0″	10′0″	9′3″	8′9″	7′9″
570	24 × 22 × 1⅞	14′9″	12′6″	11′3″	10′3″	9′3″	8′9″	7′9″
580	24 × 22 × 2	15′0″	12′9″	11′3″	10′3″	9′6″	8′9″	7′9″
590	24 × 23 × 2	15′0″	12′9″	11′6″	10′3″	9′6″	9′0″	8′0″
600	24 × 23 × 2		13′0″	11′6″	10′6″	9′6″	9′0″	8′0″
610	24 × 23 × 2		13′0″	11′6″	10′6″	9′9″	9′0″	8′0″
620	24 × 24 × 2		13′3″	11′9″	10′6″	9′9″	9′0″	8′0″
630	24 × 24 × 2¼		13′3″	11′9″	10′9″	9′9″	9′3″	8′3″
640	24 × 24 × 2¼		13′3″	11′9″	10′9″	10′0″	9′3″	8′3″
650	24 × 25 × 2¼		13′6″	12′0″	10′9″	10′0″	9′3″	8′3″
660	24 × 25 × 2¼		13′9″	12′0″	11′0″	10′0″	9′6″	8′6″
670	26 × 24 × 2¼		13′9″	12′0″	11′0″	10′3″	9′6″	8′6″
680	26 × 24 × 2⅜		13′9″	12′3″	11′0″	10′3″	9′6″	8′6″
690	26 × 24 × 2⅜		14′0″	12′3″	11′3″	10′3″	9′9″	8′6″
700	26 × 25 × 2⅜		14′0″	12′6″	11′3″	10′3″	9′9″	8′6″
710	26 × 25 × 2⅜		14′3″	12′6″	11′3″	10′6″	9′9″	8′9″
720	26 × 25 × 2⅜		14′3″	12′6″	11′6″	10′6″	9′9″	8′9″
730	26 × 26 × 2½		14′3″	12′9″	11′6″	10′6″	10′0″	8′9″
740	26 × 26 × 2½		14′6″	12′9″	11′6″	10′9″	10′0″	8′9″
750	26 × 26 × 2½		14′6″	12′9″	11′9″	10′9″	10′0″	9′0″
760	26 × 27 × 2½		14′6″	13′0″	11′9″	10′9″	10′0″	9′0″
770	26 × 27 × 2½		14′9″	13′0″	11′9″	11′0″	10′3″	9′0″
780	26 × 27 × 2⅝		14′9″	13′0″	12′0″	11′0″	10′3″	9′0″
790	28 × 26 × 2⅝		15′0″	13′3″	12′0″	11′0″	10′3″	9′3″
800	28 × 26 × 2⅝		15′0″	13′3″	12′0″	11′0″	10′3″	9′3″
810	28 × 26 × 2⅝		15′0″	13′3″	12′0″	11′3″	10′6″	9′3″
820	28 × 27 × 2⅝			13′6″	12′3″	11′3″	10′6″	9′3″
830	28 × 27 × 2⅝			13′6″	12′3″	11′3″	10′6″	9′6″
840	28 × 27 × 2⅝			13′6″	12′3″	11′6″	10′6″	9′6″
850	28 × 28 × 2⅝			13′9″	12′6″	11′6″	10′9″	9′6″
860	28 × 28 × 2⅝			13′9″	12′6″	11′6″	10′9″	9′6″
870	28 × 28 × 2¾			13′9″	12′6″	11′6″	10′9″	9′6″
880	28 × 29 × 2¾			14′0″	12′9″	11′9″	10′9″	9′9″
890	28 × 29 × 2¾			14′0″	12′9″	11′9″	11′0″	9′9″
900	28 × 29 × 2¾			14′0″	12′9″	11′9″	11′0″	9′9″
910	30 × 28 × 2¾			14′3″	12′9″	11′9″	11′0″	9′9″
920	30 × 28 × 3			14′3″	13′0″	12′0″	11′0″	10′0″
930	30 × 28 × 3			14′3″	13′0″	12′0″	11′3″	10′0″
940	30 × 29 × 3			14′6″	13′0″	12′0″	11′3″	10′0″
950	30 × 29 × 3			14′6″	13′3″	12′3″	11′3″	10′0″
960	30 × 29 × 3			14′6″	13′3″	12′3″	11′3″	10′0″
970	30 × 30 × 3⅛			14′9″	13′3″	12′3″	11′6″	10′3″
980	30 × 30 × 3⅛			14′9″	13′3″	12′3″	11′6″	10′3″
990	30 × 30 × 3⅛			14′9″	13′6″	12′3″	11′6″	10′3″
1,000	30 × 30 × 3⅛			15′0″	13′6″	12′6″	11′6″	10′3″

Table 8.3 Standard Footings

For Use With Table 8.2.

The upper figure is the total depth of the footing and the lower figure is the total reinforcement in each of two directions. $f'_c = 3{,}000$, $f_s = 24{,}000$.

	A	*B*	*C*	*D*	*E*	*F*	*G*
1'6"			1'0"	1'0"	1'0"	1'0"	1'0"
1'9"		1'0"			1'0"	1'0"	1'0"
2'0"	1'0'			1'0"		1'0"	
2'3"			1'0"		1'0"		1'0"
2'6"		1'0" 8 No. 4	1'0" 8 No. 4	1'1" 8 No. 4	1'1" 8 No. 4	1'1" 8 No. 4	1'1" 9 No. 4
2'9"	1'0"			1'1" 8 No. 4	1'1" 8 No. 4	1'1" 8 No. 4	1'1" 9 No. 4
3'0"		1'0" 8 No. 4	1'0" 8 No. 4	1'2" 9 No. 4	1'2" 9 No. 4	1'2" 9 No. 4	1'2" 9 No. 4
3'3"	1'0" 8 No. 4	1'0" 8 No. 4	1'0" 8 No. 4	1'2" 9 No. 4	1'2" 9 No. 4	1'2" 9 No. 4	1'2" 9 No. 4
3'6"				1'2" 9 No. 4	1'3" 9 No. 4	1'3" 10 No. 4	1'3" 11 No. 4
3'9"	1'0" 9 No. 4	1'0" 9 No. 4	1'0" 9 No. 4	1'3" 12 No. 4	1'3" 12 No. 4	1'3" 12 No. 4	1'3" 13 No. 4
4'0"		1'0" 10 No. 4	1'1" 11 No. 4	1'3" 12 No. 4	1'3" 12 No. 4	1'3" 13 No. 4	1'4" 14 No. 4
4'3"	1'0" 10 No. 4		1'1" 15 No. 4	1'3" 13 No. 4	1'4" 14 No. 4	1'4" 14 No. 4	1'5" 11 No. 5
4'6"		1'0" 14 No. 4	1'2" 14 No. 4	1'4" 15 No. 4	1'4" 10 No. 5	1'4" 11 No. 5	1'6" 12 No. 5
4'9"	1'0" 13 No. 4	1'1" 14 No. 4	1'2" 15 No. 4	1'4" 16 No. 4	1'5" 11 No. 5	1'5" 12 No. 5	1'7" 13 No. 5
5'0"	1'1" 13 No. 4	1'2" 11 No. 5	1'3" 12 No. 5	1'5" 12 No. 5	1'6" 13 No. 5	1'6" 14 No. 5	1'7" 10 No. 6
5'3"		1'2" 12 No. 5	1'4" 13 No. 5	1'5" 13 No. 5	1'6" 14 No. 5	1'6" 11 No. 6	1'7" 11 No. 6
5'6"	1'1" 12 No. 5	1'3" 13 No. 5	1'4" 14 No. 5	1'6" 14 No. 5	1'7" 15 No. 5	1'7" 12 No. 6	1'8" 13 No. 6
5'9"	1'2" 12 No. 5	1'3" 13 No. 5	1'4" 16 No. 5	1'6" 16 No. 5	1'7" 12 No. 6	1'8" 12 No. 6	1'9" 13 No. 6
6'0"	1'2" 14 No. 5	1'4" 15 No. 5	1'5" 17 No. 5	1'7" 17 No. 5	1'8" 14 No. 6	1'8" 17 No. 5	1'10" 14 No. 6
6'3"		1'5" 16 No. 5	1'6" 18 No. 5	1'7" 13 No. 6	1'8" 14 No. 6	1'9" 14 No. 6	1'11" 15 No. 6
6'6"	1'3" 16 No. 5	1'5" 18 No. 5	1'6" 14 No. 6	1'7" 14 No. 6	1'8" 15 No. 6	1'9" 15 No. 6	2'0" 15 No. 6
6'9"	1'3" 18 No. 5	1'6" 19 No. 5	1'7" 14 No. 6	1'8" 15 No. 6	1'9" 15 No. 6	1'10" 15 No. 6	2'0" 15 No. 6
7'0"	1'4" 19 No. 5	1'6" 19 No. 5	1'8" 15 No. 6	1'8" 17 No. 6	1'10" 18 No. 6	2'0" 18 No. 6	2'1" 20 No. 6
7'3"	1'4" 21 No. 5	1'7" 20 No. 5	1'9" 17 No. 6	1'8" 18 No 6	1'11" 18 No. 6	2'0" 19 No. 6	2'2" 21 No. 6
7'6"	1'5" 21 No. 5	1'8" 21 No. 5	1'9" 18 No. 6	1'9" 19 No. 6	1'11" 20 No. 6	2'1" 20 No. 6	2'2" 23 No. 6
7'9"	1'5" 16 No. 6	1'8" 16 No. 6	1'9" 18 No. 6	1'10" 20 No. 6	2'0" 21 No. 6	2'2" 21 No. 6	2'3" 24 No. 6
8'0"	1'6" 16 No. 6	1'9" 18 No. 6	1'10" 20 No. 6	1'10" 22 No. 6	2'1" 22 No. 6	2'2" 23 No. 6	2'4" 18 No. 7
8'3"	1'6" 18 No. 6	1'9" 18 No. 6	1'10" 21 No. 6	1'11" 23 No. 6	2'2" 25 No. 6	2'3" 25 No. 6	2'4" 22 No. 7

Table 8.3 Standard Footings (*Continued*)

	A	*B*	*C*	*D*	*E*	*F*	*G*
8′6″	1′7″ 19 No. 6	1′9″ 20 No. 6	1′11″ 22 No. 6	2′0″ 24 No. 6	2′2″ 25 No. 6	2′3″ 20 No. 7	2′5″ 22 No. 7
8′9″	1′8″ 19 No. 6	1′10″ 21 No. 6	2′0″ 23 No. 6	2′0″ 20 No. 7	2′3″ 20 No. 7	2′4″ 22 No. 7	2′6″ 23 No. 7
9′0″	1′8″ 21 No. 6	1′10″ 22 No. 6	2′0″ 18 No. 7	2′1″ 20 No. 7	2′4″ 21 No. 7	2′5″ 22 No. 7	2′7″ 24 No. 7
9′3″	1′8″ 22 No. 6	1′11″ 23 No. 6	2′1″ 19 No. 7	2′2″ 22 No. 7	2′4″ 22 No. 7	2′5″ 24 No. 7	2′8″ 25 No. 7
9′6″	1′8″ 23 No. 6	1′11″ 19 No. 7	2′2″ 20 No. 7	2′3″ 22 No. 7	2′5″ 23 No. 7	2′6″ 25 No. 7	2′9″ 27 No. 7
9′9″	1′8″ 25 No. 6	2′0″ 19 No. 7	2′2″ 21 No. 7	2′4″ 23 No. 7	2′6″ 24 No. 7	2′7″ 26 No. 7	2′10″ 28 No. 7
10′0″	1′9″ 25 No. 6	2′0″ 20 No. 7	2′3″ 23 No. 7	2′4″ 25 No. 7	2′6″ 26 No. 7	2′8″ 27 No. 7	2′10″ 29 No. 7
10′3″	1′9″ 20 No. 7	2′1″ 21 No. 7	2′3″ 25 No. 7	2′5″ 26 No. 7	2′7″ 26 No. 7	2′8″ 29 No. 7	2′11″ 30 No. 7
10′6″	1′9″ 22 No. 7	2′1″ 23 No. 7	2′4″ 25 No. 7	2′6″ 27 No. 7	2′8″ 28 No. 7	2′9″ 29 No. 7	
10′9″	1′10″ 24 No. 7	2′2″ 24 No. 7	2′4″ 27 No. 7	2′6″ 28 No. 7	2′8″ 30 No. 7	2′10″ 31 No. 7	
11′0″	1′10″ 24 No. 7	2′2″ 25 No. 7	2′5″ 28 No. 7	2′7″ 29 No. 7	2′9″ 24 No. 8	2′11″ 25 No. 8	
11′3″	1′11″ 24 No. 7	2′3″ 26 No. 7	2′5″ 29 No. 7	2′8″ 30 No. 7	2′9″ 25 No. 8	3′0″ 27 No. 8	
11′6″	1′11″ 25 No. 7	2′3″ 27 No. 7	2′6″ 30 No. 7	2′8″ 25 No. 8	2′10″ 27 No. 8	3′0″ 28 No. 8	
11′9″	2′0″ 26 No. 7	2′4″ 28 No. 7	2′6″ 31 No. 7	2′9″ 26 No. 8	2′11″ 28 No. 8		
12′0″	2′1″ 27 No. 7	2′4″ 30 No. 7	2′7″ 24 No. 8	2′9″ 27 No. 8	2′11″ 29 No. 8		
12′3″	2′1″ 28 No 7	2′5″ 31 No. 7	2′8″ 26 No. 8	2′10″ 28 No. 8	3′0″ 30 No. 8		
12′6″	2′2″ 29 No. 7	2′6″ 31 No. 7	2′10″ 27 No. 8	2′11″ 29 No. 8	3′0″ 32 No. 8		
12′9″	2′2″ 30 No. 7	2′6″ 34 No. 7	2′10″ 28 No. 8	3′0″ 30 No. 8			
13′0″	2′3″ 31 No. 7	2′7″ 34 No. 7	2′10″ 29 No. 8	3′0″ 31 No. 8			
13′3″	2′3″ 33 No. 7	2′7″ 27 No. 8	2′11″ 29 No. 8	3′1″ 32 No. 8			
13′6″	2′4″ 33 No. 7	2′7″ 29 No. 8	2′11″ 31 No. 8	3′1″ 34 No. 8			
13′9″	2′4″ 27 No. 8	2′8″ 30 No. 8	3′0″ 32 No. 8				
14′0″	2′5″ 27 No. 8	2′8″ 31 No. 8	3′0″ 33 No. 8				
14′3″	2′5″ 28 No. 8	2′9″ 31 No. 8	3′1″ 34 No. 8				
14′6″	2′6″ 29 No. 8	2′9″ 33 No. 8	3′1″ 36 No. 8				
14′9″	2′6″ 31 No. 8	2′10″ 34 No. 8	3′1″ 38 No. 8				
15′0″	2′7″ 31 No. 8	2′10″ 36 No. 8	3′1″ 40 No. 8				

uted uniformly across the full width of the section.

(e) In two-way reinforced footings, the total tension reinforcement at any section shall provide a moment of resistance at least equal to the moment computed as prescribed in (a); and the total reinforcement thus determined shall be distributed across the corresponding resisting section as prescribed for square footings in (f), and for rectangular footings in (g).

(f) In two-way square footings, the reinforcement extending in each direction shall be distributed uniformly across the full width of the footing.

(g) In two-way rectangular footings, the reinforcement in the long direction shall be distributed uniformly across the full width of the footing. In the case of the reinforcement in the short direction, that portion determined by Eq. (23-1) shall be uniformly distributed across a band-width (B) centered with respect to the center line of the column or pedestal and having a width equal to the length of the short side of the footing. The remainder of the reinforcement shall be uniformly distributed in the outer portions of the footing.

$$\frac{\textit{Reinforcement in band-width } (B)}{\textit{Total reinforcement in short direction}} = \frac{2}{(S + 1)} \qquad (23\text{-}1)$$

where S is the ratio of the long side to the short side of the footing.

2305 Shear and Bond

(a) For computation of shear in footings, see Section 1207.

(b) Critical sections for bond shall be assumed at the same planes as those prescribed for bending moment in Section 2304(b)1; also at all other vertical planes where changes of section or of reinforcement occur.

(c) Computation for shear to be used as a measure of flexural bond shall be based on a vertical section which extends completely across the footing, and the shear shall be taken as the sum of all forces acting over the entire area of the footing on one side of such section.

(d) The total tensile reinforcement at any section shall provide a bond resistance at least equal to the bond requirement as computed from the external shear at the section.

(e) In computing the external shear on any section through a footing supported on piles, the entire reaction from any pile whose center is located 6 in. or more outside the section shall be assumed as producing shear on the section; the reaction from any pile whose center is located 6 in. or more inside the section shall be assumed as producing no shear on the section. For intermediate positions of the pile center, the portion of the pile reaction to be assumed as producing shear on the section shall be based on straight-line interpolation between full value at 6 in. outside the section and zero value at 6 in. inside the section.

1207 Shear Stress in Footings

(a) The shear capacity of footings in the vicinity of concentrated loads or concentrated reactions shall be governed by the more severe of two conditions:

1. The footing acting essentially as a wide beam, with a potential diagonal crack extending in a plane across the entire width.
2. Two-way action existing for the footing, with potential diagonal cracking along the surface of a truncated cone or pyramid around the concentrated load or reaction. The footing in this case shall be designed as required in the remainder of this section.

(b) The critical section for shear to be used as a measure of diagonal tension shall be perpendicular to the plane of the slab (a vertical plane) and located at a distance $d/2$ out from the periphery of the concentrated load or reaction area.

(c) The nominal shear stress shall be computed by:

$$v = V/b_o d \qquad (12\text{-}8)$$

in which V and b_o are taken at the critical section specified in (b). The shear stress, v, so computed shall not exceed $2\sqrt{f_c'}$.

Because shear rather than moment controls the depth of a reinforced-concrete footing, the sequence of design operations is as follows:

1. The determination of shear depth is a matter of cut and try. It is determined so that, using the nomenclature of Fig. 8.3 if concrete columns are used, or Fig. 8.4 for steel columns and billets,

$$P\left[\frac{b_f^2 - (b_a + d)^2}{b_f^2}\right] \gtreqless 4d(b_a + d)f_v$$

or if we use 3,000-psi concrete ($f_v = 110$),

$$P\left[\frac{b_f^2 - (b_a + d)^2}{b_f^2}\right] \gtreqless 440d(b_a + d)$$

The first term is the footing load which produces shear, that is, that portion of the load falling outside the shear plane. The second term is the amount of shear carried by the shear planes around four vertical

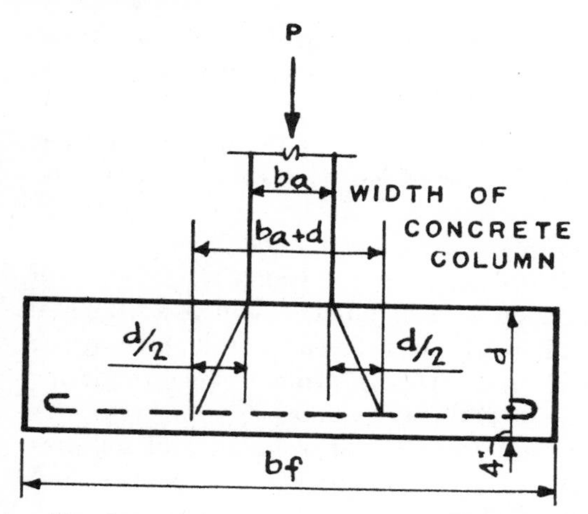

Fig. 8.3 Concrete Column on Reinforced Concrete Footings

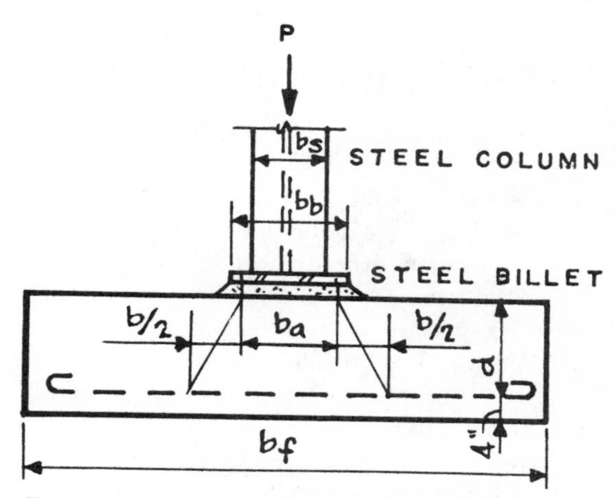

Fig. 8.4 Steel Column and Billet on Reinforced Concrete Footing

surfaces cut at $b_a + d$. As a start to cut and try for proper depth for shear, a depth d equal to b_a minus 2 in. may be assumed and correction made on the basis of results obtained.

2. Determine the load which causes moment,

$$W = P\frac{b_f - b_a}{2b_f}$$

3. The total moment on a moment section is

$$M = \left(\frac{b_f - b_a}{4}\right)W$$

Compute the area of steel required to take this moment,

$$A_s = \frac{M}{f_sjd}$$

Combining all the above equations, using 3,000-psi concrete and 20,000-psi stress in the steel, and using the load in kips,

$$A_s = \frac{P(b_f - b_a)^2}{140db_f}$$

If 24,000-psi stress is used, this becomes

$$A_s = \frac{P(b_f - b_a)^2}{168db_f}$$

4. The perimeter of bars required for bond resistance is

$$\Sigma o = \frac{W}{ujd}$$

The bond value "u" varies with the size bars used.

$$u = \frac{3.4\sqrt{f_c'}}{\text{nominal diameter}} \ngtr 350$$

Thus for 3,000-psi concrete the value of u is as follows,

For No. 3 or No. 4 bars,	$u = 350$ psi
For No. 5 bars,	$u = 300$ psi
For No. 6 bars,	$u = 250$ psi
For No. 7 bars,	$u = 214$ psi

5. The number and size of bars should be selected so that they satisfy the requirements of both area and bond as given above.

Problem 1 Assume a load of 300,000 lb with a soil pressure of 5,000 psf. The column is a 12 WF 72 and the billet is

$$\frac{300{,}000}{1{,}125} = 267 \text{ sq in.}$$

Use a 17 × 16-in. billet.

The average of base and column is 14.25. Use a trial depth of 14 in. plus 1½ in. grout, or 15½ in. in all. Then $b_a + d$ is 2.46 ft. On the basis of 14 in. to steel, or 18 in. total depth, the net soil pressure is 5,000 − 1.5(150) = 4,775 psf and the area of footing required is

$$\frac{300{,}000}{4{,}775} = 62.5 \text{ sq ft.}$$

Use 8 ft 0 in. square.

The area producing shear is $8^2 - 2.46^2$, and the shear load is

$$300{,}000 \times \frac{64 - 6.05}{64} = 272{,}000 \text{ lb}$$

The shear capacity is $29.5 \times 14 \times 440 =$ 181,700 lb. It will be found that the increase of net depth from 14 to 18½ in. will bring the two figures to 263,000 lb producing shear against an allowable shear capacity of 275,750, so the required net depth is 18½ in., or 1 ft 9 in. gross depth of footing allowing 1½ in. for grout.

From the formulae in paragraph 3 above and using 20,000-psi steel,

$$A_s = \frac{300(96 - 14.25)^2}{140 \times 17 \times 96} = 8.65 \text{ sq in.}$$

Using 24,000-psi steel,

$$A_s = \frac{140}{168} \times 8.65 = 7.22 \text{ sq in.}$$

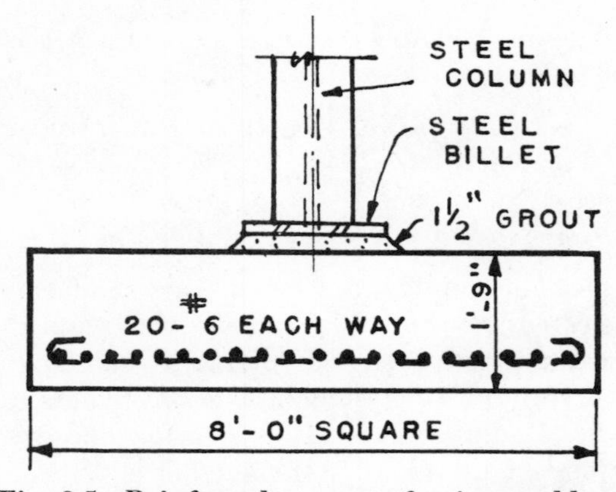

Fig. 8.5 Reinforced concrete footing problem

Using No. 6 bars, for 20 ksi = 20 bars (approx. 4¾ in. o.c.); for 24 ksi = 15 bars. For bond, using either grade of steel

$$W = 300\,\frac{96 - 14.25}{96 \times 2} = 127 \text{ kips}$$

$$\Sigma o = \frac{127{,}000}{250 \times 0.875 \times 17} = 34.3$$

$$\text{Number of bars} = \frac{34.3}{2.36} = 15 \text{ No. 6 bars.}$$

Use 20 No. 6 bars required for moment.

Although the bars provided for moment will provide sufficient bond, some authorities require that all bars in reinforced concrete footings of this type must have hooded ends.

c The design of plain concrete footings may be based on the rules and stresses quoted from the ACI Code in Art. 8.11*b*, Art. 5.12*a*, Art. 5.20*b*, and Art. 1.22.

Problem 2 Assume an 8 WF 24 steel column on a steel billet, carrying a 75,000-lb load, with an allowable soil pressure of 6,000 psf. Design a plain concrete footing, using 3,000-psi concrete.

Deducting 250 psf from the unit soil pressure for weight of footing, the required footing will be

$$\frac{75{,}000}{5{,}750} = 13.03 \text{ sq ft}$$

Use 3 ft 9 in. square.

Using a pressure of 1,125 psi under the billet, the required billet area will be

$$\frac{75{,}000}{1{,}125} = 67.6 \text{ sq in.}$$

However, in order to provide a billet under an $8 \times 6\frac{1}{2}$-in. column, the minimum size of billet should be 8×10 in. The fulcrum line would therefore be at a point $(6.5 + 8)/2 =$ 7.25 in. from the center of the column, and the width of the cantilever producing moment is

$$\frac{45 \text{ in.} - 7.25 \text{ in.}}{2} = 18.87 \text{ in.}$$

This represents 0.42 of the width of the footing, or a load of $0.42 \times 75{,}000 = 31{,}500$ lb, which will be the cantilever load.

The moment is therefore

$$M = 31{,}500 \times 9.43 = 296{,}400 \text{ in.-lb}$$

From Art. 5.12,

$$f = 88 \qquad \text{and} \qquad S = \frac{296{,}400}{88} = 3{,}368$$

Since

$$S = \frac{bd^2}{6} = \frac{45d^2}{6} = 7.5d^2$$

$$d = \sqrt{\frac{3{,}368}{7.5}} = \sqrt{449} = 21.18 \text{ in.}$$

Use 22 in.

To check this depth for shear, the area

within the shear planes is 7.25 + 22 in. = 29.25 in. square, and in a footing which is only 45 in. square, it is obvious that there is no shear problem.

d Steel-grillage footings, formerly very popular for heavy loads, have given way largely to the cheaper, simpler type of reinforced concrete spread footing, described in Art. 8.11*b*. The general rules covering the use of steel-grillage footings are that they should rest on a bed of at least 6 in. of concrete, with the grillages encased in a minimum of 3 in. of concrete. The spaces between the beams should be entirely filled with concrete. The beams are held in place usually with steel-pipe separators. The load is spread into the soil by the lowest set of beams, which are spaced uniformly across the footings. The load from the column is taken by a steel billet into the upper grillage which spreads the load into the lower grillage. At all crossings, the rules of Art. 4.10*i* relative to web shear and buckling must be observed.

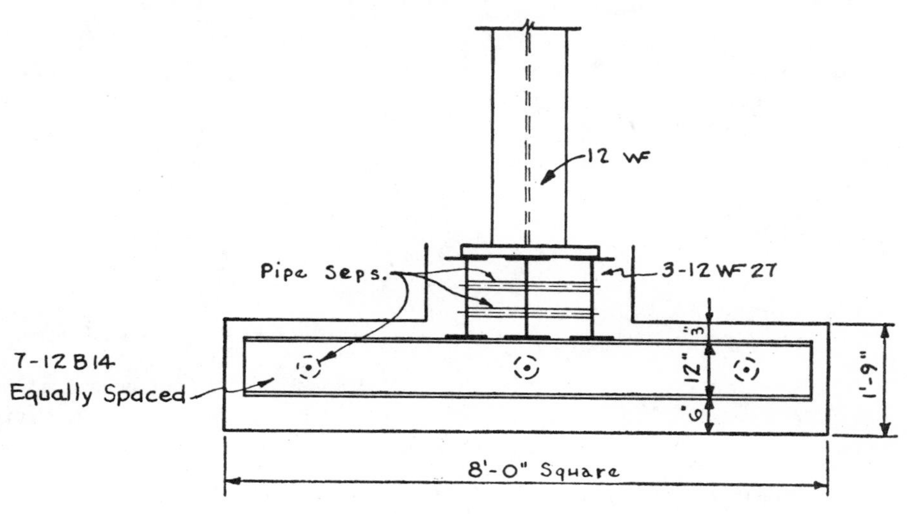

Fig. 8.6 Grillage Footing

Problem 3 Assume the same column load and soil conditions given in Art. 8.11*b*. The column spreads a load of 300,000 lb upon an 8-ft-square footing. To obtain approximate sizes, assume that the top billet will spread the load upon a 20-in. square. Then the cantilever is

$$\frac{8 \text{ ft} - 1 \text{ ft } 8 \text{ in.}}{2} = 3 \text{ ft } 2 \text{ in.} = 38 \text{ in.}$$

and the load on the cantilever is

$$\frac{96 - 20}{2 \times 96} \quad \text{or} \quad 0.395 \text{ of the load}$$

The cantilever moment is

$$0.395 \times 300{,}000 \times 19 = 2251.5 \text{ in.-kips}$$

and the section modulus required in each layer is

$$\frac{2251.5}{22} = 102.4$$

In the lower layer this can be either

Six 12 B 16.5 = 99 lb per ft
Seven 12 B 14 = 98 lb per ft
Eight 6 WF 20 = 160 lb per ft

Assuming the same section modulus in the upper layer, use three 12 WF 27. The economical grillage therefore will be three 12 WF 27 in the upper tier, and seven 12 B 14 in the lower layer if these beams will check out for web shear and buckling.

The shearing load transmitted into each upper beam is

$$\frac{0.395 \times 300{,}000}{6} = 19{,}750 \text{ lb}$$

From Table 4.4, the allowable shear is

42 kips. The reaction for $3\frac{1}{2}$-in. bearing is 28, plus 6.5 for each additional inch of bearing. The total load of 100 kips on each of three beams would require

$$L = \frac{100 - 56}{6.5} = 6.8$$

and the minimum length of billet required is $7 + 6.8 = 13.8$ in. The beams should be placed about 8 in. o.c. to allow for flow of concrete, and the billet would therefore be 18×16 in.

Referring to Fig. 7.22 for the general method, we may assume that the lever arm for the outer beam is 8 in. $- (0.80 \times 6) = 3.2$ in., and

$$M = 100{,}000 \times 3.2 = 320{,}000 \text{ in.-lb}$$

$$S \text{ required for base plate } = \frac{320}{27} = 11.85$$

Using 16 in. width,

$$S = \frac{16d^2}{6} = 2.67d^2$$

$$d = \sqrt{\frac{11.85}{2.67}} = \sqrt{4.44} = 2.12$$

Use $18 \times 16 \times 2\frac{1}{2}$ in.

At each point where the upper beam crosses the lower beam, the load transmitted is $300/21 = 14.35$ kips, and the shear on the 12-in. beam is

$$\frac{0.395 \times 300{,}000}{3} = 39.5 \text{ kips}$$

These are well within the limits of the 12 WF 27 as given in Table 4.4 for the 12 B 14.

$$\frac{0.395 \times 300{,}000}{7} = 16.92$$

which is also within the value in Table 4.4.

It will be noted that the correction of the actual cantilever load carried instead of the assumed 20×20 in. billet will not materially change the grillage. The depth required for the grillage is 6 in. of concrete + 12-in. lower tier + 12-in. upper tier, or 30 in. under the billet as against 18 in. for the reinforced footing.

8.12 Combined and cantilever footings

a Combined footings and cantilever footings are used where one of the columns is too close to the lot line or adjacent building to permit the use of a concentric square or rectangular footing. Aside from special corner conditions, etc., there are three types of combined footings which may be used to carry two columns. If we call the wall column A and the interior column B, the selection of shape of footing will be made as follows:

Rectangular type, Fig. 8.7, is used if the load on column B is greater than on column A.

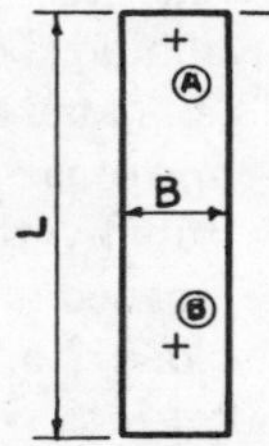

Fig. 8.7 Rectangular Footing

Trapezoidal type, Fig. 8.8, is used if the load on column A is slightly greater than on column B.

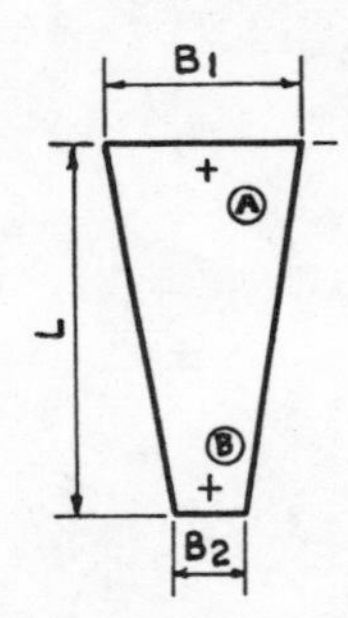

Fig. 8.8 Trapezoidal Footing

T-shape type, Fig. 8.9, is used if the load on column A is considerably greater than on column B.

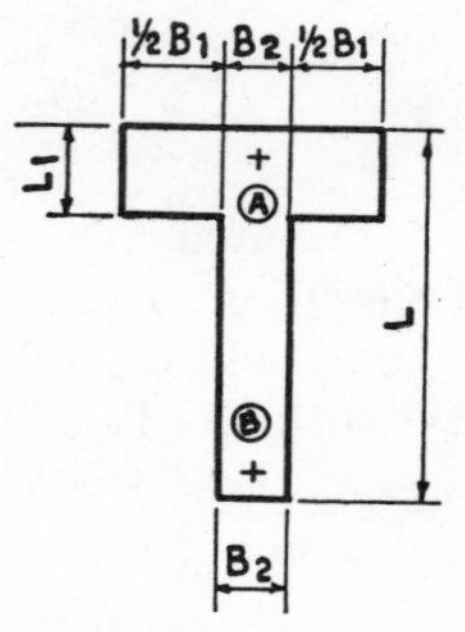

Fig. 8.9 T-Shape Footing

Starting from this basis, make an assumption of the type to be used and proceed with the design as follows:

b *For rectangular-type footing:*

1. Compute the center of gravity y of column and wall loads, using the lot line as a base. For balanced foundations, the center of gravity of the loads should coincide with the center of gravity of the footing area.

2. Compute the required area of footing. Assume the allowable soil pressure based on the nature of the subsoil. Deduct an assumed weight per square foot of footing to obtain net allowable soil pressure. Thus, assuming 8,000 psf or 4 tons, soil pressure, and footing at about 500 psf, the net allowable soil pressure is 8,000 − 500 = 7,500 psf. The total required area is the sum of the column loads plus wall load, divided by net soil pressure, or

$$A = \frac{P_a + P_b + P_{\text{wall}}}{\text{net } f}$$

3. Compute the length of the footing. For a rectangular footing the length is twice the distance to the center of gravity obtained by step 1, or $L = 2y$.

4. Compute $B = A/L$ and check footing shape obtained against interference with other footings; compute width and length to carry billets.

5. From the information thus obtained, a shear diagram may be laid out treating the entire footing as a reinforced concrete beam. From this shear diagram, compute maximum shear and maximum moment and design for depth and longitudinal steel.

6. Design cross steel if necessary to spread the load, and design stirrups. For stirrups, it is ordinarily necessary to use eight or more legs to each stirrup.

c *For trapezoidal footing:*

1. Assume the length L of footing by allowing enough clearance beyond the end of column B to carry the billet properly with 4-in. minimum coverage beyond the end of the billet. Compute the required area A as for rectangular footing, and the average width $B = A/L$.

2. Locate the center of gravity of the loads using the center line of the footing, $L/2$ from either end, as the base. Call the distance from this base to the center of gravity "e."

3. Compute $x = \dfrac{6Be}{L}$.

4. Compute $B_1 = B + x$, $B_2 = B - x$.

5. Same as step 5 for rectangular-type footing.

6. Same as step 6 for rectangular-type footing.

d *For T-type footing:*

1. Same as step 1 for rectangular-type footing.

2. Same as step 2 for rectangular-type footing.

3. Assume L as described in step 3 for trapezoidal footings. Assume B_2 similarly from the width to carry the billet for column B.

4. Compute L_1 from the formula

$$L_1 = \frac{2Ay - B_2L^2}{A - B_2L}$$

If the numerator in the formula is a negative quantity, a combined footing of this type will not be satisfactory, because the inner load is not heavy enough. Compute

$$B_1 = \frac{A - B_2L}{L_1}$$

$$B = B_1 + B_2$$

5. Same as step 5 for rectangular-type footing.

6. Design cross steel in the T, and design stirrups.

e *Combined-footing problem.* Assume a column 1 ft 2 in. from the lot line, carrying a load of 250 kips. At a distance of 18 ft from it at right angles to the lot line the inner column carries a load of 200 kips, allowable soil pressure 6,000 psf. From Table 8.2 we find that the wall column will require a 16 × 15-in. billet, and the inner column a 14 × 13-in. billet. From Art. 8.12*a* it will be noted that this requires a trapezoidal footing.

To find the center of gravity of loads, from center of wall column,

P		x		Px
250	×	0	=	0
200	×	18	=	3,600
450				3,600

$$\bar{x} = \frac{3,600}{450} = 8 \text{ ft}$$

To determine the length of the footing:

Distance from lot line	=	1 ft 2 in.
C to C of columns	=	18 ft 0 in.
Distance to end of footing	=	1 ft 0 in.
		20 ft 2 in.

Half length of footing (base line)

= 10 ft 1 in.

Deducting 500 psf for approximate weight of footing, the net bearing value is 6,000 − 500 = 5,500 psf.

$$\text{Required area} = \frac{450}{5.5} = 82 \text{ sq ft}$$

$$\text{Average width } B = \frac{82}{20.17} = 4.06$$

To compute "e," the eccentricity,

$$250[+(10.08 - 1.17)] = +2227.5$$
$$200[-(18 - 8.91)] = -1818.0$$
$$450 \qquad + 409.5$$

$$e = \frac{409.5}{450} = +0.91$$

$$x = \frac{6 \times 4.06 \times 0.91}{20.17} = 1.09$$

$$B_1 = 4.06 + 1.09 = 5.15 \text{ ft}$$
$$B_2 = 4.06 - 1.09 = 2.97 \text{ ft}$$

The increase in width per foot length is 2.18/20.17 = 0.107.

At the wall end the footing is wide enough to transmit shear along three sides, but the critical depth is across the footing on one line. At the narrow end also, the critical depth is in one line only across the footing.

Assume the net span of the footing for moment at 13½ in. less than the center-to-center distance, or

$$18 - 1.12 = 16.88 \text{ ft.}$$

The trapezoidal load carried is

$$W = 16.88 \times 4.06 \times 5.5 = 376.4$$

The width of the narrow end of the net span is

$$2.98 + (1.58 \times 0.107) = 3.15 \text{ ft} = 37.8 \text{ in.}$$

At the wide end it is

$$3.15 + (16.88 \times 0.107) = 4.95 \text{ ft} = 59.5 \text{ in.}$$

Although not strictly accurate, the result is close enough if the moments for rectangular and triangular load in Fig. 8.10 are added.

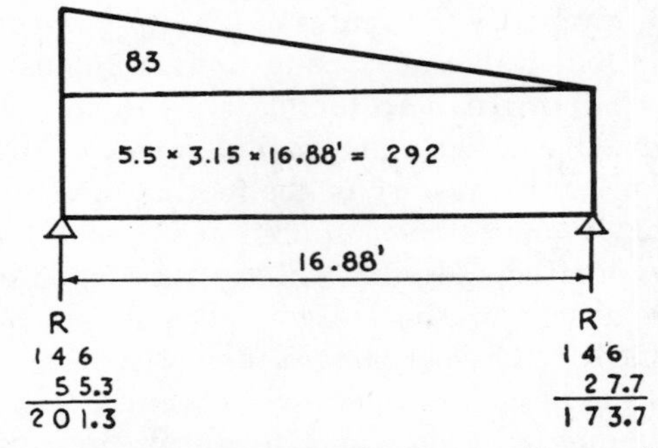

Fig. 8.10 Load Diagram for Beam—Trapezoidal Footing

From Fig. 3.13, Case 1,

$$M = \frac{292 \times 16.88}{8} = 616$$

From Fig. 3.13, Case 2,

$$M = 84 \times 16.88 \times 0.1283 = 182$$

$$616 + 182 = 798 \text{ ft-kips or } 9{,}580 \text{ in.-kips}$$

Using 3,000-psi concrete and $f_s = 24{,}000$, ($R = 201.5$) with an average width of 4.06 ft (= 48.75 in.),

$$d = \sqrt{\frac{9{,}580{,}000}{201.5 \times 48.75}} = \sqrt{975} = 31.23$$

Use 32 in. net depth, 36 in. gross depth.

$$A_s = \frac{9{,}580}{21 \times 32} = 14.25 \text{ sq in.}$$
$$= 10 \text{ No. 11 bars}$$

In accordance with Table 1.2, compute the shear at a distance d (= 32 in.) from the face of the support. At the heavy end of the footing this distance is

Lot line to center line of column	1 ft 2 in.
To face of base plate	7 in.
Distance d	2 ft 8 in.
	4 ft 5 in.

Reduction in footing width

$$= 4.42 \times 0.107 = 0.473$$

Net width of footing at this point

$$= 5.14 - 0.47 = 4.67 \text{ ft}$$

Reduction in shear load to this point is

$$4.42 \times 4.9 \times 5.5 = 119.3$$

Shear to be taken across this line

$= 201.3 - 119.3 = 82$

Shear capacity of concrete across this line

$= 56 \times 32 \times 60 = 107.5$

Therefore no stirrups are required at this end.

At the narrow end of the footing

Overhang	=	1 ft 0 in.
To face of base plate	=	6½ in.
Distance d	=	2 ft 8 in.
		4 ft 2½ in.

Increase in footing width

$= 4.2 \times 0.107 = 0.45$

Net width of footing at this point

$= 2.98 + 0.45 = 3.43$

Reduction in shear load to this point is

$4.2 \times 3.2 \times 5.5 = 73.92$

Shear to be taken across this line

$= 173.7 - 73.9 = 99.8$

Shear capacity in concrete across this line

$= 41.1 \times 32 \times 60 = 78.9$

Stress to be carried by stirrups

$99.8 - 78.9 = 20.9$

The maximum spacing of stirrups is half the net depth or 16 in. For this spacing, the stirrup dividend from Table 5.6 would be

$$16 \times 20.9 = 334{,}400$$

This would require a ⅝-in. U stirrup.

The area to a distance of 16 in. beyond this line is $1.44 \times \left(3.43 + \frac{1.33}{2} \times 0.107\right) =$ 4.57 and the reduction of load in this area is $4.57 \times 5.5 = 25.14$. The shear to be taken across this line is

$$99.8 - 25.14 = 74.66$$

The capacity across this line is in excess of 78.9 kips furnished on the line of the first stirrup. Therefore, it is advisable to use two stirrups only as shown in Fig. 8.11.

f It is sometimes desirable to use a cantilever footing instead of combined footings. There seems to be a great difference of opinion among engineers on the relative desirability of the two types.

For a cantilever footing, the wall-column footing is made rectangular and offset so that the column is eccentric on the footing. This

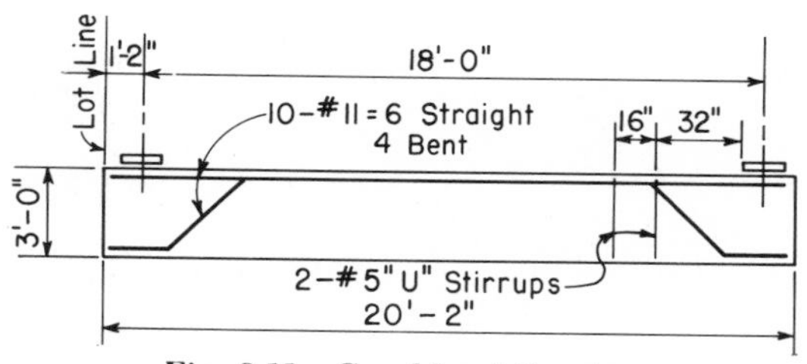

Fig. 8.11 Combined Footing

eccentricity is taken by means of a cantilever girder, or strap, which carries the column on the offset footing and uses the dead load of an adjacent column to take the uplift. The girder may be of steel or concrete, depending on the amount to be carried, but it should not bear on undisturbed earth.

The proportions shown in Fig. 8.12 are recommended by H. S. Woodward in an excellent article in *Engineering News-Record*.

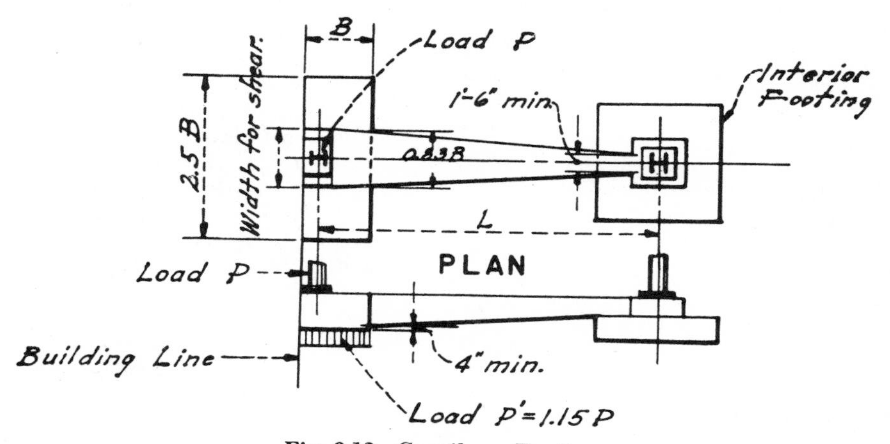

Fig. 8.12 Cantilever Footing

g *Cantilever-footing problem.* Assume the same conditions as in the combined-footing problem in Art. 8.12*e*. Using the proportions of Fig. 8.12,

$$B = \sqrt{\frac{1.15P}{2.5f}}$$

$$B = \sqrt{\frac{1.15 \times 250}{2.5 \times 5.5}} = \sqrt{20.9} = 4.58$$

Then

$$e = \frac{4.58}{2} - 1.17 = 1.12$$

$$M = 1.12 \times 250 = 280 \text{ ft-kips}$$

The uplift is

$$\frac{280}{18 - 1.12} = 16.5$$

On this basis, the factor 15 percent may be reduced to 7 percent, and

$$B = \sqrt{\frac{1.07 \times 250}{2.5 \times 5.5}} = \sqrt{19.5} = 4.42$$

Then

$$e = \frac{4.42}{2} - 1.17 = 1.04$$

$$M = 1.04 \times 250 = 260 \text{ ft-kips}$$

The uplift is

$$\frac{260}{18 - 1.04} = 15.3 \text{ kips}$$

The footing for the wall column is therefore 4 ft 5 in. by 11 ft. The footing for the interior column is not changed but remains a simple square footing for the full load as designed in Art. 8.10. The strap is designed as a simple cantilever beam for a moment of 260 ft-kips and a shear of 15.3 kips.

8.13 Pile foundations

a Foundations of bearing piles may be listed under the following classifications:

Wood piles
- Plain, untreated
- Treated

Concrete
- Cast in place
  - With shells
    - Tapered shells
    - Cylindrical shells
      - Providing sectional restraint
      - Not providing sectional restraint
    - Steel pipe
  - Without shells
    - Pedestal
    - Straight shaft
- Precast
  - Tapered
  - Parallel-sided

Composite
- Wood piles with cast-in-place concrete
- Wood piles with precast concrete

Steel H Piles

No single type of pile answers every purpose, and each type has its advantages and disadvantages. The descriptive matter in paragraphs *b* through *e* following is taken from the article "Pile Foundations," by A. E. Cummings, Director of Research, Raymond Concrete Pile Co. The article is from the *Proceedings* of the Purdue Conference on Soil Mechanics and Its Applications, sponsored by the Society for Promotion of Engineering Education.

b *Wood piles*

The plain wood pile is probably the oldest form of piling used by man. The principal advantage of the plain wood pile is its low cost. However, in spite of its low cost, it is not necessarily the cheapest type of foundation unit because of the restrictions that govern its use. In order that it will not decay, a plain wood pile must be cut off and capped below permanent ground water level. This requirement often involves deep excavation with sheeting and pumping. Another disadvantage of the wood pile is the fact that ordinarily it can support a considerably smaller load than some of the other types of pile. For a given load to be carried, this means a larger number of wood piles with larger footings and with a longer time required to do the job. Plain wood piles are subject to attack by fungi and bacteria as well as by insects and various marine animals.

In order to overcome some of the disadvantages of plain wood piles, preservative processes have been developed in which the wood is treated with various kinds of chemicals. The effectiveness of these preservatives depends on two important factors: (1) the thoroughness with which the treatment process is done and (2) the severity of the exposure conditions at the site where the treated pile is to be used. It cannot be assumed that preservative treatment will make the pile immune from attack by marine animals and insects especially in localities where these animals are very active. Neither can it be assumed that preservative treatment will entirely prevent decay in places where the pile is exposed to the air and is alternately wet and dry. However, it may be taken for granted that, under similar conditions, a properly treated timber pile will last longer than an untreated pile. Nevertheless, the engineer who proposes to use treated timber piles as part of a permanent structure should always give some consideration to the question of replacement. Under a building foundation the replacement of a pile is a very expensive process. In a railway trestle or similar structure replacement is not so expensive. Because of the cost of the treatment process, the treated pile loses some of the low cost advantage of the plain wood pile.

c Concrete piles

Concrete piles came into general use during the last few years of the past century. There are two principal types of concrete pile: cast-in-place and precast. The cast-in-place pile is formed in the ground in the position in which it is to be used in the foundation. The precast pile is cast above ground and, after it has been properly cured, it is driven or jetted just like a wood pile. The fields of usefulness of these two types of pile overlap to a large extent although there are conditions under which one or the other has a distinct advantage.

In ordinary building foundation work the cast-in-place pile is more commonly used than the precast pile. The principal reasons for this are the following. The cast-in-place pile foundation can be installed more rapidly because there is no delay while test piles are being driven to predetermine pile lengths and there is no delay waiting for piles to cure sufficiently to withstand handling and driving stresses. Of particular importance in crowded locations is the fact that a cast-in-place pile job does not require a large open space for a pile-casting yard.

In marine installations either in salt or in fresh water, the precast pile is used almost exclusively. This is due to the difficulty involved in placing cast-in-place piles in open water. For docks and bulkheads, the cast-in-place pile is sometimes used in the anchorage system. On trestle type structures such as highway viaducts, the precast pile is more commonly used. A portion of the pile often extends above the ground and serves as a column for the superstructure.

Both precast and cast-in-place concrete piles have an advantage over other kinds of piles due to the fact that the concrete pile does not have to be cut off below permanent ground water level. When the concrete has been properly mixed and placed, the concrete pile can be exposed to air or to alternate wetting and drying without danger of corrosion or decay. Another advantage of the concrete pile in comparison with some of the other kinds of pile is the fact that concrete piles usually carry higher working loads.

Precast concrete piles are always reinforced internally so that they can resist stresses produced by handling and driving. When the pile is used as an ordinary foundation pile completely submerged in soil, the stresses in the pile under the usual working loads are relatively low. The carrying capacity of the pile depends primarily on the relation between the pile and the soil and not on the strength of the pile itself. In such cases it is the handling and driving stresses which determine the design of the pile and the amount of internal reinforcement. When the precast concrete pile serves as a column or when it is a sheet pile resisting horizontal forces, the amount of reinforcement is sometimes determined by the forces that will be exerted on the pile after it is in place in the ground. Even in such cases the design of the pile should be checked against the handling and driving stresses.

The concrete in a cast-in-place pile is not subject to handling and driving stresses since the pile is poured in place in the ground. Where they are used as columns or where they are subject to horizontal forces, cast-in-place concrete piles are reinforced internally. In ordinary foundation work where they are completely submerged in soil, cast-in-place piles are not reinforced internally. The stresses in the concrete are low and the carrying capacity of the pile is determined by the relation between the pile and the soil. Under these conditions the use of internal reinforcement in a cast-in-place pile represents an unnecessary expense.

Cast-in-place concrete piles may be divided into two kinds: (1) those in which a steel shell is left in the ground and (2) those in which no shell is left in the ground. The purpose of the shell is to prevent mud and water from mixing with the fresh concrete and to provide a form to protect the concrete while it is setting and to provide restraint to the cross section. With the shell-less pile, the fresh concrete is in direct contact with the surrounding soil.

Some kinds of shell piles are formed by driving a tapered steel shell into the ground and then filling the shell with concrete. With piles of this type, the full driving resistance is maintained since the shell remains in the position into which it is driven. The Raymond pile is one of this type. A thin steel restraining shell, closed at the bottom with a steel boot, is placed on the outside of a steel mandrel or core and the shell and the core are driven into the ground. When sufficient driving resistance has been developed, driving is stopped and the core is withdrawn from the shell which is then filled with concrete. Another kind of pile which is formed in a driven tapered shell is the Union Monotube Pile. These piles are driven without an interior core and the shell thickness must therefore be great enough to withstand the stresses developed during driving.

The McArthur Pile is another kind of cast-in-place concrete pile in which a steel shell remains in the ground. The driving tube is a piece of heavy steel pipe open at both ends. A steel driving core is fitted inside this pipe and the two are driven together into the ground. When suitable driving resistance has been reached, driving is stopped and the core is removed from the tube. A thin metal shell several inches smaller in diameter than the driving tube is then dropped down inside the tube. This thin shell is then filled with concrete and the driving tube is withdrawn. With a pile of this kind the driving resistance is determined by means of the outer driving tube but the pile that remains in the ground is several inches smaller in diameter than this tube. The carrying capacity of such a pile depends largely on the point resistance since the surrounding soil may or may not come back against the shell to provide frictional resistance along the side of the pile.

Steel pipe piles are formed by driving ordinary steel pipe into the ground and filling it with concrete. Sometimes a steel shoe or boot plate is used at the bottom of the pipe and the pile is then called a closed-end pipe pile. The uses of such piles are about the same as those of other kinds of cast-in-place concrete piles. When the bottom of the pipe is left open during driving, the pile is called an open-end pipe pile. These open-end piles are usually driven to bearing on rock. The soil, which is in the pipe when driving is finished, is jetted and blown out with water and air jets and the pipe is then filled with concrete. Because of the cost of the pipe, a steel pipe pile usually costs more per foot than other types of cast-in-place concrete piles. However, when they are driven open-end to bearing on rock and then carefully cleaned out, pipe piles are usually allowed greater loads than other kinds of concrete piles. Under such conditions the high cost of the pipe is offset by the higher loading and the consequent reduction in the number of piles required to carry a given load.

The McArthur Pedestal Pile is a shell-less cast-in-place concrete pile. The driving tube is a piece of heavy steel pipe which is open at both ends. A steel core is used inside the driving tube and the two are driven together into the ground. When sufficient driving resistance has been obtained, the core is removed and a pedestal on the bottom of the pile is formed in the following manner. After driving is finished and the core has been removed, enough concrete is dumped into the tube to fill the bottom of it to a depth of five or six feet. The core is then placed back in the tube on top of this concrete and the tube is lifted about three feet. The hammer is then operated on top of the core in order to drive the concrete out of the bottom of the tube to form the pedestal. Sometimes several such charges of concrete are used in this manner to make the pedestal. After the pedestal is completed, the core is removed and the tube is filled with concrete up to the ground surface. The shaft of the pile is then formed by the same method that is used for the straight-shaft pile.

The principal advantage of piles of this type lies in the fact that they are usually the cheapest of the concrete piles. This is due to the fact that, ordinarily, the shell-less pile consists only of concrete, whereas the precast pile always has internal reinforcement and the shell types of cast-in-place pile always have a steel shell which remains in the ground. The principal criticisms of the shell-less pile are based on the following facts. It is often difficult to prevent soil and water from getting mixed with the fresh concrete and the finished piles are subject to distortion due to the soil stresses set up by the driving of adjacent piles.

Precast concrete piles may be parallel-sided or tapered. The cross sections are usually round, octagonal or square. Ordinarily, precast piles are designed for the particular job on which they are to be used and they are then made at or near the job site by the piling contractor. Sometimes they are made in pile manufacturing plants and are shipped by rail or water to the job site. Railroads and State Highway Departments often have certain standard designs for precast piles although they sometimes design special piles for a particular job.

The tapered types of precast piles are usually limited in length to about 40 feet. For longer lengths, the less flexible parallel-sided piles are used. Precast piles with diameters as great as 24 or even 30 inches are sometimes used and the maximum lengths that have been used up to the present time are well over 100 feet.

d *Composite piles*

Composite piles came into general use about 25 or 30 years ago. Their development was due to an effort on the part of engineers to make use of the advantages of different types of piles in order to produce a very long pile that was not too expensive. Ordinarily, a composite pile consists of a wood pile topped with a cast-in-place concrete pile. The concrete section extends from the permanent ground water level up to the pile cut-off elevation and the wood pile is used below water level. Such a pile takes advantage of the low cost of a wood pile, and at the same time it eliminates the necessity of the expensive sheeting, excavation and pumping that would be required to cut off and cap the wood pile at the water level. The upper end of the wood pile is sometimes cut in the form of a tenon which projects up into the concrete section. All of the various types of cast-in-place concrete piles are combined in this manner with wood piles to form composite piles. It is necessary to fasten together the two sections of the pile so that they do not separate after they are in the ground. A number of different kinds of fastenings have been developed for this purpose. Also it is necessary to provide a seal between the pile shell and the wood pile so as to exclude mud and water from the joint. Composite piles are also made with precast concrete piles superimposed on wood piles. The precast section is cast with a recess in its lower end to receive the tenon of the wood pile and various devices are used to fasten the two sections of the pile together.

e Steel H piles

A comparatively recent development in the piling industry is the use of rolled structural shapes as bearing piles. The profile usually used for this purpose is that known as the H-beam. This type of pile has proved to be especially useful for trestle structures in which the pile extends above the ground line and serves not only as a pile but also as a column. Because of the small cross-sectional area of these piles, they can often be driven into river bottoms of dense gravel into which it would be very difficult to drive a displacement pile even with the aid of jetting. By displacement pile is meant any pile with a solid cross section of considerable area. However, this ability of the H-beam pile to penetrate easily into dense materials works to its disadvantage in other kinds of soil conditions. In many cases bearing piles are used to support loads simply by friction between the pile and the surrounding soil. In other cases, piles are used primarily for compaction. Under such conditions, particularly when compaction is the principal purpose of the pile, a considerably longer H-beam pile is required to carry the same load that can be supported by a shorter displacement pile. Where the H-beam pile is exposed to air or where it is alternately wet and dry, it is subject to the same corrosive action that would occur to any other steel structure under the same conditions. In some installations the upper ends of H-beam piles have been encased in concrete from the permanent ground water level up to the pile cut-off.

f The design of pile foundations may be divided into two parts—the determination of the number of piles and their grouping, and the design of the pile cap. The previous discussion has not touched upon the allowable loads on piles. Piles are usually driven to carry their load in accordance with one of the generally accepted formulae—usually the *Engineering News* formula. The maximum loading to which they are to be driven, however, must be assumed by the engineer in order to determine the number of piles to be used. For wood piles, the usual code limitation is 500 psi of the right section of the pile at mid length. This limitation, of course, requires the assumption of sizes within the usual code limitations, and common office practice limits wood piles to a capacity of 15 to 20 tons each.

For concrete piles of any type, the permissible load is governed by the code or agency under which the designer is working. Common practice, unless otherwise specified, is to design for either 30 or 40 tons per pile, although frequently codes permit greater loads.

Rolled structural steel H piles are usually limited to 9,000 psi of section.

The spacing of piles is limited by the compressibility of the material in which they are driven and to some extent by the displacement of the pile. Ordinarily it is impracticable to drive wood piles closer than 2 ft 6 in. on centers in each direction. Although steel H piles may be driven 2 ft 6 in. on centers, if the piles are heavily loaded the rock bearing at the tip may be the controlling factor, and it is advisable to use a 3-ft spacing for steel piles as well as concrete.

Although piles may be spotted accurately when they are started, the chances are that obstructions below grade will divert them so that they will often "creep" in driving as much as 6 in. It is therefore good practice to stagger piles under a wall, and to use not less than three piles under a column or pier. This will frequently control to some extent the number of piles required since, because of this limitation, some piles may be loaded to only a fraction of their capacity.

g The general layout of pile caps is shown in Fig. 8.13. The design of pile caps is similar to the design of footings as described in Art. 8.11. The following additional requirements from the ACI Code apply:

[From Sec. 2302(*d*)] In the case of footings on piles, computations for moments and shears may be based on the assumption that the reaction from any pile is concentrated at the center of the pile.

Although, for obvious reasons, any load beyond the capacity of a single pile must be carried by a group of several piles, the capacity of a group of piles is not as great as the capacity of a single pile multiplied by the number of piles in the group. The efficiency drops with an increase in the number of piles in a row and the number of rows. There are several formulae for determination of the reduced efficiency, the most commonly accepted being the Converse-Labarre formula which is as follows

$$E = 1 - \alpha \frac{(n - 1)m + (m - 1)n}{90mn}$$

In the above formula

E = efficiency
m = number of rows of piles
n = number of piles in a row
α = numerical value in degrees of the angle whose tangent is d/s
d = pile diameter
s = pile spacing

Normally d/s may be taken as ⅓ and therefore α is 18.33.

Obviously, this formula is an approximation and any reasonable interpolation may be used with safety.

h Standard pile caps using up to 9 piles are shown in Fig. 8.13 and listed in Table 8.4. This table is based on the use of 3,000-psi concrete and a steel working stress of 24,000 psi. The net column load is based on the reduced stress of the Converse-Labarre formula given in the preceding paragraph. The loads listed for individual piles are those usually used for wood piles (30 kips), for concrete piles (60 and 100 kips), and for steel H piles (100 and 140 kips). Other values may be prorated if the spacing is maintained at 3 ft 0 in.

8.14 Caisson foundations

a Article 8.10*b* refers to caisson foundations as a fourth type of column foundation.

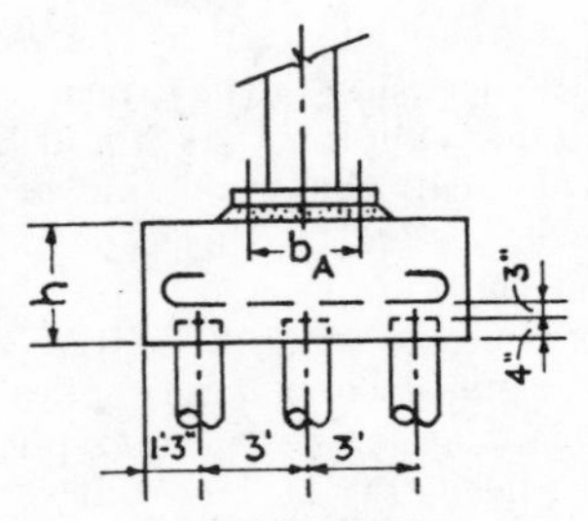

Fig. 8.13 Typical Pile Cap

Formerly, pneumatic caissons were used for practically all caisson jobs where water was encountered during excavation to rock. Now many jobs which formerly would have been pneumatic-caisson jobs are open-caisson work instead, owing to improved methods.

b In general, the pneumatic caisson is constructed as shown in Figs. 8.14 and 8.15. As the shaft is sunk, first the steel cylinder around the working chamber is filled with concrete, then the working chamber, and then the shaft; finally, the air lock is removed for use on another caisson. This question is from a state license examination: "Explain briefly the operation of a pneumatic caisson. Illustrate with a schematic diagram." The description and diagram given should be satisfactory as an answer for this problem.

c Open caissons are installed by several methods. In stiff to hard nonplastic clay, walls or pits may sometimes be excavated to rock with no support whatever. Often wood or steel sheet piling is used, being driven down as the excavating proceeds. Where the clay will stand to a depth of 4 or 5 ft, the poling-board method, or Chicago lagging method, is used. In this procedure a hole is dug about 5 ft deep and of the size required, then wood sheeting is placed and held in position by a steel internal ring or by wood supports. Then another 5-ft depth is excavated and the process repeated. This system is not workable if the soil is not stiff enough to stand.

There are several newer methods, all patented by the various foundation companies. One system—the Gow method—uses a series of telescoping cylinders in lengths of 8 ft or more, each successive cylinder being 2 in. less in diameter than the one above. The exca-

Table 8.4 Pile Caps

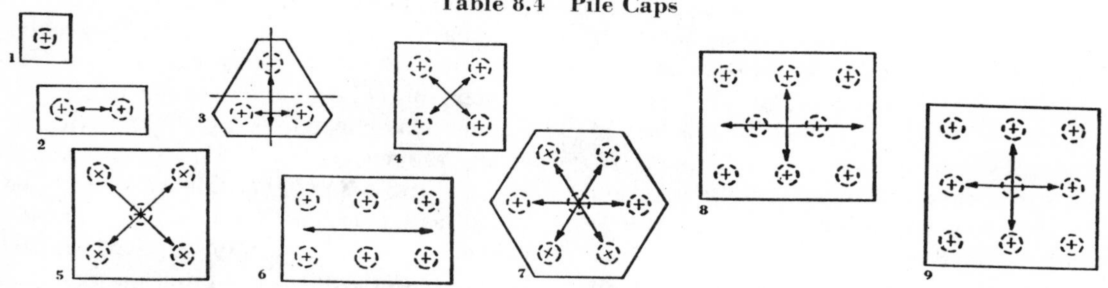

No. of piles	Pile value, kips	Net column load	Dimension b	Dimension l	Dimension h	Reinforcement Short bars	Reinforcement Long bars
1	30	28	2′6″	2′6″	1′7″		
	60	58	2′6″	2′6″	1′7″		
	100	98	2′6″	2′6″	1′7″		
	140	138	2′6″	2′6″	1′7″		
2	30	54	2′6″	5′6″	2′8″	....	5 No. 7
	60	108	2′6″	5′6″	3′5″	....	7 No. 7
	100	180	2′6″	5′6″	4′1″	....	7 No. 8
	140	252	2′6″	5′6″	4′8″	....	6 No. 9
3	30	72	See drawing		2′9″	5 No. 7	5 No. 7
	60	155			3′8″	7 No. 7	7 No. 7
	100	258			4′6″	7 No. 8	7 No. 8
	140	361			5′4″	6 No. 9	6 No. 9
4	30	108	5′6″	5′6″	2′8″	5 No. 8	5 No. 8
	60	216	5′6″	5′6″	3′5″	5 No. 9	5 No. 9
	100	360	5′6″	5′6″	4′4″	6 No. 9	6 No. 9
	140	504	5′6″	5′6″	4′8″	7 No. 9	7 No. 9
5	30	135	6′9″	6′9″	2′8″	7 No. 9	7 No. 9
	60	270	6′9″	6′9″	3′5″	6 No. 11	6 No. 11
	100	450	6′9″	6′9″	4′4″	6 No. 11	6 No. 11
	140	630	6′9″	6′9″	4′8″	7 No. 11	7 No. 11
6	30	162	5′6″	8′6″	3′5″	6 No. 6	6 No. 11
	60	324	5′6″	8′6″	4′6″	6 No. 6	8 No. 11
	100	540	5′6″	8′6″	5′6″	6 No. 6	10 No. 11
	140	756	5′6″	8′6″	6′3″	6 No. 6	12 No. 11
7	30	189	See drawing		3′5″	6 No. 8	in each direction
	60	378			4′6″	6 No. 9	
	100	630			5′6″	6 No. 10	
	140	882			6′0″	8 No. 10	
8	30	216	7′10″	8′6″	3′3″	8 No. 11	7 No. 11
	60	432	7′10″	8′6″	4′3″	11 No. 11	10 No. 11
	100	720	7′10″	8′6″	5′9″	15 No. 11	13 No. 11
	140	1008	7′10″	8′6″	6′6″	18 No. 11	15 No. 11
9	30	246	8′6″	8′6″	3′4″	11 No. 10	11 No. 10
	60	419	8′6″	8′6″	4′5″	12 No. 11	12 No. 11
	100	819	8′6″	8′6″	5′3″	15 No. 11	15 No. 11

vating is done by means of a large earth auger, and the excavation may be belled out at the base. Steel cylinders are placed as the excavation proceeds. When the bearing material is reached, the caisson is sealed, driving the bottom cylinder in to stop any inflow of water. Concrete is placed, and the cylinders withdrawn. Occasionally, if pumps cannot control the inflow of water, the caisson bottoms must be cleaned by divers and sealed by concrete tremied into the water.

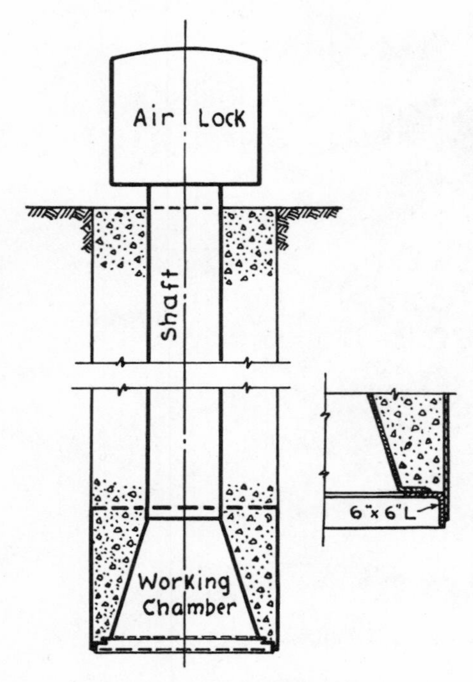

Fig. 8.14 Pneumatic Caisson

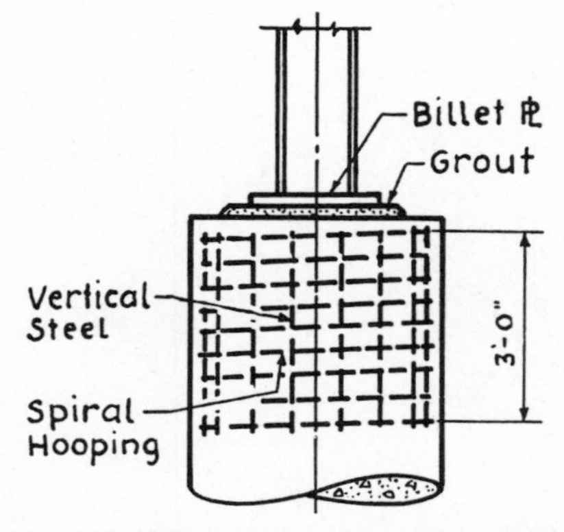

Fig. 8.15 Spiral Reinforcement—Top of Caisson

d The design of a caisson, either open or pneumatic, is a simple problem in compression. Building codes vary greatly on matters of caisson designs; the subject is frequently omitted completely. Without citing the codes from which the various clauses are taken, the following may be taken as governing good design of open caissons. For caissons or piers where the subsoil material furnishes proper lateral stability, use a maximum allowable axial compressive stress of $0.25f_c'$, but not over 850 psi. Where permanent lateral support is not provided, the allowable working stress should be reduced in accordance with the following formula, where H is the height and D the diameter.

$$f = f_c\left(1.3 - \frac{H}{20D}\right)$$

except that f may not exceed 850 psi.

Open caissons are normally limited to a minimum diameter of 2 ft, but if manual work is required, a practical minimum is 3 ft.

When constructed of reinforced concrete, and properly supported laterally, the caisson may be designed in accordance with the requirements of the ACI Code for columns as given in Art. 7.20.

Since the caisson ordinarily is cylindrical and the billet square, it is desirable to increase the bearing capacity at the top of the caisson to cut down the size of the billet and consequent overhang which requires an enlargement of the top of the caisson. This increase in bearing value is accomplished by putting spiral reinforcement in the upper 3 ft of the caisson in accordance with the design of spirally hooped columns. This procedure is shown in detail in Fig. 8.15, and may increase the bearing capacity at the top of the caisson to 1,000 psi. The billet must be designed for the special condition, however, and cannot be taken directly from the load table. The original design of a caisson does not offer any great problem, but frequently a redesign must be made in a hurry when obstacles such as boulders are encountered that cause eccentricity or a reduction in the size of the base. A common specification requires that if the stress in the caisson due to eccentricity of rock bearing increases the total design stress by over 10 percent, a like increase in strength must be provided.

Another common specification requirement is that: "The actual axis of any caisson shall not vary from its designed axis by more than 1/60th of the depth of any horizontal section. The center of any caisson at any horizontal section shall not deviate from its actual axis by more than 5 percent of the diameter. The full sectional area of any caisson must be maintained throughout." The actual stress due to eccentricity may be computed from Art. 1.60. From this formula it will be noted that the maximum eccentricity allowable without revision of design is 0.2 in. per foot of diameter. Allowable stress on a caisson may be increased by improving the quality of the mix or by putting in 1 percent or more of vertical steel, properly hooped in accordance with the column design given in Art. 7.20.

8.15 Concrete wall footings

a Offset wall footings as shown in Fig. 8.16 are often used for practical reasons even though they may not be required to spread the load. Frequently, the general excavation is carried to the elevation of the top of the footing and the wall-footing excavation made neat below this level without forming. Into this excavation the concrete is poured, providing a firm foundation on which to start the wall forming. The cheaper and surer wall forming thus provided usually will more than offset the cost of the small amount of extra concrete. For this purpose, the minimum offset a should be 4 in., and the depth 8 in. A 2 × 4 should be embedded in the top of the wet concrete to provide a key for the wall.

b If a concrete wall footing is poured for the purpose of carrying wall load, it should, if possible, be concentric with the wall above. If the wall is on the lot line so that the footing offset must be omitted on one side, the wall footing should be checked for eccentricity. Under these conditions, an offset of over half the wall thickness is useless, because when the offset is half the wall thickness, the eccentricity is one-sixth the footing width, and the soil pressure varies from zero at the edge of the offset to double the average at the outside edge.

To check the maximum soil pressure, the formula for trapezoidal loading (Art. 8.10) should be used:

$$f = \frac{P}{A}\left(1 \pm \frac{6e}{b}\right)$$

An illustration of this method is given in the following problem.

Problem 1 Assume a bearing wall with a load of 8,000 lb per running foot from a wall 16 in. thick, with a 6-in. offset inside. Compute the maximum soil pressure.

$$f = \frac{8{,}000}{1.83}\left(1 + \frac{6 \times 3}{22}\right) = 7{,}950 \text{ psf}$$

c Plain concrete is preferable for most wall footings. The depth may be governed by any one of several factors. If the offset required is not over 6 in. on either side of the wall, the depth may be made twice the offset, and computation of stresses may be omitted. This assumes a uniform stress distribution over an area enclosed by the width indicated in Fig. 8.16. If the depth

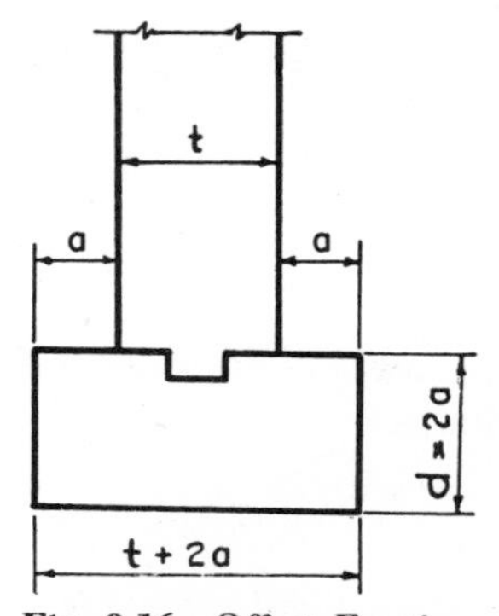

Fig. 8.16 Offset Footing

thus obtained is excessive, the proper depth may be obtained by assuming a uniform upward soil pressure on a strip of footing 1 ft wide on a cantilever span of a, and computing the depth to resist moment and shear in a plain concrete beam as described in Art. 5.12.

Problem 2 Assume a 16-in. wall with a load of 12,000 lb per running foot, bearing on a soil with a capacity of 3,000 psf. Design the footing, using a 3,000-psi plain concrete footing.

The footing width is 12,000/3,000 = 4 ft = 48 in., and the wall thickness is 16 in. The offset a is $\frac{48 - 16}{2} = 16$ in. The moment on the cantilever is

$$M = (3 \times 1.33) \times 8 = 32 \text{ in.-kips}$$

From Art. 5.12, $f_c = 88$ lb. Then the section modulus S of the footing is

$$S = \frac{32{,}000}{88} = 363$$

Since the section modulus of a rectangle = $bd^2/6$,

$$d_m = \sqrt{\frac{363 \times 6}{12}} = \sqrt{181.5} = 13.5 \text{ in.}$$

Use 14 in.

The area producing shear is 16 in. − 14 in., or 2 in. wide, which is negligible. Therefore, the depth will be controlled by moment.

d If the load is such that a plain concrete footing would be too deep for economy, reinforcement should be added in the lower face. An illustration of the calculations involved is shown in the following problem.

Problem 3 Assume a 16-in. wall with a load of 30,000 lb per foot of length, with a soil pressure of 5,000 psf. The width of footing required will be 30,000/5,000 = 6 ft. The offset a will be 28 in., and the moment on the cantilever will be

$$M = (2.33 \times 5) \times 14 = 163{,}300 \text{ in.-lb}$$

To check this first for a plain concrete footing,

$$S = \frac{163{,}300}{88} = 1{,}855$$

$$d = \sqrt{927.5} = 30.5 \text{ in.}$$

Use 31 in.

For a reinforced concrete footing, using 3,000-psi concrete, from Art. 5.11, for f_s = 24,000 lb and f_c' = 3,000. Then R is 201.5 and Rbd^2 = 163,300 or

$$d = \sqrt{\frac{163{,}300}{12 \times 201.5}} = \sqrt{67.5} = 8.2 \text{ in. net}$$

To check this depth for shear:

$$\text{Width producing beam shear} = \frac{72 - 16 + (2 \times 8.2)}{2} = 19.8$$

$$\text{Load outside the shear line} = \frac{19.8}{72} \times 30{,}000 = 8{,}040$$

This would produce a unit shear of

$$\frac{8{,}040}{12 \times 8.2} = 82 \text{ psi}$$

against an allowable limit of 60 psi.

Try 12 in. net depth:

$$\text{Width producing beam shear} = \frac{72 - (16 + 24)}{2} = 16 \text{ in.}$$

$$\text{Load in shear} = {}^{16}\!/_{72} \times 30{,}000 = 6{,}670$$

The unit shear is

$$\frac{6{,}670}{0.875 \times 12 \times 12} = 46.5 \text{ psi}$$

$$A_s = \frac{163.3}{21 \times 12} = 0.65 \text{ sq in. per lin ft}$$

Use No. 6 bars 8 in. o.c. Total depth of footing is 12 in. + 4 in. = 1 ft 4 in. (see Fig. 8.17).

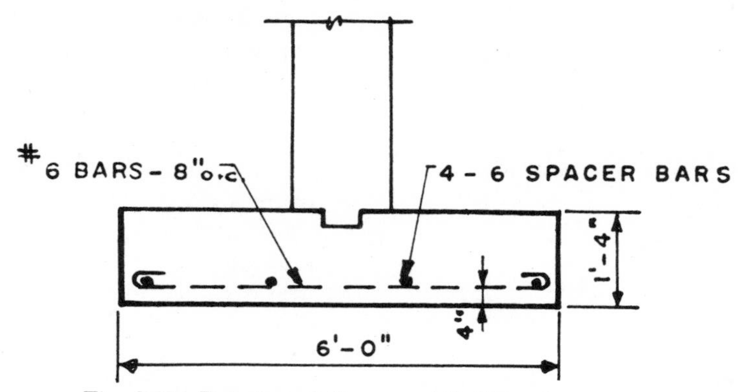

Fig. 8.17 Reinforced Concrete Wall Footing

8.20 MASONRY WALLS AND PIERS

a Masonry walls and piers include bearing walls, piers, and buttresses constructed of stone, brick, structural clay tile, architectural terra cotta, and concrete masonry units bonded together with mortar.

Aside from resistance to fire (which is discussed in Art. 1.42) and weather, the principal structural requirement of a bearing wall is resistance to vertical loads. This, however, may be combined with resistance to bending, either through eccentricity of loading, which is almost always present, or to unbalanced roof thrust or direct wind load. The discussion of this last factor—wind load—appears in Art. 10.21, since it is usually a factor which applies only to low steel buildings.

b Building codes generally set up two sets of controls for bearing walls:

1. Maximum permissible bearing load per square inch for different materials.
2. Minimum wall thickness for walls of various heights within a single story and in different stories of multistory wall-bearing buildings.

Usually the latter is the only controlling requirement that matters. If you are working

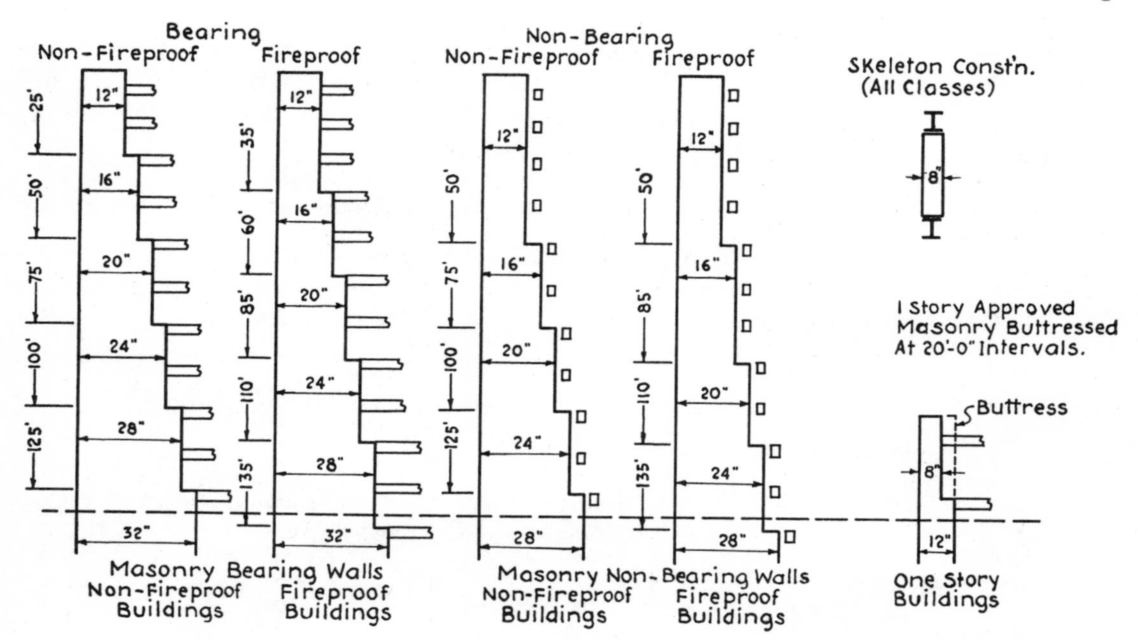

Fig. 8.18 Masonry Walls

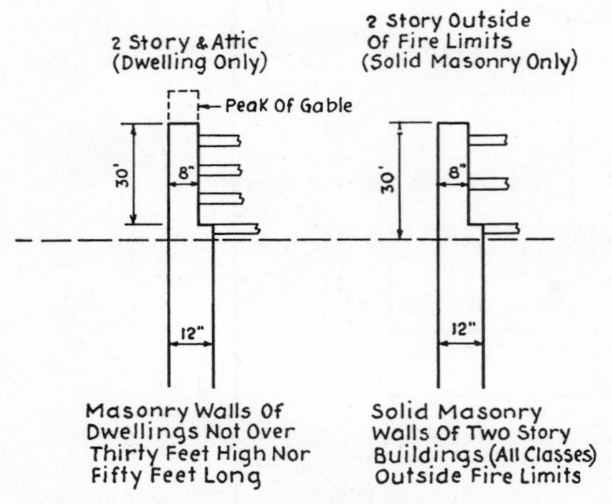

Fig. 8.18 Masonry Walls (Continued)

under a specific code, the limitations above should be ascertained before proceeding with the design. In the absence of any code requirements, the following requirements from the New York City Code may be accepted as good practice for permissible bearing loads.

The generally accepted code for the second controlling criterion given above is the building code recommended by the National Board of Fire Underwriters. These wall thicknesses and permissible heights are shown in Fig. 8.18. The maximum length of window and door openings in any wall section must not exceed 50 percent of the length unless the wall thickness is increased 4 in. for each 15 percent of added openings, and the maximum percentage of openings must not exceed 75 percent. This, however, does not preclude the use of solid masonry piers in place of columns, as described in paragraph *c* following. Piers between windows must be checked on the basis of Table 8.5.

Most building departments require that walls shall be bonded by means of header courses every fifth or sixth course, or some other type of masonry bond, if the full thickness of wall is to be considered as acting. Metal wall anchors are only given consideration as bonding facing to backup, but not for the purpose of figuring the full wall thickness in Fig. 8.18. Where metal wall anchors are used, only the thickness of backup is used as wall thickness.

If brick facing is used with tile or concrete-block backup, the bearing strength of the entire wall is taken at the lesser value rather than at the bearing value of the brick. If steel beams bear on a wall of this kind, it is common practice to design the bearing plate for a load of 250 psi and to put in a brick bearing pad of the required width, toothed into the masonry to a depth below the bearing plate equal to the width of the pad.

Solid masonry walls must be supported laterally at intervals not exceeding 18 times the wall thickness in the top story and 20 times the wall thickness elsewhere. For walls of brick and tile, these intervals become 16 and 18 times the thickness, respectively. This lateral support may be by floors, if the distance is vertical, or by crosswalls, piers, buttresses, or columns, properly bonded when the controlling distance is horizontal.

c Masonry piers may be used in place of

Table 8.5 Allowable Bearing on Masonry

Material	Using lime-cement mortar, psi	Using cement mortar, psi
Brick	250	325
Brick (with crushing strength over 4,500 psi)	$8\frac{1}{3}$% of crushing strength, not over 500	10% of crushing strength, not over 500
Clay tile, cells horizontal (on gross area)	60	70
Concrete-block masonry	$8\frac{1}{3}$% of crushing strength (usually 80)	10% of crushing strength (usually 80)
Granite	640	800
Limestone	400	500
Marble	400	500
Sandstone	250	300

columns in wall construction under the following code provisions:

1. The minimum side dimension must be 12 in.

2. The maximum height shall be 10 times the least side.

3. The maximum stresses shall not exceed those set up in Table 8.5.

d Nonbearing partitions of clay tile, gypsum block, or other masonry should be not less in thickness than 1/48th of the height of the partition. (Some codes require not less than 1/40th.) These partitions must be built solidly on the floor and wedged solidly to the underside of the slab above so that they obtain a lateral support from the slab.

e The design of masonry buttresses to resist lateral as well as vertical pressure is now seldom required, since the bending is normally taken by steel or reinforced concrete, the masonry being used for architectural purposes only. If a masonry buttress is used to resist thrust of a hammer-beam truss, a vaulted roof, or other lateral thrust, or a combined vertical and lateral load, it involves a graphical or analytical method of vector analysis combining the superimposed load with the weight of the top stone, and making the base of the top stone wide enough so that the resultant intersects the joint within the middle third of the top stone. The new resultant is used as the applied load, and is combined with the weight of the second stone in the same manner, etc., until the load is transferred to the foundation. Figure 8.19 indicates the method as applied to a buttress. The general method of each step corresponds to the method of solution

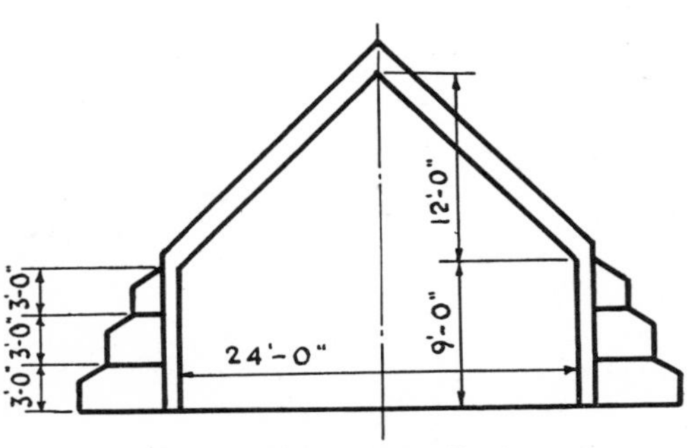

Fig. 8.19 Buttress Design

of a retaining wall footing as given in Art. 8.30*b*.

Problem Assume a church roof with the cross section shown in Fig. 8.19. The rigid frames are spaced 14 ft apart and carry a total load of 50 lb per square foot of roof area. Design a stone buttress for this rigid frame, using stone weighing 160 lb per cu ft, and buttresses 2 ft wide.

The horizontal thrust H of this type of rigid frame is

$$H = \frac{5wL^2}{32h}$$

where h is the rise of the truss. The total load wL is $50 \times 14 \times 24 = 16.8$ kips, and the vertical reaction is 8.4.

$$H = \frac{5 \times 16.8 \times 24}{32 \times 12} = 5.25$$

The top 3-ft section of the buttress weighs 2 ft 8 in. (avg ht.) $\times 3 \times 2 \times 160 =$ 2.56 kips

The total load at the bottom of the section is $8.4 + 2.56 = 10.96$, and the diversion of stress line in this height is

$$\frac{5.25}{10.96} \times 3 = 1.44$$

Center of gravity of vertical loads from inner face:

$$\begin{array}{lcl} 8.4 \times 0.33 & = & 2.8 \\ 2.56 \times 1.5 & = & 3.84 \\ \hline 10.96 & & 6.64 \end{array}$$

$$x = \frac{6.64}{10.96} = 0.605$$

$$\begin{array}{rl} \text{Diversion} & = 1.44 \\ \hline \text{Intersection of base of section} & = 2.04 \\ \text{Outer edge of middle third} & = 2.0 \end{array}$$

Therefore, the upper section is satisfactory if the pier is properly bonded to adjacent wall.

The middle 3-ft section weighs

$$\begin{array}{rl} 3 \times 4.5 \times 2 \times 160 = & 4.32 \\ \text{Load from above} = & 10.96 \\ \hline & 15.28 \end{array}$$

$$\text{Diversion} = \frac{5.25}{15.28} \times 3.0 = 1.03$$

$$\begin{array}{rl} \text{Point of application at top of section} = & 2.04 \\ \hline & 3.07 \end{array}$$

Outer edge of middle third = 3.0

Therefore, the middle section is satisfactory if the pier is properly bonded to adjacent wall.

The bottom 3-ft section weighs

$$3 \times 6 \times 2 \times 160 = 5.77$$
$$\text{Load from above} = \underline{15.28}$$
$$21.05$$

$$\text{Diversion} = \frac{5.25}{21.08} \times 3 = 0.75$$

$$\text{Point of application at top of section} = \underline{3.07}$$
$$3.82$$

Outer edge of middle third = 4.0

Therefore, the buttress is satisfactory throughout.

f Reinforced concrete building walls not subject to lateral stresses may also be classed as masonry walls. In the absence of other governing data, the following paragraphs taken from the 1963 ACI Code, Chapter 22, may be considered as good design.

2200 Notation

f_c = allowable compressive stress on concrete
f_c' = compressive strength of concrete (see Section 301)
h = vertical distance between supports
t = thickness of wall

2201 Structural Design of Walls

(a) Walls shall be designed for any lateral or other pressure to which they are subjected. Proper provisions shall be made for eccentric loads and wind stresses. Walls conforming to the provisions of Section 2202 shall be considered as meeting these requirements. The limits of thickness and quantity of reinforcement required by Section 2202 shall be waived where structural analysis shows adequate strength and stability.

2202 Empirical Design of Walls

(a) Reinforced concrete bearing walls carrying reasonably concentric loads may be designed by the empirical provisions of this section when they conform to all the limitations given herein.

(b) The allowable compressive stress using Part IV-A shall be

$$f_c = 0.225f_c'\left[1 - \left(\frac{h}{40t}\right)^3\right] \qquad (22\text{-}1)$$

When the reinforcement in bearing walls is designed, placed, and anchored in position as for tied columns, the allowable stresses shall be those for tied columns, in which the ratio of vertical reinforcement shall not exceed 0.04.

For design by Part IV-B the values from Eq. (22-1) shall be multiplied by 1.9.

(c) In the case of concentrated loads, the length of the wall to be considered as effective for each shall not exceed the center to center distance between loads, nor shall it exceed the width of the bearing plus four times the wall thickness.

(d) Reinforced concrete bearing walls shall have a thickness of at least 1/25 of the unsupported height or width, whichever is the shorter.

(e) Reinforced concrete bearing walls of buildings shall be not less than 6 in. thick for the uppermost 15 ft of their height; and for each successive 25 ft downward, or fraction thereof, the minimum thickness shall be increased 1 in. Reinforced concrete bearing walls of two-story dwellings may be 6 in. thick throughout their height.

(f) The area of the horizontal reinforcement of reinforced concrete walls shall be not less than 0.0025 and that of the vertical reinforcement not less than 0.0015 times the area of the reinforced section of the wall if of bars, and not less than three-fourths as much if of welded wire fabric. The wire of the welded fabric shall be of not less than No. 10 AS&W gage.

(g) Walls more than 10 in. thick, except for basement walls, shall have the reinforcement for each direction placed in two layers parallel with the faces of the wall. One layer consisting of not less than one-half and not more than two-thirds the total required shall be placed not less than 2 in. nor more than one-third the thickness of the wall from the exterior surface. The other layer, comprising the balance of the required reinforcement, shall be placed not less than ¾ in. and not more than one-third the thickness of the wall from the interior surface. Bars, if used, shall not be less than #3 bars, nor shall they be spaced more than 18 in. on centers. Welded wire reinforcement for walls shall be in flat sheet form.

(h) In addition to the minimum as prescribed in (f) there shall be not less than two #5 bars around all window or door openings. Such bars shall extend at least 24 in. beyond the corner of the openings.

(i) Reinforced concrete walls shall be anchored to the floors, or to the columns, pilasters, buttresses, and intersecting walls with reinforcement at least equivalent to #3 bars 12 in. on centers, for each layer of wall reinforcement.

(j) Panel and enclosure walls of reinforced concrete shall have a thickness of not less than 4 in. and not less than 1/30 the distance between the supporting or enclosing members.

(k) Exterior basement walls, foundation walls, fire walls, and party walls shall not be less than 8 in. thick.

(l) Where reinforced concrete bearing walls consist of studs or ribs tied together by reinforced concrete members at each floor level, the studs may be considered as columns, but the restrictions as to minimum diameter or thickness of columns shall not apply.

2203 WALLS AS GRADE BEAMS

(a) Walls designed as grade beams shall have top and bottom reinforcement as required by stresses. Portions exposed above grade shall, in addition, be reinforced with not less than the amount specified in Section 2202.

In accordance with the requirements stated above, the following reinforcement would be required in walls of various thickness.

For a 5-in. wall: one layer, No. 3, 9 in. o.c.

For a 6-in. wall: one layer, No. 3, 7½ in. o.c.

For an 8-in. wall: one layer, No. 4, 10 in. o.c.

For a 10-in. wall: two layers, each No. 3, 9 in. o.c.

For a 12-in. wall: two layers, each No. 3, 7½ in. o.c.

For a 14-in. wall: two layers, each No. 4, 11½ in. o.c.

For a 16-in. wall: two layers, each No. 4, 10 in. o.c.

For an 18-in. wall: two layers, each No. 4, 9 in. o.c.

g It will be noted that these requirements are specifically for "reinforced concrete walls," and must be followed when the authority under which the work is being done requires it. However, even walls which are normally classified as plain concrete walls should be provided with at least two No. 6 bars, spaced about 2 in. below the top edge and two No. 6 bars 2 in. above the footing in any wall to arrest shrinkage cracking or to assist in spanning over soft spots in the footing. The small additional cost is amply paid for by the greater freedom from cracks.

8.30 RETAINING WALLS

a The design of a retaining wall requires that it be so constructed that

1. The wall will not overturn.
2. The soil pressure under the toe will not exceed the allowable bearing value of the soil in question.
3. The wall will not move outward bodily due to thrust.
4. The wall or any of its component parts will not rupture.

b The forces acting on any wall are the horizontal thrust P_H derived from the various formulae in Art. 8.30*c* and the vertical load V of the wall and all other vertical loads on the wall. In order to avoid the danger of overturning, the line of the resultant of these two forces should pass within the middle third of the base of the wall. Referring to Art. 8.10*i*, it will be seen that if the center of pressure comes within the middle third, the pressure diagram is a trapezoid. If it comes right at the edge of the middle third, the eccentricity is one-sixth and the pressure diagram becomes a triangle, the pressure at the toe is double the normal pressure and at the heel is zero. If the center of pressure falls outside the middle third, that is, if the eccentricity exceeds one-sixth, there is an uplift at the inner edge, and the structure may or may not be stable. Therefore, good practice requires that the line of intersection shall lie within the middle third of the base.

The maximum soil pressure under the toe of the wall should not exceed the maximum soil pressure allowed by the code, or as noted in Art. 8.10*c*. This maximum soil pressure is determined by means of the formula in Art. 8.10*i*.

Resistance to sliding is furnished in two ways: by friction on the base and by direct compressive resistance of the earth in front of the toe of the wall. Friction may be taken to furnish a resistance of $0.5V$. In other words, if the total horizontal thrust is not over half the weight of the wall, the friction alone is sufficient to prevent sliding without the compressive resistance. The lateral compressive resistance at the base is resisted by a counter pressure, or "passive pressure," which is discussed in Art. 8.30*f* following.

c The fundamental law of fluid pressure states that the lateral pressure from a fluid against the wall of any container is a unit pressure equal to the weight of the fluid times the depth to the level under consideration. In other words, at depth x the unit pressure is

$$w_x = wx$$

Based on triangular loading, therefore, the loading diagram is as indicated in Fig. 8.20, and the total pressure on a strip one foot wide is

$$P_H = \frac{wh^2}{2}$$

Water, or any other liquid, is a true fluid and exerts full pressure on the wall:

$$w_x = 62.5x$$

and

$$P_H = 31.25h^2$$

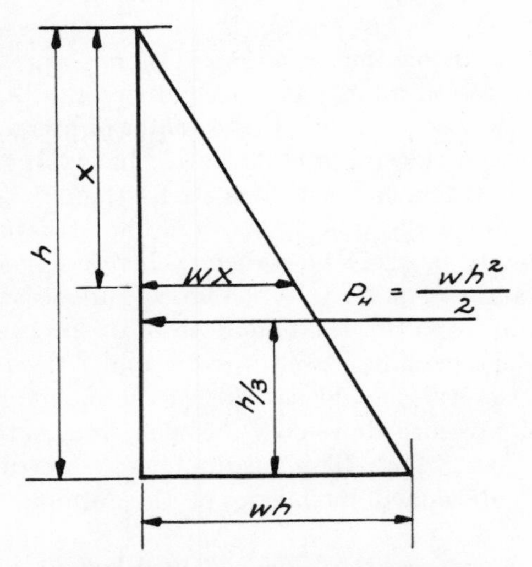

Fig. 8.20 Fluid Pressure Diagram

This law of fluid pressure is applicable also to the design of retaining walls carrying the lateral pressure from earth or other solids. The simplest method of retaining-wall computation is to convert the lateral pressure into an equivalent fluid pressure. For a level bank behind the wall, the equivalent fluid pressure may be found from the following formulae:

$$w_e = w \tan^2\left(45° - \frac{\phi}{2}\right)$$

and

$$P_H = \frac{w_e h^2}{2}$$

where w_e is the equivalent fluid pressure, w the weight of earth per cubic foot, and ϕ the angle of repose of the material. Table 8.6 gives the weight and equivalent pressures of the various materials ordinarily encountered. For ordinary clay or sand and gravel, w_e may be taken as 25 psf, or as many engineers use it, 30 psf.

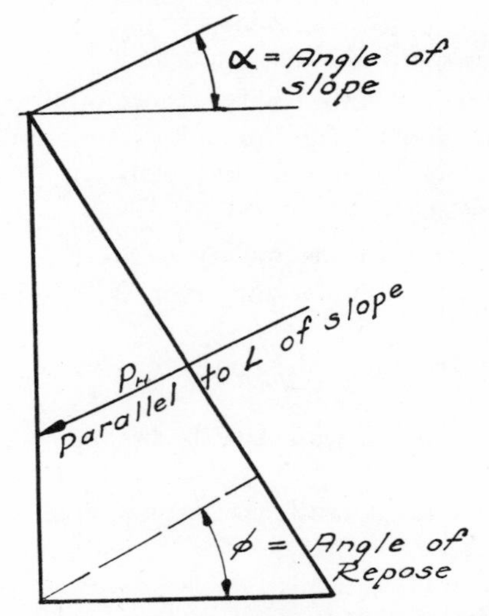

Fig. 8.21 Pressure Diagram for Sloping Surcharge

d If the bank is inclined above the top of the wall, as shown in Fig. 8.21, the equivalent fluid pressure becomes,

$$w_e = \frac{w \cos^2 \phi}{\left[1 + \sqrt{\dfrac{\sin \phi \sin (\phi - \alpha)}{\cos \alpha}}\right]^2}$$

When α, the angle of slope, is equal to ϕ, the angle of repose, which is the limiting slope,

$$w_e = w \cos^2 \phi$$

For ordinary earth or clay on natural slope, w_e becomes 64 lb, and substituting the value in the formulae of Art. 8.30*c*,

$$w_e = 64h$$
$$P_H = 32h^2$$

The pressure for other materials under this type of loading is given in Table 8.6 as "Equivalent slope pressure"

Table 8.6 Weight, Angle of Repose, and Equivalent Pressures of Various Materials

Material	Slope of repose	Angle of repose	Weight per cu ft., lb	Equivalent liquid pressure, psf	Equivalent slope pressure, psf	Passive pressure, psf
Dry ashes	1 on 1	45°	40	6.9	20.0	233
Cinders, bituminous dry	1 on 1	45°	45	7.7	22.5	262
Clay in lumps, dry	1 on $1\frac{1}{3}$	36°52′	63	15.8	40.4	252
Clay, damp, plastic	1 on 3	18°26′	110	57.1	99.0	210
Clay, sand, and gravel, dry	1 on $1\frac{1}{3}$	36°52′	100	25.0	64.0	400
Earth, dry, loose	1 on $1\frac{1}{3}$	36°52′	76	19.0	48.8	304
Earth, dry, packed	1 on $1\frac{1}{3}$	36°52′	95	23.8	61.0	380
Earth, moist, packed	1 on 1	45°	96	16.5	48.0	458
Earth, soft mud, packed	1 on 3	18°26′	115	59.7	103.5	220
Gravel, dry	1 on $1\frac{1}{3}$	36°52′	104	26.0	66.5	416
Limestone, crushed, dry	1 on 1	45°	85	14.6	42.5	494
Sand, dry	1 on $1\frac{1}{2}$	33°41′	106	30.4	73.5	370
Shale, crushed	1 on $1\frac{1}{3}$	36°52′	105	26.3	67.2	420
Slag, crushed	1 on $1\frac{1}{3}$	36°52′	72	18.0	46.1	288
Slag screenings	1 on $1\frac{1}{3}$	36°52′	117	29.3	75.0	468
Coal, bituminous	1 on $1\frac{1}{3}$	36°52′	50	12.5	32.0	200
Coal, anthracite	1 on 2	26°34′	60	23.0	48.0	157
Coke	1 on $1\frac{1}{3}$	36°52′	30	7.5	19.2	120
Iron ore	1 on $1\frac{1}{3}$	36°52′	300	75.0	192.0	1,200

e Figure 8.22 indicates the pressure diagram of a wall which retains earth pressure and is loaded with a surcharge such as would result from street traffic, railway traffic, etc. The pressures w_e and P_H in the triangles are obtained from the formulae of Art. 8.30*c*. The unit pressure on the rectangle w_{se} is obtained from the formula,

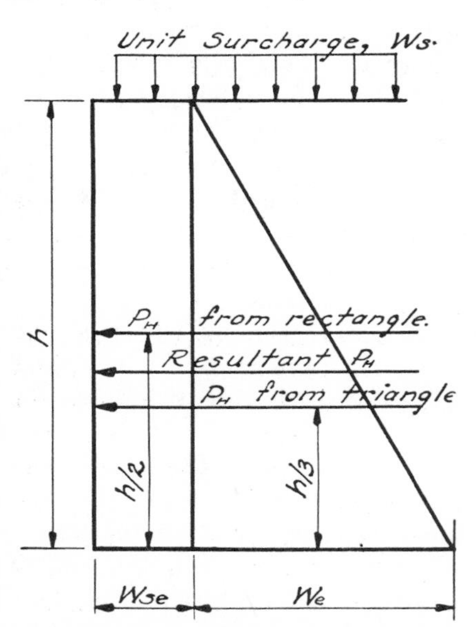

Fig. 8.22 Pressure Diagram for Ordinary Surcharge

$$w_{se} = w_s \tan^2\left(45° - \frac{\phi}{2}\right)$$

The rectangle of pressure causes a total thrust

$$P_H = w_{se}h$$

This pressure is applied at a height $h/2$. The total combined equivalent fluid horizontal pressure for ordinary earth pressure becomes

$$P_H = P_{H\triangle} + P_{H\square} = 12.5h(0.02w_s + h)$$

The point of application of the combined pressure P is

$$h_1 = \frac{h}{3}\left(\frac{w_s + 300h}{w_s + 200h}\right)$$

It may be better understood if we state that the weight of surcharge is converted into equivalent fluid earth pressure of a height equal to w_s/w and the area and center of gravity of the rectangular and triangular load diagrams are determined.

f In Art. 8.30*b* reference was made to the compressive resistance of the earth in front of the toe of the wall, or the passive pressure. The extent of the passive pressure at the maximum is

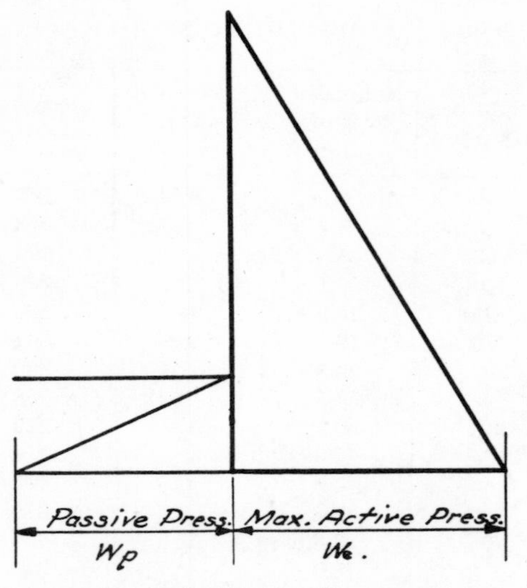

Fig. 8.23 Passive Pressure

$$w = w \tan^2 \left(45° + \frac{\phi}{2}\right)$$

Note the similarity to the formula for equivalent active pressure. The resistance due to passive pressure is used as indicated in Fig. 8.23.

g The simplest type of retaining wall is the plain concrete wall or masonry wall shown in Fig. 8.24. For the wall with no surcharge, the design consists of five steps:

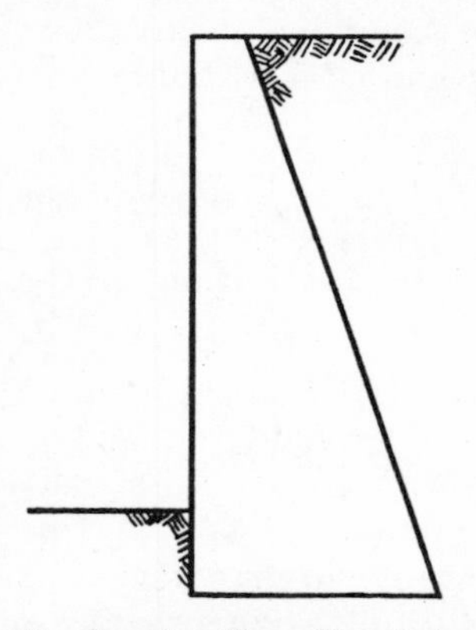

Fig. 8.24 Gravity-Type Retaining Wall

1. Compute the thrust P_H from the formulae in Art. 8.30*c* and its point of application.
2. Assume a wall on the basis that *b*, the base width, is 0.415*h* and the top width is *b*/6. Compute the weight and center of gravity of the wall.
3. Locate either graphically or analytically the point at which the resultant of the horizontal and vertical forces cuts the base. If it falls outside the middle third, vary the assumed width and revise until it comes within the middle third.
4. Check the assumed section for sliding due to thrust. In a footing of this type, it will ordinarily check satisfactorily. The frictional resistance on the base may be taken as 0.50, or in other words, if the thrust P_H is not greater than half the downward weight, the frictional resistance alone will resist sliding. In addition to this factor, the toe of the wall should be set below frost line, and when the earth has been tamped and settled against the face of the wall, it will resist a pressure equal to the passive pressure.
5. Compute the maximum soil pressure under the toe of the wall. If it exceeds the allowable soil pressure, revise the wall accordingly.

If the wall is surcharged, the width of the base will be about 0.52*h* instead of 0.415*h* as given above. If the vertical surface is on the face of the wall and the slope surface on the back, the downward load is made up of the trapezoid of masonry acting through its center of gravity and a triangle of earth acting through its center of gravity, and the resultant weight and center of gravity must be computed. This wall is not so efficient as a wall with a sloped face, and the width of base should be about 0.55*h*, or for sloping surcharge, *b* is 0.7*h*.

Tables 8.7 and 8.8 are taken from the publications of the Portland Cement Association and give recommended sizes for gravity-type walls of lower heights.

h For most retaining walls, the so-called reinforced concrete cantilever retaining wall is usually less costly than the mass-type wall described in Art. 8.30*g*. In general, there are three types in use. The most economical is an inverted T shape (Fig. 8.25) in which the footing projects beyond the face of the wall in both directions. If the wall retains earth pressure from a neighboring lot which is higher, so that all parts of the wall must be

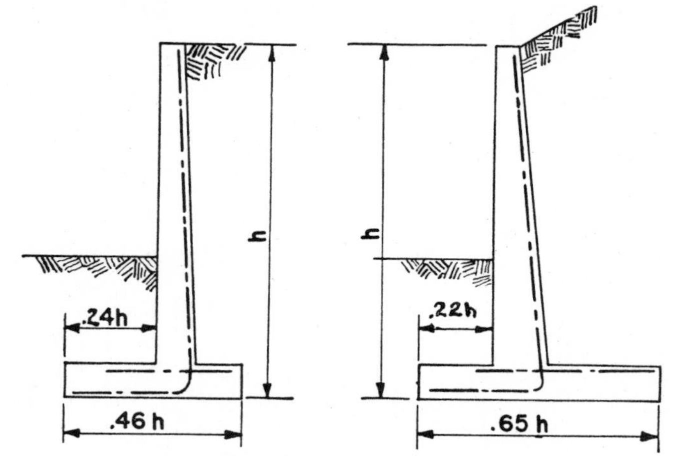

Fig. 8.25 T-Shaped Retaining Wall

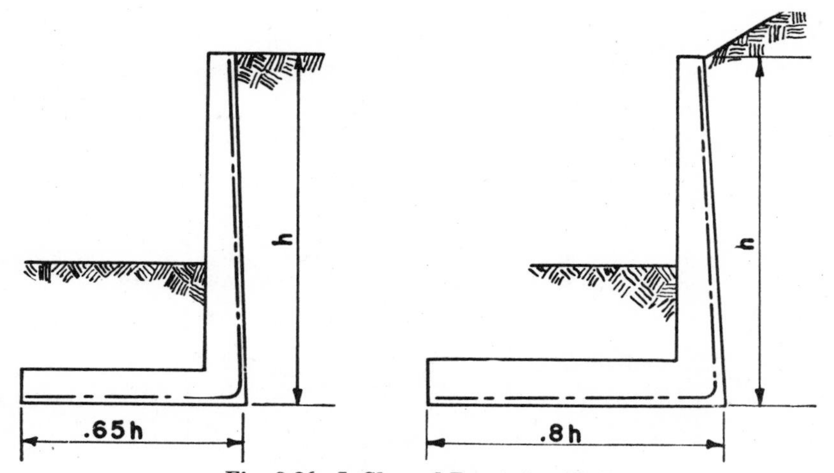

Fig. 8.26 L-Shaped Retaining Wall

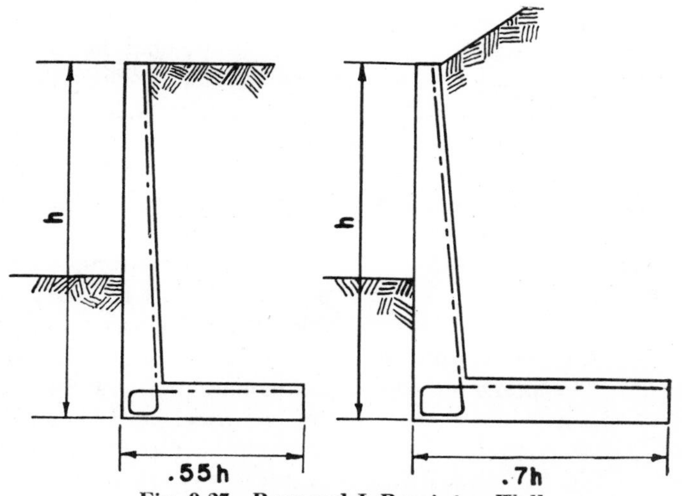

Fig. 8.27 Reversed-L Retaining Wall

Table 8.7 Vertical-faced Gravity Retaining Walls Without Surcharge

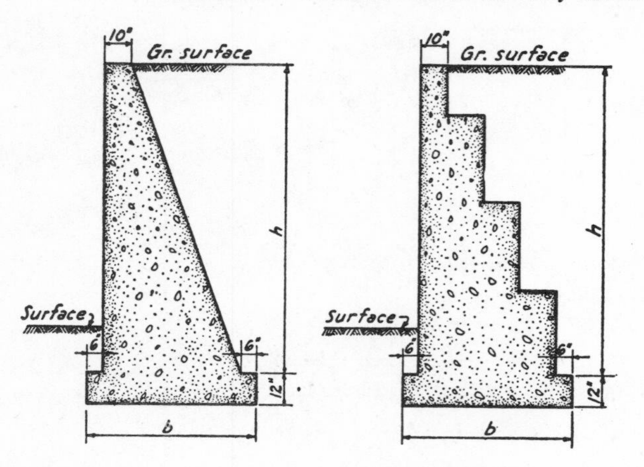

h, ft	b	Concrete, volume, cu yd per ft
3	2′3″	0.20
4	2′7″	0.27
5	3′0″	0.37
6	3′4″	0.48
7	3′9″	0.60
8	4′3″	0.76
9	4′9″	0.94
10	5′3″	1.14

Courtesy Portland Cement Association

Table 8.8 Vertical-faced Gravity Retaining Walls, Sloping Surcharge

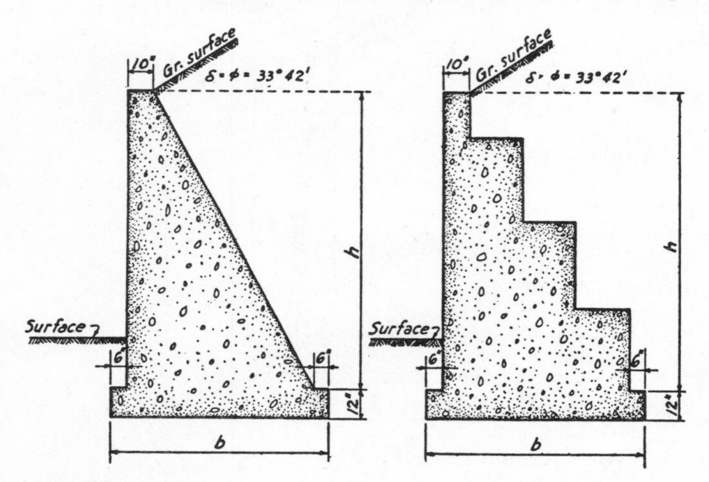

h, ft	b	Concrete, volume, cu yd per ft
3	2′6″	0.22
4	3′0″	0.32
5	3′6″	0.44
6	4′2″	0.60
7	4′10″	0.78
8	5′6″	0.99
9	6′3″	1.25
10	7′0″	1.52

Courtesy Portland Cement Association

within the lower lot, an L shape (Fig. 8.26) should be used. If the wall retains high ground, so that all parts of the wall must be on the higher lot, a reversed L (Fig. 8.27) should be used. The approximate dimensions of walls are given in the illustrations.

Tables 8.9 and 8.10 (from the Portland Cement Association) give details of design for a wall of the reversed L shape. Since the design of the wall itself has nothing to do with the design of the base, this wall design may be used for any of the three types of cantilever wall selected. The moment in the base of either the L-shaped or reversed L-shaped wall is the same as the moment in the wall. Therefore the base thickness and the steel are identical with that of the wall, except that in the L-shaped wall the base steel is in the bottom slab, and in the reversed L it is looped around and lies in the top of the slab.

The base of the T-shaped retaining wall is designed as a pair of cantilever beams loaded as indicated in Fig. 8.28. The toe of the base is loaded upward with the load on AB so that the tensile steel is in the bottom. The heel of the base is loaded downward with the weight of the earth and upward with the triangular pressure on CD, the net load being a downward load which produces tension in the top of the footing. Tables 8.9 and 8.10 are based on an f_s of 20,000. If f_s

Table 8.9 Cantilever Retaining Walls Without Surcharge

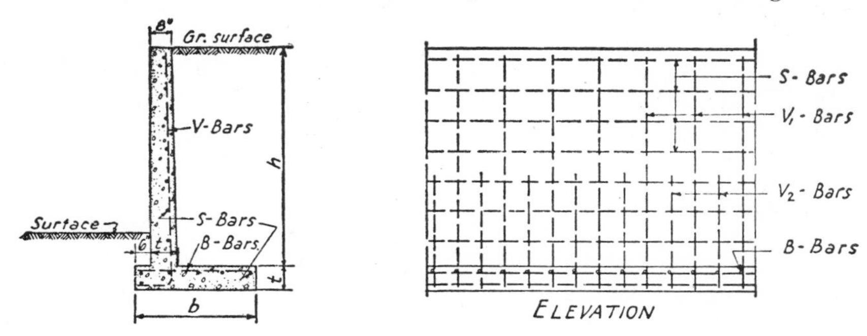

h, ft	b	t, in.	Concrete, cu yd per ft	V bars				B bars			S bars			Reinforcement, lb per ft
				Size	Spacing, in.	V_1 length	V_2 length	Size	Spacing, in.	Length	Number	Size	Spacing, in.	
5	2′9″	10	0.22	No. 2	12	6′6″		No. 2	12	2′4″	8	No. 3	12	4.5
6	3′4″	10	0.27	No. 2	7	7′6″		No. 2	7	3′0″	10	No. 3	12	6.8
7	3′10″	10	0.31	No. 3	9	8′4″		No. 3	9	3′6″	12	No. 3	12	10.4
8	4′6″	12	0.41	No. 4	12	9′8″	5′4″	No. 4	12	4′2″	13	No. 3	12	12.6
9	5′0″	12	0.46	No. 4	9	10′8″	5′4″	No. 4	9	4′8″	15	No. 3	12	16.9
10	5′6″	12	0.51	No. 5	11	11′8″	6′2″	No. 5	11	5′2″	16	No. 3	12	22.1

Courtesy Portland Cement Association

Table 8.10 Cantilever Retaining Walls, Sloping Surcharge

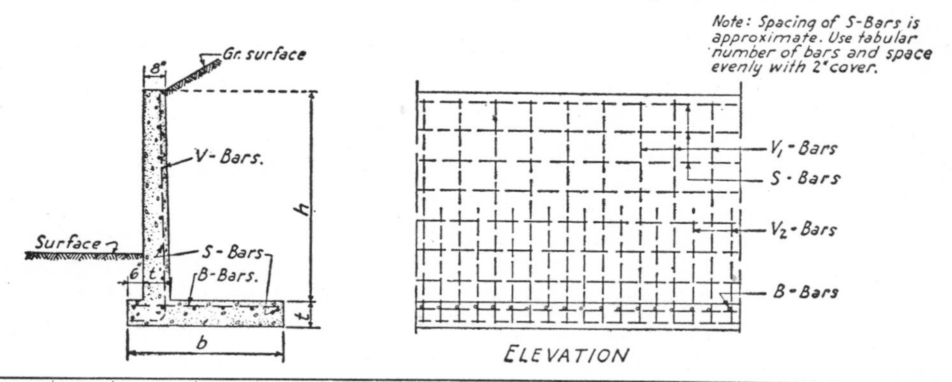

h, ft	b	t, in.	Concrete, cu yd per ft	V bars				B bars			S bars			Reinforcement, lb per ft
				Size	Spacing, in.	V_1 length	V_2 length	Size	Spacing, in.	Length	Number	Size	Spacing, in.	
5	3′6″	10	0.25	No. 3	10	6′4″		No. 3	10	3′2″	9	No. 3	12	7.7
6	4′3″	10	0.30	No. 4	10	7′4″		No. 4	10	3′10″	11	No. 3	12	13.1
7	5′0″	10	0.35	No. 4	7	8′4″		No. 4	7	4′8″	13	No. 3	12	19.8
8	6′0″	12	0.47	No. 5	9	9′8″	6′0″	No. 5	9	5′8″	15	No. 3	12	24.4
9	7′0″	13½	0.59	No. 5	7	11′0″	6′8″	No. 5	7	6′8″	17	No. 3	12	34.1
10	7′9″	15	0.71	No. 6	9	12′2″	6′10″	No. 6	9	7′4″	19	No. 3	12	40.9

Courtesy Portland Cement Association

is 24,000, the spacing given in the tables may be multiplied by 1.2.

8.31 Basement retaining walls

a Basement walls are usually called upon to resist earth pressure. Sometimes—for example, a store with the first floor level at grade—this earth pressure is on the full height of the wall, but usually—in residences, schools, and the like—it is on the lower por-

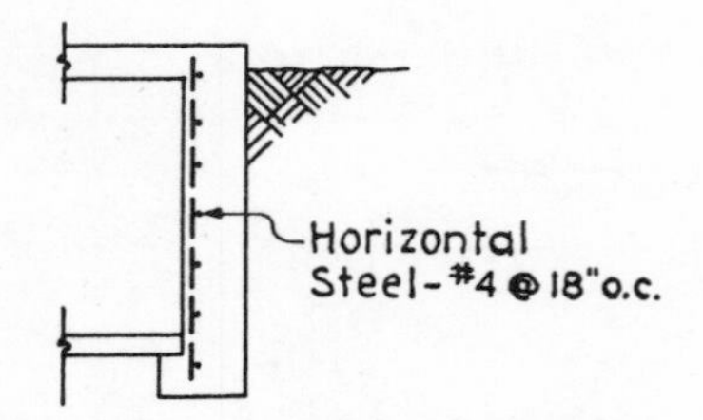

Fig. 8.29 Basement Retaining Wall

tion only. If the wall is against a driveway, or otherwise placed so that a superimposed load of trucks, railroad tracks, or the like may cause a surcharge, the walls must be designed to resist it.

The Code requirements for minimum design loads of the National Bureau of Standards contain this requirement:

4.1 *Pressure on Basement Walls.* In the design of basement walls and similar approximately vertical structures below grade, provision shall be made for the lateral pressure of adjacent soil. Due allowance shall be made for possible surcharge from fixed or moving loads. When a portion, or the whole, of the adjacent soil is below a free-water surface, computations shall be based on the weight of the soil diminished by buoyancy, plus full hydrostatic pressure.

b If the walls are properly supported at the basement and first-floor lines, they may be designed as a vertical slab. The moment resulting from earth thrust causes compression on the outer face of the wall and tension on the inner surface. The dead load of the structure on top of the wall causes compression on the entire wall. If the result of this combined bending and direct stress causes only compression, the wall is ordinarily strong enough. If the result is a net tension, the wall must be checked to determine whether reinforcement must be placed to resist this tension.

c Article 5.12*b* and Figs. 5.2 and 5.3 give a problem in the solution of a basement retaining wall of plain concrete. This problem does not allow for the effect of direct stress on the wall, and therefore has an additional factor of safety. Let us assume that we have a dead load of 1,000 lb per running foot of wall. This will distribute over the full wall thickness, and cause a uniform compression stress of 1,000/(12 × 12.5) = 6⅔ psi. The maximum tensile fiber stress on the wall is therefore 88 − 6⅔ = 81⅓ psi. If the wall above were a 12-in. brick wall, a 12½-in. thickness would be required for the foundation. However, if we wish to try a 12-in. thickness for the retaining wall,

$$f_m = \frac{46{,}600}{288} = 162$$

$$f_d = \frac{1{,}000}{144} = \underline{\quad 6.95 \quad}$$

$$\text{Net } f = 155.05, \text{ which is excessive for 12 in.}$$

Therefore a 12½-in. retaining wall is necessary.

d Earth pressure against the full height of the wall requires reinforcement as indicated in Table 8.11 for ordinary earth pressure. The values for curb or driveway wall provide for a 300-psf surcharge, the load normally to be expected from an H-20 loading on the adjacent pavement. No reduction is made for direct compression load on the wall, since the reduction is comparatively small, as was shown in Art. 8.31*c*.

Table 8.11 is based on the use of a steel fiber stress of 20,000 psi. If f_s = 24,000 psi, the spacing in the table may be multiplied by 1.2.

e The problem is more complex if the wall is concrete through part of the height, with brick above grade, as in Fig. 8.30. This wall is laid out as a beam in Fig. 8.31. The steps in the computation of this beam are as follows.

1. Compute W, total earth pressure = $15A^2$, where A is in feet.

2. Compute reactions R_A and R_B, remembering that the center of the load is $\frac{1}{3}A$ from R_A.

3. Compute x, the point of maximum moment,

$$x = \sqrt{\frac{R_B}{15}}$$

4. Design $M_{\max}$,

$$M_{\max} = R_B\left(L + \frac{2x}{3} - A\right)$$

Reduce $M_{\max}$ to inch-pounds.

Table 8.11 Basement Retaining Walls

Floor-to-floor distance, ft	Ordinary earth pressure			Curb or driveway walls		
	12-in. wall	16-in. wall	20-in. wall	12-in. wall	16-in. wall	20-in. wall
11				No. 4 at 8 in.		
12				No. 4 at 7 in.		
13	No. 4 at 8 in.			No. 5 at $8\frac{3}{4}$ in.	No. 5 at 9 in.	
14	No. 4 at $6\frac{1}{2}$ in.			No. 5 at 7 in.	No. 5 at 9 in.	
15	No. 5 at 8 in.	No. 5 at 9 in.		No. 6 at $8\frac{1}{2}$ in.	No. 5 at 8 in.	
16	No. 6 at $9\frac{1}{2}$ in.	No. 5 at 9 in.		No. 6 at 7 in.	No. 5 at 7 in.	No. 5 at 7 in.
17	No. 6 at $8\frac{1}{2}$ in.	No. 5 at $8\frac{1}{2}$ in.		No. 7 at $8\frac{1}{2}$ in.	No. 6 at 9 in.	No. 5 at 7 in.
18	No. 7 at 9 in.	No. 6 at $9\frac{1}{2}$ in.	No. 5 at 7 in.	No. 7 at 7 in.	No. 6 at 7 in.	No. 5 at $6\frac{1}{2}$ in.
19	No. 7 at $7\frac{1}{2}$ in.	No. 6 at 8 in.	No. 5 at 7 in.	No. 7 at 6 in.	No. 7 at $8\frac{1}{4}$ in.	No. 6 at $7\frac{3}{4}$ in.
20	No. 7 at $6\frac{1}{2}$ in.	No. 7 at $9\frac{1}{2}$ in.	No. 6 at 9 in.	No. 7 at $5\frac{1}{4}$ in.	No. 7 at $7\frac{1}{4}$ in.	No. 7 at $9\frac{1}{2}$ in.
Spacer bars	No. 5 at 12 in.	No. 6 at 12 in.	No. 7 at 13 in.	No. 5 at 12 in.	No. 6 at 12 in.	No. 7 at 13 in.

Curb or driveway walls are designed for surcharge of 300 psf.
(Walls up to 10 ft high do not need reinforcement.)

5. Design for M at top of concrete, $M_B = R_B B$, and reduce to inch-pounds.
6. Compute f_d, the total direct load carried at top of concrete, and reduce to pounds per square inch.
7. Compute the section modulus of the brick wall and the concrete wall for a strip 12 in. wide, using the relation $S = 2d^2$, where d is the thickness of the wall in inches.
8. Compute f_a and f_b, the bending stress at points a and b,

$$f_a = \frac{M_{\max}}{S_A}$$

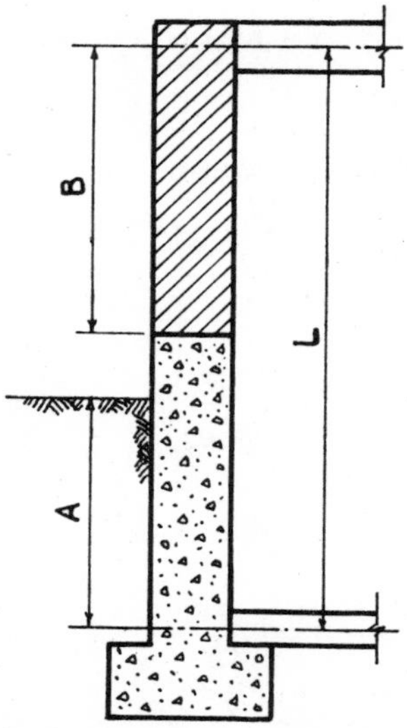

Fig. 8.30 Basement-Wall Design

$$f_b = \frac{M_B}{S_B}$$

The net tension in the concrete is $f_a - f_d$, and this is limited to 88 psi in 3,000-psi concrete (see Art. 5.12). The net tension at the top of the concrete is $f_b - f_d$, and this is limited to 10 psi (see Art. 10.33*b*).

8.32 Tunnel design

a The building designer is often called upon to design an underground passageway or tunnel between several buildings, either as a pedestrian walkway or as a pipe tunnel. This article is not intended to cover subways

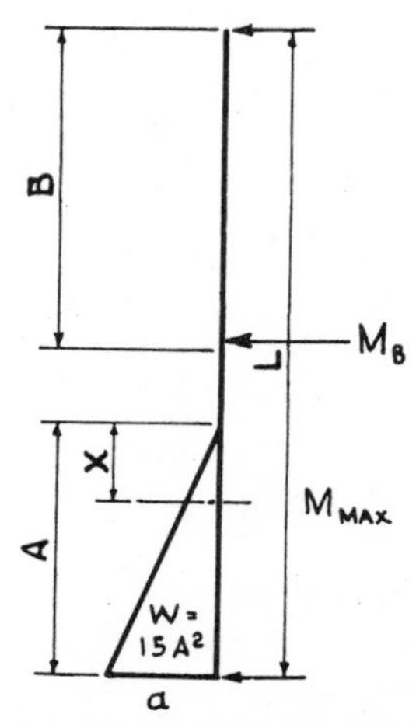

Fig. 8.31 Load Diagram—Basement Wall

or underwater tunnels, which present major problems of specialized design.

Since the ordinary tunnel of the type described here is a rectangular reinforced concrete box section, for simplicity of forming, it may be designed as a continuous beam by the method of Art. 3.21, the method of moment distribution. A tunnel of this type is shown in Fig. 8.32.

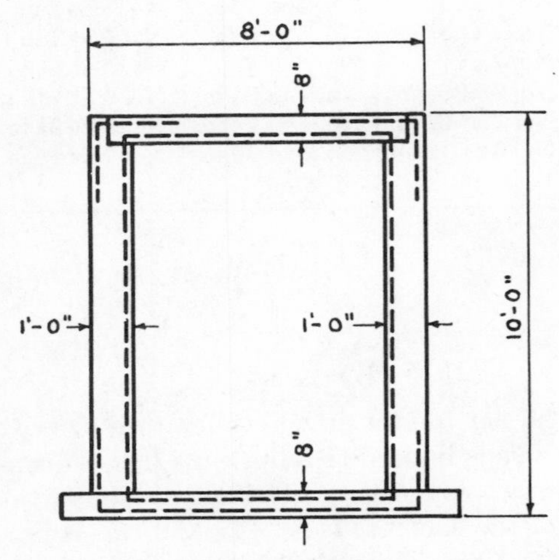

Fig. 8.32 Tunnel Section

b If we assume a tunnel of the type mentioned above, and with the dimensions shown in Fig. 8.32, overlaid with a 2-ft-thick earth fill, the applied loads and center lines will be as shown in Fig. 8.33. The tunnel shown could probably be built with an 8-in. wall instead of a 12-in. wall, but it is difficult to set two layers of reinforcement and pour an 8-in. wall of this height satisfactorily. Almost invariably a honeycombed concrete will result, and possibly a leaky wall.

In the development of Fig. 8.33, the center-line distances are used. The load on the roof of the tunnel consists of

2 ft of earth at 100	=	200 psf
⅔ ft of concrete at 150	=	100 psf
Total	=	300 psf

The load upward on the floor of the tunnel is

2 ft of earth × 8 ft wide	=	1,600 lb
2 side walls, at 10 cu ft concrete	=	3,000 lb
Roof and floor, 12 sq ft concrete	=	1,200 lb
Total	=	5,800 lb

The equivalent fluid pressure of the earth against the side walls at 30 lb per ft depth causes a trapezoidal load made up of the triangle and rectangle shown on each side of Fig. 8.33.

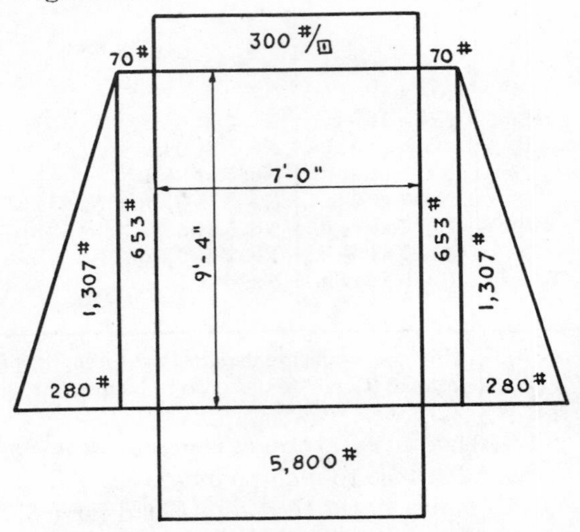

Fig. 8.33 Load Diagram—Tunnel Section

The stiffness factor $K = I/L$ gives the following for the roof and floor and the side walls. The moment of inertia varies as the cube of the thickness.

$$K_{\text{roof}} = K_{\text{floor}} = \frac{8^3}{7} = \frac{512}{7} = 73$$

$$K_{\text{side wall}} = \frac{12^3}{9.33} = \frac{1{,}728}{9.33} = \underline{185}$$

$$\text{Total} = 258$$

The distribution would be, for floor and roof,

$$\frac{73}{258} = 0.28$$

For side walls, it would be

$$\frac{185}{258} = 0.72$$

The fixed-end moments will be as follows:

$$\text{Roof fem} = \frac{(0.300 \times 7) \times 7 \times 12}{12}$$
$$= 14.7 \text{ in.-kips}$$

$$\text{Floor fem} = \frac{5.8 \times 7 \times 12}{12}$$
$$= 40.6 \text{ in.-kips}$$

$$\text{Side walls, top, fem} = \frac{0.653 \times 9.33 \times 12}{12} = 6.1$$
$$+ \frac{1.3 \times 9.33 \times 12}{15} = \underline{9.8}$$
$$\text{Total} = 15.9 \text{ in.-kips}$$

Side walls, bottom, fem $= \dfrac{0.653 \times 9.33 \times 12}{12} = 6.1$

$$+\frac{1.3 \times 9.33 \times 12}{10} = \underline{14.6}$$

$$\text{Total} = 20.7 \text{ in.-kips}$$

Since the design is symmetrical, it may be cut down the middle vertically, and a carry-over factor of opposite sign used in balancing the moment, as in the following problem.

Top	Right		Bottom
0.28	0.72	0.72	0.28
−14.7	+15.9	−20.7	+40.6
+1.8	−7.2	−14.4	−5.5
	+4.2		
		+2.1	+2.8
−0.9	−1.8	−3.5	−1.4
+0.7	+2.0		
		+1.0	+0.7
−0.4	−0.6	−1.2	−0.5
+0.3	+0.7		
		+0.4	+0.3
−0.2	−0.3	−0.5	−0.2
+0.2	+0.3		
		+0.1	+0.1
−13.2 in.-kips		−0.1	−0.1
		−36.8 in.-kips	

The positive moment is the simple span moment minus the final negative end moment. The simple-span-moment coefficient for the floor and roof slabs, ⅛, is 1.5 times the original fem, 1⁄12, or $+M = 1.5 \times 14.7 = 22.1$. The positive moment for the side walls is a combination of Cases 1 and 2 of Fig. 3.13. For the uniform load the coefficient is 0.125, and for the triangular load it is 0.128; but since they do not occur at the same point, it is a safe approximation to use ⅛ here also. Likewise, the average original fem (1⁄15 at the top and 1⁄10 at the bottom) is 1⁄12; so for the side walls we may approximate the positive moment at

$$\frac{15.9 + 20.7}{2} \times 1.5 = 27.5 \text{ in.-kips}$$

From the resultant figures, it is evident that if the equivalent fluid pressure is less than the assumed pressure, the resultant positive moment and fixed-end moments will both decrease, probably to the extent that the net positive moment in the side wall will become a negative moment. The area of steel required for computed moments is:

Roof: $+A_s = \dfrac{8.9}{24 \times 0.875 \times 6} = 0.07$ sq in./ft

At upper corner: $-A_s = \dfrac{13.2}{24 \times 0.875 \times 6} = 0.105$ sq in./ft

Side wall: $+A_s = \dfrac{3.2}{24 \times 0.875 \times 10} = 0.015$ sq in./ft

At lower corner: $-A_s = \dfrac{36.8}{24 \times 0.875 \times 6} = 0.29$ sq in./ft

Floor: $+A_s = \dfrac{24.1}{24 \times 0.875 \times 6} = 0.19$ sq in./ft

In accordance with the minimum steel requirements for reinforced concrete walls as given in the ACI Code and quoted in Art. 8.20 the minimum steel in the 8-in. floor and roof is 0.264 sq in. and in a 12-in. wall is 0.38 sq in., or 0.19 in. each face. To comply with minimum steel requirements and moment requirements, we will use:

In the roof: No. 3 rods 10 in. o.c. straight in the bottom of the slab and No. 3 rods 10 in. o.c. lapped to the corner bars in the top of the slab.

In the side wall: No. 4 rods 12½ in. o.c. in both faces, the outer bars bent around the corner to take the negative moment.

In the floor: No. 4 rods 10-in. o.c. in the top of the slab, with corner bars No. 4 at 12½ in. o.c. in the outer face to develop the negative moment.

Top Positive M	Right Approximate Positive M	Bottom Approximate Positive M
+22.1	+27.5	+60.9
−13.2	$-\left(\dfrac{36.8 + 13.2}{2}\right) = -25.0$	−36.8
+ 8.9 in.-kips	+2.5 in.-kips	+24.1 in.-kips

9

Connections

9.10 RIVETING AND BOLTING

a Until comparatively recently, the only generally accepted means of attaching metal to metal was by riveting or field bolting. The design was based on the theory that the stress was carried by bearing on the side of the connector and shear on the connectors. Recent experiments have shown that a main purpose served by the rivets or bolts is to clamp the members together, and that the greater portion of the stress in a properly riveted joint is transmitted through friction. As a result of these findings, special high-strength bolts have been developed which may be tightened to a predetermined torque or by another approved method.

As a result of this switch to high-tensile-strength bolts and to welding, riveting is rapidly being displaced as the method of attaching steel to steel. The general principles of design are similar for either rivets or bolts—either ordinary field bolts or high-strength bolts. The difference, so far as design is concerned, is in the allowable stresses used. The tables of polar section modulus and rivet groups under eccentric loading conditions apply with equal force to high-tensile bolts or to rivets.

Rivets and bolts are used in holes 1/16 in. greater in diameter than the nominal rivet or bolt diameter, and in computing the plate areas, an additional 1/16 in. is deducted to obtain the net area. Thus a 3/4-in. rivet is driven in a 13/16-in. hole and the width deducted from the plate is 7/8 in. For compression members the area of the rivet hole is not deducted; in other words, gross area is used in compression members.

The design of riveted or bolted joints is governed by the following requirements from the AISC specification:

1.5.2 *Rivets and Bolts*

1.5.2.1 Allowable unit tension and shear stresses on rivets, bolts and threaded parts (pounds per square inch of area of rivets before driving or unthreaded body area of bolts and threaded parts) shall be as given in Table 1.5.2.1.

1.5.2.2 Allowable bearing stress on projected area of bolts in bearing-type connections and on rivets.

$$F_p = 1.35F_y$$

where F_y is the yield point of the connected part. (Bearing stress not restricted in friction-type connections assembled with A325 and A354, Grade BC, bolts.)

SECTION 1.15 CONNECTIONS

1.15.1 *Minimum Connections*

Connections carrying calculated stresses except for lacing, sag bars, and girts, shall be designed to support not less than 6,000 pounds.

1.15.2 *Eccentric Connections*

Axially stressed members meeting at a point shall have their gravity axes intersect at a point if practicable; if not, provision shall be made for bending stresses due to the eccentricity.

Table 1.5.2.1

Description of Fastener	Tension (F_t)	Shear (F_v)	
		Friction-Type Connections	Bearing-Type Connections
A141 hot-driven rivets	20,000		15,000
A195 and A406 hot-driven rivets	27,000		20,000
A307 bolts and threaded parts of A7 and A373 steel	14,000		10,000
Threaded parts of other steels	$0.40F_y$		$0.30F_y$
A325 bolts when threading is *not* excluded from shear planes	40,000	15,000	15,000
A325 bolts when threading is excluded from shear planes	40,000	15,000	22,000
A354, Grade BC, bolts when threading is *not* excluded from shear planes	50,000	20,000	20,000
A354, Grade BC, when threading is excluded from shear planes	50,000	20,000	24,000

1.15.3 *Placement of Rivets, Bolts and Welds*

Except as hereinafter provided, the rivets, bolts or welds at the ends of any member transmitting axial stress into that member shall have their centers of gravity on the gravity axis of the member unless provision is made for the effect of the resulting eccentricity. Except in members subject to repeated variation in stress, as defined in Sect. 1.7, disposition of fillet welds to balance the forces about the neutral axis or axes for end connections of single angle, double angle, and similar type members is not required. Eccentricity between the gravity axes of such members and the gage lines for their riveted or bolted end connections may be neglected.

1.15.4 *Unrestrained Members*

Except as otherwise indicated by the designer, connections of beams, girders or trusses shall be designed as flexible, and may ordinarily be proportioned for the reaction shears only.

Flexible beam connections shall permit the ends of the beam to rotate sufficiently to accommodate its deflection by providing for a horizontal displacement of the top flange determined as follows:

$e = 0.007d$, when the beam is designed for full uniform load and for live load deflection not exceeding $\frac{1}{360}$ of the span

$= \dfrac{f_b L}{3{,}600{,}000}$, when the beam is designed for full uniform load producing the unit stress f_b at mid-span

where e = the horizontal displacement of the end of the top flange, in the direction of the span, in inches
f_b = the flexural unit stress in the beam at mid-span, in pounds per square inch
d = the depth of the beam, in inches
L = the span of the beam, in feet

1.15.5 *Restrained Members*

Fasteners or welds for end connections of beams, girders and trusses not conforming to the requirements of Sect. 1.15.4 shall be designed for the combined effect of end reaction shear and tensile or compressive stresses resulting from moment induced by the rigidity of the connection when the member is fully loaded.

1.15.6 *Fillers*

When rivets or bolts carrying computed stress pass through fillers thicker than $\frac{1}{4}$ inch, except in friction-type connections assembled with high strength bolts, the fillers shall be extended beyond the splice material and the filler extension shall be secured by enough rivets or bolts to distribute the total stress in the member uniformly over the combined section of the member and the filler, or an equivalent number of fasteners shall be included in the connection.

In welded construction, any filler ¼ inch or more in thickness shall extend beyond the edges of the splice plate and shall be welded to the part on which it is fitted with sufficient weld to transmit the splice plate stress, applied at the surface of the filler as an eccentric load. The welds joining the splice plate to the filler shall be sufficient to transmit the splice plate stress and shall be long enough to avoid overstressing the filler along the toe of the weld. Any filler less than ¼ inch thick shall have its edges made flush with the edges of the splice plate and the weld size shall be the sum of the size necessary to carry the splice plate stress plus the thickness of the filler plate.

1.15.7 *Connections of Tension and Compression Members in Trusses*

The connections at ends of tension or compression members in trusses shall develop the strength required by the stress, but not less than 50 percent of the effective strength of the member. Groove welds in connections at the ends of tension or compression members in trusses shall be complete penetration groove welds.

1.15.8 *Compression Members with Bearing Joints*

Where compression members bear on bearing plates, and where tier-building columns are finished to bear, there shall be sufficient rivets, bolts or welds to hold all parts securely in place.

Where other compression members are finished to bear, the splice material and its riveting, bolting or welding shall be arranged to hold all parts in line and shall be proportioned for 50 percent of the computed stress.

All of the foregoing joints shall be proportioned to resist any tension that would be developed by specified lateral forces acting in conjunction with 75 per cent of the calculated dead load stress and no live load.

1.15.10 *Rivets and Bolts in Combination with Welds*

In new work, rivets, A307 bolts, or high strength bolts used in bearing-type connections, shall not be considered as sharing the stress in combination with welds. Welds, if used, shall be provided to carry the entire stress in the connection. High strength bolts installed in accordance with the provisions of Sect. 1.16.1 as a friction-type connection prior to welding may be considered as sharing the stress with the welds.

In making welded alterations to structures, existing rivets and properly tightened high strength bolts may be utilized for carrying stresses resulting from existing dead loads, and the welding need be adequate only to carry all additional stress.

1.15.11 *High Strength Bolts (in Friction-Type Joints) in Combination with Rivets*

In new work and in making alterations, rivets and high strength bolts, installed in accordance with the provisions of Sect. 1.16.1 as friction-type connections, may be considered as sharing the stresses resulting from dead and live loads.

1.15.12 *Field Connections*

Rivets, high strength bolts or welds shall be used for the following connections:

- Column splices in all tier structures 200 feet or more in height.
- Column splices in tier structures 100 to 200 feet in height, if the least horizontal dimension is less than 40 percent of the height.
- Column splices in tier structures less than 100 feet in height, if the least horizontal dimension is less than 25 percent of the height.
- Connections of all beams and girders to columns and of any other beams and girders on which the bracing of columns is dependent, in structures over 125 feet in height.
- Roof-truss splices and connections of trusses to columns, column splices, column bracing, knee braces and crane supports, in all structures carrying cranes of over 5-ton capacity.
- Connections for supports of running machinery, or of other live loads which produce impact or reversal of stress.
- Any other connections stipulated on the design plans.

In all other cases field connections may be made with A307 bolts.

For the purpose of this Section, the height of a tier structure shall be taken as the vertical distance from the curb level to the highest point of the roof beams, in the case of flat roofs, or to the mean height of the gable, in the case of roofs having a rise of more than 2⅔ in 12. Where the curb level has not been established, or where the structure does not adjoin a street, the mean level of the adjoining land shall be used instead of curb level. Penthouses may be excluded in computing the height of structure.

SECTION 1.16 RIVETS AND BOLTS

1.16.1 *High Strength Bolts*

Use of high strength bolts shall conform to the provisions of the *Specifications for Structural Joints Using ASTM A325 Bolts* as approved by the Research Council on Riveted and Bolted Structural Joints, except that A354, Grade BC, bolts meeting the dimensional requirements of the Council's specification and tightened to their proof load, may be substituted for A325 bolts at the working stresses permitted in Sect. 1.5 and 1.6.

1.16.2 *Effective Bearing Area*

The effective bearing area of rivets and bolts shall be the diameter multiplied by the length in bearing, except that for countersunk rivets and bolts half the depth of the countersink shall be deducted.

1.16.3 *Long Grips*

Rivets and A307 bolts which carry calculated stress, and the grip of which exceeds 5 diameters, shall have their number increased 1 percent for each additional 1/16 inch in the grip.

1.16.4 *Minimum Pitch*

The minimum distance between centers of rivet and bolt holes shall be not less than 2⅔ times the nominal diameter of the rivet or bolt but preferably not less than 3 diameters.

1.16.5 *Minimum Edge Distance*

The minimum distance from the center of a rivet or bolt hole to any edge, used in design or in preparation of shop drawings, shall be that given in Table 1.16.5.

1.16.6 *Minimum Edge Distance in Line of Stress*

In bearing-type connections of tension members, where there are not more than two fasteners in a line parallel to the direction of stress, the distance from the center of the end fastener and that end of the connected part toward which the stress is directed shall be not less than

(a) for riveted connections: the area of the fastener divided by the thickness of the connected part for fasteners in single shear, and twice this distance for fasteners in double shear.
(b) for high strength bolted connections: 1½ times the distances given in (a).

The end distance may, however, be decreased in such proportion as the fastener stress is less than that permitted under Sect. 1.5.2, but it shall not be less than the distance specified in Sect. 1.16.5 above.

When more than two fasteners are provided in the line of stress the provisions of Sect. 1.16.5 shall govern.

1.16.7 *Maximum Edge Distance*

The maximum distance from the center of any rivet or bolt to the nearest edge of parts in contact with one another shall be 12 times the thickness of the plate, but shall not exceed 6 inches.

On the basis of the above specifications, Table 9.1 gives the value in tension of rivets and bolts for use in such connections as hangers as shown in Fig. 4.6 or in the tension side of a wind-bracing connection or other moment connection as used in Type I, Rigid Frame Construction.

Table 1.16.5

Rivet or Bolt Diameter (Inches)	Minimum Edge Distance for Punched, Reamed or Drilled Holes (Inches)	
	At Sheared Edges	At Rolled Edges of Plates, Shapes or Bars or Gas Cut Edges**
½	⅞	¾
⅝	1⅛	⅞
¾	1¼	1
⅞	1½*	1⅛
1	1¾*	1¼
1⅛	2	1½
1¼	2¼	1⅝
Over 1¼	1¾ × Diameter	1¼ × Diameter

* These may be 1¼ in. at the ends of beam connection angles.

** All edge distances in this column may be reduced ⅛ in. when the hole is at a point where stress does not exceed 25% of the maximum allowed stress in the element.

Table 9.1 Rivets and Threaded Fasteners in Tension

Allowable loads in kips

ASTM Designation	Tension F_t ksi	Nominal Diam., In., and Area, Sq. In.					
		5/8	3/4	7/8	1	1 1/8	1 1/4
		.3068	.4418	.6013	.7854	.9940	1.2272
A307,* A7,* A373	14	4.30	6.19	8.42	11.00	13.92	17.18
A141	20	6.14	8.84	12.03	15.71	19.88	24.54
A195, A406	27	8.28	11.93	16.24	21.21	26.84	33.13
A325	40	12.27	17.67	24.05	31.42	39.76	49.09
A354 Gr. BC	50	15.34	22.09	30.07	39.27	49.70	61.36

Note 1: Tension values are based on areas of rivets before driving and areas of unthreaded shanks of bolts and other threaded parts. See AISC Spec. Table 1.5.2.1.

Note 2: For allowable combined shear and tension loads, see AISC Spec. Sect. 1.6.3.

* Nuts shall meet specifications compatible with those of the threaded shanks.

AMERICAN INSTITUTE OF STEEL CONSTRUCTION

Table 9.2 gives the allowable load in shear for the various rivets, unfinished bolts and high-strength bolts most commonly used, and Table 9.3 the allowable load in bearing for all rivets and bolts in bearing-type connections. It will be noted in this table that the variation is based on the type of steel in the members connected and is not dependent on the connector except as to diameter of connector. In the former AISC Code,

Table 9.2 Rivets and Threaded Fasteners in Shear

Allowable loads in kips

Power Driven Shop and Field Rivets

Diam. — Area	5/8 in. — .3068 sq. in.		3/4 in. — .4418 sq. in.		7/8 in. — .6013 sq. in.	
ASTM Designation	A141	A195 A406	A141	A195 A406	A141	A195 A406
Shear F_v, ksi	15	20	15	20	15	20
Single Shear, kips Double Shear, kips	4.60 9.20	6.14 12.27	6.63 13.25	8.84 17.67	9.02 18.04	12.03 24.05

Unfinished Bolts, ASTM A307, and Threaded Parts of ASTM A7 and A373 Material

Diam. — Area	5/8 in. — .3068 sq. in.	3/4 in. — .4418 sq. in.	7/8 in. — .6013 sq. in.
ASTM Designation	A307, A7, A373	A307, A7, A373	A307, A7, A373
Shear F_v, ksi	10	10	10
Single Shear, kips Double Shear, kips	3.07 6.14	4.42 8.84	6.01 12.03

High Strength Bolts in Friction Type Connections and in Bearing Type Connections with Threads in Shear Planes

Diam. — Area	5/8 in. — .3068 sq. in.		3/4 in. — .4418 sq. in.		7/8 in. — .6013 sq. in.	
ASTM Designation	A325	A354 Gr. BC	A325	A354 Gr. BC	A325	A354 Gr. BC
Shear F_v, ksi	15	20	15	20	15	20
Single Shear, kips Double Shear, kips	4.60 9.20	6.14 12.27	6.63 13.25	8.84 17.67	9.02 18.04	12.03 24.05

High Strength Bolts in Bearing Type Connections with Threads Excluded from Shear Planes

Diam. — Area	5/8 in. — .3068 sq. in.		3/4 in. — .4418 sq. in.		7/8 in. — .6013 sq. in.	
ASTM Designation	A325	A354 Gr. BC	A325	A354 Gr. BC	A325	A354 Gr. BC
Shear F_v, ksi	22	24	22	24	22	24
Single Shear, kips Double Shear, kips	6.75 13.50	7.36 14.73	9.72 19.44	10.60 21.21	13.23 26.46	14.43 28.86

AMERICAN INSTITUTE OF STEEL CONSTRUCTION

Table 9.2 Rivets and Threaded Fasteners in Shear (*Continued*)

Allowable loads in kips

Power Driven Shop and Field Rivets						
Diam. — Area	1 in. — .7854 sq. in.		1⅛ in. — .9940 sq. in.		1¼ in. — 1.2272 sq. in.	
ASTM Designation	A141	A195 A406	A141	A195 A406	A141	A195 A406
Shear F_v, ksi	15	20	15	20	15	20
Single Shear, kips	11.78	15.71	14.91	19.88	18.41	24.54
Double Shear, kips	23.56	31.42	29.82	39.76	36.82	49.09

Unfinished Bolts, ASTM A307, and Threaded Parts of ASTM A7 and A373 Material			
Diam. — Area	1 in. — .7854 sq. in.	1⅛ in. — .9940 sq. in.	1¼ in. — 1.2272 sq. in.
ASTM Designation	A307, A7, A373	A307, A7, A373	A307, A7, A373
Shear F_v, ksi	10	10	10
Single Shear, kips	7.85	9.94	12.27
Double Shear, kips	15.71	19.88	24.54

High Strength Bolts in Friction Type Connections and in Bearing Type Connections with Threads in Shear Planes						
Diam. — Area	1 in. — .7854 sq. in.		1⅛ in. — .9940 sq. in.		1¼ in. — 1.2272 sq. in.	
ASTM Designation	A325	A354 Gr. BC	A325	A354 Gr. BC	A325	A354 Gr. BC
Shear F_v, ksi	15	20	15	20	15	20
Single Shear, kips	11.78	15.71	14.91	19.88	18.41	24.54
Double Shear, kips	23.56	31.42	29.82	39.76	36.82	49.09

High Strength Bolts in Bearing Type Connections with Threads Excluded from Shear Planes						
Diam. — Area	1 in. — .7854 sq. in.		1⅛ in. — .9940 sq. in.		1¼ in. — 1.2272 sq. in.	
ASTM Designation	A325	A354 Gr. BC	A325	A354 Gr. BC	A325	A354 Gr. BC
Shear F_v, ksi	22	24	22	24	22	24
Single Shear, kips	17.28	18.85	21.87	23.86	27.00	29.45
Double Shear, kips	34.56	37.70	43.74	47.71	54.00	58.91

AMERICAN INSTITUTE OF STEEL CONSTRUCTION

Table 9.3 Rivets and Threaded Fasteners in Bearing

Allowable loads in kips; all rivets and bolts in bearing type connections

Diam.		5/8					3/4					7/8				
F_y, ksi		33	36	42	46	50	33	36	42	46	50	33	36	42	46	50
Bearing F_p, ksi		45.0	48.5	56.5	62.0	67.5	45.0	48.5	56.5	62.0	67.5	45.0	48.5	56.5	62.0	67.5
MATERIAL THICKNESS	1/8	3.52	3.79	4.41	4.84	5.27	4.22	4.55	5.30	5.81	6.33	4.92	5.30	6.18	6.78	7.38
	3/16	5.27	5.68	6.62	7.27	7.91	6.33	6.82	7.95	8.72	9.49	7.38	7.96	9.27	10.17	11.07
	1/4	7.03	7.58	8.83	9.69	10.55	8.44	9.09	10.59	11.63	12.66	9.84	10.61	12.36	13.56	14.77
	5/16	8.79	9.47	11.04	12.11	13.18	10.55	11.37	13.24	14.53	15.82	12.30	13.26	15.45	16.95	18.46
	3/8	10.55	11.37	13.24	14.53	15.82	12.66	13.64	15.89	17.44	18.98	14.77	15.91	18.54	20.34	22.15
	7/16	12.30	13.26	15.45	16.95		14.77	15.91	18.54	20.34	22.15	17.23	18.57	21.63	23.73	25.84
	1/2	14.06	15.16				16.88	18.19	21.19	23.25		19.69	21.22	24.72	27.13	29.53
	9/16	15.82					18.98	20.46	23.84			22.15	23.87	27.81	30.52	
	5/8						21.09	22.73				24.61	26.52	30.90		
	11/16						23.20					27.07	29.18			
	3/4											29.53				
	13/16															
	7/8															
	15/16															
	1	28.13	30.31	35.31	38.75	42.19	33.75	36.38	42.38	46.50	50.63	39.38	42.44	49.44	54.25	59.06
	1 1/16															

Unit stresses F_p apply equally to conditions of single shear and enclosed bearing.
This table is not applicable to fasteners in friction type connections.

AMERICAN INSTITUTE OF STEEL CONSTRUCTION

Table 9.3 Rivets and Threaded Fasteners in Bearing (*Continued*)

Allowable loads in kips; all rivets and bolts in bearing type connections

	Diam.	1					1⅛					1¼				
	F_y, ksi	33	36	42	46	50	33	36	42	46	50	33	36	42	46	50
	Bearing F_p, ksi	45.0	48.5	56.5	62.0	67.5	45.0	48.5	56.5	62.0	67.5	45.0	48.5	56.5	62.0	67.5
MATERIAL THICKNESS	⅛	5.63	6.06	7.06	7.75	8.44	6.33	6.82	7.95	8.72	9.49	7.03	7.58	8.83	9.69	10.55
	3/16	8.44	9.09	10.59	11.63	12.66	9.49	10.23	11.92	13.08	14.24	10.55	11.37	13.24	14.53	15.82
	¼	11.25	12.13	14.13	15.50	16.88	12.66	13.64	15.89	17.44	18.98	14.06	15.16	17.66	19.38	21.09
	5/16	14.06	15.16	17.66	19.38	21.09	15.82	17.05	19.86	21.80	23.73	17.58	18.95	22.07	24.22	26.37
	⅜	16.88	18.19	21.19	23.25	25.31	18.98	20.46	23.84	26.16	28.48	21.09	22.73	26.48	29.06	31.64
	7/16	19.69	21.22	24.72	27.13	29.53	22.15	23.87	27.81	30.52	33.22	24.61	26.52	30.90	33.91	36.91
	½	22.50	24.25	28.25	31.00	33.75	25.31	27.28	31.78	34.88	37.97	28.13	30.31	35.31	38.75	42.19
	9/16	25.31	27.28	31.78	34.88	37.97	28.48	30.69	35.75	39.23	42.71	31.64	34.10	39.73	43.59	47.46
	⅝	28.13	30.31	35.31	38.75		31.64	34.10	39.73	43.59	47.46	35.16	37.89	44.14	48.44	52.73
	11/16	30.94	33.34	38.84			34.80	37.51	43.70	47.95	52.21	38.67	41.68	48.55	53.28	58.01
	¾	33.75	36.38				37.97	40.92	47.67			42.19	45.47	52.97	58.13	63.28
	13/16	36.56	39.41				41.13	44.32	51.64			45.70	49.26	57.38	62.97	
	⅞	39.38					44.30	47.74				49.22	53.05	61.80		
	15/16						47.46					52.73	56.84			
	1	45.00	48.50	56.50	62.00	67.50	50.63	54.56	63.56	69.75	75.94	56.25	60.63	70.63	77.50	84.38
	1 1/16											59.77				

Unit stresses F_p apply equally to conditions of single shear and enclosed bearing.
This table is not applicable to fasteners in friction type connections.

the allowable bearing stress was dependent on whether the connector was in an enclosed or unenclosed bearing, but this factor has been eliminated in the 1963 Code.

b There are frequent problems on rivets and riveted joints on the various license examinations. Most of the problems are quite simple, requiring only a knowledge of the tables and the general principles of joint design. Although it is common practice to use the stresses listed in the tables referred to, these stresses may be modified by the problem in hand, which should therefore be read carefully.

Problem Two plates, 18 × ¾ in. in section, are to be spliced for tension by two plates, 18 × 7⁄16 in. Using ⅞-in. rivets, design and sketch a splice for maximum efficiency, using these stresses (in psi): tension 16,000, shear 12,000, bearing 24,000. (From a state license examination.)

This brings in stresses which are not given in the tables and must be computed. The value of a ⅞-in. rivet in double shear, referring to Table 9.2 for unfinished bolts, is 1.2 × 12,030 = 14,440 psi.

The value of a ⅞-in. rivet in bearing on a ¾-in. plate at 24,000 is 0.875 × 0.75 × 24,000 = 15,760 psi. Therefore double shear controls and the rivet value is 14,440 psi.

The Code recommends that the minimum distance between centers of rivets be three diameters and the minimum edge distance for a ⅞-in. rivet in a sheared plate be 1½ in.

In an 18-in. plate, the maximum number of rivets will be 18⁄3 = 6 rivets. Net width then is 18 in. − 6 in. = 12 in. We will try this design. The net tensile value of two plates 18 × 7⁄16 (12 in. net width) = 2 × 12 × 7⁄16 in. × 16,000 = 168,000 lb. Then:

$$\frac{168{,}000}{14{,}400} = 11.65$$

Use 12 ⅞-in. rivets.

Gross area of plate = ¾ in. × 18
= 13.5 × 16,000 = 216,000

$$\frac{168{,}000}{216{,}000} = 77.8\% \text{ efficiency}$$

We will next try four rivets across in the first row next to the joint. The net tensile value of the splice plates is 2 × 14 × 7⁄16 in. × 16,000 = 196,000

$$\frac{196{,}000}{14{,}400} = 14 \text{ rivets}$$

Net in ¾-in. plate at 1st rivet = 17 × 0.75 × 16,000 = 204,000
Net in ¾-in. plate at 2d line = 16 × 0.75 × 16,000 = 192,000
Efficiency of joint 196⁄216 = 90.6% [Fig. 9.1]

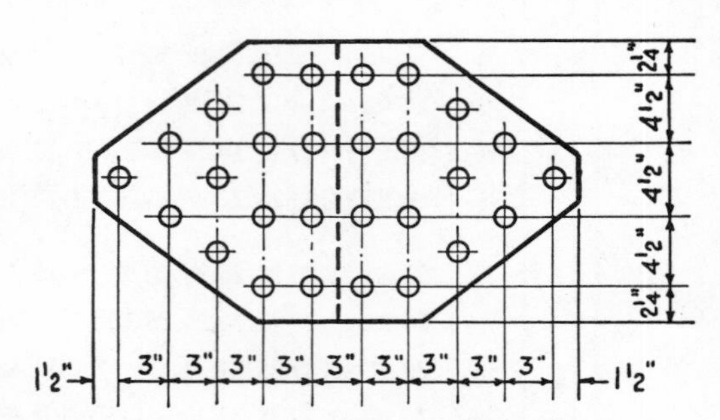

Fig. 9.1 Problem in Riveting Efficiency

c *Eccentric rivet connections.* The design of riveted joints given in Art. 9.10*a* and *b*, as well as the design of connection angle riveting given in the handbooks, is based on concentric loading, and the full value of the rivet is used to carry the load in direct shear. When the load becomes eccentric, however, the rivet group is called upon to resist bending as well as direct load and the rivets in the group are not equally stressed. In this article and in Table 9.4, the following nomenclature will be used:

n = total number of rivets in one vertical row
m = total number of vertical rows
N = total number of rivets in the group, or $N = mn$
P = total load on the connection
S = maximum load on one rivet*
C = coefficient tabulated in Table 9.4
d = distance from the center of gravity of the group to the individual rivet
b = uniform vertical rivet spacing
D = spacing of outside rows of rivets (see Table 9.4)

* This use of S is for Table 9.4 only. (Elsewhere S_p is the polar section modulus of the rivet group.)

Table 9.4 Rivet Groups Under Eccentric Application of Load

Nomenclature:
- n = total number of rivets in any one vertical row.
- P = permissible load, acting with lever arm l.
- S = permissible load on one rivet by Specification.
- C = coefficient as tabulated below.
- P = C × S; or, knowing P, required minimum $C = \frac{P}{S}$

Case I

In Table, b = 3″

n \ l	1″	2″	3″	6″	9″	12″	15″	18″	21″	24″
2	1.7	1.2	.89	.49	.33	.25	.20	.17	.14	.12
3	2.7	2.1	1.7	.95	.65	.49	.40	.33	.28	.25
4	3.7	3.1	2.6	1.5	1.1	.82	.66	.55	.47	.41
5	4.7	4.2	3.5	2.2	1.6	1.2	.98	.82	.71	.62
6	5.8	5.2	4.6	3.0	2.2	1.7	1.4	1.1	.99	.87
7	6.8	6.3	5.6	3.9	2.8	2.2	1.8	1.5	1.3	1.2
8	7.8	7.3	6.7	4.8	3.6	2.8	2.3	1.9	1.7	1.5
9	8.8	8.4	7.7	5.8	4.4	3.5	2.8	2.4	2.1	1.8
10	9.8	9.4	8.8	6.8	5.2	4.2	3.4	2.9	2.5	2.2
11	10.9	10.4	9.8	7.8	6.1	4.9	4.1	3.5	3.0	2.7
12	11.9	11.5	10.9	8.8	7.0	5.7	4.8	4.1	3.5	3.1

In general, $C = \dfrac{n}{\sqrt{\left[\dfrac{6l}{(n+1)\,b}\right]^2 + 1}}$

Case II

In Table, b = 3″ and D = 5½″ or 9½″

l	1″		2″		3″		6″		9″		12″		15″		18″		21″		24″	
n \ D	5½	9½	5½	9½	5½	9½	5½	9½	5½	9½	5½	9½	5½	9½	5½	9½	5½	9½	5½	9½
2	3.1	3.4	2.5	2.9	2.1	2.5	1.4	1.8	1.1	1.4	0.8	1.2	0.7	1.0	0.6	0.9	0.5	0.8	0.5	0.7
3	4.9	5.1	4.1	4.4	3.4	3.9	2.3	2.9	1.7	2.2	1.4	1.8	1.1	1.6	1.0	1.4	0.9	1.2	0.8	1.1
4	6.8	7.0	5.8	6.1	5.0	5.4	3.4	4.0	2.5	3.1	2.0	2.6	1.7	2.2	1.4	1.9	1.2	1.7	1.1	1.5
5	8.8	8.9	7.7	7.9	6.7	7.0	4.6	5.2	3.5	4.1	2.8	3.3	2.3	2.8	1.9	2.4	1.7	2.1	1.5	1.9
6	10.9	10.8	9.6	9.7	8.5	8.7	6.0	6.5	4.5	5.1	3.6	4.2	3.0	3.6	2.6	3.1	2.2	2.7	2.0	2.4
7	12.9	12.8	11.7	11.6	10.4	10.6	7.5	8.0	5.8	6.3	4.6	5.2	3.8	4.4	3.3	3.8	2.8	3.3	2.5	3.0
8	15.0	14.8	13.7	13.6	12.4	12.4	9.2	9.6	7.1	7.6	5.7	6.2	4.8	5.3	4.1	4.6	3.5	4.0	3.1	3.6
9	17.0	16.9	15.8	15.6	14.5	14.4	11.0	11.2	8.6	9.0	6.9	7.4	5.8	6.3	4.9	5.4	4.3	4.8	3.8	4.3
10	19.1	18.9	17.9	17.7	16.6	16.4	12.8	13.0	10.1	10.5	8.2	8.7	6.9	7.4	5.9	6.4	5.2	5.6	4.6	5.0
11	21.2	20.9	20.0	19.7	18.7	18.4	14.8	14.8	11.8	12.1	9.7	10.0	8.1	8.5	7.0	7.4	6.1	6.5	5.4	5.8
12	23.2	23.0	22.1	21.8	20.8	20.5	16.8	16.7	13.5	13.7	11.2	11.4	9.4	9.8	8.1	8.5	7.1	7.5	6.3	6.7

In general, $C = \dfrac{n}{\sqrt{\left[\dfrac{l\,(n-1)\,b}{D^2 + \frac{1}{3}(n^2-1)\,b^2}\right]^2 + \left[\dfrac{l\,D}{D^2 + \frac{1}{3}(n^2-1)\,b^2} + \frac{1}{2}\right]^2}}$

Case III

Case III, not tabulated.

In general, $C = \dfrac{n}{\sqrt{\left[\dfrac{l\,(n-1)\,b}{D^2 + \frac{1}{2}(n^2-1)\,b^2}\right]^2 + \left[\dfrac{l\,D}{D^2 + \frac{1}{3}(n^2-1)\,b^2} + \frac{1}{3}\right]^2}}$

Case IV

In Table, b = 3″

d′ = 2½″	d′ = 3″
d = 4½″	d = 5½″
d′ = 2½″	d′ = 3″
D = 9½″	D = 11½″

l	1″		2″		3″		6″		9″		12″		15″		18″		21″		24″	
n \ D	9½	11½	9½	11½	9½	11½	9½	11½	9½	11½	9½	11½	9½	11½	9½	11½	9½	11½	9½	11½
2	6.2	6.4	5.0	5.3	4.2	4.5	2.8	3.1	2.1	2.4	1.7	1.9	1.4	1.6	1.2	1.4	1.1	1.2	1.0	1.1
3	9.6	9.8	8.0	8.2	6.7	7.1	4.6	5.0	3.5	3.8	2.8	3.1	2.3	2.6	2.0	2.2	1.8	2.0	1.6	1.8
4	13.3	13.4	11.2	11.5	9.7	10.0	6.7	7.1	5.1	5.4	4.1	4.4	3.4	3.7	2.9	3.2	2.6	2.8	2.3	2.5
5	17.2	17.2	14.8	15.0	12.9	13.1	9.1	9.4	6.9	7.3	5.6	5.9	4.7	5.0	4.0	4.3	3.5	3.8	3.1	3.4
6	21.2	21.1	18.6	18.6	16.4	16.5	11.7	12.1	9.0	9.4	7.3	7.6	6.1	6.4	5.2	5.5	4.6	4.9	4.1	4.3
7	25.2	25.1	22.5	22.5	20.1	20.1	14.7	14.9	11.3	11.7	9.2	9.5	7.7	8.0	6.6	6.9	5.8	6.1	5.1	5.4
8	29.3	29.2	26.5	26.4	23.9	23.9	17.8	18.0	13.9	14.2	11.3	11.6	9.5	9.8	8.2	8.5	7.2	7.5	6.4	6.6
9	33.4	33.3	30.6	30.5	27.9	27.8	21.2	21.3	16.7	16.9	13.7	13.9	11.5	11.8	9.9	10.2	8.7	9.0	7.7	8.0
10	37.6	37.4	34.8	34.6	32.0	31.8	24.8	24.8	19.7	19.9	16.2	16.4	13.7	13.9	11.8	12.1	10.4	10.6	9.2	9.5
11	41.7	41.5	39.0	38.7	36.2	35.9	28.5	28.5	22.9	23.0	18.9	19.1	16.0	16.2	13.9	14.1	12.2	12.4	10.9	11.1
12	45.8	45.6	43.2	42.8	40.3	40.0	32.4	32.2	26.3	26.3	21.8	21.9	18.5	18.7	16.0	16.3	14.1	14.4	12.6	12.8

In general, $C = \dfrac{n}{\sqrt{\left[\dfrac{l\,(n-1)\,b}{d^2 + D^2 + \frac{2}{3}(n^2-1)\,b^2}\right]^2 + \left[\dfrac{l\,D}{d^2 + D^2 + \frac{2}{3}(n^2-1)\,b^2} + \frac{1}{4}\right]^2}}$

a = spacing between individual rows in the case of uniform spacing

l = lever arm of the applied load

All distances are measured in inches.

Using the above nomenclature, it may be stated that each rivet in the group carries an equal share of the vertical load, or P/N, but in addition, each carries a stress resisting the tendency of the entire group to rotate about the center of gravity of the group. The theoretical stress on any one rivet is the resultant of the direct and eccentric stress components, and is greatest for the rivet on the side toward the load at the distance greatest from the center of the group. The stress due to eccentricity is equal to the moment in inch-pounds divided by the polar section modulus of the group of rivets, or M/S_p.

The polar section modulus may be obtained from the formula

$$S_p = \frac{N_1 d_1^2 + N_2 d_2^2 + \cdots + N_n d_n^2}{d_{\max}}$$

Normally d is the hypotenuse of a right triangle and the two legs of the triangle are more easily obtained than the hypotenuse. We therefore substitute the squares of the two right distances for the square of the hypotenuse in the problem that follows.

Problem Assume the riveted connection shown in Fig. 9.2, which is carrying a load of 20,000 lb on a lever arm of 5 in., or 100,000 in.-lb moment. Which rivet has the maximum stress and what is the amount of the stress? Substituting the right distances for the diagonals, we may tabulate our computations as follows.

$$\begin{aligned} 8 \text{ at } 5.5^2 &= 242 \\ 8 \text{ at } 2.75^2 &= 60.5 \\ 8 \text{ at } 1.5^2 &= 18 \\ 8 \text{ at } 4.5^2 &= 162 \\ I_p &= 482.5 \end{aligned}$$

To obtain $d_{\max}$,

$$\begin{aligned} 5.5^2 &= 30.25 \\ 4.5^2 &= 20.25 \\ & 50.5 \end{aligned}$$

$$\sqrt{50.5} = 7.11$$

$$S_p = \frac{482.5}{7.11} = 67.86$$

The direct stress per rivet is

$$\frac{20{,}000}{16} = 1{,}250 \text{ lb},$$

which is a vertical stress. The stress in the corner rivet due to moment is

$$100{,}000/67.86 = 1{,}474 \text{ lb}$$

at right angles to the radial line to the corner rivets. In Fig. 9.3 the resultant forces on the outside rivets are shown.

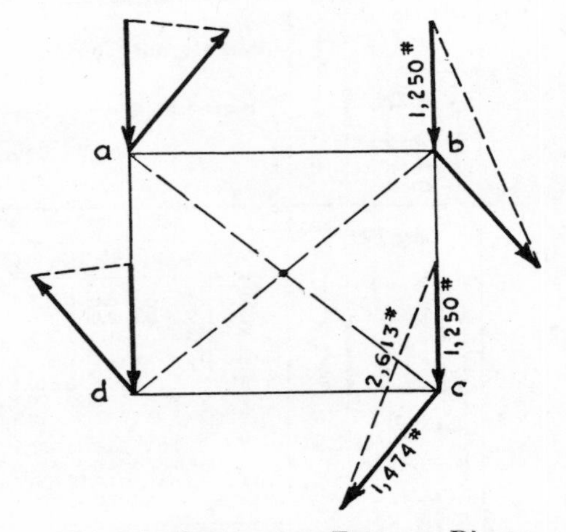

Fig. 9.3 Stresses on Extreme Rivets

There is a simple formula for obtaining the section modulus of a group of rivets with uniform spacing as follows:

1. For a single line of rivets,

$$S_1 = \frac{\text{spacing} \times N(N+1)}{6}$$

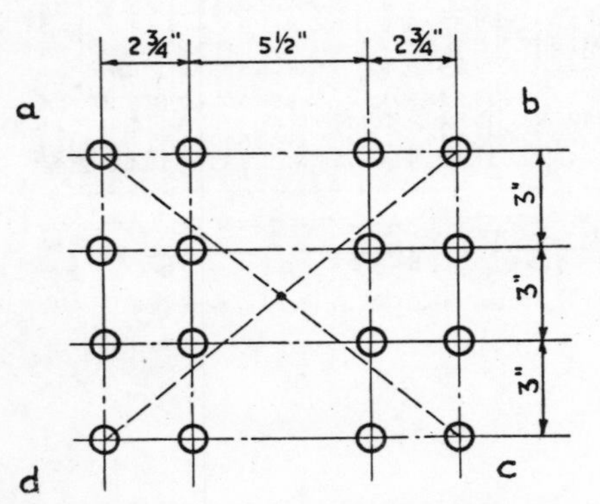

Fig. 9.2 Problem in Eccentric Riveting

2. For several lines, the polar moment of the entire group M rows wide and N rows high is

$$S_p = \sqrt{S_x^2 + S_y^2}$$

where

$$S_x = M\left[\frac{\text{spacing} \times N(N+1)}{6}\right]$$

$$S_y = N\left[\frac{\text{spacing} \times M(M+1)}{6}\right]$$

Thus for the problem above

$$S_x = S_y = 4\left(\frac{3 \times 4 \times 5}{6}\right) = 40$$

$$S_p = \sqrt{1{,}600 + 1{,}600} = \sqrt{3{,}200} = 56.6$$

d The table of rivet groups under eccentric application of load given in Table 9.4 is taken from an older AISC handbook. For any rivet group and any given lever arm of applied load, a coefficient C may be found such that C times the allowable value of one rivet equals the total load P permissible on the connection. Thus

$$P = CS$$

Or knowing P, and dividing by the allowable rivet value S, the necessary coefficient is of that magnitude or greater.

General expressions for the coefficient C are very complex, and for all except simple symmetrical cases the joint must be detailed by a cut-and-try process based on the foregoing principles and without deriving coefficients. Table 9.4 gives the coefficients C for the simplest cases occurring repeatedly in practice. Figure 9.4 shows an example of the use of the table.

In the case of eccentric brackets of the type shown in Fig. 9.5 in which the moment produces tension on the rivets, there is no exact knowledge as to the location of the neutral axis; it probably lies somewhere below the center line of the connection. Nor is there exact knowledge of the permissible combination of tension with vertical shear on the uppermost rivets. A safe and accepted method of design for brackets of the type shown is to consider the rivets to be under an eccentric loading similar to that exemplified in Table 9.4. The coefficient C

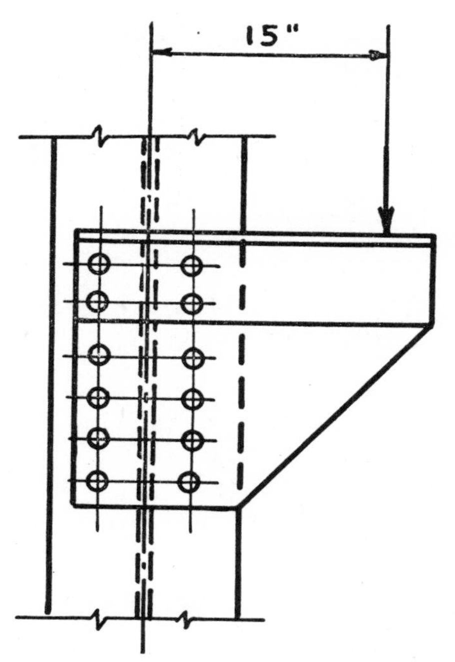

Fig. 9.4 Load on Column Bracket

for such cases will be twice the values tabulated in the table to conform with the two vertical rows of rivets.

For example, assume the following conditions: $P = 30$ kips, $l = 12$ in., ¾-in. rivets, 3-in. pitch using A141 rivets. Allow-

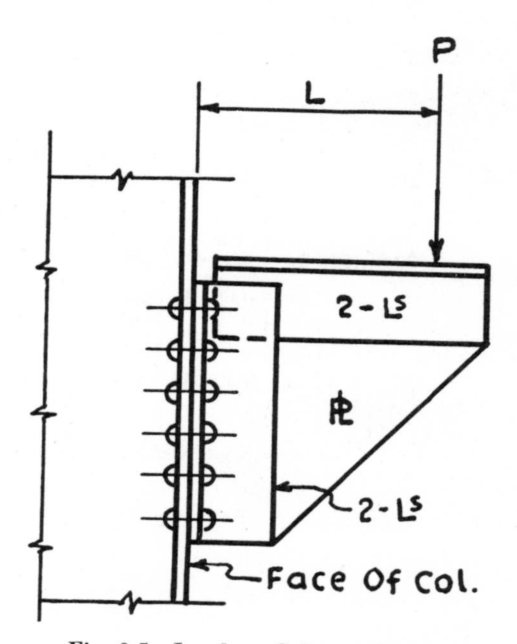

Fig. 9.5 Load on Column Bracket

Table 9.5 Net Section Moduli of Plates

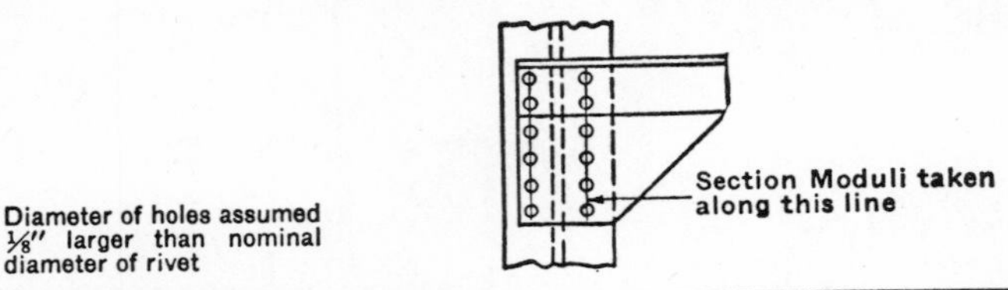

Rivets spaced 3" vertically

No. of Rivets in One Vertical Line	Depth of Plate in Inches	¾" RIVETS					⅞" RIVETS					1" RIVETS				
		Thickness of Plate, In.					Thickness of Plate, In.					Thickness of Plate, In.				
		¼	⅜	½	⅝	¾	⅜	½	⅝	¾	⅞	½	⅝	¾	⅞	1
2	6	1.2	1.8	2.3	2.9	3.5	1.7	2.3	2.9	3.4	4.0	2.2	2.7	3.2	3.8	4.3
3	9	2.5	3.8	5.0	6.3	7.5	3.6	4.8	5.9	7.1	8.3	4.5	5.6	6.8	7.9	9.0
4	12	4.4	6.3	8.7	11	13	6.2	8.2	10	12	14	7.8	9.7	12	14	16
5	15	6.8	10	14	17	20	10	13	16	19	22	12	15	18	21	24
6	18	9.6	15	19	24	29	14	18	23	27	32	17	21	26	30	34
7	21	13	20	26	33	39	19	25	31	37	43	23	29	35	41	47
8	24	17	26	34	43	51	24	32	40	48	56	30	38	45	53	61
9	27	22	32	43	54	65	31	41	51	61	71	38	48	57	67	77
10	30	27	40	53	67	80	38	50	63	75	88	47	59	71	83	94
12	36	38	58	77	96	115	54	72	90	108	126	68	85	102	119	136
14	42	52	78	104	130	157	74	98	123	147	172	92	115	138	161	184
16	48	68	102	136	170	204	96	128	160	192	224	120	150	180	211	241
18	54	86	129	172	215	259	122	162	203	243	284	152	190	228	266	304
20	60	106	160	213	266	319	150	200	250	300	350	188	235	282	329	376
22	66	129	193	257	322	386	182	242	303	363	424	227	284	341	398	454
24	72	153	230	306	383	459	216	288	360	432	504	270	338	406	473	541
26	78	180	270	359	449	539	254	338	423	507	592	317	397	476	555	634
28	84	208	313	417	521	625	294	392	490	588	686	368	460	552	644	736
30	90	240	359	478	598	718	338	450	563	675	788	422	528	633	739	845
32	96	272	408	544	680	816	384	512	640	768	896	480	600	721	841	961
34	102	308	461	614	768	922	434	578	723	867	1012	542	678	813	949	1085
36	108	344	517	689	861	1033	486	648	810	972	1134	608	760	912	1064	1216

Interpolate for intermediate thicknesses of plates.

able stress on one rivet from Table 9.1 is 8.84.

$$C = \frac{P}{2 \times S} = \frac{30}{2 \times 8.84} = 1.7$$

From Table 9.4 for l = 12 in., seven rivets are required in each of two vertical rows.

One-sided connection angles must be designed to carry the vertical load on all fasteners as well as the effect of eccentricity in the outstanding leg fasteners. In spite of this loss of efficiency, there frequently is economy, particularly in the case of filler

Table 9.6 One-Sided Connections

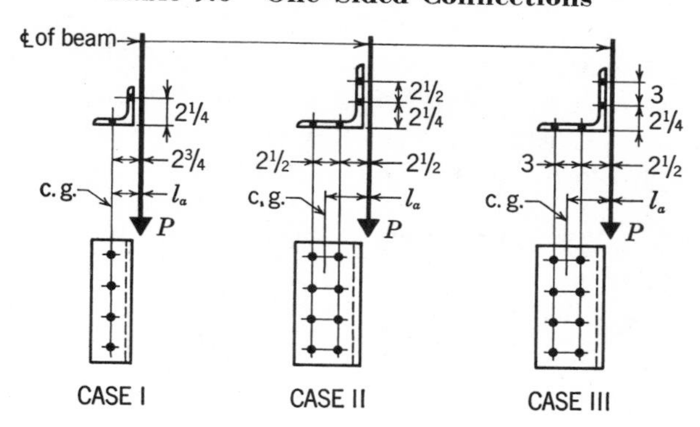

n	Coefficient C		
	Case I	Case II	Case III
1	...	0.63	0.67
2	1.41	2.05	1.99
3	2.68	4.10	3.88
4	3.92	6.64	6.31
5	5.00	9.32	8.97
6	6.00	11.83	11.60
7	7.00	14.00	14.00
8	8.00	16.00	16.00
9	9.00	18.00	18.00
10	10.00	20.00	20.00

$P = Cr_v$ or $C = P/r_v$
n = Total number of fasteners in one vertical row
C = Coefficient
P = Permissible load, kips
r_v = Allowable shear or bearing value for one fastener, kips
$l_a = l_{\text{actual}}$ = Actual arm between center line of beam and center of gravity of fasteners
$l_e = l_{\text{eff.}}$ = Effective arm between center line of beam and center of gravity of fasteners

In computation of coefficients C the actual moment arm l_a is corrected to l_e using the empirical formulas, $l_e = l_a - \left(\frac{1 + 2n}{4}\right)$ for single gage and $l_e = l_a - \left(\frac{1 + n}{2}\right)$ for double gage.

Do not exceed gages shown for web leg. Pattern of web leg fasteners may vary to suit required number of fasteners. For outstanding leg gages, other than those shown, coefficients may be interpolated from Tables XI to XIII, Part 4 of the Manual. Select angle thickness to provide sufficient gross shear capacity, or limit connection capacity to permissible shear value of angle used. Use minimum angle thickness of ⅜ in. for ¾ in. diam. and ⅞ in. diam. fasteners, and ½ in. for 1 in. diam. fasteners. It will be permissible to design a connection using combinations of leg widths as well as fastener specification and diameters.

beams, in a welded job as mentioned in Art. 9.20*c*.

The simplest method of approach to this problem is described in Table 9.6, which is taken from the AISC Manual, page 4-46, and for which several problems are given on pages 4-46 and 4-48. It is interesting to note that this is an application of Table 9.4 to a special case.

e Further problems in riveting and bolting are found in developing horizontal shear, as in riveting the flanges to plate girders or cover-plated beams, and in riveting together compound steel beams as in Art. 4.22.

The horizontal shear at any point in the web or flange may be computed from the formula given in Art. 4.22,

$$v_h = \frac{VM_s}{I}$$

where v_h is the horizontal unit shear at the point considered, V is the total shear at this point, M_s is the static moment of the area beyond the point under consideration, and I the moment of inertia of the section. This unit horizontal shear is ordinarily computed at intervals and the spacing obtained by dividing the value of one rivet by the unit shear. For example, determine the rivet spacing required to fasten a 16 × 7/8-in. plate to a 27 WF 145 at a point where the total shear is 100 kips, using 3/4-in. A141 rivets in pairs:

$$I \text{ of 27-in. beam, net} = 4{,}761$$

$$I \text{ of two } 16 \times \tfrac{7}{8}\text{-in. plates, net} = 2 \times 14 \times 0.875 \times 13.88^2 = \underline{4{,}730}$$

$$9{,}491$$

$$M_s = 14.0 \times 13.88 = 194.3$$

$$v_h = \frac{100{,}000 \times 194.3}{9{,}491} = 2{,}060 \text{ lb per in.}$$

Using two 3/4-in. rivets, the spacing may be

$$S = \frac{13{,}250 \times 2}{2{,}060} = 12.9$$

Use 12-in. spacing.

9.20 WELDING

a The subject of welding warrants much more space than can be devoted to it in this brief article. Many volumes have been devoted to it and, because of the rapid advance in welding materials and techniques, and the consequent liberalization of code requirements, welding design is advancing so rapidly that all but the most recent of these volumes are already out-of-date.

The welds most commonly used in structural welding are butt welds and fillet welds. While plug-and-slot welds are not generally accepted and the filling of such holes is not favored by the American Welding Society codes, such welds are occasionally used. Butt welds are stressed in tension or compression. The size is designated by the thickness of the thinner part of the sections joined. Fillet welds are stressed in shear for any direction of the applied load. A fillet weld is approximately an isosceles triangle of weld metal fused to the parent metal along two legs. The size is designated by the length of the shorter of these legs and the strength determined by the throat dimension, which is 0.707 times the theoretical weld size.

In a *butt weld* a long thin weld is more efficient than a short thick one of equal strength. For example, a butt weld in a 4 × 3/8 plate is equal in strength to one in a 2 × 3/4 plate but uses less weld metal. The volume of weld metal in a butt weld, neglecting reinforcement and shoulders and assuming 90-deg bevel, is equal to the width times the square of the depth. Since the strength varies as the product of width and depth, it is apparent that increasing the depth decreases the efficiency.

Fig. 9.6 Butt Weld

In a *fillet weld*, since the strength varies directly as the weld size, smaller welds are more efficient than larger ones as long as the cost per unit of volume is constant. However, welds less than 3/16 in. and welds requiring more than one pass cost more per unit of volume. Since welds larger than 3/8 in. are usually made in two or more passes, the most efficient welds are 3/16, 1/4, 5/16, and 3/8 in., respectively. While this is true of the weld itself, it may not be true of the joint, since a greater length of weld is needed for the same stress. For example a 1/4-in. weld must be half again as long as a 3/8-in. weld to carry the same stress. If the extra length of weld does not require additional detail or main material, the smaller weld is more efficient. Otherwise the cost of additional material must be considered in determining the most efficient weld size.

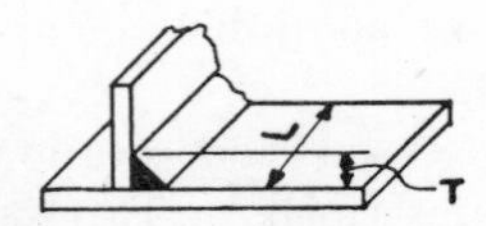

Fig. 9.7 Fillet Weld

While it is uneconomical in welded construction to transfer a stress through fillers, or across two or more joints, since each transfer requires a new weld, in riveted construction the same rivet will work at each plane of stress. For example, in a cover-plated girder the same rivets work successively through each joint from the outer cover to the flange, whereas in welded construction new welds would be required at each joint.

Where a filler is required in riveted construction, the rivets will transmit the stress across a thin filler with very little loss in efficiency. In welded construction, however, the stress must be transmitted by the filler, requiring the stress to be carried twice. Furthermore, the thickness of the filler limits the weld size; so it may be impossible to obtain sufficient welding to transmit the stress.

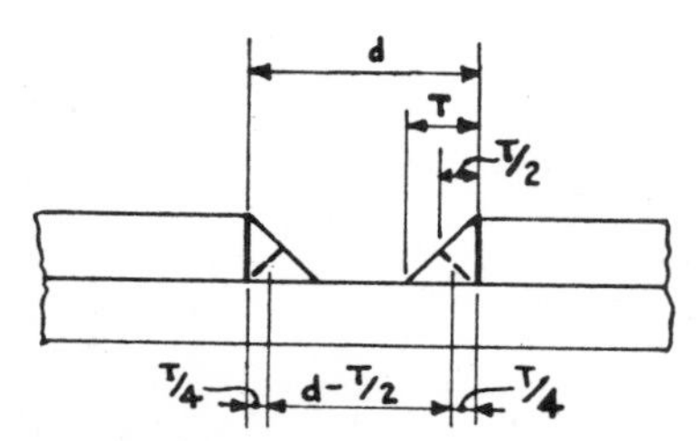

Fig. 9.8 Depth of Weld

The following sections of the 1963 AISC specification are applicable to the design and use of welding in building construction.

1.1.3 *Notations for Welding*

Note shall be made on the plans and on the shop drawings of these joints or groups of joints in which it is especially important that the welding sequence and technique of welding be carefully controlled to minimize locked-up stresses and distortion.

Weld lengths called for on the plans and on the shop drawings shall be the net effective lengths.

1.1.4 *Standard Symbols and Nomenclature*

Welding symbols used on plans and shop drawings shall preferably be the American Welding Society symbols. Other adequate welding symbols may be used, provided a complete explanation thereof is shown on the plans or drawings.

Unless otherwise noted, the standard nomenclature contained in the joint AISC-SJI *Standard Specifications for Open Web Steel Joists—Longspan or LA-Series* shall be used in describing longspan steel joists.

SECTION 1.2 TYPES OF CONSTRUCTION

In tier buildings, designed in general as Type 2 construction (that is, with beam-to-column connections other than wind connections flexible) the distribution of the wind moments between the several joints of the frame may be made by a recognized empirical method provided that either:

1. The wind connections, designed to resist the assumed moments, are adequate to resist the moments induced by the gravity loading and the wind loading at the increased unit stresses permitted therefor, or
2. The wind connections, if welded and if designed to resist the assumed wind moments, are so designed that larger moments induced by the gravity loading under the actual condition of restraint will be relieved by deformation of the connection material without over-stress in the welds.

1.5.3 *Welds* (stress in pounds per square inch of throat area)

1.5.3.1 *Fillet, Plug, Slot and Partial Penetration Groove Welds*

Stress in fillet, plug, and slot welds, tension stress transverse to the axis of partial penetration groove welds and shear in such welds, when made with A233 Class E60 series electrodes or by submerged arc welding Grade SAW-1 on all steels or with A233 Class E70 series electrodes or by submerged arc welding Grade SAW-2 on A7 and A373 steels 13,600

1.5.4 *Cast Steel and Steel Forgings*

Allowable stresses same as those provided under Section 1.5.1, where applicable.

1.14.7 *Effective Areas of Weld Metal*

The effective area of butt and fillet welds shall be considered as the effective length of the weld times the effective throat thickness.

The effective shearing area of plug and slot welds shall be considered as the nominal cross-sectional area of the hole or slot, in the plane of the faying surface.

The effective area of fillet welds in holes and slots shall be computed as above specified for fillet welds, using for effective length, the length of center-line of the weld through the center of the plane through the throat. However, in the case of overlapping fillets, the effective area shall not exceed the nominal cross-sectional area of the hole or slot, in the plane of the faying surface.

The effective length of a fillet weld shall be the overall length of full-size fillet including returns.

The effective length of a butt weld shall be the width of the part joined.

The effective throat thickness of a fillet weld shall be the shortest distance from the root to the face of the diagrammatic weld.

The effective throat thickness of a complete penetration butt weld (i.e., a butt weld conforming to the requirements of Sect. 1.23.6) shall be the thickness of the thinner part joined.

The effective throat thickness of single-V or single-bevel groove welds having no root opening and having partial penetration into their joints shall be ¼ inch less than the depth of the V or bevel groove. The effective throat thickness of single-J or single-U groove welds having no root opening and having partial penetration into their joints shall be the depth of the J or U groove. The effective throat thickness of any of these partial penetration groove welds shall be not less than $\sqrt{t_t/6}$, where t_t is the thickness of the thinner part connected by the weld.

1.15.3 *Placement of Rivets, Bolts and Welds*

Except as hereinafter provided, the rivets, bolts or welds at the ends of any member transmitting axial stress into that member shall have their centers of gravity on the gravity axis of the member unless provision is made for the effect of the resulting eccentricity. Except in members subject to repeated variation in stress, as defined in Sect. 1.7, disposition of fillet welds to balance the forces about the neutral axis or axes for end connections of single angle, double angle, and similar type members is not required. Eccentricity between the gravity axes of such members and the gage lines for their riveted or bolted end connections may be neglected.

1.15.5 *Restrained Members*

Fasteners or welds for end connections of beams, girders and trusses not conforming to the requirements of Sect. 1.15.4 shall be designed for the combined effect of end reaction shear and tensile or compressive stresses resulting from moment induced by the rigidity of the connection when the member is fully loaded.

1.15.6 *Fillers*

When rivets or bolts carrying computed stress pass through fillers thicker than ¼ inch, except in friction-type connections assembled with high strength bolts, the fillers shall be extended beyond the splice material and the filler extension shall be secured by enough rivets or bolts to distribute the total stress in the member uniformly over the combined section of the member and the filler, or an equivalent number of fasteners shall be included in the connection.

In welded construction, any filler ¼ inch or more in thickness shall extend beyond the edges of the splice plate and shall be welded to the part on which it is fitted with sufficient weld to transmit the splice plate stress, applied at the surface of the filler as an eccentric load. The welds joining the splice plate to the filler shall be sufficient to transmit the splice plate stress and shall be long enough to avoid overstressing the filler along the toe of the weld. Any filler less than ¼ inch thick shall have its edges made flush with the edges of the splice plate and the weld size shall be the sum of the size necessary to carry the splice plate stress plus the thickness of the filler plate.

1.15.9 *Combination of Welds*

If two or more of the general types of weld (butt, fillet, plug, slot) are combined in a single joint, the effective capacity of each shall be separately computed with reference to the axis of the group, in order to determine the allowable capacity of the combination.

1.15.10 *Rivets and Bolts in Combination with Welds*

In new work, rivets, A307 bolts, or high strength bolts used in bearing-type connections, shall not be considered as sharing the stress in combination with welds. Welds, if used, shall be provided to carry the entire stress in the connection. High strength bolts installed in accordance with the provisions of Sect. 1.16.1 as a friction-type connection prior to welding may be considered as sharing the stress with the welds.

In making welded alterations to structures, existing rivets and properly tightened high strength bolts may be utilized for carrying stresses resulting from existing dead loads, and the welding need be adequate only to carry all additional stress.

SECTION 1.17 WELDS

1.17.1 *Welder and Welding Operator Qualifications*

Welds shall be made only by welders and welding operators who have been previously qualified by tests as prescribed in the *Standard Code for Welding in Building Construction* of the American Welding Society, to perform the type of work required, except that this provision need not apply to tack welds not later incorporated into finished welds carrying calculated stress.

1.17.2 *Qualification of Weld and Joint Details*

Weld grooves for complete penetration welds which are accepted without welding procedure qualification under the *Standard Code for Welding in Building Construction* or the *Standard Specifications for Welded Highway and Railway Bridges* of the American Welding Society may be used under this specification without welding procedure qualification.

Weld grooves of the 60° single-V, 45° single bevel, single-J or single-U form, conforming to the details for such grooves as provided in the above AWS Standards but having partial penetration with an effective throat thickness as defined in Sect. 1.14.7 and no root opening, may be used under this specification without welding procedure qualification. However, they shall not be used in butt joints to resist tensile stress acting in a direction normal to the plane of the weld throat, except in splices or connections of columns or other members subject primarily to axial compressive stress.

Joint forms or welding procedures other than those included in the foregoing may be employed provided they shall have been qualified in accordance with the requirements of these AWS Standards.

ASTM A233 class E60 and class E70 series electrodes for manual arc welding and Grade SAW-1 or Grade SAW-2 submerged arc process may be used for welding A7, A373 and A36 steel. Only E70 low hydrogen electrodes for manual arc welding or Grade SAW-2 for submerged arc welding shall be used with A441 or weldable A242 steel, except that fillet welds or partial penetration groove welds may be made with E60 series low hydrogen electrodes and Grade SAW-1 submerged arc process.

Welding A440 steel is not recommended.

1.17.3 *Submerged Arc Welding*

The bare electrodes and granular fusible flux used in combinations for submerged arc welding shall be capable of producing weld metal having the following tensile properties when deposited in a multiple pass weld:

Grade SAW-1

Tensile strength	62,000 to 80,000 psi
Yield point, min	45,000 psi
Elongation in 2 in., min	25%
Reduction in area, min	40%

Grade SAW-2

Tensile strength	70,000 to 90,000 psi
Yield point, min	50,000 psi
Elongation in 2 in., min	22%
Reduction in area, min	40%

1.17.4 *Minimum Size of Fillet Welds*

In joints connected only by fillet welds, the minimum size of fillet weld to be used shall be as shown in Table 1.17.4. Weld size is determined by the thicker of the two parts joined, except that the weld size need not exceed the thickness of the thinner part joined unless a larger size is required by calculated stress.

1.17.5 *Maximum Effective Size of Fillet Welds*

The maximum size of a fillet weld that may be assumed in the design of a connection shall be such that the stresses in the adjacent base material do not exceed the values allowed in Sect. 1.5.1. The maximum size that may be used along edges of connected parts shall be:

1. Along edges of material less than ¼ inch thick, the maximum size may be equal to the thickness of the material.
2. Along edges of material ¼ inch or more in thickness, the maximum size shall be 1/16 inch less than the thickness of the material, unless the weld is especially designated on the drawings to be built out to obtain full throat thickness.

1.17.6 *Length of Fillet Welds*

The minimum effective length of a strength fillet weld shall be not less than 4 times the nominal size, or else the size of the weld shall be considered not to exceed one-fourth of its effective length.

If longitudinal fillet welds are used alone in end connections of flat bar tension members, the length of each fillet weld shall be not less than the perpendicular distance between them. The transverse spacing of longitudinal fillet welds used in end connections shall not exceed 8 inches, unless the design otherwise prevents excessive transverse bending in the connection.

1.17.7 *Intermittent Fillet Welds*

Intermittent fillet welds may be used to transfer calculated stress across a joint or faying surfaces when the strength required is less than that developed by a continuous fillet weld of the smallest permitted size, and to join components of built-up members. The effective length of any segment of intermittent fillet welding shall

Table 1.17.4

Material Thickness of Thicker Part Joined (Inches)	Minimum Size of Fillet Weld (Inches)
To ½ inclusive	3/16
Over ½ to ¾	¼
Over ¾ to 1½	5/16
Over 1½ to 2¼	⅜
Over 2¼ to 6	½
Over 6	⅝

be not less than 4 times the weld size with a minimum of 1½ inches.

1.17.8 *Lap Joints*

The minimum width of laps on lap joints shall be 5 times the thickness of the thinner part joined and not less than 1 inch. Lap joints joining plates or bars subjected to axial stress shall be fillet welded along the edge of both lapped parts except where the deflection of the lapped parts is sufficiently restrained to prevent opening of the joint under maximum loading.

1.17.9 *End Returns of Fillet Welds*

Side or end fillet welds terminating at ends or sides, respectively, of parts or members shall, wherever practicable, be returned continuously around the corners for a distance not less than twice the nominal size of the weld. This provision shall apply to side and top fillet welds connecting brackets, beam seats and similar connections, on the plane about which bending moments are computed. End returns shall be indicated on the design and detail drawings.

1.17.10 *Fillet Welds in Holes and Slots*

Fillet welds in holes or slots may be used to transmit shear in lap joints or to prevent the buckling or separation of lapped parts, and to join components of built-up members. Such fillet welds may overlap, subject to the provisions of Sect. 1.14.7. Fillet welds in holes or slots are not to be considered plug or slot welds.

1.17.11 *Plug and Slot Welds*

Plug or slot welds may be used to transmit shear in a lap joint or to prevent buckling of lapped parts and to join component parts of built-up members.

The diameter of the holes for a plug weld shall be not less than the thickness of the part containing it plus 5⁄16 inch, rounded to the next greater odd 1⁄16 inch, nor greater than 2¼ times the thickness of the weld metal.

The minimum center-to-center spacing of plug welds shall be 4 times the diameter of the hole.

The length of slot for a slot weld shall not exceed 10 times the thickness of the weld. The width of the slot shall be not less than the thickness of the part containing it, plus 5⁄16 inch, rounded to the next greater odd 1⁄16 inch, nor shall it be greater than 2¼ times the thickness of the weld. The ends of the slot shall be semicircular or shall have the corners rounded to a radius not less than the thickness of the part containing it, except those ends which extend to the edge of the part.

The minimum spacing of lines of slot welds in a direction transverse to their length shall be 4 times the width of the slot. The minimum center-to-center spacing in a longitudinal direction on any line shall be 2 times the length of the slot.

The thickness of plug or slot welds in material ⅝ inch or less in thickness shall be equal to the thickness of the material. In material over ⅝ inch in thickness, it shall be at least one-half the thickness of the material but not less than ⅝ inch.

1.23.6 *Welded Construction*

Surfaces to be welded shall be free from loose scale, slag, rust, grease, paint and any other foreign material except that mill scale which withstands vigorous wire brushing may remain. Joint surfaces shall be free from fins and tears. Preparation of edges by gas cutting shall, wherever practicable, be done by a mechanically guided torch.

Parts to be fillet welded shall be brought in as close contact as practicable and in no event shall be separated by more than 3⁄16 inch. If the separation is 1⁄16 inch or greater, the size of the fillet welds shall be increased by the amount of the separation. The separation between faying surfaces of lap joints and butt joints on a backing structure shall not exceed 1⁄16 inch. The fit of joints at contact surfaces which are not completely sealed by welds, shall be close enough to exclude water after painting.

Abutting parts to be butt welded shall be carefully aligned. Misalignments greater than ⅛ inch shall be corrected and, in making the correction, the parts shall not be drawn into a sharper slope than 2 degrees (7⁄16 inch in 12 inches).

The work shall be positioned for flat welding whenever practicable.

In assembling and joining parts of a structure or of built-up members, the procedure and sequence of welding shall be such as will avoid needless distortion and minimize shrinkage stresses. Where it is impossible to avoid high residual stresses in the closing welds of a rigid assembly, such closing welds shall be made in compression elements.

In the fabrication of cover-plated beams and built-up members, all shop splices in each component part shall be made before such component part is welded to other parts of the member. Long girders or girder sections may be made by shop splicing not more than three subsections, each made in accordance with this paragraph.

All complete penetration butt welds made by manual welding, except when produced with the aid of backing material or welded in the flat position from both sides in square-edge material not more than 5⁄16 inch thick with root opening not less than one-half the thickness of the thinner part joined, shall have the root of the initial layer gouged out on the back side before welding is started from that side, and shall be so welded as to secure sound metal and complete fusion throughout the entire cross-section. Butt welds made with use of a backing of the same material as the base metal shall have the weld metal thoroughly fused with the backing material. Backing strips may be removed by gouging or gas cutting after welding is completed, provided no injury is done to the base metal and weld metal and the weld metal surface is left flush or slightly convex with full throat thickness.

Butt welds shall be terminated at the ends of a joint in a manner that will ensure their soundness. Where possible, this should be done by use of extension bars or run-off plates. Extension bars or run-off plates, if used, shall be removed upon completion of the weld and the ends of the weld made smooth and flush with the abutting parts.

No welding shall be done when the ambient temperature is lower than 0°F.

Base metal shall be preheated as required to the temperature called for in Table 1.23.6 prior to tack welding or welding. When base metal not otherwise required to be preheated is at a temperature below 32°F, it shall be preheated to at least 70°F prior to tack welding or welding. Preheating shall bring the surface of the base metal within 3 inches of the point of welding to the specified preheat temperature, and this temperature shall be maintained as a minimum interpass temperature while welding is in progress. Minimum preheat and interpass temperatures shall be as specified in Table 1.23.6.

Table 1.23.6

Thickness of Thickest Part at Point of Welding	Minimum Preheat and Interpass Temperatures					
	Other Than Low-Hydrogen Welding Processes[1]			Low-Hydrogen Welding Processes[2]		
	A373 Steel	A7, A36 Steel	A441 Steel	A373 Steel	A7, A36 Steel	A441 Steel[3]
To 1″, incl.	None[4]	None[4]	Welding with this process not recommended	None[4]	None[4]	None[4]
Over 1″ to 2″, incl.	100°F	200°F		None[4]	50°F	100°F
Over 2″	200°F	300°F		100°F	150°F	200°F

Where required, multiple-layer welds may be peened with light blows from a power hammer, using a round-nose tool. Peening shall be done after the weld has cooled to a temperature warm to the hand. Care shall be exercised to prevent scaling, or flaking of weld and base metal from overpeening.

The technique of welding employed, the appearance and quality of welds made, and the methods used in correcting defective work shall conform to Section 4—Workmanship, of the *Standard Code for Arc and Gas Welding in Building Construction* of the American Welding Society.

1.24.5 *Surfaces Adjacent to Field Welds*

Unless otherwise provided, surfaces within two inches of any field weld location shall be free of materials that would prevent proper welding or produce objectionable fumes while welding is being done.

1.25.4 *Field Welding*

Any shop paint on surfaces adjacent to joints to be field welded shall be wire brushed to reduce the paint film to a minimum.

1.26.4 *Inspection of Welding*

The inspection of welding shall be performed in accordance with the provisions of Section 5 of the *Standard Code for Arc and Gas Welding in Building Construction* of the American Welding Society.

In accordance with the above specifications, the permissible stress for fillet welds is 700 lb per linear in. per 1/16 in. of thickness, or 2,800 for 1/4 in.; 3,500 for 5/16 in.; 4,200 for 3/8 in., etc. Although it is not good practice to design a structure for riveting and bolting, and to change it to a welded job on the basis of inches of weld per fastener, the general principles of use of connections and of eccentricity as stated in Art. 9.10 for rivets and bolts are applicable also to the design of welding. The text and tables in the AISC Manual on pages 4-19 through 4-31, 4-34, 4-35, 4-39 through 4-41, 4-56 through 4-65, and 4-72 through 4-76 are based on good design and practice.

Other satisfactory weld details are shown in Figs. 9.9 through 9.11.

Sometimes welds are required in places

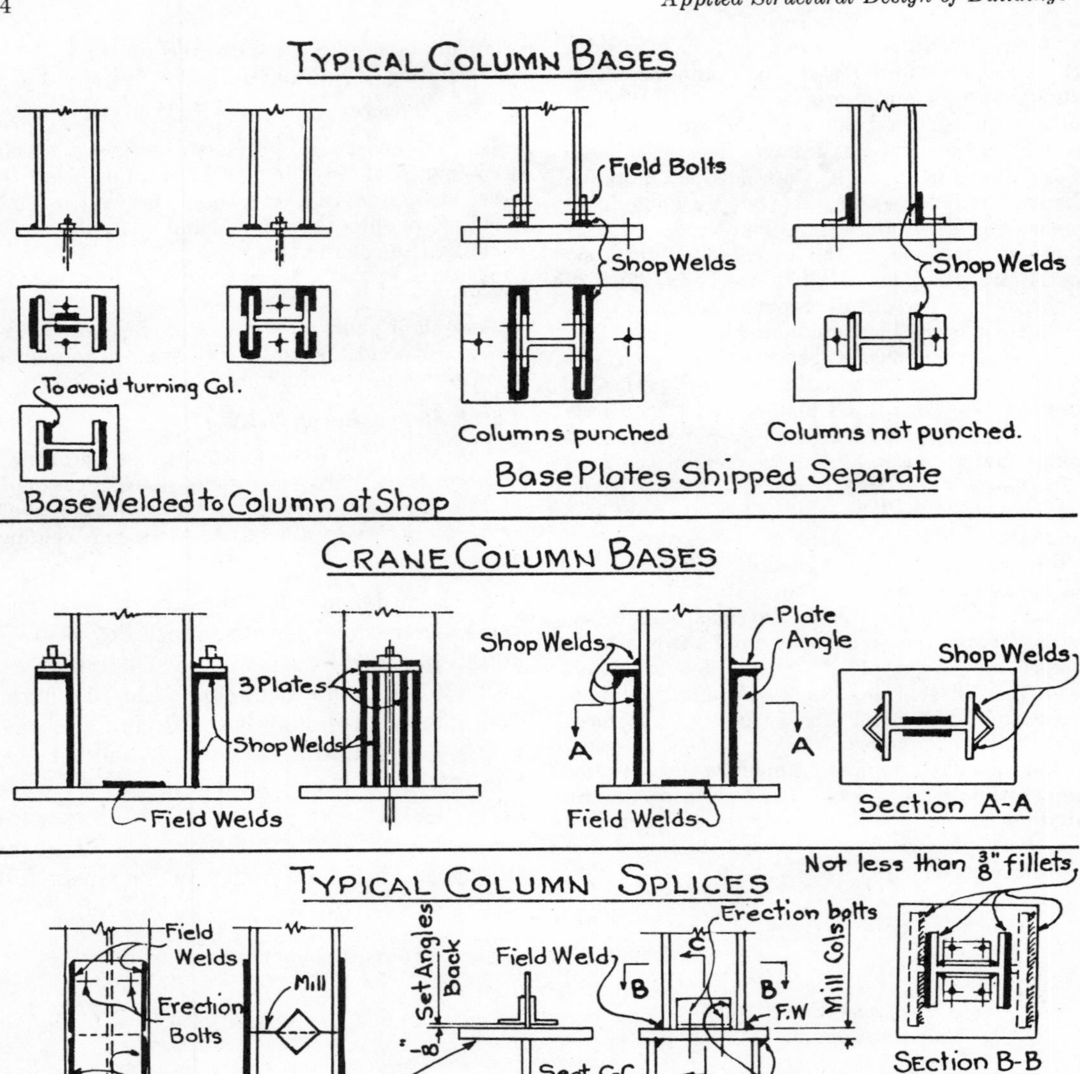

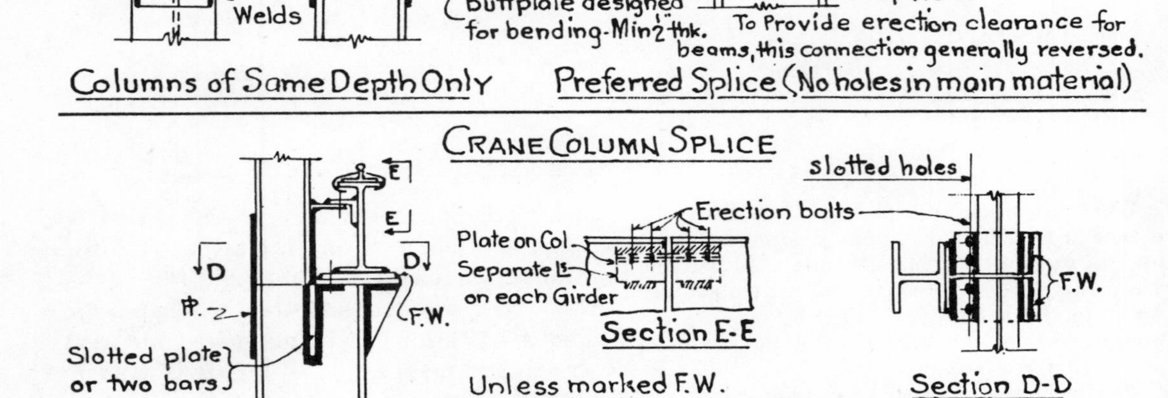

Fig. 9.9 Typical Welded-Column Details

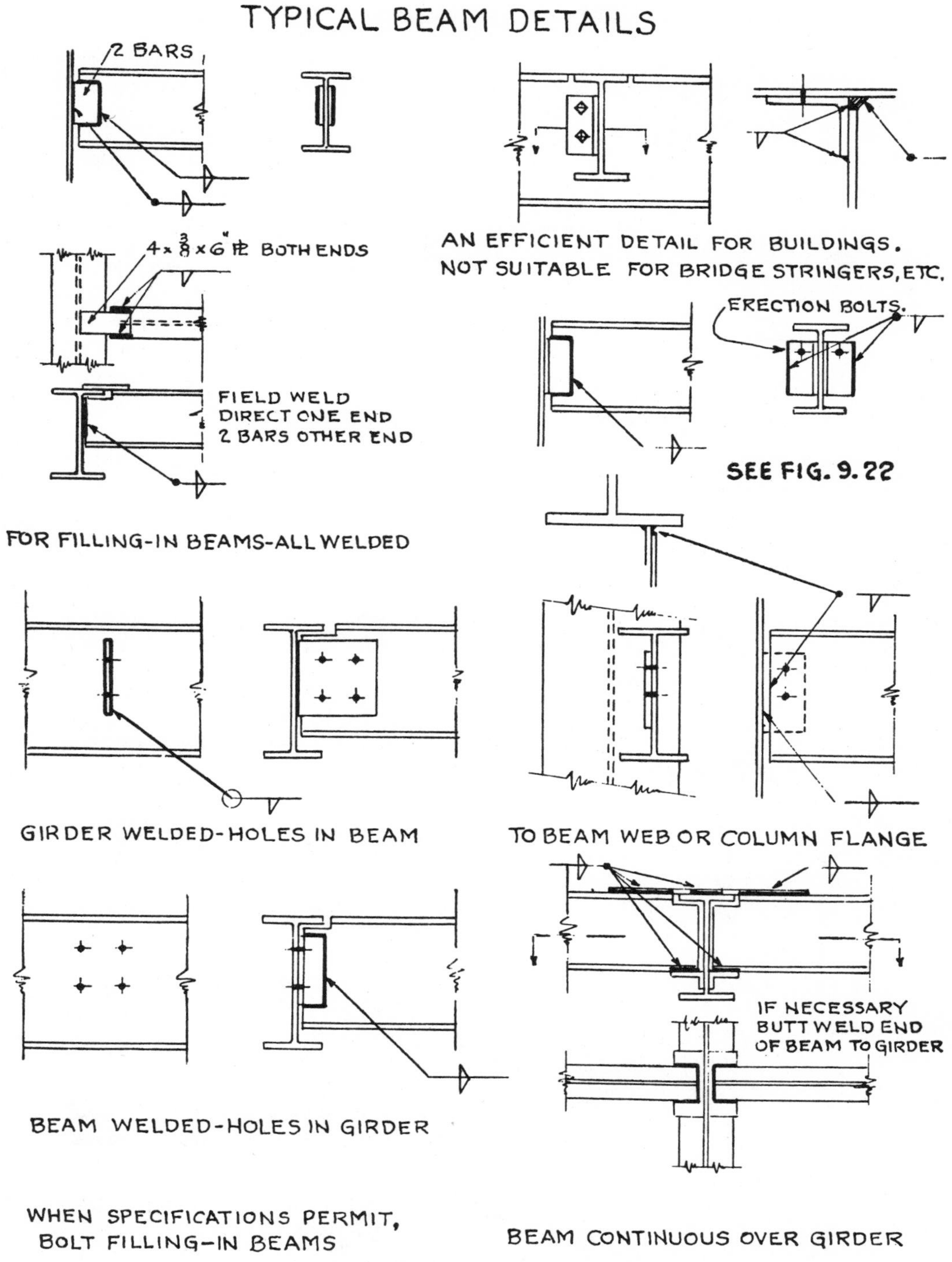

TYPICAL BEAM DETAILS

2 BARS

AN EFFICIENT DETAIL FOR BUILDINGS.
NOT SUITABLE FOR BRIDGE STRINGERS, ETC.

4 x $\frac{3}{8}$ x 6" PL BOTH ENDS

ERECTION BOLTS.

FIELD WELD
DIRECT ONE END
2 BARS OTHER END

SEE FIG. 9.22

FOR FILLING-IN BEAMS-ALL WELDED

GIRDER WELDED-HOLES IN BEAM

TO BEAM WEB OR COLUMN FLANGE

IF NECESSARY
BUTT WELD END
OF BEAM TO GIRDER

BEAM WELDED-HOLES IN GIRDER

WHEN SPECIFICATIONS PERMIT,
BOLT FILLING-IN BEAMS

BEAM CONTINUOUS OVER GIRDER

Fig. 9.10 Typical Welded-Beam Details

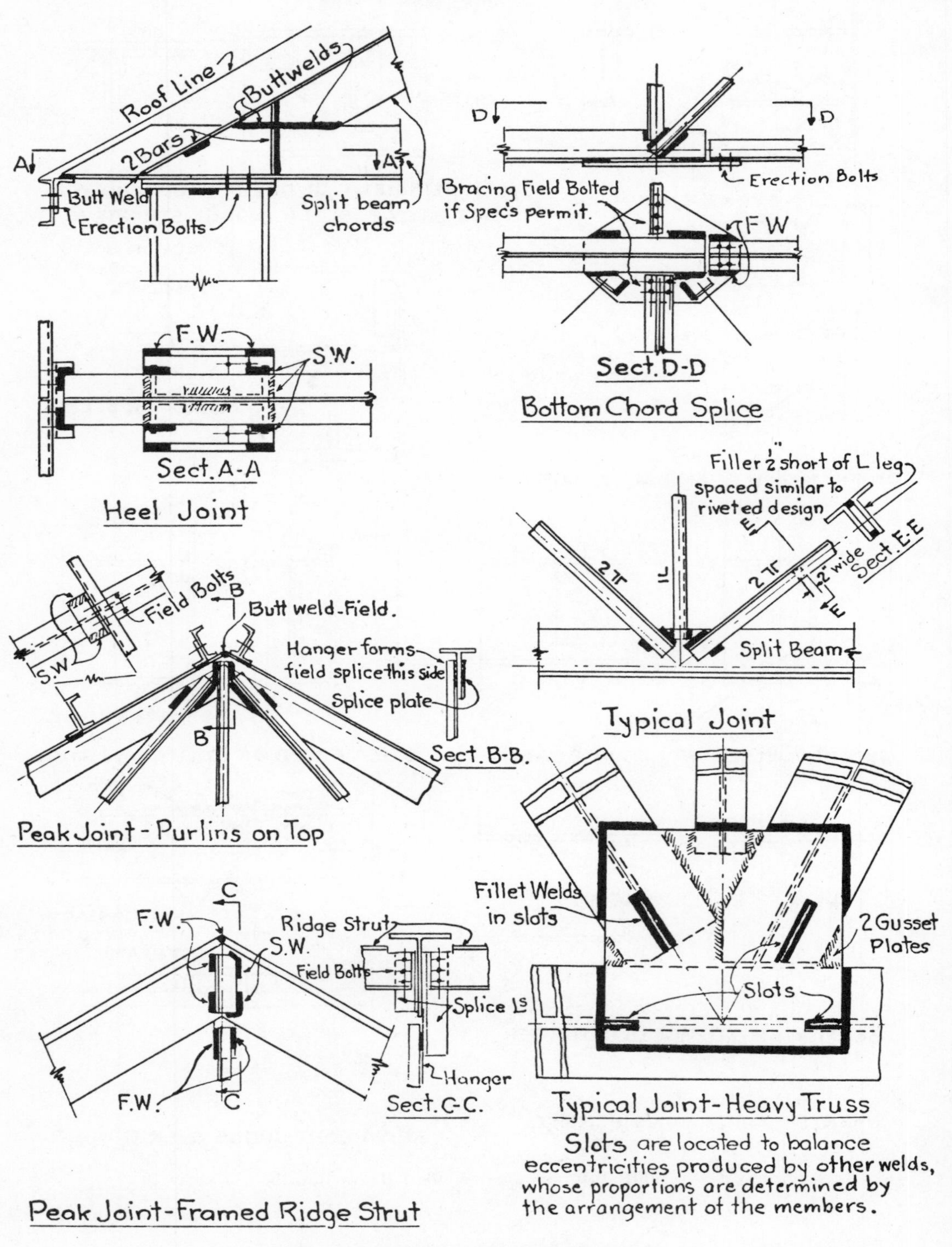

Fig. 9.11 Typical Welded-Truss Details

where no appreciable stress occurs, or where the stress is purely nominal in comparison with the thickness of the material joined. Examples are seal welds or a beam bearing on a slab or another beam. If such welds are made too small, they may crack because of the quench effect of the larger adjacent masses of metal.

The minimum fillet-weld size used should be as follows for various thicknesses of metal:

3/16-in. fillet weld up to 1/2-in. metal
1/4-in. fillet weld up to 3/4-in. metal
5/16-in. fillet weld up to 1 1/4-in. metal
3/8-in. fillet weld up to 2-in. metal
1/2-in. fillet weld up to 4-in. metal

The minimum length of a weld should be four times the weld size.

c As has been stated, to take full advantage of the economy of welding, the entire structure must be designed for welding. Moreover, costs of a welded versus a riveted structure cannot be compared on the basis of designs and prices of ten years ago since, as a result of new developments and modern codes, welded structures are on a much more competitive cost basis than they were only a few years ago. However, the designer must recognize the facts which contribute toward this economy. Having the general design in mind and partially laid out, the designer may learn much which will contribute toward economical details if he will ask for the suggestions and criticisms of a steel fabricator who is experienced in welded construction.

In general, the designer should keep in mind these basic facts with regard to welded construction.

1. The generally accepted principle today seems to be that economy can best be obtained by shop welding, and by making field connections by use of high-strength bolts, thus practically eliminating riveting. Filler beams may be connected to girders by the use of ordinary field bolting. Further economy for filler beams may be obtained by shop welding one-sided connection angles to the girders and field bolting the filler beams to these. The slight eccentricity caused by the use of a one-sided connection may be provided against by the use of connections designed in accordance with Table 9.6.

2. Shop welding is much cheaper than field welding and it is therefore desirable to use details which will allow a larger amount of shop welding with a minimum of field welding and/or high-strength bolting.

3. For moment connections at columns such as those required for rigid frames or for wind connections, the simplicity of field welding would seem to outweigh whatever other advantages high-strength bolting might offer. The economical welded-moment connection should provide that all field welding be down-pass welding. A stiffened seat may be welded to the column to permit down-pass welding of the beam to the seat. The top attachment may be by means of a blunt point slotted V-shaped plate which can be laid on top of the beam to provide for long filled welds, and an angle cut at the column into which a butt weld can be laid.

4. Adopt welding designs as nearly as possible to normal mill practice. Allowance must always be made for erection clearance but one end of the beam may be saw cut to a surface satisfactory for butt welding, thus requiring the top plate for one end only.

5. Take advantage of rigid-frame economics in all bents, i.e., in all vertical planes through columns and beams or girders, and design filler beams as simple beams.

6. In order to obtain full continuity, several methods may be resorted to. If opposite girders are of the same depth, use shop-welded stiffeners (butt welded) between the column flanges for continuity, opposite the top and bottom girder flanges.

If heavy girders are required with lighter columns, it may be more economical to run the girder completely across the lower column, and weld into the girder the stiffeners necessary to carry the column from above through the girder. If the column is relatively heavy, it is preferable to run the column through. A simple check of materials and welds required will determine this. In some cases, a shop-welded assembly of plates extending from below the floor to above the

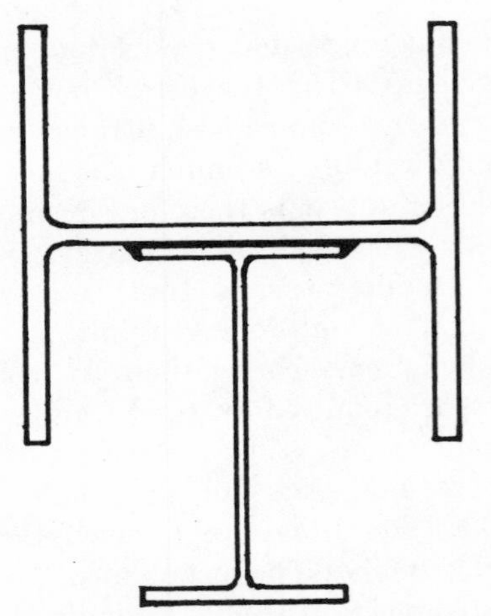

Fig. 9.12 Corner Column to Develop Moment About Both Axes

floor beam will provide an economical solution of stub columns, moment or wind connections, stub girders and column splice, and greatly simplify the field welding.

7. Where punching is necessary for bolting filler beams or for erection purposes, weld short angles to the girders, and punch the outstanding legs of the angles. This will require that only light members be handled through the punches.

8. If the design induces a heavy moment about the weaker axis of the column, as may occur in a corner column or in a column where two wind bents cross at right angles, the column may be stiffened about the weak axis by welding in a split beam to increase the lesser r (Fig. 9.12) or by welding cover plates to the edges of the flanges on one or both open sides of the column (Fig. 9.13).

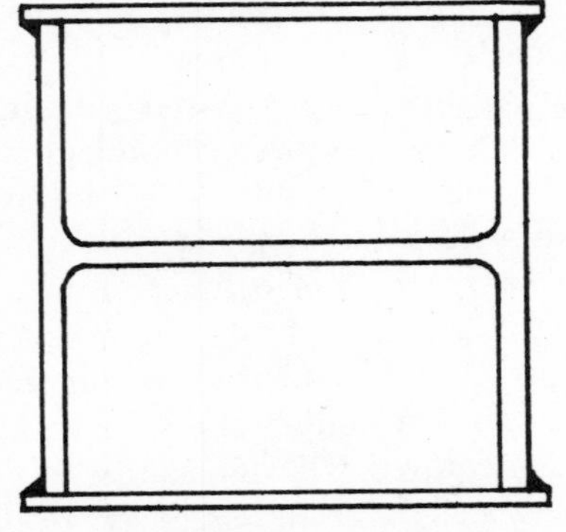

Fig. 9.13 Welded Column to Increase Least Radius of Gyration

d Welded plate girders are usually made of three plates (see Art. 4.20). The web and flange plates are designed for shear and moment in the same way as riveted girders. The flange should be so proportioned that not more than one pair of cover plates is required.

The welding required between the flange plates and the web is determined by the horizontal shear per linear inch, which equals VM_s/I, where V is the vertical shear at the section, M_s is the statical moment of that portion beyond the plane of the weld, and I is the moment of inertia of the entire section. Multiplying this shear by 12 and dividing by the allowable weld stress per linear inch gives the total inches of weld per foot.

If the factors V, M_s, or I vary throughout the length of the girder, separate calculations will be required for as many variations in the welding as the judgment of the designer dictates.

Cover plates are extended beyond the theoretical length a sufficient distance to permit continuous welding to connect for the proportion of stress in the covers. It is assumed that the stress in the girder flange without covers is the full permissible stress and that the covers will take a share of the stress based on the ratio of section modulus without covers to section modulus with covers. Between the ends tack welds are used. The total of all the welding between the end of the cover and any point should develop all the stress at that point.

The minimum size of tack weld between flange and web plates and between flange and cover plates should not be less than indicated in Art. 9.20*b*. The minimum length of weld should be about eight times the weld size and the maximum clear distance between welds not more than twenty-four times the thinnest plate for the tension flange or sixteen times the thinnest plate for the compression flange.

Continuous welds of smaller size may be used instead of intermittent welds when in the judgment of the designer it is advisable to seal the joint to prevent corrosion, or when the girder is subjected to dynamic loading.

The design of web stiffeners is similar to

riveted construction, except that bars are used instead of angles. The full area of the stiffener is considered effective to carry load. The ends are not usually milled, but fillet welds are used to carry the load from flange to stiffener and from stiffener to web. It may sometimes be cheaper to mill the stiffeners and tack-weld them to the flanges; for example, where the loads involved would require a large amount of welding and there is sufficient duplication to permit cheap milling.

Intermediate stiffeners are so proportioned that the width is approximately 90 percent of the flange projection and the thickness one-sixteenth of the width. Welds must develop the shear but should not be less than the minimum size (see Art. 9.20*b*). The minimum length of weld should be eight times the weld size, and clear distance between welds not more than sixteen times the thickness of stiffener.

The plates are usually varied in thickness only, although they may also be varied in width.

Where large shears occur near the end of long girders it may be advantageous to vary the web thickness and butt-weld the web plates.

9.30 TIMBER JOINTS

a Until timber connectors were introduced, the design of timber joints and consequently of timber trusses was a long, laborious job full of opportunities for error. Now, although much simpler than it was formerly, the design of these joints is still a much longer job than the design of steel joints. However, the design of bolted connections, of tabled fish-plate joints and the other details formerly connected with wood truss design may be completely disregarded and our study of timber joints in heavy construction may be limited entirely to the study of timber connectors and their design.

b In addition to those sections of the NLMA Code quoted in Arts. 1.23 and 6.10, the following sections of this code apply to the design of timber connector joints.

500-BASIC DESIGN CONSIDERATIONS AND LIMITATIONS

500-A. *Connector Unit.*

500-A-1. For purposes of specifying allowable connector loads herein, a connector unit shall consist of one of the following:

(a) One split ring with its bolt in single shear; or

(b) One toothed ring with its bolt in single shear; or

(c) Two shear plates used back to back in the contact faces of a timber-to-timber joint with their bolt in single shear; or one shear plate with its bolt in single shear used in conjunction with a steel strap or shape in a timber-to-metal joint.

500-A-2. In installation of connectors and bolts, a nut shall be placed on each bolt, and washers, not smaller than the size specified in Tables 9.13, 9.14, and 9.15 shall be placed between the outside wood member and the bolt head and between the outside wood member and nut. When an outside member is a steel strap or shape, the washer may be omitted, except when needed to extend bolt length to prevent metal plate or shape from bearing on threaded portion of bolt when used in conjunction with shear plates.

500-B. *Number of Connectors.*

500-B-1. Tabulated loads are for ONE connector unit with bolt in shear in any joint of any number of members.

500-B-2. For a joint assembly in which two or more connector units of the same size are used in the contact faces with the connectors concentric with the same bolt axis or in which two or more bolts are used with connectors on separate bolts, the total allowable connector load shall be the sum of the allowable connector loads given for each connector unit used. This provision applies to all conditions, except that connectors shall not be placed concentrically on the same bolt in the same timber surface, except as provided in sections 500-B-3 and 500-B-4.

500-B-3. If grooves for two sizes of split rings are cut, concentric in the same timber surface, rings shall be installed in both grooves and the total allowable load shall be the tabulated load for the larger ring only.

500-B-4. Two toothed rings in combination of the sizes 2 and 3⅜ inches, 2 and 4 inches, or 2⅝ and 4 inches may be concentric to the same bolt between the same timber surfaces. For these combinations the total allowable connector load shall be the tabulated allowable load for the larger ring plus 25 per cent of the tabulated load for the smaller ring.

500-C. *Allowable Connector Loads.*

500-C-1. The allowable connector loads given

in Tables 9.13, 9.14, and 9.15 are applicable for the connectors described in section 500-E for all conditions other than for which specific exceptions are made.

500-C-2. Where a connector joint is fully loaded to the maximum allowable load for many years, either continuously or cumulatively under the conditions of maximum design load use working stresses 90 per cent of those in Tables 9.13, 9.14, and 9.15.

500-C-3. When the duration of full maximum load does not exceed the period indicated, increase the allowable unit stresses in Tables 9.13 and 9.15 as follows:

15 percent for two months' duration, as for snow.
25 percent for seven days' duration.
33⅓ percent for wind or earthquake.
100 percent for impact.

These increases are applicable to the tabulated loads for split rings and shear plates (except that the limitations in notes in Table 9.15 apply for shear plates). These increases are not applicable to the tabulated loads for toothed rings. An increase of 20 per cent is permitted for wind, earthquake or impact on toothed rings. Otherwise the same conditions of load duration apply to connectors as apply to lumber.

500-D. *Condition of Lumber.*

500-D-1. For connectors used in lumber which is seasoned to approximately 15 per cent moisture content to a depth of three-fourths inch from the surface, before fabrication, and which will remain dry in service, the tabulated connector loads, as adjusted by section 500-C, apply for all connectors.

500-D-2. For connectors used in lumber which is fabricated before it is seasoned to a depth of three-fourths inch from the surface and which later is seasoned either before erection or while in the structure, 80 per cent of the tabulated connector loads, as adjusted by section 500-C, shall apply for all connectors.

500-D-3. For lumber partially seasoned when fabricated, proportional intermediate connector values may be used.

500-D-4. For connectors used in lumber which is fabricated in a seasoned or unseasoned condition and will remain wet in service, 67 per cent of the tabulated connector loads, as adjusted by section 500-C, shall apply for all connectors.

500-D-5. For connectors used in lumber, pressure-impregnated with fire-retardant chemicals, 80 per cent of the tabulated loads, as adjusted by section 500-C, shall apply.

500-E. *Quality of Connectors.*

500-E-1. Specifications for timber connectors and tabulated loads and modifications thereof given herein are for connectors of a quality as follows:

(a) Split-ring timber connectors manufactured from hot rolled carbon steel meeting SOCIETY OF AUTOMOTIVE ENGINEERS SPECIFICATION SAE-1010 (1943). Each ring shall form a closed true circle with the principal axis of the cross-section of the ring metal parallel to the geometric axis of the ring. The ring shall fit snugly in the precut groove. This may be accomplished with a ring, the metal section of which is beveled from the central portion toward the edges to a thickness less than at midsection, or by any other method which will accomplish equivalent performance. It shall be cut through in one place in its circumference to form a tongue and slot. (See Fig. 9.14.)

Fig. 9.14 Split Ring

(b) Toothed-ring timber connectors stamped cold from U. S. Standard 16 gage hot rolled sheet steel conforming to AMERICAN SOCIETY FOR TESTING AND MATERIALS STANDARD SPECIFICATIONS FOR CARBON STEEL A 17-29, Type A, Grade 1, and bent cold to form a circular, corrugated, sharp-toothed band and welded into a solid ring. The teeth on each ring shall be on a true circle and shall be parallel to the axis of the ring. The central band shall be welded to fully develop the strength of the band. (See Fig. 9.15.)

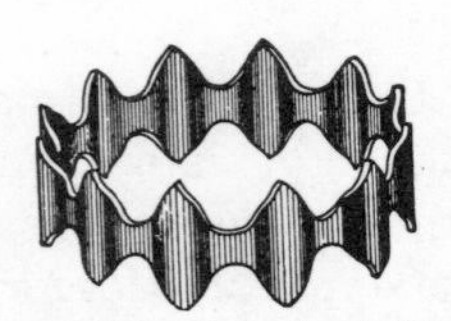

Fig. 9.15 Toothed Ring

(c) Shear-plate timber connectors:

(1) Pressed Steel Type—Pressed steel shear plates manufactured from hot rolled carbon steel meeting SOCIETY OF AUTOMOTIVE ENGINEERS SPECIFICATION SAE-1010 (1943). Each plate shall be a true circle with a flange around the edge, extending at right angles to the face of the plate and extending from one face only, the plate portion having a central bolt hole, with an integral hub concentric to the hole or without an integral hub, and two small perforations on opposite sides of the hole and midway from the center and circumference. (See Fig. 9.16.)

(2) Malleable Iron Type—Malleable iron shear

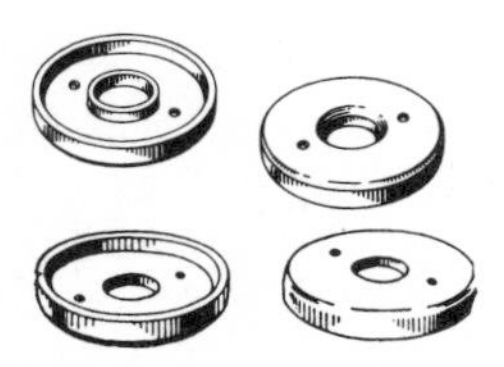

Fig. 9.16 Pressed Steel Shear Plates

plates manufactured according to AMERICAN SOCIETY FOR TESTING AND MATERIALS STANDARD SPECIFICATIONS A 47-33, Grade 35018, for malleable iron castings. Each casting shall consist of a perforated round plate with a flange around the edge extending at right angles to the face of the plate and projecting from one face only, the plate portion having a central bolt hole with an integral hub extending from the same face as the flange. (See Fig. 9.17.)

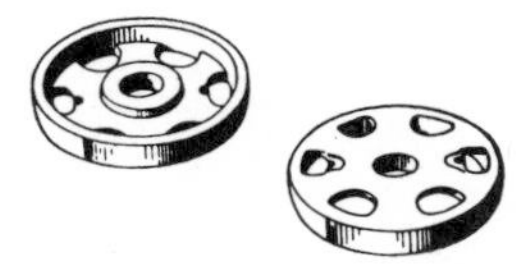

Fig. 9.17 Malleable Iron Shear Plates

500-F. *Species and Grades of Lumber.*

500-F-1. The tabulated loads for Groups A, B, C, and D, in Tables 9.13, 9.14, and 9.15 apply to stress-graded lumber of the species as listed in Table 9.12.

500-H. *Tightness of Joints.*

500-H-1. The allowable loads for connectors are based on the assumption that the faces of the members are brought into contact when the connectors are installed.

500-H-2. Tabulated connector loads and modifications thereof assume and allow for seasonal variations after the lumber has reached the moisture content normal to the conditions of service.

500-H-3. When lumber is not seasoned to the moisture content normal to the conditions of service, the joints should be drawn up by turning down the nuts on the bolts, periodically, until moisture equilibrium is reached so as to keep the adjacent faces of the members in contact.

500-I. *Side Members—Materials.*

500-I-1. Tabulated connector loads are for side members of wood.

500-I-2. If metal plates instead of wood side plates are used, the tabulated allowable connector loads for parallel-to-gain loading shall be modified in accordance with the notes on Table 9.15.

500-J. *Thickness of Lumber.*

500-J-1. Tabulated loads shall not be used for connectors installed in lumber of a net thickness less than specified in Tables 9.13, 9.14, and 9.15.

500-J-2. Loads for connectors installed in lumber of net thickness other than specified in Tables 9.13, 9.14, and 9.15 shall be obtained by straightline interpolation for thicknesses intermediate to those given.

500-K. *Edge Distance, Connectors.*

500-K-1. Edge distance, as given in Tables 9.13, 9.14, and 9.15, is the distance from edge of member to center of connector closest to the edge of the member measured perpendicular to the edge. (See Figs. 9.18 and 9.19.)

500-K-2. Edge Distance for Members Loaded Parallel to Grain (0°): The tabulated edge distance for parallel-to-grain loading, given in Tables 9.13, 9.14, and 9.15, is the minimum edge distance to be used.

500-K-3. Edge Distance for Members Loaded Perpendicular to Grain (90°): For minimum edge distance for perpendicular-to-grain loading, distinction must be made between the loaded and unloaded edges, as follows (see Fig. 9.18):

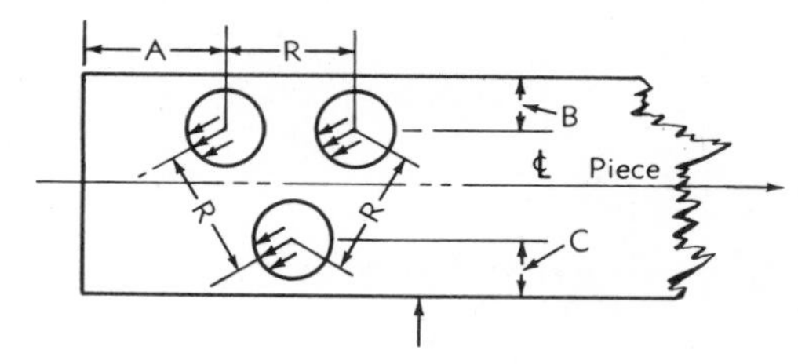

A = End distance.
B = Unloaded-edge distance.
C = Loaded-edge distance.
R = Spacing.

Fig. 9.18 End Distance, Edge Distance, and Spacing Ring Connectors

(a) Unloaded-Edge Distance. The tabulated unloaded-edge distance, given in Tables 9.13, 9.14, and 9.15 for perpendicular-to-grain loading, is the minimum distance to be used.

(b) Loaded-Edge Distance. The minimum loaded-edge distance, given in Tables 9.13, 9.14, and 9.15, is the least distance that may be used. When this distance is increased to the amount indicated in the tables, the larger load is permitted. For an intermediate loaded-edge distance, the allowable connector load is obtained by straight-line interpolation.

500-K-4. Edge Distance for Members Loaded at Angles to Grain Other Than 0° and 90°: The unloaded-edge distance and the minimum loaded-edge distance, given in Tables 9.13, 9.14, and 9.15, apply for all angles of load to grain. When

Table 9.7 Connector Spacings and End Distances for Parallel-to-Grain Loading with Corresponding Percentages of Tabulated Loads

Connector and Diameter	Spacing Parallel to Grain		Spacing Perpendicular to Grain		End Distance		
	Spacing	Percentage of Tabulated Load	Minimum	Percentage of Tabulated Load	Tension Member	Compression Member	Percentage of Tabulated Load
	(Inches)	(Per cent)	(Inches)	(Per cent)	(Inches)	(Inches)	(Per cent)
Split-Ring							
2½″	6¾	100	3½	100	5½	4	100
2½″	3½	75	3½	100	2¾	2½	62.5
4″	9	100	5	100	7	5½	100
4″	5	75	5	100	3½	3¼	62.5
Toothed-Ring							
2″	4	100	2½	100	3½	2*	100
2″	2½	75	2½	100	2	----------	66.7
2⅝″	5¼	100	3⅛	100	4⅝	2⅝*	100
2⅝″	3⅛	75	3⅛	100	2⅝	----------	66.7
3⅜″	6¾	100	3⅞	100	5⅞	3⅜*	100
3⅜″	3⅞	75	3⅞	100	3⅞	----------	66.7
4″	8	100	4½	100	7	4*	100
4″	4½	75	4½	100	4	----------	66.7
Shear-Plate							
2⅝″	6¾	100	3½	100	5½	4	100
2⅝″	3½	75	3½	100	2¾	2½	62.5
4″	9	100	5	100	7	5½	100
4″	5	75	5	100	3½	3¼	62.5

* No reduction in end distance is permitted for compression members loaded parallel to grain.

Table 9.8 Connector Spacings and End Distances for Perpendicular-to-Grain Loading with Corresponding Percentages of Tabulated Loads

Connector and Diameter	Spacing Parallel to Grain		Spacing Perpendicular to Grain		End Distance	
	Minimum	Percentage of Tabulated Load	Spacing	Percentage of Tabulated Load	Distance (Tension or Compression Members)	Percentage of Tabulated Load
	(Inches)	(Per cent)	(Inches)	(Per cent)	(Inches)	(Per cent)
Split-Ring						
2½″	3½	100	4¼	100	5½	100
2½″	3½	100	3½	75	2¾	62.5
4″	5	100	6	100	7	100
4″	5	100	5	75	3½	62.5
Toothed-Ring						
2″	2½	100	3	100	3½	100
2″	2½	100	2½	75	2	66.7
2⅝″	3⅛	100	3¾	100	4⅝	100
2⅝″	3⅛	100	3⅛	75	2⅝	66.7
3⅜″	3⅞	100	5	100	5⅞	100
3⅜″	3⅞	100	3⅞	75	3⅜	66.7
4″	4½	100	5¾	100	7	100
4″	4½	100	4½	75	4	66.7
Shear-Plate						
2⅝″	3½	100	4¼	100	5½	100
2⅝″	3½	100	3½	75	2¾	62.5
4″	5	100	6	100	7	100
4″	5	100	5	75	3½	62.5

the larger load, due to increased loaded-edge distance, is used, the required edge distance shall be determined as follows:

(a) For angles of load to grain of 45° to 90°, the tabulated loaded-edge distance at 90° shall apply.

(b) For angles of load to grain of 0° to 45°, the required edge distance shall be determined by straight-line interpolation.

500-L. *End Distance, Connectors.*

500-L-1. End distance, as given in Tables 9.7 and 9.8, is the distance measured parallel to the grain from center of connector to the square-cut end of the member. (See Fig. 9.18.)

500-L-2. If the end of the member is not cut at right angles to its length, the end distance, measured parallel to the center line of the piece from any point on the center half of the connector diameter, which is perpendicular to the center line of the piece, shall not be less than the end distance required for a square-cut member. In no case shall the perpendicular distance, from center of the connector to sloping end cut of a member, be less than the required edge distance. (See Fig. 9.19.)

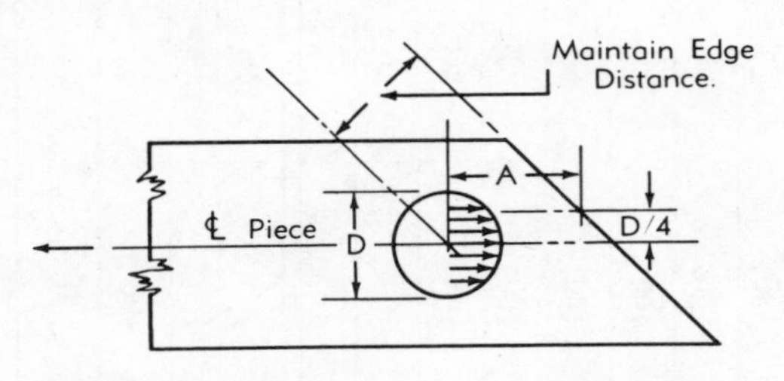

Fig. 9.19 End Distance for Member with Sloping End Cut

500-L-3. End distances for members loaded parallel to grain, and appropriate percentages of tabulated loads are listed in Table 9.7. For a distance intermediate between minimum and that required for full load, the applicable load is determined by straight-line interpolation.

500-L-4. End distances for members loaded perpendicular to grain, and the appropriate percentages of tabulated loads are listed in Table 9.8. For a distance intermediate between minimum and that required for full load, the applicable load is determined by straight-line interpolation.

500-L-5. End distances for members loaded at angles to grain, other than 0° and 90°, are as follows:

(a) Tension Member: End distances are the same for all angles of load to grain.

(b) Compression Member: For loading at angles intermediate of 0° and 90°, the end distances corresponding to full allowable load shall be determined by straight-line interpolation.

500-M. *Spacings, Connectors.*

500-M-1. Spacing, as given in Tables 9.7, 9.8, and 9.9, is the distance between centers of connectors measured along a line joining their centers. (See Fig. 9.18.)

500-M-2. Spacings for Members Loaded Parallel to Grain (0°): For parallel-to-grain loading, required spacings, parallel and perpendicular to grain, for the appropriate percentages of the tabulated loads are listed in Table 9.7. For spacing intermediate between the minimum and that required for maximum load, the allowable load is determined by straight-line interpolation.

500-M-3. Spacings for Members Loaded Perpendicular to Grain (90°): For perpendicular-to-grain loading, required spacings, parallel and perpendicular to grain, for the appropriate percentages of the tabulated loads are listed in Table 9.8. For spacing intermediate between the minimum and that required for maximum load, the allowable load is determined by straight-line interpolation.

500-M-4. Spacings for Members Loaded at an Angle of Grain Other Than 0° and 90°, and With Connector Axis at Various Angles With Grain: For tension and compression members, the spacing is measured along the connector axis which is the line joining the centers of two adjacent connectors. The spacing is determined in the following manner:

(a) For tension and compression members loaded at an angle of load to grain (θ) other than 0° and 90°, or with connector axis at an angle (ϕ) to grain other than 0° and 90°, the spacings for allowable connector loads, as provided in section 500-N, shall be determined from the following formula:

$$R = \frac{AB}{\sqrt{A^2 \sin^2 \phi + B^2 \cos^2 \phi}}$$

in which

R = spacing in inches along the connector axis required for allowable load as determined in section 500-N.

A = dimension in column (3), Table 9.9 opposite the connector in column (1) and angle of load to grain in column (2).

B = dimension in column (4), Table 9.9 opposite the connector in column (1) and angle of load to grain in column (2).

ϕ = angle of connector axis to grain.

(Graphical solution of this formula is provided in Fig. 9.20.)

(b) The dimension C in column (5), Table 9.9,

Table 9.9 Values for Use with Section 500-M-4

1 Type and size of connector	2 Angle of load to grain (θ)	3 A	4 B	5 C (75 percent value)
	(Degrees)	(Inches)	(Inches)	(Inches)
2½″ split ring or 2⅝″ shear plate	0	6¾	3½	3½
	15	6	3¾	3½
	30	5⅛	3⅞	3½
	45	4¼	4⅛	3½
	60–90	3½	4¼	3½
4″ split ring or 4″ shear plate	0	9	5	5
	15	8	5¼	5
	30	7	5½	5
	45	6	5¾	5
	60–90	5	6	5
2″ toothed ring	0	4	2½	2½
	15	3⅝	2⅝	2½
	30	3¼	2¾	2½
	45	2⅞	2⅞	2½
	60–90	2½	3	2½
2⅝″ toothed ring	0	5¼	3⅛	3⅛
	15	4¾	3⅜	3⅛
	30	4¼	3½	3⅛
	45	3¾	3⅝	3⅛
	60–90	3⅛	3¾	3⅛
3⅜″ toothed ring	0	6¾	3⅞	3⅞
	15	6	4¼	3⅞
	30	5¼	4½	3⅞
	45	4½	4¾	3⅞
	60–90	3⅞	5	3⅞
4″ toothed ring	0	8	4½	4½
	15	7⅛	4¾	4½
	30	6¼	5⅛	4½
	45	5⅜	5⅜	4½
	60–90	4½	5¾	4½

is the minimum spacing along the connector axis and will permit 75 percent of the load provided in section 500-N.

(c) For spacings intermediate between R and C, the allowable load shall be determined by straight-line interpolation between the loads for R and C spacings.

500-N. *Load at Angle to Grain.*

500-N-1. The angle of load to grain is the angle between the direction of load acting on the member and the longitudinal axis of the member.

500-N-2. For angles of load to grain other than 0° and 90°, the allowable connector load shall be determined by application of the Hankinson formula between the allowable connector loads for parallel and perpendicular-to-grain loading given in Tables 9.13, 9.14, and 9.15, except that:

(a) For Toothed Rings, allowable loads shall be determined as follows:

(1) The allowable connector loads for angles of load to grain of 45° to 90° shall be the same as the tabulated load for an angle of load to grain of 90°.

(2) For angles of load to grain intermediate of 0° and 45°, the allowable connector loads shall be determined by application of the Hankinson formula. To permit interpolation

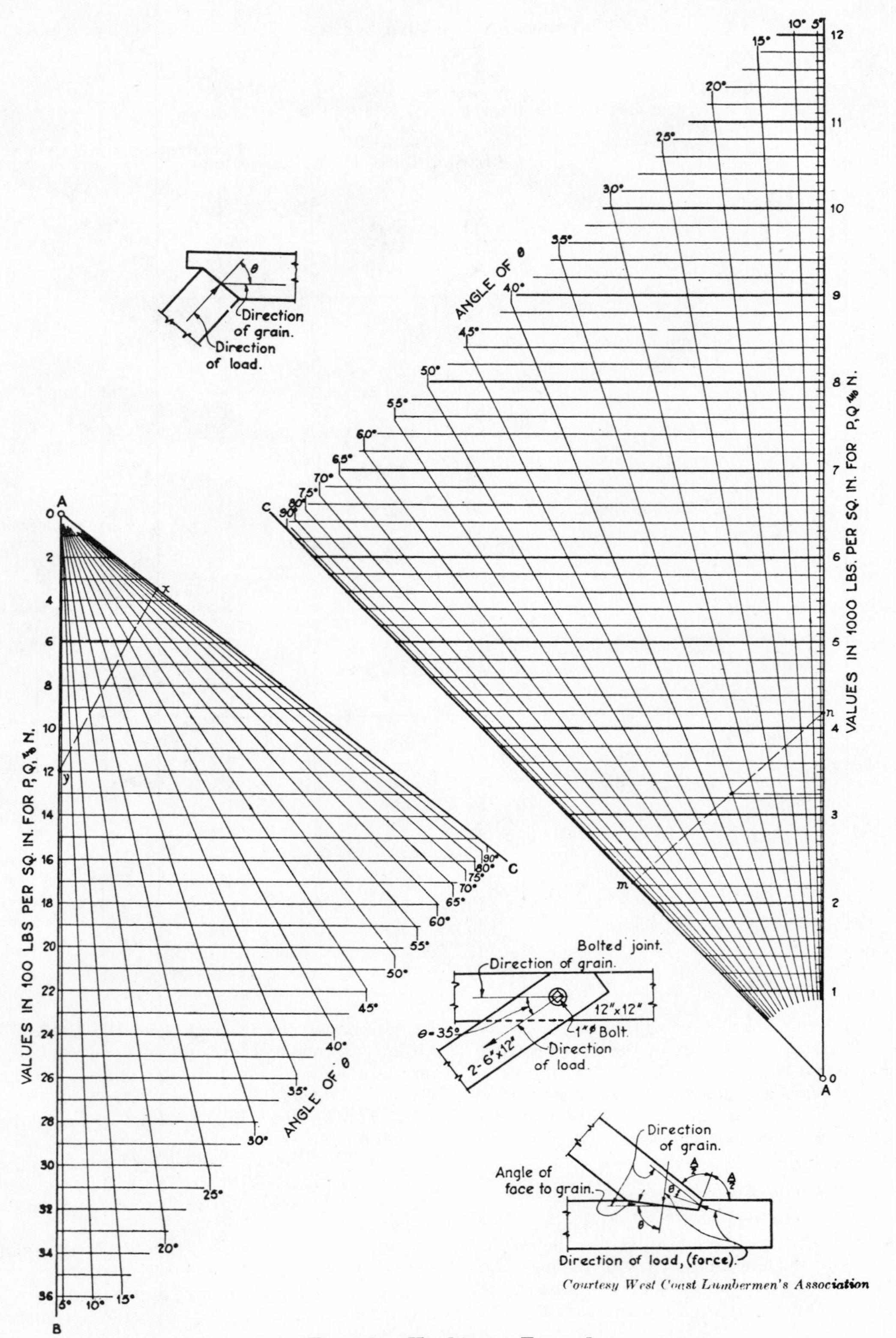

Fig. 9.20 Hankinson Formula

between 0° and 45° in this case, the actual angle shall be doubled when inserted in the Hankinson formula (Fig. 9.20).

(b) For Shear Plates, the loads determined by application of the Hankinson formula shall not exceed the limits imposed by the notes in Table 9.15.

500-O. *Inter-Relationship of Distances and Spacing.*

500-O-1. When tabulated load is reduced, due to edge distance, end distance or spacing, the reduced allowable load for each shall be determined separately, and the lowest allowable load so determined for any one connector shall apply. Such load reductions are not cumulative.

501-USE OF LAG SCREWS INSTEAD OF BOLTS WITH TIMBER CONNECTORS

501-A. *Type of Screw.*

501-A-1. The lag screw shall have a cut thread, not a rolled thread.

501-B. *Diameter of Lag Screw.*

501-B-1. The shank of the lag screw shall have the same diameter as the bolt specified for the connector.

501-C. *Hole for Shank and Threaded Portion.*

501-C-1. The hole for the shank shall be the same diameter as the shank.

501-C-2. The hole for the threaded portion of lag screw shall have a diameter equal to approximately 75 percent of that of the shank.

501-D. *Allowable Loads.*

501-D-1. Full Allowable Load. When lag screws are used with connectors, the full allowable load (the load for one connector unit with bolt) may be used for 2½-inch and 4-inch split rings and 4-inch shear plates when the minimum penetration of the lag screw into the member receiving the point, is 7 diameters for Group I woods, 8 diameters for Group II, 10 diameters for Group III, and 11 diameters for Group IV. For 2⅝-inch shear plates, the full allowable load may be used when the minimum penetration is 4 diameters for Group I woods, 5 diameters for Group II, 7 diameters for Group III, and 8 diameters for Group IV. (See Table 9.12.)

501-D-2. Reduction in Allowable Load. The allowable load for 2½-inch and 4-inch split rings and 2⅝-inch and 4-inch shear plates, when used with lag screws, shall vary uniformly from the full allowable load with penetration as specified in section 501-D-1, to 75 per cent of the full allowable load with penetration of 3 diameters for Group I woods, 3½ diameters for Group II, 4 diameters for Group III, and 4½ diameters for Group IV. When metal side plates are used with 2⅝-inch shear plates, the full allowable load may be used for the minimum penetrations specified (Fig. 9.21).

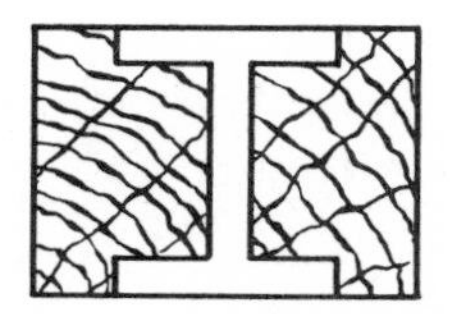

Fig. 9.21 Net Cross Section of Timber

502-NET SECTION

502-A. *Located at Critical Plane.*

502-A-1. The net section shall be determined by subtracting, from the full cross-sectional area of the timber, the projected area of that portion of the connector groove within the member and that portion of the bolt hole not within the connector groove located at the critical plane. (See Table 9.10.) Where connectors are staggered, adjacent connectors, with parallel-to-grain spacing equal to or less than one connector diameter, shall be considered as occurring at the same critical section.

502-B. *For Sawn Lumber.*

502-B-1. The stresses in the net section shall not exceed seven-eighths (⅞) of the allowable unit stress in bending for the grade of lumber used unless the provisions of section 502-B-2 are observed.

502-B-2. If knots occur at the critical section, the cross-sectional area of the knots outside the area deducted for connectors and bolts should also be deducted in determining the net section.

502-B-3. In tension and compression members, the critical or net section of a timber in square inches shall not be less than that determined by dividing the total load in pounds which is transferred through the critical section of the member, by the appropriate constant in Table 9.11.

502-C. *For Glued-Laminated Timber.*

502-C-1. Knots at or near the critical section shall be disregarded in determining the net section.

502-C-2. The required net section in tension and compression shall be determined by dividing the total load in pounds which is transferred through the critical section of the member by the appropriate allowable unit compressive stress permitted for the glued-laminated material in Part IX.

Problems in the design of timber connector joints are given in Art. 10.12.

Table 9.10 Typical Dimensions for the Timber Connectors

SPLIT RINGS

Dimensions in Inches

	2½"	4"
Split ring:		
Inside diameter at center when closed	2.500	4.000
Thickness of metal at center	.163	.193
Depth of metal (width of ring)	.750	1.000
Groove:		
Inside diameter	2.56	4.08
Width	.18	.21
Depth	.375	.50
Bolt hole:		
Diameter	9/16	13/16

	2½"	4"
Washers, standard:		
Round, cast or malleable iron, diameter	2⅝	3
Round, wrought iron (minimum):		
Diameter	1⅜	2
Thickness	3/32	5/32
Square plate:		
Length of side	2	3
Thickness	⅛	3/16
Projected area:		
Portion of one ring within member, sq. in.	1.10	2.25

TOOTHED RINGS

Dimensions in Inches

	2"	2⅝"	3⅜"	4"
Toothed ring:				
Diameter	2.000	2.625	3.375	4.000
Thickness of metal	.061	.061	.061	.061
Depth	.940	.940	.940	.940
Depth of filet (minimum)	.250	.250	.250	.250
Bolt hole:				
Diameter	9/16	11/16	13/16	13/16

	2"	2⅝"	3⅜"	4"
Washers, minimum:				
Round, cast or malleable iron (diameter)	2	2⅝	3	3½
Square plate:				
Length of side	2	2½	3	3½
Thickness	3/16	¼	¼	⅜
Projected area:				
Portion of one ring within member, sq. in.	.94	1.23	1.59	1.89

SHEAR PLATES

Dimensions in Inches

	2⅝"	2⅝"	4"	4"
Shear plate:				
Material	Pressed steel	Light gage	Malleable iron	Malleable iron
Diameter of plate	2.62	2.62	4.02	4.02
Diameter of bolt hole	0.81	0.81	0.81	0.94
Thickness of plate	0.172	0.12	0.20	0.20
Depth of flange	0.42	0.35	0.62	0.62
Circular dap — dimensions:				
A	2.63	2.63	4.03	4.03
B		1.07	1.55	1.55
C	0.81	0.81	0.81	0.94
D		0.65	0.97	0.97
E	0.19	0.13	0.27	0.27
F	0.45	0.38	0.64	0.64
G	0.25	0.14	0.22	0.22
H		0.34	0.50	0.50
I	2.25	2.37	3.49	3.49

Bolt Hole

	2⅝"	2⅝"	4"	4"
Steel strap or shapes for use with shear plates: Steel straps or shapes, for use with shear plates, shall be designed in accordance with accepted engineering practices.				
Hole diameter in straps or shapes for bolts	13/16	13/16	13/16	15/16
Bolt hole — diameter in timber	13/16	13/16	13/16	15/16
Washers, standard:				
Round, cast or malleable iron, diameter	3	3	3	3½
Round, wrought iron, minimum:				
Diameter	2	2	2	2¼
Thickness	5/32	5/32	5/32	11/64
Square plate:				
Length of side	3	3	3	3
Thickness	¼	¼	¼	¼
Projected area:				
Portion of one shear plate within member, sq. in.	1.18	1.00	2.58	2.58

9.31 Nails and screws

a Tests by the Forest Products Laboratory of the U.S. Department of Agriculture provide the best information on the holding power of nails and screws. The resistance to withdrawal of common wire nails is given by the formula $P = 1.150G^2d$, where P is the safe load per linear inch of penetration, G the specific gravity of the wood, and d the diameter of the nail in inches. Table 9.18 gives the value per inch desired from the above formula for various sizes of nails driven into the commoner kinds of seasoned wood.

Table 9.11 Constants for Use in Determining Required Net Section in Square Inches

Duration of loading	Thickness of wood member in inches	Constants for each connector load group			
		Group A	Group B	Group C	Group D
Normal	4 inches or less	2350	2000	1650	1300
	Over 4 inches	1850	1600	1300	1050
Permanent	4 inches or less	2100	1800	1500	1200
	Over 4 inches	1700	1450	1200	950
Snow	4 inches or less	2700	2300	1900	1500
	Over 4 inches	2150	1850	1500	1200
Wind or earthquake	4 inches or less	3100	2650	2200	1750
	Over 4 inches	2500	2150	1750	1400

Nails driven into bored holes of slightly less than their own diameter have a somewhat higher holding power than nails without lead holes.

b Although the above formula and Table 9.18 have been developed for the safe resistance to withdrawal for nails, it is not good policy to use nails in this manner. They should be used so that the stresses are resisted in shear instead of in direct tension. For this condition, the allowable load per common wire nail in lateral resistance when driven into the side grain of seasoned wood is determined from the formula $P = Kd^{1.5}$.

Table 9.19 gives the value of the constant K and the value per nail for various sizes of nails used in different varieties of wood. These loads apply when woods of approximately the same density are used, and where the nail penetrates the main member two-thirds of its length for softwood, or one-half for hardwood. Where metal side plates are used, the above values may be increased 25 percent. When driven into the end grain, two-thirds of the values given in Table 9.19 may be used. Nails must be so spaced and staggered as to avoid splitting the wood. It is advantageous to bore lead holes for the nails slightly smaller than the diameter of the nail.

Problem Yellow-pine roof rafters on a 45-deg slope, spanning 20 ft, spaced 16 in. o.c., and carrying a load of 50 psf are tied to prevent spreading by being nailed to the attic floor joists. What nailing is required? (Fig. 9.22).

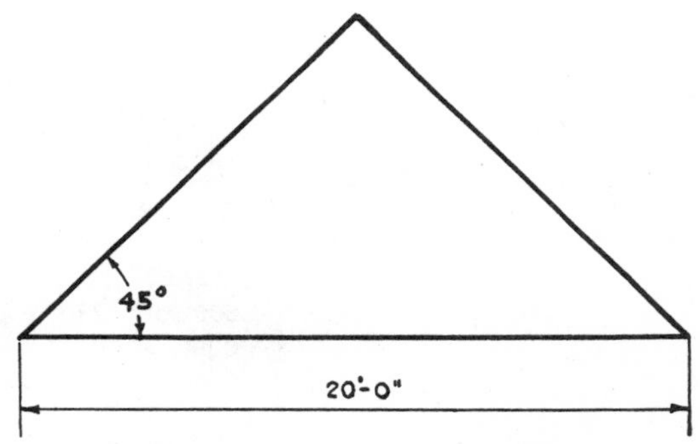

Fig. 9.22 Roof Rafters as a Truss

The load causing thrust is $10 \times 1.33 \times 50 = 666$ lb. Since the slope is 45 deg, the thrust is equal to one-half of this, or 333 lb. This will require three 20d nails. Since both rafters and floor joists will probably be nominal 2-in. material ($1\frac{5}{8}$ in. actual thickness), the nail cannot have two-thirds of its length in the main member, but it can be clinched to serve the same purpose (Fig. 9.23).

c For wood screws, the total safe lateral resistance is given by the formula $P = Kd^2$, where K is a constant as follows:

For eastern hemlock or white pine:	2,100
For Norway pine or redwood:	2,700
For Douglas fir or yellow pine:	3,300
For maple or oak:	4,000

Table 9.12 Grouping of Species for Determining Allowable Loads for Lag Screws, Nails, Spikes, Wood Screws, Drift Bolts

Group	Species of Wood	Specific Gravity (G)
I	Ash, commercial white	0.64
	Beech	.67
	Birch, sweet	.71
	Birch, yellow	.66
	Elm, rock	.66
	Hickory, true	.80
	Maple (hard), black	.62
	Maple (hard), sugar	.68
	Oak, commercial red	.66
	Oak, commercial white	.71
	Pecan	.69
II	Douglas fir, coast region	0.51
	Douglas fir	.51
	Larch	.59
	Pine, southern	.59
III	Cedar, Alaska	0.46
	Cedar, Port Orford	.44
	Cypress, southern	.48
	Hemlock, West Coast	.44
	Hemlock, western	.44
	Pine, Norway	.47
	Redwood	.42
	Sweet gum	.53
IV	Basswood	0.40
	Cedar, northern white	.32
	Cedar, southern white	.35
	Cedar, western red	.34
	Cottonwood	.37
	Fir, Balsam	.41
	Fir, commercial white	.42
	Hemlock, eastern	.43
	Pine, lodgepole	.43
	Pine, ponderosa	.42
	Pine, sugar	.38
	Pine, eastern white	.37
	Pine, western white	.42
	Spruce, Engelmann	.36
	Spruce, red	.41
	Spruce, Sitka	.42
	Spruce, white	.45
	Yellow poplar	.43

Table 9.13 Allowable Loads for One Split Ring and Bolt in Single Shear

The allowable loads below are for normal loading conditions. See other provisions of Part V for adjustments of these tabulated allowable loads.

Split-Ring diam. (inches)	Bolt diam. (inches)	Number of faces of piece with connectors on same bolt	Thickness (net) of lumber (inches)	Loaded parallel to grain (0°): Edge distance (inches)	Loaded parallel to grain (0°): Allowable load per connector unit and bolt (pounds): Group A woods	Group B woods	Group C woods	Group D woods	Loaded perpendicular to grain (90°): Edge distance (inches): Unloaded-edge (See Fig. 12)	Loaded-edge (See Fig. 12)	Loaded perpendicular to grain (90°): Allowable load per connector unit and bolt (pounds): Group A woods	Group B woods	Group C woods	Group D woods
2½	½	1	1 min.	1¾ min.	2630	2270	1900	1640	1¾ min.	1¾ min.	1580	1350	1130	970
										2¾ or more	1900	1620	1350	1160
			1⅝ and thicker	1¾ min.	3160	2730	2290	1960	1¾ min.	1¾ min.	1900	1620	1350	1160
										2¾ or more	2280	1940	1620	1390
		2	1⅝ min.	1¾ min.	2630	2270	1900	1640	1¾ min.	1¾ min.	1580	1350	1130	970
										2¾ or more	1900	1620	1350	1160
			2 and thicker	1¾ min.	3160	2730	2290	1960	1¾ min.	1¾ min.	1900	1620	1350	1160
										2¾ or more	2280	1940	1620	1390
4	¾	1	1 min.	2¾ min.	4090	3510	2920	2520	2¾ min.	2¾ min.	2370	2030	1700	1470
										3¾ or more	2840	2440	2040	1760
			1⅝ and thicker	2¾ min.	6140	5260	4380	3790	2¾ min.	2¾ min.	3560	3050	2540	2190
										3¾ or more	4270	3660	3050	2630
		2	1⅝ min.	2¾ min.	4310	3690	3070	2660	2¾ min.	2¾ min.	2490	2140	1780	1540
										3¾ or more	3000	2570	2140	1850
			2	2¾ min.	4950	4250	3540	3050	2¾ min.	2¾ min.	2870	2470	2050	1770
										3¾ or more	3440	2960	2460	2120
			2⅝	2¾ min.	6030	5160	4310	3720	2¾ min.	2¾ min.	3490	3000	2490	2150
										3¾ or more	4180	3600	3000	2580
			3 and thicker	2¾ min.	6140	5260	4380	3790	2¾ min.	2¾ min.	3560	3050	2540	2190
										3¾ or more	4270	3660	3050	2630

The values for converting screw gage to d^2 are as follows:

Ga	d^2	Ga	d^2
0	0.0036	9	0.0313
1	0.0053	10	0.0361
2	0.0074	11	0.0412
3	0.0098	12	0.0467
4	0.0125	14	0.0586
5	0.0156	16	0.0718
6	0.0190	18	0.0864
7	0.0228	20	0.1024
8	0.0269	24	0.1384

The above formula is based on a length of screw in the holding block equal to seven times the shank diameter.

9.32 Bolted wood joints

a The design of bolted wood joints does not involve simple compression on the grain, but rather complex bearing and bending, and is most easily done by the use of Table 9.20 which was compiled by the National Lumber Manufacturers Association.

In this table, P and Q are unit stresses in compression parallel and perpendicular to the grain, respectively. The safe loads given are for one bolt, acting singly or in combination, with snug-fitting bolt holes. The table is designed for side members half the thickness of the center member. If side members are thinner than one-half the main member, the safe load is decreased proportionately. Where steel plates are used for side members, the tabulated values may be increased by 25 percent for loads parallel to the grain, with no increase being allowed for loads perpendicular to the grain.

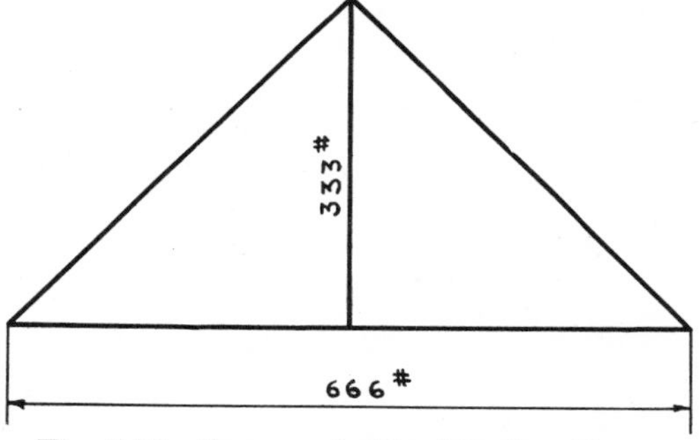

Fig. 9.23 Stresses in Roof-Rafter Truss

The woods that may be used in each group in Table 9.20 are:

Group 1. Tidewater red cypress, coast and inland Douglas fir, western larch, south-

Table 9.14 Allowable Loads for One Toothed-Ring and Bolt in Single Shear

The allowable loads below are for normal loading conditions. See other provisions of Part V for adjustments of these tabulated allowable loads

Toothed-Ring diam. (inches)	Bolt diam. (inches)	Number of faces of piece with connectors on same bolt	Thickness (net) of lumber (inches)	Loaded parallel to grain (0°): Edge distance (inches)	Loaded parallel to grain (0°): Allowable load per connector unit and bolt (pounds): Group A woods	Group B woods	Group C woods	Group D woods	Loaded perpendicular to grain (90°): Edge distance (inches): Unloaded-edge (See Fig. 12)	Loaded-edge (See Fig. 12)	Loaded perpendicular to grain (90°): Allowable load per connector unit and bolt (pounds): Group A woods	Group B woods	Group C woods	Group D woods
2	½	1	1 min.	1¼ min.	1210	1100	990	860	1¼ min.	1¼ min.	810	730	660	570
										2 or more	930	840	760	650
			1⅝ and thicker	1¼ min.	1330	1210	1090	950	1¼ min.	1¼ min.	890	810	730	630
										2 or more	1020	930	840	720
		2	1⅝ min.	1¼ min.	1210	1100	990	860	1¼ min.	1¼ min.	810	730	660	570
										2 or more	930	840	760	650
			2 and thicker	1¼ min.	1330	1210	1090	950	1¼ min.	1¼ min.	890	810	730	630
										2 or more	1020	930	840	720
2⅝	⅝	1	1 min.	1¾ min.	1820	1650	1490	1290	1¾ min.	1¾ min.	1210	1100	990	860
										2½ or more	1390	1260	1140	990
			1⅝ and thicker	1¾ min.	2270	2030	1850	1610	1¾ min.	1¾ min.	1510	1370	1240	1070
										2½ or more	1730	1570	1430	1230
		2	1⅝ min.	1¾ min.	1820	1650	1490	1290	1¾ min.	1¾ min.	1210	1100	990	860
										2½ or more	1390	1260	1140	990
			2	1¾ min.	2010	1830	1640	1420	1¾ min.	1¾ min.	1340	1220	1090	950
										2½ or more	1540	1400	1250	1090
			2⅝ and thicker	1¾ min.	2270	2030	1850	1610	1¾ min.	1¾ min.	1510	1370	1240	1070
										2½ or more	1730	1570	1430	1230
3⅝	¾	1	1 min.	2¼ min.	2360	2150	1930	1670	2¼ min.	2¼ min.	1570	1430	1290	1120
										3¼ or more	1880	1720	1550	1340
			1⅝ and thicker	2¼ min.	3180	2890	2610	2250	2¼ min.	2¼ min.	2120	1930	1740	1510
										3¼ or more	2540	2320	2090	1810
		2	1⅝ min.	2¼ min.	2360	2150	1930	1670	2¼ min.	2¼ min.	1570	1430	1290	1120
										3¼ or more	1880	1720	1550	1340
			2	2¼ min.	2590	2350	2110	1830	2¼ min.	2¼ min.	1720	1570	1410	1220
										3¼ or more	2060	1880	1690	1460
			2⅝	2¼ min.	2960	2690	2420	2100	2¼ min.	2¼ min.	1970	1790	1610	1400
										3¼ or more	2370	2150	1930	1680
			3 and thicker	2¼ min.	3180	2890	2610	2260	2¼ min.	2¼ min.	2120	1930	1740	1510
										3¼ or more	2540	2320	2090	1810
4	¾	1	1 min.	2¾ min.	2840	2590	2330	2020	2¾ min.	2¾ min.	1900	1720	1550	1340
										3¾ or more	2280	2060	1860	1610
			1⅝ and thicker	2¾ min.	3700	3360	3030	2620	2¾ min.	2¾ min.	2460	2240	2020	1750
										3¾ or more	2960	2690	2420	2100
		2	1⅝ min.	2¾ min.	2840	2590	2330	2020	2¾ min.	2¾ min.	1900	1720	1550	1340
										3¾ or more	2280	2050	1860	1610
			2	2¾ min.	3070	2790	2520	2180	2¾ min.	2¾ min.	2050	1860	1680	1450
										3¾ or more	2460	2240	2020	1740
			2⅝	2¾ min.	3470	3150	2830	2460	2¾ min.	2¾ min.	2310	2100	1890	1640
										3¾ or more	2770	2520	2270	1970
			3 and thicker	2¾ min.	3700	3360	3030	2620	2¾ min.	2¾ min.	2460	2240	2020	1750
										3¾ or more	2950	2690	2420	2100

ern yellow pine, redwood, tamarack.

Group 2. Western red cedar, Port Orford cedar, Rocky Mountain Douglas fir, western hemlock, Norway pine.

Group 3. White cedar, balsam fir, commercial white fir, eastern hemlock, white pine, Idaho white pine, Ponderosa pine, spruce.

Group 4. Commercial white ash, beech, birch, pecan, hard maple, commercial red and white oak.

For two-member joints, use one-half the value for the thinner member. For joints of over three members of equal thickness, the load varies as the number of shear planes involved, the load for each shear plane being equal to one-half the tabulated load for a piece the thickness of the member involved.

For loads neither parallel nor perpendicular to the grain, the value should be selected for each direction, and the true value obtained by Hankinson's formula as shown in Fig. 9.20.

The spacing between two rows of bolts in a line parallel to the direction of the load should be not less than 2½ times the bolt diameter for an l/d ratio of 2, and 5 times the bolt diameter for an l/d ratio of 6 or more,

Table 9.15 Allowable Loads for One Shear-Plate Unit and Bolt in Single Shear

The allowable loads below are for normal loading conditions. See other provisions of Part V for adjustments of these tabulated allowable loads. Loads tabulated below are for wood side plates.

Shear-plate diam. (inches)	Bolt diam. (inches)	Number of faces of piece with connectors on same bolt	Thickness (net) of lumber (inches)	Loaded parallel to grain (0°): Edge distance (inches)	Loaded parallel to grain (0°): Allowable load per connector unit and bolt (pounds): Group A woods	Group B woods	Group C woods	Group D woods	Loaded perpendicular to grain (90°): Edge distance (inches): Unloaded-edge (See Fig. 12)	Loaded-edge (See Fig. 12)	Loaded perpendicular to grain (90°): Allowable load per connector unit and bolt (pounds): Group A woods	Group B woods	Group C woods	Group D woods
2⅝	¾	1	1⅝ min.	1¾ min.	3370[1]	2890	2410	2080	1¾ min.	1¾ min.	1960	1680	1400	1210
										2¾ or more	2350	2020	1680	1450
		2	1⅝ min.	1¾ min.	2620	2250	1870	1610	1¾ min.	1¾ min.	1520	1300	1090	940
										2¾ or more	1820	1560	1310	1130
			2	1¾ min.	3190[1]	2730	2270	1960	1¾ min.	1¾ min.	1850	1590	1320	1140
										2¾ or more	2220	1910	1580	1370
			2⅝ and thicker	1¾ min.	3370[1]	2890	2410	2080	1¾ min.	1¾ min.	1960	1680	1400	1210
										2¾ or more	2350	2020	1680	1450
4	¾	1	1⅝ min.	2¾ min.	4750	4070	3390	2920	2¾ min.	2¾ min.	2760	2360	1970	1700
										3¾ or more	3310	2830	2360	2040
			1¾ and thicker	2¾ min.	5090[1]	4360	3640	3140	2¾ min.	2¾ min.	2950	2530	2110	1810
										3¾ or more	3540	3040	2530	2200
		2	1¾ min.	2¾ min.	3390	2910	2420	2090	2¾ min.	2¾ min.	1970	1680	1400	1250
										3¾ or more	2360	2020	1680	1410
			2	2¾ min.	3790	3240	2700	2330	2¾ min.	2¾ min.	2200	1880	1570	1360
										3¾ or more	2640	2260	1880	1630
			2⅝	2¾ min.	4440	3800	3170	2730	2¾ min.	2¾ min.	2580	2210	1840	1590
										3¾ or more	3100	2650	2210	1910
			3	2¾ min.	4830	4140	3450	2980	2¾ min.	2¾ min.	2800	2400	2000	1720
										3¾ or more	3360	2880	2400	2060
			3⅝ and thicker	2¾ min.	5090[1]	4360	3640	3140	2¾ min.	2¾ min.	2950	2530	2110	1820
										3¾ or more	3540	3040	2530	2180
4	⅞	1	1⅝ min.	2¾ min.	4750	4070	3390	2920	2¾ min.	2¾ min.	2760	2360	1970	1700
										3¾ or more	3310	2830	2360	2040
			1¾ and thicker	2¾ min.	5090	4360	3640	3140	2¾ min.	2¾ min.	2950	2530	2110	1820
										3¾ or more	3540	3040	2530	2180
		2	1¾ min.	2¾ min.	3390	2910	2420	2090	2¾ min.	2¾ min.	1970	1680	1400	1210
										3¾ or more	2360	2020	1680	1450
			2	2¾ min.	3780	3240	2700	2330	2¾ min.	2¾ min.	2200	1880	1570	1360
										3¾ or more	2640	2260	1880	1630
			2⅝	2¾ min.	4440	3800	3170	2730	2¾ min.	2¾ min.	2580	2210	1840	1590
										3¾ or more	3100	2650	2210	1910
			3	2¾ min.	4830	4140	3450	2980	2¾ min.	2¾ min.	2800	2400	2000	1720
										3¾ or more	3360	2880	2400	2060
			3⅝ and thicker	2¾ min.	5090	4360	3640	3140	2¾ min.	2¾ min.	2950	2530	2110	1820
										3¾ or more	3540	3040	2530	2180

NOTES

[1] Loads followed by "1" in the above table, exceed those permitted by Note 3, but are needed for proper determination of loads for other angles of load to grain. Note 3 limitations apply in all cases.

[2] For metal side plates, tabulated loads apply except that, for 4" shear plates, the parallel-to-grain (not perpendicular) loads for wood side plates shall be increased 18, 11, 5 and 0 per cent for groups A, B, C and D woods, respectively, but loads shall not exceed those permitted by Note 3.

[3] The allowable loads for all loadings, except wind, shall not exceed 2900 lbs. for 2⅝" shear plates; 4970 lbs. and 6760 lbs. for 4" shear plates with ¾" and ⅞" bolts, respectively; or, for wind loading, shall not exceed 3870 lbs., 6630 lbs. and 9020 lbs., respectively. If bolt threads are in bearing on the shear plate, reduce the preceding values by one-ninth.

[4] Metal side plates, when used, shall be designed in accordance with accepted metal practices. For steel, the following unit stresses, in pounds per square inch, are suggested for all loadings except wind: net section in tension, 20,000; shear, 12,500; double-shear bearing, 28,125; single-shear bearing, 22,500; for wind, these values may be increased one-third; if bolt threads are in bearing, reduce the preceding shear and bearing values by one-ninth.

where l is the length of the bolt in the main member and d the diameter of the bolt. For intermediate values, the spacing should be determined by direct interpolation. The edge margin should be at least 1½ times the bolt diameter.

In a row of bolts, the spacing must be at least four times the bolt diameter, and the end margin in tension seven times the bolt diameter for softwood and five times for hardwood; in compression the end margin should be four times the bolt diameter.

9.40 CONCRETE CONNECTIONS

a When concrete beams, girders, and columns are poured at the same time they are monolithic and present no problem of connections or joints. There are, however, several conditions under which some form of connection must be designed to carry concrete construction, as follows:

1. Where concrete beams are carried by a steel column or steel beam.
2. Where concrete beams or columns are

Table 9.16 Constants for Determining Required Net Section in Tension

Product in square inches

Type of loading	Thickness of wood member, in.	Constants for each connector load group		
		Group A	Group B	Group C
Standard	4 or less	0.00041	0.00046	0.00048
	over 4	0.00051	0.00058	0.00060
Wind or earthquake	4 or less	0.00031	0.00036	0.00037
	over 4	0.00039	0.00044	0.00046
Impact	4 or less	0.00023	0.00027	0.00028
	over 4	0.00029	0.00033	0.00035
Dead load	4 or less	0.00047	0.00053	0.00055
	over 4	0.00058	0.00067	0.00069

Table 9.17 Projected Area of Connectors and Bolts

Square inches

Split ring and bolt	Number of faces	Thickness of lumber, in.				
		$1\frac{5}{8}$	$2\frac{5}{8}$	$3\frac{5}{8}$	$5\frac{1}{2}$	$7\frac{1}{2}$
$2\frac{1}{2}$-in. ring on $\frac{1}{2}$-in. bolt	1	1.73	2.23	2.73	3.67	4.67
	2	2.64	3.14	3.64	4.58	5.58
4-in. ring on $\frac{3}{4}$-in. bolt	1	3.09	3.84	4.59	6.00	7.50
	2	4.97	5.72	6.47	7.16	9.38

Table 9.18 Resistance to Withdrawal of Common Wire Nails

Driven perpendicular to the grain in seasoned wood

Wood	Specific gravity	Resistance, lb						
		8d	10d	12d	16d	20d	30d	40d
Birch or oak	0.69	60	67	67	74	87	94	102
Douglas fir	0.51	28	32	32	35	41	44	48
Maple	0.68	57	65	65	71	84	91	99
Longleaf yellow pine	0.64	39	45	45	47	50	55	59
Shortleaf yellow pine	0.59	32	36	36	38	41	44	48
Northern white pine	0.37	15	17	17	19	22	24	26
Ponderosa pine, redwood	0.42	17	19	19	21	25	27	30

Table 9.19 Safe Lateral Resistance of Common Wire Nails

Size nail		8d	10d	12d	16d	20d	30d	40d
Length, in.		$2\frac{1}{2}$	3	$3\frac{1}{4}$	$3\frac{1}{2}$	4	$4\frac{1}{2}$	5
Diam d, in.		0.131	0.148	0.148	0.162	0.192	0.207	0.225
$d^{1.5}$		0.0474	0.0570	0.0570	0.0652	0.0841	0.0942	0.1068
Wood	**Constant K**	**Resistance, lb**						
Eastern hemlock, white pine	1,080	51	61	61	70	91	101	115
Norway pine, redwood	1,350	64	77	77	88	113	127	144
Douglas fir, yellow pine	1,650	78	94	94	107	136	155	177
Maple, oak	2,040	96	116	116	133	171	191	219

Table 9.20 Bolted Wood Joints

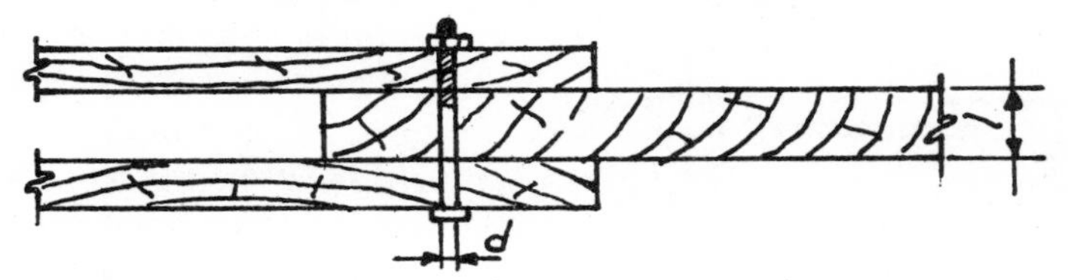

Safe loads (in pounds) with common bolts

Length, l, in.	Diam d, in.	l/d	Area $A = ld$	Group 1		Group 2		Group 3		Group 4	
				P	Q	P	Q	P	Q	P	Q
$1\frac{5}{8}$	$\frac{1}{2}$	3.3	0.81	1,000	460	780	320	620	240	1,140	660
	$\frac{5}{8}$	2.6	1.02	1,260	500	970	370	780	280	1,460	740
	$\frac{3}{4}$	2.2	1.22	1,520	560	1,180	410	940	310	1,750	830
	$\frac{7}{8}$	1.9	1.42	1,780	620	1,370	460	1,090	340	2,050	910
	1	1.6	1.63	2,030	680	1,560	490	1,260	370	2,340	1,000
$2\frac{5}{8}$	$\frac{1}{2}$	5.3	1.31	1,250	720	1,070	530	940	400	1,440	1,060
	$\frac{5}{8}$	4.2	1.64	1,850	830	1,510	600	1,250	440	2,140	1,200
	$\frac{3}{4}$	3.5	1.97	2,380	910	1,870	670	1,510	500	2,740	1,330
	$\frac{7}{8}$	3.0	2.30	2,830	1,010	2,210	730	1,760	550	3,280	1,460
	1	2.6	2.63	3,260	1,100	2,520	800	2,020	600	3,770	1,600
$3\frac{5}{8}$	$\frac{1}{2}$	7.3	1.81	1,260	960	1,090	720	980	550	1,450	1,220
	$\frac{5}{8}$	5.8	2.27	1,970	1,140	1,700	830	1,540	620	2,270	1,610
	$\frac{3}{4}$	4.8	2.72	2,810	1,260	2,360	940	2,020	700	3,240	1,840
	$\frac{7}{8}$	4.1	3.17	3,620	1,390	2,950	1,010	2,420	760	4,180	2,030
	1	3.6	3.63	4,340	1,520	3,440	1,100	2,780	830	5,020	2,210
$5\frac{1}{2}$	$\frac{5}{8}$	8.8	3.44	1,970	1,360	1,700	1,120	1,540	900	2,270	1,670
	$\frac{3}{4}$	7.3	4.13	2,830	1,810	2,460	1,390	2,220	1,040	3,260	2,340
	$\frac{7}{8}$	6.3	4.81	3,850	2,110	3,350	1,540	3,010	1,150	4,450	2,890
	1	5.5	5.50	5,020	2,300	4,340	1,680	3,860	1,260	5,780	3,320
	$1\frac{1}{8}$	4.9	6.19	6,290	2,510	5,320	1,820	4,560	1,370	7,250	3,650
$7\frac{1}{2}$	$\frac{5}{8}$	12.0	4.69	1,970	1,220	1,700	1,040	1,540	880	2,270	1,460
	$\frac{3}{4}$	10.0	5.63	2,830	1,760	2,460	1,440	2,220	1,210	3,260	2,110
	$\frac{7}{8}$	8.6	6.56	3,850	2,340	3,350	1,910	3,010	1,540	4,450	2,900
	1	7.5	7.50	5,030	2,920	4,370	2,270	3,950	1,720	5,810	3,710
	$1\frac{1}{8}$	6.7	8.44	6,400	3,400	5,560	2,500	5,000	1,870	7,340	4,500
$9\frac{1}{2}$	$\frac{3}{4}$	12.7	7.13	2,830	1,570	2,460	1,370	2,220	1,150	3,260	1,880
	$\frac{7}{8}$	10.9	8.31	3,850	2,200	3,350	1,810	3,010	1,520	4,450	2,590
	1	9.5	9.50	5,030	2,860	4,370	2,350	3,950	1,940	5,810	3,470
	$1\frac{1}{8}$	8.4	10.69	6,400	3,640	5,560	2,930	5,000	2,330	7,340	4,460
	$1\frac{1}{4}$	7.6	11.88	7,860	4,310	6,840	3,340	6,170	2,540	9,070	5,410
$11\frac{1}{2}$	1	11.5	11.50	5,030	2,690	4,370	2,230	3,950	1,910	5,810	3,200
	$1\frac{1}{8}$	10.2	12.94	6,400	3,460	5,560	2,830	5,000	2,380	7,340	4,100
	$1\frac{1}{4}$	9.2	14.38	7,860	4,220	6,840	3,460	6,170	2,840	9,070	5,170

designed to carry a future extension, requiring provision for carrying future concrete beams and slabs.

3. Where concrete columns are designed to carry additional stories to be added in the future.

b Probably the most satisfactory way of carrying a concrete-beam load on a steel column or steel beam is by means of multiple seat angles as shown in Fig. 9.24, preferably welded to the face of the column. This detail, using $3 \times 3 \times \frac{1}{4}$-in. angles the full width of the concrete beam, spaced 4 in. vertically, will provide for all the stresses to

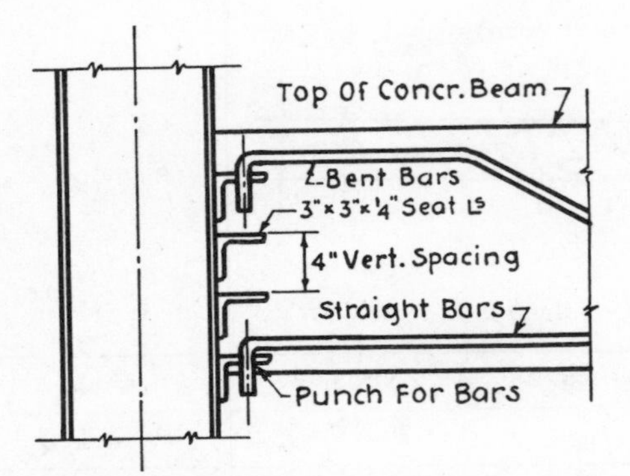

Fig. 9.24 Multiple Seat Connections

be carried into the steel column. A similar detail could be used on the face of a channel or the web of a beam.

c If concrete beams are designed to carry a future slab, a recess is usually provided in the top of the beam, as shown in Fig. 9.25.

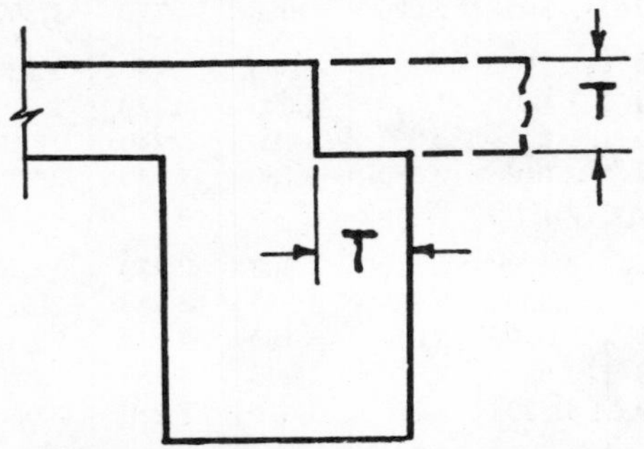

Fig. 9.25 Concrete-Beam Seat for Future Slab

d If concrete beams are designed to carry a future beam, the method used depends upon the size of beam to be added and the required shear capacity. If the original beam is twice the depth of the beam to be framed in later, a pocket can be left in the original beam for the later beam, similar to the system shown in Fig. 9.25. Another method sometimes used is to install concrete inserts in the original cast-iron or steel beam, into which at a later date bolts can be installed to carry a multiple seat connection similar to that of Fig. 9.24. If concrete inserts are used, they should be filled with grease to keep them from plugging up with concrete or from rusting. The bolts which go into the concrete inserts are in single shear, and this fact is ordinarily the governing factor in the design of such details.

e There are several details used for providing for future extension of columns. The simpler but more costly detail is to put dowels for the future bars into the original column, projecting the required length above the roof line, then pour lean concrete around them (for later removal) and flash around the column stub. As an alternate, if the bars are not too heavy, they may be bent flat against the roof so that later they may be bent back into their proper place when the addition is made. This method, although it takes less material and is easier to keep watertight, requires difficult bar bending when the addition is made.

The method preferred by the author is to use a dowel bar threaded on one end, to which is attached a sleeve nut. The open end of the sleeve nut is filled with grease to prevent concrete from filling it, and the sleeve nut is placed flush with the top of the concrete. The roofing is then level. When the addition is made, the grease is cleaned from the sleeve nut and another threaded dowel inserted to provide the dowel for future connection (see Fig. 9.26).

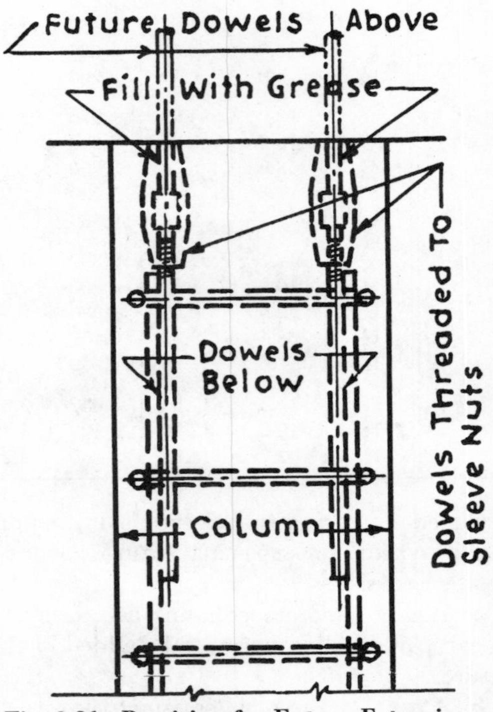

Fig. 9.26 Provision for Future Extension—Concrete Columns

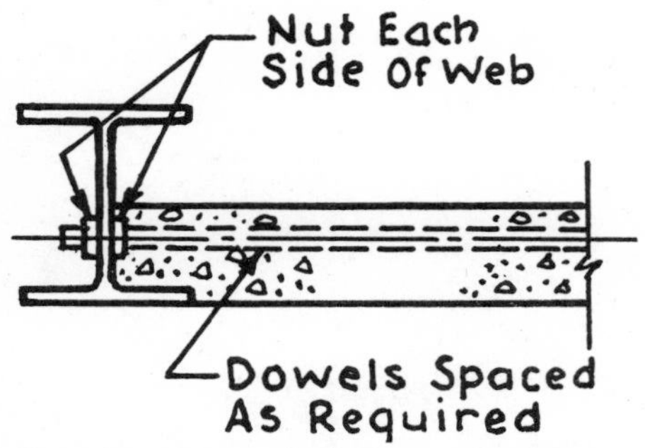

Fig. 9.27 Anchorage of Slab to Bottom Flange of Steel Beam

f Sometimes it is necessary to connect a new concrete beam or slab to an existing beam or column where no provision has been made for the connection. Provided that the existing member can be used without overstressing it, the connection may be made by installing a single seat angle for a slab or multiple seat angles for a beam, connecting them by the use of Rawl anchors or other similar lead sleeves. These sleeves are installed in holes drilled in the concrete in such a manner that when the bolt is inserted and tightened up, it is held by the lead which has been wedged into the irregularities and interstices of the concrete.

g Occasionally it is necessary to carry a concrete slab on the bottom flange of a steel beam in such a manner that the slab can be tied to the beam by dowels. These dowels are usually threaded and put through holes in the web of the steel beams with nuts on both sides of the web to lock the dowels, as shown in Fig. 9.27.

10

Complex Structures

10.10 TRUSSES

a A truss is a framework of steel, timber, or both, arranged to form a series of triangles, the bending stresses being translated into direct stresses applied concentrically to the members. The purpose of trusses is obvious —to carry loads over large rooms without use of columns. Occasionally, however, a truss may be used for decorative purposes, as in a church, or for economy where a pitched roof is required, as in some industrial plants. The ordinary pitched roof on a home employs a type of truss, the rafters forming the top chords and the attic floor joists the bottom chords.

Table 10.1 gives coefficients for the commoner type of symmetrical roof trusses, and this method of solution is perfectly satisfactory. If the truss does not come within the scope of these tables, the problem may be solved either analytically or graphically. The graphical method is faster and can be used to check results. If the diagram scale is not accurate enough, it will at least serve as a geometric representation from which to compute actual stress in members.

Usually it is uneconomical to carry bending and direct stress in a truss member. It is preferable to bring the loads into the truss at panel points only, even at the cost of some slight additional framing. This arrangement is called for more with steel than with wood trusses, and will be further discussed later. Direct stresses are always assumed to be applied on the center of gravity of the member, and these stress lines must meet at a point at the joint. Otherwise eccentric stresses will result that may set up dangerous secondary stresses.

b *Graphic analysis of a truss.* Figure 10.1 illustrates the truss analysis for a scissors truss as a means of showing this method. A graphic analysis of a truss is based on obtaining the value of the component of a stress when the value of the stress and the direction of the member along which the stress component will act are known. A graphic analysis of a truss must be accurately laid out, since a small error in angle will be magnified in proceeding through the truss.

The first step is to compute analytically the reactions on the truss and lay out the truss diagram, applying the loads at the panel points and the reactions at the end. The loads on a truss should be applied on the panel points. Loads occurring between panel points cause bending in the member, and then the member must be figured as a beam, the reactions taken to the panel points, and the truss member figured for bending and direct stress combined. In laying out a truss diagram, assign capital letters to the

Table 10.1 Trusses—Coefficients of Stresses

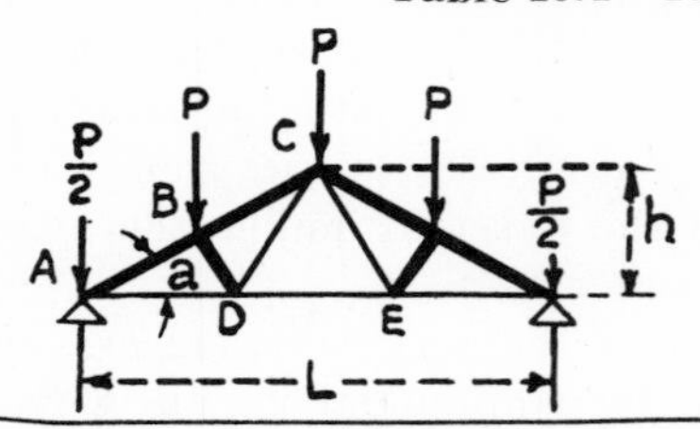

M	Pitch				
	$\frac{1}{3}$	$a = 30°$	$\frac{1}{4}$	$\frac{1}{5}$	$\frac{1}{6}$
AB	+2.70	+3.00	+3.35	+4.04	+4.74
BC	+2.15	+2.50	+2.91	+3.67	+4.43
AD	−2.25	−2.60	−3.00	−3.75	−4.50
DE	−1.50	−1.73	−2.00	−2.50	−3.00
BD	+0.83	+0.87	+0.90	+0.93	+0.95
CD	−0.75	−0.87	−1.00	−1.25	−1.50

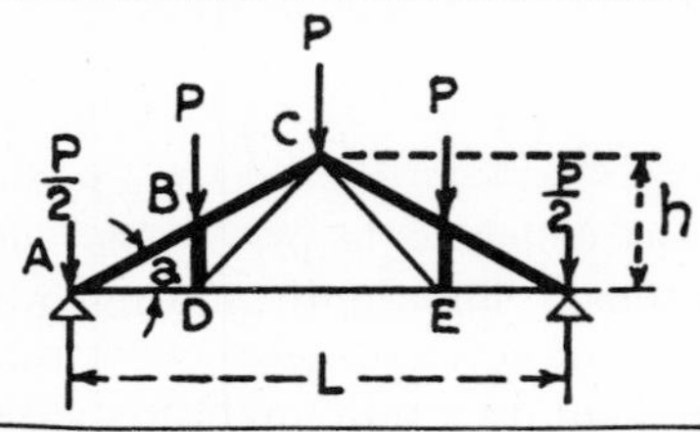

M	Pitch				
	$\frac{1}{3}$	$a = 30°$	$\frac{1}{4}$	$\frac{1}{5}$	$\frac{1}{6}$
AB	+2.70	+3.00	+3.35	+4.04	+4.74
BC	+2.70	+3.00	+3.35	+4.04	+4.74
AD	−2.25	−2.60	−3.00	−3.75	−4.50
DE	−1.50	−1.73	−2.00	−2.50	−3.00
BD	+1.00	+1.00	+1.00	+1.00	+1.00
CD	−1.25	−1.32	−1.41	−1.60	−1.80

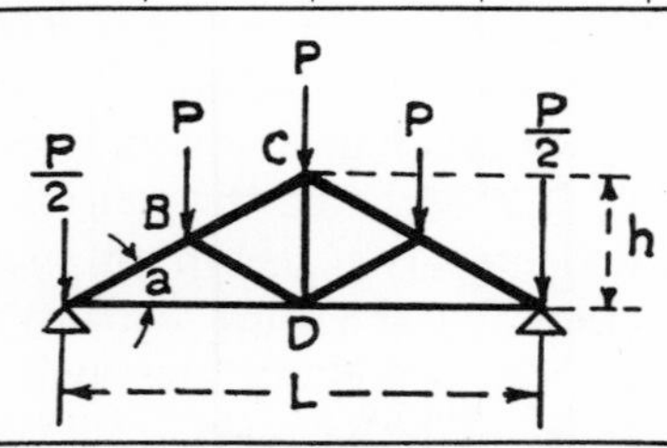

M	Pitch				
	$\frac{1}{3}$	$a = 30°$	$\frac{1}{4}$	$\frac{1}{5}$	$\frac{1}{6}$
AB	+2.70	+3.00	+3.35	+4.04	+4.74
BC	+1.80	+2.00	+2.24	+2.69	+3.16
AD	−2.25	−2.60	−3.00	−3.75	−4.50
BD	+0.90	+1.00	+1.12	+1.35	+1.58
CD	−1.00	−1.00	−1.00	−1.00	−1.00

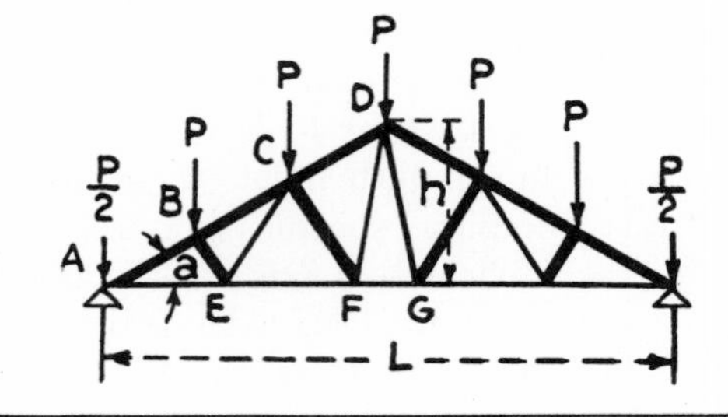

M	Pitch				
	$\frac{1}{3}$	$a = 30°$	$\frac{1}{4}$	$\frac{1}{5}$	$\frac{1}{6}$
AB	+4.51	+5.00	+5.59	+6.73	+8.00
BC	+3.95	+4.50	+5.14	+6.36	+7.70
CD	+2.77	+3.25	+3.80	+4.83	+6.00
AE	−3.75	−4.33	−5.00	−6.25	−7.50
EF	−3.00	−3.47	−4.00	−5.00	−6.20
FG	−2.25	−2.60	−3.00	−3.75	−4.70
BE	+0.83	+0.87	+0.90	+0.93	+1.00
CE	−0.75	−0.87	−1.00	−1.25	−1.50
CF	+1.25	+1.30	+1.34	+1.39	+1.50
DF	−1.04	−1.14	−1.27	−1.49	−1.70

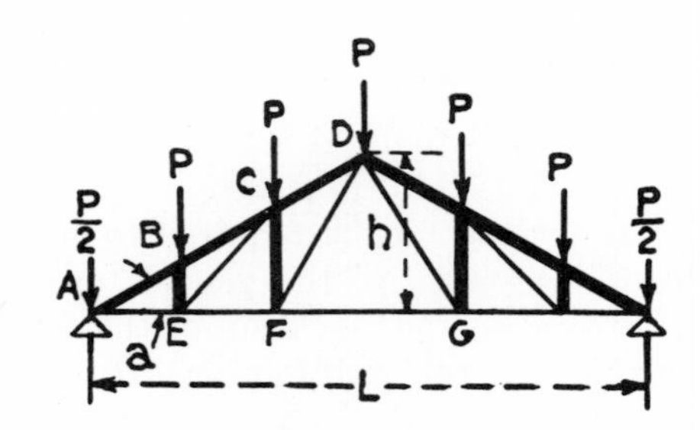

M	Pitch				
	$\frac{1}{3}$	$a = 30°$	$\frac{1}{4}$	$\frac{1}{5}$	$\frac{1}{6}$
AB	+4.51	+5.00	+5.59	+6.73	+7.91
BC	+4.51	+5.00	+5.59	+6.73	+7.91
CD	+3.61	+4.00	+4.47	+5.39	+6.32
AE	−3.75	−4.33	−5.00	−6.25	−7.50
EF	−3.00	−3.46	−4.00	−5.00	−6.00
FG	−2.25	−2.60	−3.00	−3.75	−4.50
BE	+1.00	+1.00	+1.00	+1.00	+1.00
CE	−1.25	−1.32	−1.41	−1.60	−1.80
CF	+1.50	+1.50	+1.50	+1.50	+1.50
DF	−1.68	−1.73	−1.80	−1.95	−2.12

M = Member of truss + Indicates compression
Pitch = h/L − Indicates tension
Stress in any member = coefficient multiplied by W

Table 10.1 Trusses—Coefficients of Stresses (*Continued*)

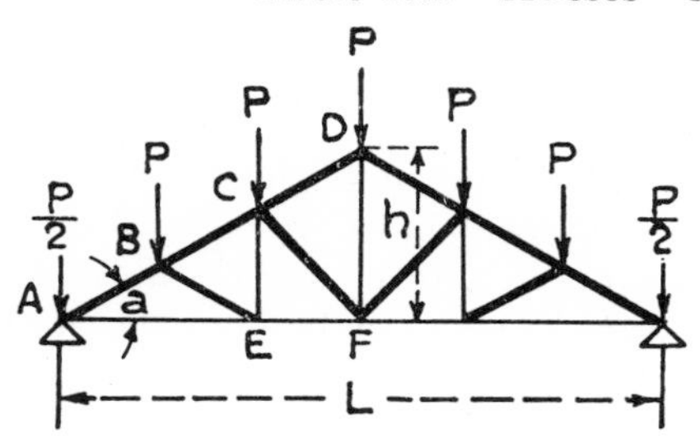

M	Pitch				
	$\frac{1}{3}$	$a = 30°$	$\frac{1}{4}$	$\frac{1}{5}$	$\frac{1}{6}$
AB	+4.51	+5.00	+5.59	+6.73	+7.91
BC	+3.61	+4.00	+4.50	+5.39	+6.32
CD	+2.70	+3.00	+3.30	+4.04	+4.74
AE	−3.75	−4.33	−5.00	−6.25	−7.50
EF	−3.00	−3.46	−4.00	−5.00	−6.00
BE	+0.90	+1.00	+1.10	+1.35	+1.58
CE	−0.50	−0.50	−0.50	−0.50	−0.50
CF	+1.25	+1.32	+1.40	+1.60	+1.80
DF	−2.00	−2.00	−2.00	−2.00	−2.00

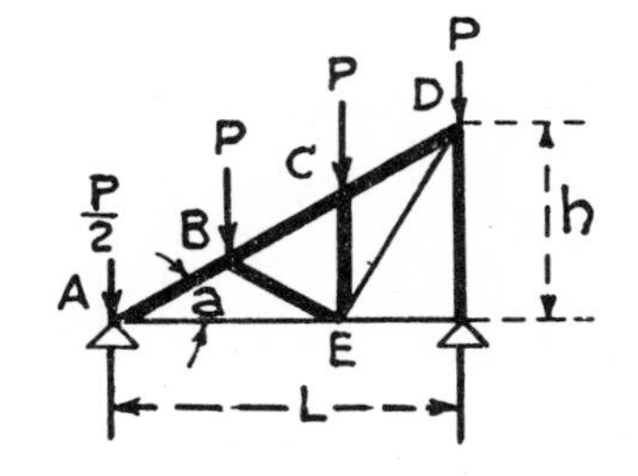

M	Pitch				
	$\frac{1}{3}$	$a = 30°$	$\frac{1}{4}$	$\frac{1}{5}$	$\frac{1}{6}$
AB	+1.80	+2.00	+2.24	+2.69	+3.16
BC	+0.90	+1.00	+1.12	+1.35	+1.58
CD	+0.90	+1.00	+1.12	+1.35	+1.58
AE	−1.50	−1.73	−2.00	−2.50	−3.00
EF	0	0	0	0	0
BE	+0.90	+1.00	+1.12	+1.35	+1.58
CE	+1.00	+1.00	+1.00	+1.00	+1.00
DE	−1.68	−1.73	−1.80	−1.95	−2.12

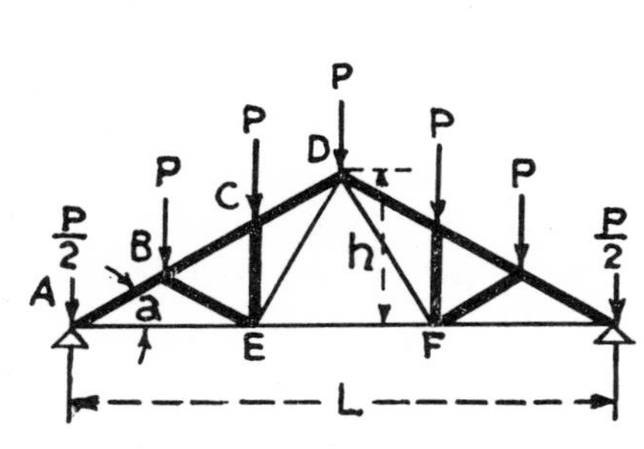

M	Pitch				
	$\frac{1}{3}$	$a = 30°$	$\frac{1}{4}$	$\frac{1}{5}$	$\frac{1}{6}$
AB	+4.51	+5.00	+5.59	+6.72	+7.91
BC	+3.54	+4.00	+4.55	+5.57	+6.64
CD	+3.40	+4.00	+4.71	+5.98	+7.27
AE	−3.75	−4.33	−5.00	−6.25	−7.50
EF	−2.25	−2.60	−3.00	−3.75	−4.50
BE	+0.93	+1.00	+1.08	+1.21	+1.34
CE	+0.93	+1.00	+1.08	+1.21	+1.34
DE	−1.50	−1.73	−2.00	−2.50	−3.00

M	Pitch				
	$\frac{1}{3}$	$a = 30°$	$\frac{1}{4}$	$\frac{1}{5}$	$\frac{1}{6}$
AB	+1.80	+2.00	+2.24	+2.69	+3.16
BC	+1.80	+2.00	+2.24	+2.69	+3.16
CD	+0.90	+1.00	+1.12	+1.35	+1.58
AE	−1.50	−1.73	−2.00	−2.50	−3.00
EF	−0.75	−0.87	−1.00	−1.25	−1.50
FG	0	0	0	0	0
BE	+1.00	+1.00	+1.00	+1.00	+1.00
CE	−1.25	−1.32	−1.41	−1.60	−1.80
CF	+1.50	+1.50	+1.50	+1.50	+1.50
DF	−1.68	−1.73	−1.80	−1.95	−2.12

M = Member of truss + Indicates compression
Pitch = h/L − Indicates tension
Stress in any member = coefficient multiplied by W

Table 10.1 Trusses—Coefficients of Stresses (*Continued*)

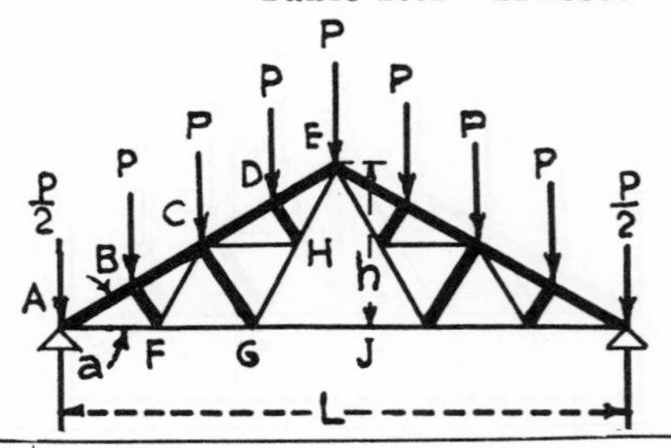

M	Pitch				
	$\frac{1}{3}$	$a = 30°$	$\frac{1}{4}$	$\frac{1}{5}$	$\frac{1}{6}$
AB	+6.31	+7.00	+7.83	+9.42	+11.07
BC	+5.76	+6.50	+7.38	+9.05	+10.75
CD	+5.30	+6.00	+6.93	+8.68	+10.44
DE	+4.65	+5.50	+6.48	+8.31	+10.12
AF	−5.25	−6.06	−7.00	−8 75	−10.50
FG	−4.50	−5.20	−6.00	−7.50	−9.00
GJ	−3.00	−3.46	−4.00	−5.00	−6.00
BF	+0.83	+0.87	+0.89	+0.93	+0.95
CF	−0.75	−0.87	−1.00	−1.25	−1.50
CG	+1.66	+1.73	+1.79	+1.86	+1.90
CH	−0.75	−0.87	−1.00	−1.25	−1.50
DH	+0.83	+0.87	+0.89	+0.93	+0.95
EH	−2.25	−2.60	−3.00	−3.75	−4.50
HG	−1.50	−1.73	−2.00	−2.50	−3.00

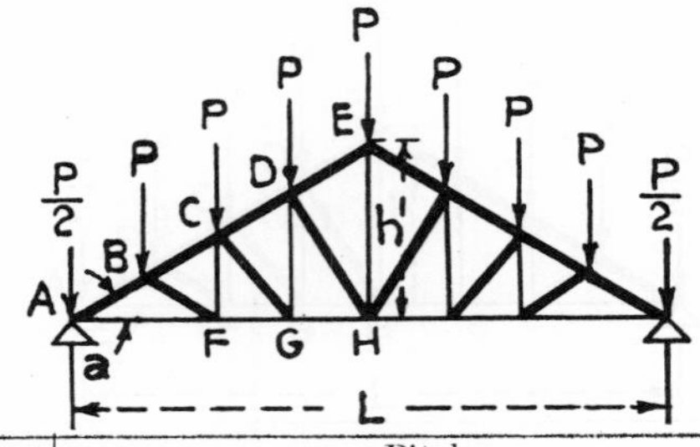

M	Pitch				
	$\frac{1}{3}$	$a = 30°$	$\frac{1}{4}$	$\frac{1}{5}$	$\frac{1}{6}$
AB	+6.31	+7.00	+7.83	+9.42	+11.07
BC	+5.41	+6.00	+6.71	+8.08	+9.49
CD	+4.51	+5.00	+5.59	+6.73	+7.91
DE	+3.61	+4.00	+4.47	+5.39	+6.32
AF	−5.25	−6.06	−7.00	−8.75	−10.50
FG	−4.50	−5.20	−6.00	−7.50	−9.00
GH	−3.75	−4.33	−5.00	−6.25	−7.50
BF	+0.90	+1.00	+1.12	+1.35	+1.58
CF	−0.50	−0.50	−0.50	−0.50	−0.50
CG	+1.25	+1.32	+1.41	+1.60	+1.80
DG	−1.00	−1.00	−1.00	−1.00	−1.00
DH	+1.68	+1.73	+1.80	+1.95	+2.12
EH	−3.00	−3.00	−3.00	−3.00	−3.00

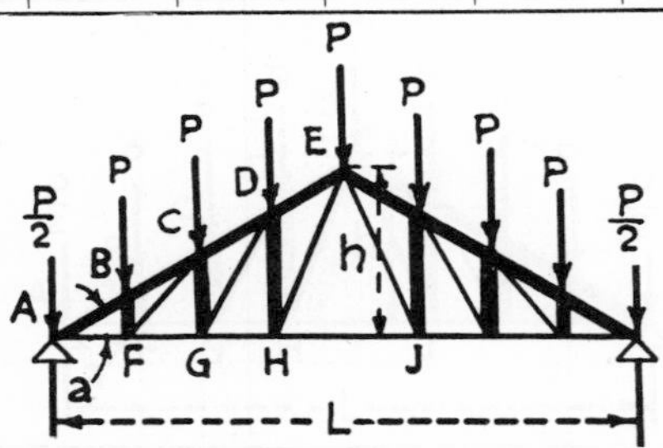

M	Pitch				
	$\frac{1}{3}$	$a = 30°$	$\frac{1}{4}$	$\frac{1}{5}$	$\frac{1}{6}$
AB	+6.31	+7.00	+7.83	+9.42	+11.07
BC	+6.31	+7.00	+7.83	+9.42	+11.07
CD	+5.41	+6.00	+6.71	+8.08	+9.49
DE	+4.51	+5.00	+5.59	+6.73	+7.91
AF	−5.25	−6.06	−7.00	−8.75	−10.50
FG	−4.50	−5.20	−6.00	−7.50	−9.00
GH	−3.75	−4.33	−5.00	−6.25	−7.50
HJ	−3.00	−3.46	−4.00	−5.00	−6.00
BF	+1.00	+1.00	+1.00	+1.00	+1.00
CF	−1.25	−1.32	−1.41	−1.60	−1.80
CG	+1.50	+1.50	+1.50	+1.50	+1.50
DG	−1.68	−1.73	−1.80	−1.95	−2.12
DH	+2.00	+2.00	+2.00	+2.00	+2.00
EH	−2.14	−2.18	−2.24	−2.36	−2.50

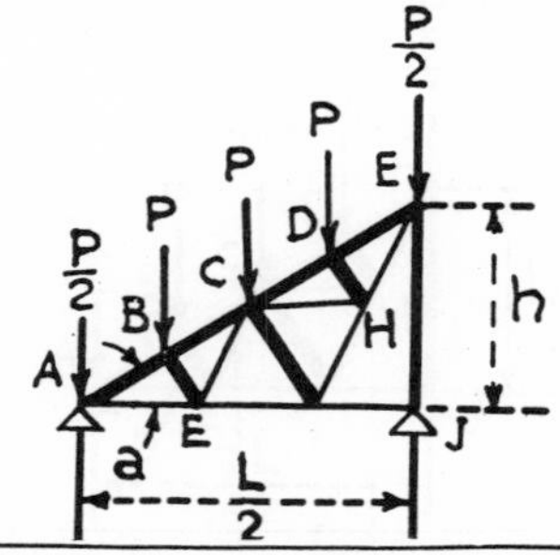

M	Pitch				
	$\frac{1}{3}$	$a = 30°$	$\frac{1}{4}$	$\frac{1}{5}$	$\frac{1}{6}$
AB	+2.70	+3.00	+3.35	+4.04	+4.74
BC	+2.15	+2.50	+2.91	+3.67	+4.43
CD	+1.59	+2.00	+2.46	+3.30	+4.11
DE	+1.04	+1.50	+2.01	+2.92	+3.79
AF	−2.25	−2.60	−3.00	−3.75	−4.50
FG	−1.50	−1.73	−2.00	−2.50	−3.00
GJ	0	0	0	0	0
BF	+0.83	+0.87	+0.89	+0.93	+0.95
CF	−0.75	−0.87	−1.00	−1.25	−1.50
CG	+1.66	+1.72	+1.79	+1.86	+1.90
CH	−0.75	−0.87	−1.00	−1.25	−1.50
DH	+0.83	+0.87	+0.89	+0.93	+0.95
EH	−2.25	−2.60	−3.00	−3.75	−4.50
HG	−1.50	−1.73	−2.00	−2.50	−3.00

M = Member of truss Pitch = h/L + Indicates compression − Indicates tension

Stress in any member = coefficient multiplied by W

Table 10.1 Trusses—Coefficients of Stresses (*Continued*)

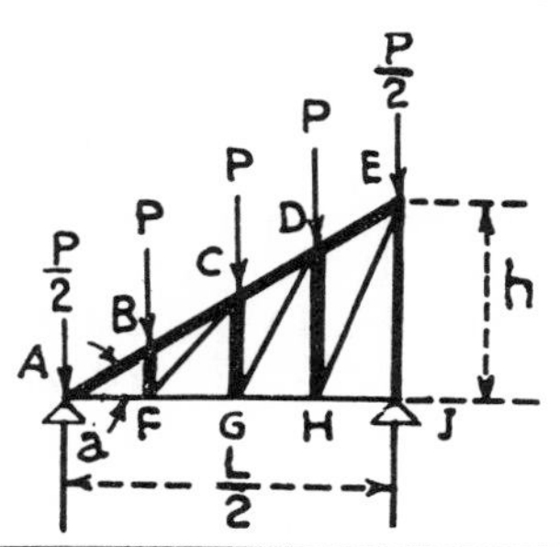

M	Pitch				
	$\frac{1}{3}$	$a = 30°$	$\frac{1}{4}$	$\frac{1}{5}$	$\frac{1}{6}$
AB	+2.70	+3.00	+3.35	+4.04	+4.74
BC	+2.70	+3.00	+3.35	+4.04	+4.74
CD	+1.80	+2.00	+2.24	+2.69	+3.16
DE	+0.90	+1.00	+1.12	+1.35	+1.58
AF	−2.25	−2.60	−3.00	−3.75	−4.50
FG	−1.50	−1.73	−2.00	−2.50	−3.00
GH	−0.75	−0.87	−1.00	−1.25	−1.50
HJ	0	0	0	0	0
BF	+1.00	+1.00	+1.00	+1.00	+1.00
CF	−1.25	−1.32	−1.41	−1.60	−1.80
CG	+1.50	+1.50	+1.50	+1.50	+1.50
DG	−1.68	−1.73	−1.80	−1.95	−2.12
DH	+2.00	+2.00	+2.00	+2.00	+2.00
EH	−2.14	−2.18	−2.24	−2.36	−2.50

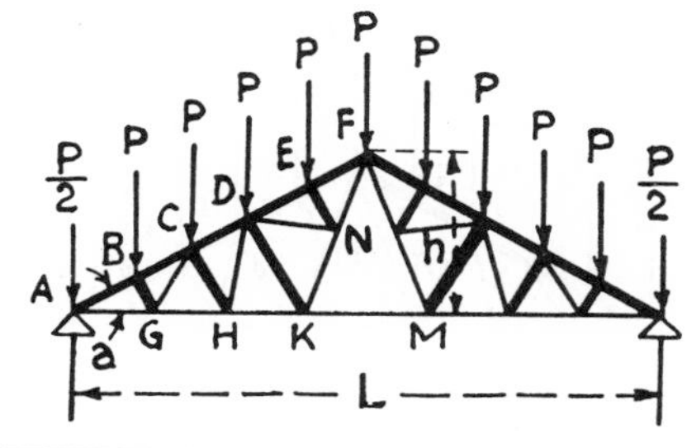

M	Pitch				
	$\frac{1}{3}$	$a = 30°$	$\frac{1}{4}$	$\frac{1}{5}$	$\frac{1}{6}$
AB	+8.11	+9.00	+10.06	+12.12	+14.23
BC	+7.56	+8.50	+9.62	+11.75	+13.91
CD	+6.38	+7.25	+8.27	+10.21	+12.17
DE	+5.62	+6.50	+7.53	+9.46	+11.38
EF	+5.06	+6.00	+7.08	+9.08	+11.07
AG	−6.75	−7.79	−9.00	−11.25	−13.50
GH	−6.00	−6.93	−8.00	−10.00	−12.00
HK	−5.25	−6.06	−7.00	−8.75	−10.50
KM	−3.75	−4.33	−5.00	−6.25	−7.50
BG	+0.83	+0.87	+0.89	+0.93	+0.95
CG	−0.75	−0.87	−1.00	−1.25	−1.50
CH	+1.25	+1.30	+1.34	+1.39	+1.42
DH	−1.04	−1.15	−1.26	−1.49	−1.71
DK	+2.08	+2.17	+2.24	+2.32	+2.37
DN	−0.59	−0.66	−0.75	−0.90	−1.06
EN	+0.83	+0.87	+0.89	+0.93	+0.95
KN	−1.77	−1.98	−2.24	−2.71	−3.18
NF	−2.35	−2.65	−2.98	−3.61	−4.24

M	Pitch				
	$\frac{1}{3}$	$a = 30°$	$\frac{1}{4}$	$\frac{1}{5}$	$\frac{1}{6}$
AB	+2.70	+3.00	+3.36	+4.04	+4.74
BC	+1.80	+2.00	+2.24	+2.69	+3.16
CD	+0.90	+1.00	+1.12	+1.35	+1.58
DE	0	0	0	0	0
AF	−2.25	−2.60	−3.00	−3.75	−4.50
FG	−1.50	−1.73	−2.00	−2.50	−3.00
GH	−0.75	−0.87	−1.00	−1.25	−1.50
BF	+0.90	+1.00	+1.12	+1.35	+1.58
CF	−0.50	−0.50	−0.50	−0.50	−0.50
CG	+1.25	+1.32	+1.41	+1.60	+1.80
DG	−1.00	−1.00	−1.00	−1.00	−1.00
DH	+1.68	+1.73	+1.80	+1 95	+2.12
EH	0	0	0	0	0

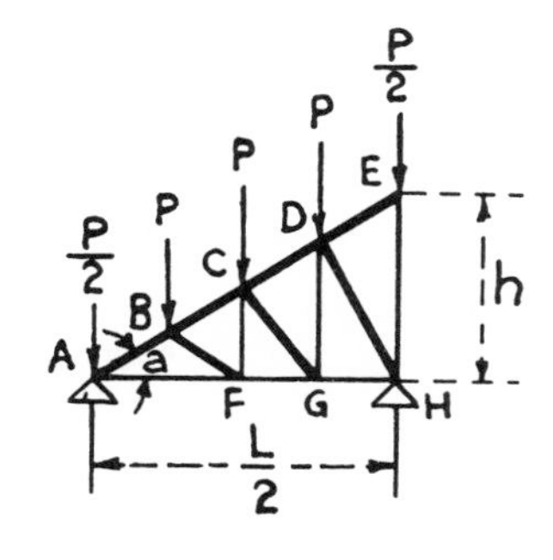

M = Member of truss   + Indicates compression
Pitch = h/L − Indicates tension
Stress in any member = coefficient multiplied by W

Table 10.1 Trusses—Coefficients of Stresses (*Continued*)

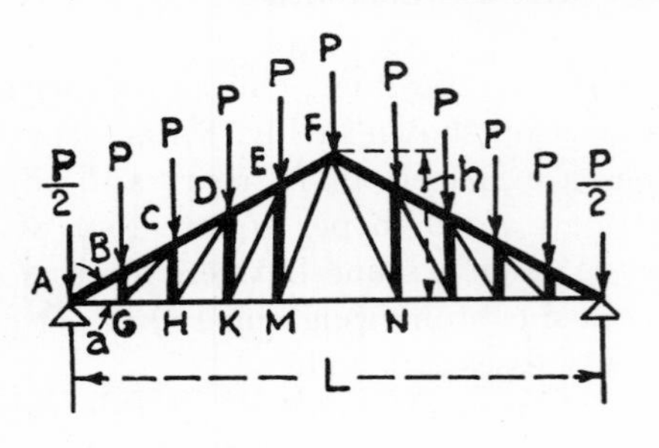

M	Pitch				
	$\frac{1}{3}$	$a = 30°$	$\frac{1}{4}$	$\frac{1}{5}$	$\frac{1}{6}$
AB	+8.11	+9.00	+10.06	+12.12	+14.23
BC	+8.11	+9.00	+10.06	+12.12	+14.23
CD	+7.21	+8.00	+8.94	+10.77	+12.65
DE	+6.31	+7.00	+7.83	+9.42	+11.07
EF	+5.41	+6.00	+6.71	+8.08	+9.49
AG	−6.75	−7.79	−9.00	−11.25	−13.50
GH	−6.00	−6.93	−8.00	−10.00	−12.00
HK	−5.25	−6.06	−7.00	−8.75	−10.50
KM	−4.50	−5.20	−6.00	−7.50	−9.00
MN	−3.75	−4.33	−5.00	−6.25	−7.50
BG	+1.00	+1.00	+1.00	+1.00	+1.00
CG	−1.25	−1.32	−1.41	−1.60	−1.80
CH	+1.50	+1.50	+1.50	+1.50	+1.50
DH	−1.68	−1.73	−1.80	−1.95	−2.12
DK	+2.00	+2.00	+2.00	+2.00	+2.00
EK	−2.14	−2.18	−2.24	−2.36	−2.50
EM	+2.50	+2.50	+2.50	+2.50	+2.50
FM	−2.61	−2.65	−2.69	−2.80	−2.92

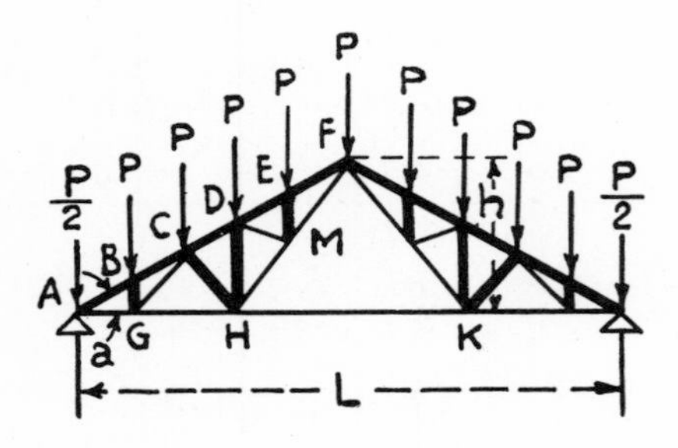

M	Pitch				
	$\frac{1}{3}$	$a = 30°$	$\frac{1}{4}$	$\frac{1}{5}$	$\frac{1}{6}$
AB	+8.11	+9.00	+10.06	+12.12	+14.23
BC	+8.11	+9.00	+10.06	+12.12	+14.23
CD	+6.31	+7.00	+7.83	+9.42	+11.07
DE	+6.91	+7.67	+8.57	+10.32	+12.12
EF	+6.91	+7.67	+8.57	+10.32	+12.12
AG	−6.75	−7.79	−9.00	−11.25	−13.50
GH	−6.00	−6.93	−8.00	−10.00	−12.00
HK	−3.75	−4.33	−5.00	−6.25	−7.50
BG	+1.00	+1.00	+1.00	+1.00	+1.00
CG	−1.25	−1.32	−1.41	−1.60	−1.80
CH	+1.25	+1.32	+1.41	+1.60	+1.80
DH	+1.50	+1.50	+1.50	+1.50	+1.50
DM	−0.53	−0.60	−0.69	−0.85	−1.01
EM	+1.00	+1.00	+1.00	+1.00	+1.00
HM	−2.92	−3.04	−3.20	−3.54	−3.91
MF	−3.89	−4.06	−4.27	−4.71	−5.21

M = Member of truss + Indicates compression
Pitch = h/L − Indicates tension
Stress in any member = coefficient multiplied by W

Table 10.1 Trusses—Coefficients of Stresses (*Continued*)

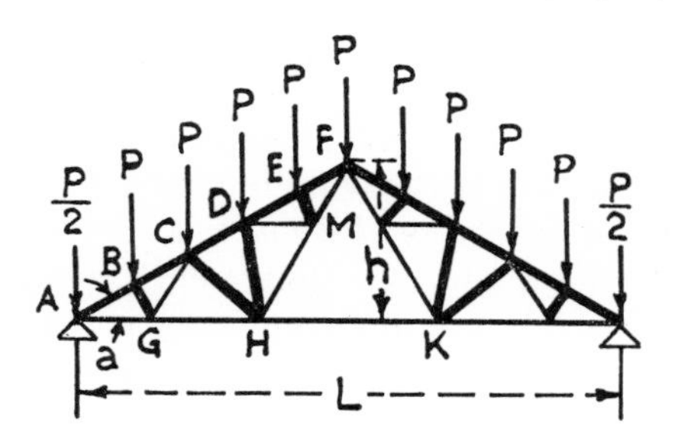

M	Pitch				
	$\frac{1}{3}$	$a = 30°$	$\frac{1}{4}$	$\frac{1}{5}$	$\frac{1}{6}$
AB	+8.11	+9.00	+10.06	+12.12	+14.23
BC	+7.56	+8.50	+9.62	+11.75	+13.91
CD	+6.00	+6.80	+7.74	+9.52	+11.32
DE	+6.45	+7.50	+8.72	+11.00	+13.28
EF	+5.89	+7.00	+8.28	+10.63	+12.97
AG	−6.75	−7.79	−9.00	−11.25	−13.50
GH	−6.00	−6.91	−8.00	−10.00	−12.00
HK	−3.75	−4.33	−5.00	−6.25	−7.50
BG	+0.83	+0.87	+0.89	+0.93	+0.95
CG	−0.75	−0.87	−1.00	−1.25	−1.50
CH	+1.31	+1.37	+1.44	+1.56	+1.66
DH	+1.31	+1.37	+1.44	+1.56	+1.66
DM	−0.75	−0.87	−1.00	−1.25	−1.50
EM	+0.83	+0.87	+0.89	+0.93	+0.95
HM	−2.25	−2.60	−3.00	−3.75	−4.50
MF	−3.00	−3.46	−4.00	−5.00	−6.00

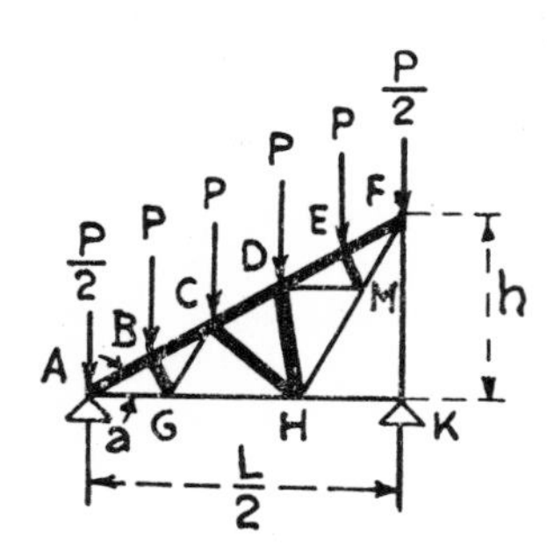

M	Pitch				
	$\frac{1}{3}$	$a = 30°$	$\frac{1}{4}$	$\frac{1}{5}$	$\frac{1}{6}$
AB	+3.61	+4.00	+4.47	+5.39	+6.32
BC	+3.05	+3.50	+4.02	+5.01	+6.01
CD	+1.50	+1.80	+2.15	+2.79	+3.42
DE	+1.94	+2.50	+3.13	+4.27	+5.38
EF	+1.39	+2.00	+2.68	+3.90	+5.06
AG	−3.00	−3.46	−4.00	−5.00	−6.00
GH	−2.25	−2.60	−3.00	−3.75	−4.50
HK	0	0	0	0	0
BG	+0.83	+0.87	+0.89	+0.93	+0.95
CG	−0.75	−0.87	−1.00	−1.25	−1.50
CH	+1.30	+1.37	+1.44	+1.56	+1.66
DH	+1.30	+1.37	+1.44	+1.56	+1.66
DM	−0.75	−0.87	−1.00	−1.25	−1.50
EM	+0.83	+0.87	+0.89	+0.93	+0.95
HM	−2.25	−2.60	−3.00	−3.75	−4.50
MF	−3.00	−3.46	−4.00	−5.00	−6.00

M = Member of truss + Indicates compression
Pitch = h/L − Indicates tension
Srtess in any member = coefficient multiplied by W

Table 10.1 Trusses—Coefficients of Stresses (*Continued*)

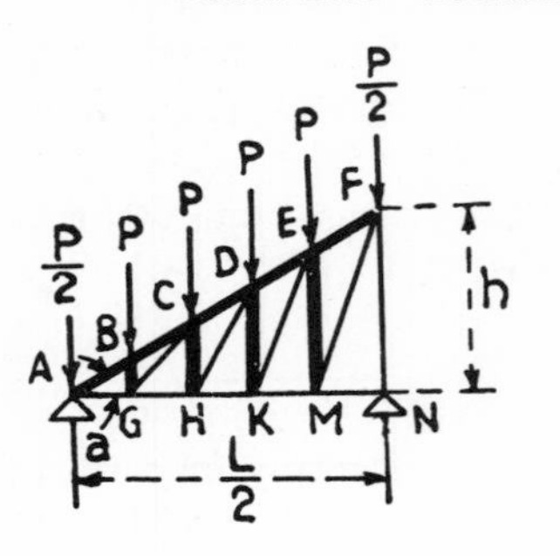

M	Pitch				
	$\frac{1}{3}$	$a = 30°$	$\frac{1}{4}$	$\frac{1}{5}$	$\frac{1}{6}$
AB	+3.61	+4.00	+4.47	+5.39	+6.32
BC	+3.61	+4.00	+4.47	+5.39	+6.32
CD	+2.70	+3.00	+3.35	+4.04	+4.74
DE	+1.80	+2.00	+2.24	+2.69	+3.16
EF	+0.90	+1.00	+1.12	+1.35	+1.58
AG	−3.00	−3.46	−4.00	−5.00	−6.00
GH	−2.25	−2.60	−3.00	−3.75	−4.50
HK	−1.50	−1.73	−2.00	−2.50	−3.00
KM	−0.75	−0.87	−1.00	−1.25	−1.50
MN	0	0	0	0	0
BG	+1.00	+1.00	+1.00	+1.00	+1.00
CG	−1.25	−1.32	−1.41	−1.60	−1.80
CH	+1.50	+1.50	+1.50	+1.50	+1.50
DH	−1.68	−1.73	−1.80	−1.95	−2.12
DK	+2.00	+2.00	+2.00	+2.00	+2.00
EK	−2.14	−2.18	−2.24	−2.36	−2.50
EM	+2.50	+2.50	+2.50	+2.50	+2.50
FM	−2.61	−2.65	−2.69	−2.80	−2.92

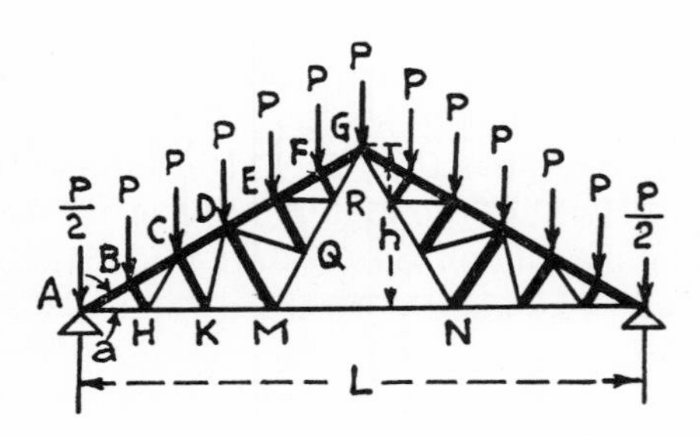

M	Pitch				
	$\frac{1}{3}$	$a = 30°$	$\frac{1}{4}$	$\frac{1}{5}$	$\frac{1}{6}$
AB	+9.92	+11.00	+12.30	+14.81	+17.39
BC	+9.36	+10.50	+11.85	+14.44	+17.08
CD	+8.18	+9.25	+10.51	+12.91	+15.34
DE	+7.63	+8.75	+10.06	+12.53	+15.02
EF	+7.70	+9.00	+10.51	+13.32	+16.13
FG	+7.14	+8.50	+10.06	+12.95	+15.81
AH	−8.25	−9.53	−11.00	−13.75	−16.50
HK	−7.50	−8.66	−10.00	−12.50	−15.00
KM	−6.75	−7.79	−9.00	−11.25	−13.50
MN	−4.50	−5.20	−6.00	−7.50	−9.00
BH	+0.83	+0.87	+0.89	+0.93	+0.95
CH	−0.75	−0.87	−1.00	−1.25	−1.50
CK	+1.25	+1.30	+1.34	+1.39	+1.42
DK	−1.04	−1.15	−1.26	−1.49	−1.71
DM	+2.50	+2.60	+2.68	+2.79	+2.85
DQ	−1.04	−1.15	−1.26	−1.49	−1.71
EQ	+1.25	+1.30	+1.34	+1.39	+1.42
ER	−0.75	−0.87	−1.00	−1.25	−1.50
FR	+0.83	+0.87	+0.89	+0.93	+0.95
MQ	−2.25	−2.60	−3.00	−3.75	−4.50
QR	−3.00	−3.46	−4.00	−5.00	−6.00
RG	−3.75	−4.33	−5.00	−6.25	−7.50

M = Member of truss — + Indicates compression
Pitch = h/L — − Indicates tension
Stress in any member = coefficient multiplied by *W*

Table 10.1 Trusses—Coefficients of Stresses (*Continued*)

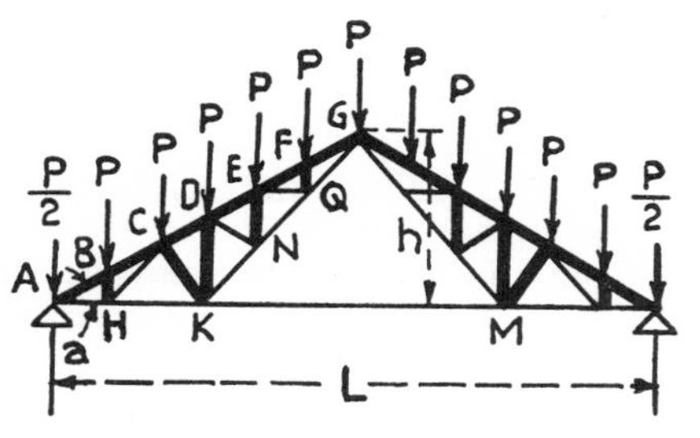

M	Pitch				
	$\frac{1}{3}$	$a = 30°$	$\frac{1}{4}$	$\frac{1}{5}$	$\frac{1}{6}$
AB	+9.92	+11.00	+12.30	+14.81	+17.39
BC	+9.92	+11.00	+12.30	+14.81	+17.39
CD	+8.11	+9.00	+10.06	+12.12	+14.23
DE	+9.01	+10.00	+11.18	+13.46	+15.81
EF	+9.92	+11.00	+12.30	+14.81	+17.39
FG	+9.92	+11.00	+12.30	+14.81	+17.39
AH	−8.25	−9.53	−11.00	−13.75	−16.50
HK	−7.50	−8.66	−10.00	−12.50	−15.00
KM	−4.50	−5.20	−6.00	−7.50	−9.00
BH	+1.00	+1.00	+1.00	+1.00	+1.00
CH	−1.25	−1.32	−1.41	−1.60	−1.80
CK	+1.25	+1.32	+1.41	+1.60	+1.80
DK	+2.00	+2.00	+2.00	+2.00	+2.00
DN	−0.90	−1.00	−1.12	−1.35	−1.58
EN	+1.50	+1.50	+1.50	+1.50	+1.50
EQ	−0.75	−0.87	−1.00	−1.25	−1.50
FQ	+1.00	+1.00	+1.00	+1.00	+1.00
KN	−3.75	−3.97	−4.24	−4.80	−5.41
NQ	−5.00	−5.29	−5.66	−6.40	−7.21
QG	−6.25	−6.61	−7.07	−8.00	−9.01

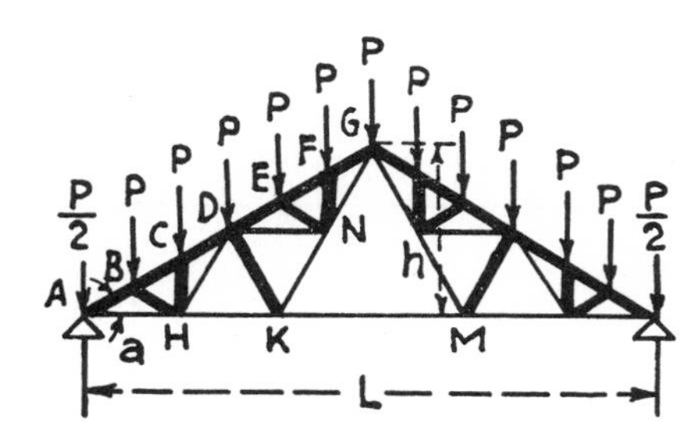

M	Pitch				
	$\frac{1}{3}$	$a = 30°$	$\frac{1}{4}$	$\frac{1}{5}$	$\frac{1}{6}$
AB	+9.92	+11.00	+12.30	+14.81	+17.39
BC	+8.95	+10.00	+11.25	+13.66	+16.13
CD	+8.81	+10.00	+11.40	+14.07	+16.76
DE	+8.25	+9.50	+10.96	+13.70	+16.44
EF	+7.28	+8.50	+9.91	+12.55	+15.18
FG	+7.14	+8.50	+10.06	+12.95	+15.93
AH	−8.25	−9.53	−11.00	−13.75	−16.50
HK	−6.75	−7.79	−9.00	−11.25	−13.50
KM	−4.50	−5.20	−6.00	−7.50	−9.00
BH	+0.93	+1.00	+1.07	+1.21	+1.34
CH	+0.93	+1.00	+1.07	+1.21	+1.34
DH	−1.50	−1.73	−2.00	−2.50	−3.00
DK	+2.50	+2.60	+2.68	+2.79	+2.85
DN	−1.50	−1.73	−2.00	−2.50	−3.00
EN	+0.93	+1.00	+1.07	+1.21	+1.34
FN	+0.93	+1.00	+1.07	+1.21	+1.34
KN	−2.25	−2.60	−3.00	−3.75	−4.50
NG	−3.75	−4.33	−5.00	−6.25	−7.50

M = Member of truss
+ Indicates compression
Pitch = h/L
− Indicates tension
Stress in any member = coefficient multiplied by W

Table 10.1 Trusses—Coefficients of Stresses (*Continued*)

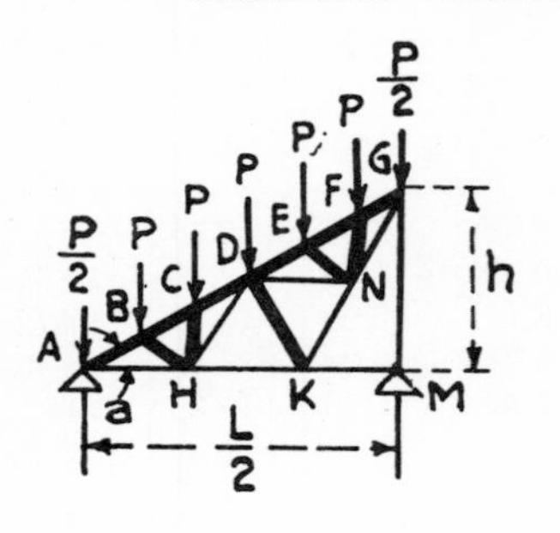

M	Pitch $\frac{1}{3}$	$a = 30°$	$\frac{1}{4}$	$\frac{1}{5}$	$\frac{1}{6}$
AB	+4.51	+5.00	+5.59	+6.73	+7.91
BC	+3.54	+4.00	+4.55	+5.59	+6.64
CD	+3.40	+4.00	+4.70	+5.99	+7.27
DE	+2.84	+3.50	+4.25	+5.62	+6.96
EF	+1.87	+2.50	+3.21	+4.47	+5.69
FG	+1.73	+2.50	+3.35	+4.87	+6.32
AH	−3.75	−4.33	−5.00	−6.25	−7.50
HK	−2.25	−2.60	−3.00	−3.75	−4.50
KM	0	0	0	0	0
BH	+0.93	+1.00	+1.07	+1.21	+1.34
CH	+0.93	+1.00	+1.07	+1.21	+1.34
DH	−1.50	−1.73	−2.00	−2.50	−3.00
DK	+2.50	+2.60	+2.68	+2.79	+2.85
DN	−1.50	−1.73	−2.00	−2.50	−3.00
EN	+0.93	+1.00	+1.07	+1.21	+1.34
FN	+0.93	+1.00	+1.07	+1.21	+1.34
KN	−2.25	−2.60	−3.00	−3.75	−4.50
NG	−3.75	−4.33	−5.00	−6.25	−7.50

M = Member of truss — + Indicates compression
Pitch = h/L — − Indicates tension
Stress in any member = coefficient multiplied by W

M	Pitch $\frac{1}{3}$	$a = 30°$	$\frac{1}{4}$	$\frac{1}{5}$	$\frac{1}{6}$
AB	+13.52	+15.00	+16.77	+20.19	+23.72
BC	+12.97	+14.50	+16.32	+19.82	+23.40
CD	+12.41	+14.00	+15.88	+19.45	+23.08
DE	+11.86	+13.50	+15.43	+19.08	+22.77
EF	+11.30	+13.00	+14.98	+18.71	+22.45
FG	+10.75	+12.50	+14.53	+18.34	+22.14
GH	+10.19	+12.00	+14.09	+17.97	+21.82
HK	+9.64	+11.50	+13.64	+17.60	+21.50
AM	−11.25	−12.99	−15.00	−18.75	−22.50
MN	−11.50	−12.12	−14.00	−17.50	−21.00
NQ	−9.00	−10.39	−12.00	−15.00	−18.00
QR	−6.00	−6.93	−8.00	−10.00	−12.00
BM	+0.83	+0.87	+0.89	+0.93	+0.95
CM	−0.75	−0.87	−1.00	−1.25	−1.50
CN	+1.66	+1.73	+1.79	+1.86	+1.90

M	Pitch $\frac{1}{3}$	$a = 30°$	$\frac{1}{4}$	$\frac{1}{5}$	$\frac{1}{6}$
CS	−0.75	−0.87	−1.00	−1.25	−1.50
DS	+0.83	+0.87	+0.89	+0.93	+0.95
ES	−2.25	−2.60	−3.00	−3.75	−4.50
SN	−1.50	−1.73	−2.00	−2.50	−3.00
EQ	+3.33	+3.46	+3.58	+3.71	+3.79
ET	−2.25	−2.60	−3.00	−3.75	−4.50
TU	−1.50	−1.73	−2.00	−2.50	−3.00
FT	+0.83	+0.87	+0.89	+0.93	+0.95
GT	−0.75	−0.87	−1.00	−1.25	−1.50
GU	+1.66	+1.73	+1.79	+1.86	+1.90
GV	−0.75	−0.87	−1.00	−1.25	−1.50
HV	+0.83	+0.87	+0.89	+0.93	+0.95
QU	−3.00	−3.46	−4.00	−5.00	−6.00
UV	−4.50	−5.20	−6.00	−7.50	−9.00
VK	−5.25	−6.06	−7.00	−8.75	−10.50

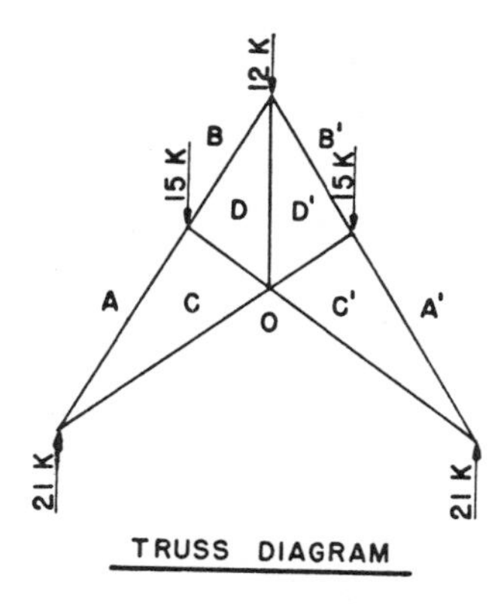

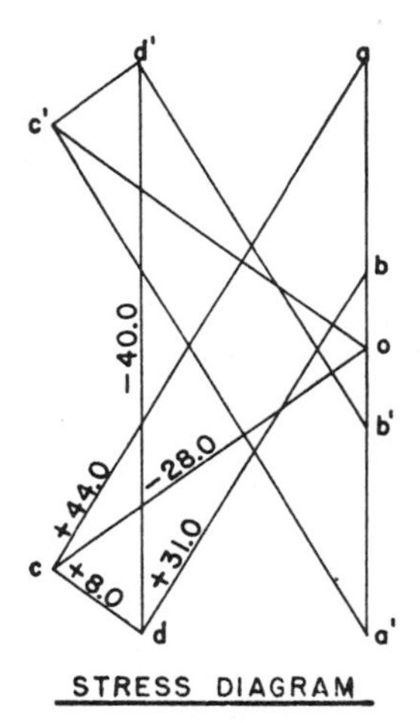

Fig. 10.1 Scissors Truss

spaces between loads or reactions, but not to panel points. A truss is made up of a series of triangles, so letter each internal triangle and each space between two external loads.

For the stress diagram, lay out at a convenient scale the applied load on the load line, going clockwise from the left end top of the truss. Use lower-case letters corresponding to the capitals on the truss diagram, but a space between loads on the truss diagram becomes a point of intersection on the stress diagram. Thus in the scissors truss, go from A to B and use lower-case letters in putting these loads on the stress diagram; then from B to B', B' to A', then back up the reaction line from A' to O, and from O back to A, closing the load line. Some texts show the load applied at the outer panel points of the truss immediately over the reaction, but this may be confusing rather than helpful. The load comes into the support, not into the truss. It has no effect whatever on the truss stresses, and is therefore omitted in the trusses that are shown.

Starting from the reaction stresses OA, we know the amount of the stress and the line of the two components AC and OC. From point a we may draw on our stress diagram, parallel with the top chord AC, a line of indefinite length. From point o, parallel with the lower chord OC, we may draw another line of indefinite length. Where these lines intersect will determine point c, and the length of the lines oc and ac will tell us the amount of stress in these two members. Having point c on our stress diagram, we look for other truss triangles where the adjacent letters in the stress diagram have been located. We have point c and the direction of the member CD and point b and the top chord direction. From point c draw an indefinite line parallel to CD, and from point b draw an indefinite line parallel to BD; the intersection of these two lines locates point d. From this we have the direction of line DD' and $B'D'$ and points b' and d' on the stress diagram; point d' may be located, then c', and if the layout was accurately made, it will check back at the point a' and close the truss.

Complete the layout of the stress diagram using the same scale as for the load line. Scale off the amount of the stresses and put them on the diagram. The next step is to determine the nature of the stresses, tension or compression. This may be done by reading from the truss diagram around any joint in a clockwise direction. Using the lower-chord center joint $ocdd'c'o$ as an example, read from the stress diagram in the same order. Reading from o to c, we read diagonally down or away from the joint, indicating that the stress is away from the joint or tensile stress. From c to d we read diagonally down against the joint—a compressive stress; from d to d' we read up or away from the joint—tensile stress; from

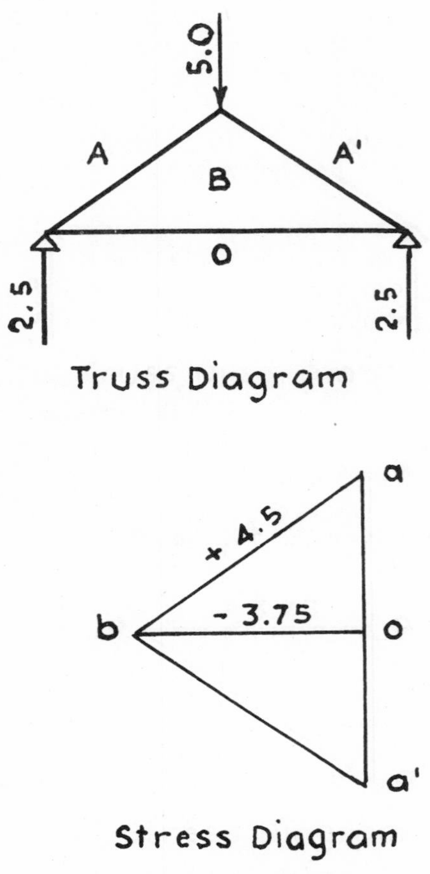

Fig. 10.2 King-Post Truss

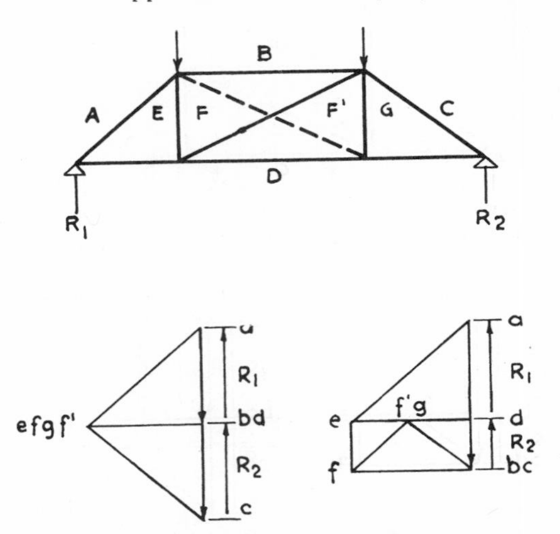

Fig. 10.3 Queen-Post Truss

d' to c' diagonally down—a compressive stress; and finally from c' to o diagonally down—a tensile stress. This simple method may be applied to any truss, no matter how complicated it may be. Although other engineers may differ in this convention, + to indicate compressive stresses and − to indicate tensile stresses are used here.

A number of the commoner forms of roof trusses and their respective stress diagrams are given in Figs. 10.2 to 10.4.

c *The Fink truss,* a common type of factory roof truss, particularly applicable to use as a steel truss because of its short compression members, requires a slight variation of procedure from that described above. In following the method described, it will be noted that after reaching point n on the stress diagram, the designer can no longer follow the normal procedure (see Fig. 10.5). Members PO and PQ are temporarily omitted here and an imaginary member OQ is installed as shown in dotted lines. This enables the designer to find point q on the stress diagram, with which the stress in members PQ and OP may be found.

The Fink truss may also be solved in a simple manner by breaking it down into component parts as shown in Fig. 10.6—four small trusses which may be easily solved graphically and the loads and stresses applied and added to a master truss which may also be solved simply. The truss shown in the truss diagram of Fig. 10.5 is broken down in Fig. 10.6 with the working members of each truss shown in solid lines. The reactions indicated at the center and quarter points of the upper part of Fig. 10.6 are added to the panel point load of the master truss in obtaining the solution. Also, the stress in the top chord of the four small trusses is added to the stresses obtained in the master truss and in any other members common to the two trusses, the stresses are added algebraically.

d Supports for roof signs and elevated tanks, radio towers, and similar structures are usually variations of a cantilever truss. Figure 10.7 shows the load and stress diagram of a truss of this kind. For the sake of comparison with the other diagrams, the truss is laid out horizontally with loads applied vertically. Normally this condition is reversed, with wind loads applied horizontally. For loads applied in the usual manner, from left to right, it will be noted that the stress diagram is laid out to the right of the load line instead of to the left as in the case of a standard truss. It is to be noted also that stresses in cantilever trusses for wind loads are subject to complete reversal, and that at the base of each leg anchorage must be provided in the foundation or structure on which it stands to carry the tensile stress or uplift under each leg. Failure to provide such proper anchorage is a common cause of failure of such cantilever trusses.

10.11 Steel trusses

a Steel trusses are not economical for roof spans below 60 ft if it is permissible to use flat-roof construction. Frequently, however, the roof pitch required (see Art. 10.40) makes flat construction undesirable, and if it is necessary to build up a two-way pitch, it is frequently advisable to take advantage of the condition and use steel trusses.

b Allowance must be made for the weight of the truss itself in selecting the dead load. Since this weight constitutes only a small portion of the total weight, a sizable error between actual and assumed weight will not materially influence the design. A simple way of approximating the weight of the truss is to compute the weight of a steel beam or girder which will carry the design load as a uniform load, and use a truss weight of 65 percent of the beam weight.

For instance, assume a 40-lb per sq ft total

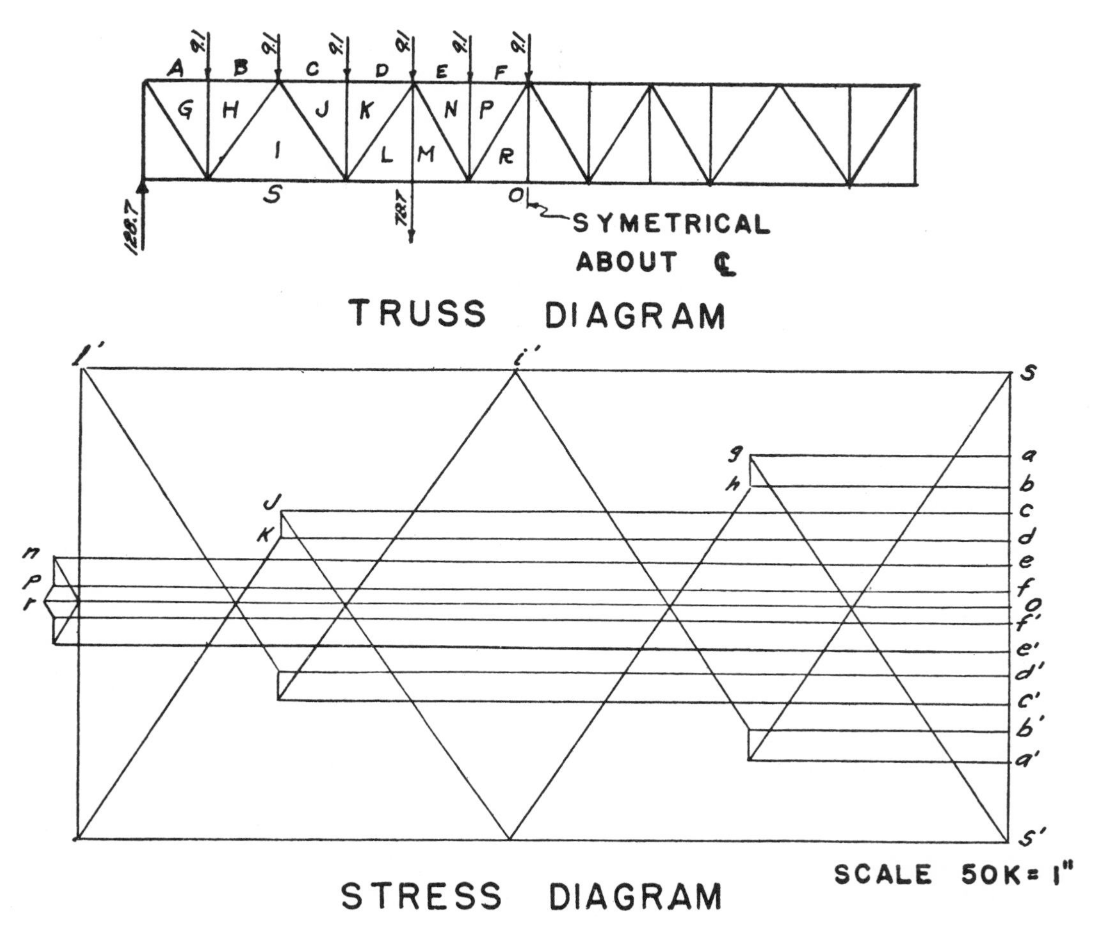

Fig. 10.4 Irregular Warren Truss with Columns Hung

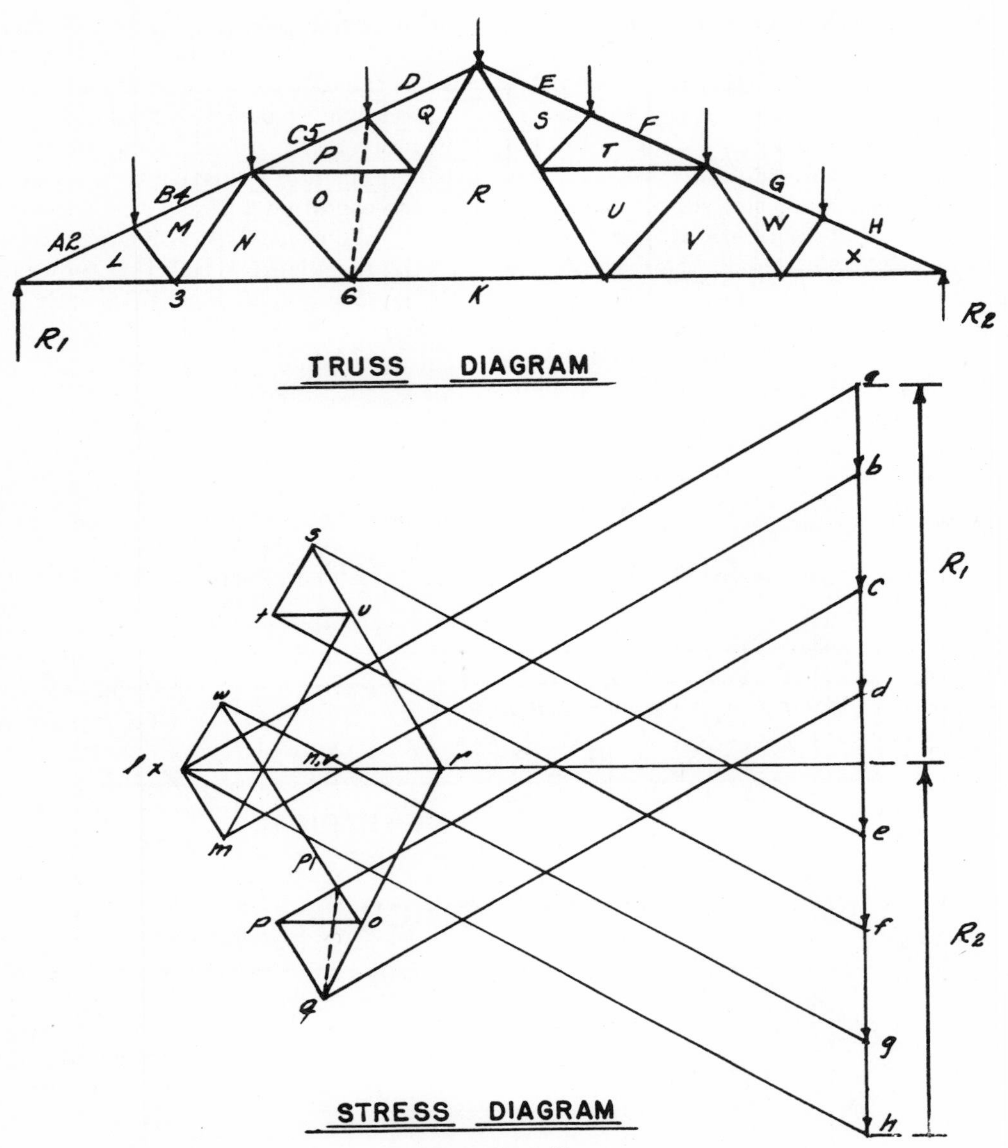

Fig. 10.5 Fink Truss

load on trusses with 60-ft span, spaced 16 ft.

$$W = 40 \times 60 \times 16 = 38.4K$$

$$S = \frac{38.4 \times 60}{16} = 144 = 24 \text{ WF } 68$$

Use a truss-weight allowance of $0.65 \times 68 = 44.2$ lb per lin ft. (Actual trusses on this span weigh 43 lb per ft for a Fink truss and 48 lb per ft for a Warren truss.)

c Stresses are computed by the methods described in Art. 10.10, either graphically or by the application of Table 10.1. The computation of truss members is usually simplified by setting the truss up as shown in Table 10.5.

The computation of member sizes is a problem in the application of simple stresses, as described in Arts. 2.10, 2.12 (as modified by Art. 7.10), and 2.13 and Table 2.2. For carrying ordinary stresses without a combination of bending and direct stress, a pair of angles is ordinarily used because they lend

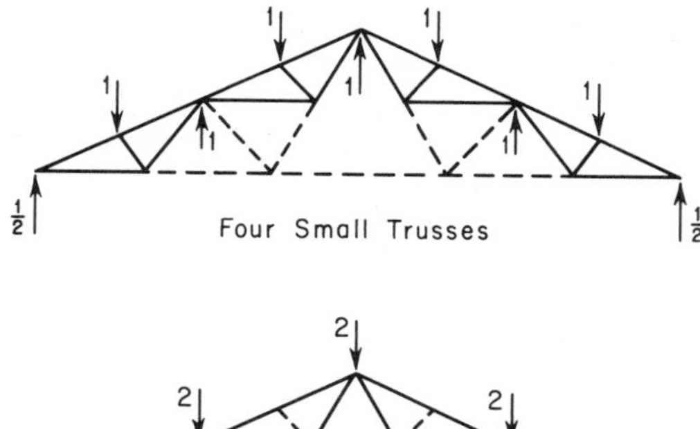

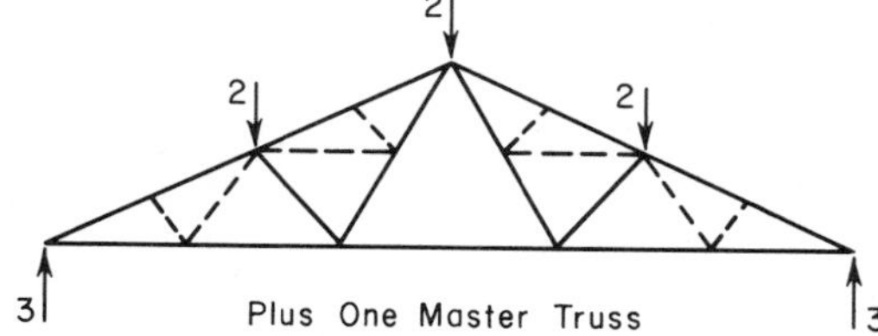

Fig. 10.6 Alternate Solution of Fink Truss

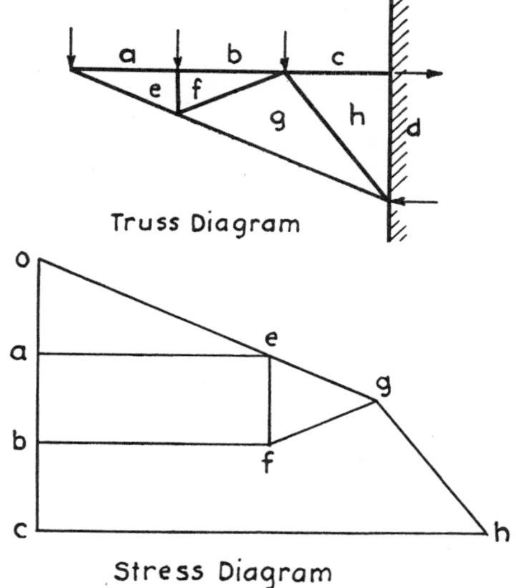

Fig. 10.7 Cantilever Truss

themselves to economical connection to gusset plates and to use as stiff compression members. When a truss member carries bending between panel points, angles are not satisfactory because they are weak in bending. In this case, the preferable member is made up of a pair of channels. Channels, however, do not lend themselves economically to compression because the radius of gyration of two channels around a ⅜- or ½-in. gusset is not so high as a pair of angles, and the added depth to the neutral axis of a channel lowers the effective depth for a given maximum total depth.

For tension members the net effective area is used. On this basis members may be selected from Table 2.2.

Compression members are designed in accordance with Art. 7.10*b*. Tables 10.2, 10.3, and 10.4 list the strength of various angles in pairs for various unbraced lengths. In the case of top-chord members, the spacing of purlins controls the unbraced length, in feet, about axis x-x, and the spacing of panel points controls the unbraced length about axis y-y. Usually they are the same.

As chord stresses are often too great to be carried economically by a pair of angles, wide-flange sections sometimes are used for truss members, with the flanges vertical. The members are usually kept to the same depth, and two gusset plates are used. Occasionally it is economical to use this type of truss even though the stresses could be carried by angle members, as in the irregular Warren truss shown in Fig. 10.4. Because columns are hung from the bottom chord, wide-flange sections lend themselves to better detail.

The selection of gusset-plate thickness to use throughout is a matter of detail. If stresses are light it is common practice to use a ⅜-in. gusset plate with ¾-in. rivets. The AISC specification requires a minimum of two rivets, which would carry a stress of $2 \times 13.24 = 26.5$ kips (see Table 9.2).

Problem Design an eight-panel compound Fink truss on 60-ft span, 15-ft rise, with loads at panel points of 20,000 lb each (including the weight of the truss). Refer to Table 10.1 for the compound Fink truss, pitch, $h/L = 15/60 = 1/4$. Since it is not economical to splice the chord members except at the center, the only chord stress it is necessary to design for is the heaviest one. Table 10.5 will use the same letters as in Table 10.1.

Assume that one truss in the building has eccentric loads on one side of the peak of the Fink truss, so that the purlins from one side are at panel points carrying 10 kips each, while from the other panel the span from peak line to eave line is divided into five equal spans carrying 8 kips each. The load

Table 10.2 Double Angle Members—Equal Legs in Compression

Size, in.	t, in.	Area, sq in.	r		Unbraced length in feet ⅜ in. back to back											
			$x-x$ axis	$y-y$ axis	Axis $x-x$						Axis $y-y$					
					4	6	8	10	12	14	4	6	8	10	12	14
8 × 8	1⅛	33.46	2.42	3.55	680	658	637	615	583	552	697	684	671	658	640	621
	1	30.00	2.44	3.53	610	591	572	553	525	496	625	613	602	590	573	556
	⅞	26.46	2.45	3.51	538	522	505	488	464	439	551	541	531	520	505	490
	¾	22.88	2.47	3.49	466	451	437	423	402	381	476	467	458	449	436	423
	⅝	19.22	2.49	3.47	391	380	368	356	339	321	400	392	384	377	366	355
6 × 6	1	22.00	1.80	2.72	436	416	397	369	338	304	451	439	427	415	398	380
	⅞	19.46	1.81	2.70	386	369	352	327	300	270	398	387	377	366	351	335
	¾	16.88	1.83	2.68	336	321	306	285	262	236	346	336	327	317	304	290
	⅝	14.22	1.84	2.66	283	271	258	241	221	200	291	283	275	267	256	243
	9/16	12.86	1.85	2.65	256	245	234	218	201	181	263	256	249	241	231	220
	½	11.50	1.86	2.64	229	219	209	195	180	163	235	229	223	216	206	196
5 × 5	⅞	15.96	1.49	2.31	312	295	271	244	214	180	340	317	304	290	275	258
	¾	13.88	1.51	2.28	272	258	237	214	188	160	286	275	264	251	238	223
	⅝	11.72	1.52	2.26	230	218	201	181	160	136	241	232	223	212	200	187
	½	9.50	1.54	2.23	186	177	163	148	131	112	196	188	180	171	161	151
	7/16	8.36	1.55	2.22	164	156	144	131	116	99	172	165	158	150	142	132
4 × 4	¾	10.88	1.19	1.88	208	189	166	140			221	210	199	185	171	155
	⅝	9.22	1.20	1.86	177	161	142	120	95		188	178	168	157	144	131
	½	7.50	1.22	1.83	144	131	116	99	79		152	144	136	127	116	105
	7/16	6.62	1.23	1.82	128	116	103	88	71		134	127	120	112	102	92
	⅜	5.72	1.23	1.81	110	100	89	76	61		117	110	103	96	88	79
	5/16	4.80	1.24	1.80	93	85	75	64	52		97	92	87	81	74	66
3½ × 3½	½	6.50	1.06	1.64	122	108	92	73			132	123	114	105	94	82
	7/16	5.74	1.07	1.62	108	96	82	65			116	108	100	92	82	72
	⅜	4.96	1.07	1.61	93	83	71	56			99	93	87	79	71	62
	5/16	4.18	1.08	1.60	79	70	60	48			85	79	73	66	59	52
3 × 3	½	5.50	.90	1.43	99	84	67				108	101	92	82	71	59
	7/16	4.86	.91	1.42	88	75	60				96	89	81	72	62	51
	⅜	4.22	.91	1.41	76	65	52				83	77	70	62	54	44
	5/16	3.56	.92	1.40	65	55	44				70	65	59	52	45	37
	¼	2.88	.93	1.38	52	45	36				56	52	47	42	36	
2½ × 2½	½	4.50	.74	1.24	76	60					87	79	70	60	49	
	⅜	3.46	.75	1.21	59	47					66	60	53	45	36	
	5/16	2.94	.76	1.20	50	40					56	51	45	38	30	
	¼	2.38	.77	1.19	41	33					46	41	36	31		
2 × 2	⅜	2.72	.59	1.02	41						51	45	37	29		
	5/16	2.30	.60	1.00	35	24					43	37	31	24		
1 × 1	¼	1.88	.61	.99	29	20					35	30	25			
	3/16	1.42	.62	.98	22	15					26	23	19			

Table 10.3 Double Angle Members—Unequal Legs in Compression

Longer leg vertical

Size, in.	t, in.	Area, sq in.	r		Unbraced length in feet 3/8 in. back to back											
			$x-x$ axis	$y-y$ axis	Axis $x-x$						Axis $y-y$					
					4	6	8	10	12	14	4	6	8	10	12	14
9 × 4	1	24.00	2.84	1.56	493	481	468	456	430	420	471	448	414	375	333	285
	7/8	21.22	2.86	1.52	437	426	415	404	389	372	425	394	363	328	289	246
	3/4	18.38	2.88	1.50	378	369	359	350	337	323	369	340	313	282	248	210
8 × 6	1	26.00	2.49	2.52	530	514	499	481	459	434	530	515	499	483	460	435
	7/8	22.96	2.51	2.50	468	454	439	426	406	384	467	453	439	425	405	383
	3/4	19.88	2.53	2.48	405	393	381	369	352	333	404	392	380	368	350	331
	5/8	16.22	2.54	2.46	341	330	320	311	297	281	340	330	319	309	294	277
8 × 4	1	22.00	2.52	1.61	448	435	421	408	389	369	444	414	384	351	314	273
	7/8	19.46	2.53	1.58	396	384	373	361	345	327	392	365	338	307	274	237
	3/4	16.88	2.56	1.56	345	335	325	314	300	284	339	315	291	264	234	201
	5/8	14.22	2.57	1.53	290	282	273	265	253	240	286	265	244	221	195	166
7 × 4	7/8	17.72	2.20	1.64	359	347	335	318	299	279	357	334	311	285	256	224
	3/4	15.38	2.22	1.62	312	301	291	277	261	243	311	290	269	246	220	192
	5/8	12.96	2.24	1.59	263	254	246	234	220	206	261	243	225	205	183	159
	9/16	11.74	2.24	1.58	239	231	223	212	200	187	236	220	204	185	165	143
6 × 4	7/8	15.96	1.86	1.71	318	304	291	271	250	226	324	304	284	262	237	211
	3/4	13.88	1.88	1.69	277	265	253	237	218	198	280	263	246	227	205	181
	5/8	11.72	1.90	1.66	234	225	215	201	185	169	237	222	207	190	171	150
	9/16	10.62	1.90	1.66	212	203	194	182	168	153	215	201	187	172	155	136
	1/2	9.50	1.91	1.65	190	182	174	163	151	137	191	179	167	153	138	121
5 × 3½	3/4	11.62	1.55	1.54	231	217	200	182	161	138	231	217	200	181	160	137
	5/8	9.84	1.56	1.51	196	184	170	154	137	118	196	183	168	152	133	113
	1/2	8.00	1.58	1.49	159	150	139	126	113	97	158	148	136	122	107	90
	7/16	7.06	1.59	1.47	141	132	123	112	100	86	139	130	119	107	93	78
5 × 3	1/2	7.50	1.59	1.25	149	141	130	119	106	92	145	132	118	101	82	
	7/16	6.62	1.60	1.24	132	124	115	105	94	82	128	117	104	89	72	
4 × 3½	5/8	8.60	1.22	1.60	165	151	133	113	91		171	162	150	137	122	106
	1/2	7.00	1.23	1.58	135	123	109	93	75		139	131	121	111	98	85
	7/16	6.18	1.24	1.57	119	109	97	83	63		123	118	107	97	87	75
	3/8	5.34	1.25	1.56	103	94	84	72	59		106	100	92	84	74	64
	5/16	4.50	1.26	1.55	87	80	71	61	50		89	84	78	70	62	53
4 × 3	5/8	7.96	1.23	1.36	153	140	124	106	85		156	144	130	115	97	
	1/2	6.50	1.25	1.33	126	115	102	88	71		127	117	105	92	77	
	7/16	5.74	1.25	1.32	111	101	90	77	63		112	103	93	81	65	
	3/8	4.96	1.26	1.31	96	88	78	67	55		96	89	80	70	58	
	5/16	4.18	1.27	1.30	81	74	66	57	47		81	75	67	58	48	
3½ × 3	1/2	6.00	1.07	1.38	113	100	85	68			118	109	99	87	75	
	7/16	5.30	1.08	1.37	100	89	76	61			104	96	87	77	65	
	3/8	4.60	1.09	1.36	87	77	66	59			90	83	75	66	56	
	5/16	3.88	1.10	1.35	73	65	56	46			75	70	63	56	47	

Table 10.3 Double Angle Members—Unequal Legs in Compression (*Continued*)

Longer leg vertical

Size, in.	t, in.	Area, sq in.	r		Unbraced length in feet 3/8 in. back to back											
			$x - x$ axis	$y - y$ axis	Axis $x - x$						Axis $y - y$					
					4	6	8	10	12	14	4	6	8	10	12	14
3½ × 2½	½	5.50	1.09	1.13	104	93	79	64			104	94	81	67		
	7/16	4.86	1.09	1.12	92	82	70	57			92	83	71	59		
	3/8	4.22	1.10	1.11	80	71	61	50			80	72	62	50		
	5/16	3.56	1.11	1.10	67	60	52	42			67	60	52	42		
3 × 2½	½	5.00	.91	1.18	91	77	61				96	87	76	64		
	7/16	4.42	.92	1.17	80	69	55				84	76	67	56		
	3/8	3.84	.93	1.16	70	60	48				73	66	58	48		
	5/16	3.24	.94	1.14	59	51	41				62	55	48	40		
	¼	2.62	.95	1.13	48	41	34				50	45	39	32		
3 × 2	½	4.50	.92	.94	82	70	56				82	71	57			
	7/16	4.00	.93	.93	73	63	50				73	63	50			
	3/8	3.46	.94	.92	63	54	44				63	54	43			
	5/16	2.94	.95	.90	54	47	38				53	45	36			
	¼	2.38	.96	.89	44	38	31				43	36	28			
2½ × 2	3/8	3.10	.77	.96	53	43					57	49	40			
	5/16	2.82	.78	.95	45	36					48	41	34			
	¼	2.12	.78	.94	37	30					39	33	27			

condition is shown in Fig. 10.8. The computation of reactions indicated in Fig. 10.9 indicates that the loads for use in the stress diagram will closely approximate the original truss designed above, so it will not be necessary to recompute the truss stresses. The top chord under consideration is a continuous beam, and the moments may be computed by the moment-distribution methods of Art. 3.21. We shall consider the ends as freely supported. Figure 10.10 shows the computation of moment.

Since the negative moment is always greatest in a continuous beam loaded in this manner, we must design for the combination of greatest direct stress and moment. Referring again to Table 10.1, the direct stress in AB is $7.83 \times 20 = 156.6$ kips, and the

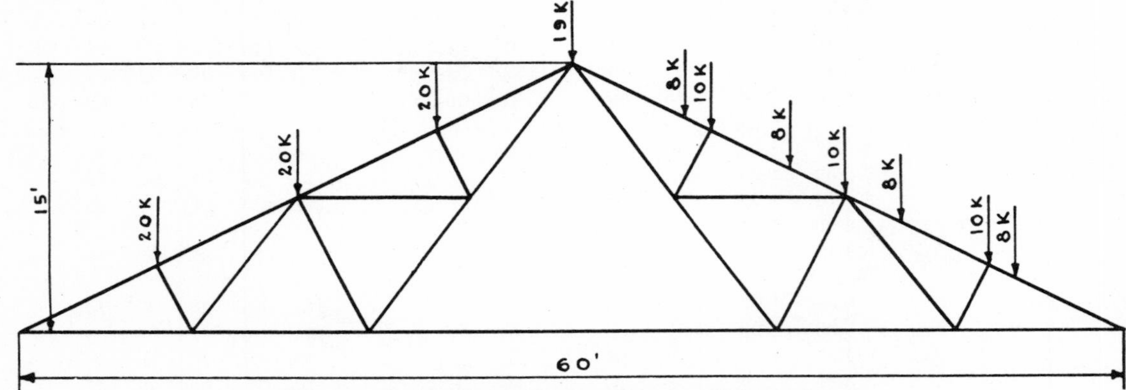

Fig. 10.8 Truss Load for Bending and Direct Stress

Table 10.4 Double Angle Members—Unequal Legs in Compression

Shorter leg vertical

Size, in.	t, in.	Area, sq in.	r		Unbraced length in feet 3/8 in. back to back											
			$x-x$ axis	$y-y$ axis	Axis $x-x$						Axis $y-y$					
					4	6	8	10	12	14	4	6	8	10	12	14
9 × 4	1	24.00	1.00	4.66	445	389	323	247			503	496	488	481	473	465
	7/8	21.22	1.01	4.63	394	346	288	222			445	438	432	425	418	412
	3/4	18.38	1.02	4.61	342	301	252	195			385	379	374	368	362	356
8 × 6	1	26.00	1.73	3.78	513	489	464	427	390	353	543	533	523	513	503	489
	7/8	22.96	1.74	3.76	453	432	411	378	345	312	479	470	461	453	444	432
	3/4	19.88	1.76	3.73	393	375	357	329	301	273	414	406	399	391	384	373
	5/8	16.22	1.77	3.72	330	315	300	277	254	231	348	342	336	329	323	313
8 × 4	1	22.00	1.03	4.10	410	362	304	237			460	452	445	437	430	420
	7/8	19.46	1.04	4.07	364	321	271	213			406	400	393	386	380	371
	3/4	16.88	1.05	4.04	316	280	237	187			353	347	341	335	330	321
	5/8	14.22	1.07	4.02	267	238	202	162			297	292	287	282	277	270
7 × 4	7/8	17.72	1.07	3.51	333	296	252	202			370	364	357	348	339	330
	3/4	15.38	1.09	3.49	290	259	222	179			320	315	309	302	294	286
	5/8	12.96	1.10	3.47	245	219	188	153			271	266	261	254	247	240
	9/16	11.74	1.11	3.46	222	199	172	140			245	241	236	230	224	218
6 × 4	7/8	15.96	1.11	2.97	302	271	233	190			330	323	316	306	295	284
	3/4	13.88	1.12	2.96	263	236	204	167			287	281	274	265	256	246
	5/8	11.72	1.13	2.92	223	200	173	143			242	237	231	224	216	207
	9/16	10.62	1.14	2.91	202	182	158	131			219	214	209	202	195	187
	1/2	9.50	1.15	2.90	181	163	142	118			196	192	187	181	175	168
5 × 3½	3/4	11.62	.98	2.48	214	187	154				237	231	224	215	205	194
	5/8	9.84	.99	2.45	182	159	131				201	195	189	181	173	163
	1/2	8.00	1.01	2.43	149	130	109	84			164	159	154	147	140	132
	7/16	7.06	1.01	2.41	131	115	96	74			144	140	136	130	123	116
5 × 3	1/2	7.50	.83	2.50	132	109	82				153	149	145	139	132	125
	7/16	6.62	.84	2.49	117	97	73				136	132	128	123	117	110
4 × 3½	5/8	8.60	1.03	1.91	160	141	119	93			174	167	158	148	130	124
	1/2	7.00	1.04	1.89	131	116	97	77			142	135	128	120	110	100
	7/16	6.18	1.05	1.89	116	102	87	69			125	120	113	106	97	89
	3/8	5.34	1.06	1.88	100	89	75	60			108	103	97	91	84	76
	5/16	4.50	1.07	1.86	85	75	64	51			91	87	82	76	70	64
4 × 3	5/8	7.96	.85	1.99	141	118	90				162	155	147	139	129	118
	1/2	6.50	.86	1.96	116	97	74				132	126	120	113	104	96
	7/16	5.74	.87	1.95	103	86	67				117	112	106	99	92	84
	3/8	4.96	.88	1.94	89	75	58				101	96	91	86	79	72
	5/16	4.18	.89	1.93	75	64	50				85	81	77	72	67	61
3½ × 3	1/2	6.00	.88	1.70	108	91	71				120	114	107	98	89	79
	7/16	5.30	.89	1.68	95	81	63				106	100	94	86	78	69
	3/8	4.60	.90	1.67	83	71	56				92	87	81	75	67	59
	5/16	3.88	.90	1.66	70	59	47				77	73	68	62	56	50

Table 10.4 Double Angle Members—Unequal Legs in Compression (*Continued*)

Shorter leg vertical

Size, in.	t, in.	Area, sq in.	r		Unbraced length in feet ⅜ in. back to back											
			$x - x$ axis	$y - y$ axis	Axis $x - x$						Axis $y - y$					
					4	6	8	10	12	14	4	6	8	10	12	14
3½ × 2½	½	5.50	.70	1.76	91	69					111	105	99	91	83	74
	7/16	4.86	.71	1.75	81	62					98	93	87	81	73	66
	⅜	4.22	.72	1.74	71	55					85	81	75	70	63	57
	5/16	3.56	.73	1.73	60	47					72	68	64	59	53	47
	¼	2.62	.75	1.45	45	35					52	48	44	39	34	28
3 × 2	½	4.50	.55	1.57	65						90	84	78	71	63	54
	⅞	4.00	.55	1.56	58						80	75	69	63	56	48
	⅜	3.46	.56	1.54	51						69	64	60	54	48	41
	5/16	2.94	.57	1.53	44						58	55	50	46	40	34
	¼	2.38	.57	1.52	35						47	44	41	37	32	28
2½ × 2	⅜	3.10	.58	1.27	47						60	55	49	42	35	
	5/16	2.82	.58	1.26	39						51	46	41	35	29	
	¼	2.12	.59	1.25	32						41	37	33	29	23	

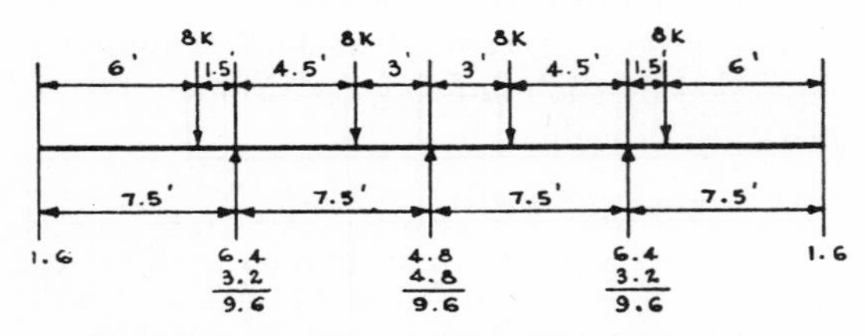

Fig. 10.9 Projected Load for Moment Computation

moment from Fig. 10.10 is 88.3 in.-kips; and in BC the direct stress is $7.38 \times 20 = 147.6$ kips, and the moment from Fig. 10.10 is 93 in.-kips. Using the member selected in Table 10.5, in AB

$$\text{Direct stress} = \frac{156.6}{10.62} = 14.75$$

$$\text{Eccentric stress} = \frac{88.3}{4.6} = \frac{19.20}{33.95}$$

which is excessive. Try two 12 channels, 20.7.

$$\text{Direct stress} = \frac{156.6}{12.06} = 12.98$$

$$\text{Eccentric stress} = \frac{88.3}{42.8} = 2.06$$

$$\frac{12.98}{17.43} = 74.6\%$$

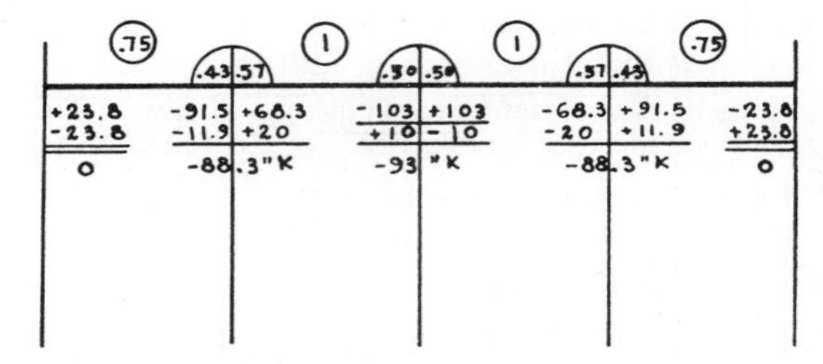

Fig. 10.10 Computation for Top Chord Moment

$$\frac{2.06}{22} = \frac{9.4\%}{84.0\%}$$

which is allowable. Therefore the top chord in question is two 12 channels, 20.7.

To compute the permissible stress in direct stress given above, the method of Art. 1.31 for radius of gyration of a compound section must be used.

$$I = 2 \times 3.9 = 7.8$$
$$2 \times 6.03 \times 0.89^2 = \underline{9.55}$$
$$17.35$$

$$r = \sqrt{\frac{17.35}{12.06}} = \sqrt{1.44} = 1.2$$

$$\frac{l}{r} = \frac{72}{1.2} = 60$$

Table 10.5 Design of Truss Members

Member	C	Stress, kips	L, ft	A	Section (all two angles), in.	Remarks
Top chord	7.83	+156	8.37		6 × 4 × 9/16 LLV	From Table 10.3
Bottom chord	7	−140		6.4	6 × 6 × 9/16	From Table 2.2
BF	0.89	+17.8	4.19		2½ × 2 × ¼ LLV	From Table 10.3
CF	1	−20		.91	2½ × 2 × ¼	From Table 2.2
CG	1.79	+35.8	8.37		3½ × 3 × ¼ LLV	From Table 10.3
CH	1	−20		.91	2½ × 2 × ¼	From Table 2.2
DH	0.89	+17.8	4.19		2½ × 2 × ¼ LLV	From Table 10.3
EH	3	−60		2.7	3 × 2½ × 5/16	From Table 2.2

The allowable stress from Table 7.1 is 17.43 which is used in the foregoing problem. It will be found that the 10 in. ⊔ 15.3 would be overstressed.

10.12 Wood trusses

a The design of wood trusses is primarily a problem in the design of truss joints. In compression members the l/d ratio will have some effect on the size of the member, but in tension members any member that will satisfactorily develop the connection will usually take the tensile stress.

Many of the older wood trusses used wood for compression members and wrought iron or steel for tension members—usually in the form of rods. The design of the joints of a wood truss involve a complicated set of computations of inclined bearing, shear, and other factors, as described in Art. 9.30.

b With timber connectors available, most trusses are designed using them as described in Art. 9.30*b*. If decorative wood trusses are used, the actual working truss may be built up of 2- or 3-in.-thick members to develop the connectors efficiently, and then boxed in with thin enveloping material to give the appearance of a solid member.

In computing the loads to be carried by the truss, an allowance must be made for the weight of the truss itself. Formulae derived from tabulated weights given by the Timber Engineering Co. are:

For a flat-top wood truss, height ⅛ to 1/10 of span, the weight of truss is $3 + 0.116S$ percent of the live plus dead load, where S is the span in feet.

For a pitched truss with ⅕ to ⅓ pitch, the weight becomes $1.6 + 0.113S$ percent.

Figures 10.11 and 10.12 illustrate the design of a wood truss using timber connectors. It has been taken from the pamphlet of the Timber Engineering Co. entitled "Designing Timber Connector Structures."

c *Procedure for the design of connector joints.* While this procedure will apply to most structures, special cases may call for variations, but with the fundamental steps in mind the variations can easily be made.

1. Determine stresses for structural members.
2. Form a tentative plan for framing the structure and select lumber species to be used.
3. Compute the sizes of the structural members.
4. Select the type and size of connectors which seem to be most suitable for the structure.
5. Determine the angle of load to grain for each member in the joint to get the ring capacity, required edge and end distances, and spacings for connectors.
6. Design those joints first which carry the greatest loads, since the sizes of the members required to space the connectors in these joints will be the determining factor in arriving at final member sizes.
7. Determine the amount of load in the joint which must be transferred between each two members to be joined and design for the largest load first. The positioning of the connectors for the smaller loads will usually not be a problem; it should, however, be checked.

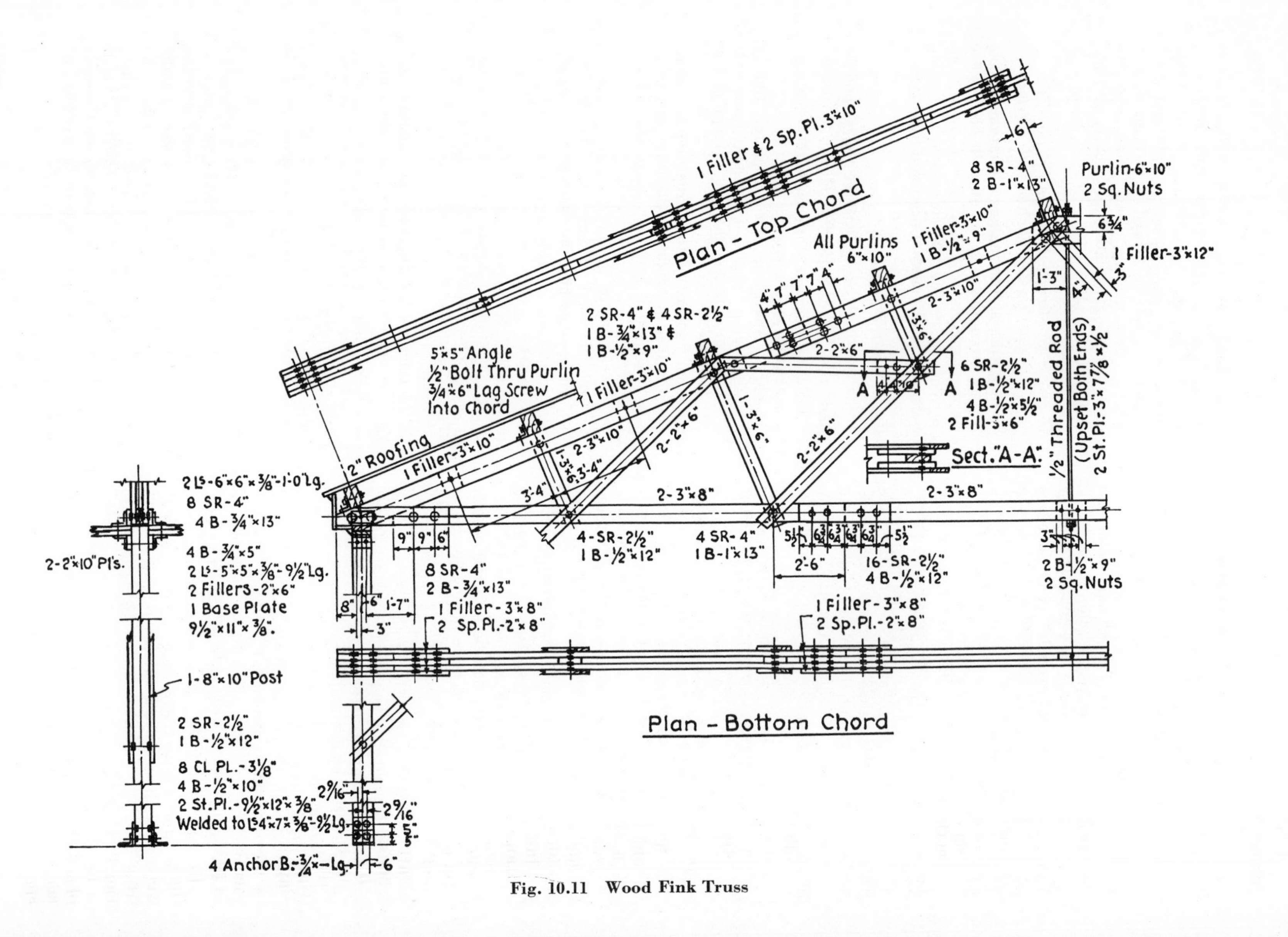

Fig. 10.11 Wood Fink Truss

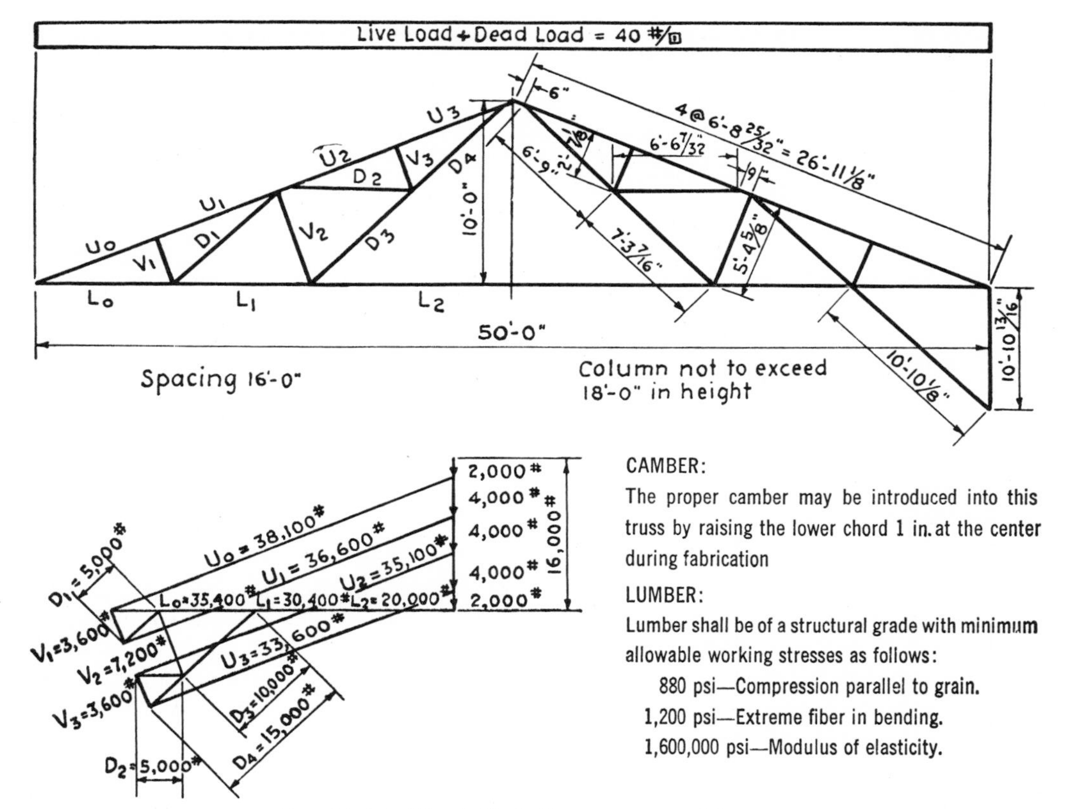

Fig. 10.12 Wood Fink Truss—Stress Diagram

8. Compute the number of connectors required to carry the loads and proceed to locate them on gage lines drawn with reference to the edges of the overlapped members. The spacings between connectors must also be checked.

9. If gage lines do not permit full spacings and the full capacities of connectors must be developed, the face widths of certain members must then be increased to provide the spacings required, or it may be that other sizes of connectors can be used without increasing the lumber size.

10. Check end distances and spacings for connectors in each member.

Problem A diagram for one-half of a 50-ft Fink roof truss with loads for each member is shown in Fig. 10.12. It is assumed that the roof loads are applied at the panel points.

Top chord. Since the greatest load is developed in the top-chord panels, the structural sizes should be designed for these loads first. To determine the size of the top chord in Fig. 10.11, it is assumed that (1) the chord will be composed of two members rather than one, thereby providing greater surface area for placing connectors in the overlapped joint members; (2) the members will be joined with connectors at the panel points to make it possible to take advantage of greater working stresses in compression due to the spaced-column principle (see Art. 7.30*a* and Fig. 10.13); and (3) lateral support will be provided at each panel point to prevent lateral buckling of the top chord.

When computing the sizes of spaced-column members as for the top chord in question, it will usually be found that a large value for d in the l/d ratio (l = length of panel in inches, d = least thickness in inches) is most suitable up to the point where the width of the member becomes too narrow to permit the placement of the necessary number of connectors. If the value for d becomes too small, greater overlapped area

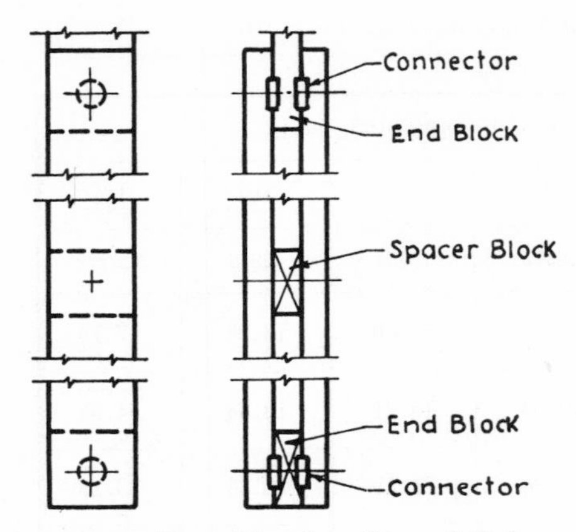

Fig. 10.13 Two-Member Spaced Columns

results but the working stresses of the lumber are rapidly reduced and the members become inefficient and uneconomical.

Reference to the top chord of the truss (Fig. 10.11) shows a panel length of 6 ft 8¾ in. and a maximum load of 38,100 lb. With members 1⅝ in. thick the l/d ratio for this panel length is 50; with lumber 2⅝ in. thick the ratio is 31; and with lumber 3⅝ in. thick the ratio is 22. Then, for lumber with a modulus of elasticity E of 1,600,000 psi and a compressive stress f_c of 880 psi, the working stresses for the different l/d ratios will be 440, 820, and 880 psi respectively. Dividing the 38,100-lb load by the values for the different l/d ratios, the 1⅝-in. thickness requires 87 sq in., the 2⅝-in. thickness requires 47 sq in., and the 3⅝-in. thickness requires 43 sq in. The 1⅝-in. thickness is evidently impractical since two chord members 28 in. wide would be necessary. Chord members 3⅝ in. thick take only a 7½-in. width to carry the load but do not furnish sufficient surface area to accommodate the connectors. Members 2⅝ in. thick require a 9½-in. face width to carry the load and provide sufficient overlapped area for placing connectors and are therefore recommended for the top chord.

Bottom chord. The bottom-chord member L_0 (Fig. 10.11) has a tension load of 35,400 lb. Assuming a lumber grade of 1,200 psi in tension or extreme fiber in bending (f), 29.5 sq in. of cross section are required. This is equivalent to approximately two pieces 3 × 6 in., nominal size, with a total net section of 29.54 sq in.

A thickness of 3 in. (2⅝ in. net) is selected primarily to maintain the same thickness as in the top-chord members, so that the splice plates will lie flat on the faces of the abutting members of upper and lower chords. The 6-in. width of lower-chord members is sufficient to accommodate the 4-in. split-ring connectors in this joint, but because of the necessity of an 8-in. nominal width to develop the load at an angle to grain in another bottom-chord joint for this truss, the bottom chord is made 8 in. wide throughout its entire length, thereby eliminating an extra splice joint in the bottom chord.

Web members in compression can best be handled as solid columns, with a relatively small l/d ratio, in contrast to web members in tension, which may be comparatively thin since their l/d is not a factor in their working stresses. This combination provides for efficient design and a symmetrically loaded joint by placing the compression members between the double chord members and the two tension members on the outside. Compression member $V2$, with a greater load than occurs in $V1$ or $V3$, is tentatively considered as having a 3-in. thickness (2⅝ in. net) to correspond to the appropriate 3-in. thickness of the center splice plates in the chord members. (The thickness of the center member of three splice plates used for joining a two-member chord should normally equal half the total thickness of the two members comprising the chord. The two outside splice pieces make up the remaining necessary section.) Using the same 880-psi grade of lumber as in the top chord and an l/d ratio of 25, the allowable working stress for member $V2$ is 680 psi. The load of 7,200 lb requires 10.6 sq in. of cross section. A 3 × 6-in. piece, with a cross section of 14.77 sq in., will accommodate this load and provide a 6-in. face for connectors.

Assuming a 2⅝-in. net thickness for compression members $V1$ and $V3$, corresponding in thickness to member $V2$, the l/d ratio is 12.3 for each of these two members. In the 880-psi grade of lumber, this gives a working stress of 880 psi, and for the 3,600-lb load approximately 4 sq in. of cross section are required. A 3 × 4-in. piece, with a net cross section of 9.52 sq in., is the smallest size possible, since the 2⅝-in. thickness must be maintained to fit between the chord members, and a 3⅝-in. width of face is necessary for a 2½-in. split ring. Therefore this size is recommended.

Tension members D3 and *D4*, with 15,000 lb maximum load and the 1,200-psi grade of lumber used in the truss, will require 12.5 sq in. of cross section. Two pieces 2 × 6 in., with a total of 18.2 sq in. of cross section, will provide more section than necessary. The other tension members, *D*1 and *D*2, with a load of 5,000 lb each, need less section than member *D*4, but 2 × 6s are recommended to keep the sizes uniform, and because smaller structural members are not generally recommended.

d *Design of timber joints.* The order in which the joints in the Fink truss (Fig. 10.11) might logically be designed would be first the heel joint, because of the large loads to be transmitted; then the peak joint, with relatively large loads and with members entering the joint from three directions; and finally the remaining joints in any order desired, since they are comparatively simple. However, to present more clearly the design principles which apply to timber-connector joints, a few typical joints will be discussed first, and then the heel and peak joints of Fig. 10.11.

Load parallel to grain

Tension joint. In the tension joint loaded parallel to the grain (Fig. 10.14), consisting of two overlapped members and split rings placed in the contact faces, assume that the

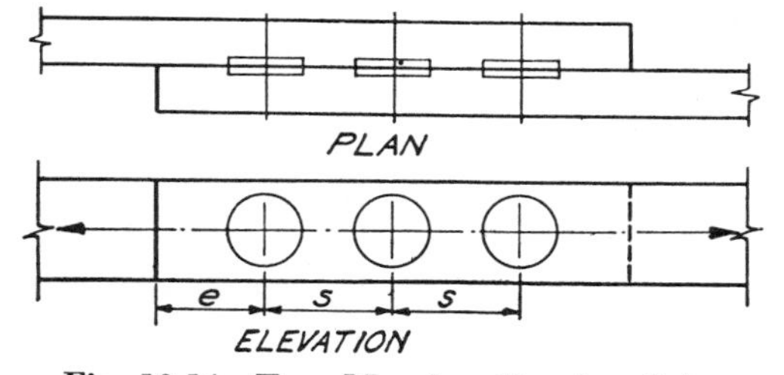

Fig. 10.14 Two-Member Tension Joint

load to be carried is 14,000 lb and that the lumber is 1,200-psi grade of group B species.

Design the joint for the following conditions:

1. Full end distances and spacings for connectors
2. Permissible reduced spacings
3. Permissible reduced end distances
4. Permissible reduced spacings and end distances

The four conditions outlined are presented to show the different combinations possible and how substandard spacings and distances may be computed for reduced loads. In actual design, it is recommended that reductions for both spacings and distances be made, rather than for one alone, thereby maintaining more uniform load reductions for all connectors.

1. *Full end distances and spacings.* Lumber Size: The 1,200-psi grade of lumber requires 11.7 sq in. of section to carry the 14,000-lb load. It will be found that a 2 × 8-in. member, with a sectional area of 12.19 sq in., and a 3 × 6-in. member, with a sectional area of 14.77 sq in., are the smallest nominal lumber sizes that will carry the load in tension.

Connector Size: Since split rings are specified for the joint, either one or two rows of 2½-in. rings or one row of 4-in. rings could be used in the 8-in. width of lumber, or one row of 4-in. rings could be used in the 6-in. width of lumber. To keep the number of bolts at a minimum, 4-in. connectors are recommended, although in a structure where 2½-in. split rings are used in other joints, it may be best to use them in this joint also.

Net Section: The net section of a member remaining after boring the bolt holes and cutting the grooves must then be checked to determine if it is adequate; it will usually be found adequate except where lumber of the higher stress grades is designed to nearly its full capacity. Checking the net section is readily done by referring to Tables 9.16 and 9.17. In Table 9.17 it will be found that the projected area of one 4-in. split ring and bolt in a member 1⅝ in. thick is 3.09 sq in., and in a member 2⅝ in. thick it is 3.84 sq in. These values subtracted from the sectional areas of the 2 × 8-in. and 3 × 6-in. lumber sizes leave 9.10 sq in. and 10.93 sq in. respectively. To determine if the net section remaining is sufficient, the actual square inches of net section required is computed from the table of constants (Table 9.16) by multiplying the load to be carried by the constant for standard loading for group B species in material less than 4 in. thick, which in this case is 14,000 × 0.00046, or 6.45 sq in. Therefore, either the 2 × 8-in. or 3 × 6-in. lumber sizes furnish adequate net section; the choice of size selected for a structure will be determined from the design as a whole.

Connectors Required: Working loads for split rings are given in Art. 9.30. A 4-in. split ring used in one face of a member 1⅝ in. or more in thickness and loaded parallel to the grain has a load capacity of 5,500 lb for group B species, and to carry the 14,000-lb load 2.55, or 3, split rings are required. The standard spacing and end distance for these connectors for load applied parallel to grain (see Art. 9.30) are 9 and 7 in. respectively. With three bolts spaced 9 in. apart and a 7-in. end distance, the joint length from the end of one piece to the end of the other piece in the joint is 7 + 9 + 9 + 7, or 32 in.

2. *Permissible reduced spacings.* Assume that it is desirable to shorten as much as possible the spacing between bolts with 4-in. split rings in the tension joint discussed above and still carry the 14,000-lb load.

The total capacity of three 4-in. split rings is 3 × 5,500 lb, or 16,500 lb, but only 14,000 lb, or 85 percent of their capacity, need be developed. Therefore, instead of three connectors carrying 300 percent of the capacity of one connector, they need carry a total of only 255 percent. Since one connector in a row with reduced connector spacing is assumed to carry 100 percent capacity, the other two connectors need to develop 77.5 percent capacity each. By interpolation of spacing requirements from full spacing of 9 in. at 100 percent capacity to 4⅞ in. at 50 percent capacity, as given in Art. 9.30, the spacing may be reduced from 9 to 7⅛ in. The total joint length from the end of one member to the end of the other member in the joint is then 7 + 7⅛ + 7⅛ + 7, or 28¼ in., a reduction of 3¾ in. from the length of the joint with full spacing and end distances.

3. *Permissible reduced end distances.* If the end distance only is to be reduced, the percentage reduction may be applied to the end ring up to the maximum of 37.5 percent allowed, which takes the minimum end distance permitted. Since the permissible reduction for the joint is 45 percent, it will be seen that the end distance may be reduced to the minimum allowed, or to 3½ in. for the 4-in. split ring. The total joint length is then 3½ in. for the 4-in. split ring. The total joint length is then 3½ + 9 + 9 + 3½, or 25 in.

4. *Permissible reduced spacings and reduced end distances.* It may sometimes be desirable to reduce both end distances and spacings. Such reductions are made by combining the two systems described under paragraphs 2 and 3 above. Assuming that the end distances are reduced, the full amount thereby reducing the capacity of the joint by 37.5 percent, it still leaves 45 − 37.5, or 7.5 percent, which can be applied to reduced spacing. The 7.5 percent divided by the two bolts with connectors (exclusive of the end bolt) gives 3.75 percent. By interpolation for reduced spacing, the 3.75 percent permits the spacing to be reduced from 9 to 8¾ in. The total joint length would then be 3½ + 8¾ + 8¾ + 3½, or 24½ in.

Compression joint. A compression joint may be designed similarly to a tension joint with all the load transferred by connectors, or it may be designed with part of the load carried by connectors and part carried by end bearing of the members with a metal plate fitted snugly between them. It is frequently found more practical to design a compression joint with connectors carrying the entire load, since splice plates provide the necessary stiffness for the members joined and a minimum amount of labor is required to fabricate the joint. The design of the joint with part of the load carried by direct end bearing offers a convenient solution, on the other hand, when using comparatively large members with limited space for splicing and where lateral stiffness is provided by means other than the joint itself—such as placement near or in a joint with lateral bracing. Compression joints in the top chords of bridge trusses offer typical examples where this method of designing joints is found. As much as 100 percent of the bearing stress of the members may be developed by end bearing provided that a metal bearing plate is placed between the ends of members and that fabrication is accurate. Normally, however, it is not practical to count on more than 50 to 75 percent, since the joint must be held together by some means. This is handled advantageously by connectors and one or more splice plates for holding the joint in line and for carrying a portion of the load; the remaining portion of the load being transferred by end bearing of members against a snugly fitted bearing plate.

Another system sometimes used is that of filling with concrete the space enclosed by the metal gusset plates and the ends of the members, the concrete forming end bearing for all members entering the joint.

The joint in the top-chord member $U2$ in Fig. 10.11 will serve to explain the two design systems. Reference to the previous text, describing the method of finding the size of the top-chord members, will show that the top chord takes two pieces of lumber 3 × 10 in. nominal. For one condition assume that the entire load of 35,100 lb is carried by connectors, and for the other condition that part of the load will be transferred by end bearing of the chord members. See Fig. 10.15, joints A and B respectively.

With the entire load transmitted by splice plates, their combined sectional area must equal at least the total sectional area of the members joined. A center piece 2⅝ in. thick and two side pieces 1⅝ in. thick provide the necessary section. Both 2½- and 4-in. diameter rings will be considered, since these are the most common sizes used in this type and size of truss. The 35,100-lb load

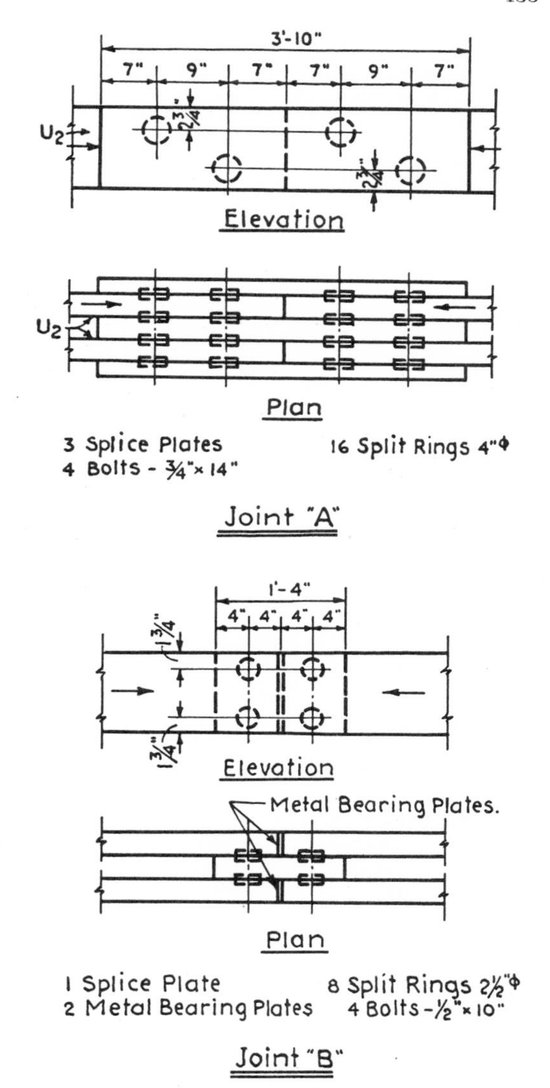

Details of compression splices: A—entire load of 35,100 lb carried by connectors; B—31 percent of load carried by connectors, 69 percent carried by end bearing of members against metal plates

Fig. 10.15 Details of Compression Splices

for group B species required 12.3 split rings 2½ in. in diameter on each side of the joint, or 6.5 split rings 4 in. in diameter. Twelve 2½-in. split rings on each side of the joint will safely carry the load and will fit into the joint conveniently. It will be necessary to use eight split rings 4 in. in diameter to keep the joint symmetrically loaded. The cost of labor for the installation of rings of either size will be about equal, the single apparent advantage of one ring size over the other being that the 2½-in. split ring takes slightly shorter splice plates. Joint *A* in Fig. 10.15 shows 4-in. split rings with end distances and spacings for full load capacity for the rings. The rings are placed off the center line to provide more even distribution of load to the members.

Figure 10.15 shows the same joint in joint *B* with the same load transmitted and member sizes as given in joint *A*, but with bearing plates between the ends of the members. The outside splice plates are omitted and 2½-in. split rings are used instead of the 4-in. split rings. In this joint 31 percent of the load is carried by the rings and 69 percent is transferred through the abutting ends of the compression members.

Load at angle to grain

Most structural joints have members entering at an angle with reference to other members, and therefore produce bearing at an angle to grain. The several examples of joints in the following discussion are all of this type, with each one bringing out certain design features.

Chord members placed between web members. This is a type of joint commonly used in the design of pitched-roof trusses. By making the compression member the same thickness as the splice plates used between the chord members, a comparatively low l/d ratio is secured, and furthermore the compression member will also fit between the chord members. The tension members in the joint may be thinner, since their l/d ratio is not a factor and they are placed outside the chord members.

A joint of this type is formed by members *L*1, *L*2, *V*2, and *D*3 of Fig. 10.11 and is detailed to larger scale in Fig. 10.16.

In this joint each of the two tension members *D*3 exerts its load of 5,000 lb (total 10,000 lb) at 45 deg to the grain in the two lower-chord members. At this angle of load to grain a 4-in. split ring in lumber 2⅝ in. thick of group B series develops a safe working load of 4,590 lb. Two of these rings, one between each chord member and each diagonal, with a ¾-in. bolt will

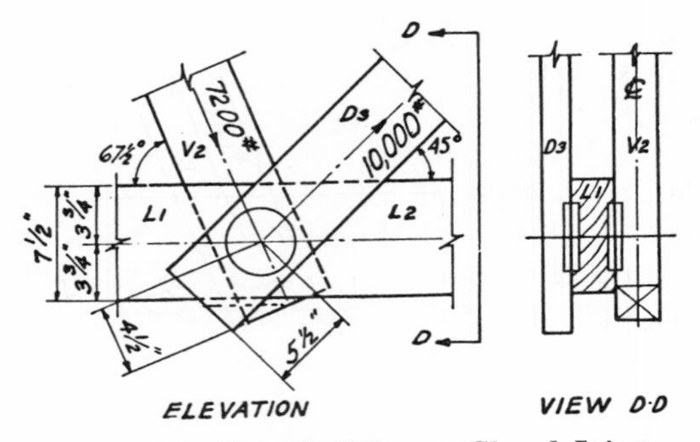

Fig. 10.16 Detail of Lower Chord Joint

develop 9,180 lb. But this is not sufficient to carry the 10,000-lb load, and therefore to increase the load capacity without increasing the size of the members to accommodate an extra bolt or more rings, a 1-in. bolt is used in place of the ¾-in. bolt, thereby increasing the capacity of the connectors by 7 percent, giving a total of 9,800 lb. This is only 2 percent under the full connector capacity required, and the joint is adequate to meet good-design requirements. Since the load in the tension member is acting upward at 45 deg, the edge distance in the chord members must be 3¾ in. on the upper side of the ring where its outside surface is in compression against the wood.

The end distance required for the 4-in. split ring in tension member *D*3, where the load is acting parallel to the grain, is determined by the load it carries. The 4,590-lb load capacity of the 4-in. split ring is increased 7 percent for the 1-in. bolt, equaling 4,910 lb, or 85 percent of the capacity of a ring with a 1-in. bolt loaded to 0 deg to grain. This percentage capacity permits a 5½-in. end distance.

Member *V*2 transmits the load to the lower chord at 67½ deg to grain, at which angle the load per ring with a 1-in. bolt is 4,185 lb plus 7 percent, or 4,480 lb, and for two rings it is 8,960 lb. This is greater than the 7,200-lb load necessary. The load in this web member is acting downward, so that the compression side of the ring is toward the bottom edge of the chord; therefore this edge distance as well as that for the upper side must be 3¾ in., making a member with a 7½-in. face necessary. It is this joint in the bottom chord which requires the widest face width and therefore determines the width of the member.

The standard end distance for a compression member with a 4-in. split ring is 5½ in., but since only 80 percent of the load need be developed in member *V*2, the distance may be reduced to 4½ in. The ends of both members *V*2 and *D*3 may be sawed off parallel to the edge of the bottom chord about ½ in. below it, as shown by the dotted lines (Fig. 10.16), which still leaves sufficient end distance.

Chord member placed outside of web members. The lower-chord joint of a flat-top Pratt truss (Fig. 10.17) has the diagonal

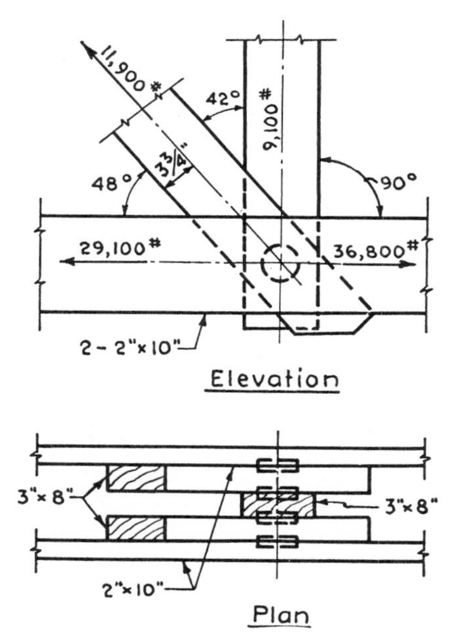

Fig. 10.17 Diagonals Between Vertical and Horizontal Members

members placed between the vertical and the horizontal members. Advantages gained by this arrangement as compared to the diagonals being placed outside the chord members are (1) the diagonal members, which must transmit the greatest load, have two faces into which connectors may be placed, whereas if they are located outside the chord members, connectors could be placed in only one face of each diagonal; (2) the bottom chord, which is in tension, does not need to be the same thickness as the top chord to assemble the truss; and (3) the working load per connector between the vertical and the diagonal with a 42-deg angle is considerably greater than it would be between the vertical and the bottom chord with a 90-deg angle of load to grain.

An analysis of the joint in Fig. 10.17 with axial loads in the members shows that where three overlapping members in a joint come together at different angles, it is the center member with connectors in both faces which has the load at an angle to the grain; the two side pieces are stressed only parallel to the grain. The side pieces (in Fig. 10.17 the horizontal chord and the vertical web member) must deliver their loads in the direction of their respective axes, thereby introducing stresses in the diagonal piece, which combine to produce the resultant force equal to the load carried by the diagonal member. Since the compression side of the ring in one face of the diagonal member applies to one edge and the compression side of the ring in the opposite face applies to the opposite edge, because of the direction of loads the diagonal must have a 3¾-in. edge distance on each side, resulting in a member with a face width of 7½ in.

The vertical member (in Fig. 10.17) with a 9,100-lb load to transfer to the two diagonal members at 42 deg will take two 4-in. split rings, one on each side to carry the load, since at this angle, in group B species and for lumber 2⅝ in. thick, one 4-in. ring will carry 4,645 lb and two will carry 9,290 lb. The two horizontal chord members must transmit a total of 7,700 lb to the diagonal members (36,800 − 29,100) at an angle to grain of 48 deg. Under the conditions presented, one ring will carry 4,540 lb and two will carry 9,080 lb, which is more capacity than necessary to carry the 7,700-lb load. Therefore, since the 4-in. split rings are sufficient to transfer the vertical and horizontal loads, they must also be sufficient to carry the load in the diagonal members.

Heel joint—Fink truss. The heel joint in the Fink truss shown in Fig. 10.18 is simplified by extending the top chord members to bear on the support, thereby transmitting directly to the support that portion of the load acting as the vertical component and eliminating the necessity of transferring this portion of the load by connectors through the bottom chord (see Fig. 10.18).

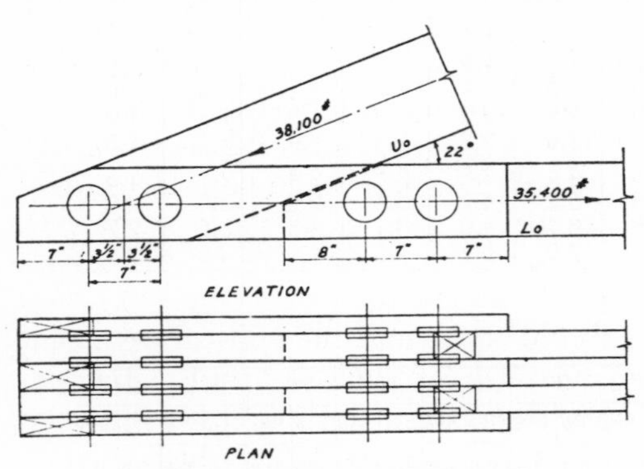

Fig. 10.18 Heel Joint of Fink Truss

In the design of the heel joint, it is apparent that split rings may be used in this timber-to-timber connection. As previously noted it is usually more efficient to use the larger connector sizes since they keep the number of bolts and connectors to a minimum and facilitate fabrication. If the most appropriate connector size for the joint is not immediately evident, the number of connectors required for each size may then be computed and one selected which fits most advantageously into the members. The loads for split rings for the heel joint in Fig. 10.18 are dependent on the 22-deg angle of load bearing to grain of the connectors in the top chord, on the 2⅝-in. lumber thickness as determined previously, and on the group B species of lumber. It is assumed that the truss will be used in a location where the lumber will reach an air-dry condition and, therefore, the moisture content of the lumber need not be considered as influencing split-ring loads. Based on the above conditions, the safe load for a 2½-in. split ring is 2,640 lb and, for a 4-in. split ring, 5,005 lb.

The 35,400-lb horizontal load which must be transferred to the bottom chord will take 13.4 split rings 2½ in. in diameter, or 7.1 split rings 4 in. in diameter. With four contacting faces into which connectors must be placed to keep the joint symmetrically loaded, the number of rings must be rounded off to 16 2½-in. rings or 8 4-in. rings. This number of rings for each size can be located in the heel joint to carry the load. It will be noted, however, that the 4-in. ring is more efficient, because the number of connectors actually required in the joint approaches nearest the number which is recommended for a symmetrically loaded joint. Also, the 4-in. size requires only 2 bolts and 8 rings as compared to 4 bolts and 16 rings of the 2½-in. size.

A further consideration in the design of a joint in which more than one bolt with connectors occurs is that if the full capacity of the connectors is not developed, as in this joint, it may be desirable to reduce their spacings and distances. The rule for permissible reduced spacing and reduced end distances may be applied, since the full capacities of the 4-in. rings are not developed. According to the rule, the 4 rings on one bolt will carry 400 percent of one ring capacity, or 400 × 5,005 lb = 20,020 lb. This amount subtracted from 35,400 lb leaves 15,380 lb for the 4 connectors on the second bolt, or 77 percent of their rated load. To develop 77 percent load capacity of the connectors in the second bolt, spacing between bolts may be reduced to 7 in. as determined by interpolation from design data given in Art. 9.30. End distance specified for the bottom-chord splice plates stressed in tension parallel to the grain is 7 in., and end distance for the top chord in compression is 5½ in. measured parallel to the grain. Edge distance, since the angle of load to grain is less than 30 deg, remains at the 2¾-in. minimum for both members and is provided by the members used.

The load to be transferred between the two bottom-chord members and the splice plates is parallel to the grain, and for the conditions presented a 4-in. split ring will carry 5,400 lb. To carry the 35,400-lb load, 6 rings are required; 8 rings, however, are recommended to maintain a symmetrically loaded joint. The diagonal cut on the end of the bottom-chord member through a point giving a 7-in.

standard end distance, measured along the center line, is not sufficient to provide the necessary 2¾-in. distance (equal to the edge distance for a 4-in. split ring) measured perpendicularly from the diagonal cut to the center of the bolt hole. Therefore, the end distance at the center line must be increased. Since the edge distance must be 2¾ in., the distance along the center line of the bottom chord is 2¾ in./sin 22 deg, or 7.2 required. In practice the end distance probably would be scaled and rounded off to 7½ or 8 in. Spacing of the connectors may be taken as a full 9 in. as specified for full-load capacity where there is sufficient area to place them along the chord, or the spacing may be reduced following the practice discussed under Tension joint, Art. 10.12, above. Connector spacing is shown as 7 in. in Fig. 10.18 to correspond to the 7-in. end distance of the splice. A check of the net section required for the members, using the same procedure as that explained for the tension joint, will show that sufficient section is provided by the size of members used.

Peak joint. The design of the peak joint in the Fink truss (Fig. 10.11) demonstrates how two bolts with connectors may be located to accommodate three members entering the joint when the angle of load to grain is less than 30 deg for each two contacting members. In this joint, as detailed in Fig. 10.19, the 15,000 lb must be transferred from the two diagonals $D4$ to the two chord members $U3$, and the 20,000-lb horizontal component must be carried through the splice to the other half of the truss. (This horizontal component is equal to the tension in the bottom chord at the center of the truss—see Fig. 10.11.) In this joint connectors are located in the two overlapped areas between the chord members and the diagonals, also between the chord members and the filler piece. The larger load of 20,000 lb should be considered first in the design of the joint, since connectors placed on the bolts used for the larger load will quite probably provide spacings and distances to develop the smaller load. Part of the thrust between the two half trusses could be carried by end bearing of the top-chord members, requiring a fitting job for the metal bearing plates. However, the two half sections of the roof truss must be securely attached at this joint and, regardless of the means for absorbing the thrust, a splice is necessary, and can perform the functions of both a tie and a splice to carry the load.

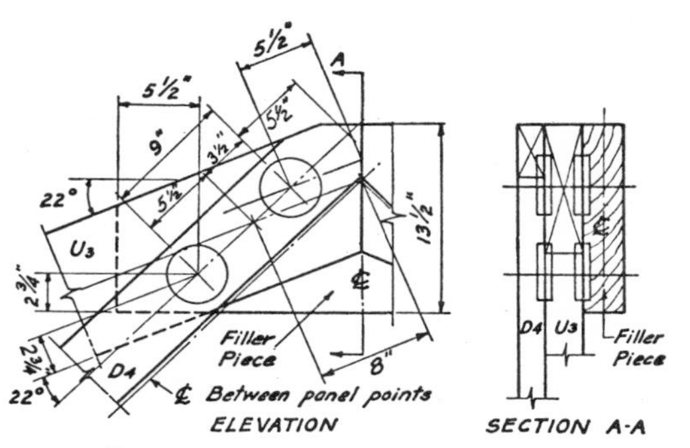

Fig. 10.19 Detail of Peak Joint

In the top-chord members a 22-deg angle of load to grain is made by the splice piece and also by web members $D4$. At this angle and with standard distances and spacings, the load per connector is 5,005 lb. With full distances and spacings, therefore, four connectors on two bolts will carry the 20,000-lb load to be transferred between the two chord pieces $U3$ and the filler piece. Since this part of the joint is in compression, the end distance for 4-in. split rings must be 5½ in. This end distance for piece $U3$ then determines the gage line for the end connector, the gage line being parallel to the end cut and measured parallel with the grain in the chord. The required 9-in. spacing and 2¾-in. edge distances for all members joined must also be met. In order to locate the two bolts with connectors within the limits of the diagonal $D4$ and still have the specified edge distance for connectors in the chord member, it is necessary to offset the diagonal so that the intersection of the center lines of the diagonal and the chord members is 8 in. to the left of the panel point. The slight eccentricity introduced into the joint by offsetting diagonal $D4$ is of little importance and may be neglected. The filler piece must be of sufficient width and length to

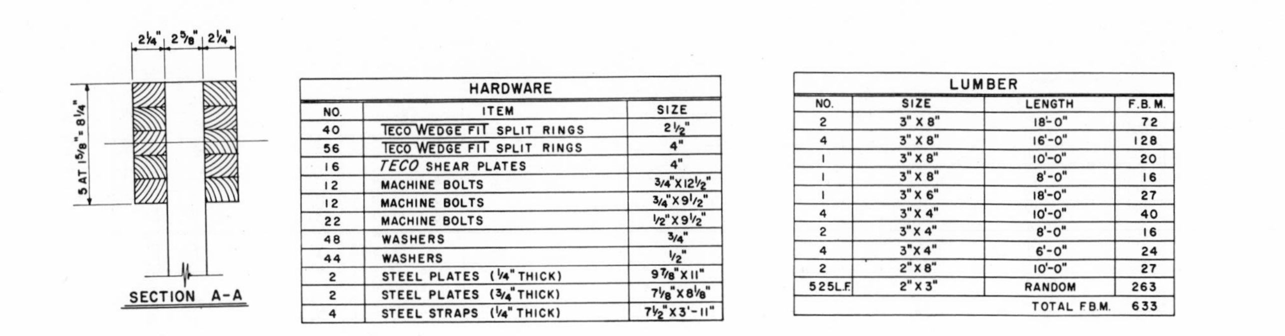

HARDWARE		
NO.	ITEM	SIZE
40	TECO WEDGE FIT SPLIT RINGS	2½"
56	TECO WEDGE FIT SPLIT RINGS	4"
16	*TECO* SHEAR PLATES	4"
12	MACHINE BOLTS	¾"X12½"
12	MACHINE BOLTS	¾"X9½"
22	MACHINE BOLTS	½"X9½"
48	WASHERS	¾"
44	WASHERS	½"
2	STEEL PLATES (¼" THICK)	9⅞"X11"
2	STEEL PLATES (¾" THICK)	7⅛"X8⅛"
4	STEEL STRAPS (¼" THICK)	7½"X3'-11"

LUMBER			
NO.	SIZE	LENGTH	F.B.M.
2	3" X 8"	18'-0"	72
4	3" X 8"	16'-0"	128
1	3" X 8"	10'-0"	20
1	3" X 8"	8'-0"	16
1	3" X 6"	18'-0"	27
4	3" X 4"	10'-0"	40
2	3" X 4"	8'-0"	16
4	3" X 4"	6'-0"	24
2	2" X 8"	10'-0"	27
525 L.F.	2" X 3"	RANDOM	263
		TOTAL F.B.M.	633

Fig. 10.20 Typical Timber Bowstring Truss

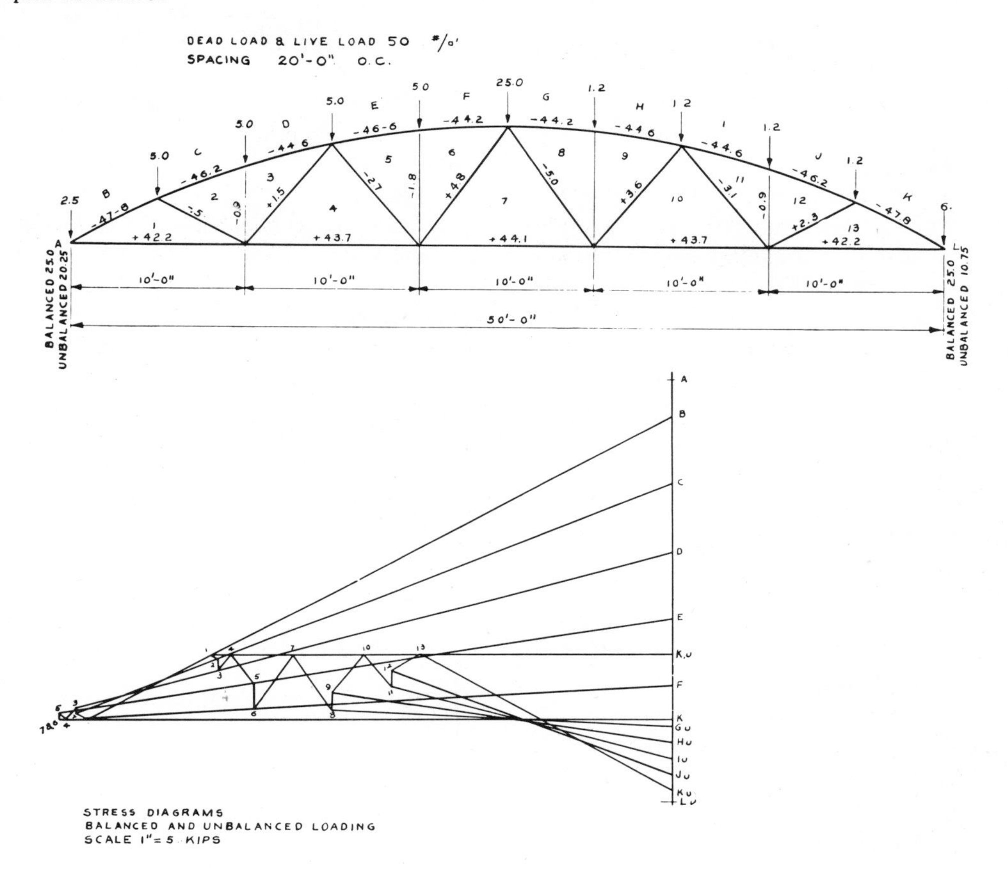

THIS TYPICAL DESIGN IS A GUIDE FOR DESIGNERS IN DEVELOPING THEIR OWN SPECIFIC DESIGNS, INCLUDING BRACING, ANCHORAGE, ETC., AS MAY BE REQUIRED.

LUMBER SHALL BE OF A GROUP "B" CONNECTOR LOAD SPECIES AND OF A STRESS GRADE WHICH QUALIFIES FOR MINIMUM ALLOWABLE WORKING STRESSES OF:

C = 1000 p.s.i COMPRESSION PARALLEL TO GRAIN
f = 1200 p.s.i. EXTREME FIBER IN BENDING
E = 1,600,000 p.s.i. MODULUS OF ELASTICITY

TOP CHORD MEMBERS SHALL BE GLUED LAMINATED LUMBER IN ACCORDANCE WITH PROVISIONS OF PART IX, NATIONAL DESIGN SPECIFICATION, PUBLISHED BY THE NATIONAL LUMBER MANUFACTURERS ASSOCIATION.

TIMBER CONNECTORS SHALL BE TECO WEDGE FIT $2\frac{1}{2}$" AND 4" DIAMETER SPLIT RINGS, 4" SHEAR PLATES AND TRIP-L-GRIP FRAMING ANCHORS AS MANUFACTURED BY THE *TIMBER ENGINEERING COMPANY*, WASHINGTON, D.C.

BOLTS SHALL BE $\frac{1}{2}$" AND $\frac{3}{4}$" MACHINE BOLTS. $\frac{1}{2}$" BOLTS SHALL HAVE 2"X2"X$\frac{1}{8}$" PLATE WASHERS, OR 2" DIAMETER CAST OR MALLEABLE IRON WASHERS, AND $\frac{3}{4}$" BOLTS SHALL HAVE 3"X 3"X $\frac{3}{16}$" PLATE WASHERS OR 3" DIAMETER CAST OR MALLEABLE IRON WASHERS. BOLTS SHALL BE TIGHT AT TIME OF ERECTION.

CAMBER OF 2" SHALL BE INTRODUCED IN THE TOP AND BOTTOM CHORDS THROUGH FABRICATION AS INDICATED.

SPLICE PLATES SHALL BE SELECTED OF THE BEST PIECES WITH A SLOPE OF GRAIN NOT STEEPER THAN 1" IN 12".

WHEN THE LUMBER IS NOT SEASONED AT THE TIME OF ERECTION TO THE MOISTURE CONTENT NORMAL TO THE CONDITIONS OF SERVICE, THE TRUSSES SHOULD BE INSPECTED PERIODICALLY AND MAINTAINED AND TIGHTENED IF NECESSARY UNTIL MOISTURE EQUILIBRIUM IS REACHED.

Fig. 10.21 Stress Diagram and Design Specification—Bowstring Truss

provide the specified edge and end distances for connectors in that piece. A splice piece 14 in. wide will provide the necessary width, and the length may be extended to meet the requirements.

The four 4-in. split rings between the chord and the web members must be investigated to determine if they develop sufficient load capacity to carry the 15,000-lb load. Since web members *D*4 are in tension, they take an end distance of 7 in. to develop full-load capacity for the connectors. The spacing of connectors is the required 9 in.; therefore the only reduction in load capacity is that due to the reduced end distance, from 7 to 5½ in. (see Fig. 10.18). It will be noted that for these diagonal members there are two diagonal end cuts, and therefore the end distance is not measured to the extreme end of the member but to a point a short distance from the end, which gives approximately the required shear area for the connector. A reduction of 1½ in. in end distance reduces the load capacity 17 percent (see Table 9.7). Therefore two connectors with full load carry 10,010 lb, and the other two connectors carry 83 percent of 10,010 lb, or 8,310 lb, which totals 18,320 lb, or 3,320 lb in excess of the 15,000 lb necessary to be developed.

For spans up to 120 or 150 ft, for such structures as bowling alleys, supermarkets, etc., the most economical method of carrying the roof is by means of timber bowstring trusses, as shown in Fig. 10.20. As will be seen by the stress diagram, Fig. 10.21, the stresses in the web members are comparatively light and therefore easily taken by one pair of connectors. The chord members may be made up of a number of plank sections, or of built-up glued, laminated members, as shown in Section *A*-*A* of Fig. 10.20.

10.13 Brackets, jib cranes, marquees

a Although widely variant in use and construction, brackets, jib cranes, and marquees are classified together because as cantilever structures they present similar problems. The bracket may support a balcony, a sign, or other structure, but it usually consists of a horizontal tension member supported from below by a diagonal compression member. A jib crane, a marquee, and some brackets have a horizontal compression member or members, and diagonal hangers above. The flagpole construction shown in Fig. 10.53 is a type of bracket.

Be sure that the structure to which a bracket or a marquee is attached can resist the resulting tension and compression. Almost always one, and frequently both, of these stresses are applied between two floors, and brick masonry is a poor material to resist bending. It is frequently necessary, even for comparatively light loads, to run a column from floor to floor as a vertical beam for attachment of the connections.

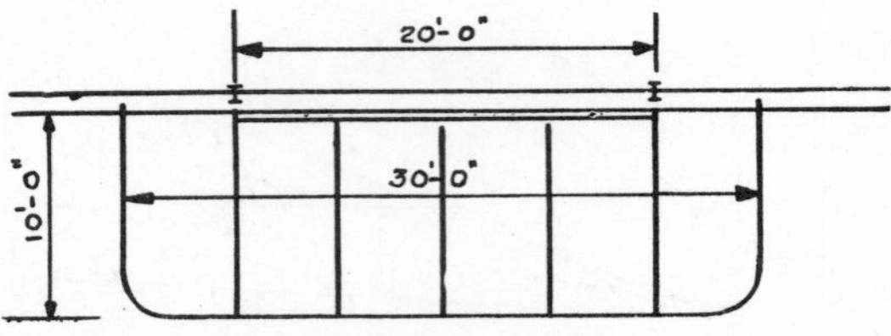

Fig. 10.22 Plan of Marquee

b Figures 10.22 and 10.23 show a marquee which carries a gypsum roof on Ts, with a dead load of 30 and a live load of 30 psf. Design the marquee.

The three center beams each carry

$$W = 60 \times 5 \times 10 = 3{,}000 \text{ lb}$$

$$S = \frac{3 \times 10}{14.22} = 2.11$$

Use 5 channel 6.7.

The beam next to the wall carries three concentrations of 1,500 lb on a 20-ft span.

$$S = 4.5 \times 0.09 \times 20 = 8.1$$

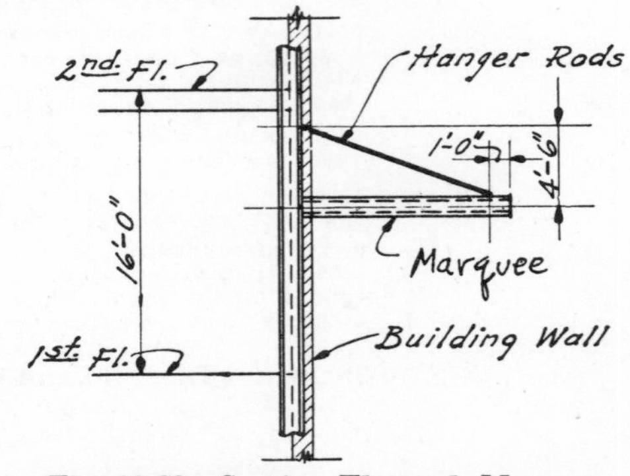

Fig. 10.23 Section Through Marquee

Use 9 channel 13.4.

The facia beam is carried by the hanger at the column line, but cantilevers out to pick up another load 5 ft beyond the support. Because of lighting, and other factors, this load will be assumed the same as interior loads.

$$M_c = 1.5 \times 5 = 7.5$$

$$\begin{aligned} M &= 5.25 \times 10 = 52.5 \\ &\quad -1.5 \times 20 = 30 \\ &\qquad\qquad\quad 22.5 \end{aligned}$$

$$S = 12.38$$

Architectural treatment suggests the use of a 12 channel 20.7 for a facia beam, bending it around to form the end beams. The bracket is made up of the beam on the column line and the hanger. The beam carries bending and direct stress. The bending is almost the same as that on the 5-in. beams, the load from the 9-in. channel coming in so close to the support as to cause practically no bending. The direct stresses in the marquee bracket are found by the stress diagram, Fig. 10.24. The beam carries a direct compression of 10.5 kips and a bending moment of 3.75 ft-kips, or 45 in.-kips.

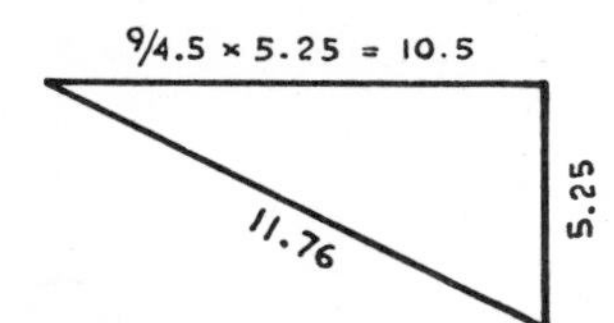

Fig. 10.24 Stress Diagram—Marquee

Try two 5 channels 6.7, back to back with a plate between to which a clevis can be pinned for the hanger. The bending factor for a 5-in. channel is

$$\frac{A}{S} = \frac{1.95}{3} = 0.65$$

The bending converted into equivalent direct stress is

$$\begin{aligned} 45 \times 0.65 &= 29.25 \\ \text{Direct stress} &= 10.5 \\ &\quad 39.75 \text{ kips} \end{aligned}$$

Since it is stiffened throughout its length by the marquee roof construction,

$$A = \frac{39.75}{21.5} = 1.85 \text{ sq in.}$$

Therefore two 5-in. channels with an area of 3.9 sq in. may be used.

The hanger carries a stress of 11.76 kips, which will require an area at the root of the thread of

$$\frac{11.76}{22} = 0.535 \text{ sq in.}$$

Use 1⅛-in. rod hanger.

The hanger should be furnished with clevises for each end, one with right-hand thread, the other with left-hand thread, for adjustment.

c *A jib crane* in its simplest form is a beam which is bracketed out from a column and carries a trolley on its lower flange by means of which loads up to several tons may be moved through an angle and a comparatively short distance. The jib crane is frequently supported on its own column, which rotates by means of pins, top and bottom, through brackets. Sometimes the brackets are on a building column, and the crane is supported by and pivots on the column brackets.

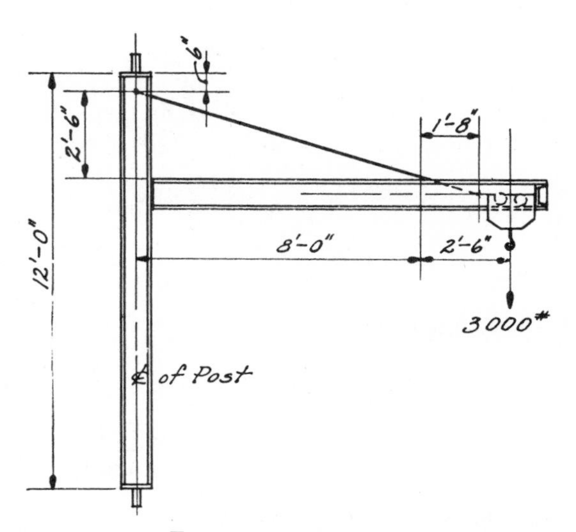

Fig. 10.25 Jib Crane

Figure 10.25 indicates a jib crane supported on its own column. Let us assume that it carries a load of 3,000 lb. Design the members. The sides of the triangle through which the stresses are figured are 2 ft 6 in., 9 ft 8 in., and 10 ft 0 in.

The maximum direct stresses in the truss will occur when the trolley is at its extreme outermost point, and the load on the truss will be

$$\frac{10.5}{9.67} \times 3{,}000 = 3{,}260 \text{ lb}$$

The direct stress in the beam is

$$\frac{9.67}{2.5} \times 3{,}260 = 12{,}600 \text{ lb}$$

The direct stress in the tie rod is

$$\frac{10}{2.5} \times 3{,}260 = 13{,}040 \text{ lb}$$

The maximum bending will be in one of two positions—the end of the cantilever or the center of the simple span.

$$M_c = 3{,}000 \times 0.67 = 2{,}000 \text{ ft-lb}$$
$$M_s = \frac{3{,}000 \times 9.67}{4} = 7{,}250 \text{ ft-lb}$$

However, when the moment is 7,250, the direct stress is reduced to

$$\frac{1{,}500}{3{,}260} \times 12{,}600 = 5{,}800 \text{ lb}$$

Therefore, the beam must be designed for either a combination of 12,600 lb direct stress with 2,000 ft-lb bending moment, or 5,800 lb direct stress with 7,250 ft-lb bending moment.

Try 8 WF 17. The bending factor B is

$$\frac{A}{S} = \frac{5}{14.1} = 0.355$$

Equivalent direct stress for the first condition is

$$P_e = 0.355 \times 2{,}000 \times 12 = 8{,}520$$
$$P_d = 12{,}600$$
$$21{,}120 \text{ lb}$$

For the second condition this becomes

$$P_e = 0.355 \times 7{,}250 \times 12 = 30{,}885$$
$$P_d = 5{,}800$$
$$36{,}685 \text{ lb}$$

The second condition therefore governs.

$$\frac{l}{r} = \frac{112}{1.16} = 96.6$$
$$f = 13{,}400$$
$$\text{Required area} = \frac{36{,}685}{13{,}400} = 2.74$$
$$\text{Area furnished} = 5$$

A spot check on the 7 I 15 shows that the l/r is in excess of 120 and it cannot be used.

The area required for the hanger is

$$\frac{13{,}040}{22{,}000} = 0.593 \text{ sq in.}$$

Use 1⅛-in. round rod.

The post is subject to a vertical load of 3,000 lb and bending as indicated in Fig.

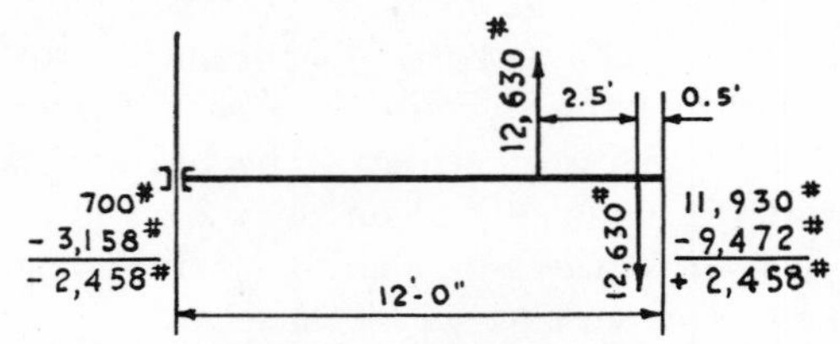

Fig. 10.26 Beam Diagram of Jib Crane

10.26. The maximum moment from this diagram is

$$2{,}625 \times 9 = 23{,}625 \text{ ft-lb}$$

Try 8 WF 24.

$$P_e = 0.339 \times 23{,}625 \times 12 = 96{,}100$$
$$P_d = 3{,}000$$
$$P = 99.1 \text{ kips}$$

The allowable load on 8 WF 24 = 101. However, the moment conditions would require the use of eq. 7(a) in Art. 7.11a. Therefore use 8 WF 31.

The pin at the top puts a pull of 2,625 on the column to which it is attached, and since this jib crane may be rotated, the pull may be against either axis.

10.20 WIND STRESSES

a In all or part of any structure, large or small, exposed to wind, wind pressure must be considered, and main structural members, anchorage, and other details must be designed accordingly. Although most structures built are satisfactorily stable and require no special investigation, there are also many which should be investigated, including (1) one-story, single-bay-width structures, such as churches, garages, and factories, (2) narrow multistory structures, even if only four or five stories high, and (3) roof structures, such as steeples, lanterns, cupolas, roof signs, high penthouses, high parapets, and roof tanks. Most building codes set up a minimum ratio of height to width, below which any investigation for wind stresses in multistory buildings may be omitted. This ratio is never less than 1½ to 1; in New York City and some other communities it is 2½ to 1.

b The most modern code for wind loads is that of the American Standards Association, and the following sections of this code are applicable to building design.*

* This material is reproduced from the American Standard Minimum Design Loads in Buildings and Other Structures, A58.1–1955, copyright 1956 by ASA, copies of which may be purchased from the American Standards Association at 10 East 40th St., New York, N.Y. 10016.

5. Wind Loads

5.1 *Minimum Design Pressures.* Buildings or other structures shall be designed and constructed to withstand the applicable horizontal pressures shown in Tables 10.6 and 10.7, allowing for wind from any direction. The height is to be measured above the average level of the ground adjacent to the building or structure.

5.2 *Exterior Walls.* Every exterior wall shall be designed and constructed to withstand the pressures specified in 5.1, acting either inward or outward.

5.3 *Roofs*

5.3.1 OUTWARD PRESSURES. The roofs of all buildings or other structures shall be designed and constructed to withstand pressures, acting outward normal to the surface, equal to 1¼ times those specified for the corresponding height zone in which the roof is located. The height is to be taken as the mean height of the roof structure above the average level of the ground adjacent to the building or other structure and the pressure assumed on the entire roof area.

5.3.2 INWARD PRESSURES. Roofs or sections of roofs with slopes greater than 30 degrees shall be designed and constructed to withstand pressures, acting inward normal to the surface, equal to those specified for the height zone in which the roof is located, and applied to the windward slope only.

5.3.3 EAVES AND CORNICES. Overhanging eaves and cornices shall be designed and constructed to withstand outward pressures equal to twice those specified in 5.1.

5.3.4 ANCHORAGE. Adequate anchorage of the roof to walls and columns, and of walls and columns to the foundations to resist overturning, uplift, and sliding, shall be provided in all cases.

5.4 *Chimneys, Tanks, and Towers*

5.4.1 CHIMNEYS. Chimneys, tanks, and solid towers shall be designed and constructed to withstand the pressures specified in 5.1, multiplied by the following factors:

Horizontal cross section	*Factor*
Square or rectangular	1.00
Hexagonal or octagonal	0.80
Round or elliptical	0.60

5.4.2 TRUSSED TOWERS. Radio Towers and other towers of trussed construction shall be designed and constructed to withstand wind pressures specified in 5.1 multiplied by suitable shape factors, and in accordance with good engineering practice.

5.5 *Signs and Outdoor Display Structures*

5.5.1 WIND PRESSURE

(*a*) For the purpose of determining wind pressures, all signs shall be classified as either open or solid. Signs in which the projected area exposed to wind consists of 70 percent or more of the gross area as determined by the over-all dimensions shall be classified as solid signs; those in which the projected exposed area is derived from open letters, figures, strips, and structural framing members, the aggregate total area of which is less than 70 percent of the gross area so determined, shall be classed as open signs.

(*b*) All signs shall be designed and constructed to withstand wind pressures of not less than the intensities (Table 10.8) applied to the projected exposed area.

5.5.2 PROJECTED EXPOSED AREA. The exposed area subjected to wind pressure shall be the total area of all parts of the sign, including structural framing projected on a plane perpendicular to the direction of the wind. Solid signs shall be designed on the basis of their gross area. In determining the stress in any member, the wind shall be assumed to blow from that horizontal direction and from that inclination from the vertical (but not to exceed 20 degrees above or below the horizontal) which produces the

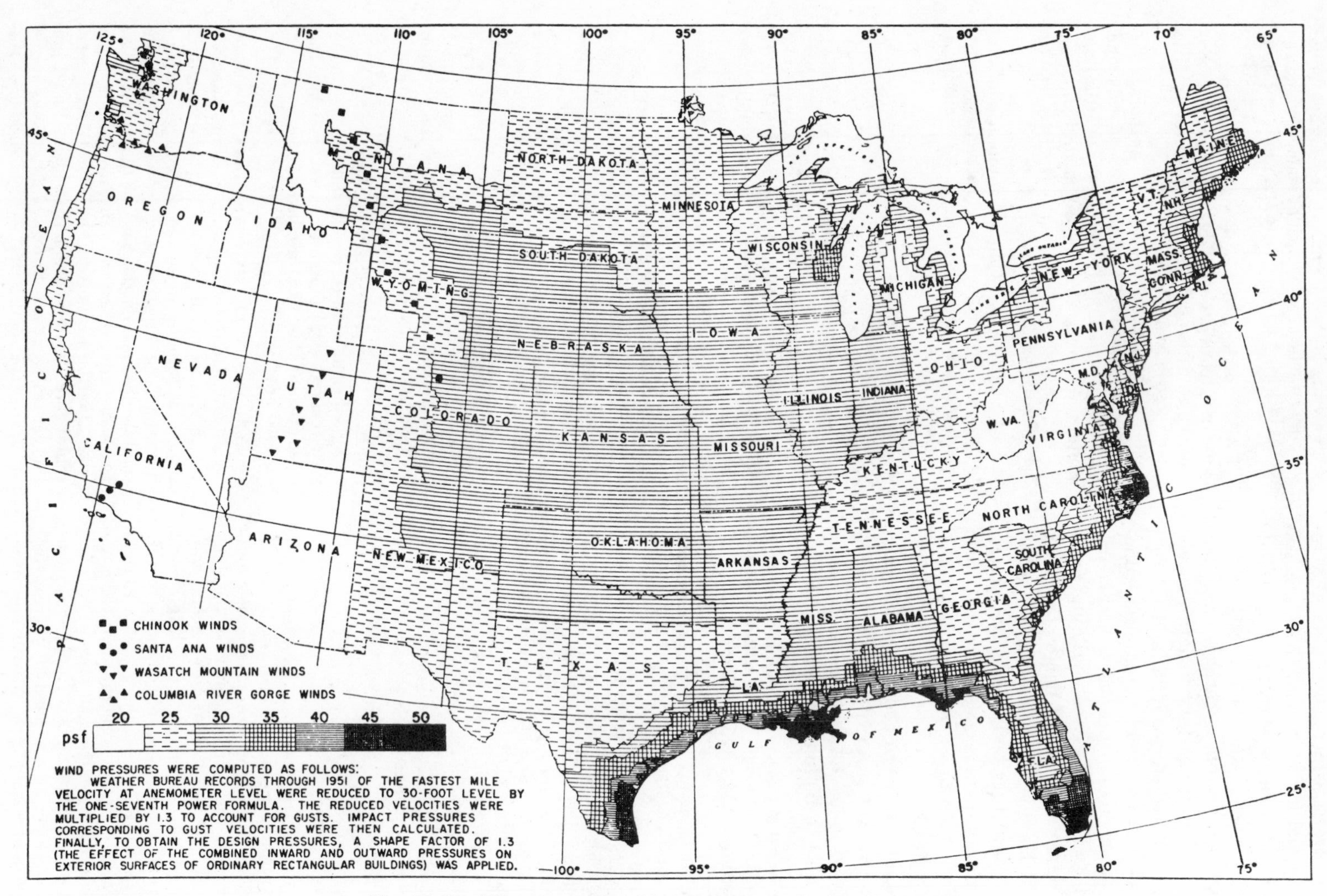

Fig. 10.27 Minimum allowable Wind Pressures (see Table 10.6)

Table 10.6 Design Wind Loads for States and Counties

This table contains a list of counties and associated wind pressures for those states along the Gulf and Atlantic Coasts, as indicated in Fig. 10.27.

State and County	lb per sq ft
Alabama	
Counties not listed	30
Baldwin	40
Clarke	35
Conecuh	35
Covington	35
Escambia	35
Geneva	35
Mobile	40
Monroe	35
Washington	35
Connecticut	
Counties except New London	25
New London	30
Delaware	
Kent	30
New Castle	25
Sussex	35
District of Columbia	20
Florida	
Counties not listed	30
Baker	25
Bay	45
Brevard	40
Broward	45
Calhoun	40
Charlotte	35
Citrus	35
Collier	45
Dade	50
Dixie	40
Escambia	40
Flagler	35
Franklin	45
Gadsen	40
Gilchrist	35
Glades	35
Gulf	45
Hendry	40
Hernando	35
Hillsborough	35
Holmes	35
Indian River	40
Jackson	35
Jefferson	40
Florida (Continued)	
Lafayette	35
Lee	40
Leon	40
Levy	35
Liberty	40
Madison	35
Manatee	35
Martin	40
Monroe	45
Okaloosa	40
Okeechobee	35
Orange	35
Osceola	35
Palm Beach	45
Pasco	35
Pinellas	35
St. Johns	35
St. Lucie	40
Santa Rosa	40
Sarasota	35
Seminole	35
Taylor	40
Volusia	35
Wakulla	45
Walton	40
Washington	40
Georgia	
Counties not listed	25
Baker	30
Brooks	35
Bryan	30
Camden	30
Chatham	35
Colguitt	30
Cook	30
Decatur	35
Early	30
Echols	30
Effingham	30
Glynn	30
Grady	35
Liberty	30
Lowndes	30
McIntosh	30
Miller	30
Georgia (Continued)	
Mitchell	30
Seminole	35
Thomas	35
Louisiana	
Counties (parishes) not listed	25
Acadia	35
Allen	30
Ascension	40
Assumption	40
Beauregard	30
Calcasieu	35
Cameron	40
East Baton Rouge	35
East Feliciana	30
Evangeline	30
Iberia	40
Iberville	35
Jefferson	45
Jefferson Davis	35
Lafayette	35
LaFourche	45
Livingston	35
Orleans	45
Plaquemines	45
Pointe Coupee	30
Saint Barnard	45
Saint Charles	40
Saint Helena	30
Saint James	40
Saint John the Baptist	40
Saint Landry	30
Saint Martin "Lower"	40
Saint Martin "Upper"	35
Saint Mary	45
Saint Tammany	40
Tangipahoa	35
Terrebonne	45
Vermillion	40
Washington	35
West Baton Rouge	35
West Feliciana	30
Maine	
Counties not listed	25
Androscoggin	30

Table 10.6 Design Wind Loads for States and Counties (*Continued*)

State and County	lb per sq ft
Maine (Continued)	
Aroostook	30
Cumberland	30
Hancock	35
Kennebec	30
Knox	35
Lincoln	35
Penobscot	30
Sagadahoc	35
Waldo	30
Washington	35
York	30
Maryland	
Counties not listed	20
Calvert	25
Caroline	30
Cecil	25
Charles	25
Dorchester	30
Kent	25
Queen Annes	25
Saint Mary's	30
Somerset	35
Talbot	25
Wicomico	35
Worcester	40
Massachusetts	
Counties not listed	25
Barnstable	40
Bristol	35
Dukes	40
Essex	35
Middlesex	30
Nantucket	40
Norfolk (main inland part)	30
Norfolk (part facing Mass Bay)	35
Plymouth	35
Suffolk	35
Mississippi	
Counties not listed	30
George	35
Green	35
Hancock	40
Harrison	40
Jackson	40
Pearl River	35
Stone	35
New Hampshire	
Counties not listed	25
Belknap	30
Hillsborough	30
Merrimack	30
Rockingham	35
Strafford	30
New Jersey	
Counties not listed	25
Atlantic	30
Cape May	35
Cumberland	30
Monmouth	30
Ocean	30
North Carolina	
Counties not listed	20
Beaufort	40
Bertie	35
Bladen	30
Brunswick	35
Camden	40
Carteret	40
Chowan	35
Columbus	35
Craven	35
Cumberland	25
Currituck	40
Dare	45
Duplin	30
Edgecombe	30
Franklin	25
Gates	35
Green	35
Halifax	30
Harnett	25
Hertford	30
Hoke	25
Hyde	45
Johnston	25
Jones	35
Lenoir	35
Martin	35
Nash	25
New Hanover	35
Northampton	30
Onslow	35
Pamlico	40
Pasquotank	40
Pender	35
Perquimans	35
North Carolina (Continued)	
Pitt	35
Robeson	30
Sampson	30
Scotland	25
Tyrrell	40
Warren	25
Washington	40
Wayne	30
Wilson	30
Rhode Island	
Bristol	35
Kent	30
Newport	35
Providence	30
Washington	35
South Carolina	
Counties not listed	25
Beaufort	35
Berkeley	30
Charleston	35
Colleton	30
Dillon	30
Dorchester	30
Florence	30
Georgetown	35
Horry	35
Jasper	30
Marion	30
Williamsburg	30
Texas	
Counties not listed	25
Aransas	40
Archer	30
Armstrong	30
Atacosa	30
Austin	30
Bailey	30
Baylor	30
Bee	35
Bowie	30
Brazoria	40
Briscoe	30
Brooks	40
Calhoun	40
Cameron	45
Camp	30
Carson	30
Cass	30
Castro	30

Table 10.6 Design Wind Loads for States and Counties (*Continued*)

State and County	lb per sq ft
Texas (Continued)	
Chambers	40
Childress	30
Clay	30
Cochran	30
Collin	30
Collingsworth	30
Colorado	30
Cooke	30
Cottle	30
Crosby	30
Dallam	30
Deaf Smith	30
Delta	30
Denton	30
DeWitt	30
Dickens	30
Donley	30
Duval	35
Fannin	30
Floyd	30
Foard	30
Fort Bend	35
Franklin	30
Frio	30
Galveston	40
Garza	30
Goliad	35
Gray	30
Grayson	30
Hale	30
Hall	30
Hansford	30
Hardeman	30
Hardin	35
Harris	35
Hartley	30
Haskell	30
Hemphill	30
Hildago	40
Hockley	30
Hopkins	30
Hunt	30
Hutchinson	30
Jack	30
Jackson	35
Jasper	30
Jefferson	40
Jim Hogg	35
Jim Wells	40
Karnes	30

State and County	lb per sq ft
Texas (Continued)	
Kenedy	45
Kent	30
King	30
Kleberg	45
Knox	30
Lamar	30
Lamb	30
La Salle	30
Lavaca	30
Liberty	35
Lipscomb	30
Live Oak	35
Lubbock	30
Lynn	30
McMullen	35
Matagorda	40
Montague	30
Montgomery	30
Moore	30
Morris	30
Motley	30
Newton	30
Nueces	45
Ochiltree	30
Oldham	30
Orange	35
Parmer	30
Polk	30
Potter	30
Randall	30
Red River	30
Refugio	40
Roberts	30
San Jacinto	30
San Patricio	40
Sherman	30
Starr	35
Stonewall	30
Swisher	30
Terry	30
Throckmorton	30
Titus	30
Tyler	30
Victoria	35
Waller	30
Webb	30
Wharton	35
Wheeler	30
Wichita	30
Wilbarger	3

State and County	lb per sq ft
Texas (Continued)	
Willacy	45
Wise	30
Yoakum	30
Young	30
Zapata	30
Virginia	
Counties and independent cities not listed	20
Counties	
Accomac	40
Brunswick	25
Caroline	25
Charles City	30
Chesterfield	30
Dinwiddie	25
Elizabeth City	35
Essex	30
Gloucester	35
Greensville	30
Hanover	25
Henrico	25
Isle of Wight	35
James City	35
King and Queen	30
King George	25
King William	30
Lancaster	35
Mathews	35
Middlesex	35
Nansemond	35
New Kent	30
Norfolk	40
Northampton	40
Northumberland	35
Prince George	30
Princess Anne	40
Richmond	30
Southampton	30
Surry	35
Sussex	30
Warwick	35
Westmoreland	30
York	35
Independent Cities	
Hopewell	25
Newport News	35
Norfolk	40
Portsmouth	40
Suffolk	35

Table 10.7 Wind Pressures for Various Height Zones Above Ground*

Height zone (ft)	Wind-pressure-map areas (lb per sq ft)						
	20	25	30	35	40	45	50
Less than 30	15	20	25	25	30	35	40
30 to 49	20	25	30	35	40	45	50
50 to 99	25	30	40	45	50	55	60
100 to 499	30	40	45	55	60	70	75
500 to 1199	35	45	55	60	70	80	90
1200 and over	40	50	60	70	80	90	100

* Reference should be made to Fig. 10.27 and that wind-pressure column in the table should be selected which is headed by a value corresponding to the minimum permissible resultant wind pressure indicated for the particular locality in Fig. 10.27.

The figures given are recommended as minimum. These requirements do not provide for tornadoes.

maximum stress in that member. No shielding effect of one element by another shall be considered where the distance between them exceeds 4 times the smaller projected dimension of the windward element.

5.6 *Other Structures.* The building official may require evidence to support the values for wind pressure used in the design of structures not specifically covered in this section.

5.7 *Shielding and Unusual Exposures*

5.7.1 ALLOWANCE. No allowance shall be made for the shielding effect of other buildings or structures.

5.7.2 HIGHER WIND PRESSURES. If the building or other structure is on a mountain, ocean promontory, or in any other location considered by the building official to be unusually exposed, higher wind pressures may be prescribed by the building official.

5.8 *Overturning and Sliding*

5.8.1 OVERTURNING. The overturning moment due to the wind load shall not exceed 66⅔ percent of stability of the building or other structure due to the dead load only, unless the building or other structure is anchored so as to resist the excess overturning moment without exceeding the allowable stresses for the materials used. The axis of rotation for computing the overturning moment and the moment of stability shall be taken as the intersection of the outside wall line on the leeward side and the plane representing the average elevation of the bottoms of the footings. The weight of earth superimposed over footings may be used in computing the moment of stability due to dead load.

5.8.2 SLIDING. When the total resisting force due to friction is insufficient to prevent sliding, the building or other structure shall be anchored to withstand the excess sliding force without exceeding the allowable stresses for the materials used. Anchors provided to resist overturning moment may also be considered as providing resistance to sliding.

5.9 *Stresses During Erection.* Provision shall be made for wind stress during erection of the building or other structure.

A5.4.2 TRUSSED TOWERS. It is believed that the use of the following factors, when applied to the wind pressures given in 5.1, will assist in safe, economical design. These factors are in accord with those recommended by Flaschbart and with the results of other wind-tunnel tests of tower models made in this country. In comparing these factors with those recommended by other authorities, it should be kept in mind that the factors given below take into account the fact that the values for wind pressures given in 5.1 include the effect of gusts, and a shape factor of 1.3 for ordinary rectangular buildings.

Table 10.8 Wind Pressures for Signs at Various Height Zones Above Ground*

Height above ground, ft	Solid signs — all types† Wind-pressure-map areas lb per sq ft							Open signs — all types‡ Wind-pressure-map areas lb per sq ft						
	20	25	30	35	40	45	50	20	25	30	35	40	45	50
Less than 30	17	22	28	28	33	39	44	23	31	39	39	46	54	62
30 to 49	22	28	33	39	44	50	55	31	39	46	54	62	70	77
50 to 99	28	33	44	50	55	61	66	39	46	62	70	77	85	92
100 to 499§	33	44	50	61	66	77	83	46	62	70	85	92	104	108

* Reference should be made to Fig. 10.27 and that wind-pressure column in Table 10.7 should be selected which is headed by a value corresponding to the minimum permissible resultant wind pressure indicated for the particular locality in Fig. 10.27.

The figures given are recommended as minimum. These requirements do not provide for tornadoes.

†Solid ground signs less than 50 ft in height shall be designed for 0.6 of the tabular values given above, but not less than 17 pounds per square foot (psf).

‡Open ground signs less than 50 feet in height shall be designed for 0.6 of the tabular values given above but not less than 23 psf.

§Design pressures on signs located 500 feet or more above ground should be determined by special analysis of conditions.

Available information indicates that tower construction should be designed to withstand the wind pressures specified in 5.1 of the standard, applied to the total normal projected area of all the elements of *one* face (excluding letters, ladders, conduits, lights, elevators, etc., which should be taken care of separately, making use of the same wind pressures but using the indicated factors for individual members) and multiplied by the following factors:

	Factor
Wind normal to one face of tower	
4-cornered, flat or angular sections, steel or wood	2.20
3-cornered, flat or angular sections, steel or wood	2.00
Wind on corner, 4-cornered tower, flat or angular sections	2.40
Wind parallel to one face of 3-cornered tower, flat or angular sections	1.50
Factors for towers with cylindrical elements are approximately 2/3 of those for similar towers with flat or angular sections.	
Wind on individual members	
Cylindrical members	
2 in. or less in diameter	1.00
Over 2 in. in diameter	0.80
Flat or angular sections	1.30

The factors for towers actually vary over a rather wide range of values and are dependent upon what is referred to as the "solidity ratio," defined as the ratio of the projected area of the elements of one face of a tower to the solid area of the face. Experimental data have shown that the factors vary according to the "solidity ratio" in somewhat the following manner, if used in conjunction with the design pressures of 5.1.

Type of tower	Solidity ratio	Factors		
		Wind normal to one face	Wind on a corner	Wind parallel to one face
4-cornered towers				
flat or angular sections	.14	2.4	2.4	
flat or angular sections	.27	1.7	1.9	
3-cornered towers				
flat or angular sections	.14	2.1		1.6
flat or angular sections	.27	1.6		1.3

c The Factory Mutual Fire Insurance Companies' bulletins present the best summary generally available for wind stresses in factories. Although these are prepared primarily for industrial buildings, they will be applicable also to churches, schools, and other buildings. Bulletin 7.13 recommends that roofs with slopes of not over 20 deg be designed for a gross uplift of 30 psf and that walls and window supports shall resist a total horizontal force of 27 psf (see Fig. 10.28). The code of the U.S. Corps of Engineers requires that the details and method of attachment shall be designed to resist pressures as follows:

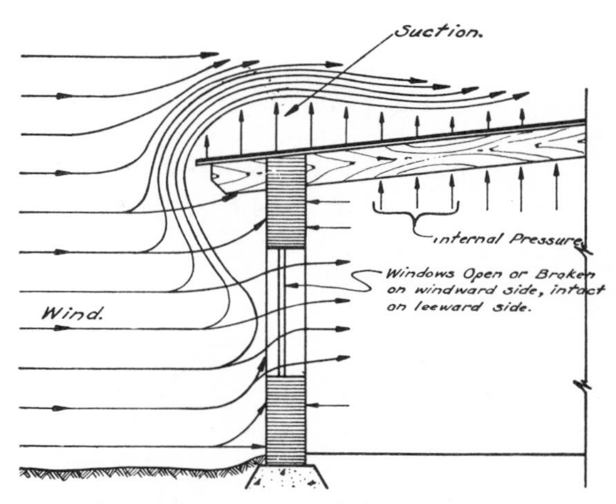

Fig. 10.28 Nature of Wind Stresses

Wind on flat roofs—suction, 12 psf

Wind on windward side of sloping roofs for slopes of under 20 deg—suction, 12 psf

For slopes between 20 and 30 deg—suction decreasing uniformly from 12 to 0 psf

For slopes between 30 and 60 deg—pressure increasing uniformly from 0 to 9 psf

For slopes above 60 deg—pressure, 9 psf

For the leeward side of all sloping roofs—suction, 9 psf

Side walls should resist pressure or suction of 10 psf

d All members subject to stresses produced by a combination of wind and other loads or by wind loads alone may be proportioned for unit stresses $33\frac{1}{3}$ percent greater than those normally specified, but this allowance should never be used to reduce the normal design for live and dead loads only. In other words, if wind does not cause stresses greater than 33⅓ percent of the original design for live and dead loads, it may be disregarded in the design of the member.

10.21 Wind bracing—low steel buildings

a As has been stated in Art. 10.20*c*, the design to resist wind on low steel-framed buildings applies to factory buildings and also to churches and other similar structures. The anchorage of steel or wood sash and its support should meet the requirements of the codes quoted. In buildings with wood siding, corrugated metal or asbestos-cement, and the like, the wind pressure is carried through the system of girts into the columns in the same manner as the application of floor loads to horizontal members. This design, although actually a wind-pressure design, is treated in Art. 10.41.

b The wind pressure on a panel is divided between the two columns at either side of the panel, and the load is divided between reactions at the top and bottom of the column. The load going into the bottom of the wall or column is transferred directly to the foundation without causing any moment on the column. The load at the top of the column is considered to be distributed across the building, being divided roughly in proportion to the stiffness factor $K = I/h$. Thus for a simple two-column bent, half the eave load goes to each column.

c The knee brace is designed as the support of a cantilever beam, to resist both cantilever load and uplift, and, since it is a sloping member, the stress is multiplied by the vector for the angle of the knee brace.

d The design for one column is on the basis of a cantilever beam. The wind load on the column is applied as a shear load at the point of contraflexure. Obviously, as a cantilever beam, the top of the column must have two points of support—either the upper and lower chords of a truss, as in Fig. 10.29, or a knee brace, as in Fig. 10.30.

The length of the cantilever is taken to the point of contraflexure as determined from the amount of fixity at the base of the column. For a completely fixed base which may be obtained by means of building the steel column into the concrete foundation or by developing the full bending moment with anchor bolts, the point of contraflexure is midway between the knee or the lower chord, whichever forms the fulcrum, and the base. Otherwise the capacity of the anchors may be deducted from the cantilever of full height, as indicated in the following problem.

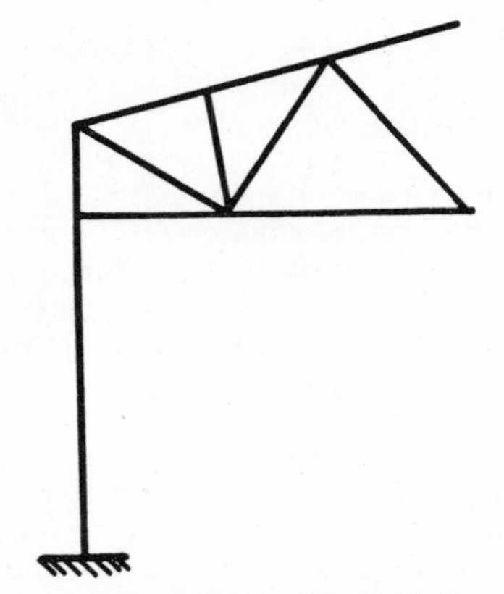

Fig. 10.29 Square End of Truss

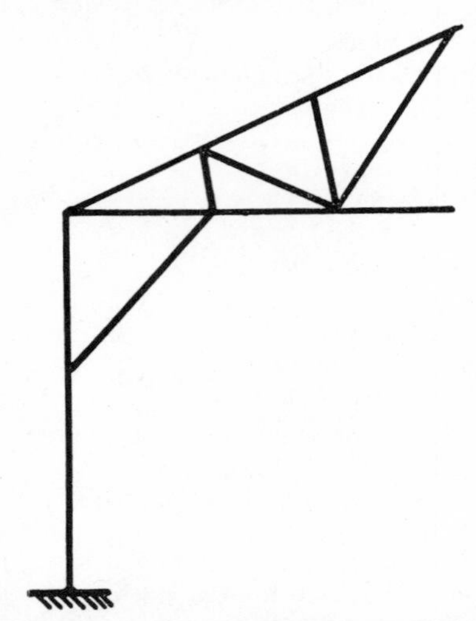

Fig. 10.30 Knee-braced End of Truss

Problem Assume wind at 10 psf acting on the structure shown in Fig. 10.31.

$$W \text{ per panel} = 10 \times 16 \times 20 = 3{,}200 \text{ lb}$$

The load at the top of the column is

$$\frac{3{,}200}{2} = 1{,}600 \text{ lb}$$

This load divided between the two columns *A* and *B* is (Fig. 10.32)

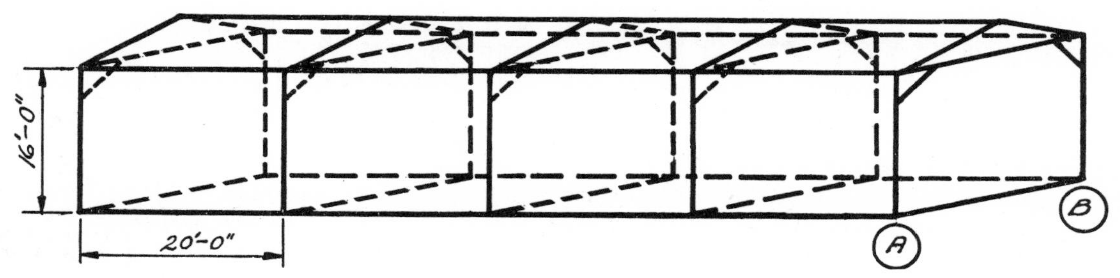

Fig. 10.31 Wind-braced Mill Building

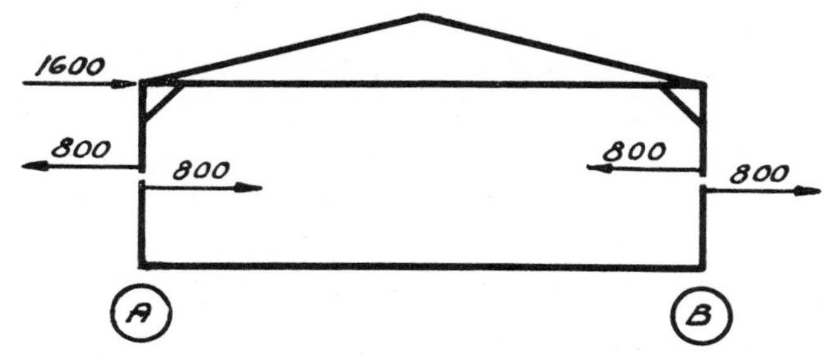

Fig. 10.32 Application of Wind Loads—Mill Building

$$\frac{1{,}600}{2} = 800 \text{ lb per column}$$

Apply a knee brace as a fulcrum 3 ft below the bottom of the truss. If there were no contraflexure—that is, if the column were free to rotate about its base—the wind moment would be

$$M = 800 \times 13 = 10{,}400 \text{ ft-lb}$$

The knee brace is designed as a strut, as shown in Fig. 10.34.

To hold 800 lb in equilibrium, taking moments about c,

$$3P_b = 9.5 \times 800$$

Then

$$P_b = \frac{9.5 \times 800}{3} = 2{,}533 \text{ lb}$$

Assuming a knee brace at 45 deg,

$$P = \sqrt{2} \times 2{,}533 = 3{,}580 \text{ lb}$$

The length of knee brace at 45 deg is $\sqrt{2} \times 3 = 4.24$ ft. The knee brace to carry this load may be selected from Table 10.4 by dividing 3,580/1.33, or 2,690 on 4.24 ft length. The divisor 1.33 allows for increased stress for wind.

If we assume an 8-in. column with one ⅞-in. anchor bolt 2 in. out from each face of the column and with the neutral axis of tension and compression at $d/7$ from the face of the column (see Fig. 10.33) which are safe

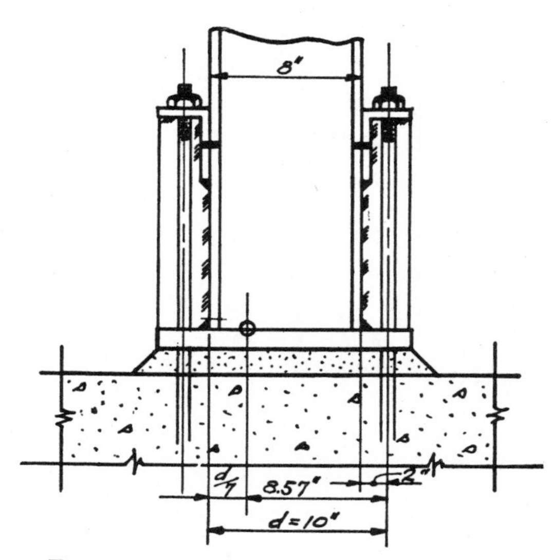

Fig. 10.33 Anchorage at Base of Column

assumptions, the value of the anchorage may be computed thus:

$$d = 8 + 2 = 10$$
$$d/7 = 10/7 = 1.43$$
$$d - d/7 = 8.57 \text{ (half lever arm of } M \text{ couple)}$$

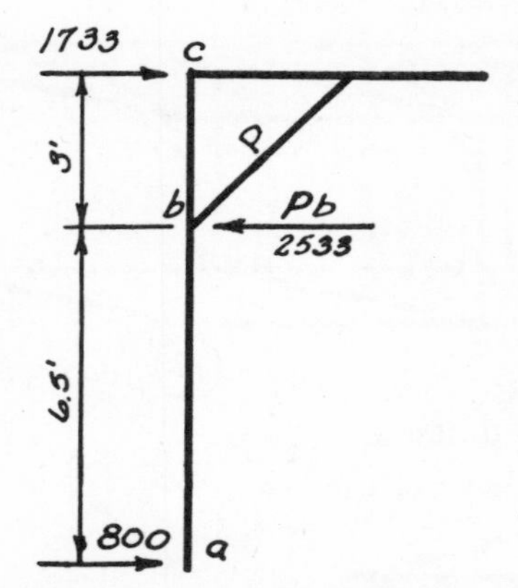

Fig. 10.34 Application of Wind Loads to Column

The value of one ⅞-in. bolt, under wind stress, is

$$0.419 \times 20{,}000 \times 1.33 = 11.2 \text{ kips}$$

$$M = 11.2 \times \frac{8.57 \times 2}{12} = 16$$

which is over half of the full moment. Therefore the point of contraflexure is at mid-height, or $M = 10{,}400/2 = 5{,}200$ ft-lb. This moment is to be applied to the columns in accordance with the method of Art. 7.10*e*.

In addition to the column stress due to moment, there is a direct stress due to wind —a compression on the leeward column, and a tension on the windward column. The amount of this direct stress is found by taking moments about the base of one column. In the problem here stated, let us assume that the truss span is 40 ft and the dead load 10 lb per square foot of roof and 8 lb per square foot of side wall. The overturning moment due to wind load is 1,600 × 16 = 25,600 ft-lb. The direct stress in columns due to wind is

$$P = \frac{25{,}600}{40} = 640 \text{ lb}$$

The dead load on the windward column is

$$\begin{aligned} \text{Roof: } 20 \times 20 \times 10 &= 4{,}000 \text{ lb} \\ \text{Side wall: } 20 \times 16 \times 8 &= \underline{2{,}560 \text{ lb}} \\ \text{Total} &= 6{,}560 \text{ lb} \end{aligned}$$

The building is, therefore, stable against overturning because the uplift is only 9.8 percent of the dead load against 66⅔ percent allowed (see Art. 10.20*b*).

In addition to the dead load, let us assume that there is a roof load of 30 psf. The load on each column is

$$\begin{aligned} \text{Dead load} &= 6{,}560 \\ \text{Live load} = 30 \times 20 \times 20 &= \underline{12{,}000} \\ \text{Subtotal} &= 18{,}560 \text{ lb} \\ \text{Wind, D.S.} &= 640 \text{ lb} \end{aligned}$$

Wind moment, equivalent

$$\begin{aligned} \text{D.S.} = 5{,}200 \times 12 \times 0.333 &= \underline{20{,}800} \\ \text{Total} &= 40{,}000 \text{ lb} \end{aligned}$$

To use Tables 7.1 to 7.4, divide by 1.33 because of wind allowance.

$$P = \frac{40{,}000}{1.33} = 30{,}000 \text{ lb}$$

Since the columns are braced the weak way by the girts, we may enter the tables at 13/2.12 = 6.1 ft unbraced height; the 8 WF 24 is adequate to carry all stresses.

e Aside from proper design of columns, anchorage, and other component parts, it is necessary also to cross-brace the entire structure properly against weaving. This is a matter of practical selection of size and location rather than of design, and is discussed more fully in Art. 10.40*d*.

f Article 10.20*c* states the requirements of some codes for anchorage of roof. The requirement of the Factory Mutual Fire Insurance Companies for a gross uplift of 30 psf seems rather severe, and it would seem satisfactory generally to use the U.S. Corps of Engineers' requirement that details for attachment be capable of resisting a net suction of 12 psf. Even this amount requires special anchorage of corrugated steel, asbestos-cement, or other lightweight roofing material —usually more anchorage than is ordinarily provided by the commonly used types, as is evidenced by the large number of light roofs lost in wind storms annually.

10.22 Wind bracing of multistory buildings

a The subject of wind bracing of high buildings may be best introduced by quoting from the final report of the Committee on

Wind Bracing of the American Society of Civil Engineers, submitted in 1940.

Proper design of wind bracing involves provision for stability, strength and rigidity. The relative importance of these three factors varies considerably, but adequate consideration must be given to each.

The frame of a building is made up, in general, of vertical members (columns) and horizontal members framed between the verticals (beams and girders). In tier buildings the horizontal members of the floors occur at fairly regular intervals throughout the height and, with floor slabs, form horizontal diaphragms uniting the columns at each floor level. The purpose of the wind bracing is to preserve this frame from undue distortion when subjected to horizontal forces.

In a building frame as described, the tendency of the panels formed by the columns and their connecting beams or girders to distort, due to horizontal thrust, is resisted by the connections at the columns and by the members in bending and direct stress. Where several columns occur in a line, they form a bent which (to a degree depending on the rigidity of the beams, their connections, and the nature of the bracing utilized) tends to resist the horizontal thrust in the manner of a vertical truss with the columns forming multiple chords and the columns, beams, and girders forming the web. The columns on the windward side of the axis of the bent become tension chords and those on the leeward side, compression chords.

In successful design, the building must be assumed to act as a unit, and not distort appreciably when subjected to the applied lateral loads.

The floors of a building must act as horizontal diaphragms for the delivery of wind load increments that arise in each story to the braced bents that are provided to receive them. Bracing planes usually run in more than one direction and the floors must possess the necessary strength and rigidity under loads in their own planes to deliver these wind loads. If the floor is incapable of serving as a horizontal distribution unit, horizontal bracing should be provided between the braced bents. Most of the customary systems of floor construction used in fireproof buildings possess the requisite strength for this purpose.

As ordinarily constructed, the walls of a modern tall building cannot be relied upon to absorb any appreciable fraction of the applied wind force. Apart from the uncertainty as to whether the walls act integrally with the frame at the outset, it is obvious that if stressed heavily under horizontal loading they would crack and cease to be a dependable element of strength. In setback buildings, the walls in the planes of the tower sides, of course, are discontinued below the roof of the widened portion of the building. Nor can wind resistance be counted upon for partitions, which, in general, are removable. For these reasons, whatever may be their role in lessening deflection and vibration, walls and partitions in buildings having a high ratio of height to width should be ignored in strength calculations, and provisions should be made in the structural frame for the entire recommended wind load.

b There are a number of acceptable methods of distributing wind loads and computing bents for resistance of wind pressures. Judging from results, any one of these methods properly applied will give satisfactory results. The Miami hurricane of 1926 furnished a full-scale test of the wind bracing of tall buildings and proved that any building designed in accordance with the general principles of good wind bracing suffered little or no structural damage, even under stresses far in excess of those used in the design. On the other hand, several tall, narrow buildings which failed did not comply with the general principles, and although quite satisfactory for carrying vertical loads, were weak against lateral pressures.

c *The cantilever method* assumes that the entire bent acts as a cantilever beam or truss fixed at the base. There are many refinements of the simple method presented here, but this method has proven quite satisfactory and is recommended for the normal range of tall buildings up to about 30 stories high—that is, a height-to-width ratio from $1\frac{1}{2}:1$ up to $5:1$. Beyond these limits, nobody but an experienced specialist on wind bracing should attempt the design.

In accordance with paragraph *a*, the loads applied against the outside wall are distributed by the floors as a diaphragm between the bents. Bents may be assumed at each column line and the wind stresses thus kept low, or, as is usually done, the bents are selected in the end walls and occasionally in interior bents where the projections of special details may be easily buried in partitions. The selection of wind-bent locations is a matter of experience.

The cantilever method assumes that (1) the direct stresses in the columns induced by wind are proportional to their distances from the center of gravity of the bent (similar to unit stresses in a beam); and (2) the point of inflection of beams and columns is at mid-span.

d The design of a wind bent by the cantilever method involves

1. The computation of coefficients for the bent.
2. The application of wind loads to the bent, and the computation of resultant shears and moments on the bent as a whole.
3. The computation of beam and column moments and direct stresses, from the coefficients found in step 1 and the shears and moments found in step 2.
4. The application of these shears and moments to the members already designed for live and dead load.

In detail, these steps are as follows:

1(*a*) Lay out to scale a cross section of one floor only of the bent, as in Fig. 10.35.

1(*e*) For each beam in the bent, compute the shear coefficient to apply at the center of the beam.

$$v_A = d_1, \quad v_B = d_1 + d_2, \quad v_C = d_1 + d_2 + d_3 \text{ (added algebraically)}$$

1(*f*) For each beam, calculate beam-moment coefficient.

$$m_A = \frac{v_A A}{2}, \quad m_B = \frac{v_B B}{2}, \quad m_C = \frac{v_C C}{2}$$

1(*g*) For each column, compute the shear coefficient.

$$v_1 = m_A, \quad v_2 = m_A + m_B, \quad v_3 = m_B + m_C, \quad m_4 = m_C$$

1(*h*) Calculate the direct stress coefficients for each beam, working from each side and using the larger for design purposes.

$$p_A = 1.0 - v_1 \qquad p_C = 1.0 - v_4$$
$$p_B = p_A - v_2 \quad \text{or} \quad p_B = p_C - v_3$$
$$p_C = p_B - v_3 \qquad p_A = p_B - v_2$$

2(*a*) Sketch the entire bent to scale, and on the right-hand side run a line of story heights (Fig. 10.37).

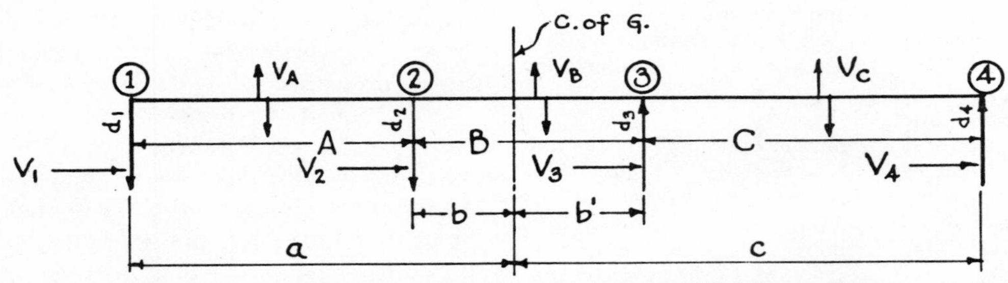

Fig. 10.35 Application of Wind Loads—Beam of Multistory Bent

1(*b*) Locate the center of gravity of the bent, assuming the area of each column as unity.

1(*c*) Compute the moment of inertia of the bent.

$$I = a^2 + b^2 + b'^2 + c^2$$

1(*d*) For each column, compute the direct-stress coefficient.

$$d_1 = \pm\frac{a}{I}, \quad d_2 = \pm\frac{b}{I}, \quad d_3 = \pm\frac{b'}{I}, \quad d_4 = \pm\frac{c}{I}, \text{ and so on}$$

It will be noted that each stress may be compression or tension, depending on the direction of the wind.

2(*b*) Compute and apply the wind loads at each story, P_R, P_8, and the rest. Set these down at the left side of the bent. Where there are several wind bents of the same cross section, it may be advantageous to compute a bent for an applied load of 1 kip per foot of height and prorate the bents from the results obtained.

2(*c*) At mid-story on the left side of the bent, indicate an arrow for the position of the shear load—the total load applied above this point—W_R, W_8, etc. Thus $W_R = P_R$, $W_8 = W_R + P_8$, $W_7 = W_8 + P_7$, etc. Set this shear value down above the mid-story arrow.

2(*d*) Compute at each mid-story point the total moment on the bent above this point, M_R below the roof, M_8 below the eighth floor, and so on, and set it down below the mid-story arrow. These moments are obtained as follows:

$$M_R = W_R \frac{h_R}{2}$$

$$M_8 = M_R + W_R \frac{h_R}{2} + W_8 \frac{h_8}{2}$$

$$M_7 = M_8 + \frac{W_8 h_8}{2} + \frac{W_7 h_7}{2}$$

Etc.

3(*a*) For each story of each column, compute and set down in parentheses the direct stress due to wind,

$$P_{1'R} = d_1 M_R \qquad P_{1'8} = d_1 M_8$$
$$P_{2'R} = d_2 M_R \qquad P_{2'8} = d_2 M_8$$
$$P_{3'R} = d_3 M_R \qquad P_{3'8} = d_3 M_8$$
$$P_{4'R} = d_4 M_R \qquad \text{Etc.}$$

[The direct-stress coefficient for each column from step 1(*d*) multiplied by the moment from step 2(*d*).]

3(*b*) For each beam in each story, compute and set down above the beam the wind moment for which the beam and the end connections are to be designed.

$$M_{AR} = m_A M_R$$
$$M_{BR} = m_B M_R$$
$$M_{CR} = m_C M_R$$

[The moment coefficient for each beam from step 1(*f*) multiplied by the roof moment from step 2(*d*).]

$$M_{A8} = m_A(M_8 - M_R)$$
$$M_{B8} = m_B(M_8 - M_R)$$
$$M_{C8} = m_C(M_8 - M_R)$$

Etc.

[The moment coefficient for each beam from step 1(*f*) multiplied by the difference in moment above and below, from step 2(*d*).]

3(*c*) Compute the moment at each floor of each column and set it down parallel with the column under the line below the floor.

$$M_{1'R} = v_1 W_R \frac{h_R}{2}$$

$$M_{2'R} = v_2 W_R \frac{h_R}{2}$$

$$M_{3'R} = v_3 W_R \frac{h_R}{2}$$

$$M_{4'R} = v_4 W_R \frac{h_R}{2}$$

$$M_{1'8} = v_1 W_8 \frac{h_8}{2}$$

$$M_{2'8} = v_2 W_8 \frac{h_8}{2}$$

$$M_{3'8} = v_3 W_8 \frac{h_8}{2}$$

Etc.

3(*d*) For each beam in each floor compute the direct stress in the beam, and set it down below the beam in parentheses.

$$DS_{AR} = p_A P_R$$
$$DS_{BR} = p_B P_R$$
$$DS_{CR} = p_C P_R$$
$$DS_{A8} = p_A P_8$$
$$DS_{B8} = p_B P_8$$
$$DS_{C8} = p_C P_8$$

Etc.

4(*a*) Check the beam sizes to carry the combined live- and dead-load moments plus the wind moment and wind direct stress (at the increased allowable stress). Figure 10.36 indicates the combination of live-, dead-, and wind-load moments in a beam of a wind bent. It will be noted that the maxi-

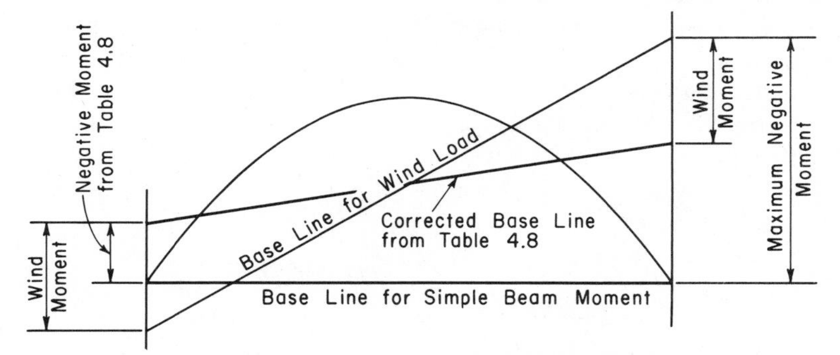

Fig. 10.36 Combination of Live, Dead, and Wind Moments on Beam

mum moment due to wind loads is at the end of the beam, whereas the maximum moment due to live and dead loads is at or near the middle of the beam, so that usually these moments are not additive. For the condition indicated in this figure, it is ordinarily necessary only to check the section modulus furnished for live and dead loads against the section modulus furnished for wind load at the increased stress. If the allowable stress increase for wind is 33⅓ percent, the required section modulus becomes $S = 0.375M$, where M is the moment in foot-kips.

Since in order to develop wind moment, it is necessary that the beam be rigidly connected to the column, a coefficient method of design for determination of the beams should be adopted—either that given in Table 4.8 or a similar method adapted from the Hardy Cross method. In making the beam design, three moment diagrams must be considered.

(*a*) Dead and live load only in accordance with the method of Art. 4.15 and Table 4.8 using full working stress.

(*b*) The moment as above with the added wind moment applied from the left, using 1⅓ times the allowable working stress, 22,000 × 1⅓ or 24,000 × 1⅓.

(*c*) The moment derived in (*a*) above with the added wind moment applied from the right, using 1⅓ times the allowable working stress.

In many instances, even though the negative moment is considerably greater than the positive, the beam size itself will be determined by condition (*a*) above, since the end connections may provide sufficient additional reinforcement to carry the moment without any increase in beam size.

It will be noted that the direct stress in the beam is so light that it may almost be ignored. The addition of about 10-kips direct load to a 16 WF 36 is approximately 1 kip per sq in., and a simple approach would be to deduct 1 kip from the allowable working stress to be used.

There are two types of end connection used to develop the end moment of the beam: (1) the simpler connection, of cap and seat connections to provide the necessary moment resistance without materially stiffening the joint; and (2) either a gusset-plate end connection or angle knee braces which will also furnish stiffness. If the bent is located in an outside wall, the stiffness may be provided by the masonry, and the cap and seat connection will be satisfactory. The knee braces or gussets may be buried in partitions.

If a knee brace or a gusset connection is used, the section modulus of the beam to be checked is that required for the combined moment at the inner end of the connection.

4(*b*) Having selected the beams to be used, design the end connections to transfer the moment to the column. Either riveted, high-strength bolted, or welded connections may be used. To aid the designer, examples of both types of connections have been designed in the following problem.

4(*c*) Having previously computed the live and dead loads and selected the necessary column to carry these loads, check the column section to carry the wind-load moments and direct stresses at the increased allowable stress.

e Applying the foregoing steps to the wind bent shown in Fig. 10.37, these results will be obtained:

1(*b*) $\bar{x}$ (from col. 1) $= \dfrac{18 + 30 + 44}{4} = 23$ ft

Therefore, $a = +23, b = +5, b' = -7, c = -21.$

1(*c*) $I = 23^2 + 5^2 + (-7)^2 + (-21)^2 = 1{,}044$

1(*d*)
$$d_1 = \pm\frac{23}{1{,}044} = 0.022$$
$$d_2 = \pm\frac{5}{1{,}044} = 0.0048$$
$$d_3 = \pm\frac{7}{1{,}044} = 0.0067$$
$$d_4 = \pm\frac{21}{1{,}044} = 0.0201$$

1(*e*)
$$v_A = 0.022$$
$$v_B = 0.0268$$
$$v_C = 0.0268 - 0.0067 = 0.0201\ (= d_4)$$

1(*f*)
$$m_A = \frac{0.022 \times 18}{2} = 0.198$$
$$m_B = \frac{0.0268 \times 12}{2} = 0.1608$$

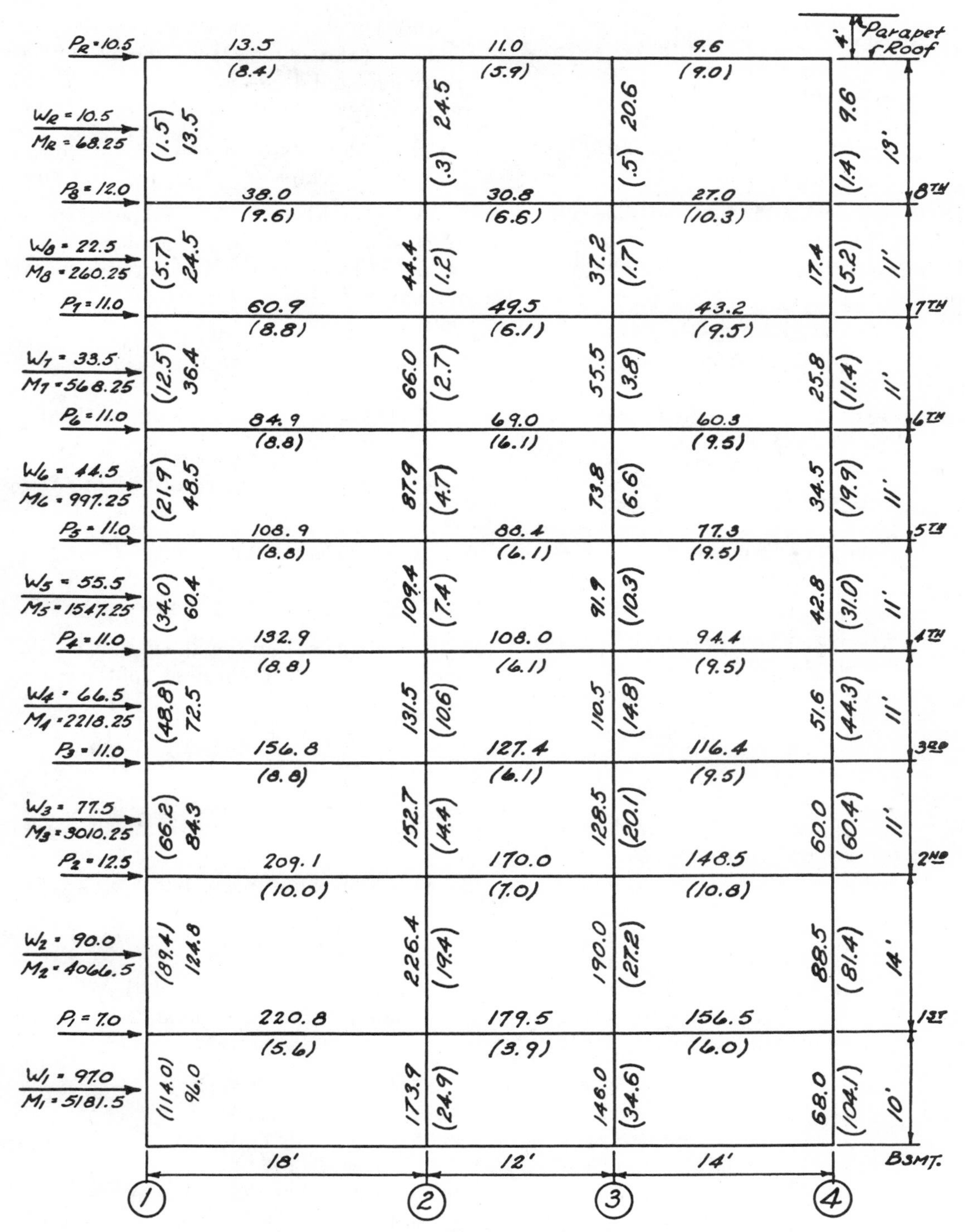

Fig. 10.37 Typical Wind Bent—Multistory

$$m_C = \frac{0.0201 \times 14}{2} = 0.1407$$

1(*g*) $v_1 = 0.198$
$v_2 = 0.198 + 0.1608 = 0.3588$
$v_3 = 0.1608 + 0.1407 = 0.3015$
$v_4 = 0.1407$

(Note that within the limits of slide-rule accuracy, the sum of these coefficients equals 1. In the above example, it totals 0.999.)

1(*h*) $p_A = 1.00 - 0.198 = 0.802$
$p_B = 0.802 - 0.3588 = 0.4432$
$p_C = .4432 - 0.3015 = 0.1417$

or

$p_C = 1.00 - 0.1407 = 0.8593$
$p_B = 0.8593 - 0.3015 = 0.5578$
$p_A = 0.5578 - 0.3588 = 0.199$

Use $p_A = 0.802$, $p_B = 0.5578$, $p_C = 0.8593$.

2(*b*) Applying a 1-kip load per foot of height,

$p_R = 4$ (= height of parapet) $+ \frac{13}{2} = 10.5$
$p_8 = \frac{13}{2} + \frac{11}{2} = 12.0$
$p_7 = p_6 = p_5 = p_4 = p_3 = 11.0$
$p_2 = \frac{11}{2} + \frac{14}{2} = 12.5$
$p_1 = \frac{14}{2} = 7$

(There is no wind applied below grade, which is assumed in this problem as first-floor level.)

2(*c*) $W_R = P_R = 10.5$ kips
$W_8 = 10.5 + 12.0 = 22.5$
$W_7 = 22.5 + 11.0 = 33.5$
$W_6 = 33.5 + 11.0 = 44.5$
$W_5 = 44.5 + 11.0 = 55.5$
$W_4 = 55.5 + 11.0 = 66.5$
$W_3 = 66.5 + 11.0 = 77.5$
$W_2 = 77.5 + 12.5 = 90.0$
$W_1 = 90.0 + 7 = 97.0$

2(*d*) $M_R = 10.5 \times 6.5 = 68.25$
$M_8 = 68.25 + 68.25 + (22.5 \times 5.5) = 260.25$
$M_7 = 260.25 + 123.75 + (33.5 \times 5.5) = 568.25$
$M_6 = 568.25 + 184.25 + (44.5 \times 5.5) = 997.25$
$M_5 = 997.25 + 244.75 + (55.5 \times 5.5) = 1{,}547.25$
$M_4 = 1{,}547.25 + 305.25 + (66.5 \times 5.5) = 2{,}218.25$
$M_3 = 2{,}218.25 + 365.75 + (77.5 \times 5.5) = 3{,}010.25$
$M_2 = 3{,}010.25 + 426.25 + (90 \times 7) = 4{,}066.5$
$M_1 = 4{,}066.5 + 630 + (97 \times 5) = 5{,}181.5$

Steps 3(*a*), 3(*b*), 3(*c*), and 3(*d*) are set down in Fig. 10.37 directly from slide-rule calculations.

4(*a*) In the bent shown in Fig. 10.37, assume that the columns below the second floor carry live- and dead-load direct stresses with negligible eccentricity, as follows:

Col. 1—261.5 kips Col. 3—257 kips
Col. 2—297 kips Col. 4—221.5 kips

The second floor beams, computed by the method of Art. 4.15 and assuming large restraint at supports would require the following moment capacity for live and dead load only:

Col. 1–2, $M = -46.1 + 35.0 - 50.2$
Col. 2–3, $M = -15.1 + 15.1 - 15.1$
Col. 3–4, $M = -30.4 + 20.9 - 28.0$

Assuming that the bent carries 0.8 of a full wind bent stress as shown in Fig. 10.37, the moments at the end of the beams will be,

Col. 1–2, $M = 0.8 \times 209.1 = 167.3$
Col. 2–3, $M = 0.8 \times 170 = 136.0$
Col. 3–4, $M = 0.8 \times 148.5 = 118.8$

It is obvious from the above that the end moments will control the design of the beams.

At Col. 1,

$$S = \left(\frac{46.1 + 167.3}{1.33}\right) 0.5 = 80.1$$

At center of beam 1–2,

$$S = 0.5 \times 35 = 17.5$$

At Col. 2, end of beam 1–2,

$$S = \left(\frac{50.2 + 167.3}{1.33}\right) 0.5 = 81.6$$

It is normally safe to assume that the moment has dropped off approximately 10 percent at the point where the beam is called upon to carry the moment without the assistance of end connections, so the beam to be used should have an S of $81.6 - 8.1 = 73.5$.

Use 16 WF 45.

For the beam from Cols. 1 to 2, the maximum S will be

$$\left(\frac{15.1 + 136.0}{1.33}\right) 0.5 = 56.7$$

Reducing this by 10 percent, $S = 56.7 - 5.7 = 51.0$

Use 16 WF 36.

For the beam from Cols. 3 to 4,

$$S = \left(\frac{118.8 + 30.4}{1.33}\right) 0.5 = 56 - 5.6 = 50.4$$

Reducing this by 10 percent, $S = 56 - 5.6 = 50.4$

Use 16 WF 36.

Undoubtedly smaller beams could have been used but at a much greater added expense in end connections.

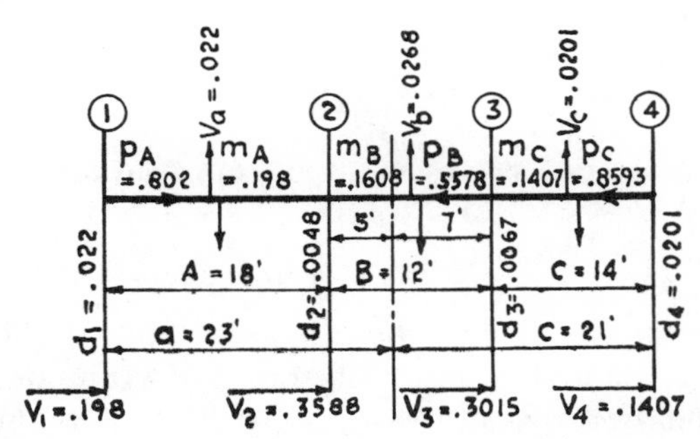

Fig. 10.38 Application of Coefficients—Beam of Multistory Bent

It will be noted that the direct stress in the beams is negligible when spread over the area of the beams above noted.

4(*b*) The moment to be provided at the ends of beams is as follows:

Cols. 1–2, 167.3
Cols. 2–3, 136
Cols. 3–4, 118.8

The stresses caused by moments in this range may be resisted by caps and seats built up from sections cut from wide-flange beams. For the beam at Col. 1–2, using the connection shown in Fig. 10.39, the moment couple between the top and the bottom of the

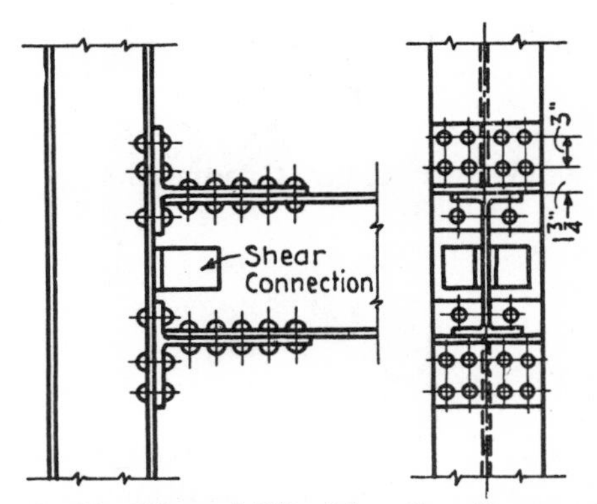

Fig. 10.39 Riveted Wind-bracing Connection

beam would produce a stress of $167.3/1.33 = 126$ kips. The value of a 7/8-in. rivet or high-strength bolt in single shear is $1.33 \times 12.03 = 16$ kips. This would require eight A354 high-strength bolts in each group between the beam and the T-type connections. The most economical connection would be cut from a 33-in. beam to make the most efficient use of materials.

In order to have eight 7/8-in. rivets in the connection shown, it will be necessary to use two rows of four each and one row of two. It would be preferable to place them as shown in Fig. 10.39.

Since the entire group of rivets in the upper or lower T connection to the face of the column would fail as a unit, the load of the eight rivets above the flange would be applied at the center of gravity of the group of rivets in the determination of the thickness of metal, as described in Art. 4.12*b*.

The distance from the fillet to the center of gravity of the eight rivets is $1\frac{3}{4} + 1\frac{1}{2} - \frac{3}{8} = 2\frac{7}{8}$ in. $= 2.88$ in. To use the method of Art. 4.12*b*, the unit load per inch is $0.8 \times 111.5/12 = 89.2/12 = 7.4$, and to use this at the increased unit stress,

$$P = \frac{7.4}{1.33} = 5.56$$

$$M = 5.56 \times \frac{2.88}{2} = \frac{20t^2}{6}$$

or

$$t = \sqrt{\frac{5.56 \times 1.44 \times 6}{22}} = 1.49 \text{ in.}$$

Use 36 WF 280.

In the same manner, for the beam from

Cols. 2 to 3, the stress developed by the moment couple is 136/1.33 = 102, or seven ⅞-in. bolts (use eight).

The thickness of the T connection is

$$P = 0.8 \times \frac{102}{1.33 \times 12} = 5.1$$

$$t = \sqrt{\frac{5.1 \times 1.44 \times 6}{22}} = 1.44$$

Use 36 WF 260.

For the beam from Cols. 3 to 4, the stress developed by the moment couple is 119/1.33 = 89.2 = six ⅞-in. high-strength bolts.

These can be put into a T-type connection using four rivets above and four below the center lines of the T. Using a 3½-in. gage on the beam, the total lever arm is 1¾ − ⅜ = 1.38 in. The load on the four rivets is 44.6/12 = 3.72, and allowing for the increased fiber stress, this is equivalent to 3.72/1.33 = 2.8.

$$t = \sqrt{\frac{2.8 \times 0.69 \times 6}{22}} = 0.56 \text{ in.}$$

Use 24 WF 68.

The translation of the connections described from a high-strength bolted connection to a welded connection is relatively simple. The moment at the end of the beams may be converted into direct stress to be taken directly from the tables by adding the dead- and live-load moment, divided by 1.33 to cover the increase in allowable stress and again by 1.33, the depth of the beam in feet or

$$\text{For beam 1–2, } \frac{50.2 + 167.3}{1.33 \times 1.33} = 123 \text{ kips}$$

$$\text{For beam 2–3, } \frac{15.1 + 136.0}{1.33 \times 1.33} = 85 \text{ kips}$$

$$\text{For beam 3–4, } \frac{30.4 + 118.8}{1.33 \times 1.33} = 84 \text{ kips}$$

Using the connection shown in Fig. 10.40 with the seat shop welded to the column, the fillet weld to the beam will be a downpass weld. The top plate will be shipped loose and will be field welded to both column and beam. The 16 WF 45 between Cols. 1 and 2 has a flange thickness of $\frac{9}{16}$ in. and a flange width of 7 in. Using ½-in. fillet weld at the bottom of the beam, the weld required

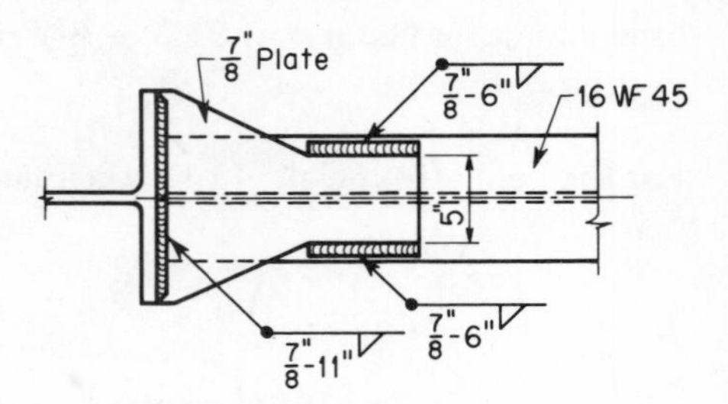

Top Plate

Fig. 10.40 Welded Wind-bracing Connection

will be $\frac{123}{5.6}$ = 22 in. or 11 in. each side of the flange. Allowing for 1 in. each side of the beam and 1½ in. for clearance of beam and ends of the weld, the seat must be a minimum of 9 in. wide and 12½ in. projection. The seat may be built up from two plates, or cut from any one of several beams. The author suggests the selection of 16 WF 88 with a flange width of 11½ in. for the seat and a T distance for the vertical weld of 13⅛ in. be used. For the cap plate, the width of the column would control the plate thickness, and this width would undoubtedly be a minimum of 12 in. Allowing 11 in. of net weld, this would require at the column, $\frac{123}{11} = \frac{11.7}{0.7} = \frac{14}{16}$, or ⅞-in. weld. This would require a ⅞-in. thick top plate. The length of the straight sides for welding to the top of the beam would be a minimum of 6 in. each side. The top plate would therefore be cut to shape indicated from a minimum of ⅞-in. plate. The method indicated would be applicable to the other connections in this line.

4(*c*) The column should be placed with the web parallel with the line of the bent to provide for the moment. Without wind stresses, the load carried by the column is

For Col. 1, 261.5 kips
For Col. 2, 217.0 kips
For Col. 3, 257.0 kips
For Col. 4, 221.5 kips

From the wind load diagram, we must add the direct load and moment assuming the bent to carry 0.8 of a unit wind bent.

For Col. 1, $DS = 71.7$, $M = 99.8$
For Col. 2, $DS = 15.5$, $M = 181.1$
For Col. 3, $DS = 21.8$, $M = 152.0$
For Col. 4, $DS = 65.1$, $M = 70.8$

Assuming a 12×12 column, $B_x = 0.217$. Equivalent direct stress for each column

For Col. 1, $0.217 \times 99.8 = 21.7$
For Col. 2, $0.217 \times 181.1 = 39.4$
For Col. 3, $0.217 \times 152.0 = 33.0$
For Col. 4, $0.217 \times 70.8 = 15.4$

The stress to be selected from the table for Col. 1 is

$$\begin{array}{r} 261.5 \\ 71.7 \\ \underline{21.7} \\ 354.9 \end{array} \qquad \frac{354.9}{1.33} = 263 \text{ for } P_{a+b} \text{ (at 22 ksi)}$$

or 261.5 for direct stress only.

Therefore, direct load controls and we should use a 12 WF 58. However, this column has only a 10-in. face and for reason of details, it would be better to use a 12 WF 65 which has a 12-in. face. In all probability the difference in weight would be overcome by economy of details.

For Col. 2,

$$\begin{array}{r} 217.0 \\ 15.5 \\ \underline{39.4} \\ 271.9 \end{array} \qquad \frac{271.9}{1.33} = 204$$

This would call for a 10 WF 49 and it would be preferable to use this column and revise the details.

For Col. 3,

$$\begin{array}{r} 257 \\ 21.8 \\ \underline{33.0} \\ 311.8 \end{array} \qquad \frac{311.8}{1.33} = 234$$

Here again the 12 WF 65 would be preferable.

For Col. 4,

$$\begin{array}{r} 221.5 \\ 65.1 \\ \underline{15.4} \\ 302.0 \end{array} \qquad \frac{302.0}{1.33} = 226$$

Here again the 10 WF 49 would be the economical section to use.

The reader's attention is called to the fact that this wind bent is not a hypothetical problem but has been selected from a design made in the author's office for a community where a constant wind pressure was required throughout the height of the building, and not a varying pressure as specified by the National Bureau of Standards.

10.30 SHALLOW BINS AND BUNKERS

a The general theory of lateral pressure of solids and the conversion of the load of ordinary level loads, sloping loads, and surcharges into equivalent fluid pressures has been discussed in Art. 8.30, Retaining Walls. A table of equivalent fluid pressures for various materials was given in Table 8.6. These rules in general will hold for the design of shallow bins, with certain modifications for sloping sides, and suspension bunkers. The designation "shallow bins" includes bins or vessels of all kinds in which the plane of rupture of the material intersects the open surface of the material, or a width, diameter, or side of square of about half the depth. If the depth exceeds this, the application of Janssen's formula will apply for granular material, and the bin will be classified as a deep bin and designed in accordance with Art. 10.31.

b For bunkers and bins in boiler rooms, two generally accepted cross sections of bins are used, the hopper-type bin (Fig. 10.41),

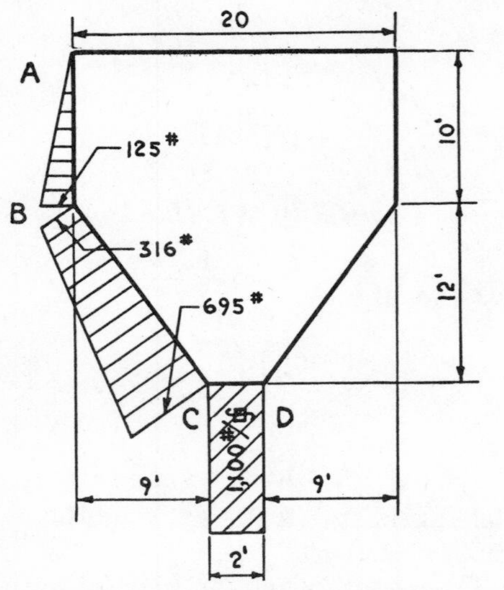

Fig. 10.41 Hopper Type Bunker

and the suspension bunker (Fig. 10.42). The hopper-type bin may be square in plan, with the same cross section both ways, or it may be rectangular in plan, with the same cross section throughout its length, and with gates at intervals. For the design of the hopper-type bin, pressure on vertical side walls may be computed by the methods of Art. 8.30 for retaining walls. Pressure on the flat bottom is the weight per square foot of the total depth of material. The pressure on the diagonal sides is the vector combination of the vertical pressure and horizontal pressure.

Problem Find the pressure on the walls of the bunker shown in Fig. 10.41 level full, using coal with a weight of 50 lb per cu ft and equivalent fluid pressure of 12.5 lb.

The pressure on vertical side AB varies uniformly from zero at the top to 125 psf at the bottom at B. On the bottom CD, the pressure is $22 \times 50 = 1{,}100$ psf.

The normal pressure on the diagonal plates is the vector sum of a vertical pressure of 1,100 and a lateral pressure of $22 \times 12.5 = 275$ psf. This may be obtained mathematically with sufficient accuracy as follows:

One square foot of sloping side carries $9/15$, or 0.6 ft, of vertical load, and $12/15$, or 0.8 ft, of horizontal pressure:

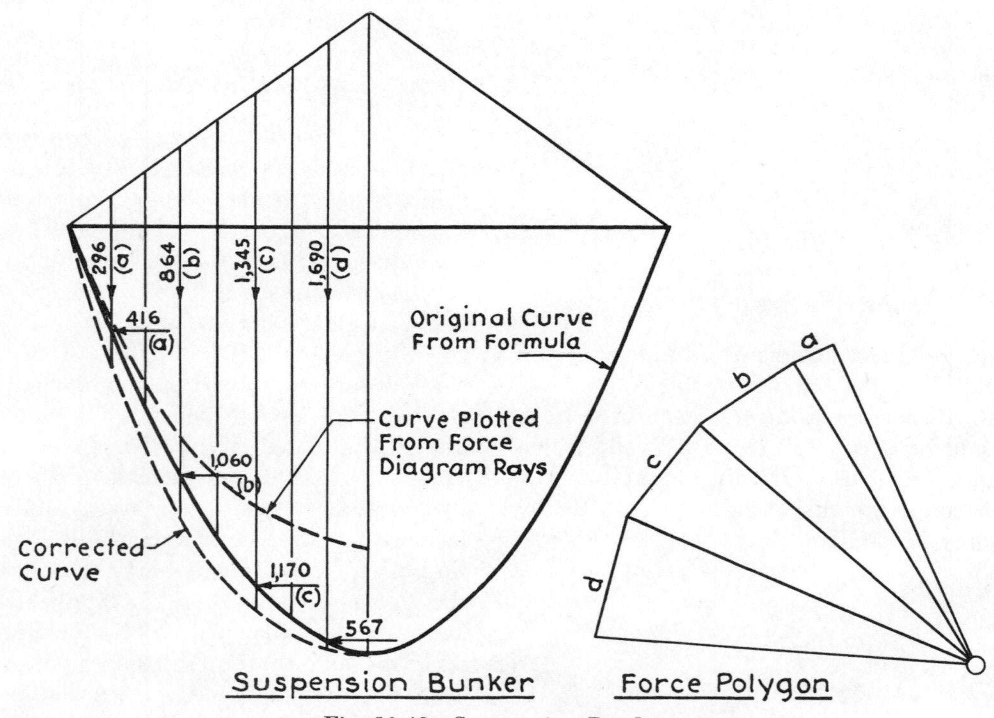

Fig. 10.42 Suspension Bunker

$$0.6 \times 1{,}100 = 660 \text{ psf}$$
$$0.8 \times 275 = 220 \text{ psf}$$

$$660^2 = 435{,}600$$
$$220^2 = 48{,}400$$
$$\sqrt{484{,}000} = 695 \text{ psf}$$

Therefore, the pressure at C normal to the sloping side is 695 psf. From C to B the pressure decreases uniformly in proportion to the total depth, so that at B this pressure is $^{10}/_{22} \times 695 = 316$ psf.

If the bunker were loaded with surcharge sloping with the angle of repose of the material, the equivalent fluid pressure would be increased in accordance with the slope pressure in Table 8.6. Moreover, the vertical loading would be increased and would be variable from the center to the edge of the bunker.

When the loads on the various plates of the bunkers have been determined, the design of plates, with the beams and the stiffeners that support them and carry the loads and stresses into the main building structure, is in accordance with ordinary moment methods if the plates are supported by beams in one direction only. Since continuity is usually developed, design is by the methods of Art. 3.21. If the stiffeners are used as supporting members from beam to beam, the method of determining the plate thickness is by one of the several flat-plate methods. The following formulae by Grashof may be used for a rectangular plate of longer dimension a, shorter dimension b, and thickness t (all in inches), with a uniform load of w per square inch.

$$f_a \text{ in the longer direction} = \frac{a^2b^4w}{2t^2(a^4 + b^4)}$$

$$f_b \text{ in the shorter direction} = \frac{a^4b^2w}{2t^2(a^4 + b^4)}$$

$$\delta = \frac{wa^4b^4}{32t^3(a^4 + b^4)E}$$

For a square plate this simplifies to

$$f = \frac{wa^2}{4t^2}$$

$$\delta = \frac{wa^4}{64t^3E}$$

These formulae are for plates which are fixed at all edges, either welded, bolted, or riveted. Plates simply supported at the edges have a fiber stress that is approximately $1\frac{1}{2}$ times as great, or comparable to that obtained by ordinary moment theory of simple or continuous beams.

c The design of the suspension-type bunker shown in Fig. 10.42 is an entirely different type of problem. Here the plates are figured to carry direct tension in the manner of a suspension bridge. The upper portion of the bin side is treated as a plate girder with the web sloping in the plane of the top of the bunker side plates. If the bunker is too narrow in proportion to its height, there is a tendency to distortion of the plates under partial load. The distortion does not come out as the balance of the bunker is filled, thereby causing a lateral deflection of the plate-girder portion of the bunker between the end connections and the central portion where the distortion takes place. Because of the rigid connections, the bunker cannot distort there. The safest bunkers have a ratio of total sag S to width B of between 5:8 and 7:8.

The theoretical equation for the curve of the bunker is

$$y = \frac{2S}{B^2}\left(3x^2 - \frac{2x^3}{B}\right)$$

where x is the horizontal and y the vertical coordinate of any point on the curve.

The weight per running foot of length for the material contained in a level full bunker is

$$W_1 = 0.625SBw$$

To this the triangle of surcharge W_2 must be added.

$$W = W_1 + W_2$$

The horizontal tensile stress at the bottom of the curve is

$$H = \frac{WB}{6S}$$

The maximum tension for which the plates must be designed is

$$T = W\sqrt{\frac{1}{4} + \left(\frac{2B}{3S}\right)^2}$$

It is not necessary to correct the tension or horizontal stress for the angle of the top girder, but it is advisable to lay out a corrected curve of the bunker to determine the angle with the vertical for the web of the girder. This is a graphical problem, as explained later.

The various steps required for the problem shown in Fig. 10.42 are partially worked out here and partially shown graphically.

Problem Assume a bunker having a 16-ft width and a 12-ft sag, carrying coal weighing 50 lb per cu ft. Assume the upper 3 ft of the bin sides to act as a plate-girder web. Figure the required thickness of plate and the slope of the plate-girder web.

$$W_1 = 0.625 \times 16 \times 12 \times 50 = 6{,}000 \text{ lb per ft}$$

For an angle of repose of 1 on $1\frac{1}{3}$, the height of triangle W_2 is $8/1.33 = 6$.

$$W_2 = \frac{6 \times 16}{2} \times 50 = 2{,}400 \text{ lb per ft}$$

$$W = 6{,}000 + 2{,}400 = 8{,}400 \text{ lb per ft}$$

$$H = \frac{8{,}400 \times 16}{6 \times 12} = 1{,}867$$

$$T = 8{,}400 \sqrt{\tfrac{1}{4} + (\tfrac{32}{36})^2}$$

$$= 8{,}400 \sqrt{1.04} = 8{,}560$$

The required net area of plate per foot is $8{,}560/22{,}000 = 0.39$ sq in., or 0.0324 per in.

Even allowing for rivet holes, etc., it is obvious that $\frac{1}{4}$-in. plate will be adequate. The direct pull on the plate girder is therefore 8,560 lb per ft.

The coordinates for laying out the curve for the bunker, as obtained from the formula, are as follows:

$x =$	1	2	3	4	5	6	7	8
$y =$	0.269	1.03	2.22	3.75	5.56	7.58	9.75	12

Divide half the width of the bunker into equal parts (four have been used), and compute the load from the area of the intercepted parts. For the trapezoidal loads it is accurate enough to apply them at the center point; for triangles, at the third point. From Table 8.6 determine the equivalent fluid pressure. In this example it is the equivalent slope pressure, or 32, which is 0.64 of the vertical pressure. The pressure on the bottom section is

$$\frac{1{,}690}{2} \times 1.03 \times 0.64 = 567 \text{ psf}$$

Similarly on the second, third, and top sections, we find horizontal pressures of 1,170, 1,060, and 416 psf. From the combination of vertical and lateral loads, loads a, b, c, and d may be laid out on a force polygon, and the rays drawn from which the curve may be constructed, as shown in dotted lines within the bunker. This curve may be corrected by prorating the position of each point in the curve in proportion to the correct center depths. The revised end slope at the upper end is more nearly the correct slope for the web of the girder.

10.31 Deep concrete bins and silos

a As stated in Art. 10.30, when the dimensional proportion of a bin carrying granular materials, such as grains, sand, or coal, reaches a ratio for H/D of approximately 2, the rules of equivalent fluid pressure no longer are applicable, and Janssen's formula governs the design. This formula involves several variables for each material, and instead of giving the formulae, Table 10.9 is presented here, from which the vertical and lateral pressures may be obtained directly.

Problem Compute the lateral pressure 30 ft below the top of a 10-ft-diameter silo for storage of anthracite coal: Table 10.9 indicates the value for anthracite is 0.52 times the value selected from col. 1, or, for $H/D = 3$,

$$\frac{L}{D} = 0.52 \times 50.24 = 26.1$$

and

$$\frac{V}{D} = 0.52 \times 119.57 = 62.1$$

Therefore $L = 26.1 \times 10 = 261$ psf, and $V = 62.1 \times 10 = 621$ psf. Since the total weight of a column of coal at this point is $30 \times 52 = 1{,}560$ lb, it is obvious that the greater part of the vertical weight is carried

Table 10.9 Pressure Factors for Various Materials in Concrete Bins

A = Area of bin, sq ft
H = Height of material in bin, ft
D = Diameter of bin, ft
V = Vertical pressure at depth H, psf
L = Lateral pressure at depth H, psf
P = Perimeter of bin, ft
$R = A/P$ = Hydraulic radius
$K = L/V$ = Ratio of lateral to vertical pressure
v' = Tangent of angle of friction of material on concrete wall

$$V = \frac{WR}{kv'}\left[1 - \left(1 \div \text{number whose common log is } \frac{Kv'h}{2.303\,R}\right)\right]$$

W = Weight of material, lb per cu ft

Values of constants and pressures

Material	W	K	v'	L/D and V/D
Wheat	50	.60	.4167	Col. 2
Cement	100	.42	.445	Col. 1
Raw mix	75	.42	.445	.75 Col. 1
Stone	100	.30	.767	Col. 3
Sand	100	.30	.52	Col. 4
Gravel	100	.30	.52	Col. 4
Anthracite	52	.42	.445	.52 Col. 1
Bituminous	50	.18	.695	Col. 5
Clinker	95	.30	.52	.95 Col. 4
Coke	30	.30	.767	.30 Col. 3

	1		2		3		4		5	
H/D	L/D	V/D	L/D	V/D	L/D	V/D	L/D	V/D	L/D	V/D
0.1	4.05	9.65	2.85	4.75	2.90	9.68	2.88	9.60	.88	4.88
0.2	7.74	18.42	5.43	9.05	5.58	18.59	5.62	18.72	1.71	9.52
0.3	11.24	26.75	7.77	12.94	7.92	26.40	8.16	27.20	2.51	13.93
0.4	14.56	34.65	9.90	16.49	10.16	33.88	10.61	35.36	3.26	18.13
0.5	17.42	41.46	11.80	19.67	12.18	40.59	12.82	42.72	3.98	22.12
0.6	20.15	47.96	13.54	22.56	13.99	46.64	14.98	49.92	4.67	25.92
0.7	22.93	54.57	15.10	25.17	15.71	52.36	16.99	56.64	5.32	29.53
0.8	25.35	60.33	16.53	27.54	17.16	57.20	18.86	62.88	5.94	32.97
0.9	27.54	65.55	17.81	29.67	18.55	61.82	20.64	68.80	6.52	36.24
1.0	29.56	70.35	18.97	31.61	19.87	66.22	22.27	74.24	7.08	39.35
1.1	31.47	74.90	20.02	33.36	20.99	69.96	23.86	79.52	7.61	42.30
1.2	33.27	79.18	20.97	34.94	22.06	73.54	25.39	84.64	8.12	45.12
1.3	34.98	83.25	21.83	36.38	23.00	76.67	26.69	88.96	8.60	47.80
1.4	36.47	86.80	22.60	37.67	23.91	79.70	27.94	93.12	9.06	50.34
1.5	37.90	90.20	23.31	38.85	24.68	82.28	29.18	97.28	9.49	52.76
1.6	39.23	93.37	23.95	39.91	25.41	84.70	30.29	100.96	9.91	55.07
1.7	40.46	96.30	24.52	40.87	26.07	86.90	31.39	104.64	10.31	57.26
1.8	41.54	98.87	25.04	41.73	26.73	89.10	32.40	108.00	10.68	59.34
1.9	42.63	101.46	25.52	42.52	27.26	90.86	33.31	111.04	11.04	61.33
2.0	43.61	103.79	25.94	43.24	27.76	92.53	34.18	113.92	11.38	63.21
2.1	44.51	105.93	26.33	43.88	28.22	94.05	35.02	116.72	11.69	64 93
2.2	45.36	107.96	26.67	44.45	28.63	95.43	35.85	119.49	12.01	66.71
2.3	46.14	109.81	27.00	45.00	29.04	96.80	36.55	121.84	12.29	68.30
2.4	46.87	111.55	27.28	45.47	29.37	97.90	37.27	124.24	12.58	69.88
2.5	47.55	113.17	27.54	45.90	29.70	99.00	37.92	126.40	12.84	71.35
2.6	48.10	114.48	27.77	46.29	29.99	99.95	38.61	128.70	13.09	72.75
2.7	48.75	116.03	27.99	46.64	30.24	100.79	39.07	130.24	13.33	74.05
2.8	49.28	117.29	28.18	46.96	30.50	101.66	39.63	132.10	13.56	75.34
2.9	49.88	118.71	28.35	47.25	30.71	102.38	40.13	133.76	13.78	76.57
3.0	50.24	119.57	28.51	47.51	30.91	103.04	40.61	135.36	13.98	77.69
3.2	51.07	121.55	28.78	47.96	31.25	104.18	41.48	138.27	14.37	79.81
3.4	51.78	123.24	29.00	48.33	31.56	105.20	42.24	140.80	14.71	81.70
3.6	52.39	124.69	29.18	48.64	31.80	105.99	42.92	143.06	15.03	83.47
3.8	52.93	125.97	29.33	48.88	31.99	106.65	43.51	145.04	15.30	85.00
4.0	53.38	127.04	29.45	49.09	32.17	107.23	44.05	146.82	15.52	86.47
4.5	54.26	129.14	29.66	49.43	32.47	108.24	45.11	150.35	16.10	89.46
5.0	54.86	130.57	29.80	49.66	32.67	108.90	45.88	152.93	16.52	91.79
6.0	55.57	132.26	29.93	49.88	32.87	109.56	46.86	156.21	17.10	95.02
8.0	56.06	133.47	29.99	49.98	32.98	109.93	47.65	158.83	17.67	98.17
10.0	56.17	133.69	30.00	50.00	33.00	109.99	47.91	159.70	17.88	99.13

into the side wall through friction, and in designing the footings this fact should be taken into consideration.

b In the design of silos, the stresses are carried by ring tension, as described in Art. 10.32, Pipe Design. The concrete not only is not given credit for carrying any of the tension, but because of concrete's low tensile value, the steel stress should be limited to 14,000 psi in design to prevent cracking of the concrete. The following method for the design of single reinforced concrete silos has been in use successfully for some years. It is theoretical, but is governed largely by practical considerations. For example, at the base of the silo, the sides are doweled to the base or floor, and therefore cannot work in ring tension, but only by retaining-wall action from the base. It is therefore satisfactory to base the thickness design on stresses at the distance above the base determined in the following paragraph. Since the concrete is only an envelope for the material and the stresses are carried by the steel, the determination of thickness of concrete wall is dependent on good coverage for the steel, and may be assumed as 6 or 8 in., depending on whether heavy jack rods are to be used for raising slip forms.

The wall will carry a cantilever moment from the base which will resist pressure through a height of

$$H = \frac{20d}{\sqrt{L}}$$

where d is half the wall thickness and L is the unit lateral pressure as derived in Art. 10.31a. Thus, for the problem given in Art. 10.31a, use 8 in.

$$H \text{ carried from the base} = \frac{20 \times 4}{\sqrt{261}} = 4.96 \text{ ft}$$

We may start figuring ring tension 5 ft above the base and develop the dowels into the base to carry the full moment capacity of the 4-in. thickness.

At a point 5 ft above the base, $H/D = 2.5$, $L/D = 0.52 \times 47.55 = 24.7$, and $L = 247$. The ring tension at this point is, $P = 10 \times 247 = 2{,}470$ lb per foot of height. This is resisted by twice the steel areas of the section (once for each wall—see Art. 10.32). Therefore

$$A_s = \frac{2{,}470}{2 \times 14{,}000} = 0.088$$

The minimum steel should be ⅜-in. round, 8 in. o.c. So the minimum steel may be used throughout the height.

The attention of the designer is called to the fact that for true fluids, such as oil, molasses, etc., and for shallow tanks or silos, whether of concrete or steel, the laws of fluid pressure still govern the design.

10.32 Pipe design

Under the internal pressure of water, steam, or any other fluid under pressure, a pipe has a tendency to rupture longitudi-

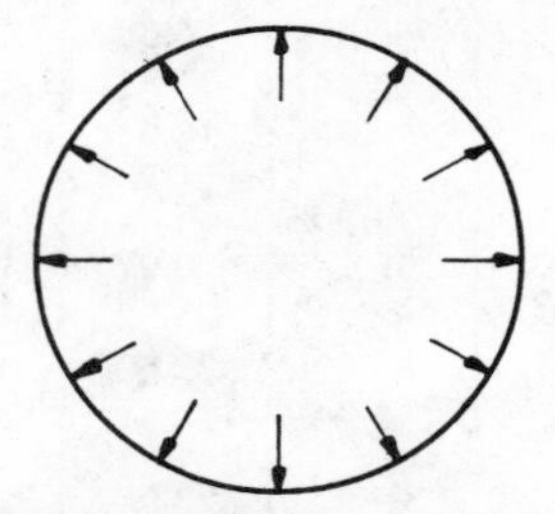

Fig. 10.43 Application of Pressure—Pipe Design

nally. Figure 10.43 indicates the method of application of pressure on the inside walls of the pipe, and Fig. 10.44 indicates the

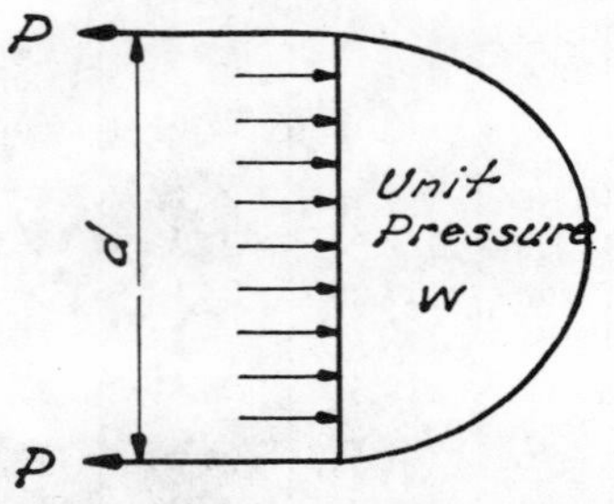

Fig. 10.44 Application of Pressure—Pipe Design

method of solution. The internal pressure is considered to be applied uniformly on an area equal to a unit length times the diameter. This is resisted by the two walls of steel or wrought iron pipe, by the hoops of wood-stave pipe, or by the reinforcement of concrete pipe. Therefore

$$P = \frac{wd}{2}$$

Problem A water supply pipe is 6 ft in diameter and built of longitudinal wood staves held by hoops of 1-in. round steel rod. What spacing of hoops is required when the pressure is 50 psi?

In this problem (from a state license examination),

$$P = \frac{50 \times 72}{2} = 1{,}800 \text{ lb per inch of length}$$

This would require a steel area of 1,800/22,000 = 0.09 sq in. per inch of length.

The area at the root of the thread of a 1-in. round rod is 0.55 sq in. (see Table 2.4), and through the body of the rod it is 0.7854 sq in. However, joints will be staggered so that two adjacent rods will never have the threaded area at the same point, and it is safe to reduce the area of one rod out of four, so that the average area to use is

$$\begin{aligned} 3 \times 0.7854 &= 2.3562 \\ 1 \times 0.551 &= \underline{0.551} \\ &\ \frac{2.9072}{4} = 0.7268 \text{ sq in.} \end{aligned}$$

The spacing may be 0.7268/0.09 = 8.07 in. Use 8-in. spacing.

10.33 Brick stacks

a The design of a brick stack is a problem combining direct compression and bending due to wind stresses. The controlling factor normally is the net tension on the windward side of the stack. Usually the lower part of a stack is lined with firebrick, but since it is free from structural brickwork, firebrick may not be counted upon to resist stresses.

Large industrial and power-house stacks present specialized problems and should be designed and built by recognized chimney-construction companies. However, since there is a tendency to smaller-area higher stacks, particularly on schools and hospitals —stacks which will be neither designed nor built by chimney specialists—it is well to recognize the structural limitations of such stacks. Frequently the chimney design is left entirely to the heating engineer, and it is not referred to the structural engineer at all. As a result, although the stack does not blow over, it may become the object of excessive maintenance costs.

b As a code to govern the design of brick stacks, the requirements of the U.S. Corps of Engineers may be used:

> Brick stacks shall be designed for a wind load of 30 psf multiplied by the shape factor listed below:
>
> | | |
> |---|---|
> | Square or rectangular | 1.00 |
> | Hexagonal or octagonal | .80 |
> | Round or elliptical | .60 |
>
> The maximum unit tensile strength in pounds per square inch shall be 10 for ordinary brickwork and 35 for radial brick construction. The maximum unit compression shall be as set forth for ordinary brickwork, and 350 for radial brick construction.

c The maximum resultant compression on the leeward edge is

$$f_c = \frac{W}{A} + \frac{M}{S}$$

and the maximum allowable stress is the compression stress of the material. For normal brickwork, this should not be over 250 psi.

The maximum resultant tension on the windward side should be

$$f_t = \frac{M}{S} - \frac{W}{A}$$

The 30-psf wind pressure required by the U.S. Corps of Engineers should definitely be used in any area subject to high winds or in any tornado section. There are many areas, however, where a 20-psf wind pressure is sufficient.

Attached chimneys or chimneys within buildings need not be investigated below their point of attachment or the roof at which they are supported.

The worst condition for a square stack results from a wind blowing against a corner of the stack, on an area based on the diagonal dimension, but using a shape factor of 0.6. The section modulus of a brick stack of this shape is

$$S = 0.1178 \frac{D^4 - D_1^4}{D}$$

where D is the side of the square, in feet.

For an octagonal stack the section modulus is

$$S = 0.1095 \frac{D^4 - D_1^4}{D}$$

where D is the depth between any two parallel sides, that is, the short diameter.

For a round stack the section modulus is

$$S = 0.098175 \frac{D^4 - D_1^4}{D}$$

where D is the outside diameter of the circle.

Using the above formulae, with a 10-psi maximum tensile stress (1,440 psf), the maximum allowable height for a plain brick stack without reinforcement may be obtained by solving the following quadratic for height x (using foot units throughout).

$$\frac{M}{S} x^2 - 120x = 1{,}440$$

A close approximation of the maximum safe height, using 30-psf wind load, may be arrived at from the empirical formulae in Table 10.10. The section modulus of various square, octagonal, and round stack sections of smaller stacks is given in foot units in Table 10.11.

d The foundation for a free-standing stack must provide for eccentricity in any direction. Applying the load under the chimney walls, compute the eccentricity *e* by combining wind and dead loads. Compute the total load on the footing, and assume a footing at least 12 in. greater in each direction than the stack size. Check the footing for soil pressure, using the eccentricity computed above, and the formula for eccentric pressures,

$$f = \frac{P}{A}\left(1 \pm \frac{6e}{b}\right)$$

Problem Check a square stack, 50 ft total height, attached up to 20 ft height. Use a 20-in. square inside stack with 1-ft walls (3 ft 8 in. square outside) and a 20-psf wind pressure.

The critical point in the masonry is at the top of the attachment, with a height of 30 ft above this point.

The load applied on the diagonal of the stack is

$$P = 0.6 \times 20 \text{ lb} \times 3.67 \sqrt{2} \times 30 = 1{,}867 \text{ lb}$$

$$M = 1{,}867 \times 15 = 28{,}000$$

$$\frac{M}{S} = \frac{28{,}000}{5.56} = 5{,}040 \text{ psf}$$

$$W = 120 \times 30 = 3{,}600$$

The net tension = (5,040 − 3,600)/144 = 10.0 psi, which is within the allowable stress.

If the stress is beyond the allowable limit, the designer has two alternatives: to increase the thickness of wall or to put in steel corner reinforcements. In the above problem, let us assume 30-psf wind load. If the wall thickness were to be increased 4 in., the 3.67 figure would become 4.33,

$$P = 0.6 \times 30 \times 4.33 \sqrt{2} \times 30 = 3{,}309$$

$$M = 3{,}309 \times 15 = 49{,}630$$

$$\frac{M}{S} = \frac{49{,}630}{9.39} = 6{,}300$$

$$W = 160 \times 30 = 4{,}800$$

The net tension is $\frac{6{,}300 - 4{,}800}{144} = \frac{1{,}500}{144} =$ 10.4 psi, which is less than 5 percent overstress and is therefore acceptable.

As an alternative design, using reinforcement with a 12-in. wall,

Table 10.10 Approximate Maximum Height of Brick Stacks

Wall thickness, in.	Square stacks	Octagonal stacks	Round stacks
8	$6.132D - 2.528$	$6.1D - 2.64$	$7.07D - 3.38$
12	$8D - 7.1$	$8.12D - 7.79$	$8.87D - 7.6$
16	$9.84D - 13.33$	$9.9D - 14.06$	$12.2D - 20.6$

Table 10.11 Section Modulus of Brick Stacks

Foot units

D	Square stacks			Octagonal stacks			Round stacks		
	8-in. wall	12-in. wall	16-in. wall	8-in. wall	12-in. wall	16-in. wall	8-in. wall	12-in. wall	16-in. wall
2′0″	.93			.865			.775		
2′4″	1.45			1.35			1.20		
2′8″	2.09	2.22		1.94	2.06		1.74	1.85	
3′0″	2.88	3.14		2.68	2.92		2.4	2.62	
3′4″	3.8	4.25	4.36	3.53	3.95	4.05	3.16	3.53	3.62
3′8″	4.86	5.56	5.77	4.51	5.16	5.37	4.04	4.62	4.8
4′0″	6.05	7.07	7.45	5.62	6.57	6.92	5.03	5.88	6.2
4′4″	7.4	8.8	9.39	6.86	8.16	8.72	6.15	7.3	7.8
4′8″	8.85	10.7	11.6	8.21	9.9	10.8	7.35	8.86	9.63
5′0″	10.5	12.8	14.1	9.56	11.9	13.0	8.72	10.6	11.7
5′4″		15.2	16.8		14.1	15.6		12.6	13.9
5′8″		17.7	20.0		16.4	18.6		14.7	16.6
6′0″			23.0			21.4			19.2

Maximum tension $= \dfrac{42{,}000}{5.56 \times 144} = 52.8$

Reduction for direct compression $= \dfrac{3{,}600}{144} = 25.0$

Net tension $= 27.8$

Referring to Fig. 10.45, we will compute the area which is in tension.

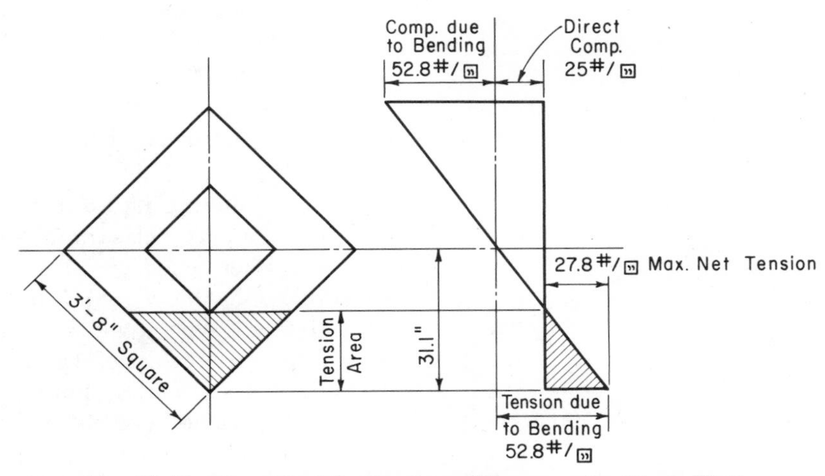

Fig. 10.45 Graphical Indication of Pressure in Stack Wall

Height of tension triangle $= \dfrac{27.8}{52.8} \times 31.1 = 16.4$

Area of tension triangle $= \dfrac{16.4 \times 32.8}{2} = 268.96$

Total tension $= 268.96 \times \dfrac{27.8}{2} = 3{,}720$

Using steel at 20,000 psi, $A_s = \dfrac{3{,}720}{20{,}000} = 0.186$

Use one No. 4 bar in each corner.

Spot checks will show that the stack does not need reinforcement within 20 ft of the top for the 20-psf load.

The stresses due to wind moment theoretically do not increase below the roof line, but the compression forces increase at the rate of 120/144 or .83 psi per foot of height. The tension reinforcement is therefore required down to the foundation.

The total weight on the footing is

$$1{,}600 \times 50 = 80{,}000 \text{ lb}$$

Because the stack is attached, eccentricity on the footing due to wind can be disregarded. Using a 12-in. overhang on all sides of the stack, the footing will be

$$3'8'' + 1'0'' + 1'0'' = 5'8'' \text{ square}$$

The soil pressure is

$$\frac{80{,}000}{5.67 \times 5.67} = 2{,}490 \text{ psf}$$

$$\text{Allowing 12-in. footing thickness} = \underline{\quad 150}$$
$$2{,}640 \text{ psf}$$

The footing should have a minimum thickness of 12 in., and no reinforcement is required for bending stresses. However, because of temperature differentials, it is advisable to use a layer of bars in the bottom and a layer in the top of the slab, each layer of ⅜-in. round, 12 in. o.c. in both directions.

10.40 MILL BUILDINGS

a Factory buildings, or as they are commonly termed, mill buildings, are treated separately because the attention of the designer should be called to a large number of design items scattered throughout this text which are particularly applicable to the design of this type of one-story building. Figures 10.46 and 10.47 indicate two of the common cross sections of mill buildings, but there are endless variations depending on such factors as use and location.

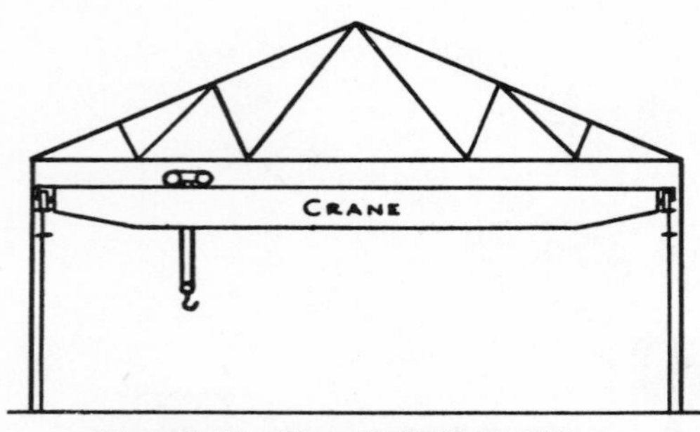

Fig. 10.46 Simple Mill Building

b The type of mill building illustrated by Fig. 10.46 is about the cheapest type of wide-span structure, and is particularly applicable to the use of corrugated steel, aluminum, or asbestos-cement roofing and siding. If corrugated roofing of any kind is used, the minimum pitch of the roof should be ⅕, or 1 on 2½, if the sheets are cemented, and ¼, or 1 on 2, if the sheets are lapped and riveted, bolted, or nailed. This will determine the slope of the truss, and may even determine whether or not trusses should be used. If tar and gravel roofing on wood, concrete, or gypsum is used, the roof may be kept practically flat, and rolled beams will be cheaper up to spans of about 60 ft.

Trusses are used on spans of from 40 to 120 ft, and since they are heavily fabricated, it is preferable to space them from 20 to 25 ft apart, and thus to increase plain rolled steel in the purlins and save fabrication in the trusses.

A number of the articles in this text that have special application to mill buildings of the type shown in Fig. 10.46 are,

Art. 3.21*g*, Method of moment distribution (the indirect Hardy Cross method)

Art. 4.14, Unsymmetrical bending in purlins

Art. 4.23, Crane-runway girders

Art. 4.40, Corrugated siding and roofing

Art. 7.11*e*, Steel columns

Art. 10.10, Trusses

Art. 10.21, Wind bracing—low steel buildings

Art. 10.41, Girts

c The building shown in Fig. 10.47 is indicative of a better class of factory buildings, required where good working conditions are desirable—that is, where heat loss from the building is to be avoided. The flat roof shown requires the use of a roof covering over the main roofing material of concrete, wood, or gypsum. This type of building is usually used with masonry side walls to correspond in heat loss and other characteristics with the roof. Since the roof is flat, or relatively flat, it is not necessary to use trusses. This particular section shows the center bay as a monitor bay for the distribution of light more uniformly over the floor area.

d In addition to proper design of all the component parts of a mill building as discussed in other articles of this text, special

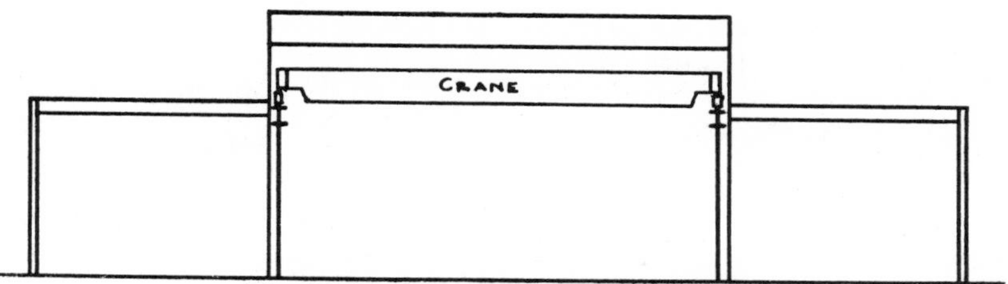

Fig. 10.47 Three-Bay Mill Building

care should be given to the bracing of mill buildings. The extent of the bracing is dependent on the type of construction and the use of the building. Obviously a building with corrugated siding and roofing, or even wood or precast concrete roofing, must be braced much more rigidly than one with masonry side walls and poured-concrete roof. Unless the columns are unusually high without support, bracing may usually be omitted in the latter type.

In Fig. 10.48 a system of bracing is indicated for a seven-bay building of the type shown in Fig. 10.46. The end bays should

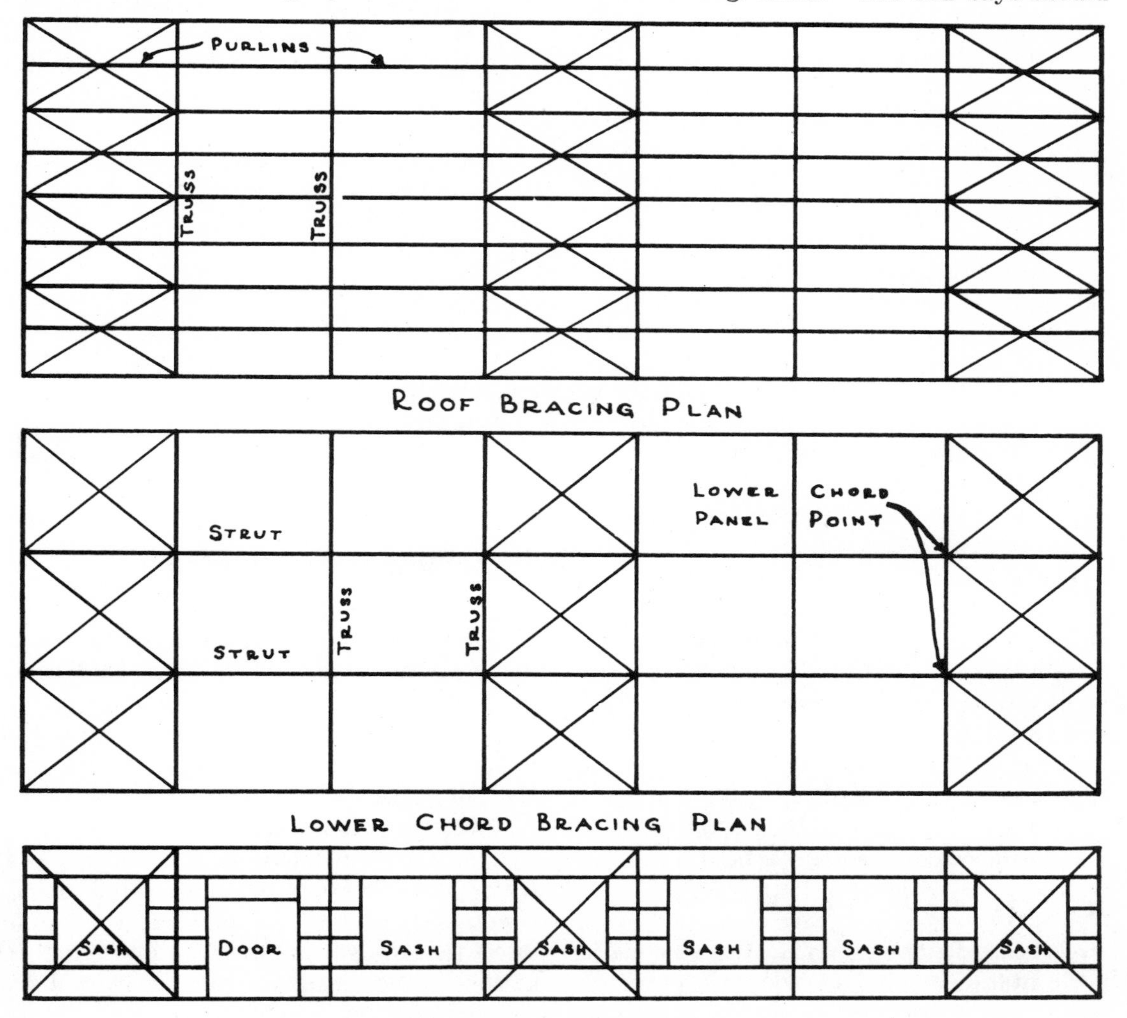

Fig. 10.48 Bracing of a Steel Mill Building

be braced, and at least every third bay inside. Thus the bracing will work out as follows:

For three- and four-bay buildings, brace end bays only.

For five and seven bays, brace end bays and center bay.

For six bays, brace bays 1, 3, and 6.

For eight bays, brace bays 1, 3, 6, and 8.

For nine bays, brace bays 1, 4, 6, and 9.

For ten bays, brace bays 1, 4, 7, and 10.

In the roof framing or upper-chord system, the purlins act as struts. Formerly much of the cross bracing was rod bracing, usually ⅞-in. rods, but nowadays there is a tendency toward light angles. The AISC Code specifies that for bracing members in tension, the least l/r ratio shall be 300.

The lower-chord bracing is made up similarly of angle cross bracing in tension and a continuous line of struts. These struts are required by the AISC specification to have a least l/r ratio of 200. A very satisfactory strut section is made up of two channels, one flatwise with the flanges turned downwards, on top of a channel with the web vertical. For this type of strut, a 6 channel 8.2 vertically with a 5 channel 6.7 horizontally is good to a span of 22 ft 10 in. Beyond this, to 26 ft, we use an 8 channel 11.5 vertically with a 6 channel 8.2 horizontally, and up to 30 ft a 9 channel 13.4 vertically with a 7 channel 9.8 horizontally.

The side-wall bracing uses the girts for strut members, and the cross bracing of angles as already noted. The designer should remember that it is not possible to operate pivoted sash in the areas covered by side-wall cross bracing. If the cross bracing is placed tight against the inside edge of the girts and connected to them, the length to use in computing the l/r ratio may be greatly reduced and smaller members may be used.

10.41 Girts

Girts are defined in Art. 1.11 as secondary horizontal members in a side wall, designed to resist wind pressure. They are ordinarily used with corrugated-steel siding and steel sash, although not necessarily. Other types of siding such as wood sheathing or asbestos-cement may be used instead of steel. (See Arts. 10.40 and 4.31 for further information on girt framing.)

Girts are ordinarily designed as angles or channels, depending on the span used. Since the main load to be carried is wind, normal to the face of the building, the main axis of the member is flatwise. To prevent deflection due to the weight of the girt and the siding, sag rods are used to carry the vertical weight back into the eave purlin, which is either vertical or at an acute angle with the vertical. The practical consideration of avoidance of dust pockets requires that as far as possible, channels be placed with the flanges downward. An end wall using girt framing is shown in Fig. 10.49. Details for framing around openings are shown in Fig. 10.50.

Problem Design a girt for the 24-ft span shown in Fig. 10.49. Assume that the building code requires us to design for a 20-psf wind load.

$$W = 24 \times 5.67 \times 20 = 2.72 \text{ kips}$$

$$S = \frac{2.72 \times 24}{14.66 \times 1.33} = 3.3$$

Use 6 channel 8.2.

To prevent sag of the girts due to their flatwise position, hang them by means of rods at the third points from the eave beam.

10.50 POLES AND POSTS

a Poles and posts are frequently subjected to stresses in such a manner that it is necessary to compute (1) the size of the pole, or (2) the method of anchorage in the ground. Figure 10.51 indicates the application of loads to such a post, and the resultant passive soil pressures.

The applied loads may be from a hanging sign or pipe indicated by P, wind on the hanging member or on a flag indicated by H_1, wind on the pole itself, indicated by H_2, or a direct pull from a wire at H_1. The determination of these loads is probably more difficult than the solution of the structure itself.

b The loads to be used in computing the stresses to apply to such poles may be de-

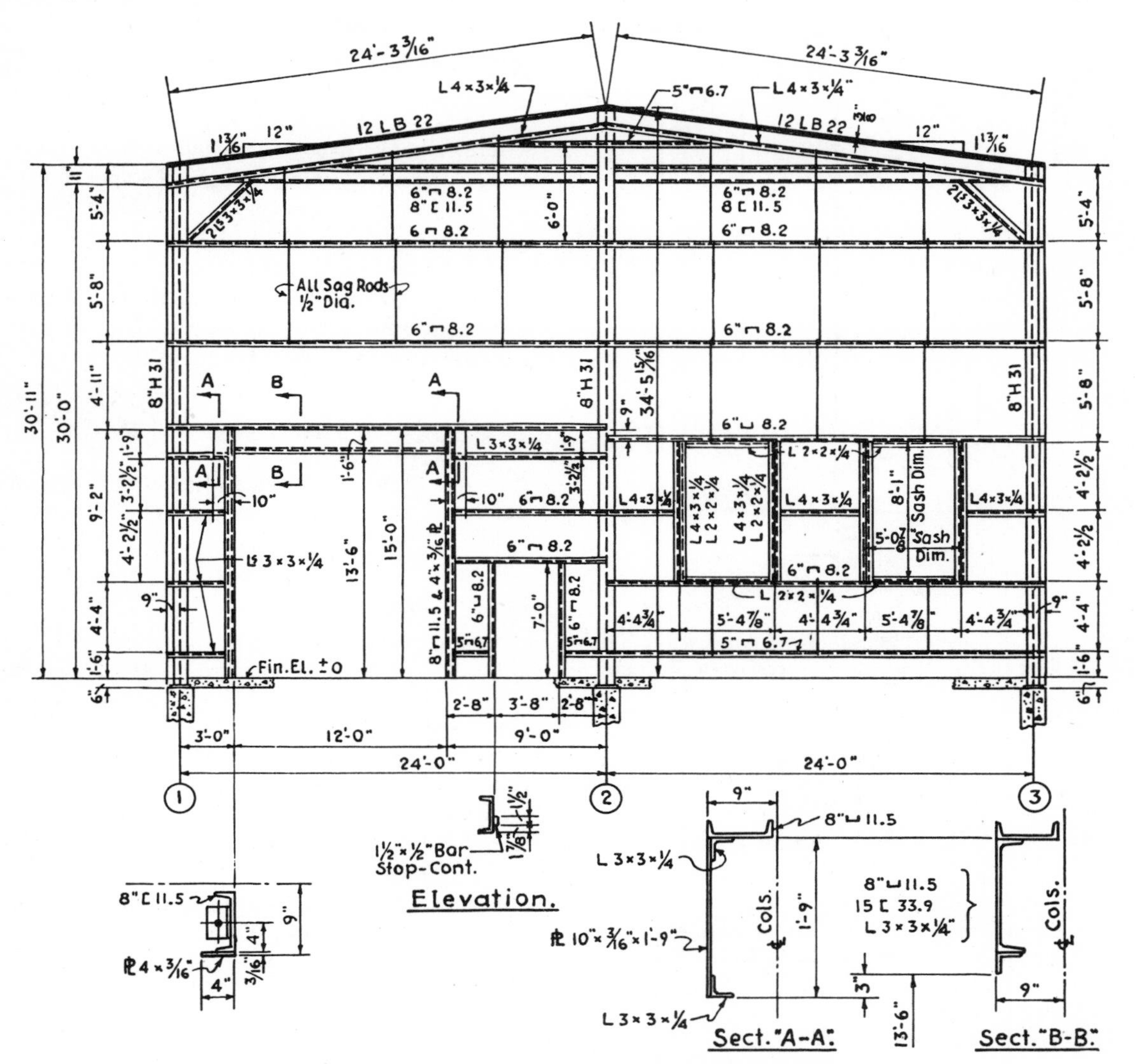

Fig. 10.49 Typical End Elevation—Mill Building Framing

rived from the following. The weight of a sign may be given, but it is well to add 50 percent to this weight for wind gusts or for weight of sleet. Pipe lines and power lines should have added to their known weight an allowance for a coating of ice 1 in. thick. For the suspension of trolley wires, the amount of sag must be known to compute the tension H_1 at the top of the pole. The horizontal tension is

$$H_1 = \frac{wL^2}{8\Delta}$$

where w is the unit weight per linear foot including the weight of ice. A safe total

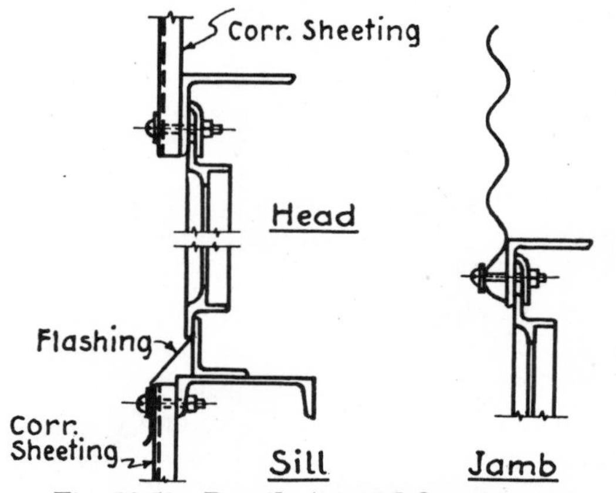

Fig. 10.50 Details Around Openings in Buildings With Corrugated Siding

load for a wire of this kind is 1.5 lb per linear foot.

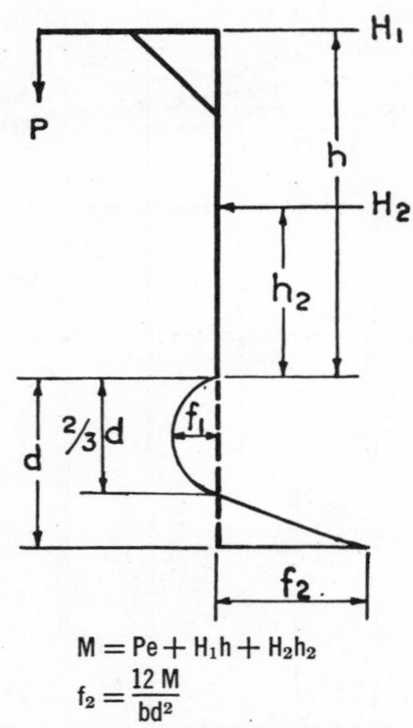

Fig. 10.51 Loads and Stresses on a Post

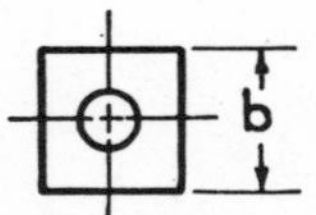

Fig. 10.52 Foundation for Post or Pole

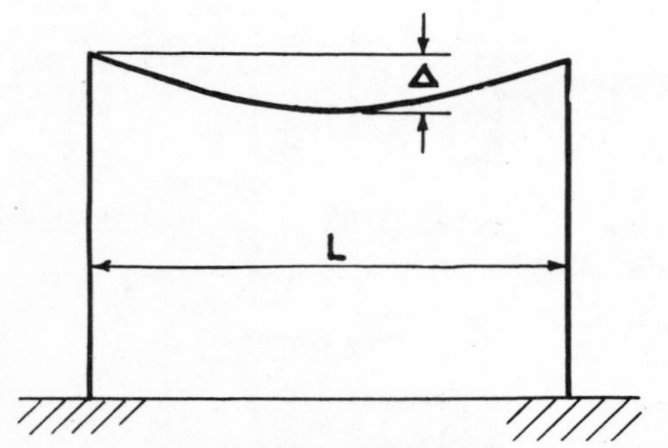

Fig. 10.53 Load on Cable Suspended from Poles

The direct pull on a pole from a power line under equal tension on both sides of the pole is zero as long as the line is intact. However, provision must be made for the stress created when the line on one side of the pole is broken.

In the design of flagpoles several U.S. government sources provide the best criteria for computing the pressure. The U.S. Navy standard formula for pressure H_1 applied at the top of the flag is

$$H_1 = 0.0003 A V^{1.9}$$

where A is the area of the flag in square feet and V is wind velocity in miles per hour. For a 60-mile maximum,

$$V^{1.9} = 2{,}390$$

For U.S. Post Office buildings, the standard requirements for flag sizes and the resulting pressure computed by the Navy formula are, for a one-story building,

$$A = 3.52 \text{ ft} \times 6.69 \text{ ft} = 23.5, \quad H_1 = 17 \text{ lb}$$

For two or more stories or for free-standing poles,

$$A = 5.19 \text{ ft} \times 9.77 \text{ ft} = 50.2, \quad H_1 = 36 \text{ lb}$$

For exceptional buildings,

$$A = 8.94 \text{ ft} \times 16.99 \text{ ft} = 151.9, \quad H_1 = 109 \text{ lb}$$

c The computation of the structure above grade is a simple matter of computation of moments as given by the moment formula in Fig. 10.51 and the resultant section modulus. Because the cantilever moment is reduced from the ground up, such poles, which are usually of pipe section, may be reduced by welding a smaller section into the lower section. The point of bending to be used in determining the diameter of the maximum pipe and the passive pressure should be taken at a point about one-third of the depth d below grade.

d The maximum allowable soil pressure f_2 is the passive pressure discussed in Art. 8.30 as a resisting pressure to movement of sheet piling or retaining walls:

$$f_p = w_e \tan^2 (45 \text{ deg} + \tfrac{1}{2}\phi)$$

where w_e is the weight per cubic foot of the material at the bottom and ϕ the angle of repose of the material. This passive pressure is given for various types of soil in Table 8.6. For normal conditions, a value of 400 psf may be used.

Problem 1 Design an outdoor steel flagpole 60 ft high to carry a flag of 50 sq ft area.

The flagpole is spliced 20 ft above grade and 40 ft above grade. Allow for a wind pressure of 10 psf on the diameter of the pole.

From Art. 10.50*b*, the pressure on the flag is 36 lb. At the first splice point the moment is $36 \times 20 = 720$. Assume 3-in.-diameter pole.

$$20 \times \tfrac{1}{4} \times 10 = 50 \times 10 = 500$$
$$720 + 500 = 1{,}220 \text{ ft-lb} = 14{,}640 \text{ in.-lb}$$

Since all moment is due to wind, all unit stresses may be increased $\frac{1}{3}$, or

$$\text{Required } S = \frac{14{,}640}{29{,}330} = 0.5$$

From Table 1.5(*a*) for 2-in. pipe,

$$S = \frac{0.666}{1.19} = 0.56$$
$$(\text{O.D.} = 2.375 \text{ in.})$$

At the next splice point, the moment is

$$36 \times 40 = 1{,}440$$

On upper 2-in. pipe,

$$20 \times \frac{2.375}{12} \times 10 = 39.4 \times 30 = 1{,}182$$

On assumed 4-in. pipe,

$$20 \times \frac{4.5}{12} \times 10 = 75 \times 10 = \underline{750}$$
$$3{,}372 \text{ ft-lb}$$
$$= 40{,}450 \text{ in.-lb}$$
$$\text{Required } S = \frac{40{,}450}{29{,}330} = 1.38$$

For 3-in. pipe,

$$S = \frac{3.017}{1.75} = 1.72$$
$$(\text{O.D.} = 3.5 \text{ in.})$$

Assume the depth in ground at 8 ft. Then the point of maximum moment = 2 ft 8 in. below grade.

The moment on the flag is

$$36 \times 62.67 = 2{,}260$$

On the upper 2-in. pipe,

$$39.4 \times 52.67 = 2{,}075$$

On the middle section,

$$20 \times \frac{3.5}{12} \times 10 = 58 \times 32.67 = 1{,}900$$

And on the lower section, try 4-in. pipe.

$$20 \times \frac{4.5}{12} \times 10 = 75 \times 12.67 = \underline{950}$$
$$7{,}185 \text{ ft-lb}$$
$$= 86{,}400 \text{ in.-lb}$$
$$\text{Required } S = \frac{86{,}400}{29{,}330} = 2.95$$

For a 4-in. pipe,

$$S = \frac{7.233}{2.25} = 3.21$$

Since stresses may be increased by $\frac{1}{3}$ for wind stresses, the allowable passive soil pressure is $1\frac{1}{3} \times 400 = 533$ psf and the formula for passive soil pressure in Fig. 10.51 becomes

$$533 = \frac{12 \times 7{,}185}{b \times 64}$$
$$b = \frac{12 \times 7{,}185}{533 \times 64} = 2.53$$

Therefore the pipe should be bedded in a concrete foundation 8 ft deep and 2 ft 7 in. square.

Problem 2 A flagpole is carried from the front of a building (Fig. 10.54). The supports BD and BE are not ordinarily designed to take any compression, so they may be rods. On the other hand, the pole from B to C carries compression, so l/r must be considered in design of this pole. Points B, D,

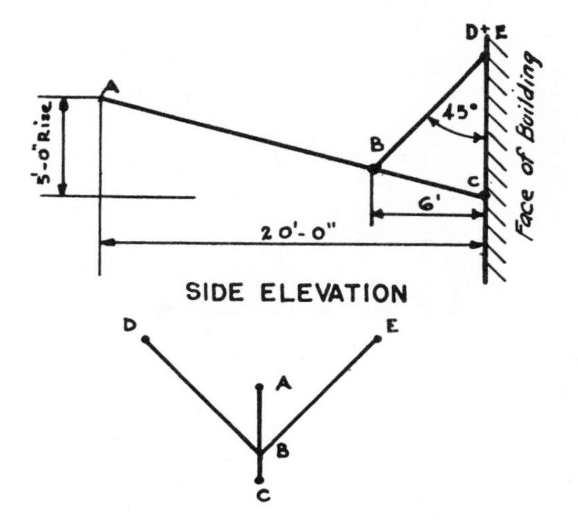

Fig. 10.54 Flagpole on Face of Building

and E should be definitely supported into the steel structure of the building, and if they occur between stories, as they ordinarily do, vertical beams should be framed from spandrel to spandrel to carry the pull or thrust.

Assume a 50-sq ft flag, $P = 36$ lb. The cantilever moment about B, assuming a 2-in. pipe pole is

$$\begin{aligned} 36 \times 14 &= 504 \\ 3.65 \times 14 \times 7 &= 358 \\ &\overline{862} \text{ ft-lb} = 10{,}350 \text{ in.-lb} \end{aligned}$$

Using a fiber stress of 29,333 psi

$$\text{Required } S = \frac{10{,}350}{29{,}333} = 0.353$$

For 2-in. pipe,

$$S = \frac{0.666}{1.19} = 0.56$$

For thrust on a 6-ft length, for $l/r \leqq 120$,

$$r = \frac{6 \times 12}{120} = 0.6$$

From Table 1.5(a) r of 2-in. pipe = 0.79

Since the wind load is horizontal and the dead load vertical, the resultant pressure will approximate the angle of the diagonal tie closely enough to disregard the variation. The load at B parallel to the front wall is

$$\begin{aligned} 36 \times 20 &= 720 \\ 3.65 \times 20 \times 10 &= 730 \\ &\frac{1{,}450}{6} = 242 \text{ lb} \end{aligned}$$

On the line of BD or BE the stress is therefore

$$\sqrt{2} \times 242 = 342$$

The area at root of thread is

$$\frac{342}{29{,}333} = 0.0117 \text{ sq in.}$$

Therefore, minimum size rod, which is ⅝-in. round, will be permissible.

11

Office Practice

11.10 STRUCTURAL CHECK LISTS

a No check list, however complete, can take the place of experience in the design and preparation of structural plans or in the checking of shop detail drawings. On the other hand, even for a thoroughly experienced designer or structural draftsman, the use of a check list frequently prevents the omission of details desirable for the benefit of bidders and calls attention to items which might otherwise be overlooked in the rush to complete the job.

No check list can be complete enough to cover every type of job. Each office must prepare its own check list as the result of experience. Included here are the various check lists used in the author's office, not in the single words on the check list, but expanded into the actual question to be answered.

b *Design information* (see also Art. 11.50):

1. Name, location, and general use of building.
2. Under what building code will the design be made?
3. Does this code have any unusual requirements or are they standard AISC, ACI, etc., codes?

c *Structural plan check list—Foundations:*

1. Are footings limited by lot lines or other restrictions, so as to require special shapes, combined footings, or other special treatment?
2. Is underpinning of adjacent structure required?
3. Does any adjacent excavation require footings to be dropped?
4. Are there old excavations on the site which will require the footings to be lowered?
5. Are stepped wall footings between different levels required?
6. Check comparative level of adjacent column footings.
7. Are all exterior footings below frost line?
8. Are retaining walls required outside the building?
9. Are building walls reinforced against earth pressure?
10. Are wall piers of sufficient area to take grillages?
11. Are area walls required?
12. Are pipe sleeves to be provided in foundation walls?
13. Do downspouts on interior columns require footings to be dropped to allow for pipe bend over the footing?
14. Are there any boiler pits, elevator pits, sump pits, or other pits, and are footings to be dropped for these pits?
15. Has provision been made in the wall design to permit installation or change of boilers or other mechanical equipment?

16. Is the waterproofing of walls or floors required?

17. Is there sufficient hydrostatic head to require special reinforcement of the basement floor?

18. Show column numbers.

19. Show column center dimensions, wall thickness, and other necessary dimensions.

20. Show elevation of bottom of all footings.

21. Show detail wall sections on the drawing.

22. Indicate scale of drawing.

23. Include general explanatory notes.

24. Show footing schedule of reinforced column footings.

d *Structural plan check list—Floor plans:*

(Go through this list for each floor.)

1. Are all slab designations noted?

2. Are all beam sizes shown?

3. Are all lintels provided for, interior as well as exterior?

4. Are radiator recess lintels provided?

5. Show column numbers.

6. Dimension column centers and special beam offsets, etc.

7. Include a general note covering beam levels, and note on the plan any exceptions.

8. Are all masonry piers of sufficient capacity for all loads?

9. Check architectural and mechanical plans for any wall chase in critical locations.

10. If there is a mail chute, it must carry through from top to bottom without any offsets. Check for interference by any beams.

11. Check interference of beams with water closet roughing.

12. Depress beams and slabs for floor-type urinals.

13. Frame for all shafts and other openings.

14. Provide double beams where floor elevation changes.

15. Note scale on drawings.

16. Show detail of floor construction.

17. Are there any beams across windows in stair wells (mid-story windows)?

18. Do beams provide sufficient stair clearance?

19. Are there any stair cantilever details to show?

20. In gypsum floors, are tie rods and end-span bracing shown?

21. In bar-joist floors, are sufficient details of general construction shown?

22. Are there any story heights over 13 ft 6 in. at elevator shafts? (If so, intermediate supports may be needed for elevator guides. Check with the elevator company.)

e *Structural-plan check list—Roof plans:*

1. Check all applicable notes from check list for floors.

2. Note high and low points and all sloped beams so as to inform steel bidders of type of detail.

3. Include general notes on levels of roof beams and slabs.

4. Locate and detail ducts, skylights, and scuttles.

5. Are there any flagpoles or roof signs to be supported by roof steel?

6. If there is an overhanging cornice, provide anchorage.

7. Indicate scale of drawings.

8. Are there any elevator or stair penthouses?

9. In an elevator penthouse, is there a secondary framing level?

10. Are elevator trolley beams required in penthouse?

11. Indicate elevation of various elevator penthouse levels.

f *Structural plan check list—Miscellaneous schedules:*

1. On column schedule, show all story heights, splice points, and top and bottom of all columns with relation to some given floor line.

2. Is a detail required for future extension of any columns, or future connections to any columns?

3. Is lintel schedule shown?

4. Is concrete-slab schedule shown?

5. Is concrete-beam schedule shown?

11.20 SURVEY

a The survey required for proper structural design covers much more than an ordinary survey made for a property transfer.

Obviously, a structural-design survey should show the exact dimensions and angles of the property, the street lines, and the north pointer. It should also show all the old buildings or remains of old buildings on the site, together with the grades of the cellars. It should locate and give the size of all trees, and locate any public-utility right of way or easement on the property. Moreover, it should include the location, height, and, if possible, depth below grade of any building on the adjacent property within several feet of the property being surveyed, with a word or two of description.

If the building is to have a full basement, grades on the entire site are not necessary except where they will enable the contractor to bid more intelligently. Of course, grades at the building walls and outside the building walls are needed. If the lowest floor is at or above grade, the site should be completely cross-sectioned, the spacing of the cross sections being dependent on whether the site is comparatively level or full of humps and hollows. The location of rubbish fills, sloughs, water courses, springs, filled-in wells or cisterns, and deep holes should be given whenever possible.

If the site is nearby, there are certain facts which may be obtained directly from the public works department, although the author's experience has been that a surveyor can do a better job of obtaining data on the location, level, and size of water mains and tees, storm-water and sanitary sewers, gas mains, width of streets, and the like. Perhaps the surveyor can also advise on regulations governing projection upon city property of such elements as steps, belt courses, cornices, and open areas.

b If the building site is adjacent to one or more old buildings, or if foundation conditions are such that there is even a remote possibility of damage to other structures or claims for damage as a result of construction operations, it may be advisable to make a complete condition survey of all adjacent buildings to uncover existing defects before new construction is begun. Most old buildings have defects of some sort and unless a careful and systematic inspection is made, the builder may be saddled with false damage claims.

The inspection report should be systematic and should cover:

1. The type and location of building, date and time of examination, owner, parties making the examination, and whom they represent.
2. Condition of exterior, including photographs of all available fronts.
3. Sketch plans of all floors with notations of all rooms as a means of identification.
4. Complete description of all rooms, starting at the roof and working down to the cellar. Take the rooms in regular order to avoid omissions.
5. Divide stair halls at floor lines as "1st floor stair hall and stair to second."
6. In each room describe floor first, then ceiling, north wall, east wall, south wall, and west wall.
7. Whenever possible, examine girders in the cellar and their bearings. Also examine all floor beams whenever possible, especially under bathrooms and kitchens. If any defect or settlement is noted, check for resultant effects in floors above.
8. Record all separations between floors and baseboards if they exist. Note floors and heads of doors and windows out of level. Note stairs out of level and stairs pulled out of the mortises in the wall string.
9. Note wainscots, chair rails, or picture moldings where they indicate any settlement. Note also separations between wood mantels and chimney breasts and hearths cracked or sunk below adjoining floors.
10. In describing cracks, record them under the wall which appears to have moved, as "East Wall, crack at intersection with south wall open 1/8 in. from base to 1 ft below ceiling, thence diagonally across east wall to ceiling."
11. Always state width of crack and length, as "hairline crack 3 ft long" or "crack 1/32 in. full width of room."
12. Describe condition of wallpaper if it

shows wrinkles or is cut diagonally across the corner, indicating a crack beneath.

13. Record water stains on walls or ceilings.

14. Where no defects are found, record "nothing noted."

Inasmuch as such a report might become an important document in court, it is well to have some representative of the owner accompany the person making the examination, and it might even pay to have a neutral party make the survey.

11.30 SOIL TESTS

a In order to determine satisfactorily the bearing capacity of the soil on which the foundation is to rest, it is necessary to have some knowledge of the materials both at the level of bearing and for some distance below it. This information is frequently available from former excavations, tests, or other data. If it is not available, or if the behavior of existing buildings has been unsatisfactory, soil samples should be taken and evaluated by a competent soils laboratory. Such laboratories are now available in practically all areas of the country to assist the designer in scientifically evaluating the subsoil condition, and in determining the most economical type of footing to use with safety.

b If in the opinion of the designer, the additional cost of such laboratory investigation is unwarranted for the structure under consideration, either test pits or soil-loading tests may be substituted under the supervision of the designing engineer.

The drilling or digging of good test holes depends on a number of factors. For example: Is your building to be two stories or fifteen stories? Is it to be built in the downtown section of a city, in a small town, or out in the open country? Are there any neighboring conditions which might occasion distrust of the subsoil? Is the building designed for light occupancy or heavy loads? All of these points and many others should influence the decision about the number of test holes required, the depth below your lowest footing level to which tests should be carried, and the type of borings that should be made.

In or near a large city, there may be a local drilling company which specializes in such borings, and which for $3 to $5 a foot will drill as deeply as required with a 6-in. casing, and take dry samples of the soil. If such test drilling service is not available, a local well driller may be employed to do such boring, but if he is so employed he should be advised not to use water to simplify his drilling. It may even be necessary for the engineer or his representative to watch so that he may be sure that no change in the natural condition of the soil was permitted to occur through ingress of rain water or drainage.

With power excavating equipment now available practically any place, it is frequently economical, if it is known that soil conditions are relatively uniform, to employ an excavating contractor to come onto the site with a back hoe or similar excavating equipment, and to dig holes to the required depth at predetermined locations for actual inspection of the soil in its native bed by the engineer.

Obviously, it is not satisfactory to go only to the level of the footings. It is the material below the footings which is of most importance. For high buildings it is advisable to carry several borings to rock, or at least 20 or 30 ft below the foundations, to be sure that there are no quicksand layers below the water line. For two- or three-story structures, it is usually satisfactory to go down about 4 to 6 ft below the footing with an open hole, then drive a bar down an extra 4 ft to be sure you are not too close to soft material. It is well to avoid location of a fairly large test hole under or near a column footing in the finished structure.

c After the test pit results have been obtained, they must still be interpreted. Here also there are factors aside from the actual test results which will influence interpretation. For example: In what season of the year were the tests made? Was it dry or wet? And finally, there is the ever-present human factor: the judgment of the man who makes the interpretation. Article 8.10 lists the recommended allowable bearing values on soils of different kinds.

d Although a soil-loading test is an expensive method of determining allowable soil pressure, it may be required by the building department of the municipality, or for some other reason, it may be an advisable precaution. Obviously, such a loading test should be made at the level where footings will be placed. The following extract from the Pacific Coast Uniform Building Code offers a good criterion to follow in making this test.

Where the bearing capacity of the soil is not definitely known or is in question, the Building Inspector may require load tests or other adequate proof as to the permissible safe bearing capacity at that particular location. To determine the safe bearing capacity of soil it shall be tested by loading an area not less than two square feet (2 sq ft) to not less than twice the maximum bearing capacity desired for use. Such double load shall be sustained by the soil until no additional settlement takes place for a period of not less than 48 hours in order that such desired bearing capacity may be used. Examination of sub-soil conditions may be required when deemed necessary.

The area specified must be a single area of 2 sq ft, and cannot be divided into four smaller areas. The platform which is built to carry the load therefore must be supported on a single central post as shown in Fig. 11.1. Since provision should be made for at least a 20,000 lb load on the platform, consideration should be given to the material used for providing the load. If it is easily obtainable, pig iron is ideal, since it takes so little space. If a steel tank of sufficient capacity is available, it may be filled as required with water. Otherwise, cement in bags, or sand in bags is generally used. Obviously the nature of the material used to make the test will influence the size of the platform that must be built. Probably the simplest way to hold the platform in horizontal position is by means of a guy cable at each of the four corners, although sometimes loose wedges against the sides are used.

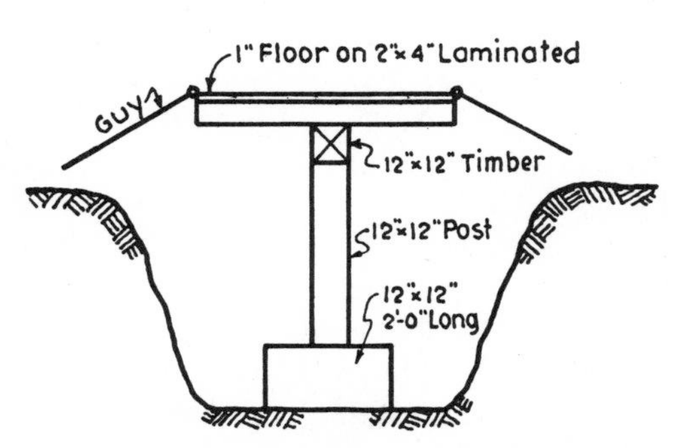

Fig. 11.1 Platform for Soil-loading Test

11.40 COST DATA

a Entire volumes are devoted to the subject of estimating and cost data, and it is impossible to give an adequate discussion in the few paragraphs that follow. There are, however, several points of general estimating practice which may be stated in this volume, including (1) a brief mention of rules of cubage and cubic foot costs and (2) the use of cost indices.

b While cubage methods are not 100 percent accurate, the approximate cost of a building may be obtained quickly by multiplying the cubage of the proposed structure by the known cubic-foot cost of a comparable building. The method of finding cubage may be shown by the following "Rules of Cubage" of the New York State Education Department.

1. All horizontal areas shall be measured from the outside of the inclosing walls at the plane of the floor.

2. (*a*) The height of principal stories shall be measured from a horizontal plane two feet below the surface of the finished floor of the first story to the underside of the ceiling of the top story in that section of the building in which the cube occurs which is being calculated.

(*b*) In sections without finished ceilings, the perpendicular dimension shall be taken from a horizontal plane two feet below the finished floor of the first story to the mean height of roof surface of the section.

3. The height of developed basement areas shall be measured from the surface of the finished basement floor to a horizontal plane two feet below the surface of the finished floor of the first story.

4. Roof areas: (*a*) Parapet roofs (flat roof type) shall be measured from the underside of the ceiling of the top story to the high point of the finished roof surface.

(*b*) Sloping ridge roofs shall be measured from the underside of the ceiling of the top story to a horizontal plane at the mean height of the sloping roof.

(*c*) Sloping roofs with deck shall be calculated to obtain actual volume of the roof section above the plane of the ceiling of the top story.

5. Other areas: (*a*) The actual volume of towers, cupolas, penthouses, connecting passageways, boiler stacks, hoist shafts, area entrances and developed sub-basements shall be obtained. All measurements shall be taken from outside face to outside face of inclosing walls and from the surface of the floor through to the mean height of roof or inclosing surface.

Note: No cube shall be included more than once. If, for example, a boiler stack is within the principal walls of the building, only that volume which has not been included in 5, 6 and 7 shall be added under this classification.

(*b*) The actual volume of porticos and covered porches shall be obtained. Only one-half of this volume shall be included in computing the total cubage of the building. Horizontal measurements shall be taken from lines marking the outside of the faces of the frieze of the cornice, or wall structure above columns. Perpendicular measurements shall be taken from mean finished grade to mean height of roof.

6. Uncovered steps, stoops, approaches, buttresses, parapets, light courts, window areas, foundation walls, pipe trenches, cisterns, septic tanks, and retaining walls shall not be calculated nor included.

7. Definitions: (*a*) The principal, or first story, is the lowest story the floor of which is wholly above finished grade to the building.

(*b*) The basement is that portion of the building immediately below the principal, or first story.

(*c*) A basement, sub-basement or a cellar shall not be termed a story.

(*d*) A sub-basement is that portion of a building which has been developed below the basement level.

(*e*) The mean height is the midpoint between the eaves and the peak of the roof.

c Obviously, cubic-foot prices are constantly changing, and consideration must be given to this fact. Cubic-foot costs of today may be compared with those of another date and current estimating costs prorated by the use of any one of a number of building-cost indices which are published regularly. Because it is easily obtained from any current issue of the magazine, included here is the *Engineering News-Record* Cost Index Table (Table 11.1). Unfortunately, this index makes no attempt to rate labor efficiency, but this is extremely difficult to do under any circumstances.

Problem Assume that a factory building 50 ft wide, 120 ft long, and 14 ft high cost $35,000 when bid in the spring of 1940. The owner wished to build another similar building 40 ft wide, 180 ft long, and 16 ft high in 1953, when the *Engineering News-Record* Building Cost Index was 430. What was the estimated cost of the construction?

The volume of the original building was 84,000 cu ft and the cost was 41.6 cents per cubic foot. The cost index in 1940 was 201.5. The proposed new building will have 115,200 cu ft. The revised cost should be

$$\frac{430}{201.5} \times 41.6 = 88.8 \text{ cents per cubic foot}$$

and the total cost of the building would be $102,297.

11.50 DIMENSIONS AND CLEARANCES

a Many of the dimensions and clearances given in this article are architectural rather than structural, but the engineer is frequently called upon to use them. Other dimensions and areas are given elsewhere in this book, particularly in Art. 11.60.

b For steel or concrete columns, with no other factors influencing spacing, 20 ft each way is a good spacing. For timber framing, the high cost of long timbers makes a shorter spacing more economical.

In general, it is cheaper to run the beams the long way of the span and the girders the short way.

For masonry buildings, most codes require that exterior columns be covered with 8 in. of masonry. In a two- or three-story building estimate on the basis of 8-in. columns; up to five stories, 10-in. columns; up to about nine stories, 12-in. columns; and above nine stories, 14-in. columns. In setting the centerline for wall columns, allow 8 in. for brick, 1 in. for splice plates and the like, and half the size of the column; thus it would be 1 ft 1 in. for an 8-in. column, 1 ft 2 in. for a 10-in. column, and so on. Some codes permit 4 in.

Table 11.1 *Engineering News-Record* Building Cost Index

1913 = 100

	1924	1925	1926	1927	1928	1929	1930	1931	1932	1933	1934
Av.	189.3	182.7	185.0	186.1	188.0	190.9	185.4	169.4	140.9	147.8	168.7

Month	1935	1936	1937	1938	1939	1940	1941	1942	1943	1944	1945
J	166.5	168.3	185.5	198.6	196.2	201.5	207.9	217.1	226.9	231.6	236.5
F	165.8	168.4	186.6	199.0	196.2	201.5	208.4	218.4	226.9	232.0	237.7
M	164.1	169.7	187.7	199.3	196.5	201.5	207.9	218.9	226.7	232.3	237.9
A	164.4	170.0	195.0	198.2	196.5	201.4	208.6	219.6	227.5	234.5	238.5
M	163.3	170.3	197.2	197.9	196.6	201.2	209.2	220.1	227.6	234.8	238.5
J	164.9	170.2	199.2	198.2	196.4	201.7	209.4	221.2	227.7	235.3	239.4
J	165.3	170.9	199.8	194.4	196.8	201.7	210.2	223.5	227.9	235.6	239.6
A	166.2	173.9	200.7	194.5	196.5	201.7	212.5	225.4	228.5	236.1	239.9
S	166.1	174.3	200.6	194.8	196.6	203.1	214.6	225.5	231.2	236.0	240.1
O	167.3	175.3	201.0	195.5	198.5	204.0	215.9	225.9	231.3	236.0	240.4
N	167.3	175.5	200.3	195.6	201.2	206.2	216.5	226.6	231.4	236.1	240.5
D	167.2	179.3	200.5	196.1	201.2	208.2	216.4	226.6	231.5	236.5	240.8
Av.	165.8	172.2	196.2	196.8	197.4	202.8	211.5	222.4	228.8	234.7	239.1

Month	1946	1947	1948	1949	1950	1951	1952	1953	1954	1955
J	242.2	294.6	333.6	354.9	356.2	393.0	405.8	425.0	436.4	458.8
F	243.9	301.6	335.5	352.9	356.5	398.2	406.2	425.2	437.1	459.4
M	245.4	303.3	334.2	352.5	360.0	399.3	407.4	424.9	436.7	459.6
A	254.4	305.2	334.6	351.4	362.8	400.1	407.9	426.4	437.5	460.6
M	258.1	304.6	333.9	348.9	364.3	401.1	410.3	426.2	438.2	462.6
J	265.3	307.4	339.3	349.3	373.4	400.8	412.5	426.2	439.7	464.8
J	267.2	308.9	342.4	349.5	377.7	400.3	414.5	435.2	444.1	467.5
A	272.3	317.8	355.5	350.9	383.1	401.2	422.4	436.9	455.5	478.0
S	272.4	322.6	355.5	352.0	392.8	400.3	424.4	436.0	454.3	479.2
O	273.0	327.3	357.1	353.0	397.4	403.4	424.8	436.0	455.3	480.0
N	274.0	329.2	355.9	352.9	390.2	404.5	426.0	436.0	456.3	479.3
D	280.0	333.1	355.6	353.2	391.4	405.7	424.9	435.5	456.8	478.7
Av.	262.4	313.0	344.0	351.8	375.5	400.6	415.6	430.8	445.7	468.8

Month	1956	1957	1958	1959	1960	1961	1962	1963	1964
J	480.5	502.0	516.8	536.2	554.4	562.9	570.9	584.5	604.2
F	483.1	502.6	515.8	537.3	555.4	562.9	572.8	585.2	604.0
M	483.1	501.4	516.1	540.3	555.2	562.9	575.0	586.0	606.3
A	485.6	501.4	516.6	542.9	556.0	565.2	575.9	586.2	607.5
M	487.4	503.0	518.6	544.4	559.6	569.0	579.4	588.3	609.5
J	488.7	504.2	521.1	548.4	561.0	570.2	579.6	590.2	612.2
J	489.2	506.6	524.3	551.7	562.6	571.7	582.8	596.2	614.7
A	491.2	516.9	525.6	554.3	562.2	570.9	585.6	601.9	616.3
S	500.1	516.1	535.0	555.9	562.8	571.3	585.8	601.7	617.1
O	499.5	516.8	536.5	555.4	561.5	570.6	584.9	604.2	616.8
N	500.5	516.8	535.0	554.2	561.1	570.8	584.0	602.2	616.7
D	500.2	516.3	534.8	552.9	561.5	570.4	584.1	603.0	
Av.	490.8	508.7	524.7	547.8	559.4	568.2	580.0	594.1	

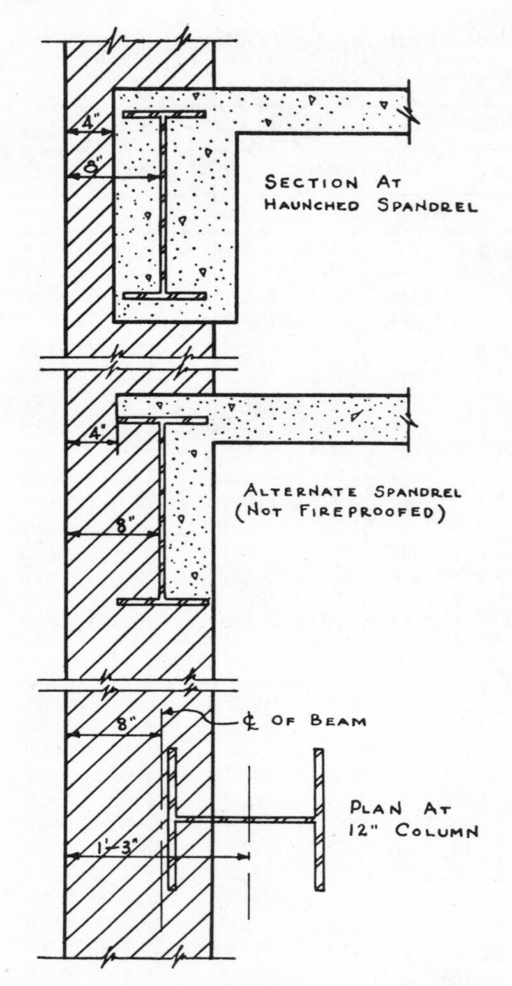

Fig. 11.2 Location of Wall Beams and Columns

of cover, which would reduce the above requirements. Consult code requirements before setting locations. Practically all codes require 8 in. of covering on beam webs, and 4 in. minimum on beam flanges in walls. Figure 11.2 indicates this requirement.

c For theater or church seating, allow 7 to 8 sq ft per person when using curved seats and 6 to 7 sq ft per person when using straight rows. These allowances are for the entire area, including aisles.

For dining-room seating, allow approximately 12 sq ft per person.

Toilet stalls are 2 ft 6 in. to 3 ft wide by 4 to 5 ft deep inside. Shower stalls are 2 ft 8 in. square to 3 ft 6 in. square. Urinal stalls are 2 ft to 2 ft 2 in. wide by 1 ft 8 in. to

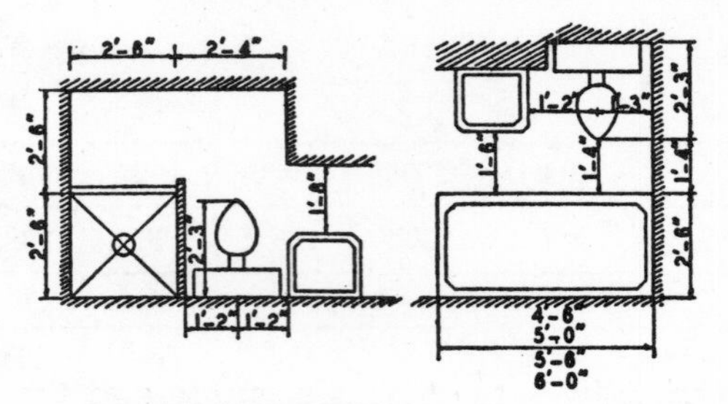

Fig. 11.3 Toilet-Room Clearances

2 ft deep (see Fig. 11.3 for further toilet-room clearances).

A telephone booth is 2 ft square inside.

Commonly accepted angles and pitches for ramps, stairs, and ladders are:

For pedestrian ramps, not over 20 deg, or $4\frac{3}{8}$ in. rise to 12 in. run. Preferred practice is 15 deg, or $3\frac{1}{4}$ in. rise to 12 in. run.

For stairs, not over 50 deg, or $14\frac{1}{4}$ in. rise to 12 in. run, with the preferred angle between 30 and 35 deg. The New York State Labor Department sets a maximum rise of $7\frac{3}{4}$ in. and minimum run of 10 in. exclusive of nosing, for each step. Vertical clearances are determined from Fig. 11.4.

Ladders should be used for slopes of over 50 deg. The preferred slope is 75 deg, with rungs ranging from $9\frac{3}{8}$ in. at 50 deg to $12\frac{3}{4}$ in. at 75 deg.

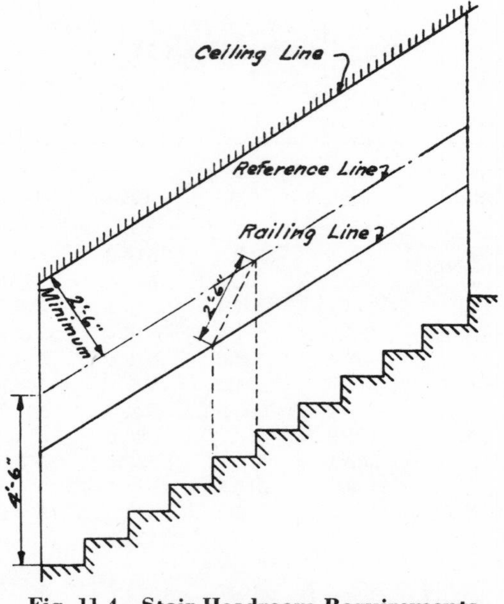

Fig. 11.4 Stair Headroom Requirements

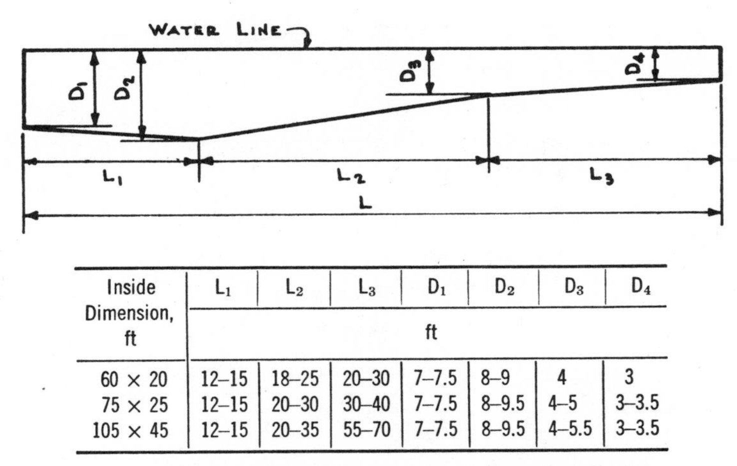

Inside Dimension, ft	L_1	L_2	L_3	D_1	D_2	D_3	D_4
	ft						
60 × 20	12–15	18–25	20–30	7–7.5	8–9	4	3
75 × 25	12–15	20–30	30–40	7–7.5	8–9.5	4–5	3–3.5
105 × 45	12–15	20–35	55–70	7–7.5	8–9.5	4–5.5	3–3.5

Fig. 11.5 Standard Rectangular Swimming Pools

d Certain indoor sports require minimum clearances and lengths, the most common requirements being as follows:

Shuffleboard requires a court 6 ft wide by 52 ft long.

Bowling alleys require a width of 11 ft 6 in. and a length of 83 ft for a pair of alleys and the ball return.

A basketball court requires a floor area of 35 by 60 ft for elementary schools, 42 by 74 ft for junior high schools, 50 by 84 ft for senior high schools, and 50 by 94 ft for colleges, with 3 ft (ideally, 10 ft) clearance on all sides of the court.

A handball court requires an area of 20 by 44 ft, with 6 ft between courts or 8 ft from the side of the court to the nearest wall or construction.

The dimensions and contours of standard swimming pools are shown in Fig. 11.5.

e Standard railroad clearances are shown in Fig. 11.6. There may be local variations which will permit less clearance for certain switching railroads, but it is advisable to

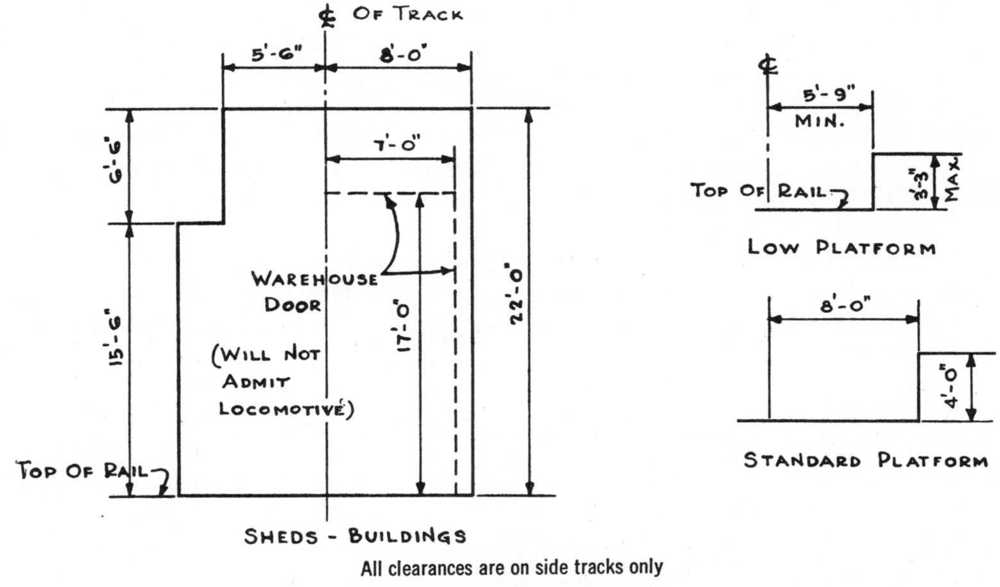

All clearances are on side tracks only

Fig. 11.6 Railroad Clearances

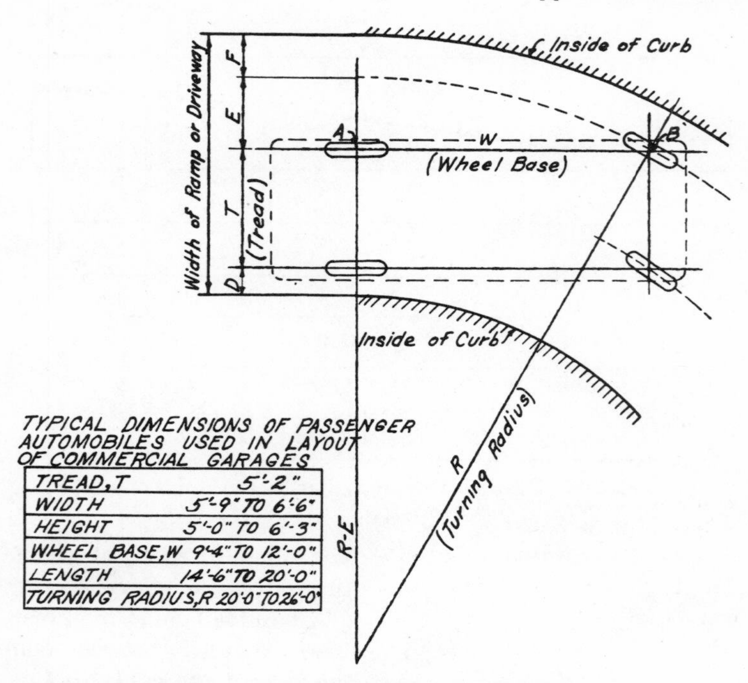

Fig. 11.7 Automobile Curved Ramps

clear any variations from these figures with the railroad which will use the track.

Parking-garage spaces and clearances are given in Figs. 11.7 and 11.8. Where ramps are used, a 12 percent grade is desirable, with a preferred maximum of 15 percent. Curved ramps should have a 10 percent lateral slope. A one-way straight ramp should be 11 to 12 ft wide flared out at the ends, including 1-ft-wide curbs 6 in. high at each side. The outside diameter of the curve should be 60 ft.

Truck garages for the accommodation of trailer trucks or even large vans should have doors 12 to 14 ft in width, depending on directness of approach, and 13 ft high.

11.60 ECONOMIC PROBLEMS OF LAYOUT

a Many of the problems of layout are architectural rather than structural, and, therefore, may be out of place in this volume. On the other hand, the problems of economics are largely engineering problems, and for the benefit of engineers who may have occasion to use such statistics, some of them are included here. Some points pertaining to these problems have been discussed in other articles, such as Arts. 1.40 to 1.42 and 5.10.

It should also be borne in mind that any economic data or comparisons will change from time to time with the fluctuation of

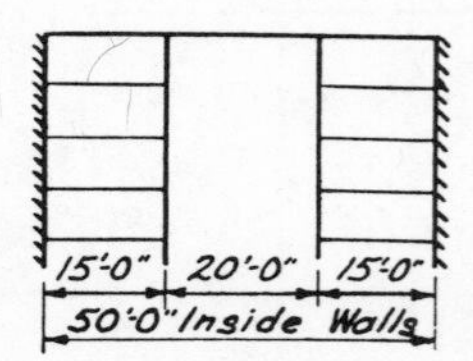

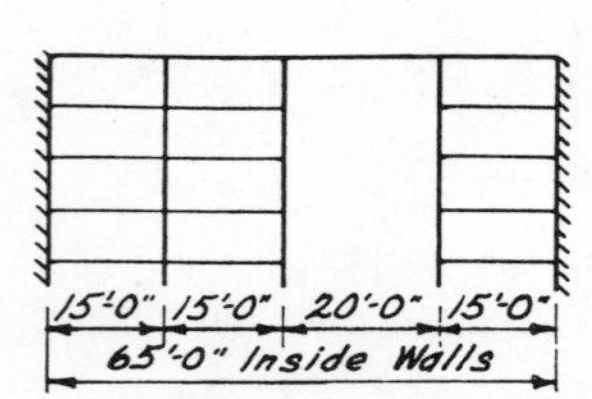

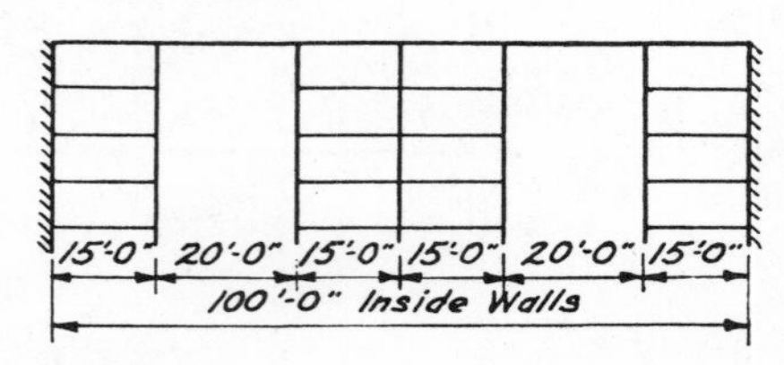

Fig. 11.8 Layout of Garage Parking Units

costs. The following approximate figures may tend to guide the engineer in his thinking, however.

b A few years ago one of the country's large industrial corporations contemplated a large expansion program utilizing reinforced concrete buildings. In preparation for this program it conducted a series of cost investigations, and the findings were as follows: Width and height have a greater effect on the cost of factory buildings than floor loads and column spacing. Width is affected by conditions of the site and by requirements for natural light. Height is affected by the intended use of the building and by land values. For buildings of the same width, on the basis of cost per square foot of floor a three-story building costs 96 percent as much as a two-story building, a four-story 89 percent, and a five-story 79 percent as much. On the basis of width alone, regardless of the number of stories from two to five, referring to the cost per square foot of floor as 100 percent for a building 60 ft wide, the cost for an 80-ft building would be 84 percent as great, for a 100-ft building would be 74 percent, and for a 120-ft building would be 66 percent.

Between limits of 16- and 25-ft spacing, using a 20-ft column spacing as 100 percent, the square-foot cost dependent on column spacing varies with the ratio $70 + 1.5s$, where s is the spacing in feet. As to live load, using a 200-psf load as 100 percent, the square-foot cost dependent on live load varies with the ratio $72 + 0.14w$, where w is the load in pounds per square foot. By applying these formulae to the ordinary range of spacings and loads, it will be seen that the effect of varying either spacing or load is relatively small.

In concrete structures with fairly heavy live loads, the rich-mix concrete shows a considerable saving over a lean-mix concrete. For each 1,000 psi of unit strength provided, a 2,000-psi concrete costs about 45 percent more than a 3,000-psi concrete. Therefore, if load conditions and design are such that concrete can be used to anything like its economical maximum, it would pay to use a rich-mix concrete. This principle does not apply without reservations, however; it is suggested as one point in the cost picture where we do not ordinarily look for economy.

These ratios, of course, are prepared on the basis of concrete-building costs, but in general the same ideas would probably apply to steel construction.

c *Office-building design.* The engineer or architect may be called upon to advise on the soundness of proposed projects as well as on their actual layout. With that idea in mind, the following is included.

In the ordinary office building there should be a reasonable expectation that the building should be 70 percent full within a year and 90 to 95 percent full in two years. A general rule used by many experts is that the ground floor of a tall building should earn enough to pay the taxes and a reasonable return on the cost of the land. Some demand taxes on the first floor also. The shape of the lot is of vital importance. If the area is too small or the frontage too narrow, it is impossible to improve it intensively enough to secure proper return. The larger the lot and the more nearly it approaches a square in shape, the lower the unit cost and the greater the efficiency of the rental space. Ordinarily the net rentable area is from 60 to 70 percent of the total lot area, and from 70 to 80 percent of the gross building area. The rentals of the first floor and basement should pay a minimum of 5 percent interest, the taxes on the land alone, and the insurance on the building.

Investigations indicate that the average requirement of office space is 3.9 sq ft per capita in Detroit, 6.5 in Los Angeles, and 6.4 in Chicago. Another study based on 12 southern cities with a population of around 300,000 indicated 4½ sq ft per capita. This is based on net rentable areas in buildings offering competitive office space.

Under normal conditions, in an office building column spacing should approximate 17 to 18 ft. This makes possible the most generally serviceable type of office unit, 17 to 18 ft in width and 25 to 30 ft in depth. In a building of 10 or more stories, the column spacing for offices should take precedence over that for stores. Offices facing an open park receive natural light to a much greater depth than those facing a narrow street. Best financial return is usually obtained by using the first floor for stores, and the available return from these stores may be the governing factor in determining the column spacing, rather than the spacing for offices. The ceiling height for offices should be 9 to 10 ft.

Each office unit should have two windows, about 4 ft 6 in. wide by 7 ft 6 in. high.

Corridors should be 6 ft 6 in. to 7 ft wide. No windows are necessary for corridors. Office doors should be 42 in. wide and doors between offices should be 30 in.

Population of office buildings may be estimated at 1 person per 125 sq ft.

d *Hotel design.* The size of hotel bedrooms has become almost standardized. Rooms facing the street are usually about 10 to 12 ft wide by 16 ft in length; rooms facing a court are 9 to 11 ft wide by 14 ft long. Corner rooms are usually 15 to 16 ft wide by 18 to 20 ft long. It is well, if possible, to make hotel room sizes in multiples of standard carpet widths for economy.

Corridors should serve rooms on both sides and there should be ample linen-closet, slop-sink, and maid-room accommodations on each floor.

For dining rooms and cafeterias, provide 20 sq ft of dining space for each bedroom. For kitchens, provide the same amount.

Private bathrooms and closet room should be provided at the corridor end of each bedroom, and one ventilating shaft provided for each two bathrooms.

An article by Professors Randolph and Sayles of Cornell University published in *Architectural Record* gives a formula for approximate cubage as follows:

$$x = 0.41y + 0.013y^2 + 0.0002y^3$$

where y = hundreds of rooms

x = millions of cubic feet

Several other formulae give approximate areas as follows:

Lobby area in hundreds of square feet = $0.373y + 0.68$

Laundry area in hundreds of square feet = $2.12y - 5.59$

e *School design.* Much of the following on school design is from the regulations of the New York State Department of Education.

Classroom dimensions should be 22 ft wide and 27 or 30 ft long, or 19 ft wide and 30 ft long.

Classroom windows should be located on one of the broad sides on the left of the pupils, grouped with a space not exceeding 1 ft between windows. The glass area should be equal to one-fifth the floor area. The window near the forward end should be 6 ft from the corner if possible, and never less than 4 ft. A room 22 by 27 ft should have five windows 3 ft 4 in. wide and 8 ft 6 in. high, or four windows 4 ft wide by 8 ft 6 in. high, with the head of the window 11 ft or more from the floor. A room 19 by 30 ft should have either five windows 3 ft 4 in. by 7 ft 10 in. or four windows 4 ft by 7 ft 10 in., with the window head 10 ft or more from the floor. Windows for other rooms may be of any stock size. For toilet and coat rooms a window 2 ft 10 in. wide by 5 ft 2 in. high is satisfactory.

In smaller buildings, well-lighted and well-ventilated coat rooms having at least 50 sq ft of floor space should be provided for each sex.

In toilet rooms, provide one lavatory for each 50 pupils, one water closet for each 25 girls, and one closet and one urinal for each 40 boys.

Main corridors should be from 9 to 14 ft wide.

Recessed wardrobes for classrooms should be 2 to 3 ft deep with sliding or folding doors full width.

f *Hospital design.* For hospitals, the Veterans' Administration has set up criteria requiring a total area ranging from 432 sq ft minimum per person for a 1,000-bed hospital, to 595 sq ft for a 200-bed hospital. Corridors should be at least 8 ft wide. Single rooms should have a minimum size of 120 sq ft. Minimum height floor to floor is 11 ft.

g *Warehouse design.* Some years ago a committee of the American Railway Engineering Association made an investigation and report on the economics of various facors of warehouse design which are worth recording here. They recommended for warehouses where the turnover of goods is fairly rapid:

1. One elevator for each 40,000 sq ft of warehouse space
2. Shipping platform 4 percent of warehouse storage floor area
3. One car length of track siding for each 17,600 sq ft of storage area
4. One foot of tailboard frontage for each 1,100 sq ft of storage space
5. Tailboard frontage of 16 ft for each car length of siding

***h** Automobile salesrooms and service station design.* Automobile sales and service stations are designed to fulfill requirements which have been established by the large manufacturers. Stalls are required for the following uses.

1. Service and repair stalls for regular customers
2. Used-car conditioning stalls
3. New-car conditioning stalls
4. Shop-storage stalls
5. Customer-reception stalls
6. Wash, polish, body, and paint stalls

The number of stalls required for the above uses is determined as follows: Take the average number of new cars sold per year for the past three years and divide by 28 for the number of service and repair stalls. Add one used-car conditioning stall for each 100 new cars sold, and one new-car conditioning stall for each 300 new cars sold. For shop storage, allow half the number of service and repair stalls. For customer reception, allow one-third the number of service and repair stalls. The number of stalls for work, polish, body, and paint must be based on past experience, since neighborhood competition is largely the governing factor.

For cars parked at right angles to the wall, allow 10 ft per car for service stalls and for customer reception, and 9 ft for shop-storage stalls. If diagonal parking is necessary add 15 percent. The stalls should be from 23 to 25 ft deep with a 25-ft aisle for diagonal parking. Lubrication racks require from 15 × 25 to 20 × 25 ft; wash racks, paint stalls, and body stalls, 15 × 25 ft; polish stalls, 12 × 25 ft. The minimum height if a lubrication hoist is used is 12 ft for passenger cars and 13 ft for trucks. Service doors should be 14 ft wide by 12 ft high for passenger cars and 13 ft high for trucks.

***i** Stair design.* Since improper stairs constitute a fruitful cause of claims against public liability insurance companies, it is well to adhere very closely to accepted practice in their design.

The rise and run of a stair are determined by the following rules:

1. The sum of the rise and run should be not less than 17 in. nor more than 18 in.
2. Twice the rise plus the run should be not less than 24 in. nor more than 25 in.
3. The product of the rise and run should be not less than 70 nor more than 75.

For important stairs, a rise of 7 in. and a run of 11 in. will give satisfactory results. For less important stairs, 7 or 7½ in. by 10 or even 9½ in. may be used, and for basement stairs in residences, even 8 in. rise. The New York State Labor Department specifies a maximum rise of 7¾ in. and minimum run of 10 in. for exit stairs. The nosing may project ¾ to 1½ in. beyond this.

The width of stairs is determined by their importance and by the number of persons that are expected to use them at one time. In general, stairs should be at least 3 ft wide. The New York State Labor laws specify a minimum of 3 ft 8 in. for exit stairs. The Building Code of the National Board of Fire Underwriters says that the width of stairs should be based on the number of people using the stairs, computed on the basis of 14 persons for each 22 in. width of stairway, plus 1 person for each 3 sq ft of hallway floor and stairway landings in the story height of each floor. Office buildings do not require as much stair area as other buildings, because the elevators are supposed to serve as proper exits. Many office buildings of large occupancy have only two 3-ft to 3-ft 8-in. stairways, and sometimes only one.

The New York City Code requires a minimum stair width of 3 ft 8 in., or a total of 22 in. width per person per 1½ treads, plus one person for each 2¾ sq ft of floor area on the landings, platforms, and halls within the stairway. In small buildings, this width may be cut to 3 ft. The occupancy is figured on the basis of one person for 10 sq ft of dance halls, restaurants, lodge rooms, and places of assembly; 15 sq ft in courtrooms and in classrooms in schools and colleges; 75 sq ft in work rooms, reading rooms, markets, showrooms, and stores; 50 sq ft in billiard rooms, bowling alleys, golf schools, and similar uses; 100 sq ft in office buildings and hospitals.

The New York State Labor law requires two enclosed fire stairways remote from each other for all buildings coming under its jurisdiction. This includes factories and loft buildings but not office buildings.

All stairways required as a legal means of exit shall run continuously from the roof to the ground floor and be connected by means of a fire enclosure to an outside exit at ground level.

j *Elevators.* The number and size of elevators are governed by:

1. The character of the building
2. The height of the building
3. The rentable area
4. The time interval between cars
5. The number of stories to be served
6. The average number and duration of stops per trip

No iron-clad rules can be given for all types of buildings, but office buildings and loft buildings have been sufficiently standardized in design to have developed some general rules. Good elevator service requires one elevator for each 20,000 to 30,000 sq ft of floor area. The elevator cabs should be at least as wide as they are deep, and, if possible, 20 to 50 percent wider. Cabs should have from 28 to 30 sq ft minimum floor area and be at least 7 ft wide. Several moderate-sized cars of high speed are preferable to a greater number of large, slow cars.

Buildings equipped with both local and express cars should have the same number of cars for each. Express cars are not necessary up to about 18 floors. Express cars will not serve as many floors as local cars, because they must go a greater distance to reach their first stop, so local cars should serve to about three-fifths of the height.

It is advisable to consult the local representative of one of the reliable elevator manufacturers to assist in planning elevators.

11.70 CHECKING STEEL SHOP DRAWINGS

a Although the detailers in the employ of the steel fabricator are responsible for sizes taken from the designer's drawings, and for correct fitting of column-and-beam details, steel shop drawings should be checked by the architectural engineer to ensure general compliance, and to pick up interferences and discrepancies that may have been missed in the original design. The following check lists will indicate the general checking that should be done.

b *Anchor-bolt plan:*

1. Dimensions and angles
   - *a.* Over-all dimensions
   - *b.* Column centers
   - *c.* Slopes of lines
2. Check column bases for:
   - *a.* Column number
   - *b.* Column turned right direction
   - *c.* Elevation
   - *d.* Billet or grillage sizes
   - *e.* Interference of billet or grillage
   - *f.* Size of pier to receive billet or grillage

c *Erection diagrams* (each checked individually):

1. Dimensions and angles
   - *a.* Over-all dimensions
   - *b.* Column centers
   - *c.* Slopes of lines
2. Check columns for:
   - *a.* Column section
   - *b.* Column number
   - *c.* Turned right direction
   - *d.* Lateral supports
3. Check beams, girders, and trusses for:
   - *a.* Section and general arrangement

   b. Location
   c. Elevation
   d. Clearance for stair wells, elevators, spandrels, fireproofing, etc.
4. Special sections
   a. Tie rods
   b. Shelf angles
   c. Special spandrels
   d. Lintels
   e. Hangers and struts

d *Beam details:*

1. Main material and over-all dimension
   a. Check for floor plan
   b. Clearance at ends
   c. Length of wall bearing
2. End connections
   a. Load capacity
   b. Eccentricity
   c. Material of connection
   d. Coped or blocked either end
   e. Square or sloping end
   f. Interferences
   g. Any countersunk rivets required
3. Connections into beam
   a. Location and size of beam connecting
   b. Elevation
   c. Beams framing opposite to use same holes
   d. Shelf angles, punching for nailers, knee braces, tie rods, etc.
4. General
   a. Size of rivets, open holes; exceptions
   b. Painting
   c. End marks, special marks, etc.
5. Special beams
   a. Hand holes for double beams
   b. Distance apart for double beams
   c. Separators or stiffeners
   d. Cover plates or web reinforcement

e *Column details:*

1. Main material
   a. Check from column schedule
2. Faces in right direction
   a. Faces noted
3. Dimensions
   a. Floor to floor
   b. Relation of ends to floor level
4. Base detail, splice detail, milled ends
5. Beam connections
   a. Elevation
   b. Size
   c. Interferences
   d. Clip angle loose
   e. Strength of seat connections
6. Miscellaneous connections
   a. Knee-brace connections
   b. Siding and girts
   c. Punch for nailers
   d. Bracing connections
7. General
   a. Column marking
   b. Size of rivets and open holes; exceptions
   c. Painting

f *Truss details:*

1. Main material
   a. Compare with design
   b. Which leg of angle outstanding
2. Main dimensions
   a. Span
   b. Elevation of underside and depth
   c. Depth on gage lines
   d. Slope of members
   e. Panel spacing
   f. Monitor width and height
3. Joints
   a. Develop stress
   b. Eliminate eccentricity in connections
4. Miscellaneous connections
   a. Purlin clips
   b. Lower lateral bracing
   c. Top lateral bracing
   d. Tie rods
   e. Holes for nailers
   f. Attachment of beams to lower chords
5. General
   a. Truss marking
   b. Mark end
   c. Size of rivets, open holes; exceptions
   d. Painting

Index